Scientific Charge-Coupled Devices

Scientific Charge-Coupled Devices

James R. Janesick

SPIE PRESS

A Publication of SPIE—The International Society for Optical Engineering
Bellingham, Washington USA

Library of Congress Cataloging-in-Publication Data

Janesick, James R.
    Scientific charge-coupled devices / James R. Janesick.
        p. cm. — (SPIE Press monograph ; PM83)
    Includes bibliographical references and index.
    ISBN 0-8194-3698-4
    1. Charge coupled devices.   I. Title.   II. Series.

TK7871.99.C45 J36 2000
621.36'7–dc21
                                                                        00-036616
                                                                        CIP

Published by

SPIE—The International Society for Optical Engineering
P.O. Box 10
Bellingham, Washington 98227-0010
Phone: 360/676-3290
Fax: 360/647-1445
Email: spie@spie.org
WWW: www.spie.org

Copyright © 2001 The Society of Photo-Optical Instrumentation Engineers

All rights reserved. No part of this publication may be reproduced or distributed in any form or by any means without written permission of the publisher.

Printed in the United States of America.

To the Universe.

Contents

	Preface	xiii
1	**History, Operation, Performance, Design, Fabrication and Theory**	**3**
1.1	Scientific CCD History	3
1.2	Operation and Performance	22
	1.2.1 Operation	22
	1.2.2 Performance Functions	25
	1.2.3 Performance Specifications	36
1.3	Architecture, Design, Photolitography and Fabrication	37
	1.3.1 Architecture	37
	1.3.2 Design and Photolithography	42
	1.3.3 Processing and Fabrication	51
1.4	CCD Theory	61
	1.4.1 MOS Capacitor	61
	1.4.2 Surface-Channel Potential Well	65
	1.4.3 Buried-Channel Potential Well	70
	References	92
	Further Reading	93
2	**CCD Transfer Curves and Optimization**	**95**
2.1	CCD Transfer Curves	95
	2.1.1 CCD Performance	95
	2.1.2 CCD Camera Performance	96
	2.1.3 CCD Camera Calibration	97
2.2	Photon Transfer	97
	2.2.1 Photon Transfer Derivation	98
	2.2.2 Photon Transfer Curve	101
	2.2.3 Camera Gain Constants	105
	2.2.4 Camera Gain Histogram	108
	2.2.5 Camera Gain Uncertainty	110
	2.2.6 Dynamic Range	113
	2.2.7 Linearity	117
	2.2.8 Flat-Field Signal-to-Noise	120
	2.2.9 Contrast Signal-to-Noise	121

		2.2.10	High-Speed Photon Transfer Generation	125
		2.2.11	Photon Transfer Simulation	130
	2.3	X-Ray Transfer		131
		2.3.1	X-ray Characteristics and Use	131
		2.3.2	Fe^{55}	132
		2.3.3	X-ray Images	139
		2.3.4	X-ray Transfer	141
		2.3.5	X-ray Histograms	143
		2.3.6	Fano-Noise-Limited Performance	147
		2.3.7	Cadmium X Rays	151
	2.4	QE Transfer		151
	2.5	CCD Clock and Bias Optimization		156
		2.5.1	Clock and Bias	156
		2.5.2	Set-point Transfer Curves	160
	References			165
3	**Charge Generation**			**167**
	3.1	Charge Generation		167
	3.2	QE Formulas		170
		3.2.1	Backside Illumination	170
		3.2.2	Frontside Illumination	173
		3.2.3	Miscellaneous QE Losses	174
		3.2.4	Monte Carlo Simulation	177
	3.3	Frontside Illumination		178
		3.3.1	Phosphor Coatings	178
		3.3.2	Virtual Phase	183
		3.3.3	Open Pinned Phase	188
		3.3.4	Thin Gate	190
		3.3.5	Transparent Gate	194
		3.3.6	Poly Hole Gate	194
	3.4	Backside Illumination		195
		3.4.1	Thinning	195
		3.4.2	Quantum Efficiency Hysteresis	196
		3.4.3	Accumulation and QE Pinning	199
		3.4.4	Self-Accumulation	201
		3.4.5	Accumulation Theory	214
		3.4.6	Passive Accumulation	227
		3.4.7	Active accumulation	256
		3.4.8	Antireflection coatings	266
	References			268

4 Charge Collection **273**

- 4.1 Charge Collection 273
- 4.2 Well Capacity 274
  - 4.2.1 Bloomed Full Well 275
  - 4.2.2 Surface Full Well 277
  - 4.2.3 Optimum Full Well 277
  - 4.2.4 Clocking Modes 280
  - 4.2.5 Full Well Transfer 280
  - 4.2.6 Full Well Data 283
  - 4.2.7 Self-induced Emission and Thermionic Emission 289
  - 4.2.8 Clocked Antiblooming 293
  - 4.2.9 Antiblooming Structures 300
  - 4.2.10 High-Speed Erasure 305
  - 4.2.11 Window Clocking 310
  - 4.2.12 Multipinned Phase (MPP) 310
- 4.3 Fixed-pattern Noise 318
  - 4.3.1 Pixel Nonuniformity 318
  - 4.3.2 Flat Fielding 321
  - 4.3.3 Fixed-pattern Sources 325
- 4.4 Charge Diffusion 332
  - 4.4.1 Charge Diffusion 332
  - 4.4.2 Measurement and Modeling Techniques 338
  - 4.4.3 Aliasing and Beating 373
- References 383

5 Charge Transfer **387**

- 5.1 Charge Transfer 387
- 5.2 Transfer Mechanisms 390
  - 5.2.1 Diffusion Drift 393
  - 5.2.2 Self-induced Drift 395
  - 5.2.3 Fringing Field Drift 396
  - 5.2.4 High-Speed Data 400
  - 5.2.5 Clock Propagation 405
  - 5.2.6 Substrate Bounce 408
- 5.3 CTE Measurement Techniques 418
  - 5.3.1 X-ray Transfer 418
  - 5.3.2 Extended Pixel Edge Response (EPER) 423
  - 5.3.3 First Pixel Edge Response 429
  - 5.3.4 Pocket Pumping 430
  - 5.3.5 Charge Injection 433
- 5.4 Traps 433
  - 5.4.1 Design Traps 434
  - 5.4.2 Process traps 439

		5.4.3	Bulk Traps	453

            5.4.3   Bulk Traps 453
            5.4.4   Radiation-induced Traps 466
            5.4.5   Proportional and Fixed Loss 466
            5.4.6   Fat-zero 469
            5.4.7   Notch Channel CCD 472
    5.5   Transfer Power 476
            5.5.1   Charge Motion Power 478
            5.5.2   Potential Power 482
            5.5.3   Reactive Power 482
    References 486

6 **Charge Measurement**   **489**
    6.1   Charge Measurement 489
    6.2   Output Amplifier Characteristics 490
            6.2.1   Operation 490
            6.2.2   Voltage Gain 498
            6.2.3   Loading 499
            6.2.4   Output Impedance 503
            6.2.5   Time Response 504
            6.2.6   Frequency Response 504
            6.2.7   Bias 507
            6.2.8   Sensitivity 509
            6.2.9   Linearity 516
            6.2.10 Temperature Characteristics 519
            6.2.11 Lightly Doped Drain 520
            6.2.12 Amplifier Luminescence 523
            6.2.13 Multistage Amplifiers 527
            6.2.14 Power Consumption 531
    6.3   Output Amplifier Noise 532
            6.3.1   Johnson Noise 532
            6.3.2   Reset Noise 537
            6.3.3   White Noise 541
            6.3.4   Flicker Noise 543
            6.3.5   Shot Noise 553
            6.3.6   Contact and Popcorn Noise 555
            6.3.7   Output Amplifier Noise Equation 555
    6.4   Correlated Double Sampling 556
            6.4.1   Correlated Double Sampling Circuit Elements 557
            6.4.2   Correlated Double Sampling Circuits 558
            6.4.3   Camera Gain Constant 561
            6.4.4   Correlated Double Sampling Transfer Function 563
    6.5   Dual Slope Processor 578
            6.5.1   Dual Slope Circuit Elements 578
            6.5.2   Dual Slope Transfer Function 579

CONTENTS   xi

    6.6 Remnant Signal and Noise 582
          6.6.1 Remnant Signal 582
          6.6.2 Remnant Noise 583
    6.7 Skipper Amplifier 585
          6.7.1 Introduction 585
          6.7.2 Operation 586
          6.7.3 Performance 588
          6.7.4 Design 592
          6.7.5 Signal-to-Noise (Extended Images) 595
          6.7.6 Signal-to-Noise (Point Images) 599
    References 602

7 Noise Sources **605**

    7.1 On-chip Noise Sources 605
          7.1.1 Dark Current 605
          7.1.2 Spurious Charge 649
          7.1.3 Fat-zero 654
          7.1.4 Transfer Noise 656
          7.1.5 Residual Image 657
          7.1.6 Luminescence 665
          7.1.7 Cosmic Rays and Radiation Interference 670
          7.1.8 Excess Charge 674
          7.1.9 Cosmetic Defects 678
          7.1.10 Blem Spillover 679
          7.1.11 Seam Noise 680
    7.2 Off-chip Noise Sources 684
          7.2.1 Light Leak 684
          7.2.2 Preamplifier Noise 684
          7.2.3 ADC Quantizing Noise 686
          7.2.4 Clock-Jitter Noise 697
          7.2.5 Electromagnetic Interference 699
          7.2.6 Grounding 706
          7.2.7 Image Cross Talk 711
          7.2.8 Noise-Reduction Techniques 714
          7.2.9 Noise-Reduction Summary 716
    References 719

8 Damage **721**

    8.1 Radiation Damage 721
          8.1.1 Introduction 721
          8.1.2 Near-Earth Radiation Environment 723
          8.1.3 Radiation Units 726
          8.1.4 Transient Events 736

	8.1.5	Ionization Damage Equivalence	749

           8.1.5   Ionization Damage Equivalence 749
           8.1.6   Ionization Damage 750
           8.1.7   Clock and Bias for Minimum Ionization Damage and
                   Control 759
           8.1.8   Ionization Damage Measurements 761
           8.1.9   Bulk Damage 773
   8.2   Electrical, Thermal and ESD Damage 837
           8.2.1   Electrical Damage 837
           8.2.2   Thermal Damage 838
           8.2.3   Electrostatic Discharge (ESD) Damage 839
   References 841

Appendixes   **847**

Glossary of CCD Terms   **871**

Index   **899**

Preface

This book began from a series of lecture notes for courses held on charge-coupled devices and digital camera systems at UCLA Extension and SPIE meetings in the mid-1980s. These sessions were well attended and met with great enthusiasm by the scientific and commercial imaging communities. The courses, and the enthusiasm, continue today after 15 years.

The courses are intended for scientists, engineers, and hardware managers involved with CCD imaging sensors and camera systems. The material details advances made in pixel count (arrays as large as 10,000 by 10,000 pixels), quantum efficiency (spectral coverage of 1 to 11,000 Å), charge transfer efficiency (99.9999% efficient per pixel transfer), read noise (less than 1 e^- rms), large dynamic range (greater than 10^6), and high-speed operation (diffusion-limited). The CCD technologies used to achieve such high levels of performance are discussed. The courses also review the electronic design of slow-scan and fast-scan CCD imaging camera systems. Applications include near-IR, visible, UV, EUV, x-ray, and particle cameras. The success of these courses prompted us to bring these notes together, along with additional detailed discussions, into a single comprehensive reference manual and tutorial. It is a timely collection, as the CCD has recently celebrated its thirtieth birthday.

This book is written for a wide audience—from the novice to the advanced CCD user. The level of the book's presentation is suitable for students in physics and engineering who have received a standard preparation in modern solid state physics and electronic circuits. Numerous examples throughout the text provide valuable exercises for students and perspective for the professional imaging engineer in terms of modern CCD performance. The text captures 30 years of experimentation with the technology, giving the less-experienced engineer the benefit of the lessons learned during the development of the CCD. Although the book focuses on scientific devices, it is also of interest to other imaging engineers who work with commercial CCDs for broadcasting and photography. Other areas of overlap include CMOS, CID, and photodiode imaging arrays. The book can be used as a reference for participants in educational short courses organized by SPIE and other educational institutions as well.

Scientific Charge-Coupled Devices contains more than 500 figures and illustrations which present experimental and modeling data products taken from many scientific CCDs in operation. The majority of these sensors are found in space imaging cameras that are currently generating new and exciting facts about the universe in which we live. The book provides hundreds of modeling equations used to support the data presented. It has been very important that theory and experiment work hand-in-hand to bring about a sensor that is nearly textbook perfect. This intimate connection also shows us the physical limitations of device performance and what potential advances might be made in the future. Also, the CCD has inspired its own language to describe its unique characteristics and operational features. Therefore, we have included a glossary of CCD terms to which the reader can refer.

The book is organized into eight chapters. Chapter 1 reviews historical aspects of the scientific CCD as taken from the author's perspective and experiences. As with most celebrated technologies, the CCD of today was not born overnight; the technical and political climates that gave rise to the CCD imaging revolution was complex and interesting. Chapter 1 also includes a review of the basics of CCD operation and performance which serve as the book's skeleton—charge generation, charge collection, charge transfer and charge measurement—as well as performance characteristics and related specifications. The chapter will also acquaint the reader with different CCD architectures and how the sensors are designed and fabricated. Presented is basic solid state CCD theory, which is necessary to understand and support the experimental findings presented in subsequent chapters. In particular the potential well, which is responsible for collecting and transferring charge, is studied in detail.

Chapter 2 introduces standard tests and absolute units used to characterize, optimize, and calibrate CCD performance, presented in the form of transfer curves. For example, an important transfer curve called photon transfer produces a multitude of critical performance data products: read noise, full well, dynamic range, linearity, signal-to-noise, etc. The chapter will also take the reader through a CCD clock and bias optimization procedure using transfer curves as a guide. The material in this section is considered advanced but it is critically important in order to achieve the high and reliable performance results that CCD camera users demand.

Chapter 3 discusses the first major operation performed by the CCD: charge generation. It is shown that the charge generation process is capable of covering an enormous wavelength range, from the IR to the hard x ray, covering more than four decades of spectral range (i.e., 1 to 11,000 Å). We will review several loss mechanisms that prevent incoming photons from interacting with the CCD. Discussions are then given to high-performance frontside-illumination CCDs whose design features reduce interaction loss. We then discuss the highest-sensitivity device available to the imaging community, the backside-illuminated CCD. Detailed studies are given on the backside accumulation process required by this technology.

Chapter 4 on charge collection explores the ability of the CCD to form an image after charge is generated. Three performance parameters associated with charge collection efficiency are elaborated: well capacity, pixel nonuniformity, and charge diffusion. The modulation transfer function (MTF) is discussed in considerable detail in quantifying charge diffusion effects and limiting performance.

Chapter 5 deals with charge transfer, the third basic CCD operation. Discussions include a review of the charge transfer efficiency (CTE) requirements for high-performance large-area arrays and physical descriptions responsible for high-speed charge movement. The chapter discusses several CTE measurement techniques in characterizing charge traps that limit CTE performance. Numerous operational, process and design solutions are given to solve CTE problems when encountered. We close the chapter with a short discussion on the power dissipated behind the charge transfer process.

Chapter 6 discusses charge measurement, the last major CCD operation. Discussions here are devoted to the sensor's output amplifier and off-chip signal processing electronics. We will describe the progress that has been achieved in the areas of design, processing and operation to achieve ultralow-noise performance. The chapter discusses other amplifier characteristics such as loading, output impedance, frequency response, sensitivity, linearity, and power consumption. The technique of correlated double sampling (CDS) is reviewed in detail, an important video processing circuit that delivers optimum S/N performance. The last section reviews a floating gate amplifier that allows subelectron noise performance.

Chapter 7 focuses on noise sources other than the CCD's output amplifier. The sources are grouped into two major categories: on-chip and off-chip. This chapter familiarizes the reader with the multitude of known noise sources, which can fundamentally be reduced below that of the noise generated by the output amplifier. Important noise sources include dark current, spurious charge, fat-zero, residual image, luminescence, cosmic rays, cosmetic defects, quantizing noise, clock jitter, electromagnetic interference, grounding noise and other sources.

Chapter 8 is on the subject of CCD damage. The majority of the chapter is devoted to the damage induced by high-energy radiation sources. Numerous design, processing, and operational solutions are given to address and alleviate the problem. Discussions on transient events produced by high-energy particles and photons are also provided. A technique that converts a complicated radiation environment into a single energy that produces the same damage is described. The last subject of this chapter is on electrical, thermal and ESD damage mechanisms.

In the course of the writing this book, many people have assisted me and offered their support. I would like to express my appreciation to the management of NASA's Jet Propulsion Laboratory for providing the rich environment for the research and investigation of CCDs from their conception; without their support this book could have not been written. I also thank Pixel Vision for the R&D environment directed toward high-speed backside-illuminated CCDs discussed in Chapter 5. I wish to express my gratitude to Tom Elliott for being my partner for nearly 15 years in generating the majority of the data products presented in this book. His artistic skills and ingenuity in conjunction with his natural curiosity about CCDs reflects upon the figures presented. Special thanks go to Jeff Pinter and Jim McCarthy, who through countless hours of CCD discussions on the black board brought excitement into the book. I have benefitted significantly from technical discussions with Stewart Collins, Taner Dosuoglu, Morley Blouke, Denis Heidtman, Dick Bredthauer, George Soli, Jim Westphal, Fred Landauer, Jim Gunn, Bob Locke, Mark Wadsworth, Michael Lesser, Bob Hlivak, Lloyd Robinson, Barry Burke, Willard Boyle, Jim Early, Albert Theuwissen, George Smith,

Gene Weckler, Rudy Dyck, Ken Klaasen, Charles Chandler, Kurt Liewer, Dave Burrows, Gorden Garmire, Paul Vu, Dick Savoye, Rusty Winzenread, Andrew Holland, David Lumb, Fred Harris, Russ Schaefer, Taher Daud, Jerry Hynecke, Dave Norris, Robert Groulx, Raymond Frost, Arsham Dingzian, Gorden Hopkinson, Bev Oke, Terrence Lomheim, Gary Sims, Walter Kosonocky, Nelson Saks, Cheryl Dale, Paul Marshall, Fred Pool, Loren Acton, Bob Lockart, Dave Swenson, John Geary, Jim Beletic, Alice Reinheimer, Larry Hovland, Sandra Faber, Bedabrata Pain, John Flemming, and Harry Marsh.

Special thanks goes to Shouleh Nikzad in writing the section of MBE delta doping and Steve Holland on his biased ohmic contact on high-resistivity substrate, both included in Chapter 3. I am further indebted to Steve Holland, Peter Hughes, Gloria Putnam, Sharon Streams, and Rick Hermann for technical and copy editing of the entire manuscript. Finally, I wish especially to thank my wife Linda Janesick, daughter Amanda, sister Barb, and Mom and Dad for their assistance in many ways.

Scientific
Charge-Coupled Devices

CHAPTER 1

HISTORY, OPERATION, PERFORMANCE, DESIGN, FABRICATION AND THEORY

1.1 SCIENTIFIC CCD HISTORY

The charge-coupled device (CCD) was invented October 19, 1969, by Willard S. Boyle and George E. Smith at Bell Telephone Laboratories. Figure 1.1(a) shows their original laboratory notebook entry describing the idea. Jack Morton, vice-president of Bell Labs Electronics Technology, was an important motivator behind the CCD. Jack was a strong proponent of the magnetic bubble memory and wondered if an analogous device could be made with a semiconductor. He encouraged Boyle and Smith to look into a potential connection. At the same time, the Picturephone was being developed using diode arrays in silicon.[1] Charge storage on individual diodes had been a problem that prompted some new research. The CCD was born with these two existing technologies in mind. As the story goes, Boyle and Smith brainstormed on the blackboard for approximately a half hour, producing the CCD structure. The first notebook drawing of the device is shown in Fig. 1.1(b), which presents a basic three-phase configuration. The original timing diagram presented also accurately describes today's CCD and is used throughout this book.

A few weeks later, a three-phase device was designed, fabricated and tested: a simple row of nine 100-μm metal plates separated by 3-μm spacings. The first gate electrode was used to inject charge into the second plate. The ninth plate was used to detect charge. Plates 2–8 were clocked to demonstrate the transfer process. The experiment was a success.

The first public announcement, which lasted only five minutes, was made at the New York IEEE convention held March 1970. The ideas were subsequently reported by Boyle and Smith in the April 1970 issue of the *Bell System Technical Journal*.[2] This paper was important because it introduced potential problems behind the invention. For example, limitations in storage time caused by dark current generation were discussed. Also, they found three important factors that limited charge transfer efficiency: surface interface state trapping, thermal diffusion and fringing fields. The bubble memory analogy was introduced by shifting digital information in the form of charge and implemented with a serial shift register with p-n inputs and outputs. Also very important was the idea that the CCD could be used for imaging using linear and array areas.

The first experimental results were reported in a paper by Amelio, Tompsett, and Smith.[3] The first technical paper was presented by Smith at the Device Research Conference in Seattle, June 1970.[4] This paper presented an 8-bit p-channel CCD with integrated input and output diodes used to inject and read charge. The

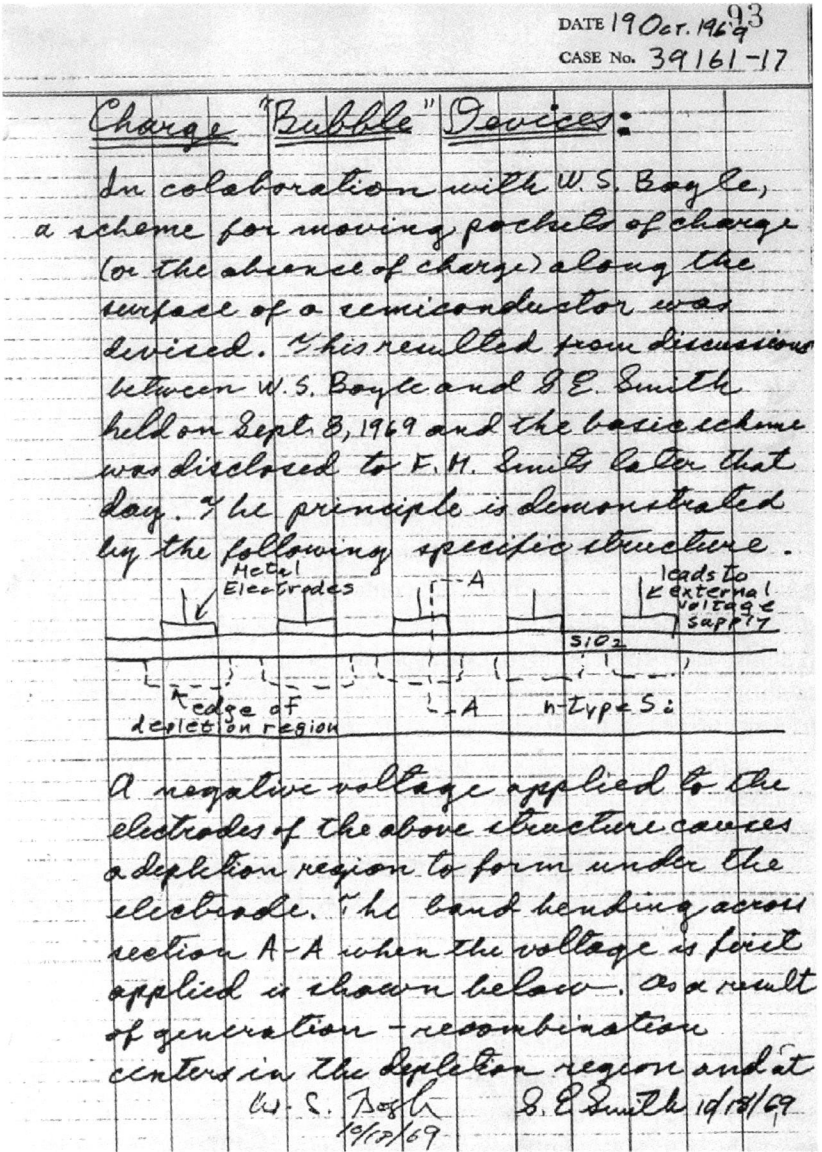

Figure 1.1(a) Boyle and Smith's original laboratory notebook entry describing the CCD concept.

signal carrier for a p-channel CCD is the hole. Today most CCDs are n-channel devices that transfer electrons. Charge-transfer performance was reported to be 99.9% at 150-kHz readout. The 8-bit device was also used to demonstrate imaging capability for the first time (i.e., first light!).

It should be mentioned that Teer and Sangster of Philips Research Laboratories were actively working on an imaging transfer device at the same time. Their device

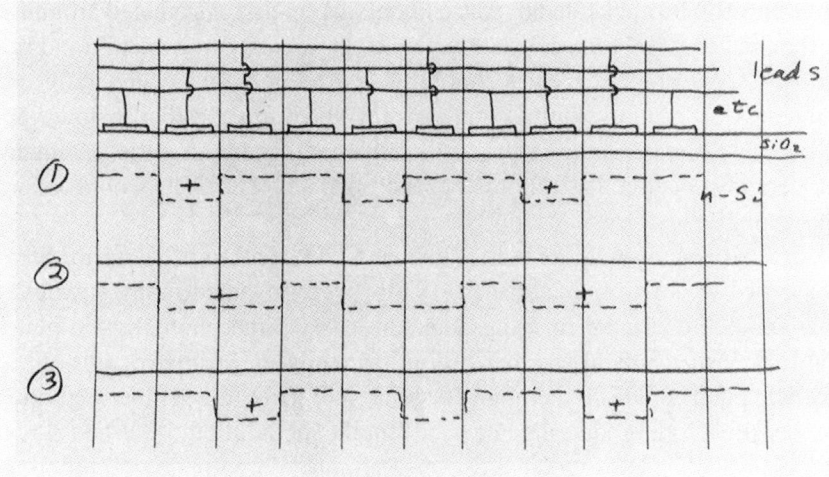

Figure 1.1(b) Boyle's and Smith's original schematic and timing diagram for a three-phase CCD.

was called the Bucket Brigade device (BBD). In comparing the devices, the BBD is essentially a circuit. The BBD was constructed by wiring discrete metal oxide semiconductor (MOS) transistors and capacitors together. The main innovation was to integrate the devices on a single substrate, making the drain of one transistor the source of the next. The CCD, on the other hand, is a functional device because signal charge is passed as minority carriers from one storage site to the next through a series of MOS capacitors. The first publication on the BB device was in 1970.[5-7] Bell Labs was unaware of the Philips BBD until that time. For a variety of reasons the BBD was never seriously considered for use as an imaging device.

The CCD idea swept through Bell Labs, and many others made contributions in the weeks that followed. For example, Dawon Kahng developed the idea of a built-in barrier under one side of the metal plate to make a two-phase device required to avoid metal bus cross-overs.[8] However, Bell's interest in imaging devices died with the Picturephone, since the company was not allowed by the government to sell devices on the open market at that time.

Although the CCD was originally conceived as a memory device, it became immediately clear to some that the CCD had potential uses that ranged far wider than simple memory applications. For example, astronomers, with their fundamental interest in the detection of photons from far-away places, could potentially benefit from a scientific CCD. An entire book could be written on the technical, political, and historical aspects of the CCD. Discussions here are limited to the author's own experience and to the three key factors that significantly influenced the development of the scientific CCD as it exists today.

Photographic film was the CCD's main competitor as an imaging detector, and therefore film acted as the first important driver on CCD history. Curiously, film was not immediately accepted by astronomers when first introduced around 1850. Over the first 30 years film did poorly, exhibiting a host of problems, and science return was sparse for the new technology. Astronomers were generally content with the measurement techniques on hand (i.e., the human eyeball). However, film quality slowly matured and by the turn of the century became the dominant tool for modern observational astronomy and astrophysics. The CCD would experience similar evolutionary trials.

Film has been a tough contender for the CCD because its performance, cost, and ease of use are well established. If film were invented today it would be a serious competitor for all imaging detectors. Very large photographic plates are available for astronomical use that can permanently map huge regions of the sky (equivalent pixel counts greater than 10^8 pixels). Besides the visible spectrum, film is also sensitive to a broad range of wavelengths including the UV and x-ray. For example, in the hands of medical doctors, similarly large photographic films can record a chest x-ray image—for example, at unity (1:1) magnification (i.e., life size)—and reveal internal features a fraction of a millimeter in size. The CCD had to match these features in order to compete.

The second historical factor was the commercial need for an electronic "solid state" imaging detector to replace tube-type detectors. The emphasis was on realizing the CCD's advantages in size, weight, low power consumption, ultralow noise, linearity, dynamic range, photometric accuracy, broad spectral response, geometric stability, reliability and durability, while attempting to match tube characteristics in format, frame rate, cosmetic quality and cost. Today we see the outcome of this effort: nearly all commercial video cameras are CCD-based (e.g., camcorders).

The third historical factor involved space imaging applications that required an electronic solid state detector. Around the same time the CCD was invented, NASA was planning two major projects that required such a detector.[9] First, the Large Space Telescope (LST, later called Hubble Space Telescope) project, initially proposed in 1965, potentially required several detectors for its focal plane. Second, unmanned space probes for solar system exploration designed by engineers and scientists at the Jet Propulsion Laboratory (JPL) required small and sensitive detectors. Past cameras, such as those used for *Surveyor*, *Ranger*, *Mariner*, *Viking*, and *Voyager*, were based on the vidicon tube technology. These tubes were highly successful at obtaining images of the planets and the moon. However, future missions—especially to the outer planets—required greater sensitivity, stability and long-term reliability than what tubes could deliver.

At first, LST experienced great difficulties in finding a satisfactory candidate sensor. In fact, more than 10 years of searching and waiting took place before a suitable detector became available, i.e., the CCD. Initially, two options became the subject of high debate: photographic film or an image tube type of detector. Film seemed to be the best choice because no development was required, it had a well-established history and it was readily available. Unfortunately, film exhibited a major drawback for the LST project. While in orbit film would slowly "fog," induced by high-energy radiation from energetic protons and electrons. Therefore, images taken on the film would need to be retrieved on a regular basis by astronauts, a requirement that LST could not afford. For this reason the LST team concluded that film was impractical to use.

An imaging tube would eliminate the need for manned maintenance since this technology, unlike film, is not vulnerable to radiation effects. Therefore, LST team members focused on imagers of this type. In 1972 several tube sensors were proposed for LST use. The SEC vidicon tube became the prime candidate. However, this tube and others like it exhibited two primary problems. First and fundamental was the tube's inability to retain an image over a long exposure time before readout commenced. Second, the lifetime of tube-type detectors was questionable for a mission as long as LST. The photocathode showed degradation with time, thereby affecting photometric accuracy. LST mission lifetime was anticipated to be at least 15 years.

In 1972 the CCD was introduced to the LST team by a member of the Bell Laboratories. The meeting immediately focused on three problems associated with the new sensor. First, the detector required significant cooling to eliminate thermal dark charge. Cooling to a temperature of 200 K was anticipated for integration periods greater than one hour. This new requirement was unfamiliar to the LST team since film and tubes could function at ambient spacecraft temperature. Second, charge transfer efficiency (CTE), which measures an inherent function performed by the CCD, was very poor in 1972. The problem would limit format sizes to arrays less than 100×100 pixels. Third, the CCD proposed was not responsive to UV light, a limitation not associated with vidicon tubes or film.

Some LST team members believed that these problems were insurmountable, and therefore were not initially attracted to the CCD. Nevertheless, these deficiencies did not discourage some CCD supporters since the new sensor exhibited many excellent performance characteristics not easy to ignore. For example, the sensor could potentially stare at an object for several hours as long as it was cooled and shielded from radiation sources. The CCD was extremely sensitive in the visible and near IR, a factor of five times higher than what the vidicon tube could deliver and nearly a factor of 100 times more sensitive than film. The CCD also exhibited excellent linearity response. That is, its output signal was exactly proportional to photon input. This characteristic was important for calibrating raw data generated by LST. Tubes and film did not exhibit this characteristic but rather showed a nonlinear response to light input. Also, CCDs did not exhibit reciprocity failure as exhibited by film. Here film becomes less sensitive as it is exposed, showing a nonlinear response. The signal output of the CCD is exactly proportional to exposure time. The CCD showed a remarkably large dynamic range, greater than 3000 at the time. It was also geometrically stable, small, rugged and consumed very little power (less than 10 mW running at LST rates). Lastly, the video generated by the CCD could be easily amplified and digitized, which is crucial to reducing scientific data such as that anticipated from LST.

While LST sensor activities were progressing, Jet Propulsion Laboratory in 1972 began an advanced program to develop a solid state detector for planetary imaging missions. The CCD appeared very attractive for that purpose. In the very beginning it was recognized that pixel counts required for such missions would be significant. Vidicon tubes, for example, developed and utilized in past JPL planetary imaging missions were based on 1024 × 1024 formats (e.g., the *Viking* and *Voyager* vidicon tubes sent to Mars and the outer planets, respectively). Therefore, early CCD development would need to focus on obtaining pixel counts this large. Initial reports from Bell Labs showed that arrays equivalent to this size would be impossible because of the charge transfer efficiency obstacle. This problem would be later solved by buried-channel CCD technology.

CCD development at JPL initially involved three U.S. manufacturers: RCA Corporation, Fairchild Semiconductor, and Texas Instruments (TI). Fairchild was a pioneer in developing and manufacturing buried-channel devices, due in part to James Early, a strong advocate of the technology invented by Boyle and Smith (refer to W. Boyle and G. Smith, U.S. Patent 3,792,322, Buried Channel Charge Coupled Devices, issued Feb. 12, 1974).[10–12] Fairchild's CCDs became commercially available in 1974 and were the first sensors to be tested at JPL. Initial tests showed a read noise floor of approximately $30\,e^-$ rms and a CTE of 0.9999 per pixel transfer. Boyle and Smith's "surface channel" CCDs exhibited CTEs of only 98% and a much higher noise floor. The Fairchild devices represented a remarkable achievement considering that the CCD was invented only a few years before.

The Fairchild 1 × 500 linear array and 100 × 100-pixel area arrays were the first devices available to the scientific community and sparked the imaging revolution that was about to occur. These early devices were a tremendous boost in getting the scientific CCD off the ground. For example, the 100 × 100 Fairchild CCD shown in Fig. 1.2(a) was incorporated into a simple camera system. In 1974 this camera was placed behind the author's 8-in. amateur telescope and pointed toward the moon, producing what may have been the first CCD astronomical image [Fig. 1.2(b)]. The 10,000 pixel image shown was displayed on a Heath Kit oscilloscope with the CCD video modulating its z-axis. The horizontal and vertical sweeps of the scope were generated by TTL counters with outputs weighted with discrete resistor ladder networks. Bit weighting in the sweeps is obvious.

Unfortunately, the architecture philosophy that Fairchild followed (i.e., interline transfer) was not optimized for scientific performance, the primary problem being its inability to achieve high quantum efficiency. With interline transfer devices, a portion of the imaging area is taken up with CCD shift registers that are made to be light insensitive. Meanwhile, RCA, also very active with CCD development, took a different approach: a full-frame, backside-illuminated CCD which in theory would achieve the highest sensitivity possible. The RCA CCD group, led by Dick Savoye, was developing the largest CCD at the time, a 512(V) × 320(H) frame transfer device intended for commercial TV applications. Unfortunately, early RCA devices were based on surface channel technology. JPL tests performed on these units showed very poor CTE performance (< 0.995) and high read noise ($100\,e^-$ rms) compared with the Fairchild CCD. Later, the RCA design was modified for buried-channel operation, but it maintained a surface-channel output am-

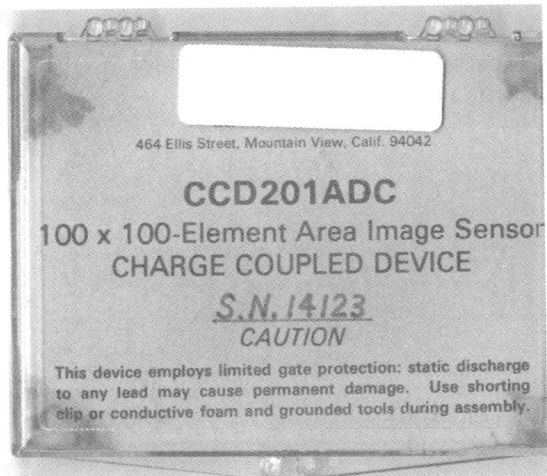

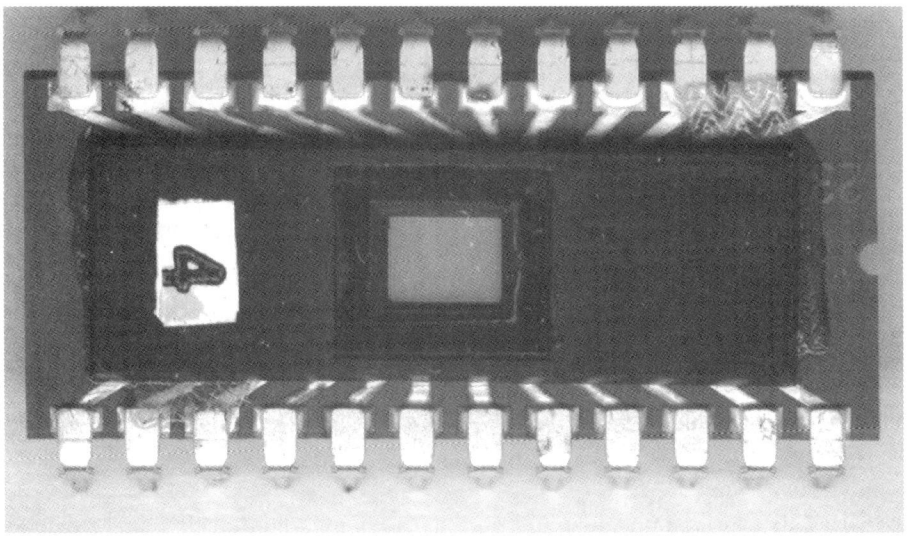

Figure 1.2(a) The 100 × 100 Fairchild CCD used to take early astronomical images.

plifier for many years. Although backside illumination was attractive, the high read noise forced JPL to pursue other avenues for low noise.

Both RCA and Fairchild focused on commercial pixel formats no larger than 512 × 320. This limited spatial resolution was unsatisfactory for most future scientific requirements. It became clear from these early studies that a custom R&D effort was necessary to combine the best attributes of all CCD technologies known at the time. JPL, under NASA contract, teamed with Texas Instruments (TI) to develop a scientific CCD sensor based on full-frame, backside illumination, buried

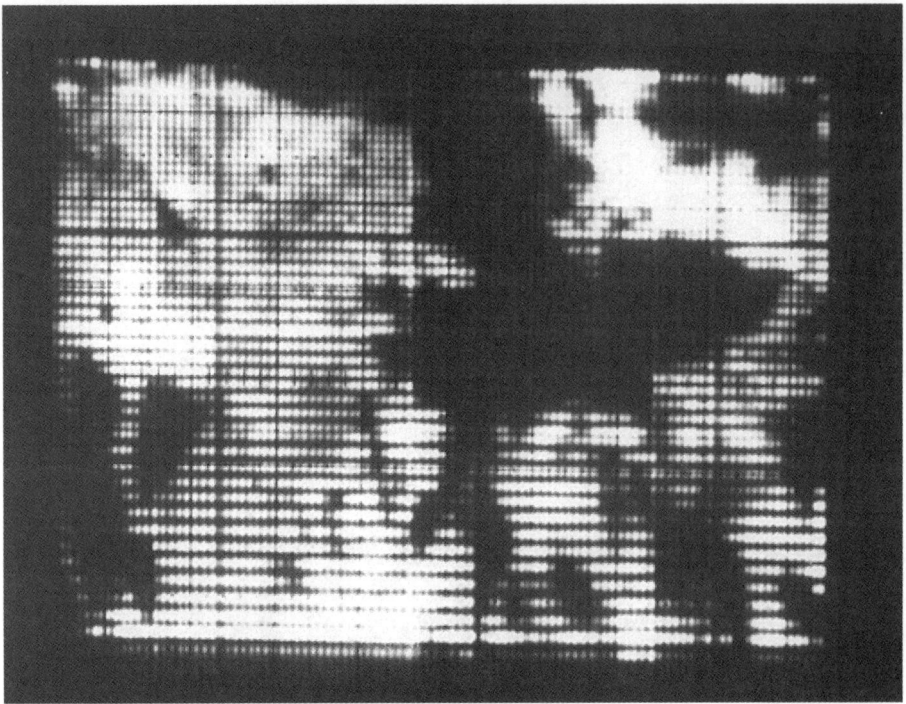

Figure 1.2(b) A CCD image of moon craters using the Fairchild 100 (V) × 100(H) CCD.

channel technology throughout, with pixel counts equivalent to or greater than the vidicon tube. Collaboration between JPL and TI progressed for more than a decade and was responsible for many significant advancements that led to the scientific CCD we have today.

In 1975 it was announced at an important LST meeting that JPL had successfully operated a 100 × 160 backside-illuminated TI CCD. The JPL team also disclosed at the same meeting that they were developing a flight 400 × 400-pixel version for the *Mariner* Jupiter-Uranus (MJU) mission. It became evident at the meeting that if the MJU CCD was sufficiently cooled, it could be used by LST too.

The 400 × 400 TI CCD became a reality the next year and characterization tests were immediately performed. The sensor exhibited exceptional performance compared with vidicon tubes in use at the time. However, process yield was a primary concern, for only a few good units had been shipped to JPL. The low yield continued and implied that sensors with pixel counts as great as 1024 × 1024 would be very difficult to build. The TI CCDs employed aluminum gates to define pixel sites, a technology very sensitive to gate shorts as discussed below. Additional CCD lots were manufactured, but it became clear that a different gate technology was required. TI then proposed using polysilicon gate technology, which is much less prone to shorts. This was the same gate technology originally used by Fairchild and RCA.

At another LST meeting in 1976 it was announced that a buried channel, backside-illuminated, polysilicon gate, 800 × 800 TI device was being developed for the Jupiter Orbiter Polar (JOP) mission (later to be called the Galileo mission). It became apparent that the CCD was being developed at a very fast pace and the sensor should be considered much more seriously by LST as a candidate sensor.

To promote further CCD interest and backing by the scientific community, the JPL team built a Traveling CCD Camera System, the first of its kind, to be used at major astronomical observatories worldwide. At the time most astronomers and engineers were not familiar with the new chip and were basically content with photographic film and tubes currently in use. JPL management recognized that scientists should become familiar with the capabilities and unique features of the CCD to help promote and support future NASA imaging flight projects. Expeditions to various observatories with a 400 × 400 backside-illuminated CCD camera system paid off because the tiny chip performed beyond anyone's expectations. New scientific discoveries were usually made each time the camera system visited a new site.

In 1976 JPL (Fred Landauer, Larry Hovland, and author) and the University of Arizona (Brad Smith) went to Mount Lemmon's 154-cm telescope with a CCD camera system to capture the first high-resolution professional images of various astronomical objects. Figure 1.3(a) shows the new CCD camera system mounted to the telescope along with two racks of support equipment on the floor of the observatory. Figure 1.3(b) shows the first CCD image taken of planet Uranus at a wavelength of 8900 Å. At that time, the planet's pole was oriented directly at earth. The dark region shows the presence of methane gas which is conspicuous at 8900 Å. The contrast was striking when the image was first taken because the CCD's sensitivity in the NIR was vastly superior to vidicon tube and photographic film type of detectors popular at this time. Figure 1.3(c) is a column trace through the center of the image showing the contrast level. The image was used to accurately determine the diameter of the planet, data required to help direct the *Voyager* spacecraft during its flyby many years later. Figure 1.3(d) shows two Polaroid CCD images of planet Saturn, taken the same night at 8300 Å and 8900 Å, respectively. Note that the ball of the planet nearly disappears at 8900 Å, again showing the presence of methane.

Later, an upgraded camera system made visits to Lick Observatory, Kitt Peak National Observatory, and Palomar Observatory, and then made a long and precarious journey to Cerro Tololo Inter-American Observatory in Chile. Amazing images were taken at each stop. Figure 1.4 shows the Traveling CCD Camera System, which housed a TI 400 × 400 × 15-μm-pixel backside-illuminated polysilicon gate CCD.

Following these visits, the demand for the CCD became intense among astronomers. Simply put, those who had access to a CCD chip would likely get a much greater scientific return from a given amount of telescope observation time,

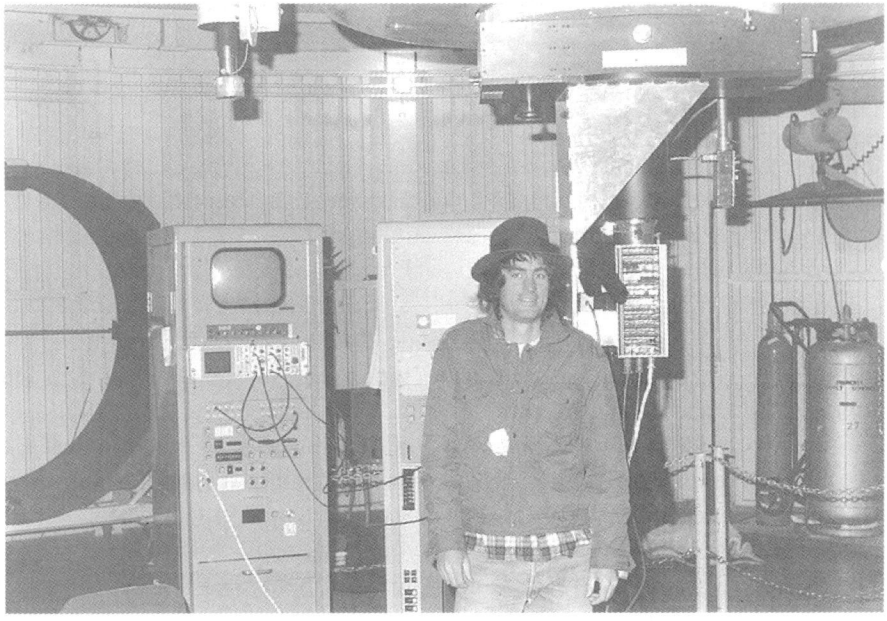

Figure 1.3(a) Author and CCD camera system used at Mt. Lemmon to take early astronomical CCD images.

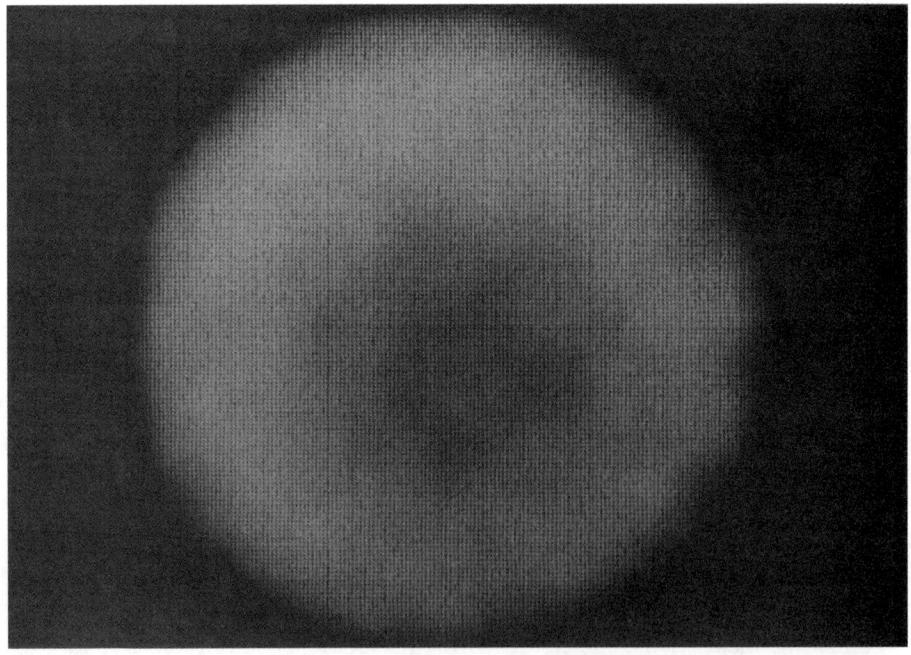

Figure 1.3(b) First CCD image of planet Uranus.

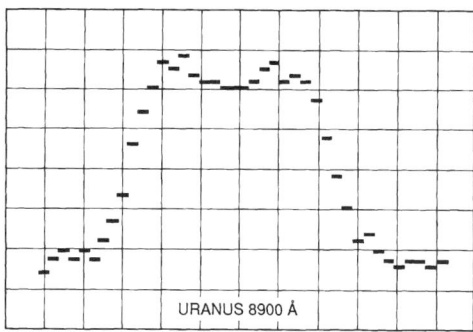

Figure 1.3(c) Column trace taken through planet Uranus showing a dark pole at 8900 Å.

Figure 1.3(d) Planet Saturn taken at 8300 Å and 8900 Å.

and with a CCD, observations could be made that were almost impossible without one. A few spare 400 × 400 TI CCDs were delivered "under the table" to the astronomical community. New CCD camera systems were fabricated at Palomar, Kitt Peak and Lick Observatories almost overnight. The first professional CCD camera system based on a 100 × 160 backside-illuminated TI CCD was designed at the California Institute of Technology (Caltech). This work was pioneered by Bev Oke with collaboration at JPL. In 1978, Dr. Oke published the first astronomical spectra of quasars taken by a CCD.[13]

Astronomy would never be the same. The CCD was beginning to revolutionize astronomical instrumentation, much as film did nearly 100 years earlier. Within a few years the CCD became the visible-light sensor of choice at all major observatories, with the possible exceptions of very large view Schmidt telescopes like the 48-in. Mt. Palomar that use large (14 × 14 in.) photographic plates. (As a side note, Kodak announced that they would probably discontinue the production of photographic plates for astronomy in early 2000.) Also, for the first time, small amateur telescopes equipped with CCDs could participate and produce new science that at one time could be generated only by larger telescopes. With a sensitivity 100 times greater than film, it was clear that more data could be produced in a shorter period of time using a CCD. The new sensor immediately set new records in seeing

Figure 1.4 A 400 × 400 "Traveling CCD Camera" that made visits to major observatories.

the most distant and dimmest objects in the universe, objects that were previously invisible to film- and tube-type detectors.

Later in 1976, NASA released an Announcement of Opportunity for the construction of an LST camera system. Proposals from the scientific and engineering community were to be submitted by March 1977. Three proposals were forthcoming. The first proposal, received from Princeton, was primarily based on an SEC vidicon and a small CCD to obtain red response, which together would cover a spectral range of 0.1 to 1 μm. Goddard proposed an intensified CCD primarily to achieve EUV sensitivity. The third submission came from the Caltech/JPL team. They proposed eight 800 × 800 × 15-μm TI CCDs arranged in two optical mosaics of 2 × 2 CCDs each, and a special phosphor coating applied directly to the CCDs to obtain UV response down to 1216 Å.

After the proposals were reviewed by NASA, the JPL/Caltech proposal was selected because it was based completely on CCDs. The system was to be called the Wide Field/Planetary Camera (WF/PC) instrument. As an engineering proof-of-concept for the WF/PC, Caltech astronomers James Gunn and James Westphal designed and constructed a camera they called "4-shooter," for ground-based imaging with the 200-in. (5-m) Hale Telescope.[14] Analogous to the wide-field channel of WF/PC, the 4-shooter used a four-sided pyramid in the focal plane to divide the field-of-view into four adjoining quadrants, sending the light from each quadrant to separate 800 × 800 TI backside-illuminated CCDs. Figure 1.5(a) shows a 1600 × 1600 image of the Crab Nebula taken with the camera system. Several years later, this prototype camera system became a reality when WF/PC I was constructed (Fig. 1.5(b)).

This book presents performance test data collected for several CCDs that have been used in spaceborne imaging applications. These sensors will be referred to by name when test data is presented. For example, the first Hubble Space Telescope Wide Field/Planetary I (WF/PC I) camera system used eight TI backside-illuminated 800 × 800 × 15-μm-pixel three-phase CCDs. Figure 1.6(a) shows a WF/PC I view of Saturn taken on August 26, 1990. Image convolution was applied because of the spherical aberration problem with Hubble's main mirror at the time.

A second generation WF/PC II camera system with corrective optics incorporated four Lockheed frontside-illuminated 800 × 800 × 15-μm-pixel three-phase CCDs. This camera systems has been highly successful and continues to return thousands of gorgeous pictures of the planets, Milky Way, and the universe beyond. Figure 1.6(b) shows the famous Hubble Deep Field image, one of the better photographs taken by the human race. The image, as described by the Hubble science team was assembled from many separate exposures (342 frames total were taken) for 10 consecutive days between December 18 and 28, 1995. Besides the classical spiral- and elliptical-shaped galaxies, the image shows a bewildering variety of other galaxy shapes. The never-before-seen dimmest galaxies are nearly 30th magnitude. Representing a narrow "keyhole" view stretching all the way to the visible horizon of the universe, the image covers a speck of sky a tiny fraction of the diameter of the full moon. Although the field is a very small sample of sky, it

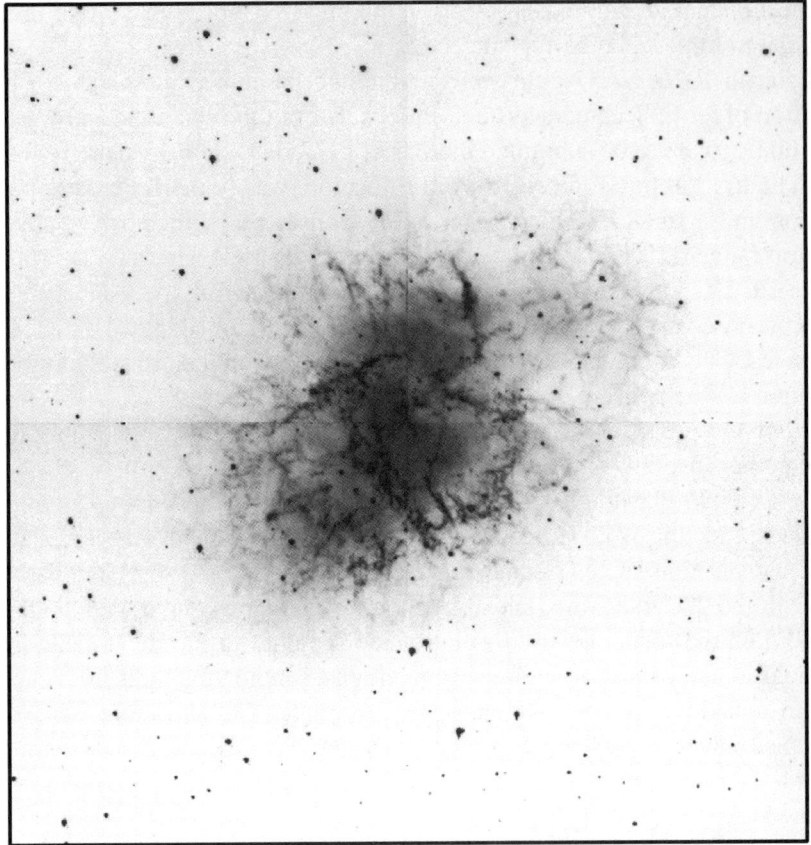

Figure 1.5(a) Crab Nebula taken with the Caltech 4-shooter CCD camera system.

is considered representative of the typical distribution of galaxies in space because the universe, statistically speaking, looks the same in all directions. Some of the galaxies may have formed less that one billion years after the Big Bang.

The Solid State Imaging (SSI) camera aboard the spacecraft *Galileo* in orbit around planet Jupiter uses a single Texas Instruments $800 \times 800 \times 15$-µm-pixel, virtual-phase CCD. Figure 1.6(c) is a full picture image of Asteroid 243 and its newly discovered moon taken on August 28, 1993. Ida, as it is called, is about 35 miles long and was about 7,000 miles away when seen en route to planet Jupiter.

The Solar X-ray Telescope (SXT) Camera System aboard the Japanese *Solar-A* spacecraft utilizes a single Texas Instruments $1024 \times 1024 \times 18$-µm-pixel VPCCD. To date (July 4, 2000) SXT has taken 961,633 full-frame images and 4,843,640 partial images of the Sun, making a grand total of 5,805,273 images since its launch in 1991. Figure 1.6(d) was taken by the SXT on the Yohkoh Satellite. It is a gray-scale image of the x-ray emission from the Sun on May 20, 2000, and is a composite image composed of a short/long-exposure pair of images to increase the dynamic range. The bright regions in this image have temperatures near 2 MK. The resolu-

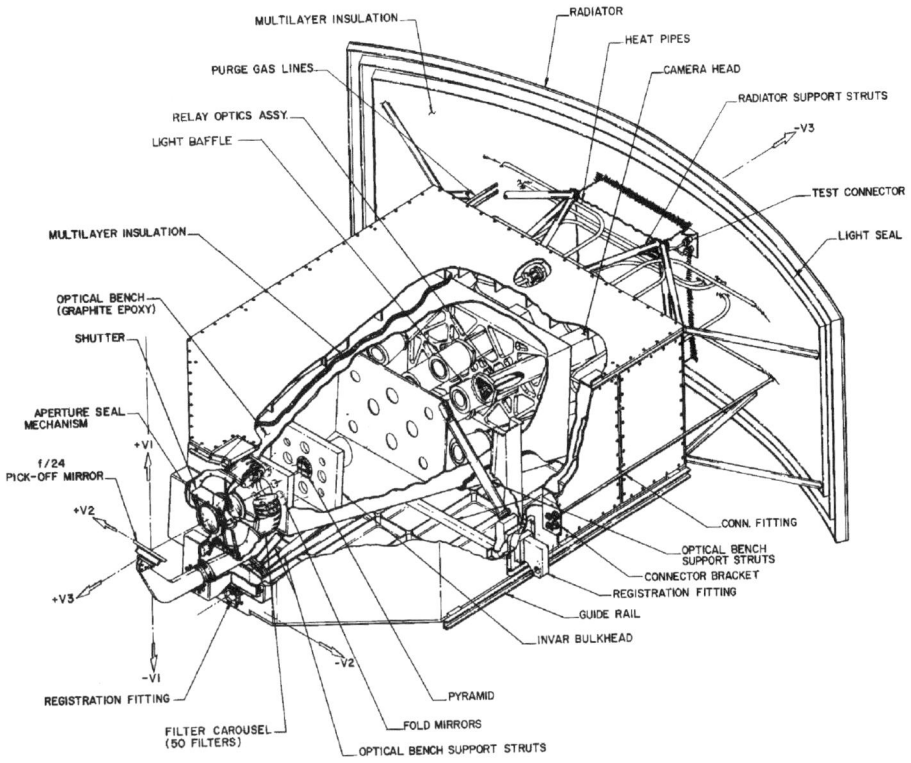

Figure 1.5(b) WF/PC I.

Figure 1.6(a) Hubble WF/PC I image of Saturn (NASA).

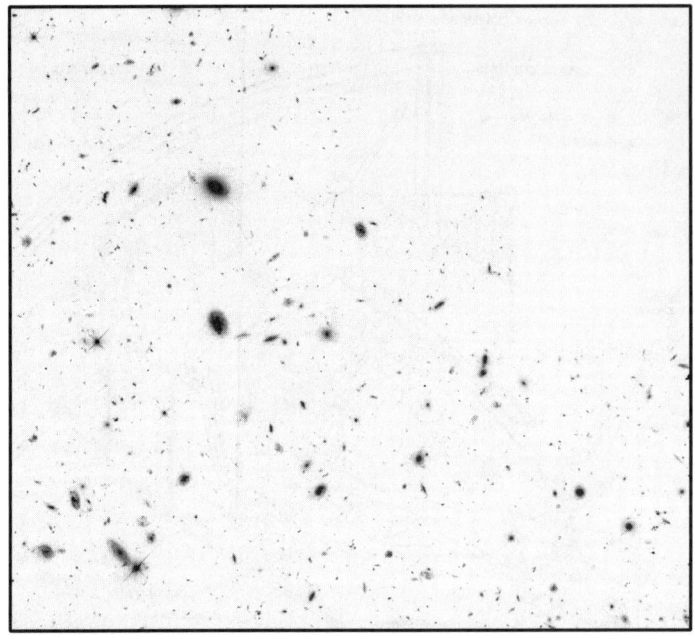

Figure 1.6(b) Hubble WF/PC II Deep Field (NASA[15]).

Figure 1.6(c) Galileo SSI image of Asteroid Ida (NASA/JPL).

tion of the original image pair was 10 arcsec or 5 arcsec/pixel. Processing included dark-current subtraction, vignette correction, and stray light removal. This image was binned to 2×2 resolution on the CCD, or 512×512 pixels.

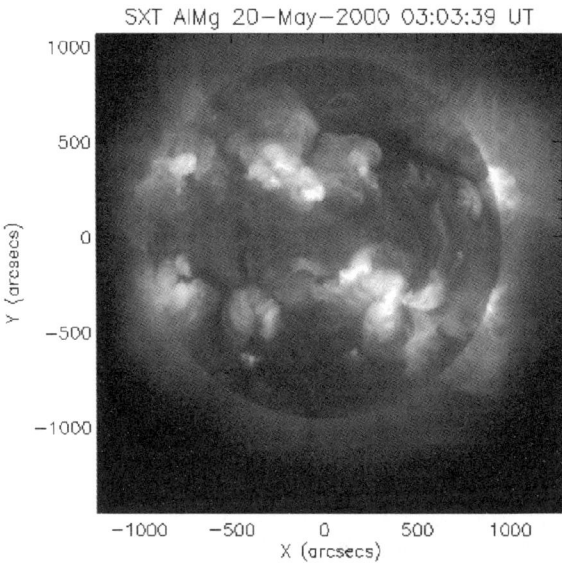

Figure 1.6(d) This solar x-ray image is from the Yohkoh mission of ISAS, Japan. The x-ray telescope was prepared by the Lockheed Palo Alto Research Laboratory, the National Astronomical Observatory of Japan, and the University of Tokyo, with the support of NASA and ISAS.

Four Lockheed 1024 × 1024 frontside-illuminated three-phase CCDs are headed toward planet Saturn and scheduled to arrive in July 2004. This mission also carries a Lockheed 512 × 512 × 23-µm-pixel CCD to probe Saturn's largest moon, Titan (NASA's Cassini/Huygens Probe Mission Descent Imaging Spectro-Radiometer (DISR)). Figure 1.6(e) shows a narrow-angle ISS test image of the Moon taken August 18, 1999, on its flyby past Earth in route to Saturn. The 80-msec exposure was through a spectral filter centered at 0.33 µm; the filter bandpass was 85 Å wide. The spatial scale of the image is ∼ 2.3 km/pixel.

The Multi Imaging Spectral Radiometer (MISR) camera system is an earth resource imaging mission. It uses several quad 1504 × 1 × 18-µm-pixel CCDs. Each of the nine MISR cameras takes images in four spectral bands (443, 555, 670, and 865 nm). Figure 1.6(f) shows an image of the South Atlantic Anomaly (SAA), a region around Earth of unusually high proton levels. The MISR cameras, designed to detect visible light, are also sensitive to energetic protons at high altitudes. With the camera cover closed, background levels of protons stand out. This map was created by processing MISR dark current images taken between February 3 and February 16, 2000. Each picture element is a square measuring 0.25 deg in latitude and longitude, and each contains thousands of pixels from the raw MISR imagery.

The Mars Orbital Camera (MOC) system on board the Mars Global Surveyor (MGS) used two Loral 3456 × 1 × 7-µm-pixel linear CCDs for the wide-angle cameras, and two Loral 2048 × 1 × 13-µm-pixel linear CCDs for a wide-angle system. Figure 1.6(g) is a view of Phobos, a moon of Mars, as viewed on August 19,

Figure 1.6(e) Cassini image of Earth's moon (Cassini Imaging Team/University of Arizona/JPL/NASA).

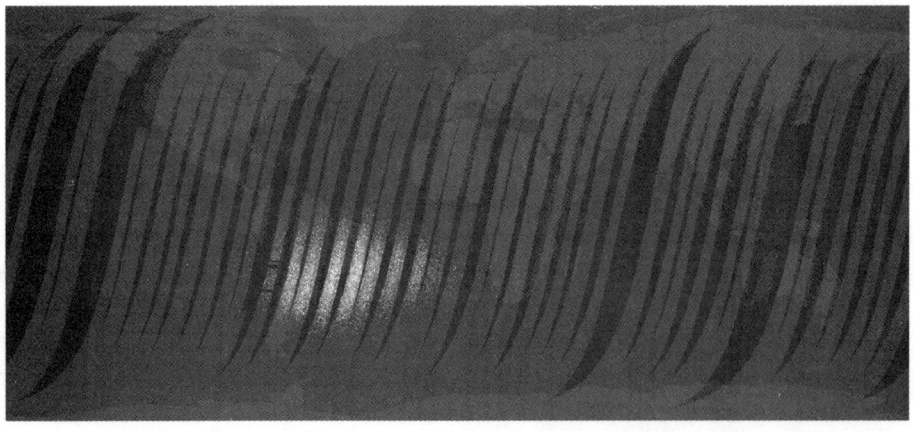

Figure 1.6(f) MISR image of the South Atlantic Anomaly (NASA).

Figure 1.6(g) MOC image of Mars satellite *Phobos* (NASA/JPL/Malin space science systems).

Figure 1.6(h) Mars *Pathfinder* image of Twin Peaks (NASA/JPL).

1998, 10 AM PDT. The MGS spacecraft was approximately 1,080 km (671 miles) from Phobos at closest approach. This image, about 8.2 km (5.1 miles) wide by 12 km (7.5 miles) tall, shows the full field-of-view. The image as shown here has a scale of 12 m (40 ft) per pixel.

The Mars *Pathfinder* uses the Cassini/Huygens Probe Descent Imaging Spectra Radiometer (DISR) Lockheed CCD mentioned above. Figure 1.6(h) shows a high-resolution image of Mars looking in the direction of Twin Peaks.

Future or new missions include *Deep Space I*, which will visit two asteroids (Braile and Borrely) and a comet, using a Reticon $1024 \times 1024 \times 9$-µm-pixel three-phase CCD. Another exciting mission recently launched by the Shuttle in July 1999 is the Advanced X-ray Astrophysics Facility (AXAF). Now called Chandra, the large x-ray telescope includes the Advanced CCD Imaging Spectrometer (ACIS) instrument. The camera is generating high-resolution x-ray images of active galaxies, supernovas, quasars, etc.; and it simultaneously measures the energy of incident x rays. Several high-performance $1024 \times 1026 \times 24$-µm-pixel CCDs fabricated by Lincoln Laboratory are used in Chandra's focal planes. The upcoming Advanced Camera System for Hubble uses two mosaiced SITe $4049 \times 4096 \times 15$-µm backside-illluminated three-phase CCDs. The sensitivity of the new camera system is approximately three times better than WF/PC II in terms of QE improvement.[15,16]

1.2 OPERATION AND PERFORMANCE

1.2.1 OPERATION

As previously mentioned, CCDs were initially conceived to operate as a memory device, specifically as an electronic analog of the magnetic bubble device. In order to function as memory, there must be a physical quantity that represents a bit of information, a means of recognizing the presence or absence of the bit (reading) and a means of creating and destroying the information (writing and erasing). In the CCD, a bit of information is represented by a packet of charges (electrons are assumed in this book, although holes would also work). These charges are stored in the depletion region of a metal-oxide-semiconductor (MOS) capacitor, an important CCD structure that is described below. Charges are moved about in the CCD circuit by placing the MOS capacitors very close to one another and manipulating the voltages on the gates of the capacitors so as to allow the charge to spill from one capacitor to the next. Thus the name "charge-coupled" device. A charge detection amplifier detects the presence of the charge packet, providing an output voltage that can be processed. Charge packets can be created by injecting charge that is from an input diode next to a CCD gate or introduced optically. Like the magnetic bubble device, the CCD is a serial device where charge packets are read one at a time.

Figure 1.7(a) Bucket analogy used to describe CCD operation.

An elegant analogy devised by Jerome Kristian and Morley Blouke, which uses the concept of a "bucket brigade" to describe CCD operation for imaging applications, is presented in Fig. 1.7(a). Their simple model is explained as follows. Determining the brightness distribution in a CCD image can be likened to measuring the rainfall at different points in a field with an array of buckets. Once the rain has stopped, the buckets in each row are moved down vertically across the field on conveyor belts. As the buckets in each column reach the end of the conveyor, they are emptied into another bucket system on a horizontal belt that carries it to a metering station where its contents are measured.

There are many ways to arrange MOS capacitors to form a CCD imager array. Conceptually, the simplest CCD is a three-phase device, the arrangement that Boyle and Smith used for their first CCD. In the three-phase device, the gates are arranged in parallel with every third gate connected to the same clock driver. The basic cell in the CCD, which corresponds to one pixel, consists of a triplet of these gates, each separately connected to phase-1, -2 and -3 clocks making up a pixel register. Figure 1.7(b) shows the layout of a three-phase CCD area array. The image-forming section of the CCD is covered with closely spaced vertical registers, also called parallel registers by some CCD manufacturers. During integration of charge, two vertical phases are biased high (e.g., phase-1 and -2), which results in a high field region that has a higher electrostatic potential relative to the lower biased neighboring gate (phase-3).The dark phases shown represent this bias condition, and it is under these gates where signal electrons collect in a pixel when the CCD is exposed to light. These phases are called collecting phases. The phase

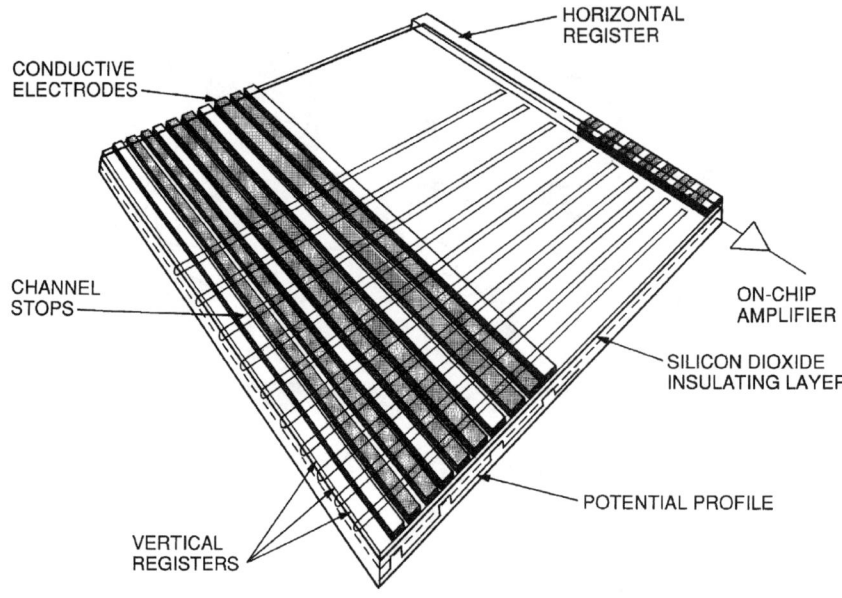

Figure 1.7(b) Primary components that make up a three-phase CCD.

that is biased at lower potential is called the barrier phase since it separates charge being collected in the pixels on either side of this phase.

Vertical registers are separated by potential barriers called channel stops which prevent the spread of the signal charge from one column into another. Similar to our bucket brigade analogy in Fig. 1.7(a), a CCD image is read out by a succession of vertical shifts through the vertical registers. For each vertical shift a line of pixels is transferred into a horizontal register (also called a serial register) which is oriented at right angles to the vertical registers. Before the next line is shifted the charge in the horizontal register is transferred to the output amplifier. The amplifier converts the charge contained in each pixel to a voltage. The device is serially read out line-by-line, pixel-by-pixel, representing the scene of photons incident on the device.

Figure 1.8 schematically shows how charge is transferred in a three-phase CCD register. At time t_1, the potential for phase-1 is held high, forming collecting wells under these gates. Charge generated in phase-1 will collect under phase-1. Electrons generated in the barrier phases (i.e., phases-2 and -3) will rapidly diffuse into phase-1 because its potential is greater. Therefore, the pixel as shown here is totally photosensitive (i.e., 100% fill factor). At t_2 charge transfer commences which involves the creation of potential wells and barriers by applying the appropriate voltages to the gates at appropriately sequenced times. During t_2 phase-2 clock goes high, forming the same well under phase-2 that exists under phase-1. The signal charge collected now divides between the two wells. Note that a potential barrier between pixels exists under phase-3 since it remains at a low bias potential. At time t_3 phase-1 is returned to ground, which forces charge to transfer to phase-2.

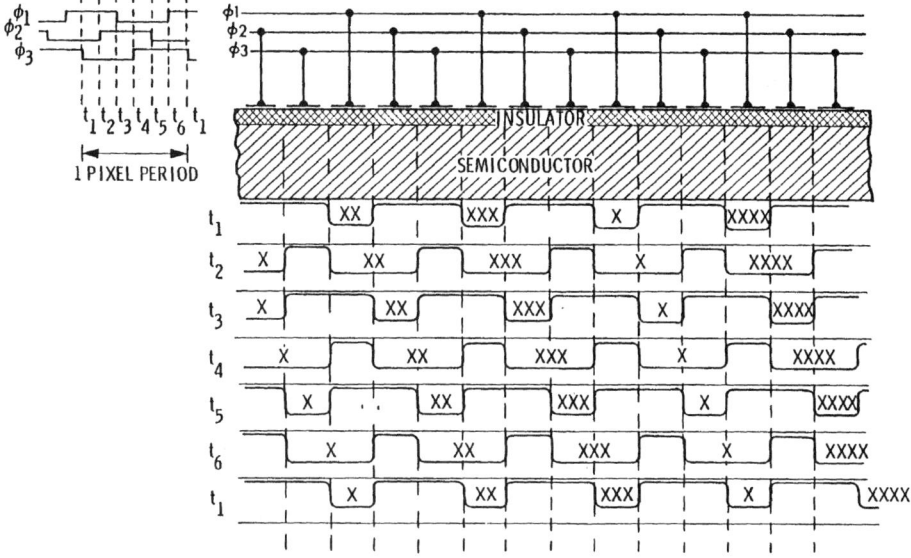

Figure 1.8 Three-phase potential well timing diagram showing signal charge being transferred from left to right.

At t_4, in the same way, charge is transferred from phase-2 to phase-3. This process continues until charge has been moved one entire pixel in one 3-phase clock cycle from left to right in the illustration. Note also that by interchanging any two of the clocks causes the charge in the 3-phase device to move in the opposite direction, an important characteristic that is beneficial in many CCD applications. For example, if charge is normally clocked in a 1, 2, 3 fashion, reversing phases-1 and -2 will cause the charge to move backwards (reversing phases-2 and -3 or phases-1 and -3 will also result in a backwards motion).

1.2.2 PERFORMANCE FUNCTIONS

Technically speaking, the CCD must perform four primary tasks in generating an image. These performance functions are called (1) charge generation, (2) charge collection, (3) charge transfer and (4) charge measurement. Each performance characteristic and its limit will be fully examined in Chapters 3, 4, 5 and 6, respectively. An introduction to each function is given below.

1.2.2.1 Charge Generation

The first operation—charge generation—is the ability of a CCD to intercept an incoming photon and generate an electric charge. Charge generation efficiency (CGE) is described by a function called quantum efficiency (QE), which is the fraction of incident photons that produces a useful charge in the silicon chip. These photons are known as interacting photons. An ideal CCD would have 100% QE at all wavelengths, but nature is rarely this kind. QE is highly dependent on wavelength, falling off steeply at the near-infrared (NIR) and blue wavelengths. Charge generation takes place in the silicon body of the CCD. When a photon interacts with this material, it creates one (or for higher-energy photons, several) free electron(s) by a physical process known as the photoelectric effect.

Interaction only occurs if the photon has enough energy to do so. This is because silicon exhibits an energy gap of approximately 1.14 eV, which is bounded by the valence and the conduction energy bands. Photon energy $E(\text{eV})$ can be converted to a wavelength using the equation

$$\lambda = \frac{12{,}390}{E(\text{eV})}, \tag{1.1}$$

where λ is the wavelength (Å).

Example 1.1

What is the wavelength of a 1.14-eV photon?
Solution:
From Eq. (1.1),

$$\lambda = 12{,}390/1.14,$$

$$\lambda = 10{,}868 \text{ Å}.$$

Most photons with energies < 1.14 eV pass through the chip; i.e., silicon is transparent in the far infrared. Some of these photons do in fact interact but do not generate signal carriers, a process called free-carrier absorption. Interacting photons with sufficient energy will excite an electron into the conduction band, creating an electron-hole (e-h) pair. The e-h pair created is free to move and diffuse in the silicon lattice structure. For high-quality silicon, the lifetime of the electron is several milliseconds before recombining.

A photon with energy of 1.1 to 3.1 eV (11,263 Å–4,000 Å) will generate a single e-h pair. This spectral range covers the near infrared (NIR) and visible spectrum (4,000–7,000 Å). Energies > 3.1 eV will produce multiple e-h pairs when the energetic conduction band electron collides with other valence electrons.[17] The average number of electrons generated for a photon with energy $E(\text{eV}) > 10$ eV is

$$\eta_i = \frac{E(\text{eV})}{E_{e\text{-}h}}, \tag{1.2}$$

where η_i is the ideal quantum yield (electrons/interacting photon) and E_{e-h} is the energy required to generate an electron-hole pair, which for silicon is 3.65 eV/e$^-$ at room temperature.

Example 1.2

Find the average number of e-h pairs generated for 10-eV (Lyman-alpha) and 5.9-keV x-ray photons.
Solution:
From Eq. (1.2),

$$\eta_i = (10)/3.65,$$

$$\eta_i = 2.74 \, e^- \text{ at } 10 \, eV,$$

$$\eta_i = 1620 \, e^- \text{ at } 5.9 \, keV.$$

The response of the CCD can potentially cover an enormous wavelength range, 1–10,000 Å (1.1 eV to 10 keV). This spectral range includes the near infrared, visible, ultraviolet (UV), extreme ultraviolet (EUV) and soft x-ray regions. CCDs fabricated with other optical materials can extend this range even further.[18] For example, small germanium CCDs were fabricated at Loral. Germanium exhibits a bandgap of 0.66 eV, which extends the IR response out to about 1.6 µm. In addition, the density of germanium is greater than silicon, pushing the x-ray response out to approximately 20 keV. The process employed in fabricating a germanium CCD is very similar to silicon processing, with the exception of the gate insulator. Unfortunately, germanium oxide is not a good insulator (e.g., it is soluble in water). Therefore, deposited SiO_2 was employed as the gate insulator for the germanium CCD. There is a severe lattice mismatch between germanium and SiO_2, which produces a high density of interface states. This leads to very high dark current and large flat-band shift problems. Nevertheless, images were taken when the devices were cooled to very low operating temperatures (< 200 K).

Charge generation in the visible region (4000–7000 Å) typically peaks near 6500 Å. At wavelengths shorter than 6500 Å, the photons begin to be absorbed by the CCD's gate structure that overlies the silicon. The chip's QE drops to only a few percent at 4000 Å. The absorption problem becomes particularly pronounced in the ultraviolet at wavelengths below 3000 Å. For example, at 2500 Å the penetration depth is only 30 Å in silicon, which is a few atomic layers. Most CCDs are inert lumps of metal in this range and regain sensitivity only at wavelengths of about 10 Å, making them useful detectors at x-ray wavelengths.

CCD manufacturers have adopted a variety of techniques to enhance blue and ultraviolet performance. One technique used for the Hubble WF/PC II CCDs was to deposit a phosphor coating directly on the gates. A photon having a wavelength shorter than 4600 Å is absorbed by the coating, which re-emits another photon at 5200 Å where the CCD is more sensitive. This down-converting wavelength technique avoids most of the gate absorption problem. However, only 50% of the 5200 Å photons are re-emitted in a direction that carries them into the silicon where they are detected, while the other 50% escape the CCD. The short-wavelength quantum efficiency of a phosphor-coated CCD is therefore approximately half that of the same CCD at 5200 Å.

Phosphor coatings are a cheap way to improve ultraviolet response. A better but much more expensive technique is known as backside illumination, whereby the image is focused on the backside of the silicon in order to bypass the gate structure. This illumination scheme results in the highest QE possible for the CCD. However, to be effective, the thick substrate layer on which the CCD is built must be completely thinned away, because only the electrons generated near the gate structure can be collected efficiently. The membrane thickness for most backside-illuminated CCDs is extremely thin, approximately 12 μm thick. Thinning is difficult, and many otherwise good chips are often destroyed in the process.

Reflection is another QE loss problem. At 4000 Å, silicon reflects about half the light falling on it, which is why most CCDs shine like mirrors. Fortunately, reflection losses can be nearly eliminated by using the same antireflection coating technique found on good camera lenses. A properly coated backside-illuminated chip looks dark, not shiny, and can achieve a peak QE over 90% in the visible range.

1.2.2.2 Charge Collection

The second step, charge collection, is the ability of the CCD to accurately reproduce an image from the electrons generated. Three parameters are included to describe this process: (1) the area or number of pixels contained on the chip, (2) the number of signal electrons that a pixel can hold (i.e., charge capacity of a pixel), and (3) the ability of the "target pixel" to efficiently collect electrons when they are generated. The last parameter is most critical because it defines the spatial resolution performance for the CCD. Spatial resolution is primarily associated with pixel size and charge diffusion characteristics, as explained in Chapter 4.

For most scientific applications, the number of pixels making up an array should be as large as possible. While early CCDs were very small, the number of pixels has grown to overwhelming proportions. For example, Fig. 1.9 shows a 4-in. wafer that holds a 2000 × 2000 × 27-μm-pixel CCD designed by Morley Blouke and his group formerly at Tektronix and currently at SITe. This was the first CCD chip to occupy the entire silicon wafer.

Today, 4096 × 4096 CCDs are fabricated at several CCD manufacturers. A 4096 × 4096 CCD can compete in number of resolution elements with 35-mm format photographic film. Such CCDs produce digital images with an enormous amount of information. For example, two football fields set side by side, including the side and goal zones, can resolve objects to about 1 inch per pixel. Looking

HISTORY, OPERATION, PERFORMANCE, DESIGN, FABRICATION AND THEORY

Figure 1.9 A wafer scale size 2 k × 2 k × 27-μm-pixel CCD imager built by Tektronix in the early eighties.

skyward, the 4096 × 4096 sensor can cover 68 arc min of the sky with 1 arc sec resolution (the moon extends about 32 arc min).

Figure 1.10(a) shows a 4000 × 4000 × 15-μm-pixel CCD built by Richard Bredthauer and his team at Ford Aerospace. (Ford was subsequently bought out by Loral and later was procured by Lockheed Martin.) Figure 1.10(b) is a dollar bill image taken with the CCD. Only a 600 × 1000-pixel portion is presented in the region of the "eye" found on the backside of the bill. The number of pixels displayed is 3.56% of the total 16,777,216 pixels. Figure 1.10(c) is a magnified view showing 300 × 400 pixels representing 0.715% of the imaging area. Figure 1.10(d) shows a 112 × 150 image which is less than 0.1% of the pixels. At this magnification individual pixels can be seen.

CCDs have been fabricated even larger than 4000 × 4000 units. For example, in 1992, JPL initiated a project to fabricate a 9000 × 7000 × 12-μm-pixel CCD for astronomical applications. The CCD, built by Philips, occupied a 6-in. wafer and contained over 63,000,000 pixels. A 9000 × 9000 × 8.75-μm-pixel CCD was fabricated by Bredthauer in 1996 on a 5-in. wafer containing a whopping 81,000,000 pixels. And a 10,000 × 10,000 × 10-μm-pixel is on the drawing board at the time of this writing.

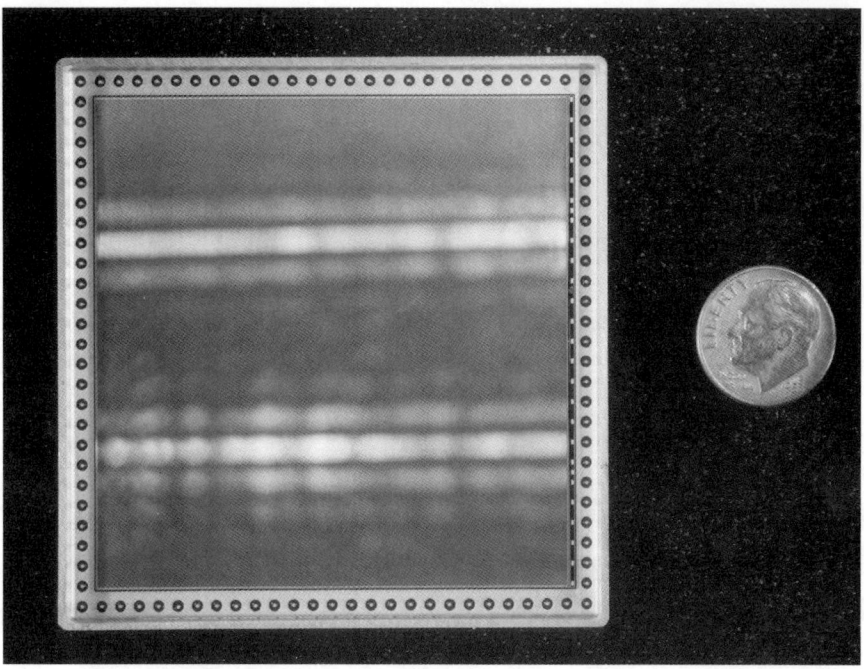

Figure 1.10(a) A 4000 × 4000 × 15-μm-pixel CCD built in the early nineties at Loral.

Figure 1.10(b) A 600 × 1000 dollar bill image taken with a 4000 × 4000 × 15-μm CCD.

Are there limits to making even larger detectors? From the manufacturer's standpoint, pixel count is ultimately limited by the production yield. Not every chip fabricated works, and the bigger the chip, the lower the yield and the higher the cost. Large CCDs fail primarily because of electrical shorts in the gates,

Figure 1.10(c) A 300 × 400 magnified view of Fig. 1.10(b).

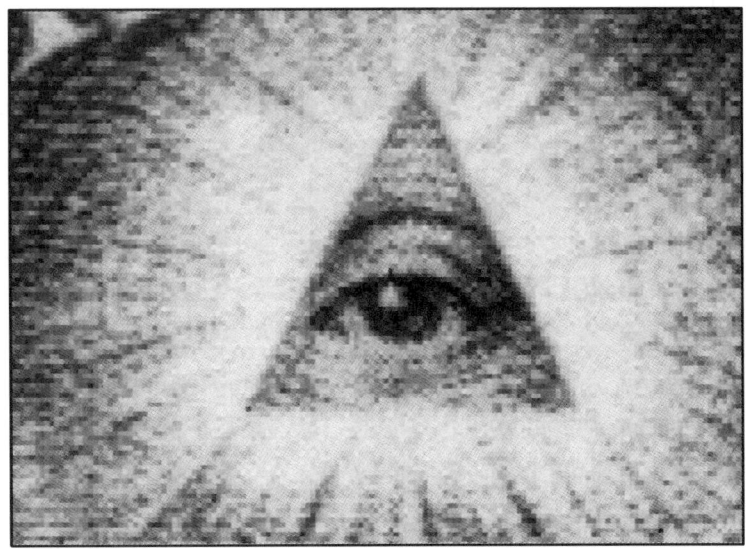

Figure 1.10(d) A 112 × 150 magnified view of Fig. 1.10(c).

a subject discussed below. So, the answer to this question is "no," assuming resources are available to build such monster devices. For example, a small-pixel 20,000 × 20,000 CCD is hypothetically possible today. However, interest and vast amounts of money would be required to build a device this size.

The cost of ultralarge CCDs is considerable because usually only one or two devices can be built per silicon wafer. A lot run typically holds 24 wafers and possibly only a few good wafers will pass the shorts test. The foundry fabrication cost for a 6-in. silicon run is approximately $100,000, depending on the manufacturer and the number of options specified for the CCD (i.e., the number of reticles or masks required for lithography). There are also costs associated with design, packaging and testing, which are expensive efforts. A custom CCD might run near $500,000 before research and development costs are concluded.

Individual off-the-shelf CCDs have price tags of more than $50,000 for ultralarge backside-illuminated arrays. For example, a class-A fully characterized 2048 × 2048 thinned chip can cost more than $70,000. Engineering grades for the same device might go for $10,000. This works out to be 0.60 and 0.23 cents per pixel, respectively.

Nevertheless, production yields and costs have improved greatly over the years. About 150 production lots were required to make 800 × 800 arrays for the Hubble WF/PC I and *Galileo* missions. In fact, the *Galileo* sensor was a one-of-a-kind CCD. Of the hundreds of chips fabricated, the sensor that now orbits Jupiter was the only one fabricated that met specifications. The sensor was found hiding in TI's lab, and was in fact an experimental CCD not intended for flight use. In contrast to all of this, only two lots were required to deliver new flight CCDs to Hubble's WF/PC II and satisfy the Cassini project.

Very large CCDs also present many difficulties for the user, primarily in data storage. For example, a 10,000 × 10,000 sensor would produce 200 million bytes of information per image, assuming 16-bit encoding. The sensor format would match that of 409 320 × 512-pixel commercial format CCDs. One night's images at an observatory from such a detector could fill a typical computer's hard drive. More important, however, is that processing time rises exponentially with pixel count. So, to use large CCDs effectively, very fast and expensive computer systems are needed. This requirement has limited the demand for large CCDs.

Also, from the camera designer's standpoint, the problem with large arrays is the amount of time required to read millions of pixels. For example, the WF/PC cameras require several minutes to read out its four 800-pixel-square arrays at 50,000 pixels/sec. At the same rate, a 10,000 × 10,000 array would require 33 minutes! Even if this read time were acceptable, long readouts degrade data because of cosmic-ray artifacts and thermally generated dark current that accumulate during the read process. To shorten readout time, large CCDs usually employ several amplifiers working in parallel. For example, the Philips CCD mentioned above uses four amplifiers. Other astronomical CCDs have employed 64 amplifiers for adaptive optics applications. Nevertheless, a price is to be paid for multiple amplifiers. Parallel readout increases the complexity of a camera's electronics and makes image processing and analysis more difficult because each amplifier exhibits its own gain, offset, linearity and noise levels.

A technique used to beat the yield problem and obtain ultralarge CCD focal planes has been to mosaic them.[19-21] For example, four 4096 × 4096 CCDs have been designed, fabricated, diced, butted and mosaicked on two edges to obtain a 8192 × 8192 format. Even larger mosaics have been configured to butt three free edges. Packaging costs are high using this approach, because very tight tolerances are required in butting the chips to keep the seam between devices to a minimum. A few pixels separation can be achieved.

Well capacity is a measure of how many electrons a pixel can hold, usually 50,000 to a 1,000,000. Again, bigger is better, since the greater the full-well capacity, the greater a CCD's dynamic range and signal-to-noise. Dynamic range is the ratio between the brightest and faintest objects that can be simultaneously discriminated. Large pixels can hold more charge but require more silicon, which in turn degrades the production and increases the price. The smallest pixel that can be manufactured is only a couple of microns in extent, set by lithography design and process rules. Also, charge capacity is an important consideration for very small pixels because full well decreases by pixel area. The advantage of small pixels is higher yield with fewer shorted devices. Production yield is roughly inversely proportional to the area used on the wafer by the CCD. Therefore, yield will be approximately four times greater for a 12-μm pixel than a 24-μm pixel, assuming the same pixel count.

Pixel size is limited due to the need to maintain high CTE with reasonable speed, as will be discussed in Chapter 5. A good compromise between CTE and well capacity for today's astronomical CCD has been the three-phase 15-μm pixel, the size used by WF/PC I and WF/PC II. Today's 15-μm pixel can hold $500,000\,e^-$, yielding a dynamic range of 250,000 assuming a readout noise of $2\,e^-$. This means that an exposure could simultaneously record two objects with a brightness difference of 13 1/2 magnitudes, the ratio between Vega (one of the brightest stars in the heavens) and the planet Pluto. Additional phases are incorporated to make super pixels. For example, a 64-phase CCD has been fabricated to obtain pixels 640 μm square. This CCD produced an enormous well capacity ($10^9\,e^-$) while simultaneously delivering good CTE characteristics at high pixel rates. A total of 128 clock drivers were required to drive the horizontal and vertical registers.

Ideally, electrons are collected by the target pixel without spreading into neighboring pixels. Diffusion into other pixels makes an image look out of focus and lowers the sensor's modulation transfer function (MTF). Confining charge to a single pixel is also an issue for sensitivity, since it is easier to distinguish a star from the background when images are well defined. Charge leakage between pixels, also called "pixel crosstalk," is caused by "field-free" material found deep in the silicon. Electrons generated farther from the gate feel a weaker electric field and thus have a higher probability of diffusing into neighboring pixels. For frontside-illuminated CCDs this problem is most apparent when imaging in the near IR because long-wavelength photons penetrate deeper into the CCD than do blue photons. CCD manufacturers reduce this effect through the proper choice of silicon material. Backside-illuminated devices, which by design have light falling on the silicon surface farthest from the gates, also experience leakage if not thinned properly. Through careful design and processing, engineers have reduced crosstalk in backside-illuminated CCDs to nearly undetectable levels.

1.2.2.3 Charge Transfer

The third operation, charge transfer, is accomplished by manipulating the voltage on a parallel sequence of gates that form a CCD register, or conveyor belt in our bucket brigade analogy. For quantitative scientific measurements such as those encountered in astronomy, it is important to lose as little of the charge as possible during the transfer process. For example, the photometric accuracy of the Hubble WF/PC II 800-pixel-square array is specified to better than 1% (< 0.01 magnitude) . This means that more than 99% of the charge collected in a pixel must survive the transfer process and reach the output amplifier. At the corner of the chip farthest from the amplifier, this means surviving 1600 pixel transfers. This places enormous demands on the CTE process—no member of the bucket brigade can afford to spill a single drop!

The first CCDs fabricated at Bell Laboratories exhibited about 99% efficiency per pixel transfer. Although this may sound good, it was in fact terrible, because the loss is compounded as the number of transfers increases. If 1% of the charge is left behind in each transfer, 63% would be lost after only 100 transfers. Fortunately, great improvements during the last 25 years have raised charge transfer efficiencies (CTE) in modern CCDs to better than 99.9999%. Only one in every million electrons is lost during a typical transfer!

Understanding how charge is lost during transfers has been very important in pushing CTE performance to the limit. Charge is transferred due to fringing fields between phases, thermal diffusion, and self-induced drift, subjects discussed in Chapter 5. The relative importance of each effect is dependent primarily on how much charge is being transferred. Most scientific CCDs move the charge slowly from pixel to pixel, so the time interval needed for diffusion and self-induced drift is not a CTE limitation in slow-scan operation—the changing fields induced by the gates have time to sweep electrons cleanly from one pixel to the next. Some scientific applications, such as Star Tracker cameras, operate at rates of 30 frames per second or greater. CCDs employed in adaptive optics and biological applications operate at thousands of frames per second. For very high speed operation, charge transfer efficiency can suffer because charge diffusion and self-induced drift require a finite time to move electrons completely from phase to phase.

While most pixels in a modern CCD transfer charge smoothly, many chips have small "traps" that prevent achieving 100% efficiency. These traps extract a fixed quantity of charge as a signal packet is clocked past the problem site. The trapped charge is then slowly released into trailing pixels as "deferred charge," so called since it appears late at the output amplifier. Traps can be created during the design and manufacture of chips, and to some degree, are also found in starting silicon material before the CCD is even made. A single trap could have disastrous effects if it occurred in a pixel near the output amplifier. Nevertheless, traps are avoidable, and through careful design and fabrication, a large chip today will contain just a few pixels with traps that siphon only one hundred electrons from a charge packet.

Energetic particles such as protons, electrons, neutrons, heavy ions, and gamma rays damage the silicon crystal lattice and create traps too. The problem is most pronounced for CCDs that fly on space missions. For example, the 800-pixel-square array currently circling Jupiter on the *Galileo* spacecraft has been severely damaged by energetic protons and neutrons. Engineers had anticipated this problem, however, and compensated for it in the camera design. The spectacular images returned by *Galileo* are evidence that they did their job well. For ground-based astronomy, radiation problems prevail but concern us little. Although constant bombardment by cosmic rays and natural radiation causes damage and CTE loss, even the highest radiation levels would take a 100 years or more to cause a noticeable effect. Since the CCD is only 30 years old, no one has complained.

Bulk traps are due to impurities or lattice defects within the silicon. These typically trap a single electron during charge transfer and set limits on the ultimate CTE performance. They can be reduced by building chips with the best silicon material available. Current limits place CTE at 99.99999%.

1.2.2.4 Charge Measurement

The last major operation to occur during CCD imaging is the detection and measurement of the charge collected in each pixel. This is accomplished by dumping the charge onto a small capacitor connected to an output MOSFET amplifier. This point on the CCD is called the "sense node" or "output diode." The MOSFET amplifier is the only active element that requires power. The CCD array of pixels itself is passive in that it is made up of MOS capacitors. The output amplifier generates a voltage for each pixel proportional to the signal charge transferred.

CCD designers have worked diligently to make the output capacitor extremely small. The tinier it is, the higher the amplifier's gain and the greater the output signal. For example, a 50 fF capacitance produces a gain of 3.2 µV per electron. Thus, if 1000 electrons are transferred to the sense node, 3.2 mV appear at the transistor's output, which is a respectable signal for digital conversion. Sensitivities as high as 25 µV per electron have been achieved. However, such sensitivities limit the dynamic range of the amplifier and linearity performance. A popular sensitivity for the scientific CCD has been about 2–4 µV per electron.

Besides making the sensitivity as high as possible, engineers make sure there is minimal noise generated by the output MOSFET. This is very important because reducing noise by a factor of two doubles the CCD's sensitivity. The effect is the same as if a telescope's aperture were increased by 41%, assuming signal detection is truly read-noise-limited. Building a quieter camera is easier and cheaper than building a bigger telescope. Although there are many unwanted sources of noise involved with CCD imaging, all can be reduced to zero. For example, there is thermal noise associated with the dark current arising from electrons spontaneously generated within the silicon. Dark current can be controlled and eliminated by cooling the chip. The WF/PC II CCDs are cooled to $-90°C$ for this purpose and generate less than $3.6\,e^-$ of dark current per hour per pixel.

The only fundamental source of noise that cannot be completely eliminated comes from the output amplifier. It is generated by the random fluctuations in the current that flows through the transistor. The noise produced is similar to the "snow" sometimes seen on weak television pictures. Early CCDs had about 30 e$^-$ rms of amplifier noise associated with the signal measured for each pixel. Today's high-performance CCDs have slightly less than 2 e$^-$ of noise, a spectacular improvement. The improvement did not come easily, however. Literally millions of dollars have been spent optimizing the CCD amplifier. Design efforts striving for even lower noise amplifiers continue today in pushing the noise level below 1 e$^-$.

Noise levels this low are achieved not just by careful chip design but also by the design of electronics used to process the CCD signal. The raw output noise that accompanies a CCD signal is typically more than 100 e$^-$. This noise can be reduced almost a hundredfold by additional off-chip amplification and digital filtering.

1.2.3 PERFORMANCE SPECIFICATIONS

The introductory discussions above acquaint the reader with CCD performance characteristics, which have been grouped into the four basic CCD functions (i.e., charge generation, collection, transfer and measurement). CCD performance is further specified by three major categories associated with the architecture of the CCD: (1) vertical register performance, (2) horizontal register performance and

Table 1.1 Vertical register performance.

Performance Parameter	Specification	Transfer Curve or Image	Conditions
Quantum efficiency 3000 Å 4000 Å 5000 Å 7000 Å 9000 Å	> 0.4 > 0.6 > 0.75 > 0.50 > 0.20	QE Transfer	backside illumination
Charge capacity	> 175,000 e$^-$ > 90,000 e$^-$	Photon Transfer	partially inverted MPP
Global CTE	> 0.99999	X-ray Transfer	1620 e$^-$ (Fe^{55})
Local CTE	< three 100 e$^-$ traps	Bar target image	1000 e$^-$ signal
Line transfer time	1 μsec	Full Well Transfer	at vertical full well
Pixel nonuniformity	< 2%	Photon Transfer	
Global dark current	< 0.5 nA/cm^2 < 20 pA/cm^2	Dark Current Transfer	partially inverted MPP (300 K)
Local dark current	< 1 nA/pixel	Dark current image	300 K
Dark current nonuniformity	< 10%	Photon Transfer	
MTF	> 0.45	Modulation Transfer	Nyquist, $\lambda = 4000$ Å

Table 1.2 Horizontal register performance.

Performance Parameter	Specification	Transfer Curve	Conditions
Global CTE	> 0.99999	X-ray Transfer	1620 e$^-$ (Fe^{55})
Charge capacity	> 700,000 e$^-$	Photon Transfer	noninverted
Pixel transfer time	< 10 ns	Full Well Transfer	at vertical full well
Output summing well charge capacity	> 10^6 e$^-$	Photon Transfer	

Table 1.3 Output amplifier performance.

Performance Parameter	Specification	Transfer Curve	Conditions
Read noise	< 3 e$^-$ rms	Photon Transfer	sample time = 4.0 μsec
	< 30 e$^-$ rms		sample time = 0.04 μsec
Sensitivity	> 5 μV/e$^-$	Photon Transfer	at 10 M pixels/sec
Video time constant	τ < 5 nsec		C_L = 5 pF
Nonlinearity	< 1%	Linearity Transfer	over dynamic range

(3) output amplifier performance. As an example of this organization, Tables 1.1–1.3 tabulate major performance parameters for a backside-illuminated 12-μm-pixel three-phase CCD. For those readers new to the CCD, it will be important to return to these tables as these parameters are reviewed in future chapters. Also included is the CCD transfer curve responsible in quantifying the parameter (transfer curves are discussed in Chapter 2) and the critical operating condition for the parameter tested (for example, operating temperature is an important condition in measuring dark current).

1.3 ARCHITECTURE, DESIGN, PHOTOLITOGRAPHY AND FABRICATION

1.3.1 ARCHITECTURE

While the idea of charge coupling is well defined, it has been implemented by CCD manufacturers in many ways. There are three common scientific CCD architectures. The first is a simple "linear" shift imager, as shown in Fig. 1.8. An image for this detector is recorded by slowly scanning the scene vertically past the imager while rapidly reading out the shift register in the horizontal direction. The second format is an area array employed as a "full-frame" imager, the format shown in Fig. 1.7. For this architecture, the scene is imaged onto the device, integrated, and then, with the shutter closed, the entire array of pixels is read out. Finally, to achieve very high frame rates and provide electronic shuttering, several manufacturers use a "frame transfer" layout. In this case, the full-frame imager is divided into two halves and independently clocked. The upper half of the array contains the "imaging registers," while the lower half contains the "storage registers." The scene is imaged onto imaging registers for a specified integration time. After integration, the image is then rapidly transferred to an optically opaque masked storage region. This is accomplished by shifting lines of pixels as fast as possible without horizontal readout. Once all lines are safely stored into the storage region, the pixels are all read out while the vertical clocks in the imaging region are frozen to integrate up the next field.

This book assumes progressive scan readout. That is, each line is read out in order without interlacing common to many commercial CCD imagers. With interlacing operation individual frames are read out every other line, similar to the way images are presented to the viewer on a television set. That is not to say that interlacing is not used in scientific applications. For example, three different fields can be taken with a three-phase CCD by changing the integrating phase for each field taken. For example, for the first frame phase-1 is used to collect charge, leaving phases-2 and -3 as barriers. The second frame would use phase-2 as the collecting phase and the third frame phase-3. The three frames are read into a computer and then later assembled into a composite image. This operating mode improves spatial resolution in the vertical direction only.

A common CCD array architecture arrangement based on three-phase technology is shown in Fig. 1.11(a). Example architecture specifications for a CCD are listed in Tables 1.4–1.6. As before, architecture features are broken down into three categories: vertical and horizontal registers and output amplifier. The pixels of the array are arranged in a square format with two horizontal shift registers, one at the top and the other at the bottom of the array. The vertical section is segmented into two sections. This format operation leads to some interesting imaging possibilities. The device can be employed as a full-frame imager; alternatively, it may be used in the frame-store mode, using the upper section for imaging and the lower as the memory, with readout through the lower horizontal register. Readout from both horizontal registers can also be achieved with the appropriate off-chip analog multiplexing, to achieve twice the data rate.

The vertical registers for the device in Fig. 1.11(a) could have also been divided into quadrants allowing for "split-frame transfer." Here imaging takes place in the center two quadrants, leaving the upper and lower sections for frame store. This architecture increases the vertical transfer rate by a factor of two. The horizontal register itself can be split in multiple sections, allowing charge to transfer to two or more amplifiers. This architecture arrangement is used for ultrahigh-speed CCDs that use multiple readout amplifiers. The highest speed architecture that is possible uses split frame transfer with output amplifiers in every column, top and bottom.

The two-amplifier design shown in Fig. 1.11(a) is also beneficial in improving device yield. For instance, the upper or lower amplifier may exhibit better performance in terms of CTE or read noise. Also, cosmetics play a role in the selection process. For example, the lower amplifier can be selected if a defect is near the top of the array. A second amplifier can also be used for backup purposes in case the first amplifier stops working. However, for flight applications there has been considerable debate in taking this design philosophy. The circuits used in switching from one amplifier to the other are subject to reliability concerns. It is for this

Table 1.4 Vertical register architecture.

Feature	Specification	Comments
Backside illumination	yes	
Frame transfer	yes	
Image columns	512	
Image lines	512	
Storage columns	512	
Storage lines	512	
Image/frame store interface lines	16	lines between image and storage regions
Full frame	optional	
Split frame transfer	optional	
Phases/pixel	three-phase	
Pixel shape	12 μm	
Pixel pitch	12 μm	
Gate overlap	< 1 μm	
Fill factor	100%	
Side dark columns	yes	two each side
Metal channel stop strapping	yes	every 50 lines
Poly gate strapping	yes	
Metal gate strapping	yes	every twelfth column
Array transfer gate	yes	top and bottom
Anti-blooming	yes	active gate lateral AB
Substrate contact	yes	topside
Light shield	yes	
MPP	optional	
Notch	optional	
Super notch	yes	

Table 1.5 Horizontal register architecture.

Feature	Specification	Comments
Horizontal registers	2	top and bottom
Horizontal channel width	40 μm	
Poly gate strapping	yes	
Metal gate strapping	yes	
Barrier phase during vertical transfer	phase 1	
Collecting phases during vertical transfer	phases 1 or 2 or both	
Notch	optional	
Super notch	yes	
Output summing well	2	for each amplifier
Output transfer gate	2	for each amplifier
Extended pixels	16 pixels	for each amplifier

Table 1.6 Output amplifier architecture.

Feature	Specification	Comments
Amplifiers	2	
Dual stage	yes	
1st stage geometry	3(L) × 20(W) μm	
2nd stage geometry	3(L) × 120(W) μm	
LDD	yes	
Drain voltage, V_{DD1}		for each 1st stage amplifier
Drain voltage, V_{DD2}		for each 2nd stage amplifier
On-chip load		for each 1st stage amplifier
Reset switch		for each amplifier
Reset switch geometry	2(L) × 8(W) μm	

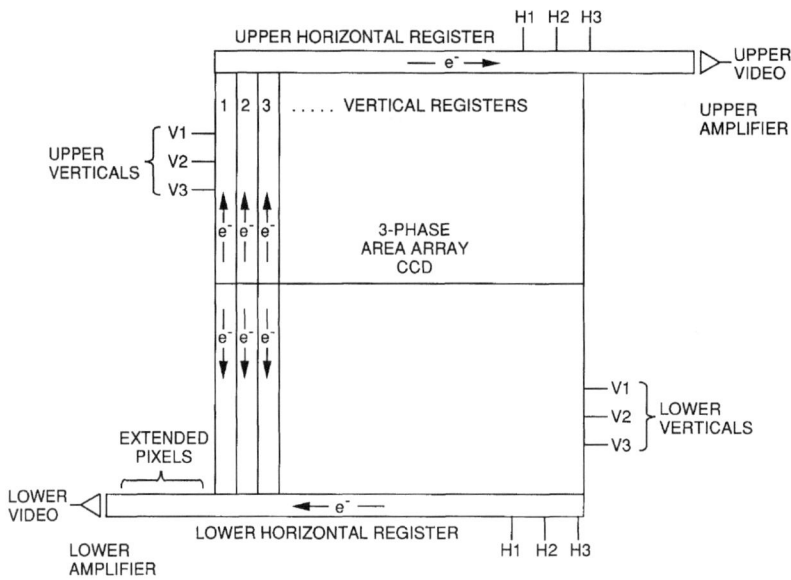

Figure 1.11(a) A full-frame transfer format showing dual vertical and horizontal registers with two amplifiers.

reason that WF/PC, Cassini and other flight cameras only use one amplifier even though multiple amplifiers were located on the CCD. The other problem with two amplifiers is associated with the time it takes to test and screen CCDs. Test time is approximately proportional to the number of amplifiers that need to be characterized because slight differences in performance are noted between amplifiers (e.g., gain, charge transfer direction, cosmetics, etc.).

HISTORY, OPERATION, PERFORMANCE, DESIGN, FABRICATION AND THEORY

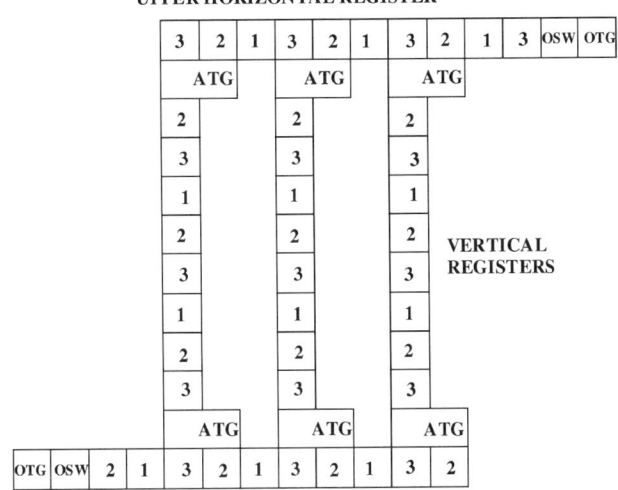

Figure 1.11(b) Functional diagram for the CCD shown in Fig. 1.11(a).

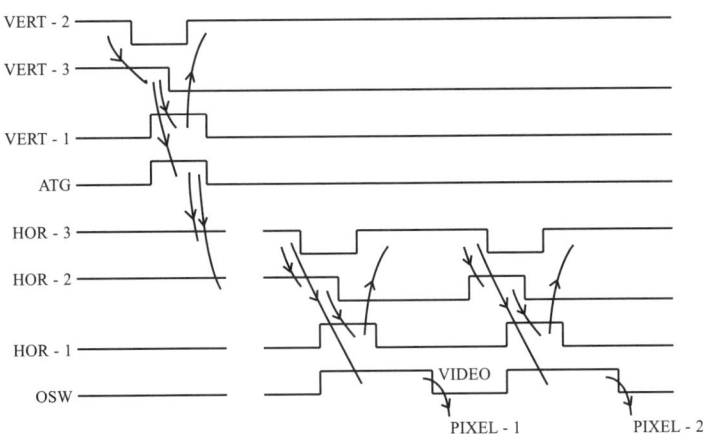

Figure 1.11(c) Timing diagram for the CCD shown in Fig. 1.11(a).

Figures 1.11(b) and 1.11(c) present a three-phase "functional diagram," and a "clock timing diagram." The two diagrams are very important to the camera designer. The interface between the vertical and horizontal registers is a particularly critical clocking region and are separated by the array transfer gate (ATG). Note that, in this example, either horizontal phases-3 and/or -2 can accept charge from the vertical registers. They are the collecting phases and are held high when a line of charge is dumped into the horizontal register. Horizontal phase-1 in this case acts as a barrier and is held low during this time. The interface between the output summing well (OSW) and the last horizontal phase is also a critical clocking region (the purpose of the output summing well is to perform pixel-binning, to be discussed below). For the lower horizontal register, phase-2 must dump charge into the output summing well, whereas the upper horizontal register involves phase-3. Unfortunately for the camera engineer, CCDs are designed differently and functional diagrams are not universal.

1.3.2 DESIGN AND PHOTOLITHOGRAPHY

1.3.2.1 Masks

Photolithography processes used in fabricating CCDs require the precise position of many doped regions and interconnection patterns. [22,23] These regions include ion implants (commonly referred to as implants) and diffusions, contact cuts for gates, metallization and protective oxide through which connections can be made to bonding pads. Two different types of mask (or reticle) sets are used in the photolithography process: (1) "contact" and "off-contact" masks and (2) "projection" masks. In contact printing, the mask is in intimate contact with the silicon wafer. With an off-contact reticle there is a small space (a few microns) between the mask and silicon wafer. The space significantly extends the lifetime of the mask. Contact printing can achieve CCD design features of a couple microns. Projection aligners optically project the image of the mask onto the silicon at a distance achieving improved design features. Projection printing can also be stepped across the wafer in making very large CCD arrays on large wafers (a process called "stitching"). Projection printing can hold design features to 1 µm or better, and therefore is used when very small pixel or CCD structures are fabricated.

1.3.2.2 Design Example

A minimum of seven masks are required to fabricate a three-phase CCD. Figures 1.12(a)–1.12(l) are the reticles used in the lithography process to form the signal channel, gates, contacts, bond pads, etc., for the Cassini and Hubble WF/PC II CCDs. The region shown illustrates the output amplifier, reset switch, sense node, output transfer gate (OTG), output summing well (OSW), extended pixels, array transfer gate (ATG), channel stops, and portions of the vertical and horizontal registers. Layer names vary, depending on the manufacturer's nomenclature. For our discussions the seven masks are defined as (1) Field, (2) Poly-1, (3) Poly-2, (4) Poly-3, (5) Contacts, (6) Metal and (7) Pad. An optional light shield layer is included for frame-transfer operation. Each layer is discussed in the order it is used in the fabrication process.

Field Reticle [Fig. 1.12(a)]
The field reticle defines the signal (buried) channel and where electrons will collect and transfer. The illustration shows four columns separated by 12 µm (i.e., the

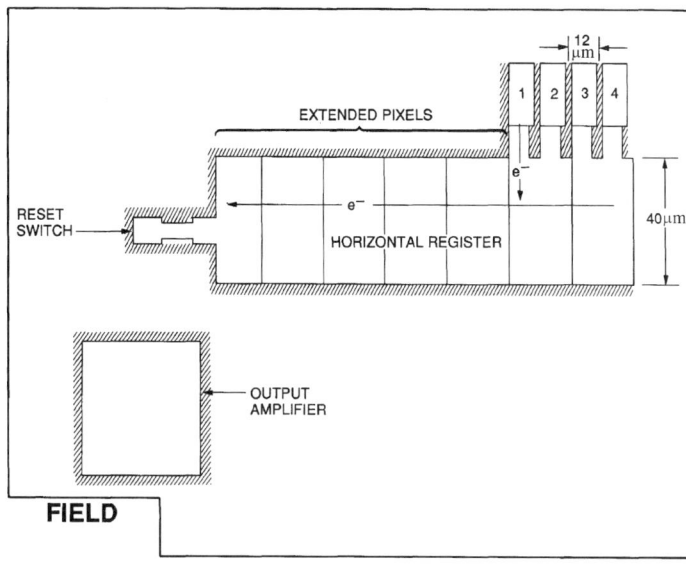

Figure 1.12(a) Field mask.

pixel pitch). The crosshatched regions between the columns represent the channel stops and are 2 µm wide. The columns interface with the horizontal register, which has a width of 40 µm. The extra width for this register is for "pixel-binning" or "pixel summation," where multiple lines can be summed on-chip. This allows the effective integration area of a pixel to be varied to a larger integral multiple of vertical pixels. For example, a 1024 × 1024 CCD can be transformed into a 512 × 512 device by summing lines of charge into the horizontal register and pixels into the output summing well in a 2 × 2 manner. This clocking technique is used in situations involving low-contrast, diffuse scenes where high spatial resolution is not critical. The large width for the horizontal register is also required for high-speed operation, a very important feature described in Chapter 5. Eight "extended pixels" exist between the first column and the output amplifier. Since there are no vertical registers above the extended region, these pixels represent zero signal level and are often sampled for offset information (discussed in Chapter 2). At the very end of the horizontal register are the sense node and reset transistor, which occupy the signal channel. The output amplifier is in its own p^+ moat isolated from the signal channel through field doping.

Poly-1 Reticle [Fig. 1.12(b)]
The poly-1 mask defines phase-1 for both the horizontal and vertical registers and output summing well as shown. Charge held in the last phase of the horizontal register transfers into a twice-normal width, independently clocked output summing well to provide for pixel binning. The increased storage capacity under this gate is designed to bin several pixels horizontally. CCDs fabricated without an independently clocked output summing well cannot do pixel binning properly. Here,

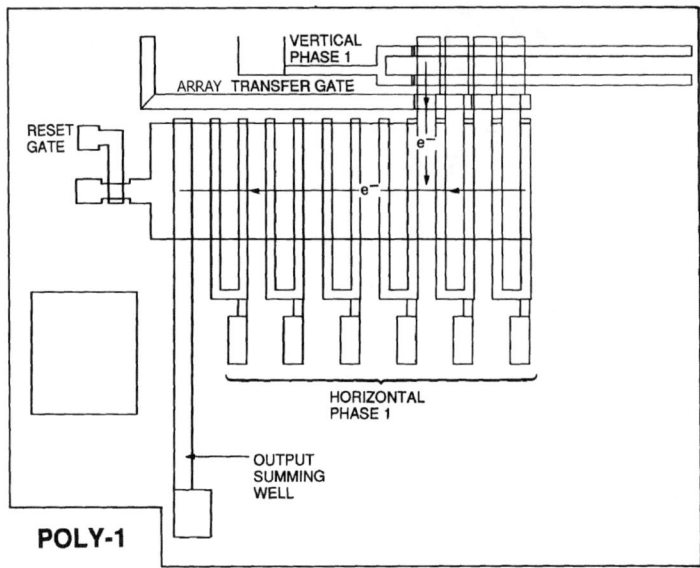

Figure 1.12(b) Poly-1 mask.

charge must be summed directly in the sense node, which results in some signal-to-noise degradation (the sense node, where charge is converted to a voltage, has not been formed at this point in the process). Pixel binning is discussed further in Chapter 4. The output summing well is also used to electrically inject charge into the horizontal register, a clocking scheme discussed below.

The array transfer gate is also defined with poly-1 in this example design. This gate is the last phase of the vertical register and directs charge into the horizontal register. The transfer gate is used as a barrier phase to prevent charge from leaking from the horizontal register back into array while the horizontal register is clocked (an effect called transfer gate tunneling). CCDs fabricated today do not usually include an independently clocked transfer gate. Instead the last vertical phase in the array is considered the transfer gate and is held in a low state when the horizontal registers are shifting charge.

Also defined with poly-1 is the reset gate, which is responsible for resetting the sense node to a reference voltage (V_{REF}) before charge is transferred onto the sense node. The V_{REF} connection is yet to be made at this stage of the design.

Poly-2 Reticle [Fig. 1.12(c)]
A poly-2 mask defines phase-2 for both the horizontal and vertical registers. It also forms the output transfer gate. The output transfer gate is a nonclocked gate that plays many roles. For example, the gate serves to reduce the clock feed-through signal generated by the output summing well when charge is dumped onto the sense node. The output transfer gate is also used to electrically inject charge into the horizontal register.

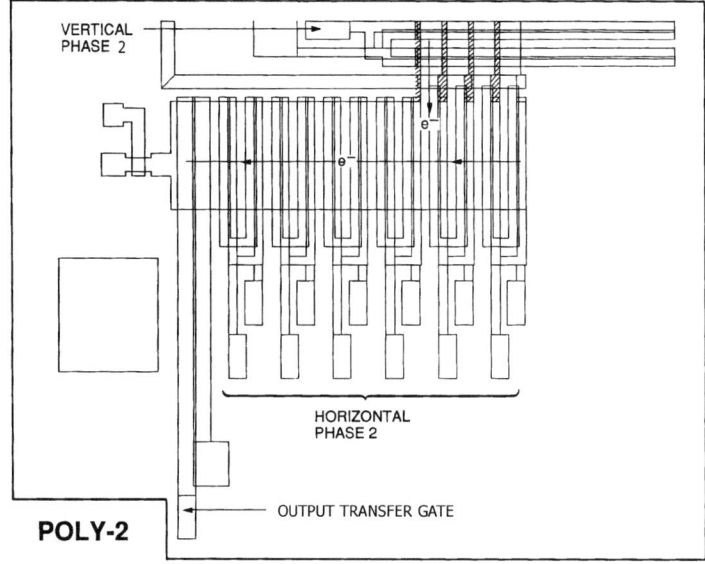

Figure 1.12(c) Poly-2 mask.

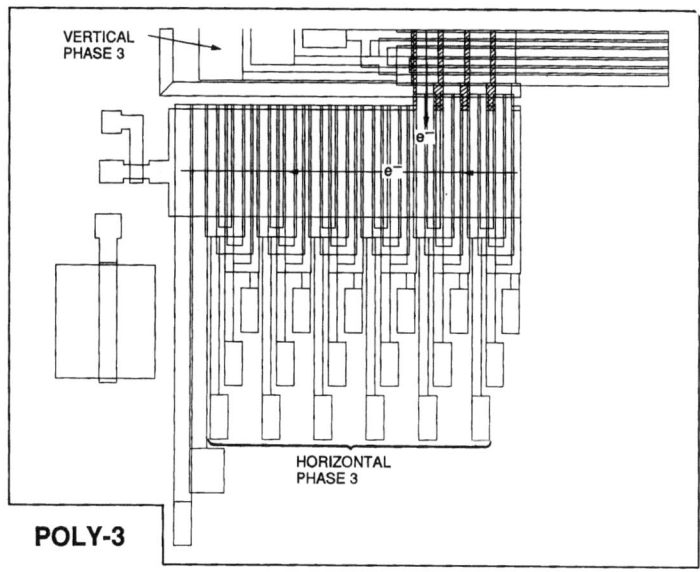

Figure 1.12(d) Poly-3 mask.

Poly-3 Reticle [Fig. 1.12(d)]
The poly-3 mask defines phase-3 for both the horizontal and vertical registers. Poly-3 also forms the gate for the output MOSFET amplifier.

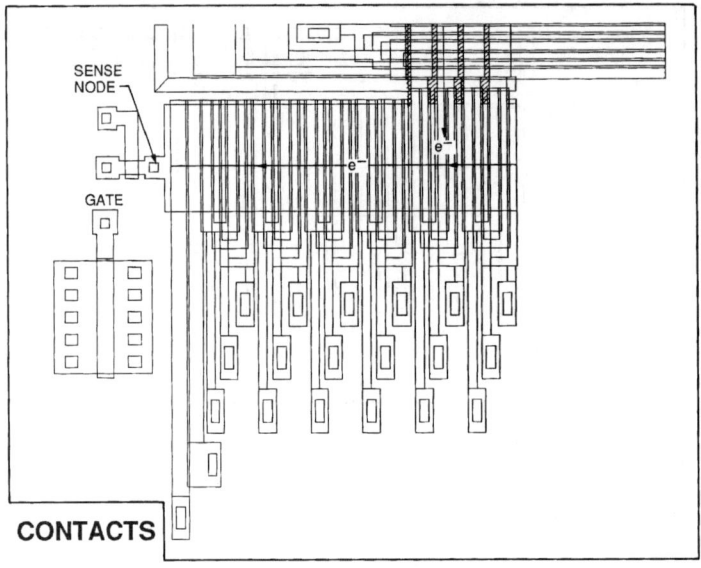

Figure 1.12(e) Contact mask.

Contact Reticle [Fig. 1.12(e)]
Each poly layer is oxidized after being patterned to isolate each layer electrically. The contact mask is used to etch away oxide in specific spots, to make electrical connections to the poly gates and diffusions with subsequent metal deposition. For example, a contact will need to be made at the sense node in connecting it to the gate of the output amplifier.

Metal Reticle [Fig. 1.12(f)]
The metal mask interconnects the horizontal and vertical three phases, connects to the source and drains of the amplifier, and connects the sense node to the gate of the amplifier.

Pad Reticle [Fig. 1.12(g)]
The pad mask etches the oxide over the metal bond pads. Bond wires are later attached to the pads and pins of the CCD package.

Light Shield Reticle [Fig. 1.12(h)]
For frame-transfer operation, a light shield is deposited over the storage regions of the CCD. For this design there are dual-frame storage regions for redundancy.

Figure 1.12(i) shows a scanning electron microscope image of the region we have discussed, after device fabrication. The vertical registers and column stops are seen along with the horizontal register, extended pixels and amplifier region. Figure 1.12(j) is a magnified view of the output amplifier region, showing the output summing well, output transfer gate, the reset MOSFET and sense node. The

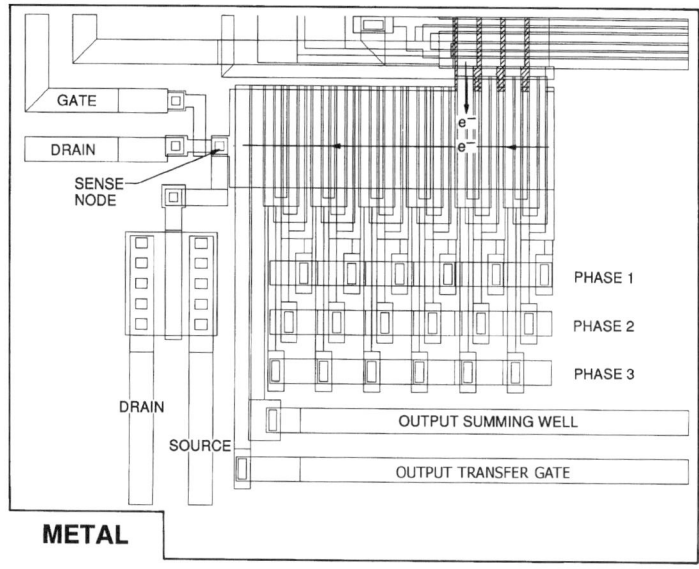

Figure 1.12(f) Metal mask.

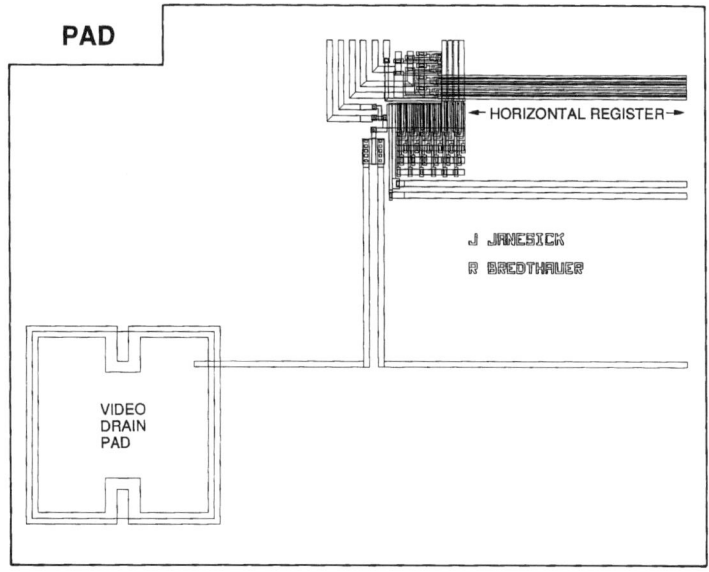

Figure 1.12(g) Pad mask.

sensitivity of the sense node is approximately 3 μV per electron. The voltage at the sense node is impressed on the gate of the output MOSFET, which changes the current flow through the device between the source and drain. Connecting an

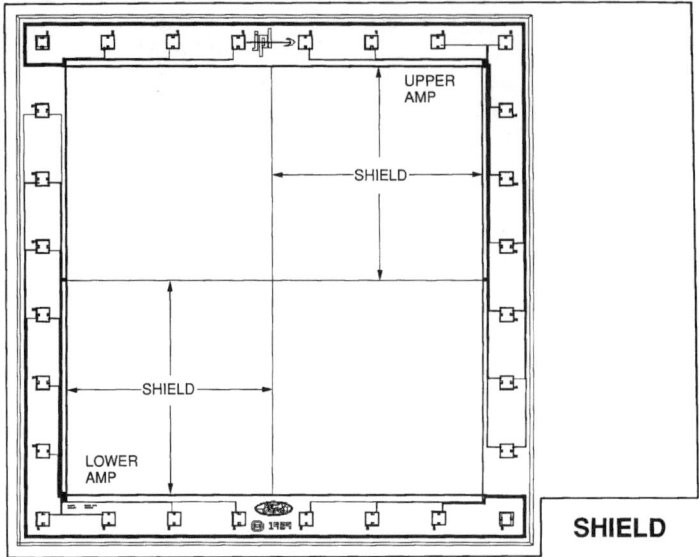

Figure 1.12(h) Light shield mask.

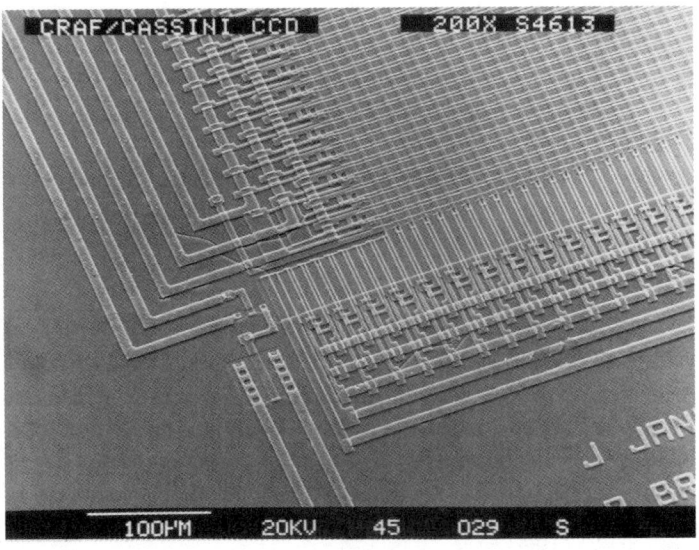

Figure 1.12(i) SEM photograph of the amplifier region.

external load resistor on the source produces a voltage change approximately equal to the voltage change on the sense node.

Figure 1.12(k) is the final device after the CCD is bonded into a package. Very small 1-mil aluminum wires are seen between the bond pads and package pins.

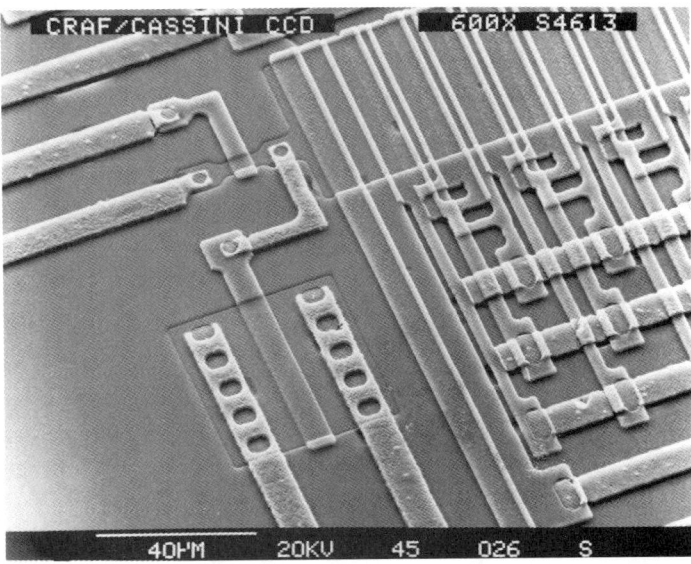

Figure 1.12(j) Magnified view of the amplifier region.

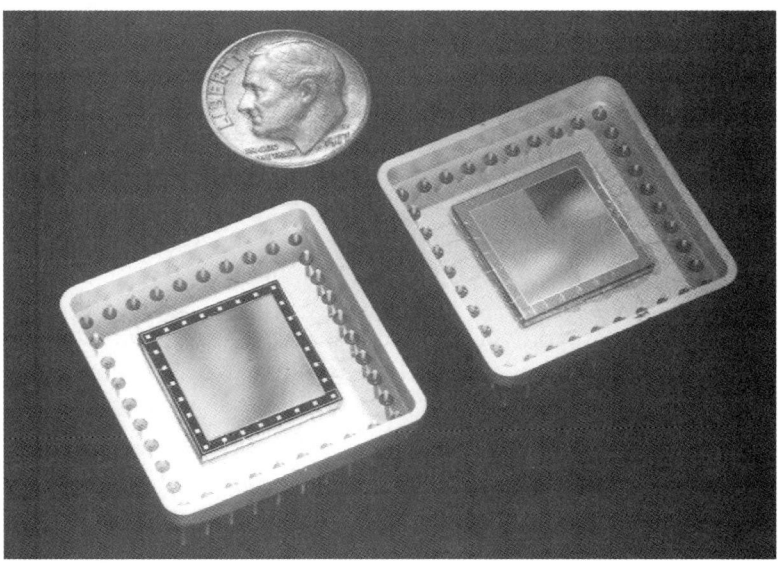

Figure 1.12(k) Packaged CCDs.

Both full-frame and frame-transfer CCDs are shown. Figure 1.12(l) shows the Whirlpool galaxy generated by the CCD using Palomar's 200-in. telescope (first light!).

Figure 1.12(l) Image of the Whirlpool galaxy (first light for the CCD).

1.3.2.3 Optional Masks

High-performance scientific CCDs may have up to twice the number of masks as those described above for specialized functions. Some of these design/process options include

1) Substrate contact mask

    This mask provides a ground contact on the top of the device. Backside-illuminated CCDs typically have this feature for grounding purposes.

2) Multipinned phase (MPP) mask

    This mask and corresponding process is employed to suppress surface dark current generation.

3) Notch channel mask

    This mask provides a small channel within the main channel to achieve the highest CTE possible. CCDs used in high-radiation environments will normally have this feature included.

4) Lightly doped drain (LDD) mask

    This mask improves output amplifier sensitivity (e^-/V) and drain/source breakdown characteristics.

5) High-speed channel stop (HSCS) mask
   This mask allows the channel stop regions to be doped independently for high-speed performance.

6) Super notch mask
   This mask defines the buried channel and isolates the signal channel from highly doped field regions (e.g., channel stops).

7) Antiblooming (AB) mask
   This mask provides a design structure to preventing blooming.

8) Amplifier doping mask
   This mask is used to dope the output amplifier independently from the array for low-voltage operation.

9) Frame/split frame transfer mask
   This light shield mask is used for frame transfer and split frame transfer operation.

10) High-speed metal strap mask
    This mask is used to strap the poly gates and channel stop regions with a second layer of metal for high-speed operation.

These optional CCD features will be discussed throughout this book.

1.3.3 PROCESSING AND FABRICATION

1.3.3.1 Silicon Wafers

CCDs are fabricated on silicon wafers that range from 4 to 8 in. in diameter with 12-in. wafers on the horizon. Epitaxial silicon is normally used in the fabrication of a CCD. An epitaxial layer of approximately 10–20 μm in thickness is grown on top of a thick, highly doped substrate layer (< 0.01 Ω-cm resistivity). The epitaxial layer during growth is usually pre-doped with boron between 10–100 Ω-cm for n-channel CCDs. The four CCD functions discussed above (i.e., charge generation, collection, transfer and measurement) all take place in the epitaxial layer. In fact, the last three functions take place within the first micron from the surface.

The thick substrate layer below the epitaxial layer serves several purposes. First, it supports the epitaxial layer to allow it to be processed. Substrate thickness for 4-in. material is approximately 500 μm thick and increases in thickness with diameter. Second, the substrate acts as a good electrical ground plane. As discussed below, displacement currents flow in and out of the substrate as the device is clocked. The importance of the substrate layer becomes critical for high-speed applications where a low-impedance ground path is required. Third, in that the substrate is highly doped (approximately 0.01 Ω-cm), it is essentially optically dead.

Photoelectrons generated in the substrate region recombine quickly with the holes provided by the dopant. This characteristic is important to achieve high CCE and good spatial resolution among pixels.

Silicon quality is extremely critical for the CCD. Impurities such as gold or lattice imperfections can have a profound effect on CCD performance. The third function, charge transfer, is especially vulnerable to epitaxial quality. CCDs are sometimes required to transfer very small charge packets through several inches of silicon without loss. Silicon requirements are discussed more fully in Chapter 5.

1.3.3.2 LOCOS Process

New silicon processing techniques were being developed for computer microelectronic circuits (e.g., RAMS, DRAMs, microprocessors chips, etc.) at the same time the scientific CCD was evolving. These competitive technologies have had a major impact on CCD development and production of large-area-array CCDs. For example, a process called "localized oxidation of silicon (LOCOS)" developed at Phillips Laboratory has significantly increased the yield to allow ultralarge-area arrays to be fabricated. The LOCOS process is based on the fact that silicon nitride can be used as a mask against thermal oxidation, which is advantageous when fabricating CCDs. In addition, there are certain enchants (e.g., H_3PO_4) that remove silicon nitride but not silicon dioxide, and vice versa. In the LOCOS process the gate insulator is based on a dual insulator system (e.g., typically 500 Å of silicon dioxide and 500 Å of silicon nitride). This insulating system is high yielding in terms of shorts, compared with a single layer of silicon dioxide (1000 Å) used by some CCD manufacturers. Most large-area-array CCD manufacturers use an oxide/nitride insulator.

CCD processing involves precise positioning (ideally, 1 µm or less) of implants, diffusions, gates, metallization, etc. The number of processing steps for the simplest CCD may require 80 individual procedures and take several weeks for a device to be fabricated. Presented below is a list of the major process steps in fabricating a three-phase buried-channel CCD with the masks described above (Fig. 1.12).

1) Starting silicon wafer material: p, {100} orientation, 10–100 Ω-cm epitaxial (10–20 µm) on p^+ substrate 0.01 Ω-cm 525 µm thick
2) Gate silicon dioxide (500 Å)
3) Gate silicon nitride deposition (500 Å)
4) Photo-resist coat

5) **— Field Oxide Mask —**
6) Resist develop
7) Etch silicon nitride within field oxide regions
8) Etch silicon dioxide within field oxide regions

9) Field implant (e.g., boron—2×10^{13} cm^{-2})
10) Resist strip
11) Field oxide (10,000 Å)
12) Buried channel implant (e.g., phosphorus—1.6×10^{12} cm^{-2})
13) Polysilicon-1 deposition (5000 Å)
14) Polysilicon-1 doping (phosphorus)
15) Photo-resist coat

16) — **Poly-1 Mask** —
17) Resist develop
18) Etch polysilicon-1
19) Resist strip
20) Polysilicon-1 oxidation (1000 Å)
21) Polysilicon-2 deposition (5000 Å)
22) Polysilicon-2 doping (phosphorus)
23) Photo-resist coat

24) — **Poly-2 Mask** —
25) Resist develop
26) Etch polysilicon-2
27) Resist strip
28) Photo-resist coat
29) Polysilicon-2 oxidation (1000 Å)
30) Polysilicon-3 deposition (3000 Å)
31) Photo-resist coat

32) — **Poly-3 Mask** —
33) Resist develop
34) Etch polysilicon-3
35) Resist strip
36) Nitride etch within source/drain regions
37) Oxide etch within source/drain regions
38) Polysilicon-3 and source/drain doping (phosphorus)
39) Polysilicon-3 oxidation (700 Å)
40) High-temperature anneal in 4% H_2 in N_2 (i.e., forming gas)
41) Photo-resist coat

42) — **Contact Mask** —
43) Resist develop
44) Oxide etch contact regions
45) Resist strip
46) Reflow glass
47) Aluminum metallization (10,000 Å)
48) Photo-resist coat

49) — **Metal Mask** —
50) Resist develop
51) Aluminum etch
52) Resist strip
53) SILOX deposition (1 μm)
54) Photo-resist coat

55) — **Pad Mask** —
56) Resist develop
57) Oxide etch
58) Resist strip

Figure 1.13(a) is a cross-section that shows dimensions for the various layers used in fabricating a CCD. Figure 1.13(b) is a cross-sectional view of a three-phase CCD showing an active signal column and two channel stop regions beside it.

1.3.3.3 Shorts, Cosmetics and Yield

The size limit of an area-array CCD is determined by the number of gate shorts and image defects that result during the fabrication of the sensor. A single fatal short (the CCD's Achilles' heel) can make the device worthless. Shorts are usually induced by small dust particles that settle on the silicon wafers or reticles during processing. Although the dust problem can be reduced by special air filters, the laboratory environment can never be made perfect. Therefore, shorts will always be experienced to some degree. The fabrication of large three-phase CCDs ($>1024 \times 1024$) usually requires a class-100 clean room environment or better

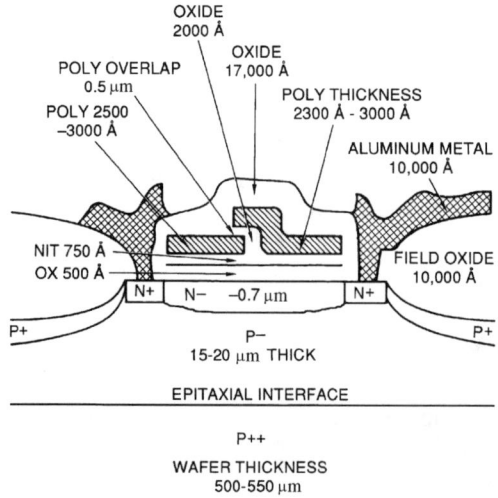

Figure 1.13(a) Dimensions for the various layers used in fabricating a simple CCD.

(class-100 is defined as having not more than 100 particles of a certain size [e.g., 0.5 μm] per cubic foot of air). Some CCD manufacturers maintain class-100 facilities where human intervention during critical process steps is not allowed. Instead, mechanical robots help with the production.

Several different types of shorts that plague the CCD have been identified and characterized. Three common shorts are (1) the interlevel short, (2) the intralevel short and (3) the substrate short. Figure 1.14 is a scanning electron microscope cross-section image of a WF/PC I three-phase 15-μm pixel that shows where short types 1 and 3 occur. Poly-1 is the smaller poly layer on the left-hand side of the image. The phase to its right is phase-2, followed by phase-3. The dark silicon dioxide layer seen in the image between each gate and silicon serves as the gate insulator (i.e., SiO_2). It is in these regions where shorts occur. The interlevel short takes place between clock phases of different poly levels. For example, phase-1 might be shorted to phase-2. This short was especially serious for early aluminum gate backside-illuminated CCDs made at TI. Aluminum gates were replaced by polysilicon-doped gates, which are much less conductive. In turn, the

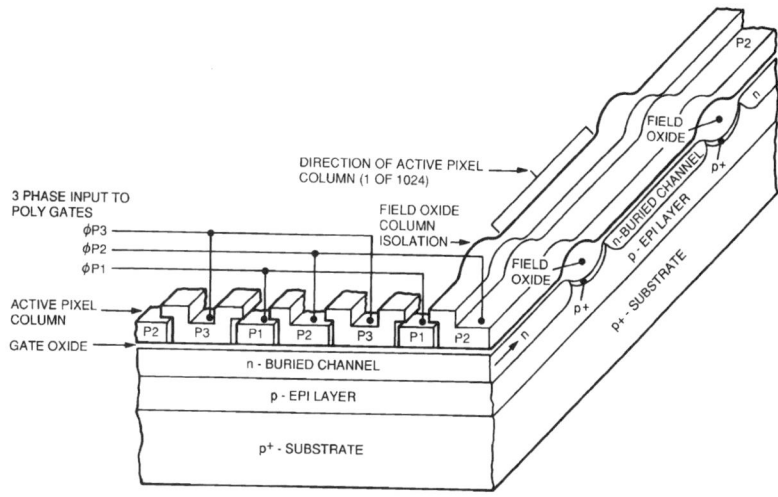

Figure 1.13(b) Cross-section showing signal channel and channel stop regions.

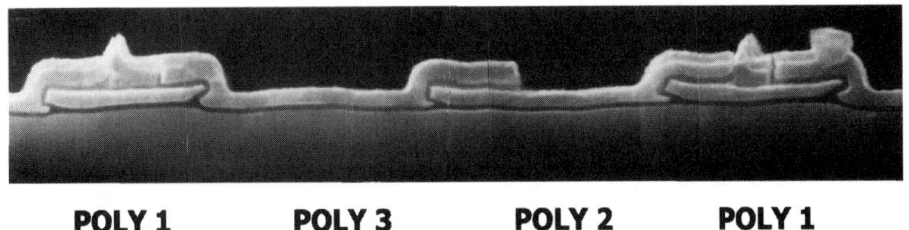

Figure 1.14 SEM cross-section image of a WF/PC I 15-μm pixel.

interelectrode short became less catastrophic. In fact, high-impedance interlevel shorts (>100 kΩ) can exist without significantly influencing performance. Several shorted CCDs are functioning quite well at the world's largest observatories. Polysilicon also has the advantage of being semitransparent to incoming visible photons, allowing for frontside illumination as discussed in Chapter 3. Today, nearly all scientific CCDs are based on polysilicon gate electrodes.

Note in Fig. 1.14 that phases-2 and -3 overlap phase-1 because phase-1 was patterned first. Overlapping gates, without gaps between phases, are crucial to good CTE performance. A gap of only 1 μm will significantly degrade performance. For example, 0.2-, 0.3- and 0.4-μm gaps will produce potential bumps in the signal channel of approximately 0.1, 0.3 and 0.7 V, respectively, for a standard scientific CCD process. These small barriers can be overcome by fringing fields between phases. However, as full well is approached, these fields disappear, making the potential bumps active. Therefore, full well performance will ultimately be the main parameter affected by the presence of gaps. It is interesting to note that there are always gaps present between phases for multiphase CCDs because of the finite thickness of the interelectrode insulator. Fortunately, these gaps (approximately 1000 Å, the thickness of the gate dielectric) are sufficiently small and will not influence CTE performance.

Triple polysilicon technology, used in fabricating three-phase CCDs, is a high-yielding process and has permitted very large area arrays to be fabricated in labs of class-100 or higher. Two-level poly gate processing, used in making two- and four-phase CCDs, is more sensitive to shorts of another type. Although two-level poly gate technology requires fewer process steps than does three-level, the process has one serious drawback: the intralevel short. An intralevel short originates when a "bridge" of poly forms between adjacent phases on the same poly level. These defects are called "poly stringers" and are unetched areas where poly crosses steps, such as over a field oxide edge. They come about because the poly is thicker over steps and inadequate etching can leave a stringer. Four-phase CCDs typically use the first level of poly for phases-1 and -3 and the second level for phases-2 and -4. An intralevel short that occurs on the first level, for example, would short phases-1 and -3. The intralevel short manifests itself as poor CTE performance for all lines above the troubled site (refer to Chapter 5). Triple poly gate processing inherently eliminates the intralevel short problem since each poly level is intentionally connected and bussed in forming phases-1, -2 and -3. An intralevel short for a three-phase CCD would affect only a localized pixel.

The most serious short for all CCD technologies is the substrate short. The substrate short develops between the gate and signal channel, usually because of improper gate dielectric definition. A substrate short of any value (e.g., Meg Meg Meg ohms) usually implies disaster to the CCD. The short generates a high-leakage current which usually saturates the CCD. Substrate shorts as high as 10 MΩ are screened by simple dc ohmic measurement; substrate shorts above this level are found by imaging the array.

Shorts can exhibit latent characteristics. That is, the impedance of the short can suddenly appear and/or worsen with time. The best way to locate latent shorts is by

Figure 1.15 An 800 × 800 flat-field image showing cosmetic problems.

"burning" in the CCD. This is accomplished by clocking the CCD for an extended period of time with clock levels higher than normal. For example, flight CCDs are burned in for hundreds of hours before being installed into the flight instrument. The sensors are often temperature cycled to accelerate latent failure promoted by mechanical stress. This technique has paid off because CCDs that fail the burn-in test could have otherwise failed in flight.

A common array size is the 1024 × 1024 CCD. Today, the "shorts yield" for this format size is exceptionally high compared to the past. Several CCD manufacturers are capable of producing chips this size with > 85% shorts yield. This does not mean that 85% of the sensors will be acceptable for sale. Cosmetics also enter into the yield equation. For example, Fig. 1.15 shows an 800 × 800 image with several bright "column blemishes" that generate high-leakage currents. Also shown in the image are several dark columns. These defects are called "blocked channels" and represent regions where charge cannot transfer properly through a pixel. Nevertheless, "cosmetic yield" can also be very high for a 1024 × 1024 CCD. Manufacturers and foundries have demonstrated consistent cosmetic and shorts yields of >70%. This figure can be used to estimate yields for larger CCDs, by using the equation

$$Y_N = Y_{sm}^N, \tag{1.3}$$

where Y_{sm} is the shorts and cosmetic yield for a small array and Y_N is the expected yield for an array that is N times larger in area. It is obvious from this equation that a CCD manufacturer must demonstrate very high yields for smaller arrays before they can consider fabricating ultralarge arrays.

Example 1.3

Determine the expected yield for a 4096 × 4096 × 15-µm-pixel array. Assume a 1024 × 1024 × 15-µm-pixel yield of 75%.
Solution:
From Eq. (1.3),

$$Y_{16} = 0.75^{16},$$

$$Y_{16} = 0.01.$$

It should be pointed out that Eq. (1.3) is approximate. Strictly speaking, IC yield goes with Poisson statistics (i.e., yield proportional to e^{-DA}, where D is defect density and A is area). This model is pessimistic due to defect clustering, and more refined models are used to account for this. Of course, one defect kills a CCD while other defects are tolerated (i.e., substrate versus interlevel shorts).

1.3.3.4 Sandbox

A "Sandbox lot" is a fabrication technique developed at JPL to reduce CCD development and production costs. In doing so, the approach enables the development of several custom CCDs that would normally be unaffordable. These savings are realized by combining several devices on a single development lot, thus dividing the production costs among several groups. For example, Fig. 1.16 shows Sandbox I with six different devices arranged on a 4-in. wafer. The CCD formats onboard are (1) 4096 × 4096 × 9-µm-pixel, three-phase, 4-amp CCD; (2) 1024 × 1024 × 18-µm-pixel, six-phase, 4-amp CCD; (3) 4096 × 3072 × 9-µm, three-phase, 4-amp CCD; (4) 128 × 64 × 36-µm, 64-amp, six-phase CCD; (5) 2048 × 1024 × 9-µm-pixel, three-phase, 4-amp CCD and (6) an unusual 16-pixel "circulator" CCD. Device (5) is currently in space aboard *Deep Space I*. Device (6) is shown in Fig. 1.17(a). The device was designed to interrogate a single electron (just for fun). To do this, a Skipper nondestructive floating gate amplifier takes a sample each time the electron makes a complete revolution around the circuit. Thousands of samples can be taken from the same pixel, thereby reducing the noise below 0.1 e$^-$ rms to detect the small particle. At the center of the array is a diode which can be slightly forward biased to emit photons and generate charge in the signal channel

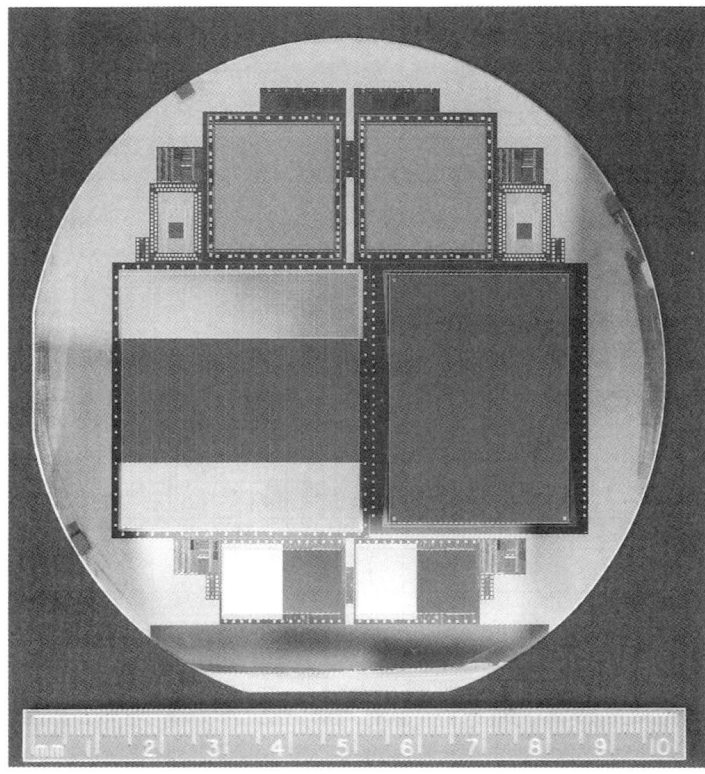

Figure 1.16 Sandbox I wafer showing six different CCD designs.

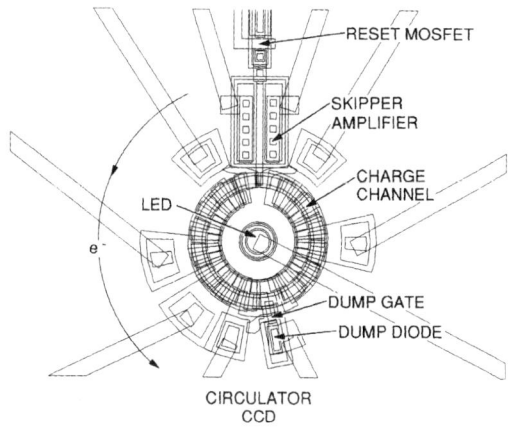

Figure 1.17(a) Sandbox I Circulator CCD used to detect a single electron.

if desired. A dump gate and drain are provided to erase charge from the channel. Refer to Chapter 6 for Skipper operation.

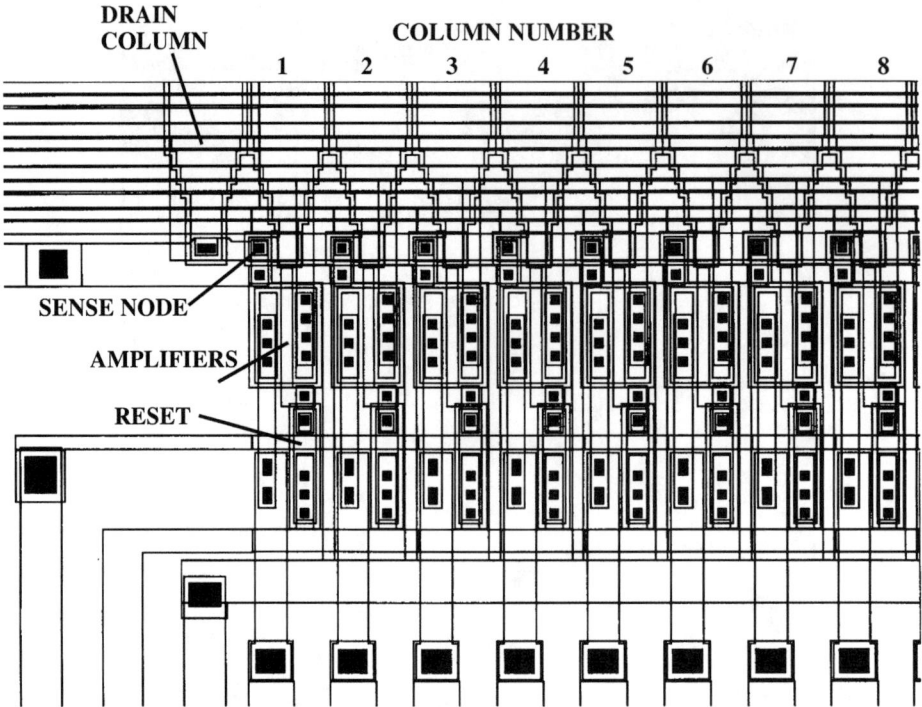

Figure 1.17(b) Magnified view of adaptive optics CCD showing output amplifiers in each column.

Figure 1.17(b) shows a magnified view of the 128 × 64 CCD. This floating gate amplifier CCD is used for wavefront adaptive optics applications and has been used at Mt. Wilson with great success. The CCD is intended for very fast framing (> 2500 frames/sec) and low-noise performance (3 e^- rms). In order to achieve both specifications, there is an output amplifier for each column for multiple amplifier readout.

The large CCD on the left side of Fig. 1.16 is called the Cinema CCD. The CCD is intended to replace photographic film used in cinematic moving picture cameras. The sensor offers several new design challenges. For example, the CCD will be read out at 60 frames/sec at an effective pixel rate of a half billion pixels per second. To achieve such rates, the CCD is divided up into 32 sections for parallel readout [refer to Fig. 1.17(c)]. The array itself is split into four 1024 × 4096 quadrants forming two image and two storage sections (i.e., split frame transfer). The split image is transferred toward the top and bottom of the array. From each storage region there are 16 horizontal registers, each with a 3-stage output amplifier. Each channel reads at approximately 16 Mpixels/sec and is digitized to 12 bits. It is anticipated that a 2- or 3-chip approach will be employed to obtain color information. A magnified view of the output amplifier region is presented in Fig. 1.17(d).

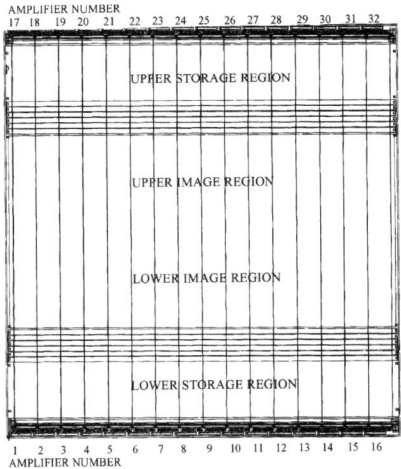

Figure 1.17(c) Cinema CCD design layout.

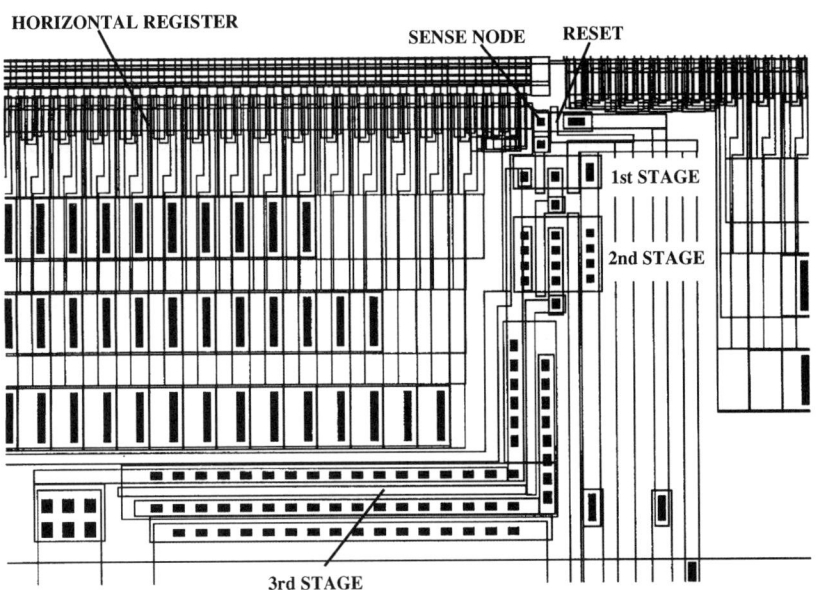

Figure 1.17(d) Magnified view of amplifier region of Cinema CCD.

1.4 CCD THEORY

1.4.1 MOS CAPACITOR

As mentioned above, the fundamental building block of the CCD is the MOS capacitor. This element is the backbone behind the second and third functions of the CCD (i.e., charge collection and charge transfer). We will begin our theoretical review by explaining how a simple p-MOS capacitor works. The insight gained in these discussions will give better understanding of CCD characteristics and performance limitations noted in other chapters of this book. The MOS capacitor will be compared first to a surface-channel CCD. Although surface CCD technology is rarely employed today, it is conceptually the easiest to analytically describe and to study. We will then work our way to the n-p MOS capacitor which is analogous to the buried-channel CCD. This CCD technology is almost universally fabricated and will deliver the highest performance possible. The difference between surface and buried channel will become apparent in these discussions.

1.4.1.1 Accumulation

A p-MOS capacitor consists of p-type (e.g., boron-doped) silicon substrate, a gate dielectric, (e.g., thermally grown silicon dioxide approximately 1000 Å thick) and a conductive gate that is deposited (usually doped polysilicon). A cross-sectional view is shown in Fig. 1.18. When a negative voltage is applied to the gate, majority carriers (i.e., holes) will accumulate at the silicon-silicon dioxide (Si–SiO$_2$) interface. The layer of holes forms immediately (in a few nanoseconds) and represents a highly conductive layer at the interface. The potential throughout the silicon layer is at ground potential. The gate capacitance associated with accumulation is

$$C_{OX} = \frac{\varepsilon_{OX}}{d}, \qquad (1.4)$$

where C_{OX} is the oxide capacitance (F/cm^2), ε_{OX} is the permittivity of SiO$_2$ (3.45×10^{-13}) and d is the thickness of the gate insulator (cm). Note that Eq. (1.4) describing the MOS capacitor has exactly the same form as the equation describing a simple metal parallel plate capacitor, since the oxide layer is situated between two conductive layers.

Example 1.4

Find the capacitance of a MOS capacitor in the accumulation state. Assume an oxide thickness of $d = 1000$ Å (10^{-5} cm). Also find the capacitance associated with a dual-gate silicon dioxide/silicon nitride insulator often used in fabricating CCDs. Assume an oxide and nitride thickness of 500 Å each and a permittivity of 6.63×10^{-13} F/cm for silicon nitride.
Solution:
From Eq. (1.4),

$$C_{OX} = 3.45 \times 10^{-13} / 10^{-5},$$

$$C_{OX} = 3.45 \times 10^{-8} \text{ F/cm}^2.$$

HISTORY, OPERATION, PERFORMANCE, DESIGN, FABRICATION AND THEORY

For the dual insulator, the capacitance for 500 Å of silicon nitride is

$$C_{NIT} = 6.63 \times 10^{-13}/5 \times 10^{-6},$$

$$C_{NIT} = 1.32 \times 10^{-7} \text{ F/cm}^2.$$

The capacitance for 500 Å of silicon dioxide is

$$C_{OX} = 3.45 \times 10^{-13}/5 \times 10^{-6},$$

$$C_{OX} = 6.9 \times 10^{-8} \text{ F/cm}^2.$$

Adding these capacitances in series,

$$C_{DUAL} = (1.32 \times 10^{-7})(6.9 \times 10^{-8})/[(1.32 \times 10^{-7}) + (6.9 \times 10^{-8})],$$

$$C_{DUAL} = 4.53 \times 10^{-8} \text{ F/cm}^2.$$

Unless otherwise noted, capacitance calculations in this book will assume an oxide insulator.

1.4.1.2 Depletion

Conversely, when a positive bias is applied to the gate, holes are driven away from the surface, leaving behind uncompensated negatively charged acceptor (boron) atoms. This region is called the depletion region since the region is depleted of mobile carriers—in this case, holes. This bias condition is shown in Fig. 1.18, where potential (V) is plotted as a function of distance into the silicon. The number of holes driven away equals the number of positive charges on the gate electrode, i.e.,

$$Q_i = qN_A x_d, \tag{1.5}$$

where Q_i is ionized acceptor charge concentration beneath a depleted MOS gate (C/cm^2), x_d is the depletion region depth (cm) and N_A is the acceptor doping concentration (atoms/cm^3).

Example 1.5

Find the ionized charge concentration for a MOS structure when depleted to 3 μm. Assume $N_A = 10^{15}$ cm^{-3}.

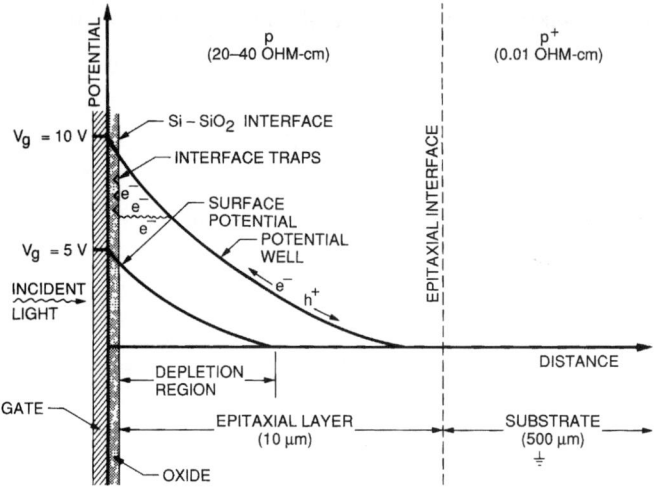

Figure 1.18 Surface channel potential well.

Solution:
From Eq. (1.5),

$$Q_i = (1.6 \times 10^{-19}) \times 10^{15} \times (3 \times 10^{-4}),$$

$$Q_i = 4.8 \times 10^{-8} \, \text{C/cm}^2.$$

This equals a total ion concentration of $3 \times 10^{11}/\text{cm}^2$.

The depletion region is nonconductive and acts as an insulator with a capacitance of

$$C_{DEP} = \frac{\varepsilon_{SI}}{x_d}, \tag{1.6}$$

where C_{DEP} is the depletion capacitance (F/cm^2), and ε_{SI} is the permittivity of silicon (1.04×10^{-12}).

Example 1.6

Find the depletion capacitance for a MOS capacitor depleted to a depth of 7 μm.
Solution:
From Eq. (1.6),

$$C_{DEP} = 1.04 \times 10^{-12} / (7 \times 10^{-4}),$$

$$C_{DEP} = 1.5 \times 10^{-9} \, \text{F/cm}^2.$$

The net gate capacitance C_T relative to substrate is, in the depletion state, the series combination of the oxide capacitance, C_{OX}, and depletion capacitance, C_{DEP}; i.e.,

$$C_T = \left(\frac{1}{C_{OX}} + \frac{1}{C_{DEP}}\right)^{-1}. \tag{1.7}$$

Example 1.7

Find the series oxide and depletion capacitances. Assume $C_{OX} = 3.45 \times 10^{-8}$ F/cm^2 and $C_{DEP} = 1.5 \times 10^{-9}$ F/cm^2 from previous examples.
Solution:
From Eq. (1.7),

$$C_T = \left[1/(3.45 \times 10^{-8}) + 1/(1.5 \times 10^{-9})\right]^{-1},$$

$$C_T = 1.43 \times 10^{-9} \, \text{F/cm}^2.$$

Note that the depletion capacitance is less than the oxide capacitance and dominates C_T.

Oxide and depletion capacitances calculated above will be added to other capacitances associated with a pixel for a total capacitance. This overall capacitance is used to calculate power dissipation by the CCD and clock driver requirements to transfer charge (refer to Chapter 5).

1.4.2 SURFACE-CHANNEL POTENTIAL WELL

The potential voltage generated in the silicon for a MOS capacitor in the depleted state is found by solving Poisson's differential equation,

$$\frac{d^2V}{dx^2} = \frac{\rho}{\varepsilon_{SI}}, \tag{1.8}$$

where V is the voltage in the silicon (V), and ρ is the charge density [i.e., $p + n + N_A + N_D$, where p is the number of free holes, n is the number of free electrons, N_A is the number of localized fixed ionized acceptors (atoms/cm^3) and N_D

is the number of fixed ionized donors (atoms/cm^3)]. Most free carriers within the depletion region will be swept away because of the electric field generated. Therefore, carrier concentrations n and p will be negligibly small in comparison to the impurity concentration over the space charge region. This assumption is called the depletion approximation, and therefore $\rho = N_A$ for a p-channel device. The origin (i.e., $x = 0$) for x is taken at the Si–SiO$_2$ interface (cm).

Assuming the boundary conditions $V = 0$, $dV/dx = 0$ at $x = x_d$ and integrating Eq. (1.8) with respect to x, the voltage drop as a function of distance in the silicon layer is

$$V = \frac{qN_A}{2\varepsilon_{SI}}(x - x_d)^2. \tag{1.9}$$

Note from this equation and Fig. 1.18 that the potential voltage is greatest at the Si–SiO$_2$ interface (i.e., $x = 0$), thus forming a potential well where electrons can collect and transfer.

The surface voltage, V_S, at $x = 0$ is

$$V_S = \frac{qN_A}{2\varepsilon_{SI}}x_d^2. \tag{1.10}$$

The electric field at the surface is found by differentiating Eq. (1.10) with respect to x, yielding

$$E_S = -\frac{qN_A}{\varepsilon_{SI}}x_d. \tag{1.11}$$

Example 1.8

Find the surface potential and electric field strength at the Si–SiO$_2$ interface for the MOS capacitor described in Example 1.5.
Solution:
From Eq. (1.10),

$$V_S = (1.6 \times 10^{-19}) \times 10^{15} \times (3 \times 10^{-4})^2 / (2 \times 1.04 \times 10^{-12}),$$

$$V_S = 6.89\,\text{V}.$$

From Eq. (1.11),

$$E_S = -(1.6 \times 10^{-19}) \times 10^{15} \times (3 \times 10^{-4})/(1.04 \times 10^{-12}),$$

$$E_S = -4.6 \times 10^4\,\text{V/cm}.$$

HISTORY, OPERATION, PERFORMANCE, DESIGN, FABRICATION AND THEORY 67

The voltage drop across the gate oxide and silicon relative to substrate potential (0 V) is

$$V_G = V_{OX} + V_S, \qquad (1.12)$$

where V_G is the gate voltage (V) and V_{OX} is the gate oxide voltage drop (V). This equation is equivalent to

$$V_G = E_S d + V_S. \qquad (1.13)$$

Substituting Eqs. (1.10) and (1.11) into Eq. (1.13) yields

$$V_G = \left(\frac{q N_A x_d}{\varepsilon_{ox}}\right) d + \left(\frac{q N_A}{2 \varepsilon_{SI}}\right) x_d^2. \qquad (1.14)$$

Solving for the depletion depth x_d,

$$x_d = -\frac{\varepsilon_{SI}}{C_{ox}} + \sqrt{\left(\frac{\varepsilon_{SI}}{C_{ox}}\right)^2 + \frac{\varepsilon_{SI} V_G}{2 q N_A}}. \qquad (1.15)$$

The depletion depth increases by the square root of V_G and decreases with the square root of doping concentration. Note that this is true if the first term inside the square root is small compared to the second term, which is usually the case. If instead C_{OX} were very small, then the first term under the square root would dominate, and changing V_G or N_A would have little effect on the depletion depth.

Example 1.9

Find the depletion depth generated for a doping concentration of $N_A = 10^{15}$ cm^{-3} and an applied gate voltage of $V_G = 10$ V. Also find the gate capacitance (i.e., depletion and oxide). Assume $C_{OX} = 3.45 \times 10^{-8}$ F/cm^2.
Solution:
From Eq. (1.15),

$$x_d = -(1 \times 10^{-12})/(3.45 \times 10^{-8})$$
$$+ \left\{ [(1 \times 10^{-12})/(3.45 \times 10^{-8})]^2 \right.$$
$$\left. + (1.04 \times 10^{-12} \times 10)/[(2 \times (1.6 \times 10^{-19}) \times 10^{15})] \right\}^{1/2},$$

$$x_d = 1.5 \, \mu m.$$

From Eq. (1.6),

$$C_{DEP} = 10^{-12}/(1.5 \times 10^{-4}),$$

$$C_{DEP} = 6.6 \times 10^{-9}\,\text{F/cm}^2.$$

From Eq. (1.7),

$$C_T = [1/(3.45 \times 10^{-8}) + 1/(6.6 \times 10^{-9})]^{-1},$$

$$C_T = 5.5 \times 10^{-9}\,\text{F/cm}^2.$$

1.4.2.1 Well Capacity

Signal electrons collect at the Si–SiO$_2$ interface when a positive gate voltage is applied. For charge neutrality to exist it is required that the charge on the gate be equal to the sum of signal charge collected in a potential well and the net lattice acceptor charge N_A. For a fixed gate potential, the number of charged acceptor atoms decrease as free electrons collect. The depletion region also becomes smaller:

$$x_d = -\frac{\varepsilon_{Si}}{C_{OX}} + \sqrt{\left(\frac{\varepsilon_{Si}}{C_{OX}}\right)^2 + \frac{\varepsilon_{Si} V_Q}{2qN_A}}, \quad (1.16)$$

where V_Q is the effective voltage drop at the gate induced by the charge collected in the potential well (V), a quantity defined by

$$V_Q = V_G - \frac{qQ}{C_{OX}}, \quad (1.17)$$

where Q is the signal charge collected at the Si–SiO$_2$ interface (e$^-$/cm^2).

Example 1.10

Find the new depletion depth when 3×10^{11} e$^-$/cm^2 of charge is stored in the potential well described in Example 1.9.
Solution:
From Eq. (1.17),

$$V_Q = 10 - (3 \times 10^{11}) \times (1.6 \times 10^{-19})/(3.45 \times 10^{-8}),$$

$$V_Q = 8.6\,\text{V}.$$

From Eq. (1.16),

$$x_d = (-1.04 \times 10^{-12})/(3.45 \times 10^{-8}) \\ + \left\{[(1.04 \times 10^{-12})/(3.45 \times 10^{-8})]^2 \right. \\ \left. + (1.04 \times 10^{-12} \times 8.6)/[2 \times (1.6 \times 10^{-19}) \times 10^{15}]\right\}^{1/2},$$

$x_d = 1.375\,\mu\text{m}$.

Therefore, the depletion width shrinks from 1.5 to 1.375 µm due to the charge collected. The surface potential and the corresponding electric field also are reduced.

Charge capacity of a MOS capacitor is defined as the amount of charge required to bring the surface potential to 0 V (i.e., $V_S = 0$ V, $Q_i = 0$, $x_d = 0$). In that the substrate and gate potentials are fixed, signal electrons at the Si–SiO$_2$ interface are shared between the oxide and depletion capacitances (i.e., parallel capacitors in respect to the signal). Therefore, a change in surface potential of V_S due to Q is

$$\Delta V_S = -\frac{Q}{C_{OX} + C_{DEP}}. \qquad (1.18)$$

Note that C_{OX} remains fixed as electrons collect at the surface, whereas C_{DEP} increases with Q. However, C_{OX} usually dominates over C_{DEP}, yielding the approximation

$$Q = C_{OX} \Delta V_S. \qquad (1.19)$$

Equation (1.19) can be used to estimate the well capacity of a MOS capacitor.

Example 1.11

Find the number of electrons that can be stored in a 4×8-µm^2 region for the MOS capacitor described in Example 1.10, assuming $C_{OX} = 3.45 \times 10^{-8}$ F/cm^2.

Solution:
From Eq. (1.10) the surface potential for an applied gate voltage of 10 V and empty well is

$$V_S = \left(1.6 \times 10^{-19}\right) \times 10^{15} \times \left(1.5 \times 10^{-4}\right)^2 / \left(1.04 \times 10^{-12}\right),$$

$$V_S = 3.44\,\text{V}.$$

From Eq. (1.19) full well is estimated as

$$Q = \left(3.45 \times 10^{-8}\right) \times 3.44,$$

$$Q = 1.19 \times 10^{-7}\,\text{C/cm}^2.$$

The number of electrons stored in a $4 \times 8\text{-}\mu m^2$ region is

$$N = (4 \times 10^{-4}) \times (8 \times 10^{-4}) \times (1.19 \times 10^{-7})/(1.6 \times 10^{-19}),$$

$$N = 2.4 \times 10^5 \, e^-.$$

It is important to note that the quantity of charge that can be stored is directly proportional to doping concentration and to the applied gate potential.

1.4.3 BURIED-CHANNEL POTENTIAL WELL

Discussions above have described the operation of a surface-channel CCD, in which charge is stored and transferred along the surface of the semiconductor. As mentioned earlier, a major problem exists with surface-channel CCDs because signal charge is trapped at the Si–SiO$_2$ interface, severely limiting CTE performance. Early in the development of the CCD, different approaches were attempted to passivate and reduce the density of interface states. However, CTE requirements were too demanding even for the best of processes, especially for large-area-array scientific CCDs.

To avoid the surface-state problem, the buried-channel CCD was invented. In a buried-channel CCD, charge packets are confined to a channel (i.e., a potential well) that lies beneath the surface. In contrast to surface-channel operation, CTE performance for buried-channel CCDs is remarkably high. As demonstrated

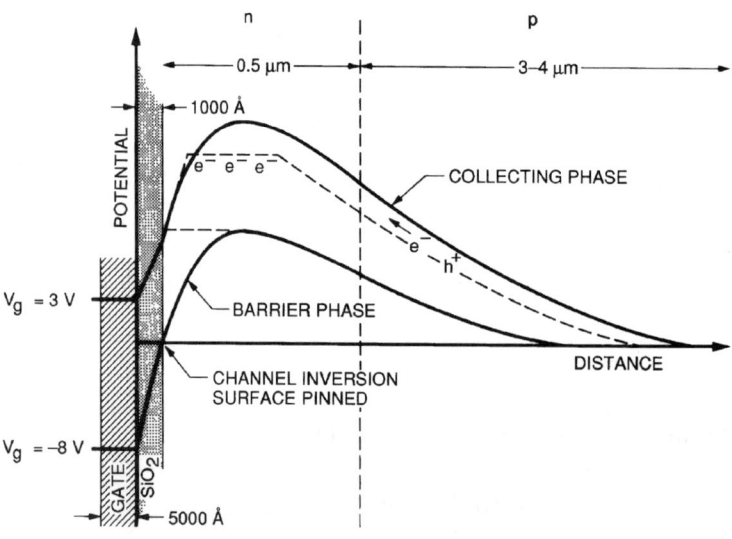

Figure 1.19 Buried-channel potential well.

in Chapter 5, CTEs > 99.9999% per pixel transfer are routinely achieved. Therefore, scientific CCDs are totally based on buried-channel technology.

Figure 1.19 presents a cross-sectional view of a buried-channel potential well showing a thin doped n-region that forms the buried-channel. In comparison to a surface-channel structure (Fig. 1.18), the n-layer reshapes the potential well to form a potential maximum (V_{max}) below the Si–SiO$_2$ interface and above the n-p junction. Two buried-channel potential wells are shown in Fig. 1.19 for applied gate voltages of -8 V and 3 V. The collecting well is at a higher potential than the barrier well and is where electrons would collect in a pixel.

Figure 1.20 shows spreading resistance measurements of the n-channel and epitaxial layer for the WF/PC II CCD. Doping is initially specified in the number of ions that impinge on the surface of the silicon per unit area (cm^{-2}). The range of these ions into the silicon is very short depending on their incident energy (refer to Appendix H). The dopant is then thermally driven into the epitaxial layer, the final depth depending on the drive temperature and processing time. The net doping is usually specified in concentration per unit volume (cm^{-3}). The two specifications

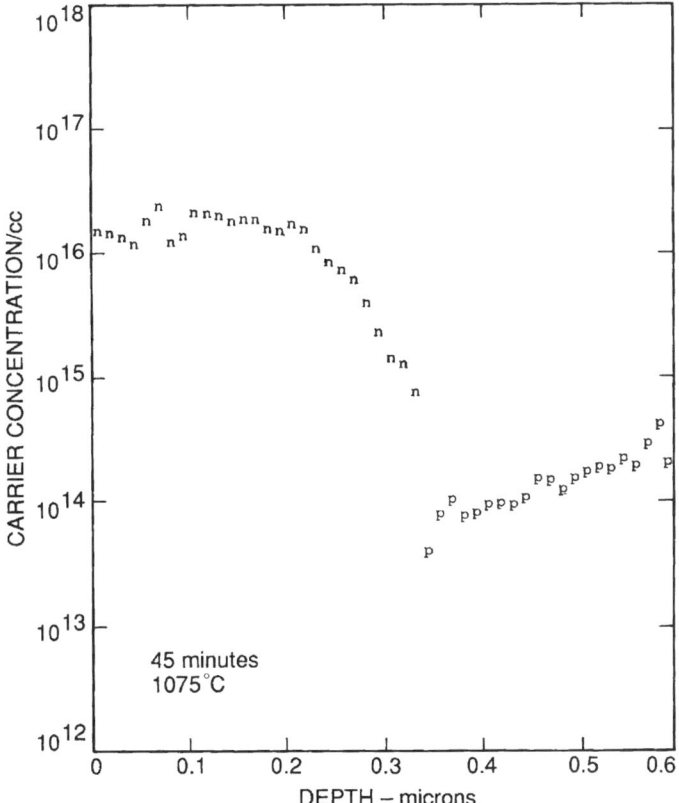

Figure 1.20(a) Buried-channel doping distribution for a 45-minute 1075°C drive.

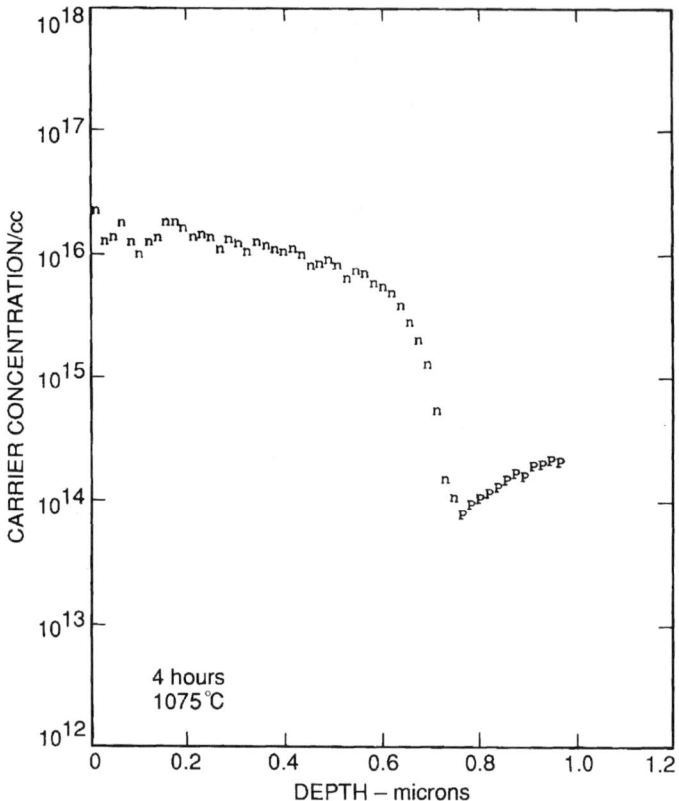

Figure 1.20(b) Buried-channel doping distribution for a 4-hour 1075°C drive.

are related through

$$N_A = \frac{N_{AREA}}{t}, \qquad (1.20)$$

where N_A is the doping concentration per unit volume (cm^{-3}), N_{AREA} is the doping concentration per unit area (cm^{-2}) and t is the depth of the channel after the thermal drive (cm).

Example 1.12

Determine the donor impurity concentration for an ion implant of 1.6×10^{12} phosphorus atoms/cm^2. Assume that the channel is driven 1 μm.
Solution:
From Eq. (1.20),

$$N_A = 1.6 \times 10^{12}/10^{-4},$$

$$N_A = 1.6 \times 10^{16} \, \text{atoms/cm}^{-3}.$$

Figure 1.20(a) shows a phosphorus-doped n-channel that reaches a depth of approximately 0.3 µm into the silicon after a 45-minute, 1075°C channel drive is applied. Figure 1.20(b) shows an extended temperature drive of four hours, which takes the channel to a depth of 0.6 µm. Channel drive affects clocking potentials and must be carefully controlled at the manufacturer, a subject related to an important process parameter called the "effective threshold voltage," to be discussed below and in Chapter 4. A shallow channel will result in a lower threshold voltage and low clock voltages, usually a desirable processing result. Deep-channel CCDs are fabricated to increase fringing fields between phases for high-speed applications. The depth of channel is also controlled by the ion energy incident on the silicon. Appendix H1 gives the range of phosphorus ions in silicon as a function of energy. Appendices H2 and H3 are for boron and arsenic, impurities that are commonly used to implant the CCD.

1.4.3.1 Depletion

The n-buried channel must be completely depleted of majority carriers (electrons) to distinguish them from signal electrons when generated. This condition is assumed to have already taken place in Fig. 1.19 in producing the potential well. To understand how depletion is first achieved in the CCD we will first review the physics behind a simple junction diode, a device very closely related to the buried-channel CCD.

Figure 1.21 shows a cross-section of an n-p diode. Recall that an n-type doped semiconductor (e.g., silicon doped with phosphorus) exhibits a large concentration of mobile electrons. Similarly, a p-type semiconductor (e.g., silicon doped with boron) contains a large concentration of holes. At room temperature every donor (n-type, column V impurity) and acceptor (p-type, column III impurity) atom thermally ionizes by kT (recall kT at room temperature is about 0.025 eV). This is because the energy level of donor states is very close to the conduction band, whereas acceptor states are close to the valence band (note that we call p-impurities acceptors because they can accept an electron from the valence band, which is equivalent to supplying a hole to the valence band). For example, the ionizing energy of phosphorus in silicon is much smaller than the silicon bandgap, approximately 0.05 eV from the conduction edge. Similarly, boron exhibits an ionization of 0.05 eV from the valence edge. Therefore, under the condition of complete ionization, the concentration of electrons in n-material is equal to N_D, and the concentration of holes in p-material is N_A. At very cold operating temperatures (<70 K) this process begins to stop. This condition is called "freeze-out" and is the point where the CCD will cease to function. We will discuss freeze-out further in Chapter 7.

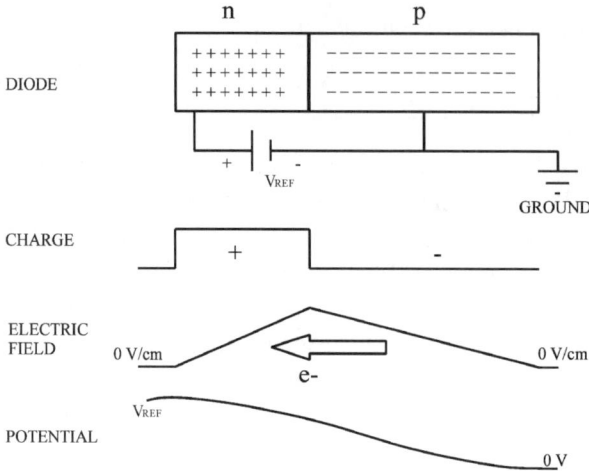

Figure 1.21 Charge, electric field and potential diagrams for a simple n-p junction diode that is reverse biased.

Before bias is applied to the diode, mobile holes diffuse from the p-side and move toward the junction where they recombine with mobile electrons on the n-side. A small depletion region appears naturally, until enough fixed lattice charges (negative on the p-side and positive on the n-side) are exposed to create an electric field. For silicon, the potential difference that forms between the two regions is about 0.7 V. The direction of this field is from the n to p side and acts to repel mobile charges from the junction and is strong enough to inhibit further drift of mobile charges toward the junction. Forward biasing the junction generates an electric field that is in the opposite direction to the built-in field. Once greater than 0.7 V, mobile charges can recombine at the junction, and current begins to flow because there is no depletion region.

Reverse biasing the junction as shown in Fig. 1.21 strengthens the electric field, further repelling mobile charges from the junction and increasing the size of the depletion region. What remains in the semiconductor are fixed lattice charges located at the dopant locations. For every donor atom that is charged, a corresponding number of acceptor atoms needs to be charged to conserve charge neutrality. If the applied voltage is great enough, full depletion will result in the n-region (this occurs in the CCD, as we shall see). The fixed lattice charge will generate an electric field and corresponding potential gradient as shown. If a free electron is generated by the photoelectric effect, the carrier will be swept to the highest potential seen, which is at the n-contact. Holes from the photoelectric effect move in the opposite direction toward the p-contact.

The maximum depletion depth, x_p, in the p-region is found by

$$x_p = \left[\frac{2\varepsilon_{SI} N_D}{q N_A (N_A + N_D)}(V_{REF})\right]^{1/2}, \qquad (1.21)$$

where V_{REF} is the reference voltage that reverse biases the diode (V), N_D is the n-donor concentration (atoms/cm^3), and N_A is the p-acceptor concentration (atoms/cm^3).

If the doping of the n and p regions are such that $N_D \gg N_A$, which will be the case for buried-channel CCDs, the above equation reduces to

$$x_p = \left[\frac{2\varepsilon_{SI}}{qN_A}(V_{REF})\right]^{1/2}. \tag{1.22}$$

The impurity concentration of the p-layer is normally specified as resistivity (Ω-cm). Resistivity and doping concentration are approximately related through

$$R_R = \frac{1}{q\mu_{SI}N_A}, \tag{1.23}$$

where R_R is the resistivity (Ω-cm), and μ_{SI} is the electron mobility for silicon (cm^2/V-sec). Mobility is a function of doping density, but is approximately constant over the doping range of 10^{14}–10^{15}, which covers the resistivity range of approximately 10–100 Ω-cm. If we assume a mobility of 500 cm^2/V-sec the Eq. (1.23) reduces to

$$R_R = \frac{1.25 \times 10^{16}}{N_A}. \tag{1.24}$$

Example 1.13

Determine the depletion depth, x_p, given a resistivity of 10 Ω-cm (1.2×10^{15} cm^{-3}) p-material. Assume $V_{REF} = 10$ V.
Solution:
From Eq. (1.22),

$$x_p = \left\{[2 \times (1.04 \times 10^{-12}) \times 10]/[(1.2 \times 10^{15}) \times (1.6 \times 10^{-19})]\right\}^{1/2},$$

$$x_p = 3.3 \, \mu\text{m}.$$

The buried-channel structure shown in Fig. 1.19 can be thought of as an n-p junction with a gate and insulator at the exterior surface of the n-side. As in the case of a reverse-biased diode, a depletion region forms around the n-p junction when biased with a reference voltage (V_{REF}). For small values of V_{REF}, the uncompensated lattice charges in the (small) depletion region at the junction give rise to an increase in potential from the p-side to the n-side, analogous to that shown in Fig. 1.21 for the reverse-biased diode. In the n-material (and also in the p-material) beyond the depletion region, the potential remains constant, since there are no uncompensated charges present. However, as the surface is approached in the n-material, the situation becomes analogous to an n-type MOS structure. If the applied gate voltage is negative relative to the channel potential, the majority carriers (electrons) will be repelled away from the surface, thereby creating a surface depletion region of uncompensated donor ions. These positive lattice charges induced by the gate cause the potential in the n-material to fall abruptly as the surface is approached.

For a fixed gate voltage, consider next what happens as these two depletion regions (junction and surface) begin to meet as V_{REF} is increased. Since V_{REF} reverse-biases the junction, increasing V_{REF} will increase the size of the depletion region at the n-p junction. Meanwhile, increasing V_{REF} will cause the fixed gate voltage to appear more negative with respect to the channel potential in the nondepleted n-material, and therefore the gate-induced depletion region at the surface will also increase in size. Finally, a reference voltage will be reached for which the two depletion regions merge into one, extending from the surface completely through the n-material and deep into the p-material. The point where the two depletion regions meet in the buried channel is where the maximum channel potential, V_{max}, is locally found. The potential is exactly equal to V_{REF} when the two regions merge.

Figure 1.22(a) provides more detail of the CCD output region and the way depletion first takes place. V_{REF} is connected, through the reset switch, to the output diode (or sense node) which is imbedded in the n-buried channel. Also shown is the output transfer gate and output summing well, each biased at V_{OTG} and V_{OSW}, respectively. The dotted lines show the three individual depletion edges as they begin to form in action to V_{REF} (gate, n-junction p-junction depletion). Here, the output transfer gate is biased less than the output summing well, to illustrate a depletion width difference.

As indicated above, $V_{max} = V_{REF}$ when gate and n-junction depletion regions meet. Increasing V_{REF} beyond this condition does not affect V_{max} or the p-depletion edge. Any increase in V_{REF} only influences depletion in the n^+ output diode and surrounding p^+ material. When the buried channel is fully depleted, the p-depletion edge and the position of V_{max} are solely determined by the gate voltage. For example, if the gate voltage increases, less depletion will come from the gate side. This in turn will require more depletion on both sides of the n-p junction. On the other hand, if the gate voltage is lowered, depletion will come more from the gate side, moving the potential maximum toward the n-p junction. The p-depletion depth will also be reduced. Figure 1.22(b) plots depletion depth as a function of V_{REF} for a family of gate voltages. The curved envelope shown is described by Eq. (1.21). Note that the depletion depth levels out at a specific gate voltage. This voltage is

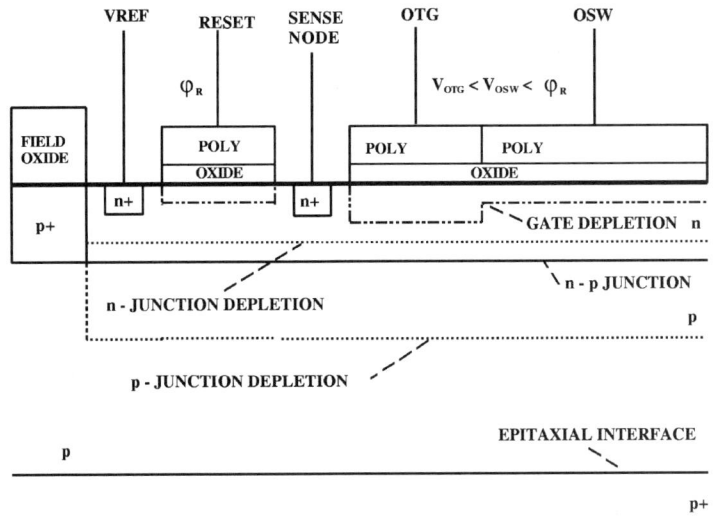

Figure 1.22(a) Output region showing gate and n-p junction depletion regions as they form.

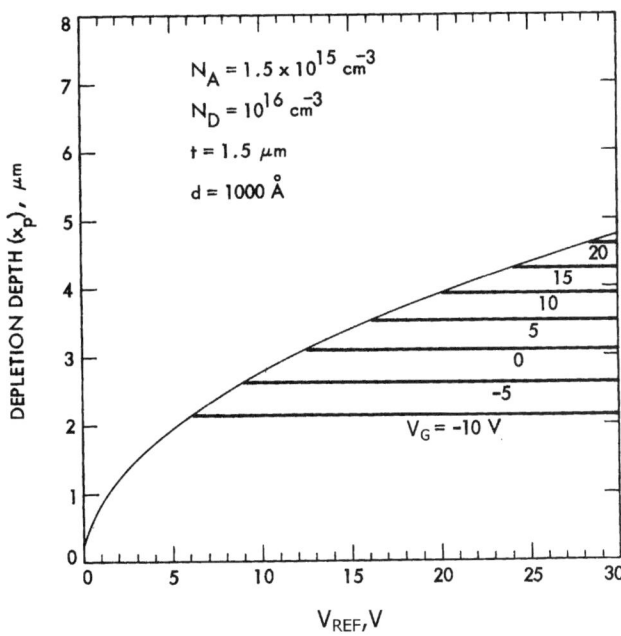

Figure 1.22(b) Depletion depth as a function of reference and gate voltage.

where the gate and junction depletion regions meet within the channel at V_{max}. The reference voltage loses control of depletion at this point leaving the gate potential to determine the extent of the p-depletion region.

Depletion must first take place under the output transfer gate before other phases of the array can be depleted of majority carriers. In other words, bias conditions must be such that the potential maximum under the output transfer gate must be less than V_{REF}. Once the output transfer gate is depleted, the output summing well and horizontal and vertical registers can be depleted simply by clocking them. For example, clocking the output summing well more negative than the output transfer gate causes electrons from its potential well to transfer through the output transfer gate onto the output diode. Once the output summing well is depleted it can act as a sink for electrons for the next horizontal phase. By clocking the CCD sequentially from gate to gate, depletion will propagate along the channel until the entire horizontal register is depleted of charge. For area array CCDs this charge transfer process continues up through the vertical registers, as they are clocked, until the entire sensor is depleted. As long as the clocking speed is fast enough (and/or the temperature is low enough) to prevent the channel from being supplied with thermally generated carriers, the channel will remain depleted and potential wells exist to collect signal charge.

It should be noted that when the output summing well is in a high state, it does not have proper static conditions for depletion as does the output transfer gate. It is in the state of nonequilibrium. That is, it immediately begins to undeplete as thermal dark current forces its potential to the level of the output transfer gate. At this point charge diffuses over the output transfer gate to the sense node until it reaches an equilibrium state. It is important to note that all clocked gates are in the state of nonequilibrium when the CCD is in use.

One can also consider the output diode equivalent to the "quasi-Fermi level" for the electrons set by V_{REF}. As long as the channel potential under the output transfer gate is less than the reference voltage, electrons will see a barrier and not move backward into the array. This does not mean that the potential within the array cannot be greater than V_{REF}. Again, the device will deplete as long as the phases are clocked more negatively than the output transfer gate.

1.4.3.2 Effective Threshold

The relationship between the applied gate voltage, V_G, and the channel maximum potential, $V_{\max}$, in the silicon is

$$V_{EFF} = V_{\max} - V_G, \qquad (1.25)$$

where V_{EFF} is the effective threshold voltage (V). Again, it is assumed that full n-channel depletion has been established when using this equation.

Equation (1.25) is used extensively to make clock adjustments and achieve optimum performance (refer to Chapter 2, Section 2.5). The effective threshold is one of the most important parameters that is measured when characterizing CCDs. Also, V_{EFF} is a critical process parameter for the CCD manufacturer. One knows a lot about the process through the effective threshold value.

HISTORY, OPERATION, PERFORMANCE, DESIGN, FABRICATION AND THEORY 79

The voltage V_{EFF} increases with channel doping and how far the n-dopant is driven into the silicon thermally. Values for V_{EFF} range from 6 to 18 V depending on processing details and charge capacity required. We will refer to V_{EFF} often in this book.

Example 1.14

Find the maximum potential within the CCD channel when $V_G = 5$ V and $V_{EFF} = 10$ V.
Solution:
From Eq. (1.25),

$$V_{\max} = 10 + 5,$$

$$V_{\max} = 15 \text{ V}.$$

1.4.3.3 Charge Injection

From discussions above, it is possible to electrically inject charge into the CCD using the output transfer gate and output summing well gate in the following manner. For charge injection to occur, V_{REF} is first set at a fixed potential (e.g., $V_{REF} = 12$ V). Next, the output transfer gate and output summing well are both biased together such that $V_{\max} > V_{REF}$. For example, setting both gates at 6 V produces a channel potential of $V_{\max} = 13$ V, assuming $V_{EFF} = 7$. Under this condition, charge from the output diode is injected under the output transfer gate and output summing well. Equilibrium is established instantaneously forcing the channel potentials under the gates to equal the output diode. Then, the output transfer gate is biased negatively to $V_G = -4$ V, setting $V_{\max} = 3$ V under this gate. This action confines charge to the output summing well which is still biased at $V_G = 6$ V. Charge from the output summing well can then be injected into the horizontal register by clocking the phase next to the output summing well high. The charge packet can be further transferred down the horizontal register and up into the array by clocking the vertical registers backward. The entire CCD can be filled with charge, if desired.

The quantity of charge electrically injected can be controlled precisely by the potential difference between the output diode and the channel potential under the output summing well as

$$S_{INJ}(\text{e}^-) = \frac{(V_{\max} - V_{REF})C_{OSW}}{q}, \quad (1.26)$$

where $S_{INJ}(\text{e}^-)$ is the charge injected (e$^-$) and $V_{\max}$ and C_{OSW} are the channel potential (V) and capacitance (F) of the output summing well, respectively.

Example 1.15

Find the amount of charge that is electrically injected, assuming $V_{REF} = 12$ V, $V_{\max} = 14$ V ($V_G = 4$ V), $C_{OSW} = 8.32 \times 10^{-15}$ F, and $V_{EFF} = 10$ V.
Solution:
From Eq. (1.26),

$$S_{INJ} = (8.32 \times 10^{-15}) \times (14 - 12)/(1.6 \times 10^{-19}),$$

$$S_{INJ} = 100{,}000\,e^-.$$

As a note, charge injection was a required feature for the CCD when it was used as a digital memory device in the early seventies. Although there were no lead solid state competitors for the CCD in the imaging field, there were many different memory technologies under development at that time (e.g., MOS memory, magnetic bubbles, disks, drums, tape, etc.). The CCD was forced to compete in price and performance with these available technologies. Several companies invested heavily and fabricated CCD memories in this competition race (e.g., Fairchild, Intel, Bell Northern, Texas Instruments). Unfortunately for the CCD, the dynamic NMOS RAM, its toughest competitor, won out. Charge injection is also required when the CCD is employed in sampled data signal-processing applications (e.g., multiplexing, recursive and transversal filtering, analog delay, correlation). However, these CCD applications did not mature as hoped, because other digital technologies were superior. Today, charge injection is rarely used in CCD imaging applications, although a few imaging applications will be discussed in this book. Charge injection input diodes and gates were included in nearly every early CCD design but for the most part have disappeared. We will be injecting charge into the CCD to determine the effective threshold voltage as described in the next section.

1.4.3.4 Output Transfer Gate Transfer

The effective threshold can be readily measured from an important transfer curve called the Output Transfer Gate Transfer. The curve is presented in Fig. 1.23(a). These data can either be generated by a computer or simply measured directly at the output of the CCD with an oscilloscope as the sensor is constantly reading out. The curve plots V_{REF} as a function of V_{OTG} for those points where charge injection takes place. As discussed above, charge injection occurs when the channel potential under the output transfer gate becomes equal to V_{REF}. This condition can be reached by increasing V_{OTG} or decreasing V_{REF}. It should be noted that the negative level of the precharge pulse must also be adjusted with V_{OTG} when this measurement is performed to maintain a barrier between the sense node and V_{REF} when the switch is off.

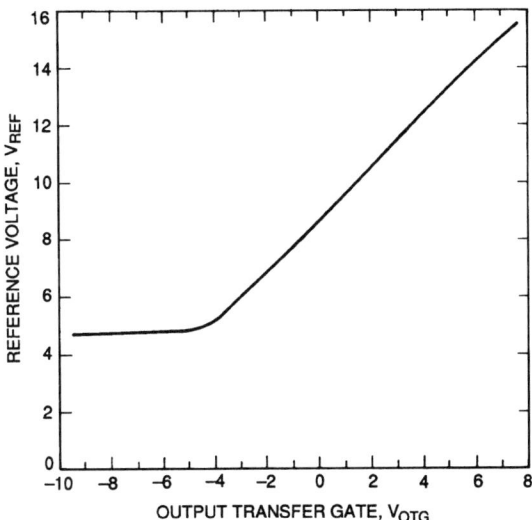

Figure 1.23(a) Output Transfer Gate Transfer.

The charge injection break point defines V_{EFF} through the equation

$$V_{EFF} = V_{REF} - V_{OTG/CI}, \qquad (1.27)$$

where $V_{OTG/CI}$ is the output transfer gate voltage when charge injection takes place(V).

The charge injection effect is shown in Fig. 1.23(b) where output signal from the CCD is measured as a function of V_{REF} for different V_{OTG}. For example, the dramatic increase in signal that occurs when $V_{OTG} = 3$ V and $V_{REF} = 11.5$ V indicates the onset of charge injection. The charge injection breakpoints seen in this figure are used to generate the output transfer gate transfer curve in Fig. 1.23(a).

Example 1.16

Find the effective threshold voltage for the CCD characterized in Fig. 1.23(a).
Solution:
From Eq. (1.27),

$$V_{EFF} = 8.5 - 0,$$

$$V_{EFF} = 8.5 \, \text{V}.$$

Note that V_{EFF} is a constant because of the linear relationship between V_{REF} and V_{OTG}.

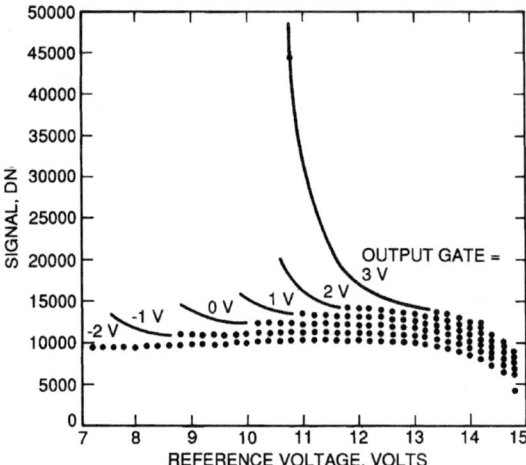

Figure 1.23(b) Reference voltage required to cause charge injection for different OTG potentials.

For several subtle reasons, the effective threshold is actually slightly lower than what the output transfer gate provides. For example, when the reset switch is turned off, the reference level on the sense node decreases by the precharge feed-through clock level. The reset clock feed-through makes the effective threshold slightly larger than it really is. The amount of reset feed-through generated on the sense node is

$$V_{RFT} = \frac{C_{GS}}{C_S + C_{GS}} V_{PC}, \tag{1.28}$$

where V_{RFT} is the reset feed-through level on the sense node (V), C_S is the sense node capacitance (F), C_{GS} is the gate to source capacitance of the reset MOSFET switch (F) and V_{PC} is the reset clock amplitude (V).

The reset pulse is always seen in the output video and can be the largest signal generated from the CCD. Figure 1.24 shows the output amplifier and reset switch and the capacitances involved in producing the reset feed-through pulse. Note the reset pulse is generated by the capacitance voltage divider action of C_{GS} and C_S. Also illustrated is the video waveform seen at the output of the CCD. Labeled are the reset feed-through pulse (V_{RFT}), the reference level at the time the reset switch is on and the video transition at the time the charge is dumped onto the sense node. The figure also shows the output summing well clock feed-through pulse, which is usually small enough to neglect in this analysis.

Example 1.17

Find the precharge feed-through amplitude assuming the following parameters: $C_S = 10^{-14}$ F, $C_{GS} = 1 \times 10^{-15}$ F, $V_{PC} = 10$ V.

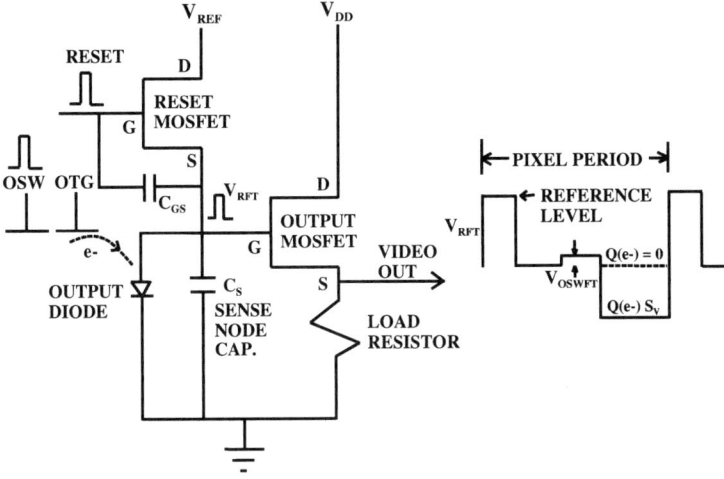

Figure 1.24 Output amplifier region showing video and feed-through signals.

Solution:
From Eq. (1.28),

$$V_{RFT} = 10 \times (1 \times 10^{-15})/[10^{-14} + (1 \times 10^{-15})],$$

$$V_{RFT} = 0.91 \text{ V}.$$

The amplitude of the reset feed-through pulse can be measured at the output of the CCD less the gain of the output amplifier, which is near unity.

Other factors also add to the threshold level measured. For example, signal on the sense node will make the sense node more negative, leading to a higher threshold value. This problem is best taken care of by running the vertical clocks backward or by cooling the CCD when V_{EFF} is measured. Also, a notch channel, if employed in the horizontal register, will add to the threshold level measured (refer to Chapter 5 on notch technology). Notch channels run about 1–2 V deeper than the main channel, thereby increasing the channel potential under the output transfer gate by this amount. Thermal energy of the electrons on the sense node will cause charge injection prematurely. The effect increases the threshold by about $50kT$, depending on measurement sensitivity and operating temperature (refer to Chapter 3 on thermal barrier heights).

With these factors included, the "true" effective threshold as measured by the manufacturer is

$$V_{EFF} = V_{REF} - V_{OTG/CI} - V_{RFT} - V_{NOC} - 50kT - S_V S(e^-), \quad (1.29)$$

where V_{NOC} is the notch built-in potential (V), S_V is the sense node sensitivity (V/e$^-$), $S(e^-)$ is the charge on the sense node (e$^-$), T is temperature (K) and k is Boltzmann's constant (eV/K).

Example 1.18

Determine the true effective threshold for the measurement performed in Example 1.16. Assume $V_{NOC} = 0$ V (i.e., no notch), $S_V S(e^-) = 0$ (vertical registers running backwards), $V_{RFT} = 1$ V and $50kT = 0.83$ V ($T = 200$ K).
Solution:
From Eq. (1.29),

$$V_{EFF} = 8.5 - 1 - 0.83,$$

$$V_{EFF} = 6.67 \text{ V}.$$

1.4.3.5 Inversion and Pinning

A very important condition develops when a CCD gate is driven negatively. As the gate voltage is lowered, the surface potential decreases until it becomes equal to the substrate potential (i.e., $V_S = V_{SUB}$). This condition is shown in Fig. 1.19 when $V_G = -8$ V. In this bias state, holes from the channel stop region are attracted and collect at the Si–SiO$_2$ interface. This condition is called the "inverted state" because minority carriers (holes) populate the n-channel. Additional gate voltage will attract more holes, "pinning" and maintaining $V_S = 0$ V independent of how negative the gate voltage that is applied. Under the pinned state the potential well is not influenced by the gate voltage, because the thin layer of holes at the surface is conductive and shields the silicon layer from any change in gate voltage applied. When inverted, any change in gate bias goes directly across the gate insulator as a voltage drop (i.e., V_{OX}). The output transfer gate transfer curve presented in Fig. 1.23(a) shows where the device inverts (i.e., V_{INV}). The onset of inversion takes place when $V_{OTG} = -4$ V, producing the flat response shown for gate voltages lower than this. Inversion has a profound effect on many CCD performance parameters, as discussed throughout this book.

1.4.3.6 Buried-Channel Model

This section presents analytical equations that describe the buried-channel potential well.[24] Equations and the potential curves derived are important to explain many performance characteristics and limitations for the CCD in upcoming chapters. Assuming full depletion—a necessary condition discussed above—the potential distribution within the gate insulator and silicon region can be analytically obtained. In the derivation that follows, an idealized doping distribution for the

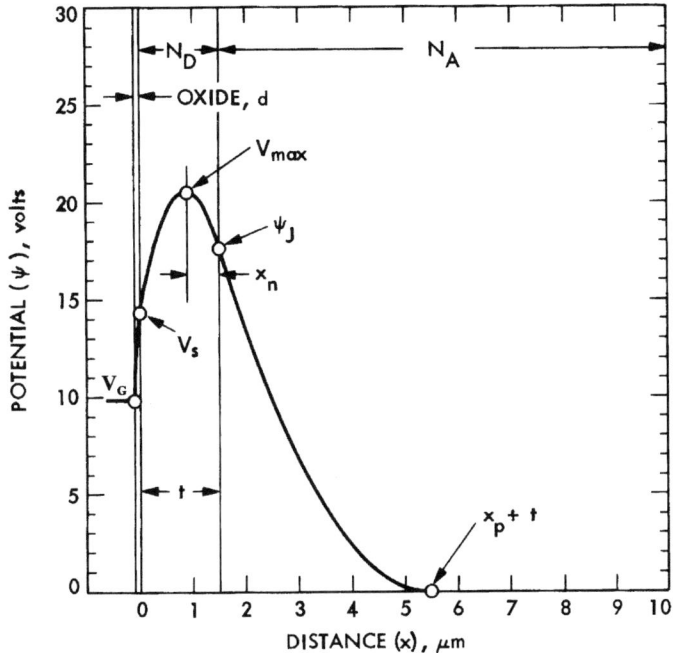

Figure 1.25 Modeling parameters for a buried-channel potential well.

buried-channel layer is assumed—that is, a "box" distribution where the impurity concentration is constant throughout the n-layer. In practice, the doping of the buried channel is not uniform because the channel is ion-implanted followed by a high temperature drive. This results in a "Gaussian" profile for the n-channel (refer to Fig. 1.20). Analysis and calculation of the potentials for a real distribution require numerical techniques. There are also many good commercial modeling programs that give such solutions (e.g., PISCES two-dimensional simulation modeling software). The box distribution model described below allows us to find explicit equations, making calculations easy to do. The model is sufficiently accurate to describe most physical characteristics exhibited by the CCD that are useful for our purposes. For the simple analysis, only a one-dimensional potential profile is calculated in the direction perpendicular to the silicon surface, similar to surface-channel analysis above. Important variables to the derivation are shown in Fig. 1.25.

The potential distribution in the gate and silicon is found by solving Poisson's differential equations, assuming depletion approximation ($\rho = qN_A$ and $\rho = -qN_D$) as

$$\frac{d^2V}{dx^2} = 0, \qquad -d < x < 0, \qquad (1.30)$$

$$\frac{d^2V}{dx^2} = \frac{-qN_D}{\varepsilon_{SI}}, \quad 0 < x < t, \tag{1.31}$$

$$\frac{d^2V}{dx^2} = \frac{qN_A}{\varepsilon_{SI}}, \quad t < x < t + x_p, \tag{1.32}$$

where t is the n-channel depth (cm), qN_D is the charge density of ionized donor (phosphorus) atoms and x_p is the depletion width of the p-region (cm). The boundary conditions for the above differential equations are

$$V(x = -d) = V_G - V_{FB}, \tag{1.33}$$

where V_{FB} is the flat-band voltage associated with fixed charge in the oxide layer (V).

$$K_{OX}\frac{dV}{dx}(x = 0^-) = K_{SI}\frac{dV}{dx}(x = 0^+), \tag{1.34}$$

where K_{OX} and K_{SI} are the dielectric constants for oxide (3.9) and silicon (11.8).

$$V(x = 0^-) = V(x = 0^+), \tag{1.35}$$

$$\frac{dV}{dx}(x = t^-) = \frac{dV}{dx}(x = t^+), \tag{1.36}$$

$$V(x = t^-) = V(x = t^+), \tag{1.37}$$

$$V(x = t + x_p) = 0. \tag{1.38}$$

Solutions to the differential equations for the three regions are

$$V_1 = V_G - V_{FB} - E_{OX}(x + d), \quad -d < x < 0, \tag{1.39}$$

where E_{OX} is the electric field across the oxide (V/cm).

$$V_2 = V_{\max} - \frac{qN_D}{2\varepsilon_{SI}}(x - x_n)^2, \quad 0 < x < t, \tag{1.40}$$

where $V_{\max}$ is the channel's potential maximum and x_n is the distance to the potential maximum from the n-p junction (cm).

$$V_3 = \frac{qN_A}{2\varepsilon_{SI}}(x - t - x_p)^2, \quad t < x < t + x_p. \tag{1.41}$$

The maximum potential, V_{max}, of the channel is

$$V_{max} = V_J\left(1 + \frac{N_A}{N_D}\right), \tag{1.42}$$

where V_J is the junction potential,

$$V_J = \frac{2(V_G - V_{FB} + V_{IMP}) + V_{OX} - \{V_{OX} + [V_{OX} + 4(V_G - V_{FB} + V_{IMP})]\}^{1/2}}{2}, \tag{1.43}$$

where V_{IMP} is a voltage associated with the n-layer given as

$$V_{IMP} = qN_Dt\left(\frac{d}{\varepsilon_{OX}} + \frac{t}{2\varepsilon_{SI}}\right), \tag{1.44}$$

and V_{OX} is the voltage drop across the oxide,

$$V_{OX} = \frac{2qN_At^2}{\varepsilon_{SI}}\left(1 + \frac{\varepsilon_{SI}d}{\varepsilon_{OX}t}\right)^2. \tag{1.45}$$

The location of the potential maximum is

$$x_n = t - \frac{x_p N_A}{N_D}, \tag{1.46}$$

where x_p is the depletion depth within the p-region given as

$$x_p = \sqrt{\frac{2V_J\varepsilon_{SI}}{qN_A}}. \tag{1.47}$$

The surface potential is given by

$$V_S = V_{max} - \frac{qN_D}{2\varepsilon_{SI}}x_n^2. \tag{1.48}$$

Example 1.19

Using the equations above, calculate V_{max}, V_J, V_{OX}, x_n, x_p, and V_S, assuming the following CCD parameters:

$N_A = 5 \times 10^{14}\,\text{cm}^{-3}$,

$N_D = 2 \times 10^{16}\,\text{cm}^{-3}$,

$t = 5 \times 10^{-5}\,\text{cm}\ (0.5\,\mu\text{m})$,

$d = 1 \times 10^{-5}$ cm,

$V_G = 6$ V.

Solution:
From the equations above,

$$V_{IMP} = (1.6 \times 10^{-19}) \times (2 \times 10^{16}) \times (5 \times 10^{-5}) \\ \times [(1 \times 10^{-5})/(3.45 \times 10^{-13})] \\ + (5 \times 10^{-5})/(2 \times 1.04 \times 10^{-12}),$$

$V_{IMP} = 8.47$ V,

$$V_{OX} = [2 \times (1.6 \times 10^{-19}) \times (5 \times 10^{14}) \\ \times (5 \times 10^{-5})^2]/(1.04 \times 10^{-12}) \\ \times \left\{1 + [(1.04 \times 10^{-12}) \times (1 \times 10^{-5})]/ \right. \\ \left. [(3.45 \times 10^{-13}) \times (5 \times 10^{-5})]\right\}^2,$$

$V_{OX} = 1$ V,

$V_J = 2 \times (6 + 8.47) + 1 - \{1 \times [1 + 4 \times (6 + 8.47)]\}^{1/2}$,

$V_J = 11.15$ V,

$V_{max} = 11.5 \times [1 + (5 \times 10^{14})/(2 \times 10^{16})]$,

$V_{max} = 11.43$ V,

$$x_p = (2 \times 11.15 \times 1.04 \times 10^{-12})/[(1.6 \times 10^{-19}) \\ \times (5 \times 10^{14})]^{1/2},$$

$x_p = 5.4 \times 10^{-4}$ cm (5.4 μm),

$x_n = 5 \times 10^{-5} - (5.4 \times 10^{-4}) \times (5 \times 10^{14})/(2 \times 10^{16})$,

$x_n = 3.65 \times 10^{-5}$ cm (0.365 μm),

$$V_S = 11.43 - [(1.6 \times 10^{-19}) \times (2 \times 10^{16}) \\ \times (3.65 \times 10^{-5})]^2/(2 \times 1.04 \times 10^{-12}),$$

$V_S = 9.39$ V.

The corresponding electric fields through each region are calculated by differentiating Eqs. (1.39), (1.40) and (1.41), yielding,

$$E_{OX} = -\frac{qN_D x_n}{\varepsilon_{OX}}, \quad -d < x < 0, \tag{1.49}$$

$$E_n = -\frac{qN_D}{\varepsilon_{SI}}(x - x_n), \quad 0 < x < t, \tag{1.50}$$

$$E_p = \frac{qN_A}{\varepsilon_{SI}}[x - (t + x_p)], \quad t < x < t + x_p. \tag{1.51}$$

Figures 1.26(a) and 1.26(b) are potential and electric field plots using the equations above for different V_G. Note that the electric field is zero at $V_{\max}$ where signal electrons collect. The field is positive on the left-hand side of $V_{\max}$ which directs electrons into the silicon. The field is negative on the right-hand side of $V_{\max}$ which directs charge to the surface. Also note that the electric fields extend deeper into the p-side of CCDs, although they are weaker. Some important observations worth noting for discussions given in future chapters are

1. For the typical doping parameters, depletion comes primarily from the gate side down, especially for low gate voltages.
2. As N_D increases, x_n decreases and $V_{\max}$ moves closer to the n-p junction, because less depletion comes from the junction side compared to the gate side.

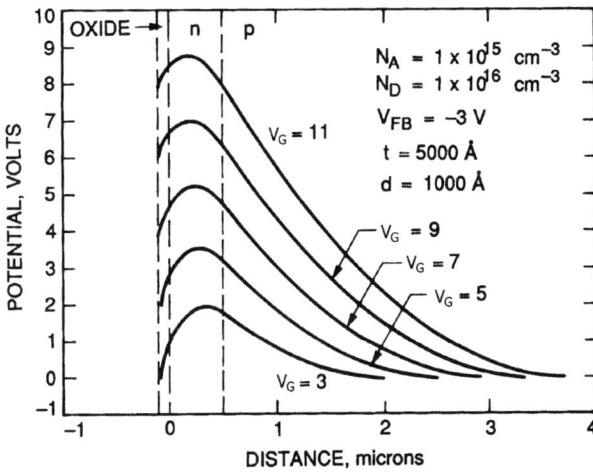

Figure 1.26(a) Buried-channel potential well modeling results.

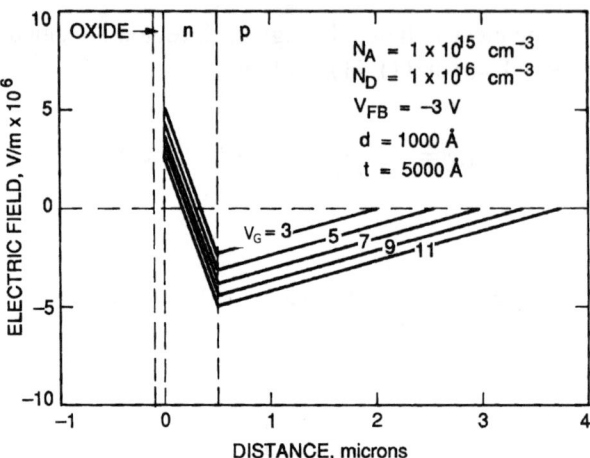

Figure 1.26(b) Buried-channel electric field modeling results.

3. As N_D increases V_{max} increases, because a higher applied V_{REF} is required to deplete channel.
4. As V_G increases, V_{max} increases.
5. As V_G increases, x_n decreases and V_{max} moves closer to the surface, because more depletion comes from the junction side. At some high gate voltage $V_{max} = V_S$ and the buried CCD acts as a surface-channel device.
6. As the oxide thickness increases, V_{max} increases.

1.4.3.7 Effective Storage Capacitance

The effective storage capacitance is defined as the capacitance between the gate electrode and the centroid of the signal charge packet. For an empty potential well, the effective capacitance is the series combination of the oxide and depletion capacitance of width $t - x_n$,

$$C_{EFF}^{-1} = \frac{d}{\varepsilon_{OX}} + \frac{1}{\varepsilon_{SI}}(t - x_n), \tag{1.52}$$

where C_{EFF} is the effective capacitance (F/cm^2), d is the dielectric thickness (cm), t is the channel depth (cm), and x_n is the location of V_{max} (cm).

Example 1.20

Calculate the effective capacitance for the following parameters: $d = 10^{-5}$ cm, $t = 5 \times 10^{-5}$ cm and $x_n = 3.65 \times 10^{-5}$ cm.

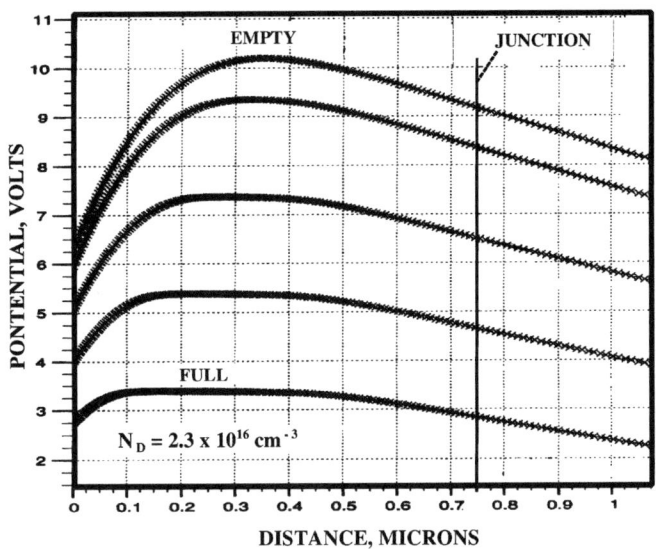

Figure 1.27 Channel potential as signal is added to a potential well.

Solution:
From Eq. (1.52),

$$C_{EFF}^{-1} = 10^{-5}/(3.45 \times 10^{-13})$$
$$+ [(5 \times 10^{-5}) - (3.65 \times 10^{-5})]/(1.04 \times 10^{-12}),$$
$$C_{EFF}^{-1} = (2.90 \times 10^{7}) + (1.3 \times 10^{7}),$$
$$C_{EFF}^{-1} = 4.2 \times 10^{7},$$
$$C_{EFF} = 2.38 \times 10^{-8} \text{ F/cm}^{2}.$$

When electrons are generated, they are stored at the potential maximum $V_{\max}$. In doing so, a portion of the well becomes "undepleted." Figure 1.27 is a two-dimensional computer simulation showing potential plots for various signal levels up to full well. Note that the addition of charge to the potential well causes the potential maximum to decrease and broaden toward the surface. Also recognize that there is not much movement of the depletion edge toward the n-p junction. As signal charge is generated, the depletion region from the surface becomes smaller, increasing the effective capacitance according to

$$C_{EFF}^{-1} = \frac{d}{\varepsilon_{OX}} + \frac{1}{\varepsilon_{SI}}\left(t - x_n - \frac{Q}{2N_D}\right), \qquad (1.53)$$

where Q is the electron density (e$^-$/cm^2).

Example 1.21

Determine the effective capacitance for a charge density of $Q = 1000\,e^-/\mu m^2$. Assume, from the example above, $N_D = 2 \times 10^{16}$ cm^{-3}, $t = 5 \times 10^{-5}$ cm, $x_n = 3.65 \times 10^{-5}$ cm.
Solution:
$1000\,e^-/\mu m^2$ translates to $1 \times 10^{11}\,e^-/cm^2$. Therefore, $Q/2N_D = 2.5 \times 10^{-6}$ cm.

From Eq. (1.53),

$$C_{EFF}^{-1} = 10^{-5}/(3.45 \times 10^{-13}) + \left[(5 \times 10^{-5}) - (3.65 \times 10^{-5}) - (2.5 \times 10^{-6})\right]/(1.04 \times 10^{-12}),$$

$$C_{EFF} = 2.53 \times 10^{-8}\,F/cm^2.$$

The effective capacitance will be used to estimate the CCD's charge capacity, as discussed in Chapter 4.

REFERENCES

1. M. Crowell and E. Labuda, "The silicon diode array camera tube," *Bell Syst. Tech. J.* Vol. 48, pp. 1481–1528 (1969).
2. W. Boyle and G. Smith, "Charge coupled devices," *Bell Syst. Tech. J.* Vol. 49, pp. 587–593 (1970).
3. G. Amelio, M. Tompsett and G. Smith, "Experimental verification of the charge coupled device concept," *Bell Syst. Tech. J.* Vol. 49, pp. 593–600 (1970).
4. M. Tompsett, G. Amelio and G. Smith, "Charge coupled 8-bit shift register," *Appl. Physics Ltrs.* Vol. 17, No. 3, p. 111 (1970).
5. F. L. J. Sangster, "The bucket brigade delay line, a shift register for analog signals," *Philips Tech. Rev.* Vol. 31, pp. 97–110 (1970).
6. F. Sangster, "Integrated MOS and bipolar analog delay lines using bucket-brigade capacitor storage," *Dig. Tech. Papers* Vol. XIII, pp. 74–75, 185 (1970).
7. F. Sangster and K. Teer, "Bucket brigade electronics—new possibilities for delay, time axis conversion and scanning," *IEEE J. of Solid State Circuits* Vol. SC-4, No. 3 (1969).
8. D. Kahng and E. H. Nicollian, "Monolithic semiconductor apparatus adapted for sequential charge transfer," U.S. Patent No. 3,651,349.
9. R. Smith and J. Tatarewicz, "Replacing a technology: the large space telescope and CCDs," *Proc. of the IEEE* Vol. 73, No. 7 (1985).

10. R. Walden, R. Krambeck, R. Strain, J. Mckenna, N. Schryer and G. Smith., "The buried channel charge coupled device," *Bell Syst. Tech. J.* Vol. 51, pp. 1635–1640 (1970).
11. L. Esser, "Peristaltic charge coupled device: a new type of charge transfer device," *Electron Lett.* Vol. 8, pp. 620–621 (1972).
12. C. Kim, J. Early and G. Amelio, "Buried channel charge coupled devices," presented at *NEREM*, Nov. 1972.
13. B. Oke, "Seyfert galaxies and quasars," *J. of the Royal Astronomical Soc. of Canada* Vol. 72, No. 3, p. 121 (1978).
14. J. Gunn, M. Carr, G. Danielson, E. Lorenz, R. Lucinio, V. Nenow, J. Smith, J. Westphal, D. Schneider and B. Zimmerman, *Optical Engineering* Vol. 26, p. 779 (1987).
15. R. Williams and the Hubble Deep Field Team (STScI) and NASA, presented to the *187th Meeting of the American Astronomical Society* in San Antonio, Texas, on January 15, 1996.
16. B. E. Burke, J. A. Gregory, M. W. Bautz, G. Y. Prigozhin, S. E. Kissel, B. B. Kosicki, A. H. Loomis and D. J. Young, "Soft-x-ray CCD imagers for AXAF," *IEEE Trans. Elec. Dev.* Vol. 44, No. 10, pp. 1633–1642 (1997).
17. F. Wilkinson, A. Farmer and J. Geist, "The near ultraviolet quantum yield of silicon," *J. Appl. Phys.* Vol. 54, No. 2, pp. 1172–1174 (1983).
18. D. Schroder, "Two-phase germanium charge-coupled device," *Appl. Phys. Lett.* Vol. 25, No. 12 (1974).
19. G. Burley, S. Chapman and G. Walker, "A precisely aligned CCD mosaic," *Publications of the Astronomical Society of the Pacific* Vol. 108, pp. 1024–1027 (1996).
20. J. Geary, G. Luppino, R. Bredthauer, R. Hlivak and L. Robinson, "A 4096 × 4096 pixel CCD mosaic imager for astronomy applications," *Proc. SPIE* (1991).
21. B. Burke, R. Mountain, D. Harrison, M. Baultz, J. Doty, G. Ricker and P. Daniels, "An abuttable CCD imager for visible and x-ray focal plane arrays," *IEEE Trans. on Electron Devices* Vol. 38, No. 5 (1991).
22. A. J. Theuwissen *et al.*, "Solid state sensor arrays: development and applications II," *Proc. SPIE* Vol. 3301, p. 37 (1998).
23. P. S. Heyes, P. J. Pool and R. Holtom, "Solid state sensor arrays: development and applications," *Proc. SPIE* Vol. 3019, p. 201 (1997).
24. C.-K. Kim, "Physics of charge coupled devices," In *Charge-Coupled Devices and Systems*, M. Howes and D. Morgan, Eds., John Wiley and Sons, pp. 51–53 (1979).

FURTHER READING

C. Sequin and M. Tompsett, *Charge Transfer Devices*, Academic Press Inc. (1975).
R. Melen and D. Buss, Eds., *Charge-Coupled Devices: Technology and Applications*, IEEE Press (1977).

M. Howes and D. Morgan, Eds., *Charge-Coupled Devices and Systems*, John Wiley and Sons (1979).

S. Sze, *Physics of Semiconductor Devices*, John Wiley and Sons (1981).

J. Janesick, Ed., "Charge-coupled-device and charge-injection-device theory and application," *Optical Engineering* Vol. 26, No. 10, pp. 963–1067 (1987).

J. Janesick, Ed., "Charge-coupled-device manufacturer and application," *Optical Engineering* Vol. 26, No. 9, pp. 827–939 (1987).

J. Janesick, Ed., "Charge-coupled-device characterization, modeling, and application," *Optical Engineering* Vol. 26, No. 8, pp. 685–806 (1987).

C. Buil, *CCD Astronomy*, William-Bell, Inc. (1991).

L. Ristic, Ed., *Sensor Technology and Devices*, Artech House (1994).

A. Theuwissen, *Solid-State Imaging with Charge-Coupled Devices*, Kluwer Acad. Publishers, Dordrecht, Netherlands, ISBN 0-7923-3456-6 (4th print) (1995).

W. Leigh, *Devices for Optoelectronics*, Marcel Dekker, Inc. (1996).

J. Janesick and M. Blouke, "Sky on a chip: the fabulous CCD," *Sky and Telescope* Vol. 74, No. 3, pp. 238–242 (1987).

J. Kristian and M. Blouke, "Charge coupled devices in astronomy," *Scientific American* Vol. 247, No. 66 (1982).

Chapter 2

CCD Transfer Curves and Optimization

2.1 CCD Transfer Curves

Traditionally, CCD performance has been measured using commercial industry standards, which usually fall short of the requirements for many scientific digital applications. Also, scientific and commercial communities have been at odds with each other because each group uses different performance standards and units. This situation has created confusion as groups attempt to compare performance and data products. Commercial performance is often based on relative and photometric units (volts, lux, foot-candles, lumens, etc.), making it very difficult to know if absolute performance is being achieved. In contrast, scientific CCD camera systems are usually predicated on calibration standards that are based on absolute and physical radiometric units (photons, electrons, amps, watts, ergs, etc.). For example, it is physically meaningless to quote a noise or full well specification in relative units such as volts. However, these specifications become completely meaningful if specified in absolute units of electrons.

In characterizing the capabilities of the CCD for scientific work, numerous absolute test tools have been developed that allow performance to be expressed in absolute units. Test results from these techniques are often presented in the form of a transfer curve. For example, the spectral sensitivity for the CCD is graphed as absolute sensitivity as a function of wavelength, an important transfer curve called QE transfer. Different transfer curves describe different CCD and camera performance parameters measured under various test conditions. Transfer curves have been organized into three different working categories: (1) CCD performance, (2) CCD camera performance and (3) CCD camera calibration.

2.1.1 CCD Performance

The first category of transfer curves, CCD performance, primarily serves the CCD manufacturer and customer. The curves form an important data baseline to aid in development of new CCD designs and fabrication processes. The curves are also used in the characterization and screening processes that grade sensors for quality and eventual component sale. The curves also form the basis of data sheets supplied to the customer.

Performance transfer curves are presented in two main groups, referred to as "mainstream" and "custom." Mainstream includes approximately a dozen curves that are universally applied by the manufacturer and custom designer. This chapter will review in considerable depth three very important mainstream performance transfer curves: photon transfer, x-ray transfer, and QE transfer. These three curves

Table 2.1 CCD performance parameters and transfer curves.

CCD Function	Performance Parameter	Transfer Curve or Image	Units
Charge Generation			
	Interacting QE	QE Transfer	interacting photons/incident photons
Charge Collection			
	Pixel nonuniformity	Photon Transfer	% nonuniformity
	Well capacity	Photon Transfer	e^-/pixel
	Charge collection efficiency	MTF	% modulation
Charge Transfer			
	Global CTE	X-ray Transfer	per pixel transfer
	Local CTE	Bar target image	
Charge Measurement			
	Sensitivity	Photon Transfer	Volts/e^-
	Linearity	Linearity Transfer	% nonlinearity
	Read noise	Photon Transfer	rms e^-
	Global dark current	Dark Current Transfer	e^-/sec/pixel
	Local dark current	Dark current image	

alone supply approximately 90% of the data required to make judgements about CCD performance. Table 2.1 tabulates these mainstream transfer curves into our four CCD operational categories: (1) charge generation, (2) charge collection, (3) charge transfer and (4) charge measurement.

There are literally hundreds of custom performance transfer curves to select from. Many of these curves will be presented in this book. Custom curves are generated under special testing conditions defined by the application at hand. For example, QE stability may be characterized in a form of a transfer curve under specific operating conditions. Such data was taken for the Hubble CCDs where 1% absolute photometric accuracy was required over a specified operating period.

2.1.2 CCD Camera Performance

The second group of transfer curves used to optimize CCD camera performance are listed in Table 2.2. For example, clock levels, dc bias voltages and clock wave shaping applied to the CCD are critical parameters characterized. The same curves define operating windows to maintain stable and reliable performance. Procedures presented at the end of this chapter will show how this optimization process is accomplished. Analog CCD signal processing circuits are also included in this group of transfer curves. For example, camera gain, offset, sample time, and bandwidth are critical parameters defined to achieve maximum signal-to-noise. These signal-processing parameters are discussed and specified in Chapter 6.

Table 2.2 CCD camera parameters and transfer curves.

Function	Performance Parameter	Transfer Curve	Units
CCD clock, bias and operating windows			
	Array threshold	OTG Transfer	
	Amplifier threshold	V_{DD} Transfer	Volts
	Inversion	OTG Transfer	Volts
	Well capacity	Photon Transfer	e^-/pixel
	Sensitivity	Photon Transfer	V/e^-
Clock wave-shaping			
	Well capacity	Photon Transfer	e^-/pixel
Signal processing			
	Offset	Photon Transfer	e^-
	Gain	Photon Transfer	e^-/DN
	Sample-to-sample time	Photon Transfer	sec
	CCD dominant time contant	Photon Transfer	sec

Table 2.3 CCD camera calibration parametrs and transfer curves.

Performance Parameters	Transfer Curve	Units
Signal-to-noise	Photon Transfer	Signal/Noise
Dynamic range	Photon Transfer	Full well/Noise
Linearity	Linearity Transfer	% nonlinearity
Quantum efficiency	QE Transfer	Interacting photons/incident photons
MTF	Modulation Transfer	% modulation
Bright cosmetics	Dark current image	
Dark cosmetics	Bar target image	
Camera Gain Constant	Photon Transfer	e^-/DN

2.1.3 CCD CAMERA CALIBRATION

The third group of transfer curves used to calibrate a CCD camera system are tabulated in Table 2.3. Groups that procure off-the-shelf camera systems will find these curves essential. It is assumed that the two previous groups of transfer curves have been fully exercised before calibration takes place (i.e., Tables 2.1 and 2.2).

2.2 PHOTON TRANSFER

The photon transfer technique has proven to be one of the most valuable CCD transfer curves for calibrating, characterizing, and optimizing performance.[1,2] As described below, the photon transfer method is used to evaluate numerous CCD parameters in absolute terms. These parameters include read noise, dark current, quantum yield, full well, linearity, pixel nonuniformity, sensitivity, signal-to-noise, offset, and dynamic range. In addition, photon transfer is usually the first test performed in determining the overall health of a new CCD camera system, because all camera hardware and software must be in perfect operating order. This includes the CCD, its clock and bias levels, signal processor, analog-to-digital converter (ADC), camera/computer interface circuits and the software behind data acquisition and image processing. Well-behaved photon transfer data imply that a camera system is capable of taking precise scientific measurements (e.g., absolute measurement errors typically less than 1% can be achieved). Also when camera problems do exist, photon transfer is usually the first diagnostic test tool called upon to identify subtle problems such as nonlinearity and noise issues. Photon transfer also gives—almost magically—the conversion constant used to convert relative digital numbers (DN) generated by the camera into absolute physical units of electrons. This very important conversion constant is referred to as the "camera gain constant," expressed in e^-/DN. The next section will show how photon transfer curves are generated and how this constant is found.

2.2.1 Photon Transfer Derivation

A schematic representation of a typical CCD camera is shown in Fig. 2.1. The camera can be described with five transfer functions, three that are related to the CCD (interacting QE, quantum yield and output amplifier sensitivity) and two associated with the off-chip signal processing circuitry (signal chain gain and ADC gain). The input to the camera is given in units of incident photons. The final output signal of the camera is achieved by encoding each pixel's signal into a digital number $S(DN)$, typically using 8, 12 or 16 bits.

The output signal, $S(DN)$, that results for a given exposure period is given by

$$S(\text{DN}) = P\text{QE}_I \eta_i S_V A_{CCD} A_1 A_2, \qquad (2.1)$$

where $S(DN)$ represents the average signal for a group of pixels (DN), P is the average number of incident photons per pixel, QE_I is the interacting QE (interacting photons/incident photons), η_i is the quantum yield (the number of electrons generated, collected and transferred per interacting photon), S_V is the sensitivity of the sense node (V/e^-), A_{CCD} is the output amplifier gain (V/V), A_1 is the gain of the signal processor (V/V), and A_2 is the gain of the ADC (DN/V).

Example 2.1

Given Eq. (2.1), determine the average output signal $S(DN)$ for a CCD camera with the following parameters.

$P = 1000$ photons/pixel
$QE_I = 0.5$
$\eta_i = 2\,e^-$/interacting photon

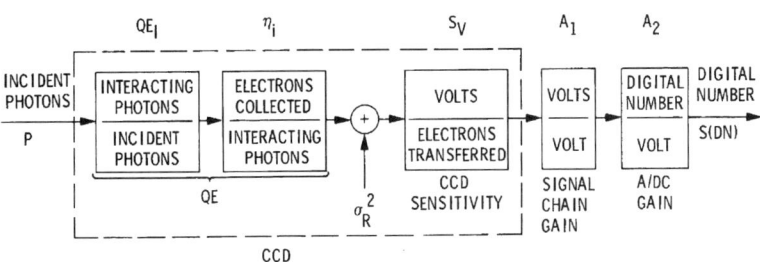

Figure 2.1 Block diagram of a typical CCD camera showing individual transfer functions.

$$S_V = 10^{-6}\,\text{V/e}^-$$
$$A_{CCD} = 0.8\,\text{V/V}$$
$$A_1 = 100\,\text{V/V}$$
$$A_2 = 2^{16}\,\text{DN/V}.$$

Solution:
From Eq. (2.1),

$$S(\text{DN}) = 1000 \times 0.5 \times \left(2 \times 10^{-6}\right) \times 0.8 \times 100 \times 2^{16},$$

$$S(\text{DN}) = 5240\,\text{DN}.$$

As mentioned above, to convert the output signal DN into fundamental physical units, it is necessary to find the appropriate factors in converting DN units to either interacting photons or signal electrons generated. The constants which accomplish this conversion are defined by

$$K = \frac{1}{S_V A_{CCD} A_1 A_2} \qquad (2.2)$$

and

$$J = \frac{1}{\eta_i S_V A_{CCD} A_1 A_2}, \qquad (2.3)$$

where K and J are camera gain constants given in units of e^-/DN and interacting photons/DN respectively. Note that these equations are related through the quantum yield, η_i, by

$$\eta_i = \frac{K}{J}. \qquad (2.4)$$

It is possible to determine the constants J and K by measuring each transfer function indicated in Fig. 2.1 separately and then combining these results as in Eqs. (2.2) and (2.3). However, it is difficult to measure each CCD parameter precisely (QE_I, η_i, and S_V). Instead, the photon transfer technique will determine J and K to any desired precision without knowledge of the individual transfer functions.

To evaluate the constant K, the CCD is stimulated with photons that generate single e-h pairs. Recall from Chapter 1 that wavelengths > 4000 Å generate single e-h pairs (i.e., $\eta_i = 1$). Wavelengths < 4000 Å are more energetic and generate multiple e-h pairs. Assuming a unity quantum yield, Eq. (2.1) can be rewritten as

$$S(\text{DN}) = \frac{P_I}{K}, \qquad (2.5)$$

where P_I is the number of interacting photons per pixel [i.e., $P_I = (QE_I)(P)$].

The constant K can now be determined by relating the output signal to its variance, $\sigma_S^2(\text{DN})$. The variance of Eq. (2.5) is found using the propagation of errors formula,

$$\sigma_S^2(\text{DN}) = \left[\frac{\partial S(\text{DN})}{\partial P_I}\right]^2 (\sigma_{P_I}^2) + \left[\frac{\partial S(\text{DN})}{\partial K}\right]^2 (\sigma_K^2) + \sigma_R^2(\text{DN}), \qquad (2.6)$$

where we have added in quadrature the output amplifier read noise floor variance $\sigma_R^2(\text{DN})$, as indicated in Fig. 2.1.

Performing the required differentiation on Eq. (2.5) and assuming that the constant K has a negligible variance (i.e., $\sigma_K^2 = 0$), the variance in $S(\text{DN})$ is found:

$$\sigma_S^2(\text{DN}) = \left(\frac{\sigma_{P_I}}{K}\right)^2 + \sigma_R^2(\text{DN}). \qquad (2.7)$$

Using $\sigma_{P_I}^2 = P_I$ (photon statistics) and Eq. (2.5),

$$K = \frac{S(\text{DN})}{\sigma_S^2(\text{DN}) - \sigma_R^2(\text{DN})}. \qquad (2.8)$$

Equation (2.8) is a very important expression and can be used, without any further calibration, to convert output noise and signal measurements in DN directly into units of electrons. We will demonstrate below how useful this result is.

For wavelengths longer than 4000 Å, the constants K and J are equal since the quantum yield is unity. The constant J is also found by relating the output signal given by Eq. (2.3) to its variance. Using the propagation of errors formula and assuming that the quantum yield has no variance, we find

$$J = \frac{S(\text{DN})}{\sigma_S^2(\text{DN}) - \sigma_R^2(\text{DN})}, \quad (2.9)$$

where J gives the number of interacting photons/DN.

Equations (2.8) and (2.9) form the basis for the photon transfer technique. By simply measuring the mean signal and its variance both for visible photons and for photons at any other specific wavelength of illumination (i.e., $\lambda < 4000\,\text{Å}$), the values K and J can be determined. Once the constants K and J are known, the quantum yield for photons at the wavelength under consideration can be determined through Eq. (2.4).

Example 2.2

Find the constant K when a CCD is stimulated with visible light. Assume the following measurement parameters: $S(\text{DN}) = 4000\,\text{DN}$, $\sigma_S^2(\text{DN}) = 300\,\text{DN}$ rms, and $\sigma_R^2(\text{DN}) = 10\,\text{DN}$.
Solution:
From Eq. (2.8),

$$K = 4000/(300 - 10),$$

$$K = 13.79\,e^-/\text{DN}.$$

2.2.2 PHOTON TRANSFER CURVE

The photon transfer curve illustrated in Fig. 2.2 is a response from a CCD that is uniformly illuminated at different levels of light. We plot noise or the standard deviation, $\sigma_S(\text{DN})$, as a function of average signal, $S(\text{DN})$, for a group of pixels contained on the CCD array. Data is plotted on a log–log scale in order to cover the large dynamic range of the CCD. Three distinct noise regimes are identified in the plot. The first regime, the read noise floor $\sigma_R(\text{DN})$, represents the random noise measured under totally dark conditions. This noise is ultimately limited by on-chip amplifier noise, but can represent any other noise sources that are independent of the signal level (e.g., shot noise generated by dark current).

As the illumination to the CCD is increased, the noise becomes dominated by the shot noise of the signal, shown in the middle region of the curve. Shot noise is the noise associated with the random arrival of photons on the CCD. Some pixels intercept more photons than others, which accounts for the variance seen in pixel values. Since the plot shown is on log coordinates, the shot noise is characterized by a line of slope 1/2. This specific slope arises because the uncertainty in the

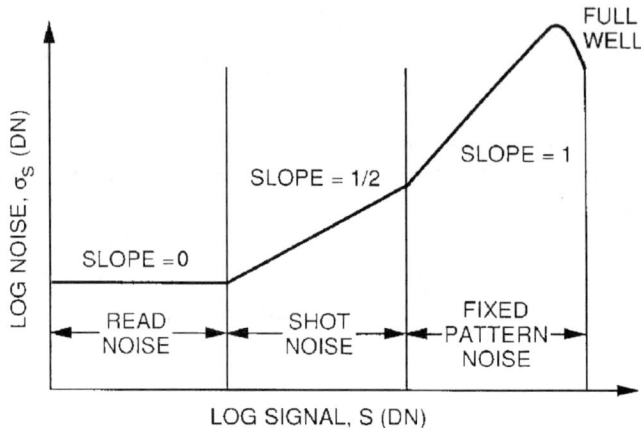

Figure 2.2 Total noise photon transfer curve illustrating three noise regimes over the dynamic range of the CCD.

quantity of charge collected in any given pixel is proportional to the square root of the number of incident photons (as governed by Poisson's statistics).

The third regime is associated with fixed-pattern or pixel nonuniformity noise that results from sensitivity differences among pixels. Pixel nonuniformity is a manifestation of processing variations and photomask alignment errors when the CCD is fabricated. This problem generates pixels with different responsivities. Pixel nonuniformity noise is proportional to signal and consequently produces a characteristic slope of unity on our plot.

The onset of full well is observed at some illumination level within the fixed-pattern noise regime and is seen as a break in the slope of one. At this point the charge spreads between pixels (i.e., blooming), smoothing and lowering the noise component as seen in the plot. This action causes the noise value to suddenly decrease, making full well obvious (the signal component remains high).

Figure 2.3(a) shows a photon transfer curve generated by a backside-illuminated $800 \times 800 \times 15$-μm CCD using a 20×20 subarray of pixels that are uniformly illuminated with visible light (7000 Å). We refer to the curve as "classical" because pixel nonuniformity has been removed, a simple computer process step that will be described in this section. The curve is generated by varying the exposure sequence beginning with a dark frame (i.e., CCD not exposed to light). Then the exposure time is increased and signal level is measured in DN units until saturation is reached. Exposure periods are typically increased logarithmically to cover the entire dynamic range of the CCD. The exposure period is the time period when a shutter is open or how long the CCD is exposed to the light source. Usually the integration period is fixed during a photon transfer sequence. In this manner, only the exposure time is varied, allowing the total frame time period (integration + read out) to be fixed. Figure 2.3(b) shows another photon transfer curve which exhibits a much greater dynamic range.

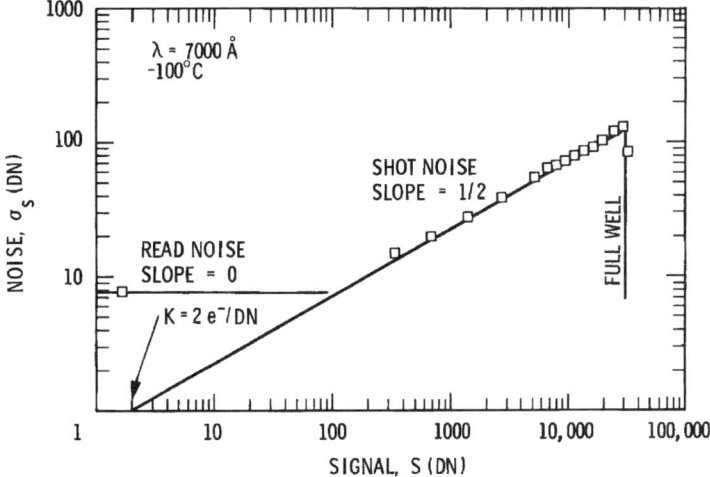

Figure 2.3(a) Photon transfer curve generated by a WF/PC I 800 × 800 CCD with pixel nonuniformity removed.

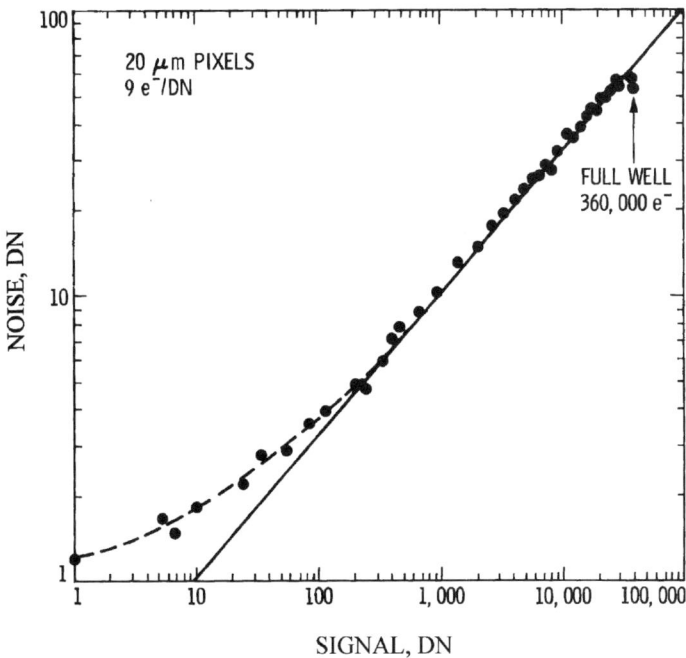

Figure 2.3(b) Photon transfer for a CCD that exhibits a wide dynamic range.

It should be mentioned that a simple LED placed in front of the CCD can be pulsed to expose the sensor, again leaving the integration period fixed. The signal charge generated will be proportional to the time the LED is "on." Also, an

LED can be turned "on" and "off" very quickly, allowing for very short exposure times (on the order of a microsecond). In contrast, mechanical shutters are limited to exposure times no shorter than 1 msec. For these shuttered systems, a 10-sec integration period may be required to cover a signal dynamic range of 10^4, which is often encountered in CCD work. On the other hand, the LED stimulus would only require a maximum of 10 msec, thus shortening the time to take data.

The abscissa $S(\text{DN})$ is proportional to the exposure period or the average number of incident photons and photo-generated charge per pixel element. The 400 pixel values are averaged, yielding $S(\text{DN})$, after a fixed average electrical offset is subtracted from each pixel. That is,

$$S(\text{DN}) = \frac{\sum_{i=1}^{N_{pix}}[S_i(\text{DN})]}{N_{pix}} - S_{OFF}(\text{DN}), \tag{2.10}$$

where $S_i(\text{DN})$ is the signal value of the ith pixel (DN), N_{pix} is the number of pixels contained in the subarray, and $S_{OFF}(\text{DN})$ is the average offset level (DN).

The offset, $S_{OFF}(\text{DN})$, represents the camera's output DN level in total absence of signal electrons. Subtracting a precise offset is an important requirement in measuring signal levels. Offset can sometimes be determined from a dark exposure using pixels from the same pixel region where signal is measured. However, the offset information on the array may be corrupted with dark current or other sources of charge (e.g., light leak, spurious charge, etc.). The best location on the CCD for a zero signal level is in the "extended pixel region" of the horizontal register (see Fig. 1.12). Extended pixels are found at the beginning of a line or are purposely created by overclocking the horizontal register. In the latter case, those pixels following the last video pixel represent a good measurement of zero charge. Extended pixels should therefore always accompany video pixels as they are stored in a computer. In the case of the Wide Field/Planetary Camera, one extended pixel for every video line of data is sent back to Earth for offset information. It is very important to track the offset level because it can change from frame to frame. Sensitive image-processing programs (e.g., flat fielding and nonlinearity measurements) require an offset accuracy of better than 1 DN. Note from Eq. (2.10) that the offset level does not need to be subtracted pixel by pixel. Only a single difference from the average signal level is required, thus saving computing time.

Noise data for the ordinate are found by calculating the standard deviation of the 400 pixels after pixel-to-pixel nonuniformity is removed. Pixel nonuniformity is eliminated by simply differencing, pixel by pixel, two identical images taken back to back at the same exposure level. The variance, $\sigma_S^2(\text{DN})$, of this differenced frame is given by

$$\sigma_S^2(\text{DN}) = \frac{\sum_{i=1}^{N_{pix}}[S_i(\text{DN}) - S(\text{DN})]^2}{2N_{pix}}. \tag{2.11}$$

Note that a factor of two must be included in the denominator of Eq. (2.11) because the noise power increases by two within the differenced frame. In other words, when two identical frames are either subtracted (as in this case) or added, the random noise component (rms or standard deviation) of the resultant frame increases by the square root of two.

To improve statistics for photon transfer data, three or more exposures at the same signal level can be taken. $S(\text{DN})$ is then found by taking the average of the three mean levels measured. The noise, $\sigma_S(\text{DN})$, is determined by first differencing frames 1 & 2, then frames 2 & 3, and lastly frames 1 & 3, and generating three standard deviations. These values are then averaged together, yielding $\sigma_S(\text{DN})$. This is the method used in generating the photon transfer curves in Fig. 2.3. One can also take a larger region of interest from the CCD (that is, greater than 20 × 20 pixels) to improve statistics, as long as the region selected does not contain blemish artifacts (multiple frames were required when early CCDs exhibited cosmetic problems).

It should also be mentioned that dark current, if it exists, does not need to be subtracted from the signal level. This is because dark charge only adds to the signal charge measured. It is not important in the photon transfer technique how charge is generated, as long as the source exhibits shot noise characteristics, which is the case for dark current (refer to Chapter 7). In fact, a photon transfer curve can be generated using only thermally generated dark current; i.e., no light source is required. However, if total noise is plotted as in Fig. 2.2 without frame differencing, dark nonuniformity will dominate the noise term with very little shot noise seen. Pixel-to-pixel dark current nonuniformity for CCDs is typically 10% rms of the average signal level, compared to pixel-to-pixel sensitivity nonuniformity, which is approximately 1%.

2.2.3 CAMERA GAIN CONSTANTS

The conversion factors K and J can be found graphically or precisely through Eqs. (2.8) and (2.9) by measuring $S(\text{DN})$ and $\sigma_S^2(\text{DN})$ at a signal level where $\sigma_S^2(\text{DN}) \gg \sigma_R^2(\text{DN})$. The graphical approach will be examined in this section to obtain approximate values. For example, the constant K for the photon transfer curve shown in Fig. 2.3(a) is found by extending the slope 1/2 line back to the x axis intercept corresponding to $\sigma_S(\text{DN}) = 1$ DN. The intercept on the signal axis represents the value K [i.e., $K = S(\text{DN})$]. This remarkable result can be proven by taking the logarithm of Eq. (2.8), yielding

$$\log[K] = \log[S(\text{DN})] - \log[\sigma_S^2(\text{DN}) - \sigma_R^2(\text{DN})]. \tag{2.12}$$

We next assume $\sigma_R(\text{DN}) = 0$ since the slope 1/2 curve is shot noise limited. Then we let $\log[\sigma_S(\text{DN})] = 0$ since $\sigma_S(\text{DN}) = 1$. Substituting these values reduces the above equation to $K = S(\text{DN})$, the desired result.

Example 2.3

Determine graphically the camera gain constant K for the photon transfer curve shown in Fig. 2.3(a). Then determine the read noise and full well for the sensor in units of electrons. Also find the read noise and full well for the photon transfer curve shown in Fig. 2.3(b).

Solution:

Figure 2.3(a):

The signal intercept for the shot noise curve is approximately $K = 2.0\,e^-/\text{DN}$ as shown. Therefore, the 8 DN rms of noise corresponds to a read noise of $16\,e^-$ rms. Saturation occurs at 30,000 DN, which corresponds to a charge capacity of $60,000\,e^-$.

Figure 2.3(b):

The camera gain constant in Fig. 2.3(b) is $9\,e^-/\text{DN}$. The 40,000 DN saturation level corresponds to a full well of $360,000\,e^-$. The read noise of 1 DN is $9\,e^-$ rms.

The same approach can be used in determining the constant J when the quantum yield is greater than 1. Since S_v, A_1, and A_2 are constant for a given CCD camera system, a decrease in K can be directly attributed to an increase in the quantum yield (i.e., multiple electron/hole pairs per interacting photon). For example, Fig. 2.4 shows three photon transfer curves that are generated by a backside-illuminated 800 × 800 CCD. The curves show the response of the CCD when exposed to light of the following wavelengths: (1) red light (7000 Å), (2) UV light (1216 Å) and (3) soft x rays (2.1 Å). The corresponding intersections on the signal axis using the graphical approach for these wavelengths are, respectively, $K = 2.3\,e^-/\text{DN}$, $J = 0.77$ and $J = 1.62 \times 10^{-3}$ interacting photons/DN.

Example 2.4

Determine the quantum yield for the three photon transfer curves presented in Fig. 2.4.

Solution:

7000 Å (1.77 eV):

Stimulating the CCD with 7000-Å light generates single electron-hole pairs yielding a signal intercept value of $K = 2.3\,e^-/\text{DN}$.

1216 Å (10.2 eV):

At this wavelength, multiple electron-hole pairs are generated, causing the photon transfer curve to shift to a smaller signal intercept value (i.e., from 2.3 to 0.77). The quantum yield is found from Eq. (2.4).

CCD TRANSFER CURVES AND OPTIMIZATION

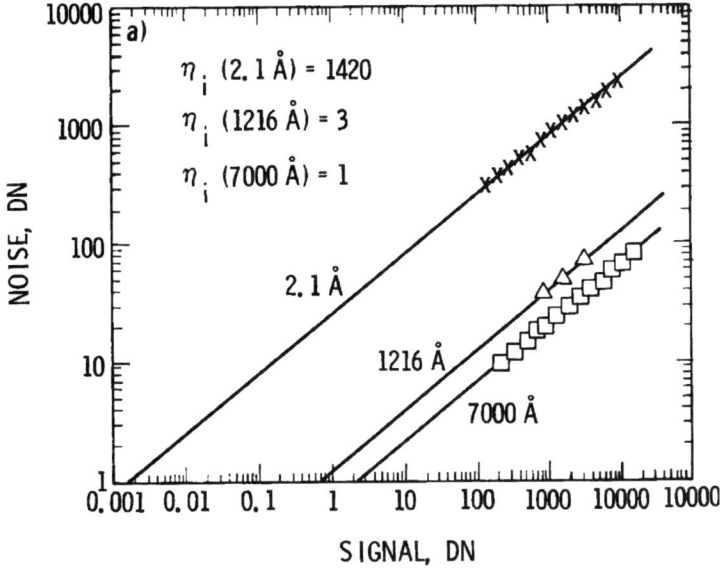

Figure 2.4 Three photon transfer curves taken at 7000 Å, 1216 Å and 2.1 Å.

$$\eta_i = 2.3/0.77,$$

$$\eta_i = 2.99 \, e^-/\text{interacting photon}.$$

2.1 Å (5.9 keV):

The intercept shifts to 1.6×10^{-3} when the CCD is stimulated with soft x rays. The quantum yield is

$$\eta_i = 2.3/(1.62 \times 10^{-3}),$$

$$\eta_i = 1420 \, e^-/\text{interacting photon}.$$

The quantum yield for x-ray interaction always falls short of the ideal quantum yield, η_i (i.e., $5900/3.65 = 1620 \, e^-$). This result is because photogenerated charge generated by the x-ray photon is collected in more than one pixel, thus reducing the measured quantum yield (referred to as split x-ray events). This effect is associated with charge collection efficiency, discussed in Chapter 4.

2.2.4 CAMERA GAIN HISTOGRAM

The graphical method of determining K and J above is accurate to within 10–20%, depending upon how well the data points on the "slope 1/2 line" are fitted to a straight edge. The constant K can be determined precisely by repeatedly measuring S(DN) and σ_S^2(DN) and inserting these values into Eq. (2.8). If many different 20×20 subarrays over the CCD are interrogated, a "camera gain histogram" for K, as shown in Fig. 2.5, can be generated. This is accomplished by storing two flat-field frames taken near full well to assure that read noise is negligible. The two frames are then subtracted pixel by pixel to remove pixel nonuniformity, to obtain σ_S^2(DN). Then, a 20×20 subarray is randomly selected from the stored data and Eq. (2.8) is applied, yielding a value for K. A different 20×20 region is then randomly selected and the process repeated for another K value. As values of K are generated in this fashion, they are plotted in histogram form. The distribution is normal and weighted about the mean value of K. The histogram technique allows one to measure the camera gain constant to any desired level of precision.

By incorporating many 20×20 pixel subarrays into the histogram, those regions on the device that contain blemish artifacts (e.g., column blemishes, blocked channels, etc.), which would give an erroneous value of K, can be eliminated. Areas on the CCD that are not well behaved are usually recognized as data points outside the main histogram peak (three of these points are seen in Fig. 2.5).

A similar histogram can be generated to measure the quantum yield, η_i (i.e., K/J). In this case, four frames of data are required: two visible frames to determine K and two frames at a specific wavelength of interest (< 4000 Å). For example, Fig. 2.6 presents two photon transfer histograms for a frontside-illuminated

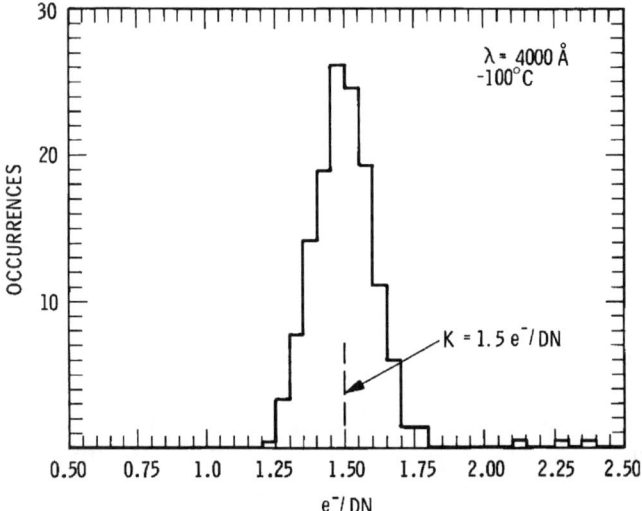

Figure 2.5 Camera gain histogram derived from photon transfer data.

CCD stimulated with 7000 Å light and 2.1 Å soft x rays. The histograms are generated by employing Eqs. (2.8) and (2.9) as before. The quantum yield for soft x rays is $790\,e^-$ (K/J), less than the ideal response ($1620\,e^-$) for reasons explained above.

Figure 2.7(a) shows quantum yield measurements taken in the UV with a backside-illuminated CCD. These data were produced by generating histograms for selected wavelengths as in Fig. 2.6. For example, Fig. 2.7(b) shows two histograms taken at 6400 Å and 2400 Å and the shift in e^-/DN that occurs between these two wavelengths.

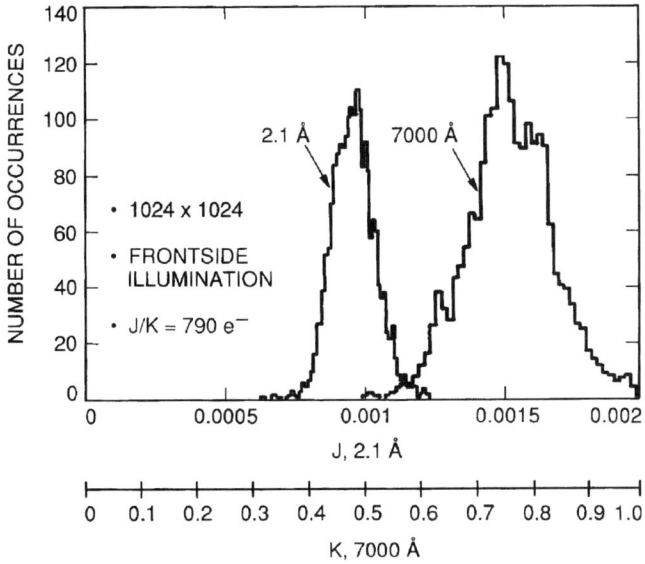

Figure 2.6 Two camera gain histograms generated at 7000 Å and 2.1 Å.

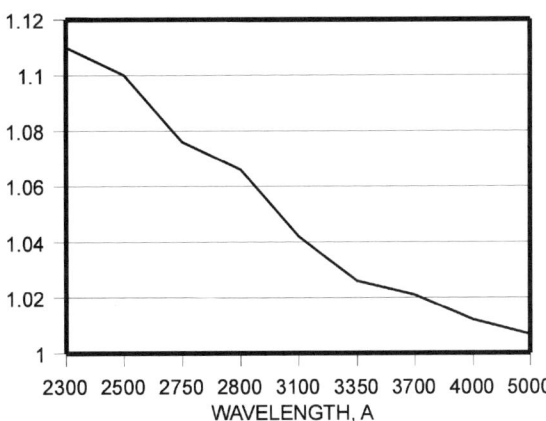

Figure 2.7(a) Quantum yield as a function of wavelength.

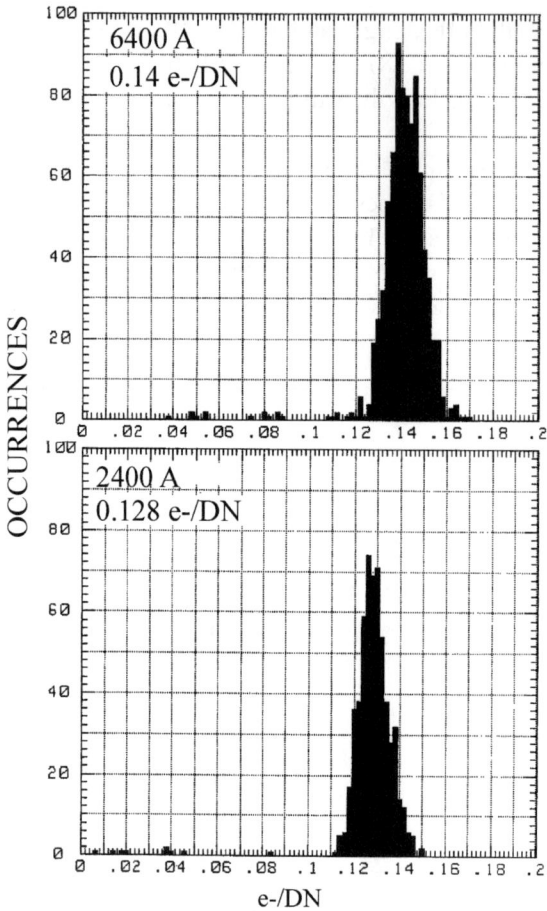

Figure 2.7(b) Quantum yield histograms used in Fig. 2.7(a).

2.2.5 CAMERA GAIN UNCERTAINTY

The variance of K is found by applying the propagation of errors formula to Eq. (2.8) rewritten in the form

$$K = \frac{S(\text{DN})}{N(\text{DN})^2}, \qquad (2.13)$$

where $N(\text{DN}) = \sigma_S(\text{DN})$. Note that we have assumed that read noise, σ_R, is negligible.

Applying the propagation of errors formula to this equation yields

$$\sigma_K^2 = \left[\frac{\partial K}{\partial S(\text{DN})}\right]^2 (\sigma_S^2) + \left[\frac{\partial K}{\partial N(\text{DN})}\right]^2 (\sigma_N^2). \qquad (2.14)$$

CCD TRANSFER CURVES AND OPTIMIZATION 111

Assuming photon statistics (i.e., $K\sigma_S(\mathrm{DN}) = [K\,S(\mathrm{DN})]^{1/2}$), the uncertainty in the signal measured is

$$\sigma_S^2 = \frac{S(\mathrm{DN})}{K N_{pix}}. \tag{2.15}$$

The uncertainty in noise measured is

$$\sigma_N^2 = \frac{N(\mathrm{DN})^2}{2 N_{pix}}. \tag{2.16}$$

Performing the differentiation required to evaluate Eq. (2.14) and substituting these two equations yields

$$\sigma_K^2 = \frac{1}{N(\mathrm{DN})^4} \frac{S(\mathrm{DN})}{K N_{pix}} + \frac{4 S(\mathrm{DN})^2}{N(\mathrm{DN})^6} \frac{N(\mathrm{DN})^2}{2 N_{pix}}. \tag{2.17}$$

Applying Eq. (2.13) simplifies this equation into

$$\sigma_K^2 = \frac{1}{N_{pix}} \left[K \left(\frac{1}{S(\mathrm{DN})} + 2K \right) \right]. \tag{2.18}$$

As long as $\sigma_S(\mathrm{DN}) \ll S(\mathrm{DN})$, which will be true for reasonable signal levels, we can use the approximation $1/S(\mathrm{DN}) \ll 2K$, reducing Eq. (2.18) to the final result,

$$\sigma_K = \left[\frac{2}{N_{pix}} \right]^{1/2} K. \tag{2.19}$$

It is interesting to note that the uncertainty of K is independent of signal level as long as the read noise is negligible and the signal level is large compared to the shot noise. Equation (2.19) has been experimentally confirmed with good precision.

Example 2.5

Determine the uncertainty (rms) of the camera gain constant when sampling a 20×20 pixel region, when $K = 5\,\mathrm{e}^-/\mathrm{DN}$.
Solution:

$$\sigma_K = 5 \times (2/400)^{1/2},$$

$$\sigma_K = 0.35\,\mathrm{e}^-/\mathrm{DN\,rms}.$$

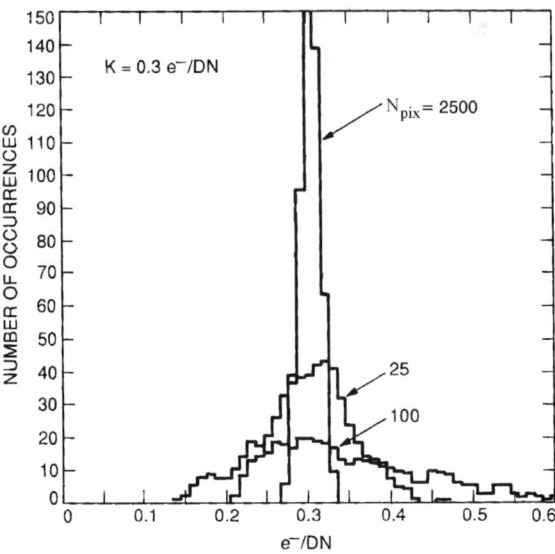

Figure 2.8(a) Camera gain variance as a function of pixels sampled.

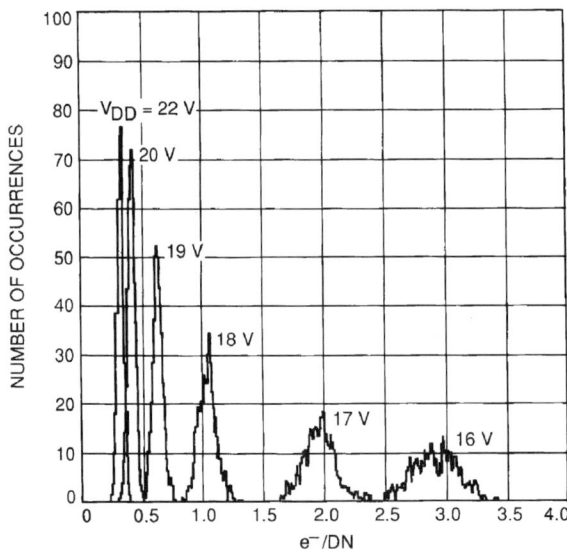

Figure 2.8(b) Camera gain histograms taken at different V_{DD} potentials.

Figure 2.8(a) presents three camera gain histograms showing that the variance of K decreases with increasing N_{pix}. Figure 2.8(b) plots camera gain histograms for different drain voltages (V_{DD}) to the on-chip amplifier. As V_{DD} is lowered, the CCD gain, A_{CCD}, decreases, causing K to increase and the histogram width to broaden.

2.2.6 Dynamic Range

Well capacity and read noise specify the dynamic range of the CCD using

$$\mathrm{DR} = \frac{S_{FW}(\mathrm{e}^-)}{\sigma_R(\mathrm{e}^-)}, \qquad (2.20\mathrm{a})$$

where $S_{FW}(\mathrm{e}^-)$ is full well.

Dynamic range can also be expressed in decibels (dB) as

$$\mathrm{DR} = 20 \log\left(\frac{S_{FW}(\mathrm{e}^-)}{\sigma_R(\mathrm{e}^-)}\right). \qquad (2.20\mathrm{b})$$

Example 2.6

Apply Eq. (2.20) to Example 2.3 and find the dynamic range for the CCD characterized in Fig. 2.3(a). Also, express dynamic range in decibels.

Solution:
From Eq. (2.20a),

$$\mathrm{DR} = 60{,}000/16,$$

$$\mathrm{DR} = 3{,}750.$$

From Eq. (2.20b),

$$\mathrm{DR} = 20 \times \log(60{,}000/16),$$

$$\mathrm{DR} = 71.48\,\mathrm{dB}.$$

The enormous dynamic range that the CCD exhibits cannot usually be represented properly when images are printed and displayed. For example, photographic film at its "spatial resolution limit" (approximately 5 μm) can only resolve 10 to 20 shades of gray. In comparison, a 5-μm pixel CCD can differentiate hundreds of levels at its spatial limit (i.e., Nyquist). The dynamic range for film and the CCD are compared in Fig. 2.9(a) using several images of a globular star cluster field. Each 8-bit image shown is derived from a single 16-bit CCD exposure. For example, the image shown in the upper right-hand corner uses the most significant 8 bits, with the least significant 8 bits dropped. Note that only the brightest stars that are near full well are seen. The picture presented in the lower left-hand corner uses the least significant 8 bits. The central region of the cluster is saturated because of

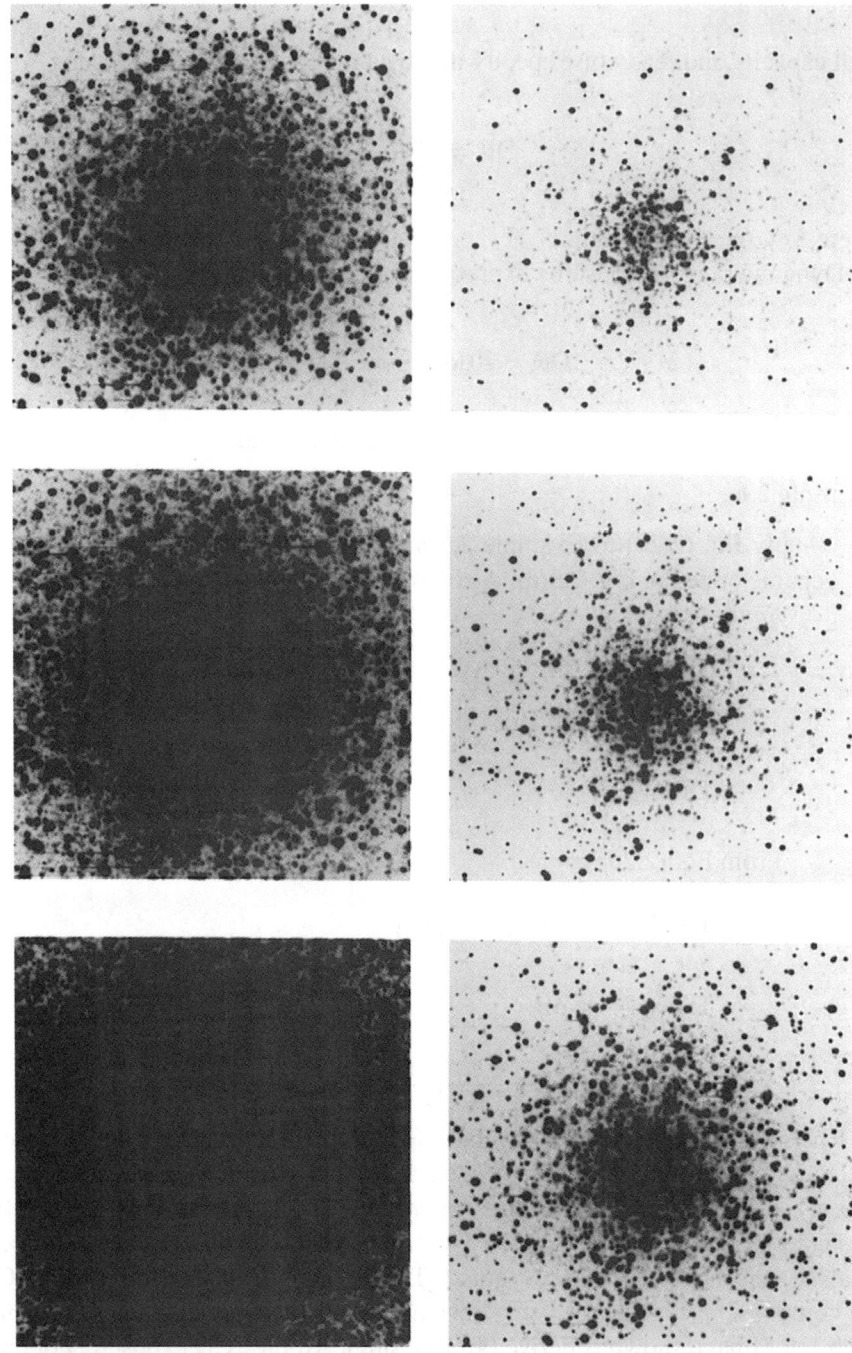

Figure 2.9(a) A globular star cluster image demonstrating the wide dynamic range of a CCD.

CCD TRANSFER CURVES AND OPTIMIZATION

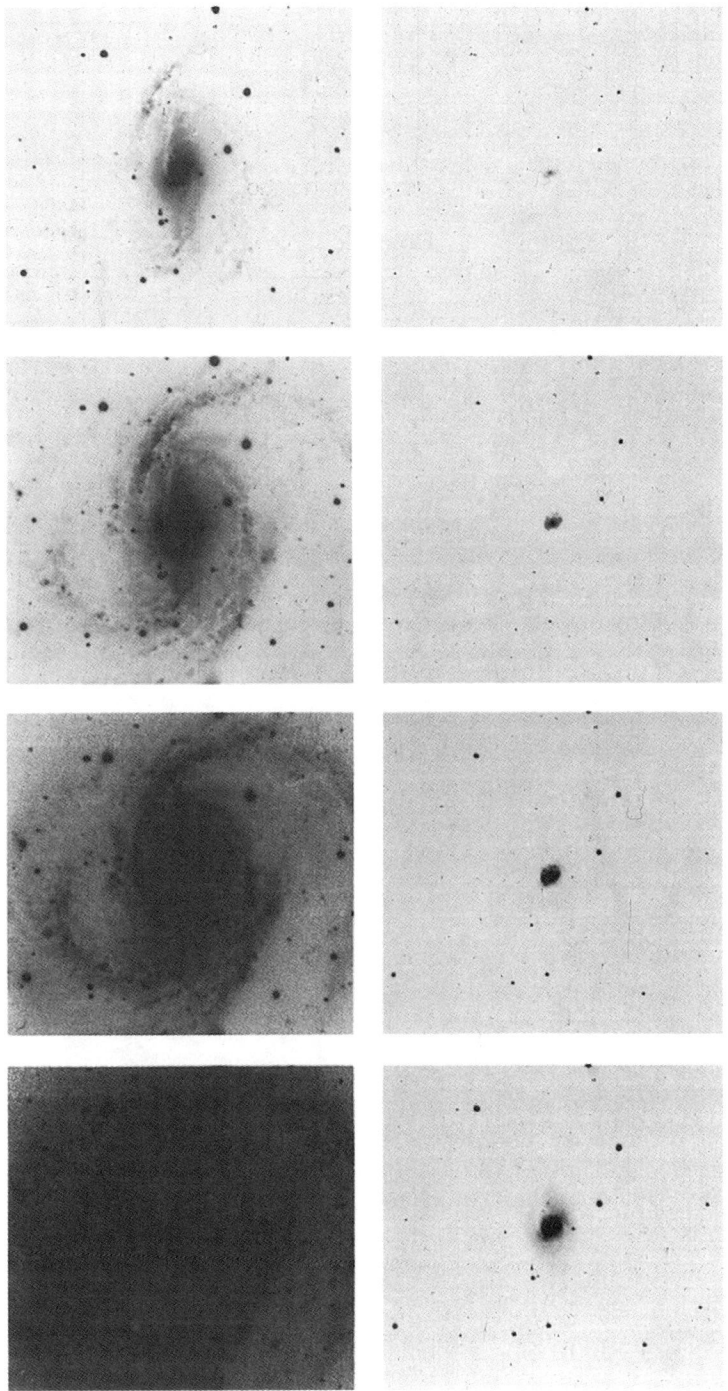

Figure 2.9(b) Galaxy image processed as Fig. 2.9(a).

Figure 2.10(a) Gobular star cluster showing saturation within center region.

Figure 2.10(b) Same gobular star cluster printed logarithmically.

the limited dynamic range of the film. Figure 2.9(b) shows a picture of a galaxy, again demonstrating the dynamic range of the CCD.

CCD TRANSFER CURVES AND OPTIMIZATION

Figure 2.10 demonstrates a compression technique that allows the CCD's full dynamic range to be better presented on film. Figure 2.10(a) shows a globular cluster star field, showing the least significant 8 bits. Figure 2.10(b) was produced by using a logarithmic gain adjustment on each pixel value, compressing information to match the dynamic range of the film. Now stars can be seen without saturating the central region of the cluster.

2.2.7 LINEARITY

Nonlinearity is a measurement of the camera gain constant as a function of signal. Ideally there should be no dependence. Any gain change with signal is important to scientific applications where absolute measurements are taken. For example, in astronomy, a star's brightness is usually determined by comparing its brightness to a "standard star" whose brightness is already known. The error (or nonlinearity) for this measurement is equal to the gain (i.e., e^-/DN) ratio of the stars signal levels. The nonlinearity specification for WF/PC is < 1% over the dynamic range of the camera system. This means that the absolute brightness for an unknown star will be accurate to within 1%.

The same nonlinearity error is measured using photon transfer data by plotting signal as a function of exposure time. In general, the output signal, $S(\text{DN})$, as a function of exposure time can be expressed in a power series as

$$S(\text{DN}) = C(t_E)^\gamma, \tag{2.21}$$

where C is a proportionality constant, t_E is the exposure time (sec), and γ is a measure of the linearity of the device. A gamma of unity signifies that the CCD output signal is proportional to exposure time.

Figure 2.11(a) presents a linearity transfer curve of $S(\text{DN})$ as a function of exposure time on log-log coordinates. The straight line of a unity slope indicates good linearity (i.e., $\gamma = 1$). As can be seen, linearity is exceptional for the CCD as compared to other types of imaging sensors (e.g., film, vidicon tubes). To measure nonlinearity at the 1% level, a different transfer curve is used. Figure 2.11(b) presents the "linearity residuals" which expand the results of Fig. 2.11(a) using the equation

$$LR = 100 \left(1 - \frac{S_M(\text{DN})/t_{EM}}{S(\text{DN})/t_E} \right), \tag{2.22}$$

where LR is the linearity residual at a specific signal level (%). The variables $S(\text{DN})$ and t_E are the signal (DN) and exposure time (sec) measured in the photon transfer process, and t_{EM} is the exposure time required to produce a signal (sec), S_M, at mid-scale in the plot (DN).

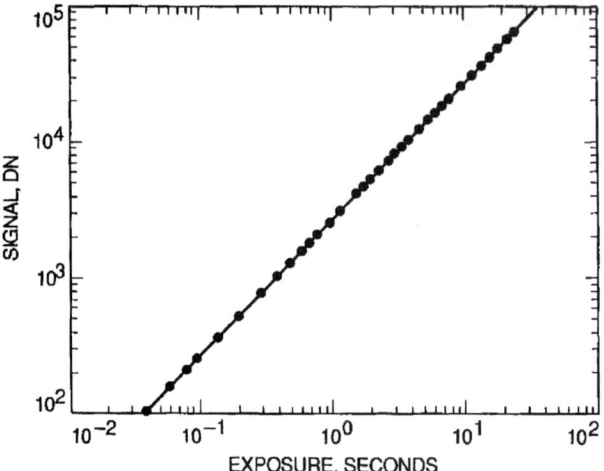

Figure 2.11(a) Linearity transfer curve.

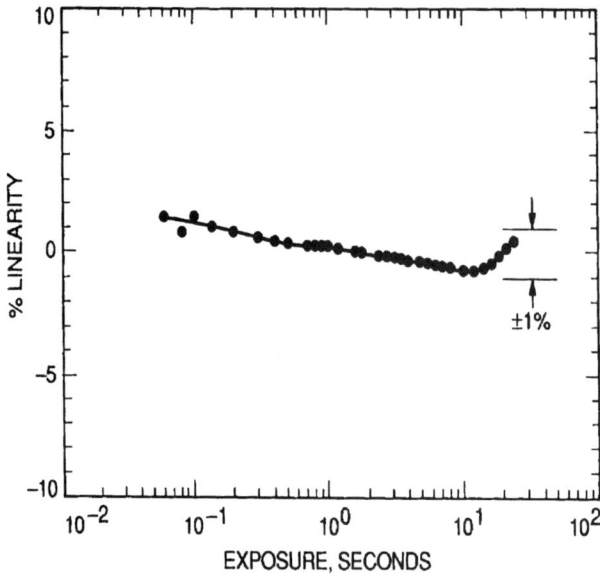

Figure 2.11(b) Linearity residual transfer curve that exhibits nonlinearity < 1% over the sensor's dynamic range.

Example 2.7

Find the linearity residual at an output signal of 4900 DN for an exposure period of 0.5 sec. Assume a mid-scale level of 10^4 DN at 1-sec exposure time.

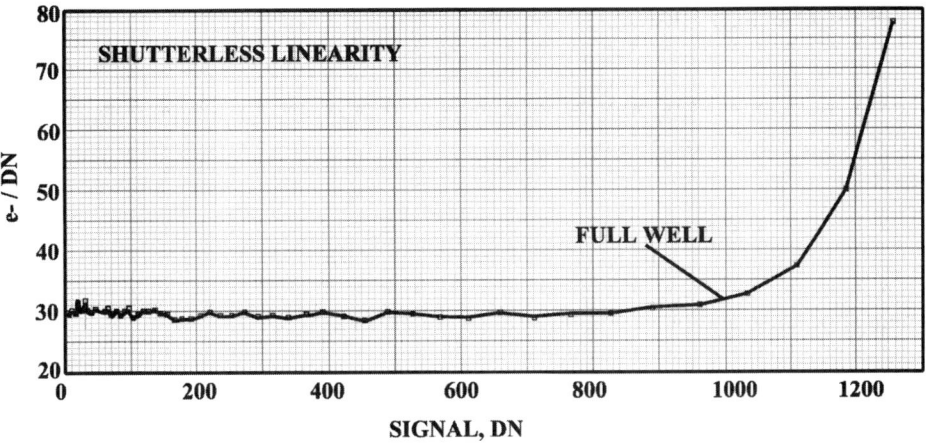

Figure 2.11(c) Linearity transfer curve based on the shutterless photon transfer method.

Solution:
From Eq. (2.22),

$$LR = 100 \times \{1 - 10^4/[(4.9 \times 10^3)/0.5]\},$$

$$LR = -2.04\%.$$

It should be mentioned that good linearity performance may still be seen beyond full well conditions. For example, linearity is well behaved when the average signal is measured about a small overexposed point source that has bloomed. Only when bloomed charge escapes the region being averaged will linearity suffer, because charge is conserved. This situation is different compared to a flat-field exposure where blooming takes place over the entire array. Here, the onset of nonlinearity and full well will occur at the same signal level. The camera gain constant (e$^-$/DN) can be plotted against signal to circumvent setup problems like this. This is because gain (e$^-$/DN) is dependent on both signal and noise [i.e., Eq. (2.8)]. Linearity measured in this manner is very sensitive to charge sharing between pixels. This measurement technique, combined with the shutterless photon transfer method described below, represents a powerful calibration tool in characterizing gain and linearity simultaneously without exposure time information. Figure 2.11(c) shows a linearity transfer curve based on this method.

2.2.8 FLAT-FIELD SIGNAL-TO-NOISE

The flat field signal-to-noise (S/N) ratio is also provided by the photon transfer curve. In equation form S/N is

$$\text{S/N} = \frac{S(e^-)}{[\sigma_{shot}(e^-)^2 + \sigma_{PN}(e^-)^2 + \sigma_R(e^-)^2]^{1/2}}, \quad (2.23)$$

where shot noise and pixel nonuniformity are given as

$$\sigma_{shot}(e^-) = [\eta_i S(e^-)]^{1/2}, \quad (2.24)$$

and

$$\sigma_{PN}(e^-) = P_N S(e^-), \quad (2.25)$$

where the constant P_N is the fraction of pixel nonuniformity.

Assuming that pixel nonunifomity is removed by flat fielding (refer to Chapter 4) and read noise is negligible, S/N reduces to

$$\text{S/N} = \left(\frac{S(e^-)}{\eta_i}\right)^{1/2}. \quad (2.26)$$

Example 2.8

Calculate the maximum S/N performance that is achieved at full well. Assume a well capacity of $10^6\, e^-$ for three different wavelengths: 7000 Å ($\eta_i = 1$), 1216 Å ($\eta_i = 3$) and 2.1 Å ($\eta_i = 1620\, e^-$).

Solution:
From Eq. (2.26) the S/N is

$\lambda = 7000\,\text{Å}\ (\eta_i = 1),$

$$\text{S/N} = (10^6/1)^{1/2},$$

$$\text{S/N} = 1000,$$

$\lambda = 1216\,\text{Å}\ (\eta_i = 3),$

$$\text{S/N} = (10^6/3)^{1/2},$$

$$\text{S/N} = 577,$$

$\lambda = 2.1 \text{ Å } (\eta_i = 1620\, e^-)$,

$\text{S/N} = (10^6/1620)^{1/2}$,

$\text{S/N} = 25$.

The S/N for visible light is 40 times better than for x rays at full well. This is because the number of electrons generated for each photon is 1600 times greater, which results in more shot noise at the same average signal level. An x-ray image compared to a visible light image would therefore look grainier.

S/N can also be calculated by using DN values directly:

$$\sigma_{shot}(e^-) = [K\eta_i S(\text{DN})]^{1/2}, \tag{2.27}$$

and

$$\sigma_{PN}(e^-) = P_N S(\text{DN}) K, \tag{2.28}$$

and

$$\sigma_R(e^-) = \sigma_R(\text{DN}) K. \tag{2.29}$$

Substituting these equations into Eq. (2.23) yields

$$\text{S/N} = \frac{K S(\text{DN})}{\{\eta_i S(\text{DN}) K + [P_N S(\text{DN})]^2 + [\sigma_R(\text{DN})]^2\}^{1/2}}. \tag{2.30}$$

Again, assuming pixel nonuniformity is removed and read noise is negligible, the resultant S/N is

$$\text{S/N} = \left(\frac{K S(\text{DN})}{\eta_i}\right)^{1/2}. \tag{2.31}$$

2.2.9 CONTRAST SIGNAL-TO-NOISE

High signal-to-noise is important in detecting and measuring low-contrast images. For example, astronomical objects such as the sun and planets are relatively bright objects, but the surface detail presented is very low contrast. It is not uncommon in scientific applications to ask the CCD to detect contrast levels less than 1%, levels not distinguishable with the human eye or photographic film. In theory, any level

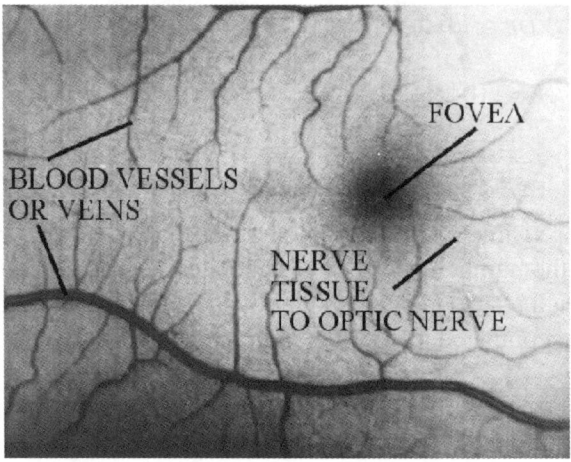

Figure 2.12(a) Low-contrast image of retina.

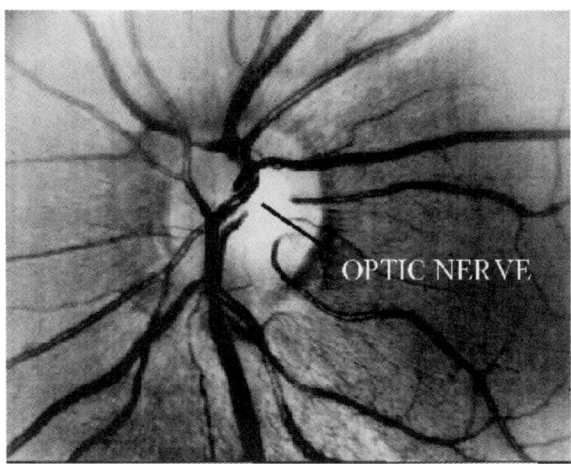

Figure 2.12(b) Low-contrast image of optic nerve.

of contrast can be seen by the CCD as long as pixel nonuniformity is removed and the charge capacity is high enough for the signal-to-noise required.

Figures 2.12(a) and 2.12(b) are low-contrast images taken of the author's eye ball. The images were generated by a MPP 1024 × 1024 CCD working at room temperature. Figure 2.12(a) shows the retina, composed of the fovea and nerve tissues surrounding the region leading up to the optic nerve, shown in Fig. 2.12(b). The most prominent structures seen are either blood vessels or veins. Although the average signal of the image is high (80,000 e^-), the modulation or information contained is relatively low ($< 5\%$ of the average signal).

The S/N for the contrast contained in an image with pixel nonuniformity removed (i.e., shot noise limited) and unity quantum yield is

CCD TRANSFER CURVES AND OPTIMIZATION

$$(S/N)_{CON} = \frac{\sigma_{CON}(e^-)}{S_{AVG}(e^-)^{1/2}}, \qquad (2.32)$$

where $(S/N)_{CON}$ is the contrast S/N and $\sigma_{CON}(e^-)$ is the rms "signal contrast," i.e., signal that is superimposed on an average signal level of $S_{AVG}(e^-)$.

The contrast is expressed as a fraction of the average signal:

$$\sigma_{CON}(e^-) = C_{CON} S_{AVG}(e^-), \qquad (2.33)$$

where the constant C_{CON} is that fraction of the average signal.

Substituting Eq. (2.33) into Eq. (2.32) yields

$$(S/N)_{CON} = C_{CON} S_{AVG}(e^-)^{1/2}. \qquad (2.34)$$

Maximum $(S/N)_{CON}$ is achieved at full well as

$$(S/N)_{CON} = C_{CON} S_{FW}(e^-)^{1/2}. \qquad (2.35)$$

Example 2.9

Find the $(S/N)_{CON}$ for a 1% contrast signal that rides on an average signal level of $S_{AVG} = 10^3 \, e^-$. Also, find the $(S/N)_{CON}$ improvement when the exposure time is increased by a factor of 1000 (i.e., $S_{AVG} = 10^6 \, e^-$). Assume that the image is shot noise limited.

Solution:
From Eq. (2.34),

$$(S/N)_{CON} = 0.01 \times (1000)^{1/2},$$

$$(S/N)_{CON} = 0.32.$$

At this exposure level, modulation is hidden in the shot noise of the signal. The S/N for a 1000× increase in signal is

$$(S/N)_{CON} = 0.01 \times (1000 \times 1000)^{1/2},$$

$$(S/N)_{CON} = 10.$$

The modulation in this case would be clearly seen. In theory, $(S/N)_{CON}$ is only limited by the charge capacity of the CCD.

Example 2.10

What is the full well requirement to detect 0.1% modulation, assuming $(S/N)_{CON} = 3$.
Solution:
From Eq. (2.35),

$$S_{FW}(e^-) = (3/0.001)^2,$$

$$S_{FW}(e^-) = 9 \times 10^6 \, e^-.$$

Figures 2.13(a) and 2.13(b) show images of a ten dollar bill that demonstrate the high S/N capability of the CCD. Figure 2.13(a) is taken near full well ($100{,}000 \, e^-$) with a contrast modulation of approximately 30%. Neglecting the $5 \, e^-$ read noise floor and removing pixel nonuniformity, a $(S/N)_{CON} = 95$ is achieved. In Fig. 2.13(b) the light flux is reduced by a factor of 100, reducing the $(S/N)_{CON}$ to 9.5. Although three low-level column blemishes emerge, S/N degradation is only slightly apparent in the image as printed.

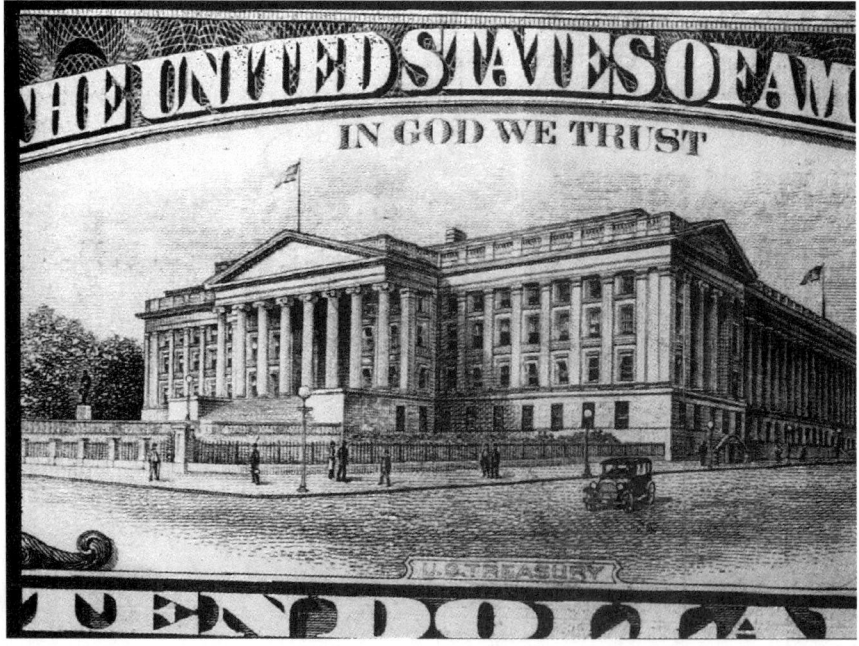

Figure 2.13(a) $100{,}0000 \, e^-$ image of a ten dollar bill exhibiting a $(S/N)_{CON} = 95$.

Figure 2.13(b) 1000 e^- image of a ten dollar bill with $(S/N)_{CON} = 9.5$.

2.2.10 HIGH-SPEED PHOTON TRANSFER GENERATION

It is clear from the above discussions that a significant amount of time is required to generate a photon transfer curve because of the number of images involved. This problem becomes more pronounced when large CCD arrays are characterized. For example, several hours would be required to generate a photon transfer curve for a 4096 × 4096 pixel CCD that is read out slow-scan. Methods used to decrease readout and processing time are (1) pixel binning, (2) shutterless photon transfer, and (3) window clocking.

2.2.10.1 Pixel Binning

Pixel binning is used to reduce readout time when generating photon transfer data. For example, Fig. 2.14 show five-dollar-bill images taken to demonstrate the pixel binning process and the corresponding improvement in readout speed. Figure 2.14(a) shows normal 1 × 1 readout, whereas Fig. 2.14(b) is based on 7 × 7 binning with readout times of 20 and 0.45 sec, respectively. Therefore, binned image is read out 44 times faster than normal readout.

Flat-field photon transfer responses are similar for binned and unbinned data. For example, the read noise and camera gain constant e^-/DN should be the same. Full well for the vertical registers will scale with the number of pixel binned, as

Figure 2.14(a) A full resolution 1024 × 1024 image of a five dollar bill.

long as the horizontal register and output summing well can sum the charge properly. However, for a given exposure period, the flat-field S/N is greater for the binned image because the signal increases proportionally to the number of pixels binned, whereas the noise increases by the square root. In equation form,

$$(S/N)_{BIN} = [P_H P_V S(e^-)]^{1/2}, \qquad (2.36)$$

where $(S/N)_{BIN}$ is the flat-field S/N after pixel binning is performed, P_H is the number of pixels summed in the horizontal direction, P_V the number summed in the vertical direction and $S(e^-)$ is the average flat-field signal level before binning is performed. The equation assumes shot-noise-limited performance and unity quantum yield.

Example 2.11

Determine the $(S/N)_{BIN}$ for a flat-field image for 1×1 readout and 7×7 binning. Assume an average signal level of $100{,}000\,e^-$/pixel.
Solution:
From Eq. (2.36),

$$(S/N) = (10^5)^{1/2},$$

CCD TRANSFER CURVES AND OPTIMIZATION

Figure 2.14(b) The same image of a five dollar bill with 7 × 7 pixel summing employed.

$$(S/N) = 316 \quad 1 \times 1,$$

$$(S/N)_{BIN} = 2214 \quad 7 \times 7.$$

2.2.10.2 Shutterless Photon Transfer

The shutterless photon transfer method is based on only a single frame of data. As the name implies, the curve is generated without a mechanical shutter or pulsed light source. Instead, the CCD is continuously exposed to light and read out without an integration or shuttering cycle. With this method, illumination to the CCD is typically a defocused spot of light above the CCD (e.g., LED). Figure 2.15(a) illustrates the video output signal when the light source is centered onto the array. After the light is adjusted to slightly above full well and video is stabilized, several lines (at least 400 lines for good statistics) are read and stored into a computer. Then $S(DN)$ and $\sigma_S(DN)$ are calculated for each column on the array. As before, the offset is found within the extended pixel region of the horizontal register. Continuous readout and exposure eliminates fixed-pattern noise because charge is averaged over the region of pixels being exposed. Therefore, frame differencing as performed before is not required to obtain shot noise data.

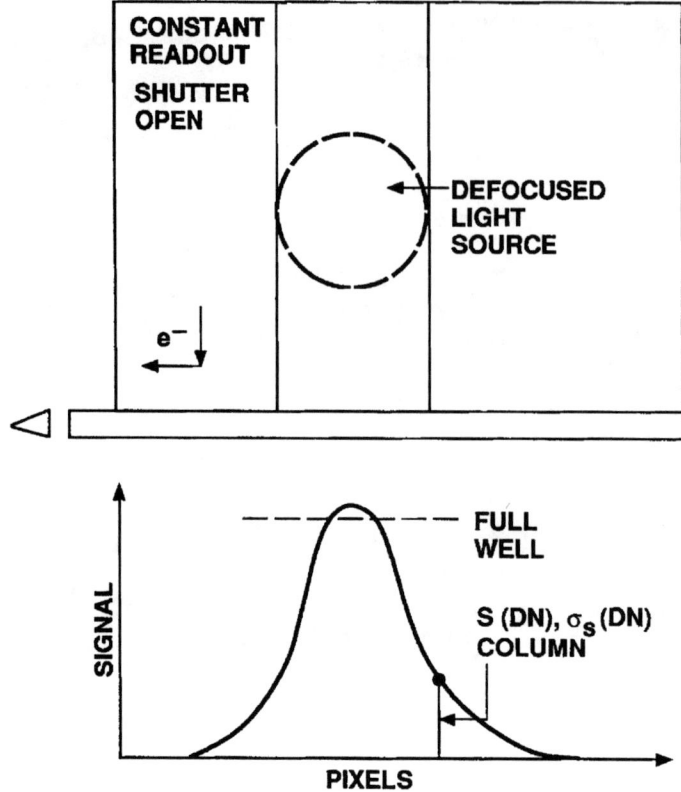

Figure 2.15(a) Configuration used to generate a shutterless photon transfer curve.

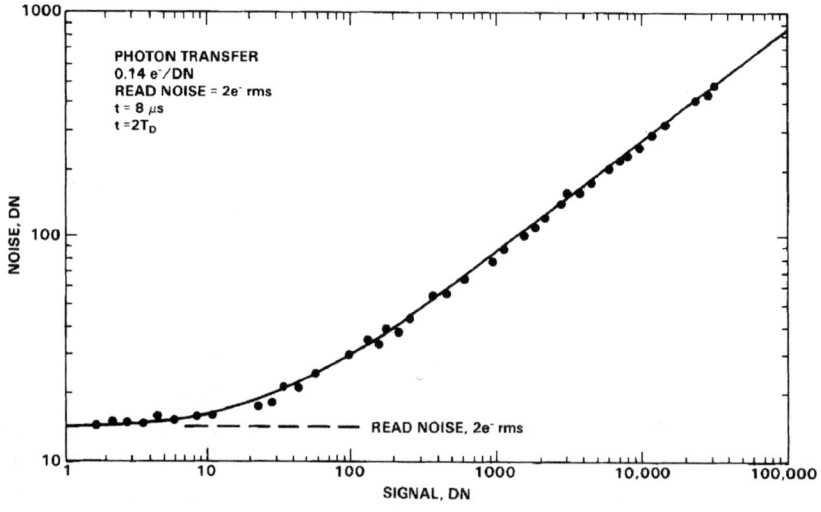

Figure 2.15(b) Shutterless photon transfer curve.

CCD TRANSFER CURVES AND OPTIMIZATION 129

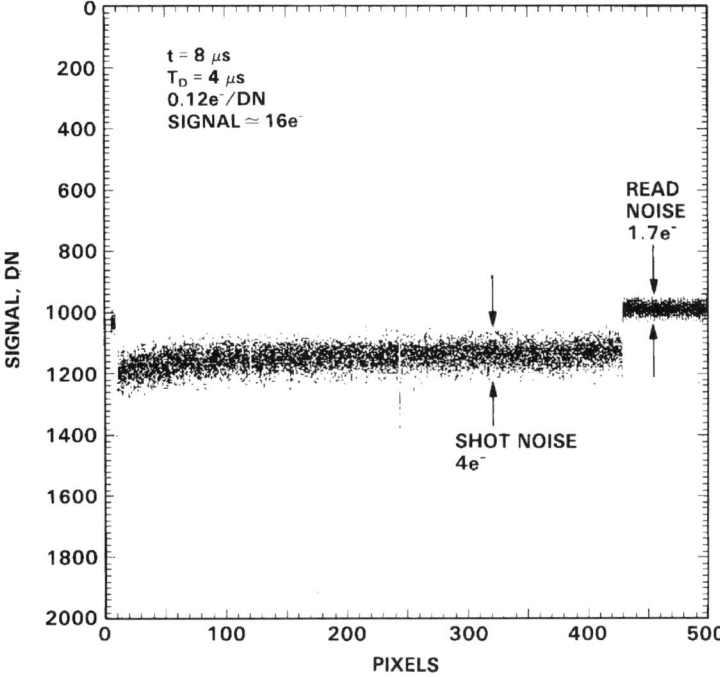

Figure 2.15(c) A 16 e⁻ raw line trace taken from the sensor characterized in Fig. 2.15(b).

Figure 2.15(b) shows a low-light-level shutterless photon transfer curve generated by a 420 × 420 CCD using the arrangement shown in Fig. 2.15(a). All data points shown are derived from one frame of data. A camera gain constant of $K = 0.14\,e^-/DN$ and a read noise floor of $2\,e^-$ rms are measured from the plot. Figure 2.15(c) shows a $16\,e^-$ flat-field video line trace taken with the same CCD. A read noise of $1.7\,e^-$ measured within the extended pixel region. The read noise is sufficiently low to see the $4\,e^-$ rms shot noise component associated with a $16\,e^-$ signal level.

It would seem that linearity data would be lost when using the shutterless technique because there is no exposure period involved. As mentioned above, linearity can be measured by calculating the gain (i.e., e^-/DN) for each column and plotting the values as a function of signal. Ideally, e^-/DN gain should be signal-independent. A change in the camera gain constant implies a nonlinearity problem [refer to Fig. 2.11(c)].

2.2.10.3 Window Clocking

Window clocking is a powerful technique often used to generate photon transfer curves very rapidly while maintaining exposure information. The method interrogates a small subarray on the CCD, similar to random access. This is accomplished by rapidly clocking the vertical registers up to a region of interest without waiting for complete horizontal readout. Normal horizontal readout only resumes when the desired region on the array is reached vertically. At that time, several lines of video are stored into the computer for photon transfer processing. After these data are collected, the high-speed vertical clocking resumes until the CCD is fully read out. The process is repeated at different exposure times to produce photon transfer and linearity transfer curves. The high-speed clocking technique can also be applied in the horizontal direction for additional speed. Window clocking is discussed further in Chapter 4.

2.2.11 Photon Transfer Simulation

Pixel values that make up read noise, shot noise and pixel nonuniformity exhibit normal frequency distributions (i.e., Gaussian) with a mean and a variance. Therefore, one can model and simulate photon transfer curves mathematically. Theoretical curves are often plotted along with experimental data to note discrepancies in solving camera and noise problems. Random numbers that follow a Gaussian distribution can be generated by the following four-step procedure.

Step 1: Generate a pair of numbers from 0 to 1 using a uniform random number generator. Call these U_1 and U_2.
Step 2: Compute

$$x = -\ln U_1. \tag{2.37}$$

Step 3: If

$$e^{-(x-1)^2/2} > U_2, \tag{2.38}$$

then accept x as the first Gaussian number; otherwise reject it.

Step 4: Randomly choose a negative or positive value for noise simulation.
Step 5: Repeat Step 1.

The numbers generated will have a normal distribution with sigma of unity. Scaling is performed by generating numbers with the desired sigma. For example, noise values for a $10\,e^-$ rms read noise floor are generated by choosing a pair of numbers that vary between 0 and 10. Shot noise would be handled by choosing a pair of numbers that range between 0 and the square root of signal. Pixel nonuniformity numbers would vary between 0 and $P_N \times$ signal. All numbers generated would ride on an electrical offset to maintain a positive DN or electron level. Figure 2.16 shows a simulated photon transfer plot with pixel nonuniformity included. The plot assumes $K = 1\,e^-/\text{DN}$, a read noise of $4\,e^-$ rms and a pixel nonuniformity of 1%. The three noise regimes, read (slope 0), shot (slope 1/2) and pixel nonuniformity (slope of unity), are clearly seen.

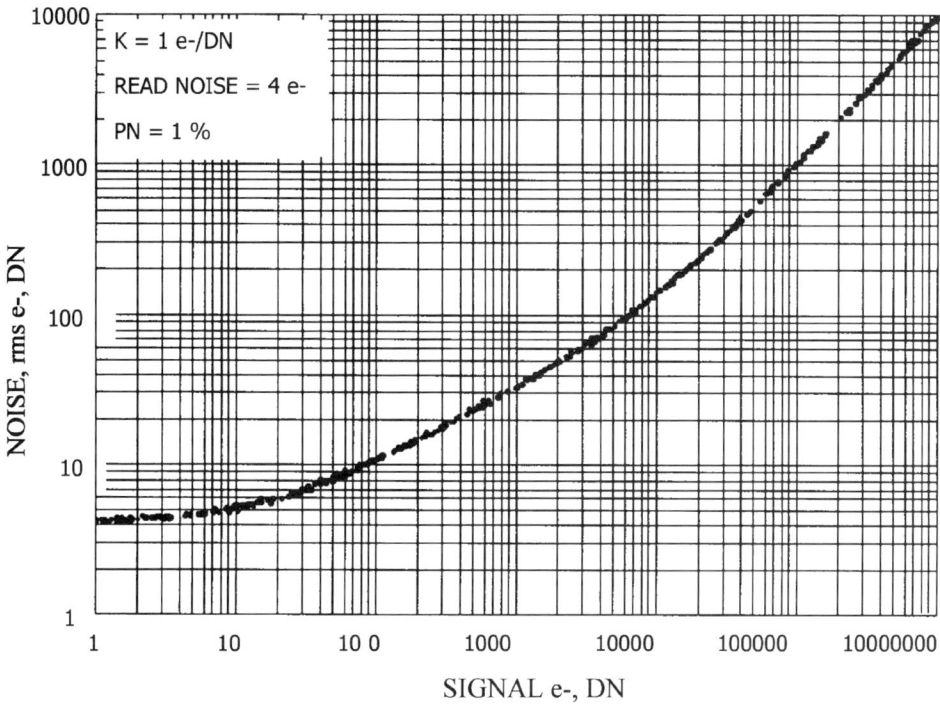

Figure 2.16 Simulated photon transfer plot with pixel nonuniformity.

2.3 X-Ray Transfer

A visible or near-infrared photon, interacting with silicon, generates a single e-h pair. It would be instructive to observe the fate of such an electron as it is drawn into the potential well, transferred through the vertical and horizontal registers and converted to an output voltage. Current CCD technology does not permit accurate measurement of such small charge units. Instead, CCD characterization has traditionally employed relatively large numbers of photons, which are carefully measured to investigate such properties as QE, charge collection efficiency (CCE) and charge transfer efficiency (CTE). The resulting data often represent averages over a large number of pixels and are limited in their accuracy by the intensity, focus and stability of the source. To circumvent these requirements, several alternative calibration techniques that use x-ray illumination to supplement visible light testing have been developed.

2.3.1 X-ray Characteristics and Use

Soft x-ray photons (1 to 100 Å) have much higher energy than do visible light photons. Absorbed by silicon, this additional energy generates multiple e-h pairs in

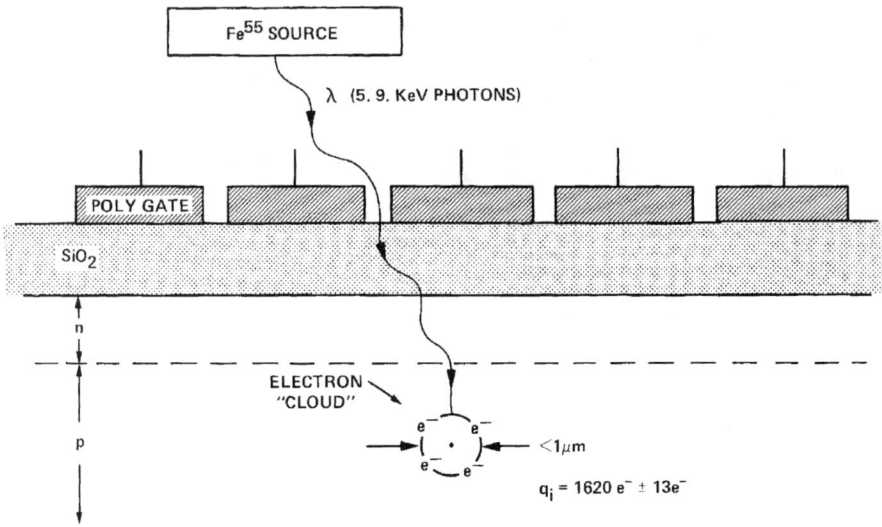

Figure 2.17 Illustration showing an Fe^{55} x-ray event interacting with a CCD.

the CCD [refer to Eq. (1.2)]. In contrast to the visible light case, the electrons are generated in a very small cloud diameter, essentially a near-perfect point source, as illustrated in Fig. 2.17. For example, a 5.9-keV photon generates 1620 e^- contained within a diameter of only 0.4 µm (FWHM). An equivalent number of focused visible photons would generate 1620 e-h pairs throughout an appreciable width and depth of the pixel. Also, their number could not be measured with an accuracy that is comparable to the uncertainty in charge created by an x-ray photon.

The shape and size of individual x-ray events will be displayed and measured below. Ideally, the charge from a single x-ray photon would be confined to a single pixel (referred to as the target pixel), and the surrounding pixels would contain no charge. In practice, the photon is sometimes absorbed below the CCD's depletion layer, in a field-free region. Charge generated there diffuses into neighboring pixels, an indicator of degraded CCE performance. Also, imperfect charge transfer causes some of the charge from the target pixel to "lag" during successive transfers so that an x-ray event exhibits a "tail" of deferred charge. The size and shape of this tail is a sensitive indicator of CTE performance. X-ray stimulation is therefore extremely valuable in characterizing the CCE and CTE.

2.3.2 Fe^{55}

The Fe^{55} soft x-ray source has become a standard in measuring CTE characteristics for the CCD. Figure 2.18 shows how K_α and K_β manganese (Mn) x rays originate from an Fe^{55} x-ray source and how these photons interact with the CCD and generate e-h pairs. An Fe^{55} atom is inherently unstable and decays into a Mn atom when a K-shell electron is quantum mechanically absorbed by the nucleus

CCD TRANSFER CURVES AND OPTIMIZATION

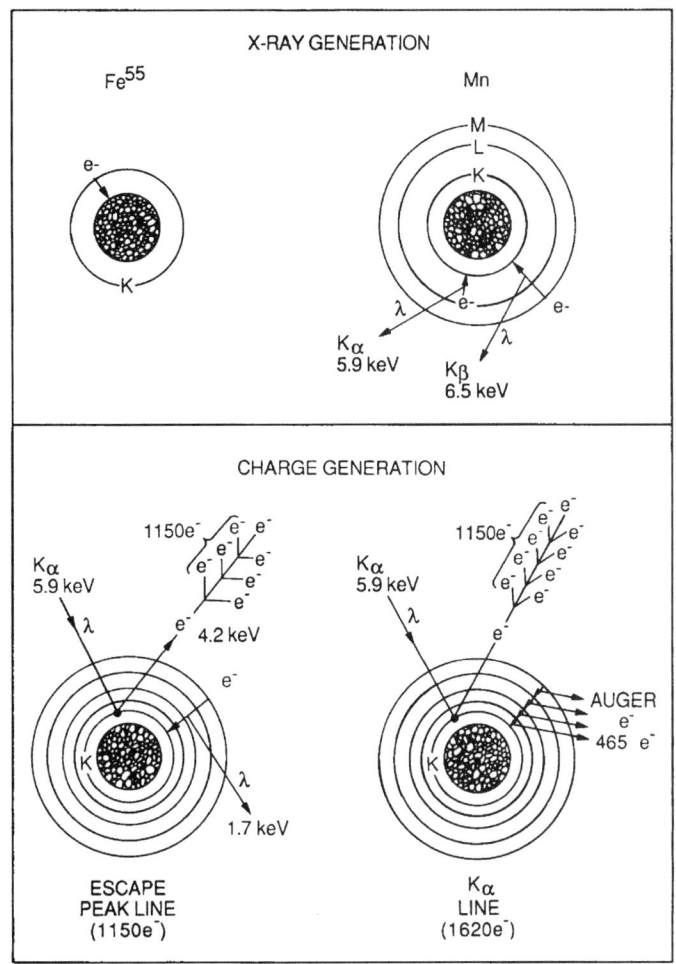

Figure 2.18 Illustration showing manganese x-ray generation and interaction with CCD.

(the half-life of this process is 2.7 years). An x ray is generated when an electron drops from either the L- or M-shell to fill the empty K-shell. This action in turn produces either a K_α or K_β x ray, respectively (the production of a K_α x ray is more likely than a K_β x ray by approximately 7 to 1).

The photoelectric absorption of a Mn x ray by the CCD ejects an electron with energy E-b, where E is the energy of the x ray and b is the binding energy for K-shell electrons (1.78 keV silicon). The free energetic electron produces an ionizing trail of electron-hole pairs (e-h) through inelastic collisions with orbital electrons of other silicon atoms. The binding energy used to fill the K-shell again is either converted into another x ray (a silicon x ray), or creates additional e-h pairs through a mechanism referred to as the Auger process. In the latter case, the atom returns to the ground state when electrons from the outer shells drop to inner ones as free electrons are ejected from the atom. The energy therefore appears largely as kinetic energy of ejected Auger electrons rather than energy in the form of an x-ray photon. The Auger process occurs with much higher probability than the x-ray emission process. If a silicon x ray is produced, it will likely "escape" from the original point of interaction. The net signal generated under these circumstances is only $1128\,e^-\,[(5900-1780)/3.65]$, creating a K_α "escape peak" charge packet. The K_β primary line and escape peak are generated in a similar fashion. The silicon x ray that is produced may be absorbed by the CCD at some other pixel location resulting in an x-ray charge packet of $487\,e^-\,(1780/3.65)$ or escape from the sensor entirely.

In summary, an Fe^{55} x-ray source produces five different signals in the CCD: K_α ($1620\,e^-$), K_β ($1778\,e^-$), K_α escape peak ($1133\,e^-$), K_β escape peak ($1291\,e^-$), and a silicon line ($487\,e^-$). It should be mentioned that several other weak x-ray lines are possible; however, their presence is relatively low compared to the five primary x-ray lines listed above. For example, it is possible for an electron to drop from the M-shell to the L-shell, producing a L_α x ray.

Figure 2.19 plots absorption length as a function of wavelength (nm) and energy (keV) for visible/near IR and soft x-ray photons, respectively. Absorption length is that distance where 63% of the photons are absorbed (i.e., $1-e^{-1}$). Note that a Mn x-ray photon penetrates the same distance into the CCD as a 900-nm near-IR photon: approximately 30 µm. Therefore, Mn x rays stimulate the photo-

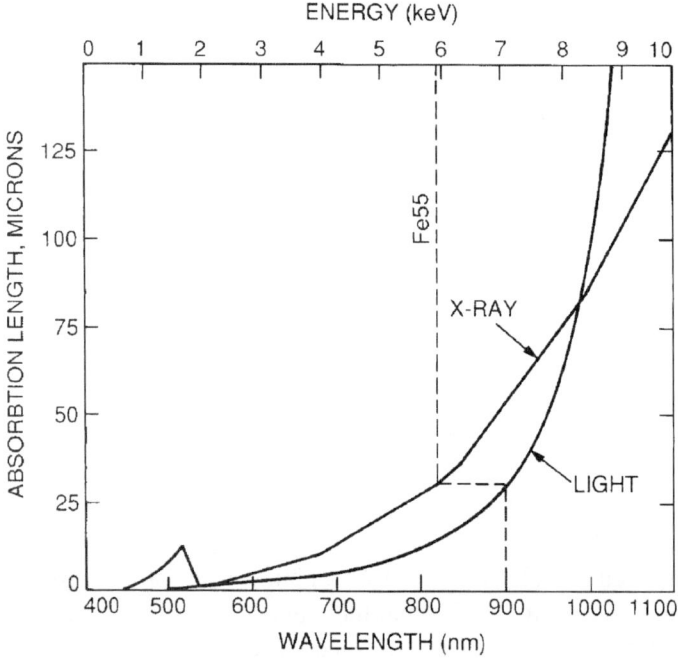

Figure 2.19 X-ray absorption length compared to visible and near-IR photons.

sensitive volume of most CCDs in depth fairly uniformly. This characteristic is important in characterizing charge collection performance as discussed in Chapter 4.

The strength of an x-ray source is specified in a unit called "curie." One curie (Ci) = 3.7×10^{10} disintegrations per second. A 100-µCi source is satisfactory for most laboratory CCD characterization tests. Sources as strong as 50 mCi are commercially available with a state license. An Fe^{55} source is convenient to use in the laboratory, as it is physically small (roughly the size of a dime) and can be placed a few inches from the CCD to provide uniform illumination, as illustrated in Fig. 2.20. An opaque metal window can be located between the CCD and x-ray source to adjust intensity outside the camera under vacuum situations. The window must be sufficiently thin to transmit Mn x rays but thick enough to maintain vacuum conditions. A half-inch diameter Be window 3 mils in thickness is used for most x-ray test results in this book, although more absorbant, thin aluminum windows are satisfactory for strong x-ray sources.

In practice, a variety of x-ray lines and energies are used in testing the CCD. As indicated above, an Fe^{55} source offers a pure flux of Mn photons at 5.9 and 6.5 keV. Other x-ray energies and signal strengths can be produced by "fluorescing" selected target elements with high-energy alpha particles. For example, alpha particles that interact with Si atoms eject K-shell electrons, which in turn will liberate x rays characteristic of the target material fluoresced. Cm^{244}, a 5.2-MeV alpha emitter, is a common radioactive source utilized for this purpose. Example target materials include fluorine (0.68 keV), sodium (1.04 keV), aluminum (1.49 keV), sulfur (2.31 keV), chlorine (2.62 keV), calcium (3.69 keV), and copper (8.0 keV). Table 2.4 is a list of x-ray energies, and the charge generated determined by Eq. (1.2) as a function of atomic number. Both K_α and K_β lines are tabulated.

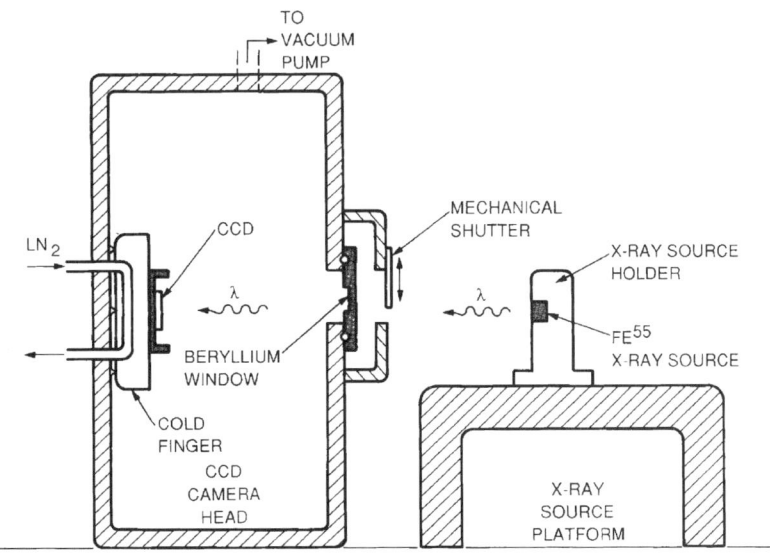

Figure 2.20 Experimental setup used to stimulate the CCD with Fe^{55} x rays.

Example 2.12

> Determine the quantity of charge generated by a 2.6-keV chlorine x ray.
> *Solution*:
> From Table 2.4, a chlorine x ray generates $718\,e^-$. This quantity can also be calculated from Eq. (1.2).

An experimental setup using Cm^{244} is shown in Fig. 2.21(a). In this arrangement, the source is contained inside a vacuum head because the alpha particle would rapidly lose energy if it traveled through the air (its ionizing path is only a few inches). Also, a high vacuum is required when low-energy x rays (< 1 keV) are produced, if they are to reach the CCD.

It is worth noting that the range of a 5-MeV alpha particle in silicon is extremely short (approximately 30 μm), and therefore, the particle deposits most of its energy in the epitaxial layer of the detector. Thus, it is very important to minimize the number of alpha particles that interact directly with the CCD in order to prevent radiation (i.e., silicon lattice) damage. The alpha source is pointed away from the CCD—as shown in Fig. 2.21(a)—for that reason. Radiation damage induced by sources like this are discussed in Chapter 8.

Example 2.13

> Calculate the charge induced by a 5-MeV alpha particle, assuming complete energy transfer within a pixel element and $E_{e-h} = 3.65$ eV.
> *Solution*:
> From Eq. (1.2),
>
> $$S(e^-) = (5 \times 10^6)/3.65,$$
>
> $$S(e^-) = 1.37 \times 10^6\,e^-.$$

An Fe^{55} x-ray source can also be used to fluoresce target materials and generate other x-ray energies. For example, Mn x rays directed toward an aluminum target will generate aluminum x rays (i.e., 1.5-keV x rays). The Mn x ray ejects an Al K-shell electron which in turn produces an Al x ray when the K-shell fills with an M-shell electron. X-ray energies using an Fe^{55} source are limited to energies no greater than 5.9 keV and no less than approximately 1000 eV. Below 1000 eV, x rays do not efficiently leave the target since their absorption depth is much smaller than the penetration depth of a Mn x ray. In contrast, alpha particles (e.g., Cm^{244}) do not penetrate deep into the target and will efficiently produce

Table 2.4 Primary x-ray lines listing photon energy and charge generated

Atomic#	Element	Lines K_α (keV)	Signal K_α (e^-)	Lines K_β (keV)	Signal K_β (e^-)
4	Be	0.109	29.863		
5	B	0.183	50.137		
6	C	0.277	75.890		
7	N	0.393	107.671		
8	O	0.525	143.836		
9	F	0.677	185.479		
10	Ne	0.848	232.329	0.858	235.068
11	Na	1.041	285.205	1.071	235.068
12	Mg	1.253	343.288	1.302	356.712
13	Al	1.487	407.397	1.557	426.575
14	Si	1.740	476.712	1.806	494.795
15	P	2.014	551.781	2.139	586.027
16	S	2.308	632.329	2.464	675.068
17	Cl	2.622	718.356	2.816	771.507
18	Ar	2.958	810.411	3.190	873.973
19	K	3.314	907.945	3.590	983.562
20	Ca	3.692	1011.507	4.013	1099.452
21	Sc	4.090	1120.548	4.461	1222.192
22	Ti	4.511	1235.890	4.932	1351.233
23	V	4.952	1356.712	5.427	1486.849
24	Cr	5.415	1483.562	5.947	1629.315
25	Mn	5.890	1613.699	6.490	1778.082
26	Fe	6.494	1779.178	7.058	1933.699
27	Co	6.930	1898.630	7.649	2095.616
28	Ni	7.478	2048.767	8.265	2264.384
29	Cu	8.048	2204.932	8.905	2439.726
30	Zn	8.639	2366.849	9.572	2622.466
31	Ga	9.252	2534.795	10.271	2813.973
32	Ge	9.887	2708.767	10.983	3009.041
33	As	10.544	2888.767	11.727	3212.877
34	Se	11.222	3074.521	12.496	3423.562
35	Br	11.924	3266.849	13.292	3641.644
36	Kr	12.650	3465.753	14.113	3866.575
37	Rb	13.396	3670.137	15.000	4109.485
38	Sr	14.166	3881.096	15.833	4337.900
39	Y	14.958	4098.082	16.737	4585.479
40	Zr	15.770	4320.548	17.662	4838.904
41	Nb	16.615	4552.055	18.623	5102.192
42	Mo	17.479	4788.767	19.608	5372.055
43	Te	18.367	5032.055	20.169	5649.041
44	Ru	19.279	5281.918	21.656	5933.151
45	Rh	20.216	5538.630	22.723	6225.479

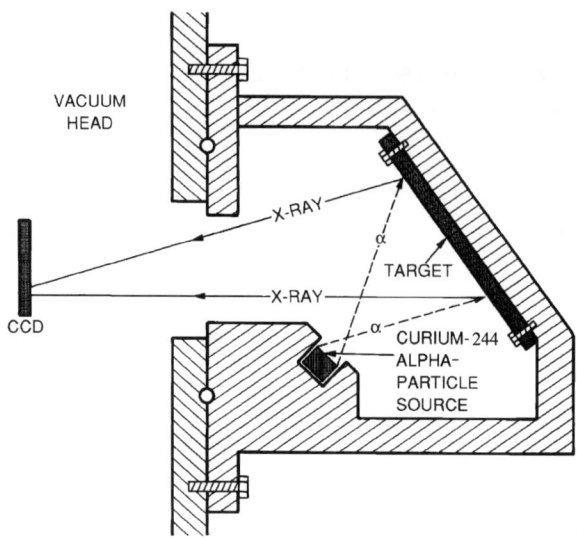

Figure 2.21(a) Apparatus used to fluoresce x rays from a target material.

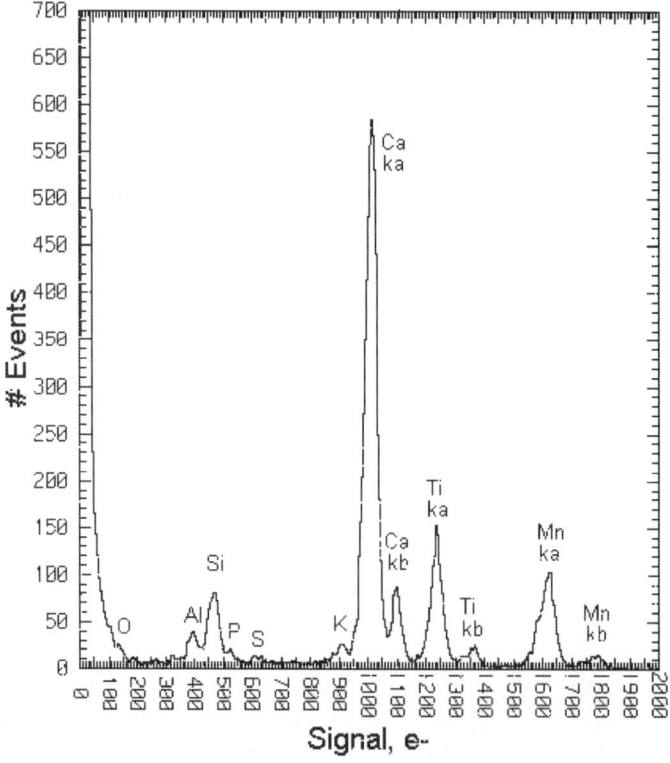

Figure 2.21(b) Fe^{55} x-ray histogram.

CCD TRANSFER CURVES AND OPTIMIZATION

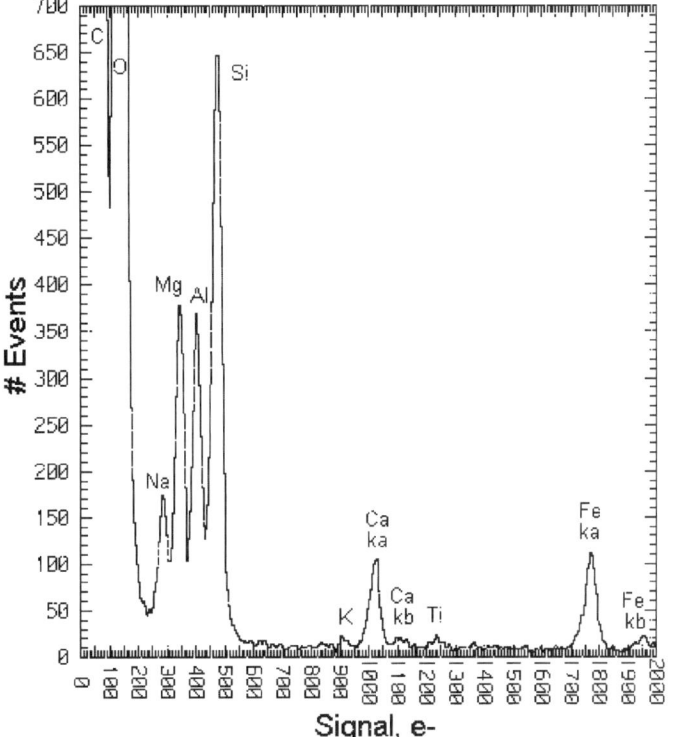

Figure 2.21(c) Cm^{244} x-ray histogram.

x rays below 1000 eV. This is the main advantage of using an alpha source over Fe^{55} when very low energy x rays are required. Figures 2.21(b) and 2.21(c) show Fe^{55} and Cm^{244} histograms generated by a thin-gate CCD with a 3-electron read noise floor. A basalt target is used to fluoresce x-ray events in both cases. It is quite apparent that the Cm^{244} is much more efficient in generating low-energy x rays than the Fe^{55} source.

2.3.3 X-RAY IMAGES

Figure 2.22(a) shows a flat field of Mn x-ray events taken by a 520 × 520 CCD. As mentioned above, each K$_\alpha$ photon that interacts with the CCD produces 1620 e$^-$. Figure 2.22(b) shows a magnified view of the events. Events are represented by the computer by displaying a "dot" for each electron measured. The dots shown are randomly positioned within each pixel in which they are found. A completely dark pixel indicates 1620 e$^-$ or more are measured. Read noise electrons are seen in all pixels (6 e$^-$ rms for this CCD). Note that the charge collected for some x-ray events ends up in more than one pixel. The amount of charge splitting depends

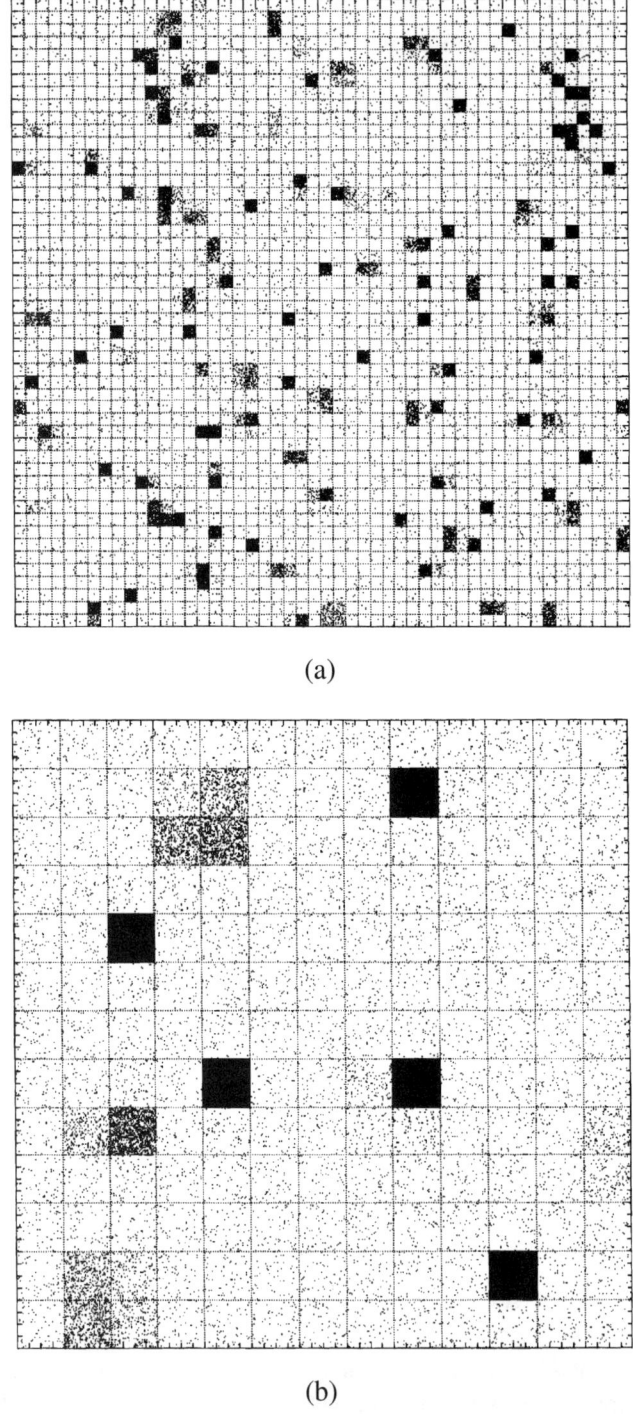

Figure 2.22 A flat-field image of Fe^{55} x-ray events. Each dot represents a measured electron and is randomly positioned within the pixel in which it was detected.

upon where in the pixel the photon is absorbed. Photons that interact within the frontside depletion region, which defines the potential well, are typically seen as the ideal event and are called "single pixel events" [five of these events are seen in Fig. 2.22(b)]. Photons that are absorbed below the depletion region in field-free material create charge clouds that thermally diffuse outward until reaching the sharply defined potential wells at the lower boundary of the pixel well. At that point, the charge cloud may split between two or more pixels. Events of this type are called "split events" [three of these event types are seen in Fig. 2.22(b)]. Occasionally, a split event will be created even though the photon interacts within the depletion region. This is because the initial electron cloud has a finite diameter (about 0.4 µm for a Mn photon interaction) and will split if it falls directly on the boundary of two or more pixels. Summing the charge produced within the affected pixels of a split event may or may not yield the expected amount (i.e., $1620\,e^-$). Most often, charge that is summed is less than what theory would predict, signifying that some carriers are lost to recombination during the diffusion process (carriers are also lost in the channel stops and at the epitaxial interface of the CCD where traps are found). Events of this type are referred to as "partial events."

2.3.4 X-RAY TRANSFER

Figure 2.23 shows a horizontal x-ray transfer plot. The trace was produced by stimulating a 520×520 CCD with a flat field of Mn x rays and stacking (summing)

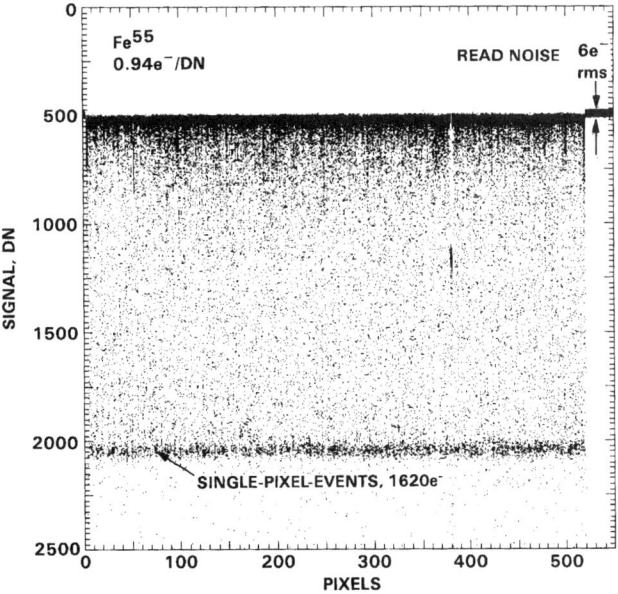

Figure 2.23 A horizontal Fe^{55} x-ray transfer plot.

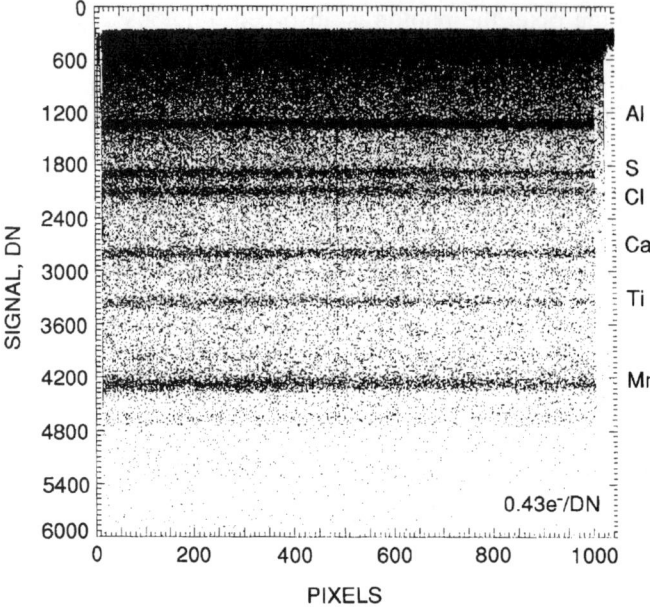

Figure 2.24 X-ray transfer plots generated by a strong Fe^{55} source that fluoresces x rays from the various target materials.

approximately 500 lines together into the composite trace presented. The vertical signal axis is expressed in DN units, measured after the video signal is digitized to 16 bits. These arbitrary units can be directly converted to electrons using a camera gain constant of $K = 0.94\,e^-/DN$ found through photon transfer. The strong signal level seen at 2200 DN represents the single-pixel-event line. Signals below this level are the result of split and partial events. The CCD was purposely over-scanned in the horizontal direction to measure the read noise floor ($6\,e^-$ rms here) and obtain offset information. Note also that the camera gain constant e^-/DN can be determined from an x-ray response and should agree with photon transfer. For example, the x-ray signal measured in Fig. 2.23 is approximately 1700 DN (i.e., single-pixel event line (2200 DN) minus the offset (500 DN)). The camera gain constant is therefore $1620\,e^-/1700\,DN = 0.95\,e^-/DN$.

Figure 2.24 shows multiple x-ray event line traces generated by fluorescing a target with a strong 10-mCi Fe^{55} x-ray source as described above. The target employed was composed of aluminum with small traces of sulfur, chlorine, calcium, and titanium all detected in the plot. Mn x rays are also observed by the CCD as background events (the Fe^{55} source was not totally shielded from the CCD). Multiple x-ray event line traces as shown in Fig. 2.24 are used to characterize CTE as a function of charge packet size.

Some interesting target materials can be fluoresced. For example, the x-ray stacking response shown in Fig. 2.25 was generated by exposing a freshly picked

CCD TRANSFER CURVES AND OPTIMIZATION 143

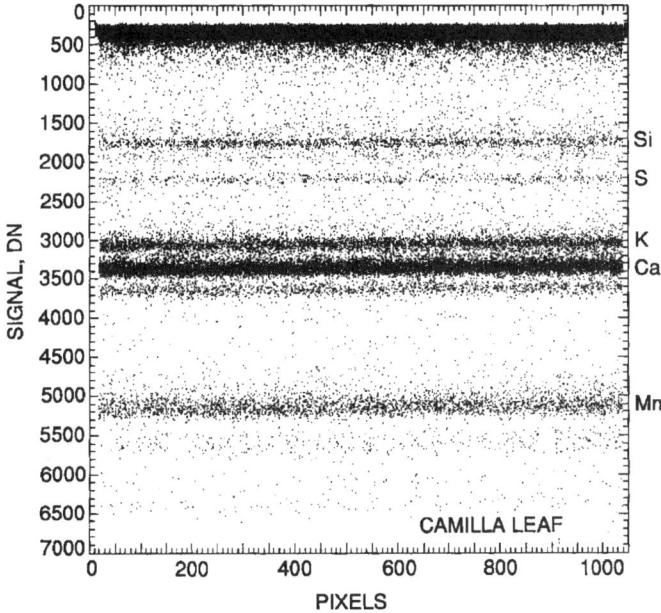

Figure 2.25 An x-ray transfer plot using a Fe^{55} source to fluoresce x rays from a camellia leaf.

camellia leaf to Mn x rays. The leaf is apparently rich in calcium and potassium, with small traces of sulfur and silicon. The strong line above Ca$_\alpha$ is the Ca$_\beta$ line. Low-energy x rays such as carbon and oxygen are not seen because the x rays are absorbed by the target. To generate these x rays, an alpha source must be used.

2.3.5 X-RAY HISTOGRAMS

X-ray events can also be displayed in histogram form. A histogram of x-ray events would ideally contain a narrow peak centered at the signal level that corresponds to the x-ray energy (e.g., 1620 e$^-$ for a 5.9-keV x ray). A histogram is produced by measuring the charge in every pixel and sorting the results into signal bins. For example, Fig. 2.26 is a very early x-ray histogram that showed for the first time clear separation of the K$_\alpha$ and K$_\beta$ lines and indications of escape peaks.[3] The data were generated by a 1024 × 1024 virtual-phase CCD (VPCCD) developed for the SXT and AXAF missions.

Figure 2.27(a) is an Fe^{55} x-ray histogram that separates x-ray signals into bins of 10 DN. The histogram shows a gain of 0.7 e$^-$/DN. Both the K$_\alpha$ (1620 e$^-$) and the K$_\beta$ (1778 e$^-$) lines are seen in the histogram. Split and partial events are indicated in the histogram and represent over half of the events measured. Figure 2.27(b) shows only single-pixel events. Here, events where more than two pixels are involved with the interaction are discarded by the computer. A signal

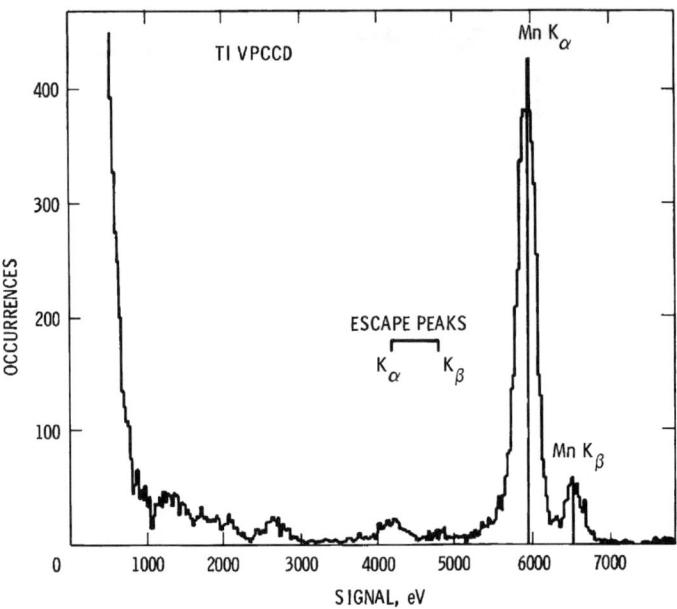

Figure 2.26 An early Fe^{55} x-ray histogram taken with a virtual-phase CCD.[3]

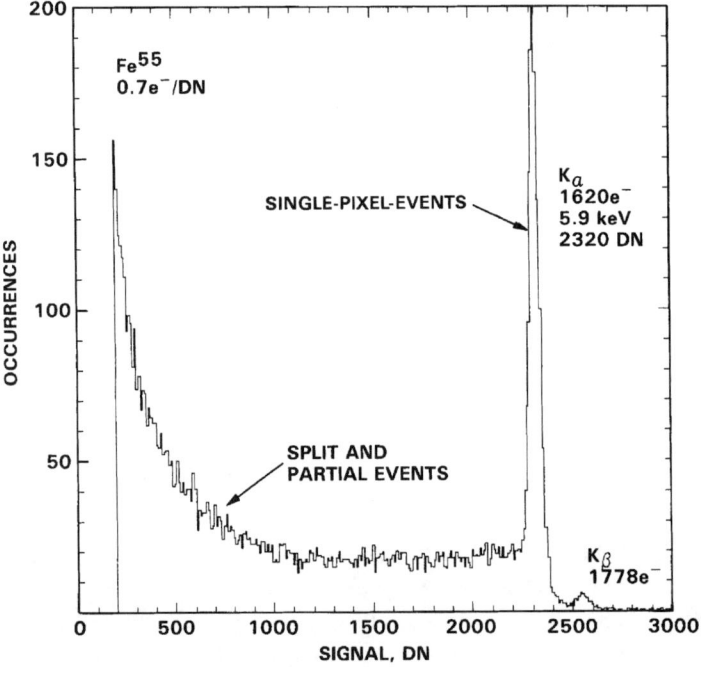

Figure 2.27(a) A Fe^{55} x-ray histogram showing all event types.

CCD TRANSFER CURVES AND OPTIMIZATION 145

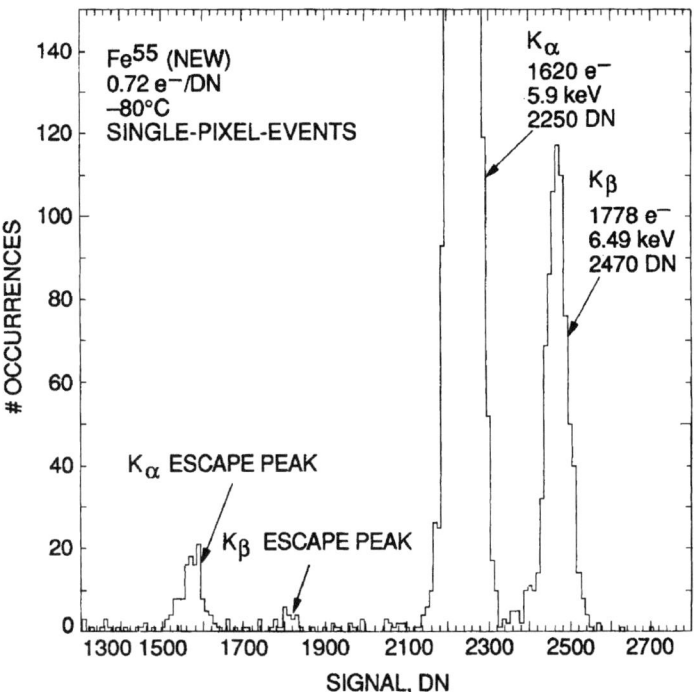

Figure 2.27(b) A Fe^{55} x-ray histogram that plots only single-pixel events.

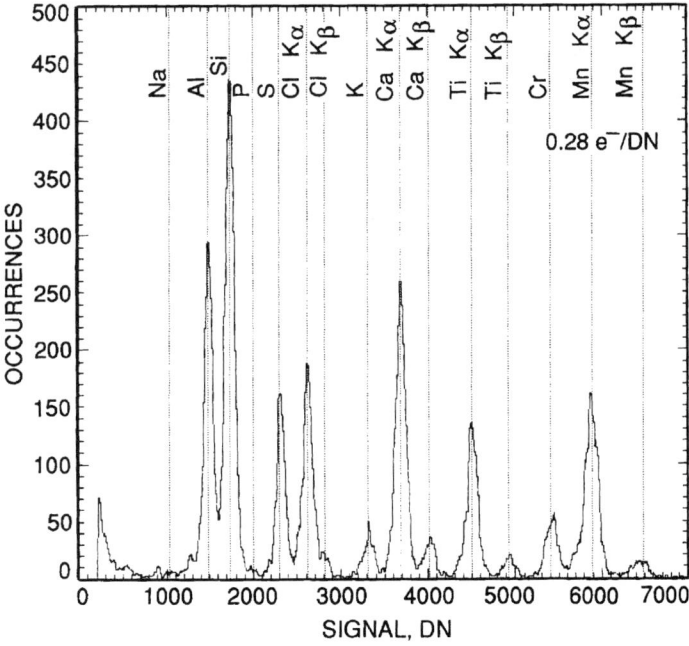

Figure 2.27(c) X-ray histogram showing many x-ray lines.

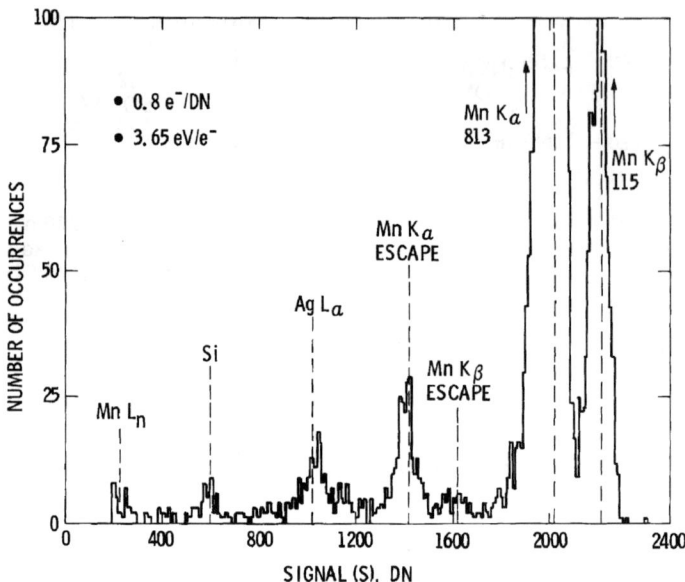

Figure 2.28 An early Fe^{55} x-ray histogram generated by a backside-illuminated CCD.

discriminator of three times the read noise is used to decide if an event is split or not. Note that the K_α and K_β lines are completely separated. These x-ray lines are separated by 600 eV, equivalent to only 165 e^- (refer to Table 2.4). Also, the K_α and K_β escape peaks not seen in Fig. 2.27(a) become apparent. Figure 2.27(c) is a histogram showing various x-ray lines for a target source. Only single-pixel events are used in the plot.

It is interesting to note that the K_β line shown in Fig. 2.27(a) is weak in contrast to what theory predicts. Here, the K_β line is down 40 times compared to the K_α line (i.e., $K_\alpha = 200$ DN and $K_\beta = 5$ DN). Recall from above that the K_α to K_β intensity ratio should be approximately 7 to 1. This ratio is about correct for Fig. 2.28 (i.e., $K_\alpha = 813$ DN and $K_\beta = 115$ DN). Other x-ray histograms and stacking plots shown in this book also show significant variance in the K_α and K_β lines depending on the x-ray source utilized. It has been theorized that Cr is present in some of these sources. Its K-edge is at 5.947 keV (Table 2.4), and therefore it would transmit the K_α line but would absorb the K_β line at 6.49 keV.

Figure 2.28 shows a single-pixel-event x-ray histogram generated by a WF/PC I CCD. The response was the first x-ray response generated by a 6-µm thinned, fully depleted backside-illuminated CCD. Although the noise level for the CCD is high (11 e^-), the sensor still achieves good separation of the Mn K_α and K_β lines. In addition, both escape peaks are seen, including the silicon absorption peak. Also, an Ag L_α peak is discerned because of Ag present in the source (i.e., silver solder was used) as well as the Mn L_α line. The partial event floor of this histogram is very small ($< 0.1\%$), indicating that virtually all the signal charge from each event measured is collected.

2.3.6 FANO-NOISE-LIMITED PERFORMANCE

It is important to note in Fig. 2.23 that the peak-to-peak width of the single-pixel-event line is greater than the peak-to-peak width of the read noise floor. This result indicates that the statistical variation in charge generated by the Mn x rays is greater than the uncertainty in charge measurement (i.e., read noise). As indicated by Eq. (1.2), it takes 3.65 eV of energy on the average to produce a single e-h pair. If all the energy of an x-ray photon were used to produce e-h pairs directly, there would be no statistical variation in the amount of charge generated by the x-ray event. However, a finite amount of energy is transferred to the silicon lattice by non–e-h processes (e.g., thermal), giving rise to a small statistical difference in the number of e-h pairs actually generated. This uncertainty is characterized by the Fano factor, originally formulated by U. Fano in 1947 to describe the uncertainty of the number of ion pairs produced in a volume of gas following the absorption of ionizing radiation.

The Fano factor relates the measured full width half maximum (FWHM) of a histogram and the average energy expended to produce an e-h pair by the relation

$$\text{FWHM(eV)} = 2.355 \left[E(\text{eV}) F_a E_{\text{e-h}} \right]^{1/2}, \quad (2.39)$$

where $E(\text{eV})$ is the photon energy (eV), and F_a is the Fano factor. Experimentally, the Fano factor has been determined to be about 0.1 for silicon.

Example 2.14

Find the theoretical limiting energy resolution for a CCD that is stimulated with 5.9-keV x-ray photons. Express the results in units of FWHM (eV) and rms eV.

Solution:
From Eq. (2.39),

$$\text{FWHM(eV)} = 2.355 \left(5.9 \times 10^3 \times 0.1 \times 3.65 \right)^{1/2},$$

$$\text{FWHM(eV)} = 109 \, \text{eV}.$$

Note that FWHM is 2.355 times greater than the rms value, assuming a Gaussian distribution. Therefore, energy resolution expressed in rms eV is 109/2.355, or 46.3 eV. This is equivalent to $12.7 \, e^-$ rms. The read noise is $6 \, e^-$ in Fig. 2.23, considerably less than the Fano noise produced.

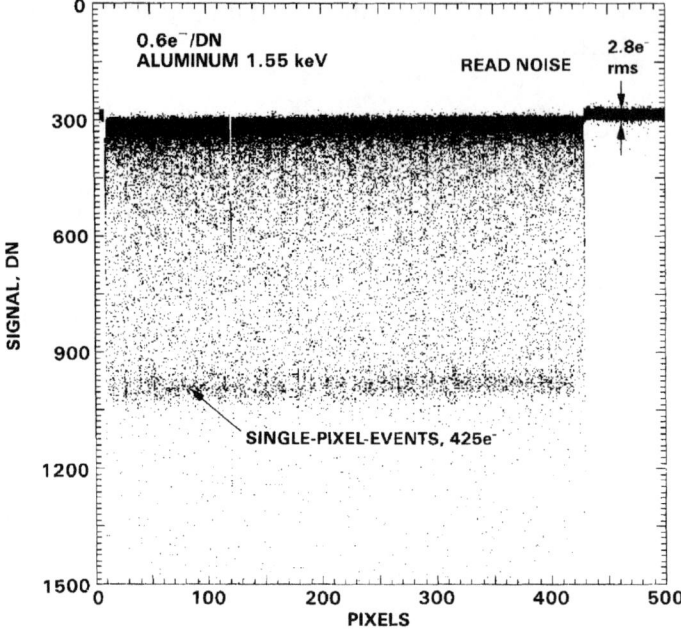

Figure 2.29(a) Horizontal aluminum x-ray transfer plot generated by Fano-noise-limited 420 × 420 CCD.

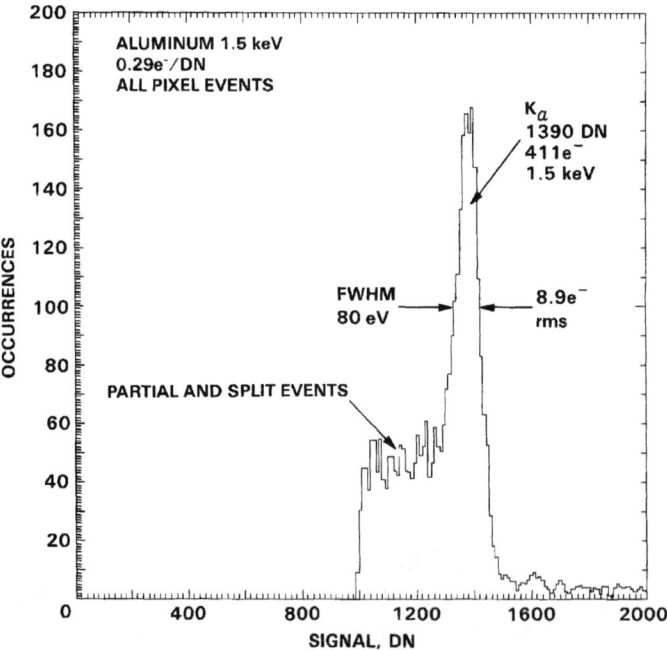

Figure 2.29(b) X-ray histogram for the 420 × 420 CCD.

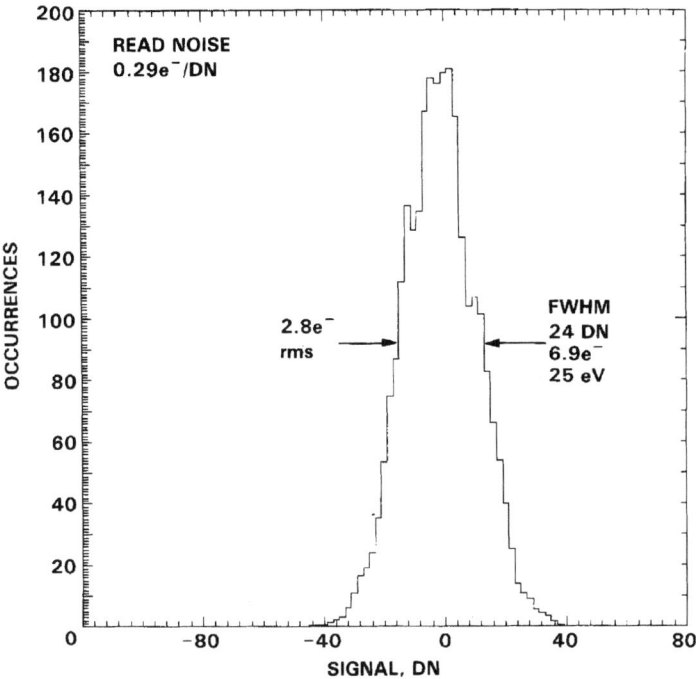

Figure 2.29(c) Read noise histogram showing a 2.8 e⁻ rms noise floor.

"Fano-noise-limited performance" implies the read noise is less than the Fano noise. With noise floors of $< 2\,e^-$ being achieved today, the CCD is capable of such performance over the entire soft x-ray regime (i.e., 100–10,000 eV), including a portion of the EUV spectrum (10–100 eV).[4]

Figure 2.29(a) shows a horizontal x-ray transfer plot generated by a 420 × 420 CCD in response to aluminum 1.5-keV x rays (425 e⁻) produced by a curium source. The weak K_β Al line cannot be resolved because the separation of the K_α and K_β lines is only 70 eV (see Table 2.4) and the K_β line is very weak. Note that the width of the K_α single-pixel-event line is greater than read noise (2.8 e⁻), indicating the sensor is Fano-noise-limited at this energy. Figure 2.29(b) shows an x-ray histogram for the same CCD with an elevated camera gain (0.29 e⁻DN). An energy resolution of 8.9 e⁻ rms (80 eV FWHM) is measured. Figure 2.29(c) shows a corresponding read noise histogram with a standard deviation of 2.8 e⁻. These data can be used to estimate the Fano factor using the formula

$$F_a = \frac{3.65(L_W^2 - \sigma_R^2)}{E(\text{eV})}, \qquad (2.40)$$

where L_W is the x-ray spectral line width (rms e⁻).

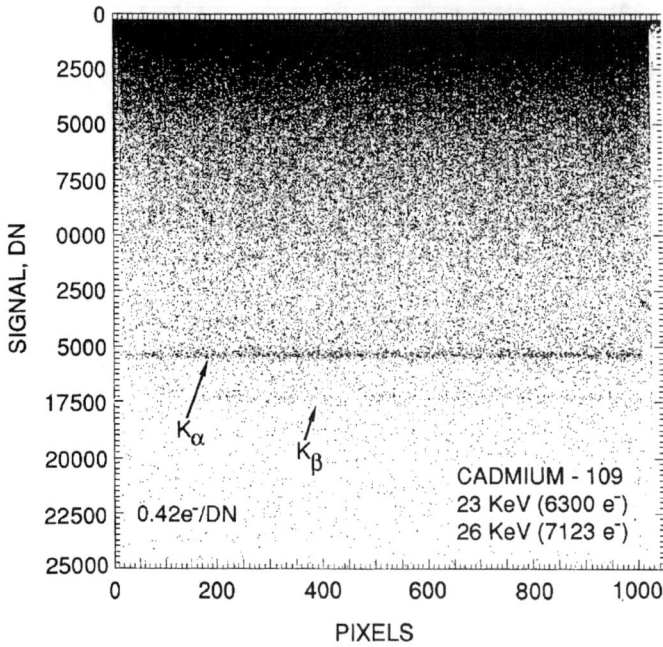

Figure 2.30(a) A horizontal x-ray transfer plot generated by a Fano-noise-limited 1024 × 1024 CCD.

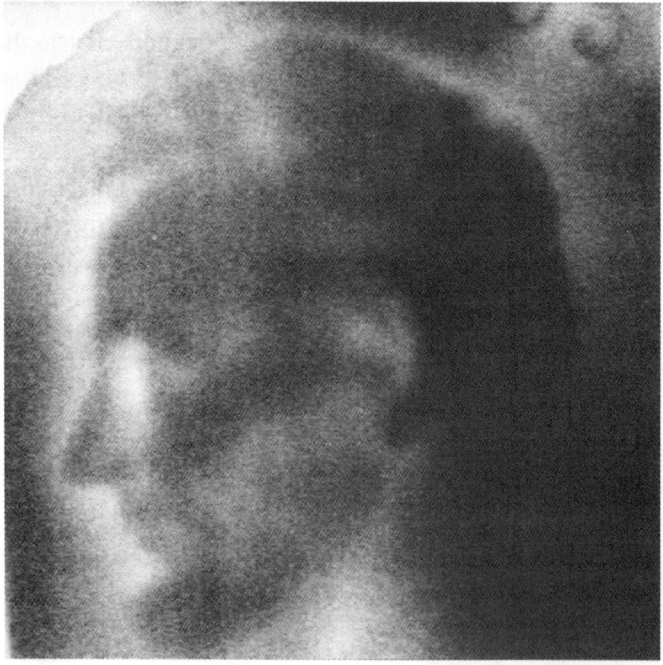

Figure 2.30(b) Cadmium x-ray image of a penny taken with a 1024 × 1024 CCD.

CCD TRANSFER CURVES AND OPTIMIZATION

Example 2.15

Estimate the Fano factor from the histogram shown in Fig. 2.29(b).
Solution:
From Fig. 2.29(b), $L_W = 8.9$ e⁻rms and $\sigma_R = 2.8$ e⁻rms.
From Eq. (2.40),

$$F_a = 3.65 \times (8.9^2 - 2.8^2)/(1.5 \times 10^3),$$

$$F_a = 0.17.$$

The Fano factor as measured by the CCD is usually higher than the accepted value of 0.1. There are at least three reasons for the discrepancy. First, a small but finite CTE loss will cause the x-ray histogram to broaden since less charge will be contained in the target pixel. Second, imperfect charge collection efficiency due to partial and split events will shift the x-ray line to lower energies, widening the peak. Third, K_α and K_β lines that are not resolved will broaden the peak.

2.3.7 CADMIUM X RAYS

Often, CCDs are protected with an optical window that prevents low-energy x rays from reaching the CCD. For these sensors, x-ray characterization can be performed using a hard x-ray Cd^{109} source. This source of x rays generates 23-keV (K_α, 6300 e⁻) and 26-keV (K_β, 7123 e⁻) events. Figure 2.30(a) shows a horizontal x-ray transfer plot of cadmium x-ray events generated by a CCD with a thick quartz window. The absorption length for cadmium x rays is very long (several hundred microns) and most photons pass completely through the CCD. Therefore, longer exposures may be required compared to Fe^{55} stimulation. Figure 2.30(b) shows a cadmium x-ray image of a penny which was positioned on top of the CCD during a test exposure. The image demonstrates that cadmium x rays are very penetrating.

2.4 QE TRANSFER

Figure 2.31 shows a QE test configuration using a calibrated silicon photodiode to cover a spectral range of 2,500 to 11,000 Å. Cadmium telluride photodiodes are typically used to cover the 1200 to 2500-Å range. The diode and CCD are in close proximity to intercept the same number of photons. Illumination can be provided by a 100-W quartz (or xeon) lamp in conjunction with a diffuser to scatter light uniformly on both detectors. Interference filters are used to filter light and measure

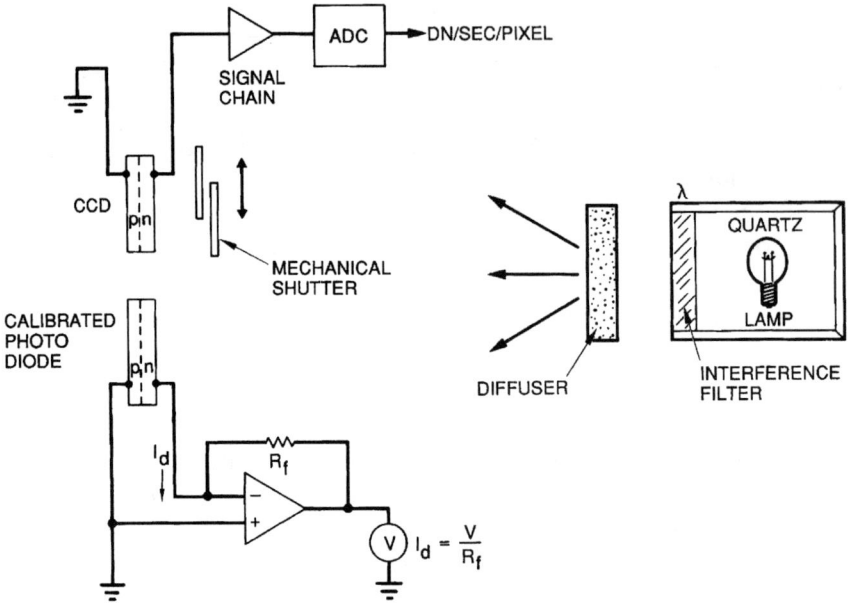

Figure 2.31 Apparatus used in making absolute QE measurements.

QE as a function of wavelength. QE of the CCD is determined by comparing the sensitivity of the CCD and standard diode using

$$\text{QE} = \frac{A_D S_{CCD} \text{QE}_D}{P_S S_D}, \qquad (2.41)$$

where A_D is the active area of the diode (cm^2), P_S is the active area of a pixel (cm^{-2}), S_{CCD} is the average signal generated by the CCD (e$^-$/pixel/sec), S_D is the photosignal generated by the diode (e$^-$/sec), and QE_D is the QE of the diode supplied by the manufacturer.

The CCD signal, S_{CCD}, is found by measuring the average signal for a small group of pixels through the equation

$$S_{CCD} = \frac{S(\text{DN})K}{t_E}, \qquad (2.42)$$

where t_E is the exposure time (s). The signal from the diode, S_D, is defined by

$$S_D = \frac{I_d}{q}, \qquad (2.43)$$

where I_d is the photocurrent measured by an amp meter (A). Note that the diode is constantly illuminated by the light source, whereas the CCD is shuttered.

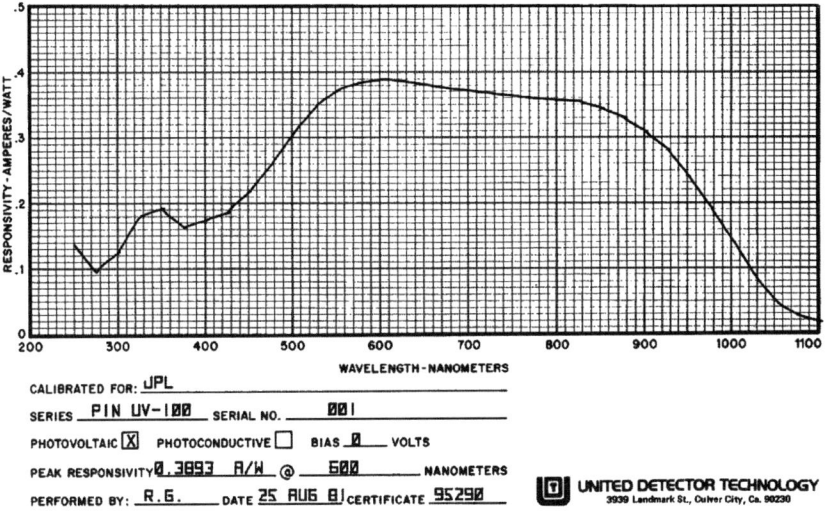

Figure 2.32 Standard photodiode responsivity plot used in making QE measurements.

The term QE_D in Eq. (2.41) is the QE of the standard diode. These data are often supplied in the form of spectral responsivity, R_e (ampere/watt or A/W), as a function of wavelength. This quantity can be converted to QE_D by the equation,

$$QE_D = \frac{12390 R_e}{\lambda}, \qquad (2.44)$$

where λ is the test wavelength (Å).

Figure 2.32 shows a typical response generated by United Detector Technology silicon diode. The responsivity covers the near UV (2500 Å) to the near IR (11,000 Å). The response of the diode (A/W) is converted to QE values through Eq. (2.44).

Example 2.16

Find the QE for the photodiode response supplied in Fig. 2.32 at 5000 Å.
Solution:
A responsivity of 0.3 A/W at 5000 Å is found from Fig. 2.32. Using Eq. (2.44), the QE of the diode is

$$QE_D = 12390 \times 0.3/5000,$$

$$QE_D = 0.75.$$

Example 2.17

Find the interacting QE for a CCD, assuming the following test data and parameters:

$$S(\text{DN}) = 25{,}000\,\text{DN}$$
$$K = 10\,\text{e}^-/\text{DN}$$
$$A_D = 1\,\text{cm}^2$$
$$P_S = 4 \times 10^{-6}\,\text{cm}^2$$
$$\text{QE}_D = 0.75$$
$$I_d = 10 \times 10^{-9}\,\text{A}$$
$$t_E = 1.5\,\text{sec}$$
$$\lambda = 5000\,\text{Å}.$$

Solution:
From Eq. (2.42),

$$S_{CCD} = 25{,}000 \times 10/1.5,$$
$$S_{CCD} = 1.67 \times 10^5\,\text{e}^-/\text{sec}.$$

From Eq. (2.43) the signal for the diode is

$$S_D = (10 \times 10^{-9})/(1.6 \times 10^{-19}),$$
$$S_D = 6.25 \times 10^{10}\,\text{e}^-/\text{sec}.$$

From Eq. (2.41) the QE for the CCD is

$$\text{QE} = 1 \times (1.67 \times 10^5)(0.75)/[(4 \times 10^{-6})(6.25 \times 10^{10})],$$
$$\text{QE} = 0.5.$$

Quantum efficiency can also be determined when the CCD is configured as a photodiode as shown in Fig. 2.33(a). The QE can be measured without external clocking or bias to the CCD. Only two connections are required when operating the CCD in the photodiode mode. The first connection is made to the drain of the reset MOSFET (i.e., V_{REF}) which is connected to the signal channel. The second connection is made to the substrate as shown (i.e., ground). The photocurrent generated in the signal channel is measured by the same amp meter used by the standard photodiode. The QE of the CCD is found by comparing photocurrents of the sensor and the calibrated photodiode through the equation,

$$\text{QE} = \frac{\text{QE}_D I_{CCD} A_D}{I_d A_A}, \qquad (2.45)$$

CCD TRANSFER CURVES AND OPTIMIZATION   155

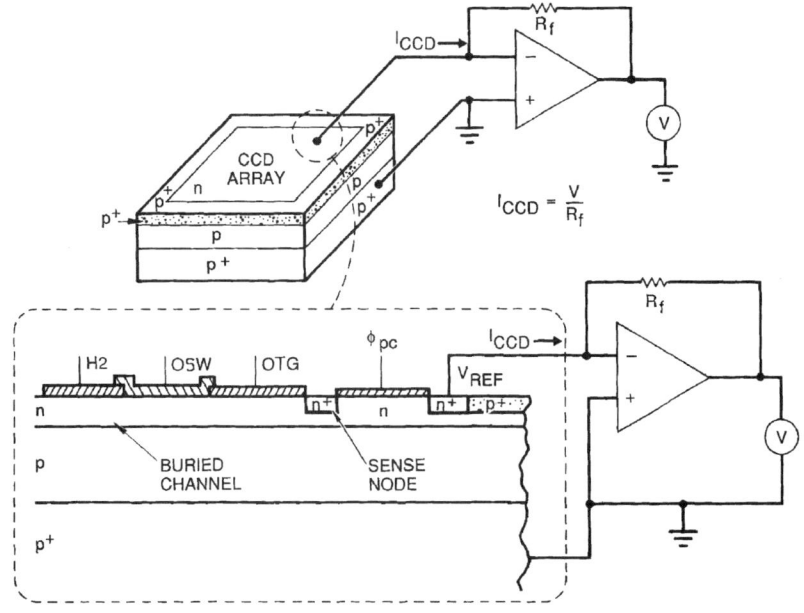

Figure 2.33(a) CCD as a photodiode used to make QE measurements.

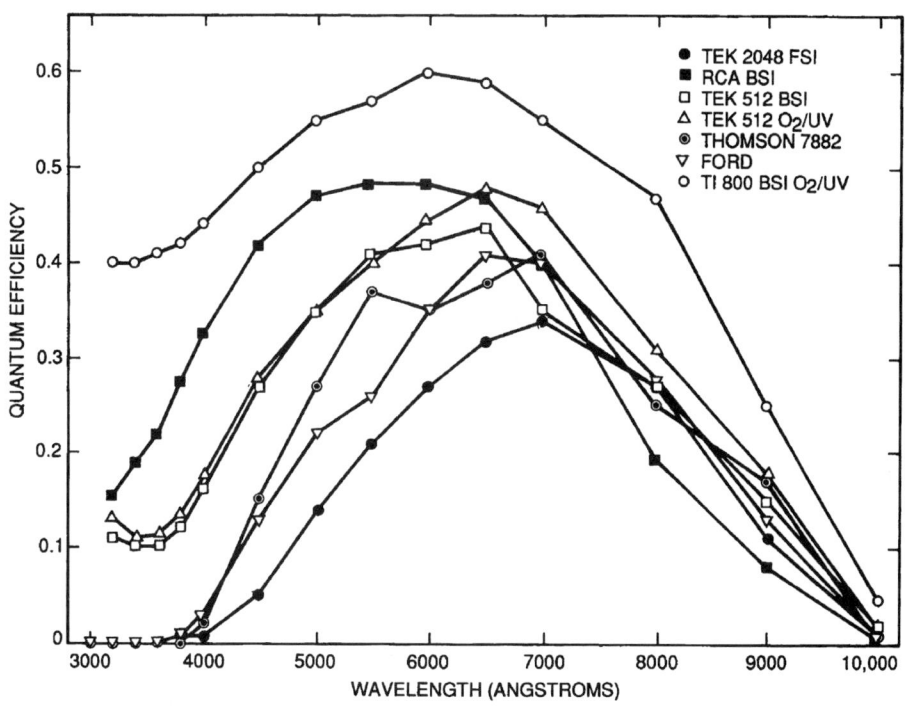

Figure 2.33(b) QE transfer curves.

where I_{CCD} is the photocurrent of the CCD (A) and A_A is the active area of the CCD array (cm^2) that includes all imaging pixels and horizontal registers.

Example 2.18

Find the QE of a CCD at a wavelength of 5000 Å, assuming the following parameters:

$$A_A = 4.19 \, \text{cm}^2$$
$$I_d = 5 \times 10^{-8} \, \text{A}$$
$$I_{CCD} = 10 \times 10^{-8} \, \text{A}$$
$$QE_D = 0.6$$
$$A_D = 1 \, \text{cm}^2.$$

Solution:
From Eq. (2.45),

$$QE = 0.6 \times 10^{-7} \times 1/\left[\left(5 \times 10^{-8}\right) \times 4.19\right],$$

$$QE = 0.286.$$

Figure 2.33(b) shows several QE transfer curves taken using the photodiode approach. QE measurements in the near IR are more difficult to perform for two reasons. First, the operating temperature must be stable or the QE will vary as the measurement is performed. Second, because near-IR response for the CCD is low, the wavelength of the light must be very pure. Blue light that accompanies red light will produce a QE higher than it actually is. For example, interference filters used in making QE measurements often exhibit small pinholes, allowing blue light through to the CCD. Near-IR laser stimulation with a blue-blocking filter is advantageous here. Such measurements will be presented in Chapter 3.

2.5 CCD Clock and Bias Optimization

2.5.1 Clock and Bias

This section will review advanced techniques used to adjust CCD clocks and bias voltages for optimum performance. For the reader inexperienced with CCDs, it would be helpful to return to this section after fully reviewing CCD characteristics in upcoming chapters. For example, spurious charge and residual image are important noise problems that bound and define the vertical register clock levels. If these noise sources are unfamiliar to the reader, it would be best to return to this section after understanding these and other important CCD characteristics.

CCD TRANSFER CURVES AND OPTIMIZATION

Unfortunately, it has been standard practice to determine CCD clock and bias voltages by trial and error methods without providing much justification for the final settings found. For example, one technique involves examining a test target while adjusting the clocks for the "best-looking" image. Such subjective procedures usually yield far from optimum performance and a bias condition that may not be stable. The optimization method described in this section relies heavily on transfer curves to precisely set clock and bias levels for optimum and reliable performance. From this technique, we will also show that CCD drive voltages can be derived from simple setup equations that are based on a few parameters measured by the user. For multiphase CCDs, seven setup parameters are required:

1) the effective threshold voltage of the array, V_{EFF}
2) the effective threshold of the output amplifier, V_{AEFF}
3) the inversion voltage, V_{INV}
4) the optimum full well level $S_{FW}(e^-)$
5) the optimum full well clock voltage, V_{FULL}
6) the output amplifier sensitivity, $S_V(V/e^-)$
7) the desired operating window.

The effective threshold, V_{EFF}, and the inversion breakpoint, V_{INV}, are found from the output transfer gate transfer curve (Section 1.4.3.4). Optimum charge capacity $S_{FW}(e^-)$ and its corresponding clock voltage, V_{FULL}, are provided by photon transfer. Output amplifier sensitivity, $S_V(V/e^-)$, is also derived from photon transfer (Section 6.2.8). If the array and output amplifier are doped at the same concentration, then $V_{AEFF} = V_{EFF}$. If they are different, then V_{AEFF} is found through V_{DD} transfer (see Section 6.2.11). The "operating window," V_{OW}, is defined by the user to allow variation in clock and bias voltages without significant performance degradation (i.e., the CCD remains in specification within this range). An operating window of $V_{OW} = 4$ V (i.e., ±2 V) about the set point is assumed in discussions that follow. Setup parameters are usually available from the CCD manufacturer and data sheets. However, verification of the parameters listed above is recommended.

Table 2.5 lists CCD voltages in the order they are to be selected. Also given is the corresponding voltage symbol and a setup equation. Figure 2.34 shows the output region of the CCD to aid in the setup process. Here we show the clocking and bias levels for each gate and the corresponding V_{max} within the signal channel.

Selection begins with the negative level of the horizontal register and the output summing well, both driven as negative as possible without inverting the channel. Inverting these gates could potentially generate spurious charge, an unwanted source of shot noise that is generated when a phase is inverted. Therefore, as indicated in Table 2.5, the negative level is set out of inversion to $-V_H = -V_{OSW} = V_{INV} + V_{OW}/2$.

Table 2.5 CCD setup clock and bias voltage equations.

Clock or Bias	Voltage Symbol	Set Point Equation
− Horizontal clock	$-V_H$	$V_{INV} + V_{OW}/2$
− Output summing well clock	$-V_{OSW}$	$V_{INV} + V_{OW}/2$
Output transfer gate	V_{OTG}	$V_{OSW} + V_{OW}/2$
Reference	V_{REF}	$V_{EFF} + V_{OTG} + V_{OW}/2 + S_V S_{FW}(e^-)$
− Vertical clock	$-V_V$	$V_{INV} - V_{OW}/2$
+ Horizontal clock	$+V_H$	V_{FULL}
+ Output summing well clock	$+V_{OSW}$	V_{FULL}
+ Vertical clock	$+V_V$	V_{FULL}
+ Pre-charge clock	$+V_{PC}$	$> V_{REF} - V_{EFF} + V_{OW}/2$
− Pre-charge clock	$-V_{PC}$	$< V_{OTG}$
Output drain	V_{DD}	$> V_{AEFF} + V_{REF} + V_{OW}/2$

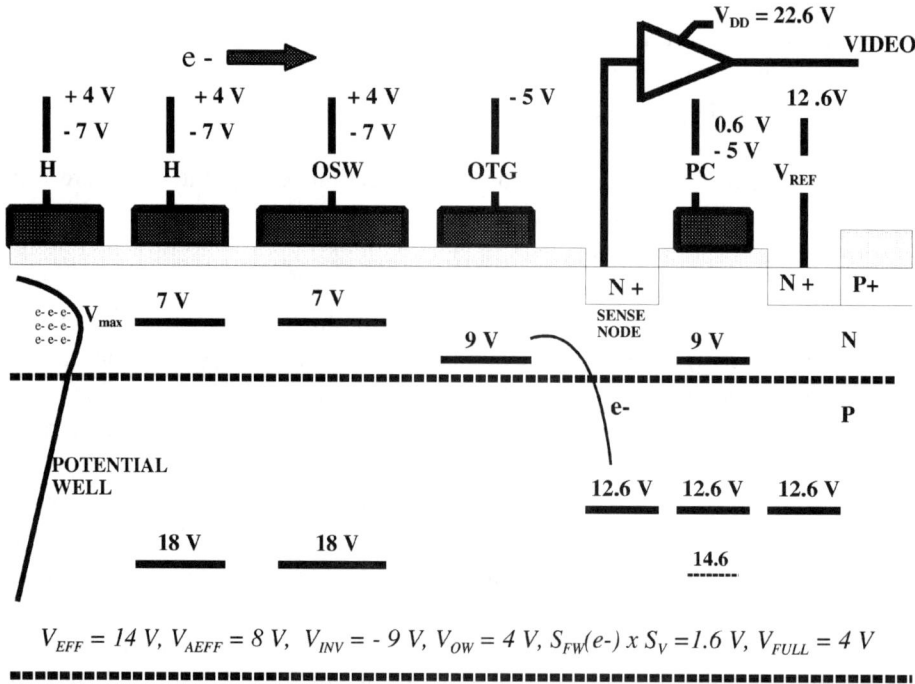

$V_{EFF} = 14\ V,\ V_{AEFF} = 8\ V,\ V_{INV} = -9\ V,\ V_{OW} = 4\ V,\ S_{FW}(e-) \times S_V = 1.6\ V,\ V_{FULL} = 4\ V$

Figure 2.34 Output amplifier region showing optimum clocking and bias levels.

The output transfer gate must be biased higher in potential than the negative level of the output summing well in order for charge to transfer through the region. The output transfer gate is therefore set at $V_{OTG} = -V_{OSW} + V_{OW}/2$.

The sense node voltage must always be greater than the output transfer gate or charge injection will occur. Recall from Chapter 1 that the voltage on the sense node becomes more negative when charge is dumped onto it. For example, at full well the sense node changes by $S_V(\text{V}/\text{e}^-) \times S_{FW}$ (e$^-$) volts in the direction of charge injection. Therefore, additional voltage margin must be applied to V_{REF} to accomodate this voltage change. Voltage requirement on the sense node is therefore $V_{REF} = V_{EFF} + V_{OTG} + V_{OW}/2 + S_V(\text{V}/\text{e}^-)S_{FW}(\text{e}^-)$.

We will assume that it is desirable to set the negative vertical clock level into inversion for low dark current generation, no residual image, and maximum charge capacity (subjects discussed in upcoming chapters). Therefore, $-V_V = V_{INV} - V_{OW}/2$.

The positive clock level for the horizontal and vertical registers and output summing well are adjusted to V_{FULL} for maximum charge capacity (i.e., $+V_V = +V_H = V_{OSW} = V_{FULL}$). As discussed in Chapter 4, the positive clock level normally does not have an operating window because only one voltage achieves the highest charge capacity possible. This is always the case for multiphase CCDs. Virtual and two-phase CCDs typically exhibit a wider voltage range where full well is constant with $+V_V$.

The precharge clock voltage applied to the reset transistor is the least critical of all CCD voltages. Its positive level is set to $+V_{PC} > V_{REF} - V_{EFF} + V_{OW}/2$ to assure that the switch is "on." The negative level is set to $-V_{PC} < V_{OTG}$ to assure that there is a good barrier between the sense node and V_{REF} when the switch is "off."

The output drain voltage to the output MOSFET amplifier, V_{DD}, is the last voltage adjustment to be made. Its set point is

$$V_{DD} = V_{AEFF} + V_{REF} + \frac{V_{OW}}{2}, \quad (2.46)$$

where V_{AEFF} is the effective threshold of the amplifier. This bias condition assures that the MOSFET amplifier is "pinched-off" and is operating in the linear gain region of the device (refer to Chapter 6). This condition comes from the fact that pinch-off requires full depletion in the drain region, which occurs when $V_{DD} > V_{AEFF} + V_{REF}$. Today it is common to see the amplifier lightly doped (i.e., low V_{AEFF}) for a low operating voltage, and the array highly doped (i.e., high V_{EFF}) for high charge capacity.

Example 2.19

Find the operating clock and bias voltages for a CCD with the following set-up parameters. Also find the maximum channel potential, $V_{\max}$, for each gate voltage.

Effective threshold: $V_{EFF} = 14$ V
Inversion: $V_{INV} = -9$ V,
Vertical full well: $S_{FW}(\text{e}^-) = 200,000\,\text{e}^-$
Optimum full well voltage: $V_{FULL} = 4$ V

Table 2.6 Clock and bias voltage settings for Example 2.19.

Clock or Bias	Voltage Symbol	Set Point	Potential Maximum, V_{max}
− Horizontal clock	$-V_H$	$V_{INV} + V_{OW}/2$ $-9 + 2 = -7$ V	$-7 + 14 = 7$ V
− Output summing well clock	$-V_{OSW}$	$V_{INV} + V_{OW}/2$ $-9 + 2 = -7$ V	$-7 + 14 = 7$ V
Output transfer gate	V_{OTG}	$V_{OSW} + V_{OW}/2$ $-7 + 2 = -5.00$ V	$-5.00 + 14 = 9$ V
Reference	V_{REF}	$V_{EFF} + V_{OTG} + V_{OW}/2$ $+ S_V FW$ $14 - 5 + 2 + 1.6 = 12.6$ V	12.6 V
− Vertical clock	$-V_V$	$< V_{INV} - V_{OW}/2$ $< -9 - 2 = -11$ V	$-9 + 14 = 5.00$ V
+ Horizontal clock	V_H	$V_{FULL} = 4$ V	$4 + 14 = 18$ V
+ Output summing well clock	$+V_{OSW}$	$V_{FULL} = 4$ V	$4 + 14 = 18$ V
+ Vertical clock	V_V	$V_{FULL} = 4$ V	$4 + 18 = 18$ V
+ Pre-charge clock	$+V_{PC}$	$> V_{REF} - V_{EFF} + V_{OW}/2$ $> 12.6 - 14 + 2 = 0.6$ V	$< 0.6 + 14 = 14.6$ V
− Pre-charge clock	$-V_{PC}$	$< V_{OTG}$ < 5 V	$< -5 + 14 = 9$ V
Output drain	V_{DD}	$V_{AEFF} + V_{REF} + V_{OW}/2$ $8 + 12.6 + 2 = 22.6$ V	

Output amplifier sensitivity: $S_V(V/e^-) = 8 \times 10^{-6}$ V/e$^-$
Operating window: $V_{OW} = 4$ V
Amplifier threshold: $V_{AEFF} = 8$ V.

Solution:

Using the setup equations given in Table 2.5 and the illustration shown in Fig. 2.34, Table 2.6 is generated.

2.5.2 SET-POINT TRANSFER CURVES

Each clock and bias operating window is bounded by performance problems that degrade read noise, full well, linearity or CTE when exceeded. Table 2.7 lists clock and bias voltages that run a multiphase CCD and the problems that develop when operating outside the window range. Figure 2.35 shows a collection of setup tranfer curves used to clock and bias the CCD optimally. For example, Fig. 2.35(a) is called $-V_H$ transfer and defines where the negative clock level is set for the horizontal register and output summing well. The curve plots read noise as a function of $-V_H$. The curve is normally generated with the vertical clocks reversed to eliminate dark current that would disturb the noise measured in the horizontal register (cooling the CCD could also be employed).

CCD TRANSFER CURVES AND OPTIMIZATION

Table 2.7 Operating window limits for CCD clock and bias voltages.

Clock or Bias	Window Extreme	Problem
− Horizontal clock	$-V_H + V_{OW}/2$	Full well
	$-V_H - V_{OW}/2$	Noise (spurious charge)
− Output summing well clock	$-V_{OSW} + V_{OW}/2$	CTE
	$-V_{OSW} - V_{OW}/2$	Noise (spurious charge)
Output transfer gate	$V_{OTG} + V_{OW}/2$	Noise (charge injection)
	$V_{OTG} - V_{OW}/2$	CTE
Reference	$V_{REF} + V_{OW}/2$	Noise (pinch-off luminescence)
	$V_{REF} - V_{OW}/2$	Noise (charge injection)
− Vertical clock	$-V_V + V_{OW}/2$	Full well and noise (dark current, residual image)
	$-V_V - V_{OW}/2$	CTE (speed reduction)
+ Horizontal clock	$V_H + V_{OW}/2$	Full well
	$V_H - V_{OW}/2$	Full well
+ Output summing well clock	$+V_{OSW} + V_{OW}/2$	Full Well
	$+V_{OSW} - V_{OSW}/2$	Full well
+ Vertical clock	$V_V + V_{OW}/2$	Full well
	$V_V - V_{OW}/2$	Full well
+ Pre-charge clock	$+V_{PC} + V_{OW}/2$	Noise (reset droop)
	$+V_{PC} - V_{OW}/2$	Reset won't turn on
− Pre-charge clock	$-V_{PC} + V_{OW}/2$	Reset won't turn off
	$-V_{PC} - V_{OW}/2$	No window requirement
Output drain	$V_{DD} + V_{OW}/2$	Noise (breakdown)
	$V_{DD} - V_{OW}/2$	Low gain and nonlinearity

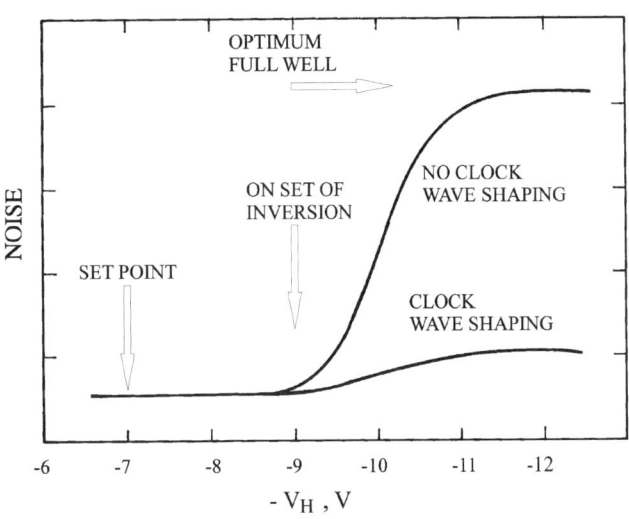

Figure 2.35(a) Read noise as a function of $-V_H$.

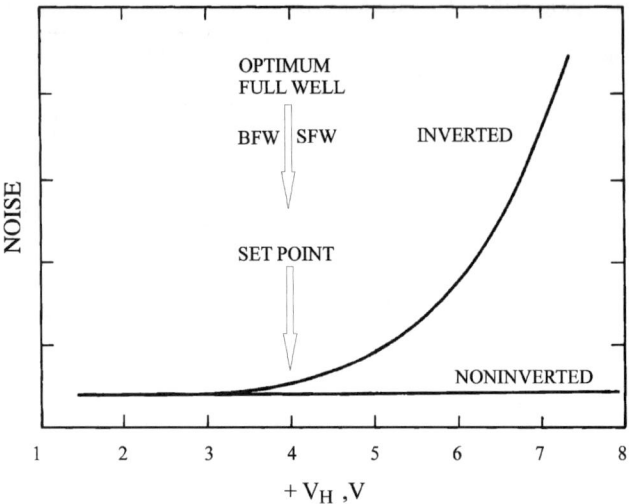

Figure 2.35(b) Read noise as a function of $+V_H$.

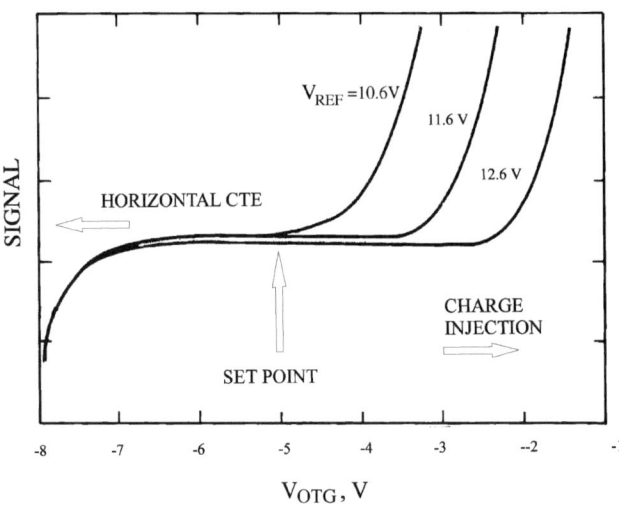

Figure 2.35(c) Signal as a function of output transfer gate V_{OTG}.

Noise should remain be constant with $-V_H$ until inversion is reached. At that point spurious charge is likely to be generated, resulting in a noise increase (see Chapter 7 for discussions on spurious charge generation). Therefore, the horizontal register is normally biased out of inversion as shown. Spurious charge is very sensitive to clock wave shaping, also illustrated in the plot.

Figure 2.35(b) plots noise as a function of $+V_H$. As discussed in Chapter 4, for multiphase CCDs there is a specific positive clock voltage for optimum full well. This is where $+V_H$ is set. Spurious charge is very dependent on $+V_H$, but, as

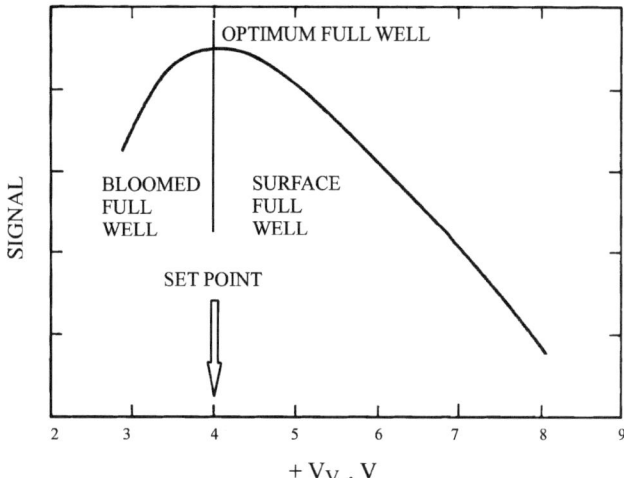

Figure 2.35(d) Full well signal as a function of $+V_V$.

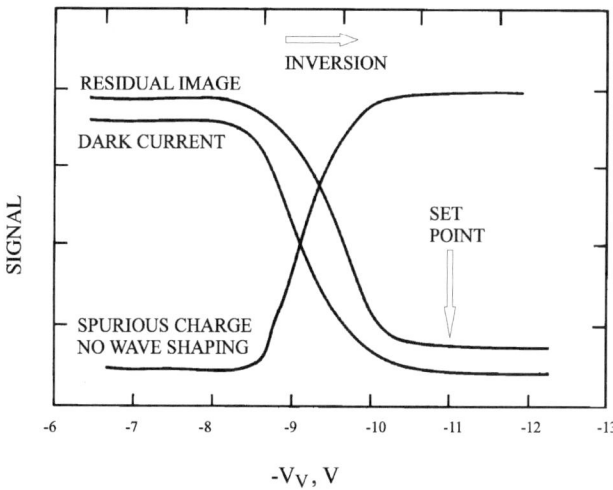

Figure 2.35(e) Full well signal as a function of $-V_V$.

mentioned above, spurious charge can be controlled by not clocking the horizontal register into inversion.

Figure 2.35(c) plots signal as a function of output transfer gate, V_{OTG}. This curve is normally generated with the vertical registers moving forward to fill the horizontal register with some charge (dark current or photon induced). V_{OTG} is bounded by charge injection and horizontal CTE. If set too high, charge from the sense node will be injected into the horizontal register as explained in Chapter 1. If set too low, improper charge transfer between the output summing well and sense node will occur.

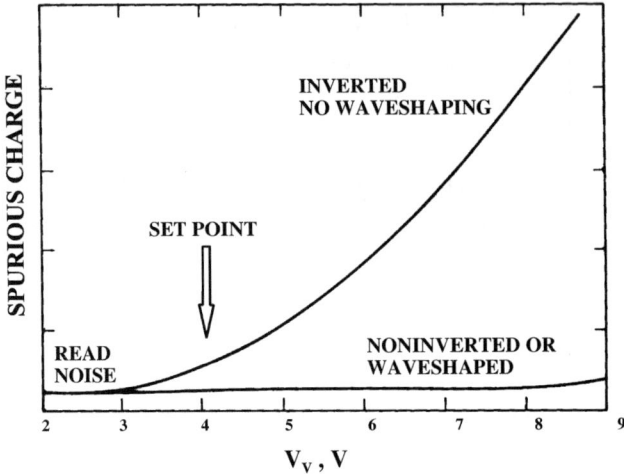

Figure 2.35(f) Spurious charge signal as a function of $+V_V$.

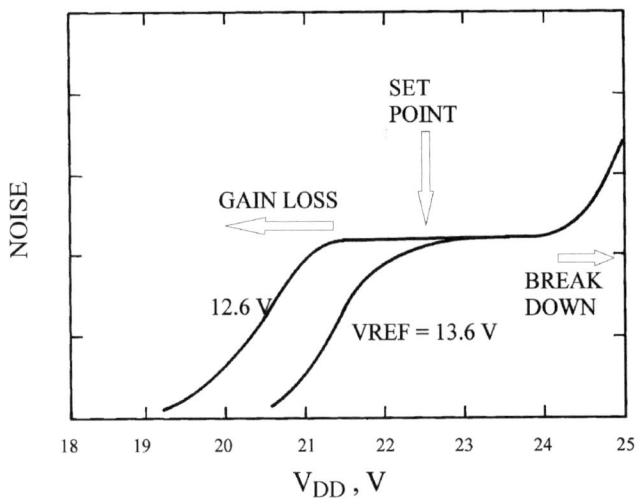

Figure 2.35(g) Noise as a function of $+V_{DD}$.

Figure 2.35(d) plots full well as a function of $+V_V$. Chapter 4 describes how this transfer curve is generated (referred to as full well transfer). The set point shown delivers optimum full well (the same clock level as $+V_H$).

Figure 2.35(e) plots signal as a function of $-V_V$. As discussed in Chapter 7, inverted clocking eliminates residual image and significantly reduces dark current. In addition, maximum full well is achieved. Therefore, it is highly desirable to drive the vertical registers inverted as shown. However, inverted operation implies the sensor is vulnerable to spurious charge generation. As discussed in Chapter 7, spurious charge can be controlled by clock wave shaping.

CCD TRANSFER CURVES AND OPTIMIZATION

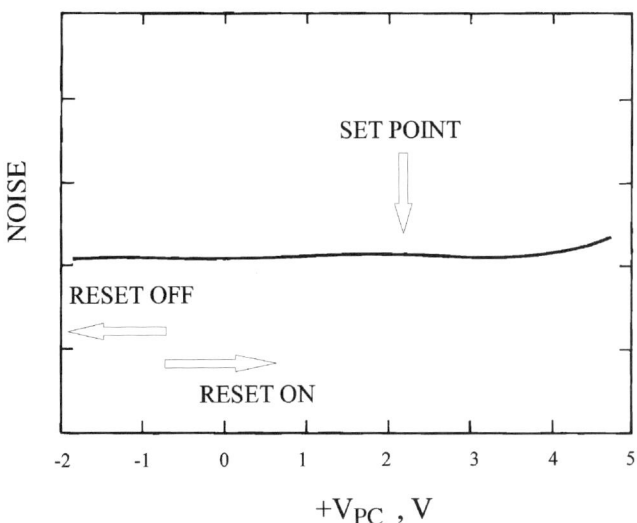

Figure 2.35(h) Noise as a function of $+V_{PC}$.

Figure 2.35(f) plots spurious charge signal as a function of $+V_V$. In that the vertical registers are normally inverted, wave shaping is usually required to control spurious charge generation for these registers.

Figure 2.35(g) plots noise as a function of $+V_{DD}$. This voltage is bounded by amplifier gain loss and breakdown. As discussed in Chapter 6, breakdown is very dependent on the output amplifier MOSFET design and fabrication process. Therefore, V_{DD} is set just high enough for good gain and linearity performance.

Figure 2.35(h) plots noise as a function of $+V_{PC}$. This voltage is not critical as long as the reset switch is sufficiently turned on and off. A similar transfer curve could be generated for $-V_{PC}$ in turning off the reset switch.

REFERENCES

1. J. Janesick, "CCD characterization using the photon transfer technique," *Proc. SPIE* Vol. 570, *Solid State Imaging Arrays*, K. Prettyjohns and E. Derenlak, Eds., pp. 7–19 (1985).
2. J. Janesick, K. Klaasen and T. Elliott, "Charge-coupled-device charge-collection efficiency and the photon-transfer technique," *Optical Engineering* Vol. 26, No. 10 (1987).
3. R. Stern, K. Liewer and J. Janesick, "Evaluation of a virtual phase charge-coupled device as an imaging x-ray spectrometer," *Rev. Sci. Instrum.* Vol. 54, No. 2 (1983).
4. J. Janesick, T. Elliott, R. Bredthauer, C. Chandler and B. Burke, "Fano-noise-limited CCDs," *Proc. SPIE* Vol. 982, pp. 70–94 (1988).

CHAPTER 3

CHARGE GENERATION

3.1 CHARGE GENERATION

The photoelectric effect is responsible for generating signal charge when incident photons interact with the photosensitive volume of the CCD. The challenge for CCD manufacturers is to let incoming photons interact. Quantum efficiency is the performance parameter that tells us how well they have accomplished this task, and is defined as

$$QE = \eta_i QE_I, \qquad (3.1)$$

where QE_I is the interacting QE and η_i is the quantum yield. An interacting photon is one that is absorbed by the CCD and generates one or more e-h pairs. The definition above may produce QE values greater than unity because multiple electron-hole pair generation is possible. Interacting QE is always less than unity and is usually the QE value specified by the CCD manufacturer.

Example 3.1

Calculate QE performance assuming a quantum yield of 3 e$^-$/interacting photon and an interacting QE of 50%.
Solution:
From Eq. (3.1),

$$QE = 3 \times 0.5,$$

$$QE = 1.5.$$

For frontside-illuminated CCDs, incident photons must pass through the poly gates before they can interact with the epitaxial layer. As shown below, gate absorption represents a significant QE loss mechanism, especially for photons with short absorption lengths. Reflection loss also degrades QE performance over a major portion of the working spectrum. Transmission loss is important for very long and very short wavelength photons that penetrate the epitaxial layer without interaction.

This chapter will discuss different CCD designs and processes that the manufacturer employs to reduce these three loss mechanisms. For example, the highest sensitivity that can be achieved by the CCD is accomplished by removing the substrate layer to allow direct illumination of the epitaxial layer from the rear of the device (i.e., backside illumination). Absorption loss for this CCD technology is

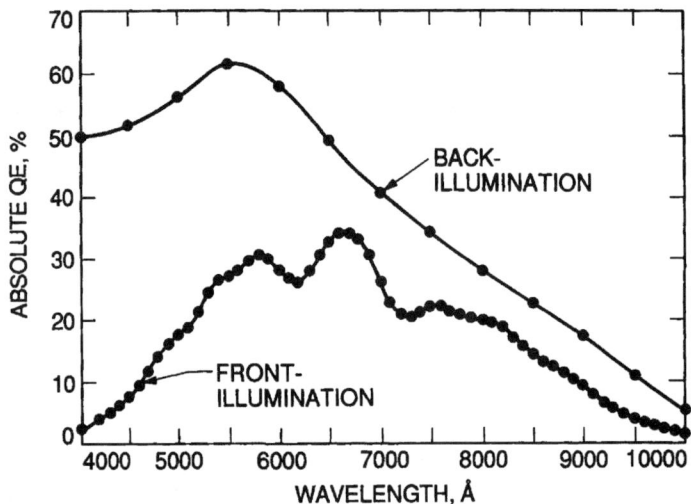

Figure 3.1(a) QE performance for a CCD illuminated from the front and back.

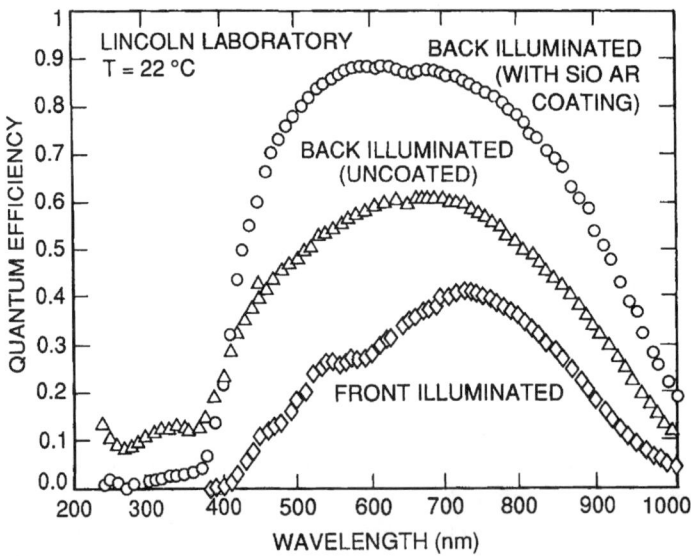

Figure 3.1(b) Frontside QE compared with backside QE with and without antireflection coating.

CHARGE GENERATION

eliminated. Reflection loss can also be reduced significantly by depositing antireflection coatings onto this rear surface. In theory, this only leaves transmission loss for the CCD.

Figure 3.1(a) shows QE transfer curves for a thinned CCD illuminated from the front and back. The CCD was thinned to 7 μm, a thickness that is approximately equal to a photon absorption length of 7500 Å. The backside response is solely limited by reflection loss for wavelengths shorter than 5500 Å, whereas the frontside response is limited by reflection and gate absorption. The main loss mechanism longward of 5500 Å for both illumination schemes is transmission loss. Figure 3.1(b) shows QE plots for frontside illumination and backside illumination CCDs with and without an antireflection coating. The difference in responses is striking and is why backside illumination is so popular in the scientific community.

Figure 3.2 presents QE transfer data collected from a 10-μm back-illuminated CCD without an antireflection coating. The top wavelength scale is given in units of angstroms, whereas the bottom energy scale is given in units of electron volts. The spectral range covers the near-IR (7,000–10,000 Å), visible (4000–7000 Å), the UV (1000–4000 Å), the extreme UV (100–1000 Å) and the soft-x-ray (1–100 Å) ranges. This early QE plot showed that the CCD could be used outside the visible range. Today the CCD is employed throughout the 1–10,000-Å range in a host of scientific imaging instruments.

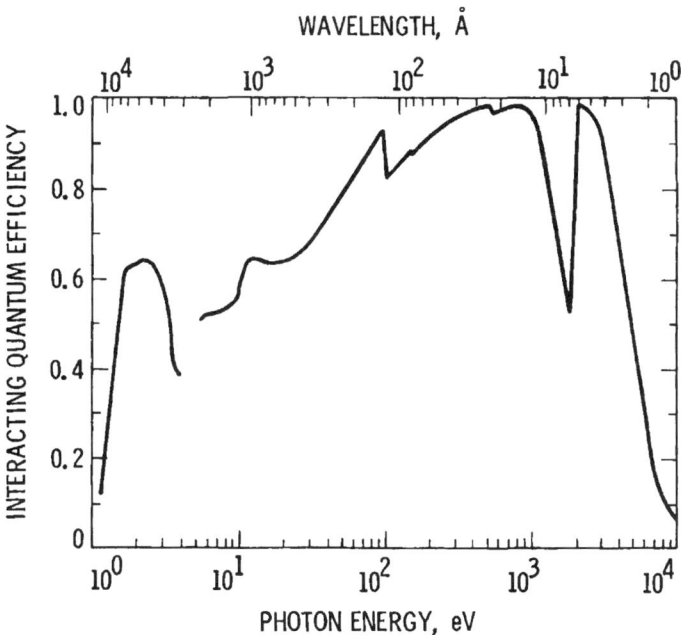

Figure 3.2 Early backside-illuminated QE response that covers 1–10,000 Å.

3.2 QE Formulas

This section will give some simple formulas used to estimate QE performance. Backside and frontside illumination formulas will be presented. We will also discuss other QE loss mechanisms not included in the formulas. Lastly, equations will be given for determining QE performance using the Monte Carlo technique.

3.2.1 Backside Illumination

The interacting QE for a backside-illuminated CCD is estimated by

$$\text{QE}_{BI} = \text{CCE}(1 - R_{REF})(1 - e^{-x_{EPI}/L_A}), \qquad (3.2)$$

where R_{REF} is the reflection coefficient for silicon, x_{EPI} is the epitaxial thickness (cm), CCE is the charge collection efficiency or the ability of the CCD to collect all photoelectrons generated within the thinned membrane (ideally this should be 100%), and L_A is the photon absorption length at the wavelength of interest (cm). The equation assumes unity quantum yield and no antireflection coatings.

Reflection and absorption silicon characteristics are shown in Fig. 3.3. Figure 3.3(a) plots reflectivity as a function of wavelength from 2,000 to 10,500 Å.[1] Note that reflectivity peaks in the UV at approximately 2700 Å. At this wavelength, more than 70% of the photons are reflected from the surface. Also note that the reflection coefficient is essentially constant for wavelengths greater than 5500 Å

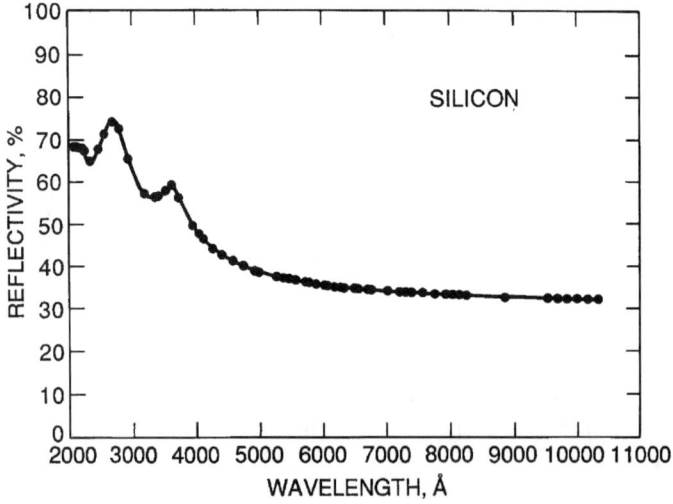

Figure 3.3(a) Silicon reflection characteristics.

CHARGE GENERATION 171

(approximately 30%). Figure 3.3(b) plots absorption length as a function of wavelength. The absorption length is shortest at 2700 Å, where photons only penetrate 50 Å. This is a very small depth considering that the silicon atom lattice spacing is 5.4 Å. Figure 3.3(c) plots absorption length over the entire spectral range in which

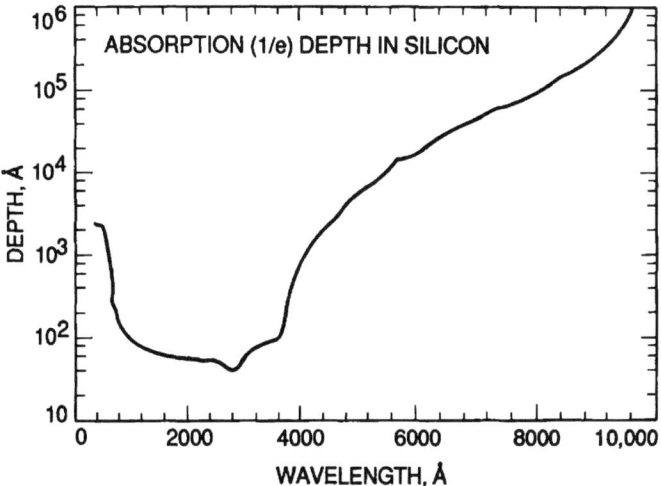

Figure 3.3(b) Silicon photon absorption characteristics.

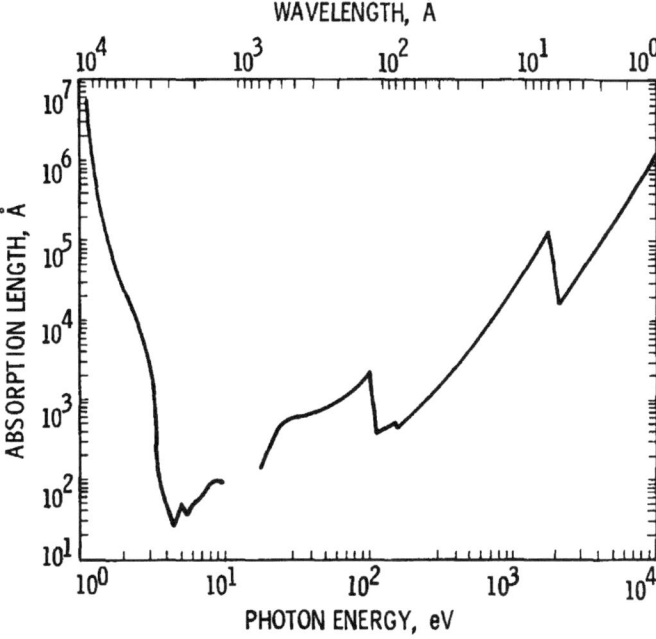

Figure 3.3(c) Silicon photon absorption characteristics that cover 1–10,000 Å.

silicon responds. At the end of the soft-x-ray regime (i.e., 1 Å, 10 keV) the QE falls off rapidly, because the probability of an x-ray photon interacting with a K-shell electron rapidly decreases. Silicon is therefore transparent for x-ray energies greater than 10 keV (i.e., hard x rays). Table 3.1 tabulates absorption length for UV, visible, and near-IR wavelengths at room temperature.

Example 3.2

Estimate the QE for a backside-illuminated CCD at a wavelength of 4000 Å. Assume a photosensitive thickness of 10 μm and a CCE of unity.
Solution:
From Fig. 3.3(a), the reflectivity at 4000 Å is 50%. From Table 3.1, the absorption length is 0.2 μm (2000 Å, 2×10^{-5} cm).
From Eq. (3.2),

$$QE_{BI} = (1 - 0.50) \times [1 - \exp-(10/0.2)],$$

$$QE_{BI} = 0.5 \ (50\%).$$

QE performance at 4000 Å is often used as a figure of merit for the CCD. The theoretical limit calculated is useful to remember as we go along.

Figure 3.4 shows theoretical QE plots using Eq. (3.2) for different membrane thicknesses. Note that the near-IR QE drops when the absorption length is about one-third of the thickness of the membrane. For example, a 10-μm thickness corresponds to an absorption length of 3.3 μm or a wavelength of approximately 6500 Å.

Table 3.1 Silicon absorption depth as a function of wavelength.

λ (Å)	L_A (μm)	λ (Å)	L_A (μm)	λ (Å)	L_A (μm)
3000	0.0065	6200	2.5773	9,400	43.4782
3250	0.00851	6400	2.9851	9,600	54.3478
3500	0.0115	6600	3.4722	9,800	70.4225
3750	0.0286	6800	4.0650	10,000	93.4579
3870	0.0667	7000	4.7619	10,200	126.5822
4000	0.2000	7200	5.5555	10,400	175.4386
4200	0.2976	7400	6.4935	10,600	250.0000
4400	0.4310	7600	7.6336	10,800	370.3703
4600	0.5814	7800	8.9285	11,000	582.3529
4800	0.7752	8000	10.5263	15,000	595.2380
5000	0.9804	8200	12.7388	20,000	454.5454
5200	1.2048	8400	15.3846	25,000	833.3333
5400	1.4265	8600	18.6915		
5600	1.6666	8800	22.9985		
5800	1.9305	9000	28.5714		
6000	2.2472	9200	35.0877		

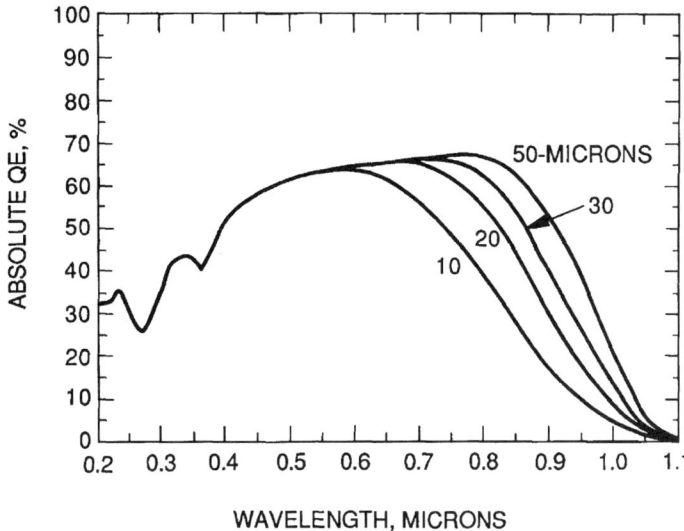

Figure 3.4 Theoretical QE performance for a backside-illuminated CCD with different membrane thicknesses.

3.2.2 FRONTSIDE ILLUMINATION

The QE for a front-illuminated CCD is more complex to analyze because of gate absorption loss. . The QE sensitivity for a multiphase CCD is estimated by

$$\mathrm{QE}_{FI} = \mathrm{CCE}(1 - R_{REF})e^{-x_{POLY}/L_A}(1 - e^{-x_{EPI}/L_A}), \qquad (3.3)$$

where x_{POLY} is the poly gate thickness (cm). This equation assumes a quantum yield of unity and that reflection and absorption length characteristics for silicon also apply to polysilicon gates.

Example 3.3

Estimate the QE performance for a frontside-illuminated CCD at 4000 Å. Assume CCE = 1, $x_{EPI} = 10\,\mu\mathrm{m}$, and $x_{POLY} = 0.5\,\mu\mathrm{m}$.
Solution:
From Eq. (3.3),

$$\mathrm{QE}_{FI} = (1 - 0.50)\exp-[0.5/(0.2)] \times [1 - \exp-(10/(0.2)],$$

$$\mathrm{QE}_{FI} = 0.04\ (4\%).$$

Compare this QE performance with that of a backside-illuminated CCD in Example 3.2.

3.2.3 MISCELLANEOUS QE LOSSES

The QE equations above only include absorption, reflection, and transmission losses. There are other many other losses (and gains) worth discussing here.

3.2.3.1 Poly Gate Overlaps

The poly gates overlap each other, reducing QE performance (refer to Fig. 1.14). Poly overlaps range from 0.5 to 2.5 µm, depending on CCD manufacturer design rules. The QE problem is more evident for small-pixel devices because the overlap is a greater fraction of the pixel area. For example, the QE at 4000 Å for a multiphase 7.5-µm pixel can be 50% lower compared with a 15-µm pixel.

3.2.3.2 Substrate Sensitivity

Charge generated in the substrate will diffuse into the epitaxial region, increasing the near-IR QE. The diffusion length for common substrates (0.01 Ω-cm) is approximately 5 µm. The p^+/p dopant gradient at the epitaxial interface also generates an electric field that directs photoelectrons to the front of the device. This small built-in field prevents carriers from diffusing from the epitaxial layer into the substrate region where charge would recombine.

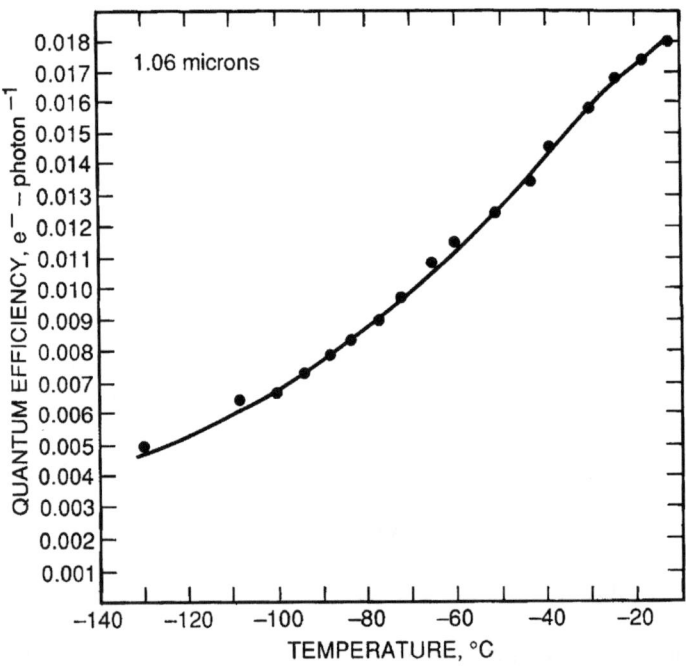

Figure 3.5 Near-IR response (10,600 Å) as a function of operating temperature.

3.2.3.3 Operating Temperature

QE performance is dependent on operating temperature. As the temperature is reduced, the photon absorption length increases because the silicon bandgap increases.[2-5] This characteristic is especially important in the near IR because phonons help energize free electrons into the conduction band. Figure 3.5 shows QE data generated by a front-illuminated Galileo CCD laser stimulated at 1.06 μm as a function of operating temperature. At this wavelength, QE changes by a factor of 3.6 over an operating temperature range of 100 degrees.

The QE temperature dependence is also seen in Fig. 3.6 for a backside-illuminated CCD. The responses shown are normalized to measurements taken at −20°C. Again, the near-IR QE decreases with decreasing temperature because more photons pass through the membrane. Notice that the visible sensitivity increases slightly with decreasing temperature. This is because photons penetrate beyond a surface dead region that usually exists for backside-illuminated CCDs.

3.2.3.4 Channel Stop Sensitivity

Recombination takes place when charge is generated in the channel stop region. The diffusion length of a carrier made within a channel stop is approximately 5 μm. Although the channel stop is small in extent, its effect on QE can be noticeable depending on incident wavelength. Figure 3.7(a) shows a 4000-Å flat-field response taken from a frontside-illuminated thin-gate $1024 \times 1024 \times 18$-μm-pixel CCD (thin-gate technology will be discussed below). The experimental device was

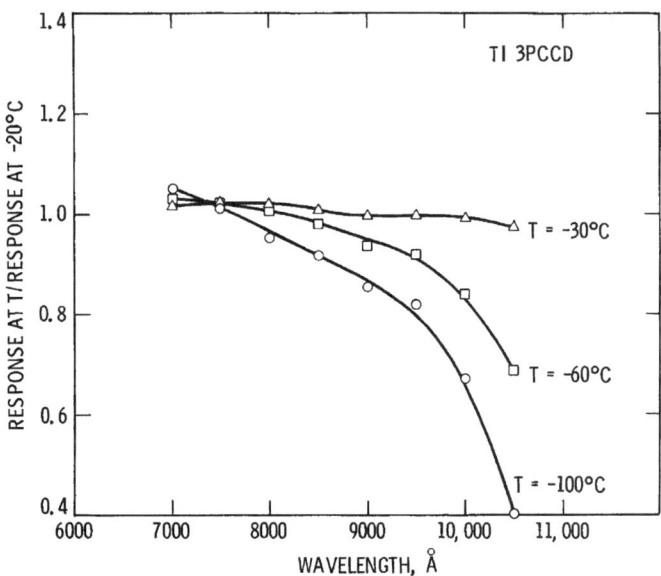

Figure 3.6 Near-IR response at different operating temperatures.

fabricated with channel stop widths that vary from 2 to 16 µm across half the array. Note that the QE drops as the channel stop becomes a larger portion of the signal channel. Figure 3.7(b) plots QE for the same CCD as a function of channel stop width. QE loss within the channel stop is worse for short wavelengths because recombination is greatest. Wavelengths longward of 4500 Å pass through the channel

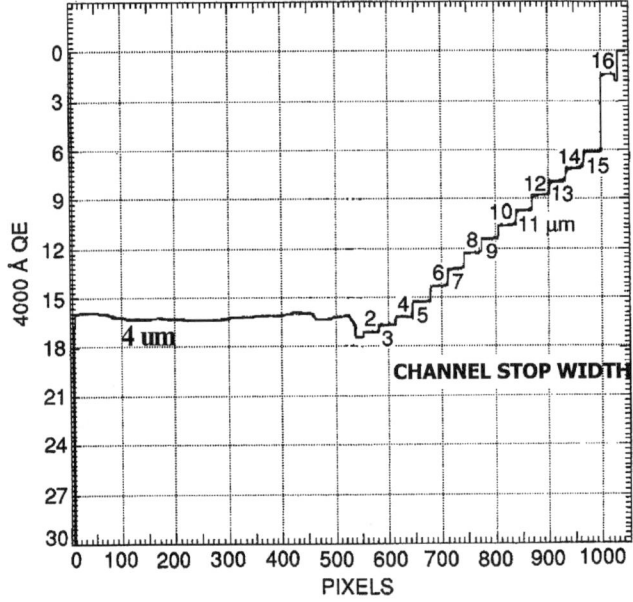

Figure 3.7(a) Flat-field 4000-Å response as a function of channel stop width.

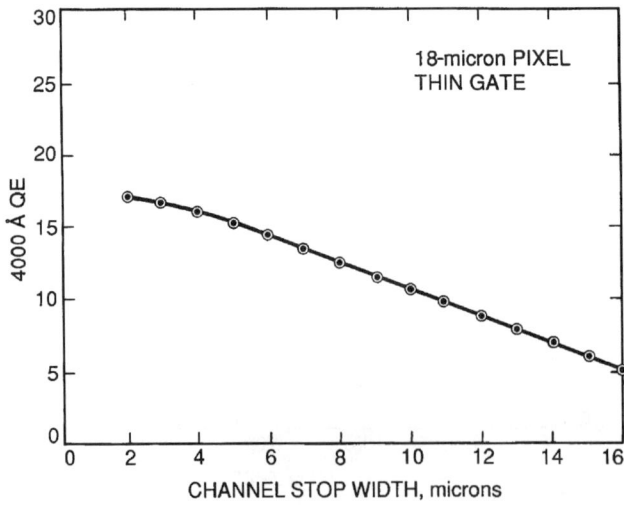

Figure 3.7(b) 4000-Å QE as a function of channel stop width.

CHARGE GENERATION

stop and show very little loss (less than 1%). For backside-illuminated devices, the channel stop has little effect on QE for any wavelength.

3.2.4 MONTE CARLO SIMULATION

Figure 3.8 shows a Monte Carlo simulation of incident visible photons (0.4–0.6 μm) on a backside/frontside CCD. For frontside, we leave half of the 5000-Å poly gate off of the surface to show gate absorption characteristics more clearly. Note that the polysilicon gates absorb most incoming photons at the blue end of the spectrum. Gate absorption becomes negligible for wavelengths longer than 6000 Å. The main loss mechanism beyond this wavelength, for both backside and frontside illumination, is transmission loss through the assumed 10-μm epitaxial layer. The simulation does include reflection loss, which is important to both illumination schemes.

Photon absorption as shown in Fig. 3.8 can be simulated by generating random numbers with an exponential probability distribution $x_i = -L_x \log_e U_i$ where x_i is the absorption depth for photon $i = 1, 2, 3, \ldots, L_x$ is the characteristic absorption length, and $U_i =$ uniform random number between 0 and 1.

Example 3.4

Generate five photon events that have a characteristic absorption length of 2 μm.
Solution:
First generate five random numbers between 0 and 1 and the correspond-

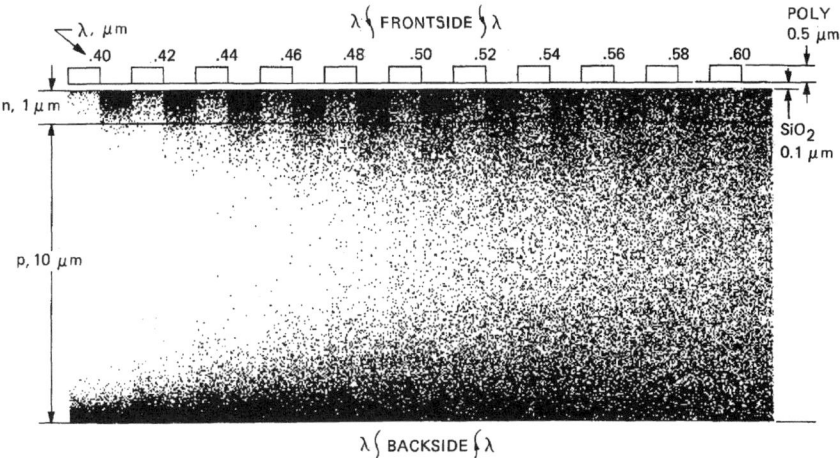

Figure 3.8 Monte Carlo photon absorption simulation for front and back illumination.

ing logarithm,

$$U_1 = 0.2, \log_e = -1.61,$$
$$U_2 = 0.6, \log_e = -0.51,$$
$$U_3 = 0.3, \log_e = -1.2,$$
$$U_4 = 0.5, \log_e = -0.69,$$
$$U_5 = 0.1, \log_e = -2.3.$$

Next, multiply the above results by $-2.0\,\mu m$. The penetrating depth for each photon is

$$x_1 = 3.22\,\mu m,$$
$$x_2 = 1.02\,\mu m,$$
$$x_3 = 2.4\,\mu m,$$
$$x_4 = 1.38\,\mu m,$$
$$x_5 = 4.6\,\mu m.$$

3.3 FRONTSIDE ILLUMINATION

Different CCD processes and designs have been employed to reduce gate absorption loss. This section will discuss six custom design schemes to reduce this loss: (1) phosphor coatings, (2) virtual phase, (3) open pinned phase (OPP), (4) thin gate, (5) transparent gate and (6) poly hole.

3.3.1 PHOSPHOR COATINGS

Phosphor coatings are used to improve and extend QE response for front-illuminated CCDs. Phosphors are wavelength converters that convert short-wavelength light (e.g., UV) into the visible spectral region where the CCD is more sensitive. Lumigen (also spelled liumogen, lumogen) is a phosphor that has been used extensively on JPL flight CCDs (e.g., the Hubble WF/PC II and Cassini CCDs).[6-8] The use of lumigen originated with another type of phosphor, called coronene, used on WF/PC I sensors. Although WF/PC I CCDs were backside illuminated, the dead layer on the backside required assistance from a phosphor coating to obtain UV sensitivity (WF/PC was required to respond to Lyman-alpha-light at 1216 Å). The backside problem for WF/PC I is explained further below.

Lumigen absorbs wavelengths $< 4800\,\text{Å}$ and isotropically generates new photons at approximately 5300 Å. Figure 3.9 shows the yellow-green emission spectra of lumigen. The quantum yield of lumigen is 100%. That is, for every incident photon a visible photon is generated. Phosphor coatings like lumigen can increase UV QE performance by manyfold for conventional front-illuminated CCDs. As shown

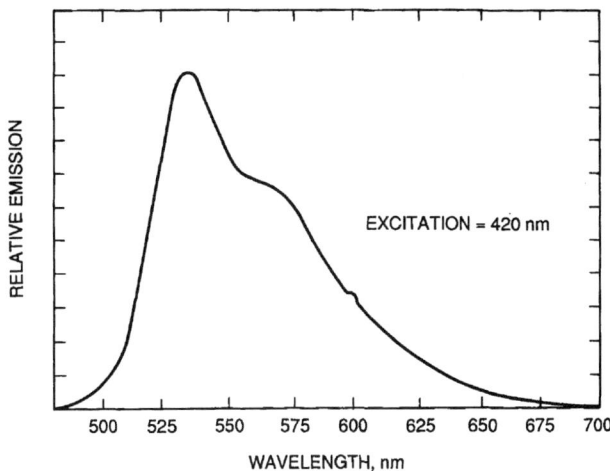

Figure 3.9 Luminescence spectra for lumigen.

in Fig. 3.3(b), the absorption length at 2500 Å is extremely short (50 Å), whereas at 5500 Å, where lumigen fluoresces, the absorption length is much longer (1.5 µm).

The proper chemical name for lumigen is 2,2′dihydroxy 1,1′naphthalazine [$C_{22}H_{16}N_2O_2$] with molecular weight = 340. The synonym for lumigen is "pigment yellow 101 (C.I. #48052)"; its trade name is Lumigen Yellow S 0790, and lumigen belongs to the azo methine chemical family. Its melting point at 760 mm Hg is 295°C and it exhibits a specific gravity of 1.34. It is insoluble in water and most solutions. It is odorless and is readily purchased as a yellow powder in a very pure (99.99%) form. Lumigen can be removed from the CCD surface by using trichloroethylene. Lumigen is the same hydrocarbon used in phosphorescent "highlighter pens." Writing on a CCD with a highlighter pen surprisingly extends the CCD's QE performance into the far UV.

For optimum performance, the lumigen coating is applied directly to the CCD by vacuum sublimation. Ideal results are achieved when the CCD is maintained at an elevated temperature (approximately 70°C) during the deposition process to stabilize crystal growth thereafter. If applied at room temperature, the QE will slowly increase as lumigen crystals form naturally. Figure 3.10 shows two SEM photographs of lumigen samples immediately following deposition and after the coating was aged for 24 hours at 100°C. Note that the lumigen coating is homogeneous in structure when first deposited [Fig. 3.10(a)], but transforms to large crystals after a high-temperature bake [Fig. 3.10(b)]. Note also that there are voids between the crystals after the temperature cycles. This sample was coated with a thin layer of lumigen, approximately 1500 Å. Depositing an adequate amount of lumigen to reduce this spacing during growth is important. We have found that a thickness of 6000 Å is optimum (thickness applied to WF/PC II sensors).

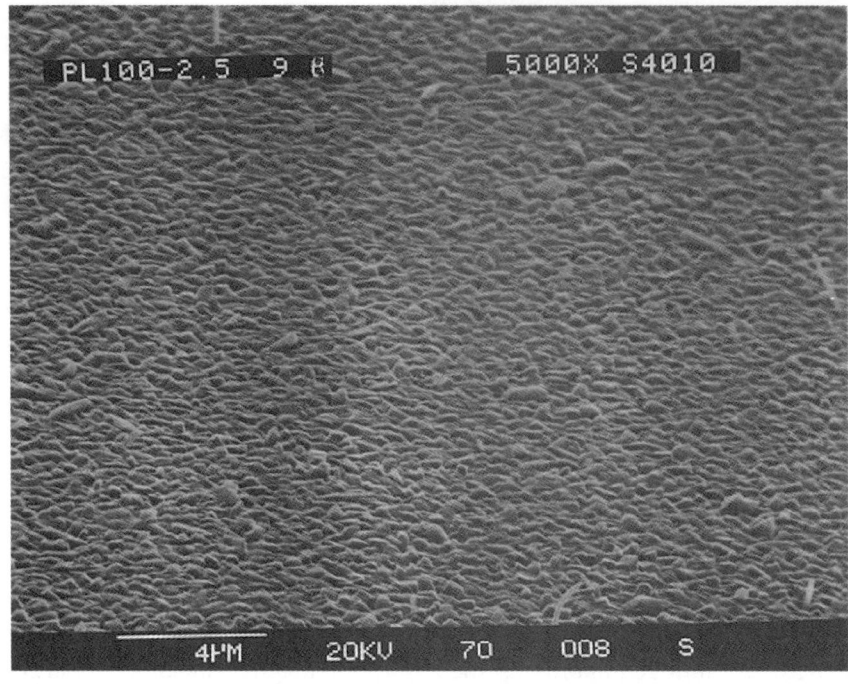

Figure 3.10(a) Lumigen before high-temperature anneal.

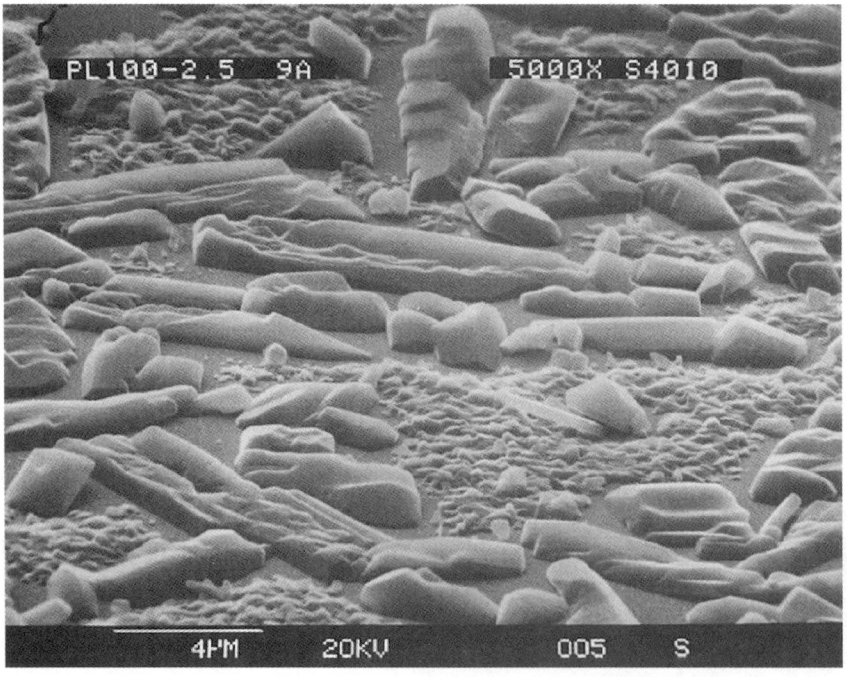

Figure 3.10(b) Lumigen after 70°C anneal.

Based on its chemical structure and high melting point, lumigen is a stable material. However, lumigen coatings can evaporate under high vacuum or temperature conditions. For example, vapor pressure calculations show that evaporation will begin to occur at $+50°C$ at 4.3×10^{-8} torr. However, CCDs such as those used on Hubble can be hermetically sealed and backfilled with an inert gas to increase the partial pressure and stability of the phosphor layer. Also, the operating temperature is never allowed to go higher than 300 K for Hubble.

Lumigen shows no significant signs of phosphorescence. Also, lumigen does not significantly degrade optical characteristics such as MTF and visible and near-IR QE performance. Longward of 4600 Å, the phosphor is essentially transparent. However, two subtle characteristics are worth noting. First, lumigen acts as an antireflection coating, influencing QE slightly. The second effect is illustrated in Fig. 3.11. The highly stretched computer-generated image shows a 4000 Å knife-edge response taken from a CCD that has been partially coated with lumigen. Note that the edge for the lumigen region extends in several pixels under the light shield compared with the uncoated region. This effect is caused by scattered light emitted from the phosphor because the lumigen layer is not in intimate contact with the silicon surface. For front-illuminated CCDs, the layer is deposited on top of the poly gates and oxide protective layers, at a distance of a few microns from the signal channel. For backside-illuminated sensors, this problem is less severe because the phosphor is in intimate contact with the silicon layer (unless the CCD incorporates an antireflection coating).

Figure 3.12 shows a QE response taken from a WF/PC II CCD before and after lumigen deposition. The sensitivity for the UV after coating is approximately 13% and remains constant for wavelengths down to approximately 500 Å. At this wavelength the phosphor becomes transparent to incident photons (to go shorter, the coating can be made thicker). Note also that the UV sensitivity is approximately

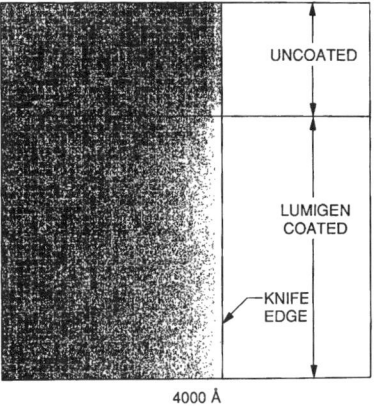

Figure 3.11 Lumigen light-scattering characteristics in response to a knife-edge stimulus.

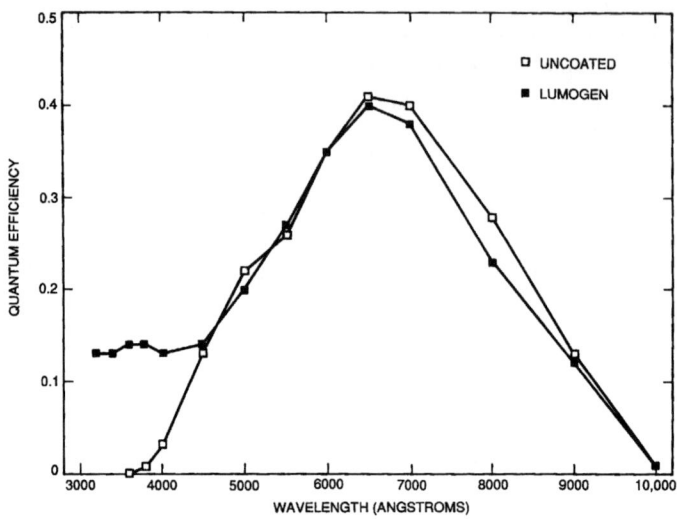

Figure 3.12 QE response of a Hubble WF/PC II CCD with and without lumigen phosphor coating.

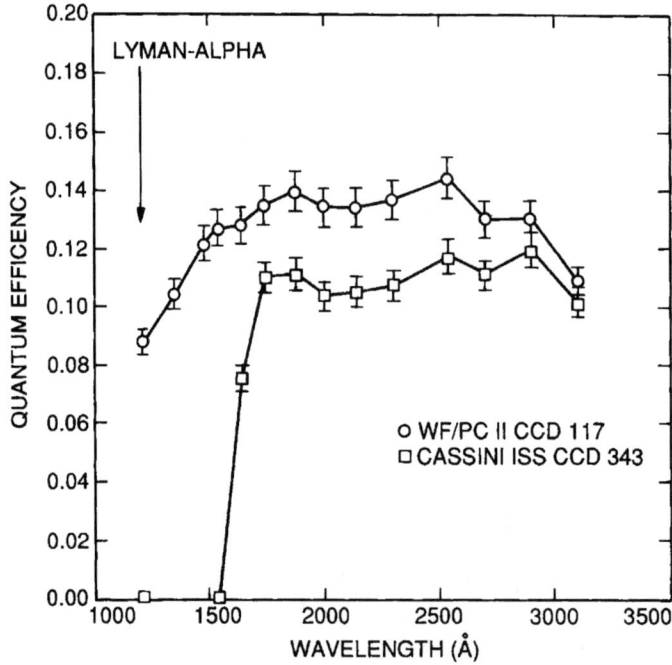

Figure 3.13 Far-UV lumigen response for Hubble WF/PC II and Cassini CCDs.

half that achieved at 5300 Å, because half the photons emitted by the phosphor travel away from the CCD rather than toward it.

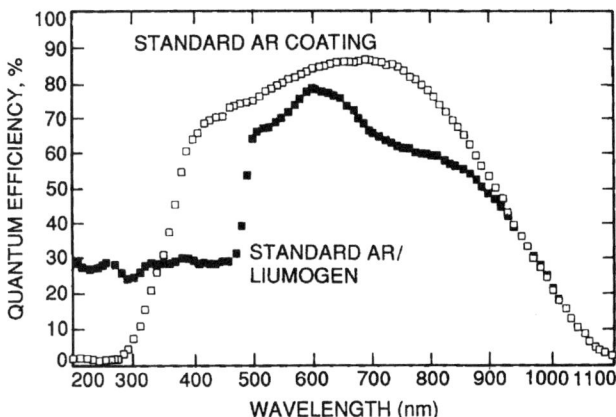

Figure 3.14 QE response for a backside-illuminated CCD with and without lumigen.

Figure 3.13 shows UV QE plots taken by WF/PC II and Cassini CCDs coated with lumigen. The falloff seen at Lyman-α (1216 Å) for the WF/PC II device is due to absorption in a magnesium fluoride window above the sensor. The QE for the Cassini CCD drops off before the WF/PC II device because it has a quartz window that becomes opaque at 1800 Å.

Figure 3.14 shows the QE of a backside-illuminated CCD with and without the lumigen coating. Note that the phosphor is essentially transparent at wavelengths longward of 5000 Å. Lumigen is applied because the antireflection coating absorbs incoming UV photons shortward of 3500 Å. The combination of lumigen and backside thinning yields a device sensitive down to 500 Å with 30% QE performance in the UV.

3.3.2 Virtual Phase

As mentioned above, conventional multiphase illuminated CCDs traditionally exhibit poor sensitivity in the UV because of the absorbing polysilicon gate layers. To bypass this problem, CCD manufacturers have been forced to either thin and back-illuminate the sensor or deposit UV-sensitive organic phosphor coatings as discussed above. Virtual-phase CCD technology, however, has partially resolved the frontside QE dilemma by leaving half the pixel "open" by employing a "virtual electrode."[9–11] The technology allows photons to enter this open region without significant absorption. For example, Fig. 3.8 shows a pixel structure that is half open.

Figure 3.15 compares the construction of a two-phase pixel and a virtual-phase CCD pixel. The illustration shows that a two-phase CCD can be operated with a single clock line phase-1 in this case. The clocked phase is driven above and below a dc potential applied to the phase-2. For a virtual-phase CCD, the potentials in the transparent region are fabricated into the silicon, using implants as shown in the lower illustration.

Figure 3.16 is a detailed cross-sectional view of a virtual-phase pixel, showing four distinct implanted regions. The "clocked phase" is made up of regions 1 and 2. Region 1 is called the "clocked barrier" (CB) and region 2 the "clocked well" (CW). Region 1 is n-doped similarly to a conventional buried-channel CCD. Region 2 receives additional n-dopant, producing a greater potential there. As illustrated in Fig. 3.15, this gives rise to a potential step between regions 1 and 2.

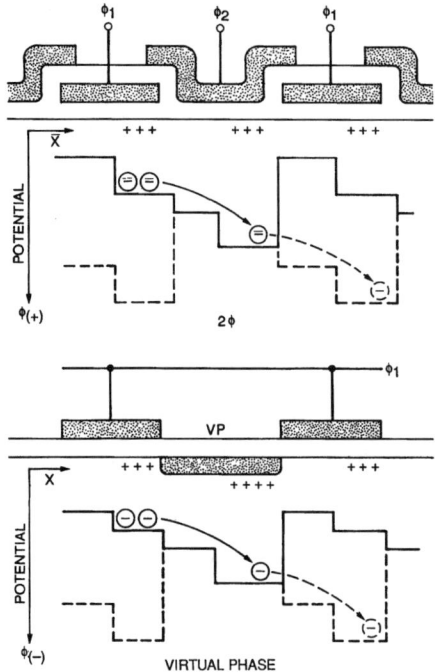

Figure 3.15 Two-phase compared to virtual-phase CCD operation.

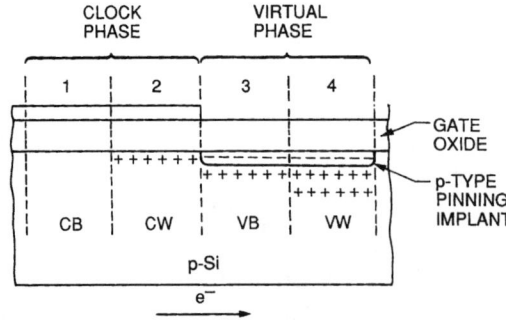

Figure 3.16 Virtual-phase pixel showing four implanted regions.

Regions 3 and 4 make up the "virtual phase." Region 3 is the "virtual barrier" (VB), which is defined by another n-implant. Region 4, the "virtual well" (VW), uses an even heavier dose. Therefore, a step in potential exists between regions 3 and 4 as shown. The potentials defined in the virtual phase are next "pinned" to ground potential. This is accomplished by a shallow but a concentrated p-type implant made at the surface. This layer is in turn connected to neighboring channel stop regions, which pins the surface potential to ground potential. This last step is most critical to the technology. If the virtual region is not rigidly pinned, the potentials at the edges will swing with the clocked phase, degrading CTE and full well performance.

Charge integration and transfer processes for the virtual phase are illustrated in Fig. 3.17. The illustration plots measured potential maxima (i.e., V_{max}) for each region as a function of gate bias. Charge integration takes place either under the clocked phase when the applied gate voltage is high, or under the virtual phase when biased low. Charge transfer from the clocked phase to the virtual phase is accomplished by lowering the gate voltage below $-V_G$. Charge transfer between the virtual-phase and clocked phase occurs when the gate voltage increases above $+V_G$. The values of $+V_G$ and $-V_G$ determine the necessary voltage swing of the gate for complete charge transfer. For this Galileo virtual-phase CCD, a swing of approximately 10 V is required.

The virtual-phase CCD is a unique technology in that a single poly gate is used to collect and transfer charge. Because of this attribute, the device is reliable, as the structure is less susceptible to inter- and intralevel shorts found in overlapping gate technologies. Therefore, the technology is high yielding, a virtue confirmed in practice. The device operates inherently inverted, resulting in low dark current generation, with no residual image problems. Only three clock lines (i.e., one vertical,

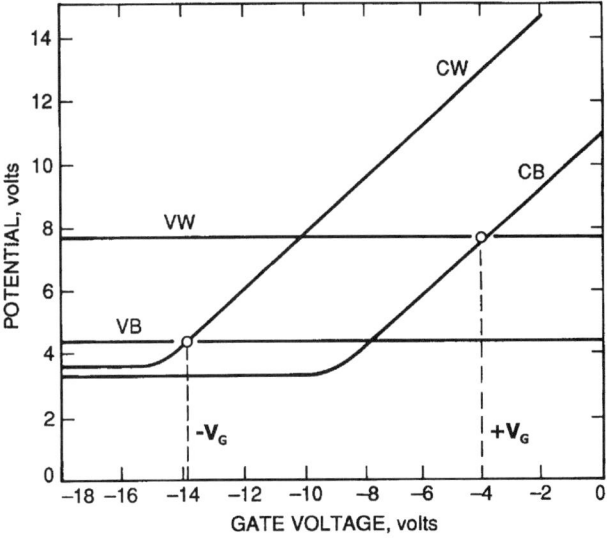

Figure 3.17 Channel potentials for a VP CCD.

one horizontal, and one reset clock) are required for CCD operation, simplifying drive electronics as compared to multiphase CCDs.

In many respects, virtual-phase CCDs are more difficult to fabricate than multiphase sensors. For example, a minimum of six implants are required to fabricate a virtual-phase pixel, five that define regions 1–4 and one implant for the channel stops. For comparison, only two implants are required for a three-phase pixel (i.e., the n-channel and channel stops). The various implants required must be precisely aligned, while multiphase implants do not require such precision. Texas Instruments of Japan (TIJ) is the only manufacturer of virtual-phase CCDs. In contrast, there are about a dozen CCD manufacturers currently fabricating scientific multiphase CCDs.

With virtual-phase technology, there is only one poly gate over half of the pixel, and, therefore it is possible to achieve significant short-wavelength sensitivity. Figure 3.18 shows a QE response generated by a Galileo 800 × 800 virtual-phase CCD demonstrating high QE well into the UV. For example, at 4000 Å the sensitivity is 20%, which is greater than lumigen-coated multiphase devices can achieve (about 15% for the same pixel size). The sudden drop in QE at 1800 Å is caused by the absorbing properties of the protective glass (SiO_2) layer that covers the CCD. This layer can be etched and removed for extended performance into the EUV.

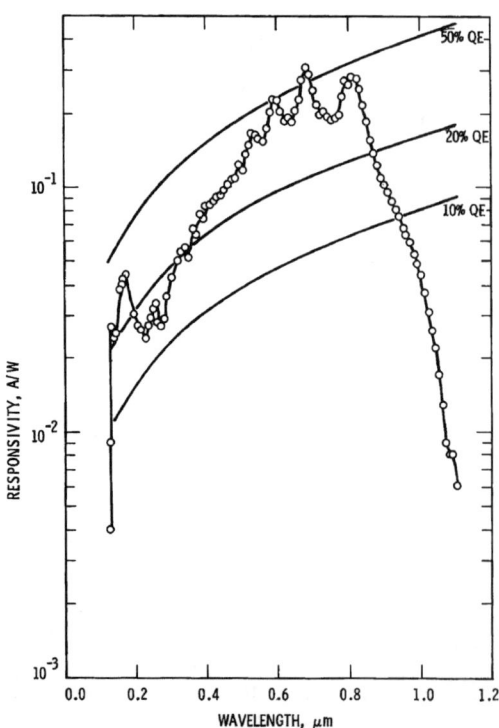

Figure 3.18 QE response for a Galileo VP CCD.

Figure 3.19 presents QE data taken in the x-ray region for a 1024 × 1024 virtual-phase SXT CCD. The low-energy response is limited by a 5000-Å glass layer that covers the entire device. Again, this layer can be removed for extended soft x-ray response.

Single-clock operation offered by virtual-phase CCDs is an advantage for high-speed applications because only one clock needs to be applied to the array. However, the inflexibility of this virtue has often proved to be a disadvantage for low-signal-level applications. The problem has been traced to "spurious potential pockets" which trap charge in the signal channel, a subject discussed fully in Chapter 5. Small misalignments of the implants can create trapping problems. For example, if the implant used in defining the VW region encroaches under the edge of the CB region, the CW would exhibit a higher channel potential, creating a potential pocket. There are many more possibilities for trap creation in processing virtual-phase CCDs. It has been difficult to completely suppress the trap problem for low-light-level applications. For this reason, virtual-phase CCDs are mainly being used in high-light applications (e.g., Galileo and SXT). In contrast, the potential wells for multiphase CCDs are created by gates that can be adjusted by the user for optimum CTE performance.

From Fig. 3.17 it can be seen that the virtual-phase CCD must be clocked in and out of inversion to transfer charge from the clocked region into the virtual region. This makes the technology vulnerable to spurious charge generation, an important clocking noise source discussed in Chapter 7. In contrast, multiphase and two-phase CCDs can be operated inverted or noninverted—the choice is left to the user. Therefore, virtual-phase CCDs are generally noisier compared to multiphase sensors.

Another inherent problem with virtual-phase (and two-phase) CCDs is that signal charge can only be transferred in one direction. For many scientific applica-

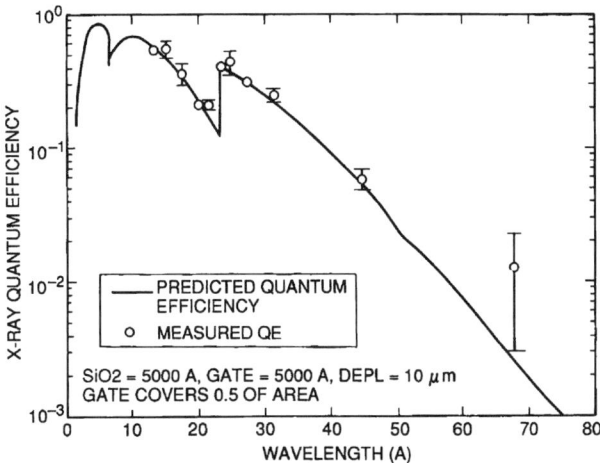

Figure 3.19 X-ray QE response for a SXT VP CCD.

tions, bidirectional clocking is useful for parallel amplifier readout. This mode of operation can be accommodated by multiphase CCDs.

3.3.3 Open Pinned Phase

As mentioned above, virtual-phase CCDs typically have limited usage in low-signal-level applications, because of CTE and noise problems. To avoid these difficulties, a CCD technology called open pinned phase (OPP) was invented.[12] The OPP CCD unites multiphase and virtual-phase CCD technologies by taking the best features of each technology and combining them for low noise and high CTE. The OPP CCD retains multiphase clocking flexibility; also, OPP complies with the three-phase fabrication process.

Figure 3.20 shows a cross-sectional view of an OPP pixel. Three sections are illustrated: two clocked regions (defined as phases-1 and -2) and an open region. The OPP CCD is fabricated using the same process steps as a three-phase CCD in defining phases-1 and -2. The third level of polysilicon is left off; instead, two implants are incorporated. The first implant dopes the channel with additional phosphorus, thereby defining the potential in the region. The second implant is a concentrated but very shallow implant of boron that pins the surface potential. This is the same pinning implant as virtual phase. The pinning implant acts as a virtual gate, maintaining a fixed surface potential within the region. Both implants are self-aligned by poly levels one and two. As with virtual phase, it is important that the pinning implant be connected to the channel stop region. If not, the potential within the open region will not stay pinned as adjacent phases are clocked. This process can be accomplished by using implanted channel stops.

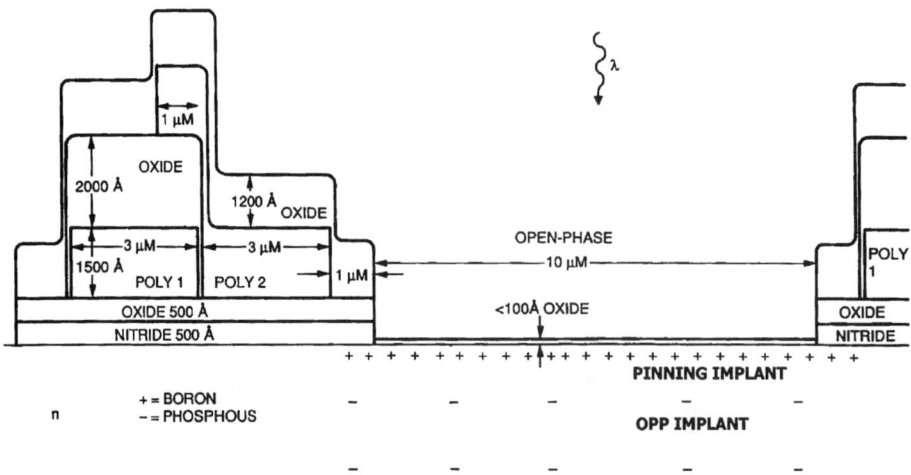

Figure 3.20 Cross-section of an OPP pixel.

CHARGE GENERATION

The OPP CCD is clocked very much like a three-phase CCD. During charge integration, phases-1 and -2 are held low, inverting the surface and forcing charge into the open-phase region. Charge transfer is initiated in a normal fashion by first taking phase-1 high, moving charge from the open phase to phase-1. Phase-2 remains pinned and acts as a barrier phase during this time, to assure that no charge flows backward from the open phase. In the next clock cycle, phases-1 and -2 are both biased high where signal charge occupies both phases. Phase-1 is then inverted, forcing charge into phase-2. The cycle is finished when phase-2 is lowered, moving charge back into the open-phase region. Bidirectional clocking is also possible by reversing phases 1 and 2.

As mentioned above, the virtual-phase CCD can exhibit CTE and spurious charge noise problems related primarily to the horizontal register. To avoid these difficulties the OPP CCD uses a conventional three-phase horizontal register (i.e., the third level of polysilicon is employed). Designed this way, the horizontal register can be clocked noninverted, circumventing the spurious charge problem. The vertical registers are inverted during charge integration and readout for low dark current generation. The OPP region is always inverted because of the pinning implant.

Figure 3.21 shows data generated by a linear OPP MISR CCD demonstrating good QE performance in the UV.

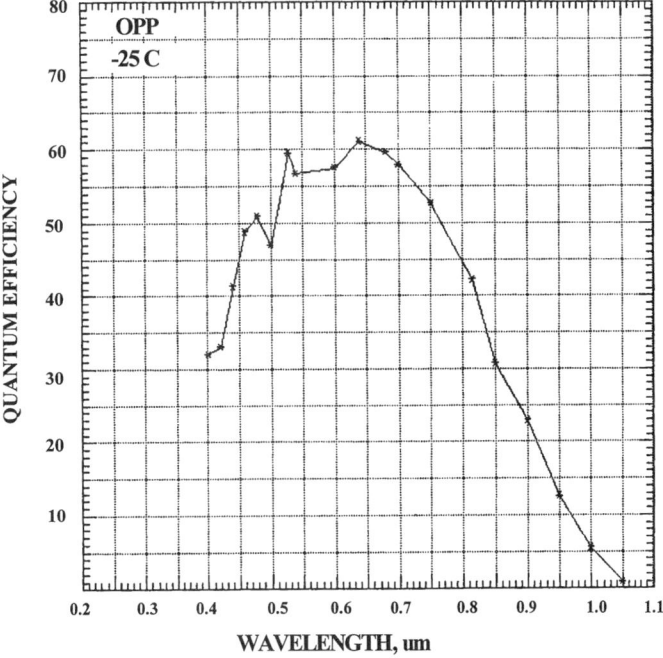

Figure 3.21 QE response for an OPP pixel.

3.3.4 THIN GATE

The poly gates can be made thinner to reduce gate absorption loss. However, gate conductivity becomes a problem for poly layers thinner than about 1500 Å. Also, the poly lines must be bussed together with metal lines on the sides of the array. The oxide that covers the poly lines must be etched in order to make contact to metal. Etching very thin poly layers would likely thin completely through the poly layer, resulting in a poor contact.

Thin-gate CCD technology was invented to avoid these problems.[13,14] Although many different schemes can be implemented, Fig. 3.22 shows a cross-sectional view of a thin-gate pixel that has been built by several CCD manufacturers. The pixel is fabricated in the following manner. First, poly gates-1 and -2 are deposited and patterned as in conventional three-phase technology. In the next step, poly-3 is patterned directly on top gates-1 and -2. At this point in the fabrication process poly-3 is not oxidized, but instead a very thin fourth level of polysilicon is deposited over the whole array. In that poly-3 is not oxidized, poly-4 is electrically connected to poly-3. In this way, poly-3 acts as a clock bus to drive poly-4. Contacts and metal buses outside the array are made to poly-1, -2, and -3 as usual. Scratch protection glass is then deposited over the entire CCD. For x-ray sensitivity this oxide layer can be etched back within the imaging portion of the array. The thin-gate CCD is clocked as a conventional three-phase sensor.

Figure 3.23(a) is a scanning electron microscope cross-sectional photograph taken of a linear thin-gate MISR CCD. Poly-3 is shown in the upper left-hand corner of the image. It is connected to a 400-Å thin gate (i.e., poly-4) which covers the photo-gate region labeled in the figure. For a linear array, the photo gate is where incident photons interact and where signal charge is collected. The magnified view also shows the gate oxide layer (300 Å), the nitride layer (600 Å) and a poly-2 gate (4000 Å). Poly-2 acts as a dump gate in taking signal charge from the photo-gate region for electronic exposure control. A scratch

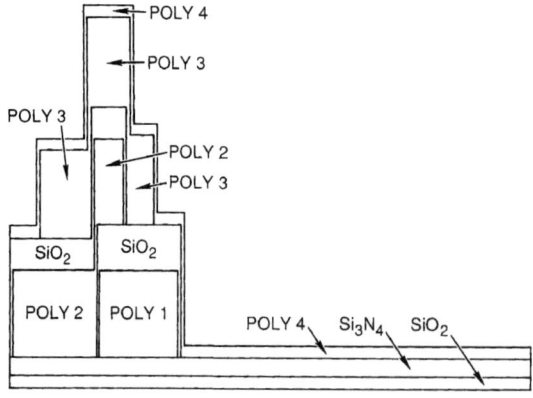

Figure 3.22 Cross-sectional illustration of a thin-gate pixel showing poly-4 layer.

oxide is also included on top of the thin gate for protection. Figure 3.23(b) is a visible and near-IR QE curve for a MISR thin gate CCD. The interference seen is caused by the thin gate. The sensitivity and interference can be improved by making the thin gate thinner. Gates as thin as 200 Å have been fabricated.

Figure 3.24(a) shows the layout of a thin-gate x-ray CUBIC three-phase 1024 × 1024 × 18-μm-pixel CCD.[15,16] The 400-Å thin-gate region occupies a 12 × 18-μm

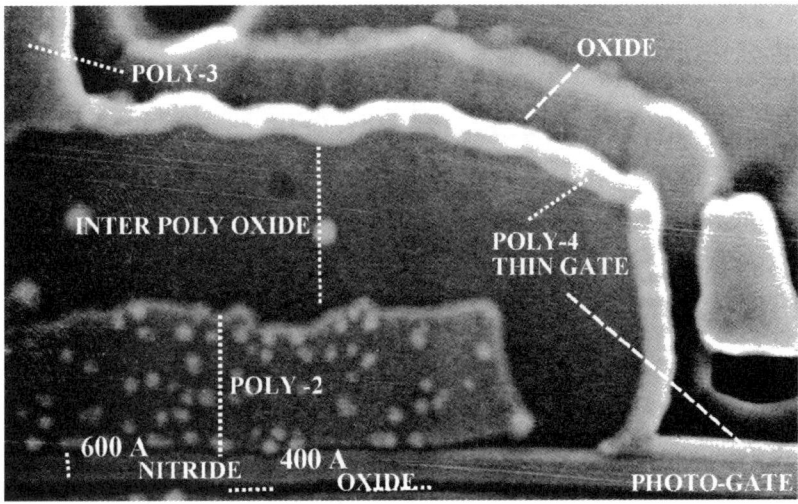

Figure 3.23(a) SEM cross-sectional photograph of an MISR pixel showing a 400-Å thin gate.

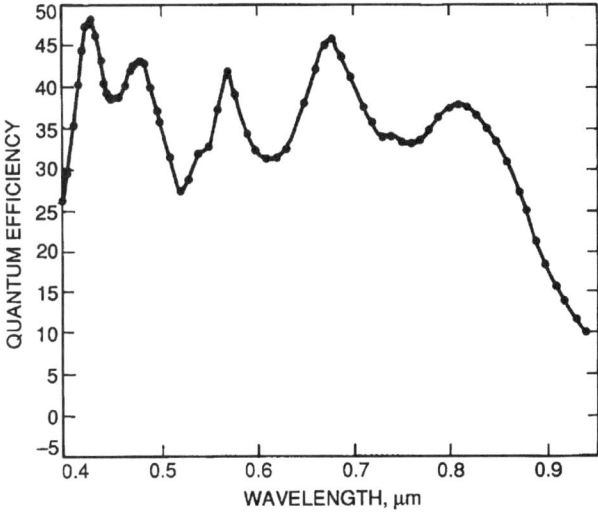

Figure 3.23(b) QE response for an MISR thin-gate CCD.

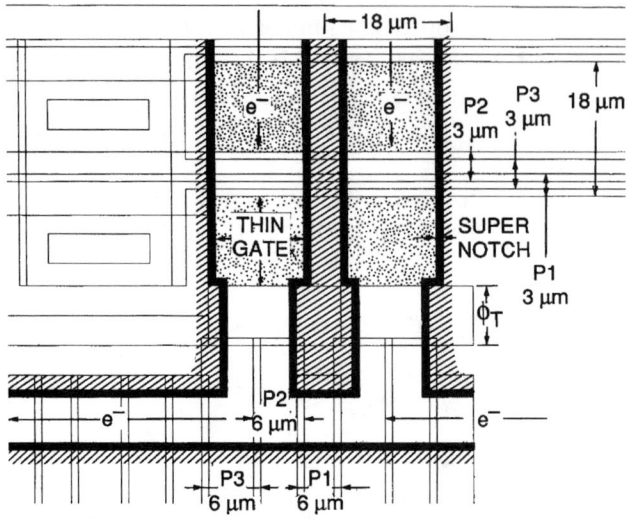

Figure 3.24(a) CUBIC thin-gate 18-μm pixel with super notch.

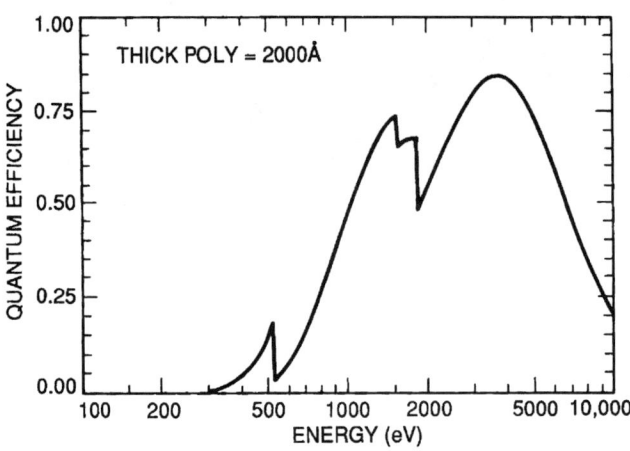

Figure 3.24(b) Predicted x-ray response for a conventional three-phase 2000-Å poly gate CCD.

region. Poly phases-1, -2, and -3 cover a 3 × 18-μm area and are 4000 Å thick. The horizontal register for the CCD uses conventional three-phase gates (i.e., poly-4 is not incorporated in the horizontal register). The super notch shown isolates the n-channel from the channel stop regions to increase radiation hardness for the CCD (refer to Chapter 8 for discussions on super notch technology).

Figure 3.24(b) shows the expected x-ray sensitivity for a conventional three-phase CCD based on 2000-Å-thick poly gates and an oxide overcoat layer. The x-ray sensitivity is limited to approximately 500 eV. Figure 3.24(c) shows the sen-

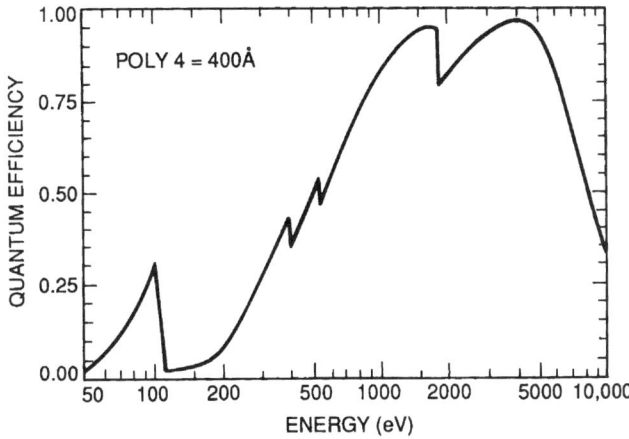

Figure 3.24(c) Predicted x-ray response for a 400-Å thin-gate CUBIC CCD.

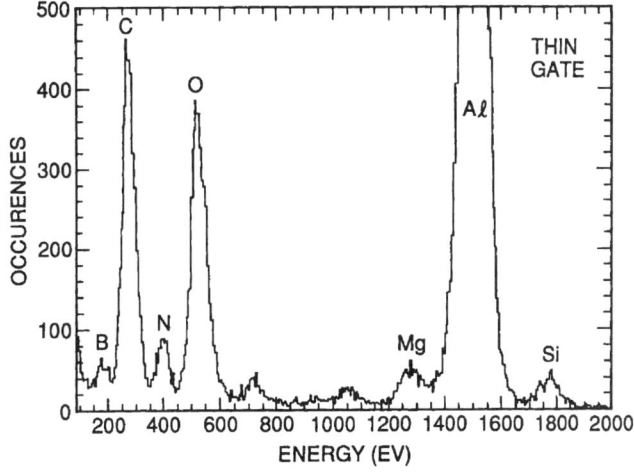

Figure 3.24(d) X-ray spectra taken from a thin-gate CUBIC CCD.

sitivity improvement for the CUBIC 400-Å thin-gate CCD. The thin gate extends the response to approximately 200 eV with substantial sensitivity even below the silicon edge. Figure 3.24(d) shows experimental x-ray histogram data generated by the sensor. Performance is exceptional where carbon, oxygen, nitrogen, and boron x-ray lines are clearly seen. The CUBIC CCD also employed a Skipper floating-gate amplifier, which allows the charge in each pixel to be measured multiple times and achieve subelectron read noise (refer to Chapter 6 for Skipper amplifier technology). The noise floor of Fig. 3.24(d) is only $0.9\,e^-$ rms with 16 samples/pixel taken. The energy resolution at 277 eV is 38 eV (FWHM) with a corresponding QE of 15%.

3.3.5 Transparent Gate

Transparent gates are also utilized to reduce gate absorption loss.[17,18] A new class of full-frame two-phase CCDs was introduced in 1999 by Kodak in which the second poly-gate electrode is replaced by an optically transparent conducting gate material made of indium-tin-oxide or ITO.[19] The technology maintains the simplicity and high yield of front-illuminated CCDs while at the same time providing significant improvements in shorter-wavelength QE. Figure 3.25 shows the QE performance of the ITO Kodak CCD as a function of wavelength from 3,700 to 11,000 Å. The ITO process, compared to an all poly process also shown, is clearly superior at wavelengths less than 7500 Å, with an improvement of $\sim 10\times$ at 4000 Å (from 3% to 30%). For wavelengths shortward of 3000 Å, the ITO response drops rapidly. The bandgap of ITO is approximately 3.75 eV, which corresponds to an absorption edge of 330 nm. For example, at 2500 Å the absorption length is approximately 500 Å, much less than the thickness of the ITO gate.

3.3.6 Poly Hole Gate

CCD manufacturers can also leave small hole openings in the poly gate to reduce gate absorption loss. For example, Fig. 3.26 shows a four-phase 8-μm-pixel CCD design that incorporates poly holes near the channel stop regions. The channel potential in the open regions is controlled by a pinning implant similar to what is

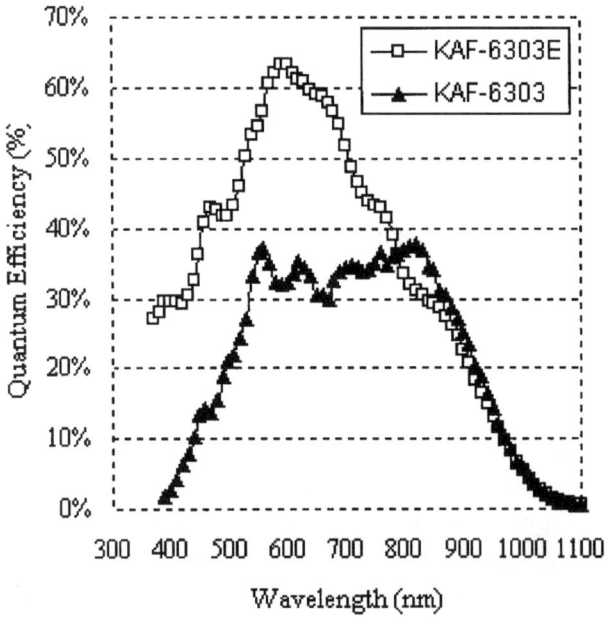

Figure 3.25 QE response for a Kodak transparent-gate CCD.

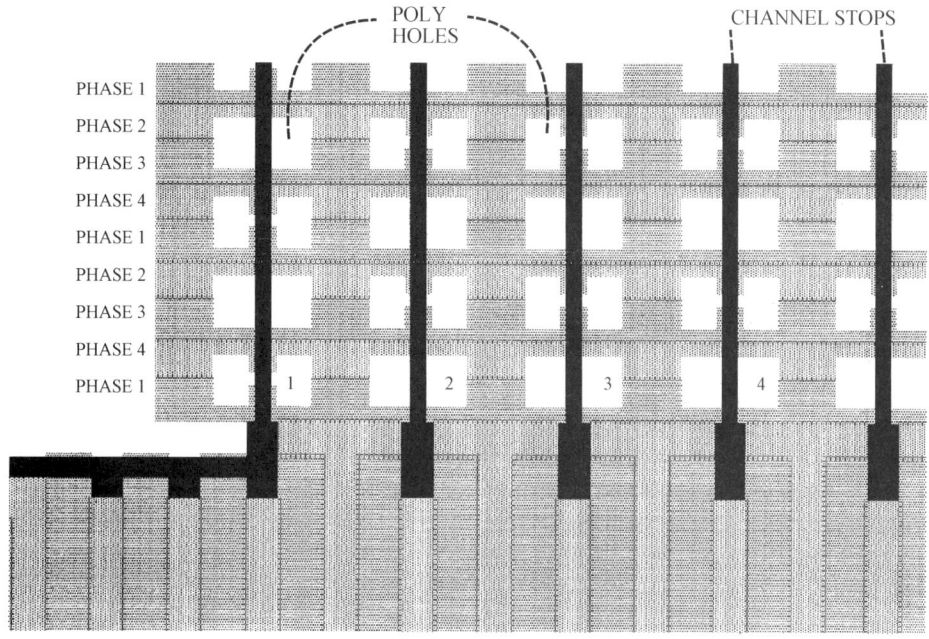

Figure 3.26 8-μm pixel poly-hole CCD design.

required for virtual-phase and OPP CCD technologies. The boron pinning implant compensates some of the buried-channel phosphorus implant, resulting in a channel potential less than the potential of the barrier phases. Therefore, signal charge does not transfer within the polyholes. The main drawback of poly-hole fabrication is a reduction of full well performance.

3.4 BACKSIDE ILLUMINATION

3.4.1 THINNING

Various methods are used to thin CCDs for backside illumination. The earliest reported work on the performance of thinned scientific CCDs occurred at Texas Instruments (TI) in the early seventies, when work began on the Hubble WF/PC I CCDs.[20] RCA was developing a thinning process at the same time for commercial use.[21–23] Today many groups are involved with thinning CCDs for both uses. This section discusses the TI process because other thinning techniques are closely related to the original process.

The wafers to be thinned are usually first "back lapped," either chemically or mechanically, to remove some of the substrate layer. Then, an acid etch is employed using a mixture of corrosive acids. For example, a typical recipe will consist of nitric acid (HNO_3), hydrofluoric acid (HF), and acetic acid (CH_3COOH).[24-28] The reaction at the silicon surface is an oxidation-reduction reaction, followed by the dissolution of the silicon dioxide formed according to the chemical equations

$$Si + 4HNO_3 = SiO_2 + 4NO_2 + 4H_2O, \qquad (3.4a)$$

$$SiO_2 + 6HF = H_2SiF_6 + 2H_2O. \qquad (3.4b)$$

Note that HNO_3 acts as the oxidizing agent and HF as the agent to remove oxidation products; together they work to consume silicon. The acetic acid is used primarily as a diluent.

One specific recipe used is 1 part per volume HF, 3 parts per volume HNO_3, and 8 parts CH_3COOH. This ratio, 1–3–8, is a selective etch. That is, it readily etches highly doped p-silicon material at a rate of approximately 0.08 to 0.1 mils/min (2–2.54 μm/min). However, when reaching the epitaxial interface, thinning dramatically slows by a factor of roughly 100. The reason for this is that the oxidation rate and consumption of silicon are much greater for highly doped material (i.e., substrate = 0.01 Ω-cm versus epitaxial = 50 Ω-cm). Therefore, the epitaxial interface is important in the thinning process and used as an etch stop.

Removing all p^+ substrate material to passivate and accumulate the surface in subsequent process steps is important, as shown below. Also, the CCD should be thinned to the front depletion edge to eliminate as much backside field free material as possible for good MTF performance. To achieve these requirements a final etch bath is used, a potassium permanganate ($KMnO_4$) and hydrofluoric mixture that removes the remaining p^+ silicon. $KMnO_4$ oxidizes silicon more quickly and therefore is less selective. The etchant thins the epitaxial layer to the final device thickness. After thinning, the silicon surface appears as an "orange peel." Figure 3.27(a) shows an SEM image of the backside surface of a WF/PC I CCD. The region shown is approximately 18 × 24 μm, about the size of a pixel.

3.4.2 Quantum Efficiency Hysteresis

After thinning, a native SiO_2 layer immediately begins to grow on the backside of the chip. The native oxide reaches a thickness of approximately 25 Å after several months' time, until its growth essentially stops. The properties associated with the silicon-native oxide interface dramatically affect the sensor's QE performance. The Si–SiO_2 interface exhibits surface states that range in density from 10^{11} to 10^{13} states/cm^2, depending on how the oxide grows. Some of these interface states are electrically neutral when filled with electrons (i.e., states below the Fermi level) and positively charged when empty (i.e., states above the Fermi level). The positive nature of the states will result in the formation of a depletion region within the p-silicon epitaxial layer. In turn, the depletion region forms a "backside potential

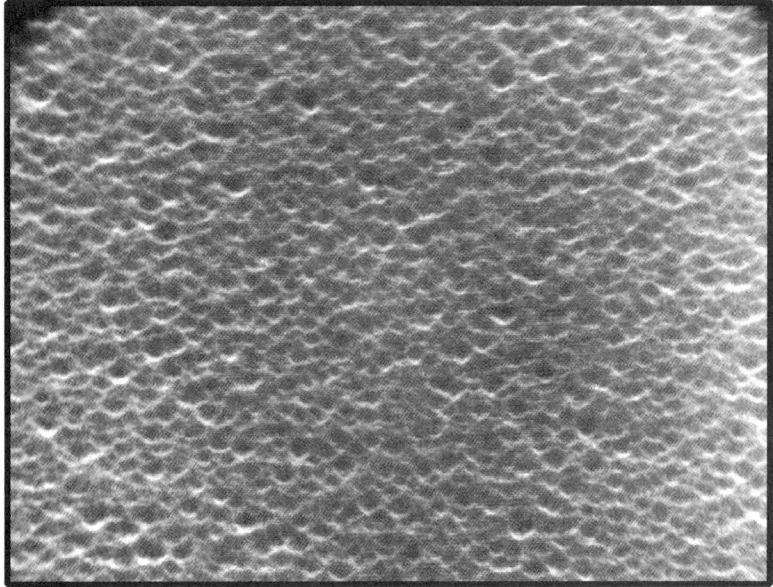

Figure 3.27(a) Greatly magnified SEM image of backside surface.

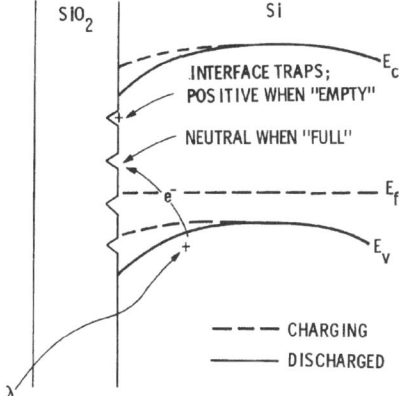

Figure 3.27(b) Backside well induced by positive charge at the Si–SiO$_2$ interface.

well" as shown in Fig. 3.27(b). The backside well can extend into the silicon several microns immediately after thinning.

The electric field generated by the backside depletion region is in the direction to attract signal electrons into the backside well. Signal charge generated during an exposure will collect at the back of the CCD and neutralize some of the positive interface states. In turn, the size of the backside well will decrease in size and cause the QE to increase. During readout these trapped electrons will eventually recombine, causing the backside well to increase in size again and the QE to de-

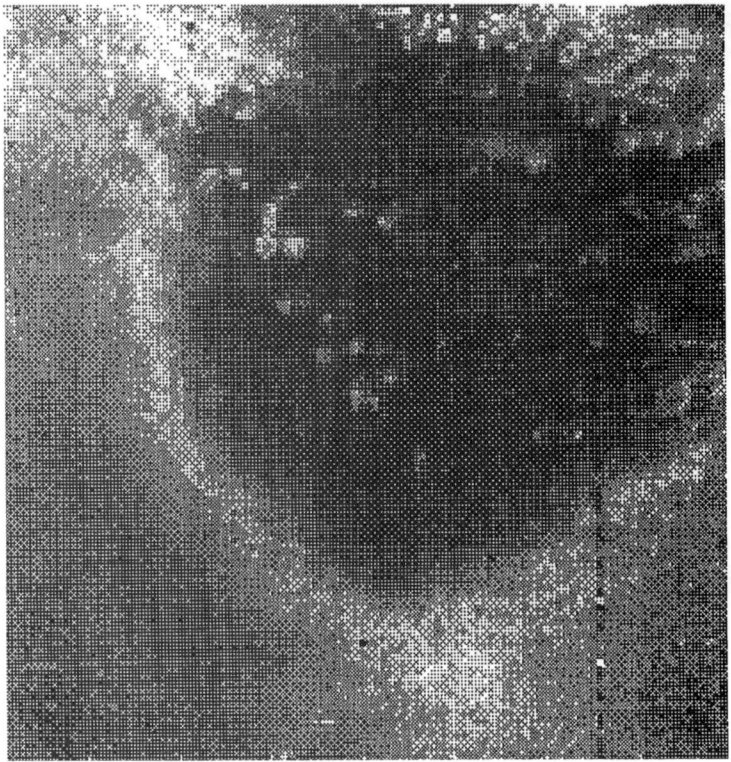

Figure 3.28(a) 4000-Å QEH image.

crease. This characteristic of charging and discharging the Si–SiO$_2$ interface with signal charge is called "QE hysteresis" or QEH. QEH has plagued the backside-illuminated CCD since thinning was first employed.

QEH behavior is shown in Fig. 3.28(a) for a recently thinned backside-illuminated CCD in response to a flat-field source of 4000-Å light. The image was generated by dividing, pixel by pixel, two frames imaged back to back with identical exposure times. Ideally, the processed image should show no modulation and contain only signal shot noise (i.e., QE should remain constant between the two images). However, the white regions seen correspond to an increase in QE for the second image. The image shows that QE for some areas changes as much as 200%.

As one might expect, QEH is highly temperature dependent and is more noticeable at cold operating temperatures. As mentioned, QEH is caused by the charging and discharging of interface states. These states exhibit a vast range of emission time constants. At very cold temperatures ($< -120°C$), the emission time constants for the traps can be much longer than an exposure/readout cycle producing stable QE performance. Effectively, the traps remain filled during a sequence of exposures. For example, Fig. 3.28(b) plots signal as a function of exposure number

CHARGE GENERATION

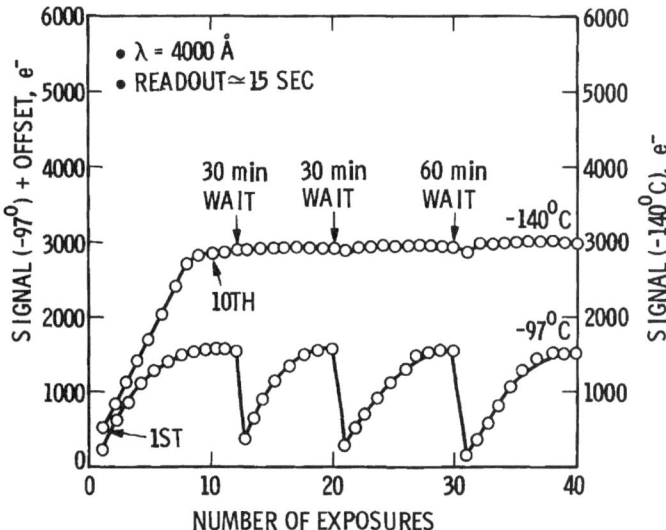

Figure 3.28(b) QEH characteristics at two different operating temperatures.

in response to 4000-Å light. Readout time between exposures was approximately 15 sec for this experiment. Data were taken at −97°C and −140°C. Note that the QE sensitivity at −97°C increases as interface states trap signal electrons, collapsing the backside well. At −140°C the sensitivity also increases, exceeding the QE at −97°C because the traps remain filled. After 12 exposures are taken, the sensor is allowed to stabilize, under dark conditions, for 30 min. During this time the traps are allowed to discharge, causing the backside well to form again. When the exposures are resumed after 30 min, the QE again increases in the −97°C experiment. However, at −140°C the sensitivity remains stable over the 30-min period. This behavior is attributed to the long emission time constants involved at this temperature (i.e., the traps never discharge). Note that a 60-min wait is also attempted, where a slight drop in QE is observed at −140°C but which recovers after the first exposure.

3.4.3 ACCUMULATION AND QE PINNING

The backside well must be eliminated if high QE performance and good stability are to be achieved for the backside-illuminated CCD. Accumulation is a process used by CCD manufacturers after thinning to accomplish this task. Accumulation is electrically the opposite of depletion and can occur when a net negative charge exists at the surface of the CCD (as opposed to a positive one). Under this condition, holes are not repelled as in the case of depletion, but attracted to the surface. A field is generated by the diffusion of holes from the regions of high concentration to regions of low concentration. The gradient of charges, which is more concentrated at the surface, generates an electric field that accelerates signal electrons toward the front of the CCD.

Optimum accumulation implies a very special condition referred to as "QE pinning." When pinned, the internal QE for the CCD is 100% at all wavelengths (i.e., all photoelectrons are collected). Also, QEH is automatically eliminated under the pinned state because photoelectrons are not allowed to diffuse to the surface and cause interface charging problems. The cross-section presented in Fig. 3.29 shows a backside-illuminated CCD in the QE-pinned state. It is this ideal band diagram that all backside CCD manufacturers hope to achieve. The backside accumulation layer shown directs photoelectrons into the frontside depletion region, which in turn collects in the buried channel. The distance between the accumulation and depletion regions is minimized for high charge collection efficiency (i.e., good MTF). This is accomplished by selecting the proper epitaxial resistivity that defines the depletion depth, which should be approximately equal to the epitaxial thickness.

Unfortunately, there is not one accumulation process universally applied by CCD manufacturers. Several different techniques are employed, each with its own set of advantages and disadvantages. Two general methods will be described in this chapter. The first category is called "passive accumulation," which includes those methods that provide negative charge to the backside, to induce an accumulation layer. Passive techniques leave the silicon lattice intact. These methods are very effective in QE pinning the CCD but are more vulnerable to environmental conditions. The second category is called "active accumulation," which includes those methods that involve doping the backside with boron atoms in forming the accumulation layer. Active accumulation methods are more stable because the accumulation region is built into the silicon. However, active approaches rearrange

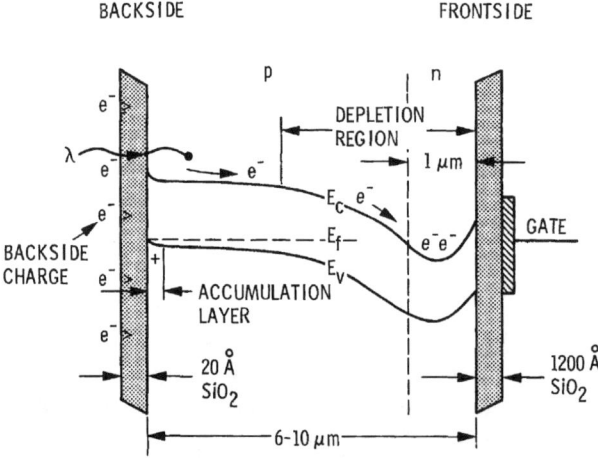

Figure 3.29 Energy band diagram showing a backside-illuminated CCD in the QE-pinned state.

3.4.4 SELF-ACCUMULATION

Self-accumulation was originally employed on the first CCDs thinned at TI and flown in the Hubble WF/PC I cameras. Here the CCD is carefully thinned to intentionally leave a small amount of p^+-substrate material on the backside to serve as an accumulation layer. The method resulted in many QE problems and dramatically demonstrated the importance of a good accumulation layer. Although self-accumulation is not employed today, the method gives insight into backside physics, on which modern accumulation schemes are based. A short review of the physics behind self-accumulation is therefore worth some discussion.

Figure 3.30(a) presents a schematic cross-section of a WF/PC I CCD normal to the surface before thinning. The CCD shown is built on a 10-Ω-cm, p-type, 10-μm epitaxial layer residing on a 0.01-Ω-cm, p^+-type substrate. The thickness of the epitaxial layer and the p^+ diffusion from the substrate depend on how the epitaxial layer was grown and on the subsequent high-temperature processing during CCD fabrication. For some thin epitaxial CCDs, the p^+ diffusion may extend nearly to the front depletion edge. For other CCDs the two edges may be widely separated, an undesirable result for field-free reasons discussed in Chapter 4. Figure 3.30(b) is a spreading resistance measurement taken from a WF/PC I CCD of impurity concentration as a function of distance from the backside surface. Note that the substrate autodoping extends 5 μm.

As mentioned above, thinning terminates close to the epitaxial interface. Ideally, for self-accumulation the stopping point should be about midway within the p^+ diffusion region, leaving some p^+ material for the accumulation layer (this requirement will be further explained below). The acid agitation technique employed by TI is called "window thinning." Here, only the active pixel area was thinned to the epitaxial interface, leaving a thick border of silicon around the thinned region (refer to Fig. 3.31(a) that shows a SEM cross-section). This method of thinning produced many interesting problems. For example, as the thinned membrane (< 10 μm) was freestanding, it was under a great amount of stress induced by the poly gates and oxide layers on the frontside. The tension pulled on the membrane in different directions, resulting in a warped surface (called the "potato chip factor"). Figure 3.31(b) shows a top view of a WF/PC I CCD that is thinned to approximately 8 μm. One can see the warping pattern in the membrane. The four holes in the CCD mount stabilize the pressure on both sides of the CCD. Figure 3.31(b) shows the inside of the package and the frontside of the CCD. Bond wires are attached from the pads of the CCD to the gold package traces.

The peak-to-peak flatness variation for the WF/PC I CCDs averaged about 50 μm over the array. Optically this did not present a problem to Hubble; however, ground-based astronomers complained about the flatness when they were utilized

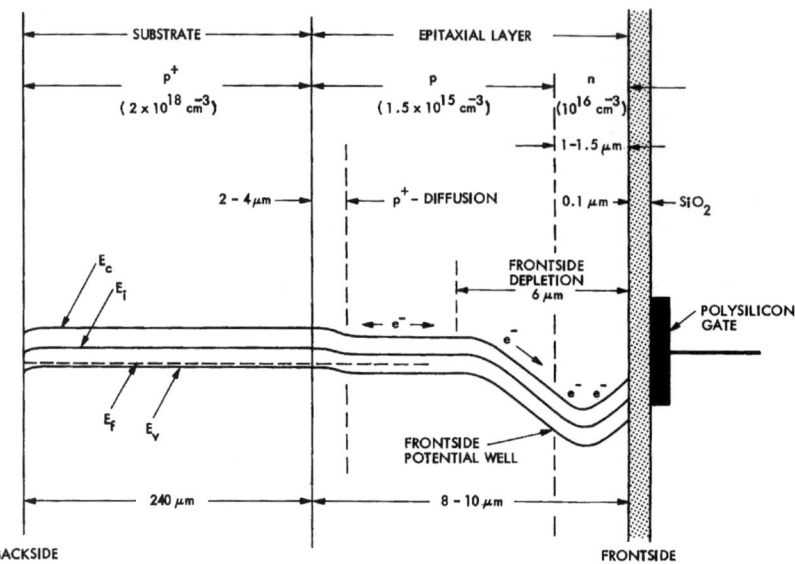

Figure 3.30(a) Cross-section of substrate and epitaxial layers before thinning is performed.

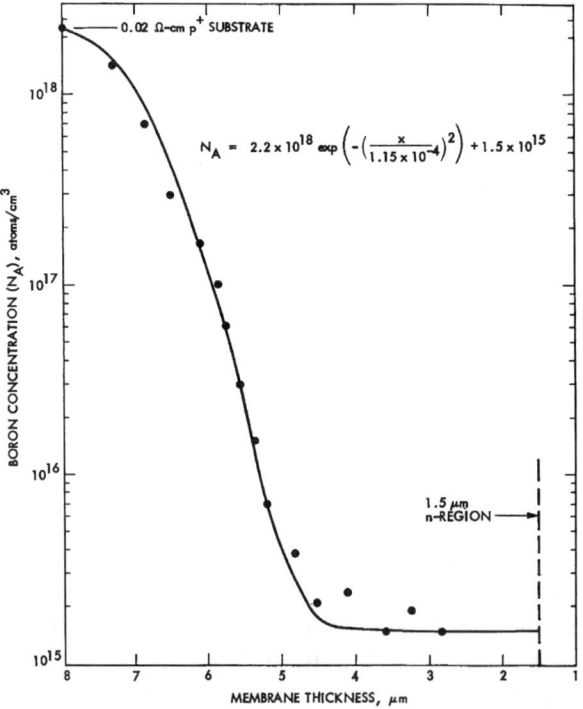

Figure 3.30(b) Doping profile at the epitaxial interface.

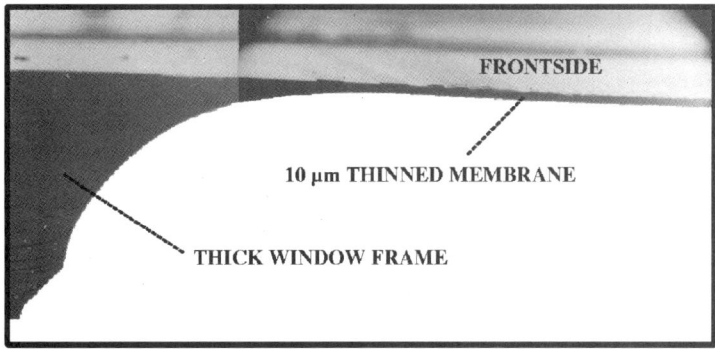

Figure 3.31(a) Side view of window thinning membrane.

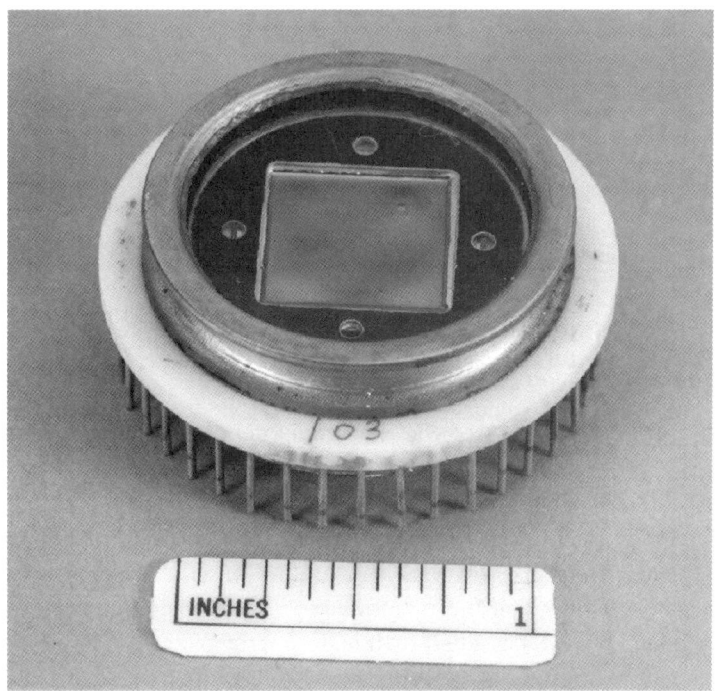

Figure 3.31(b) Top view of a WF/PC I CCD.

in fast ($< f/5$) camera systems. In fact, attempts were made to correct the flatness by introducing a conjugate lens to flatten the focal plane and gain more depth of field. These attempts were futile because when the sensor was cooled, the membrane usually buckled into a new position (called the "beer can effect").

The main concern about the freestanding membrane was its natural resonant frequency, which was about 1500 Hz. This frequency was measured by mechanically jarring the CCD while looking at the output video. Unfortunately, this fre-

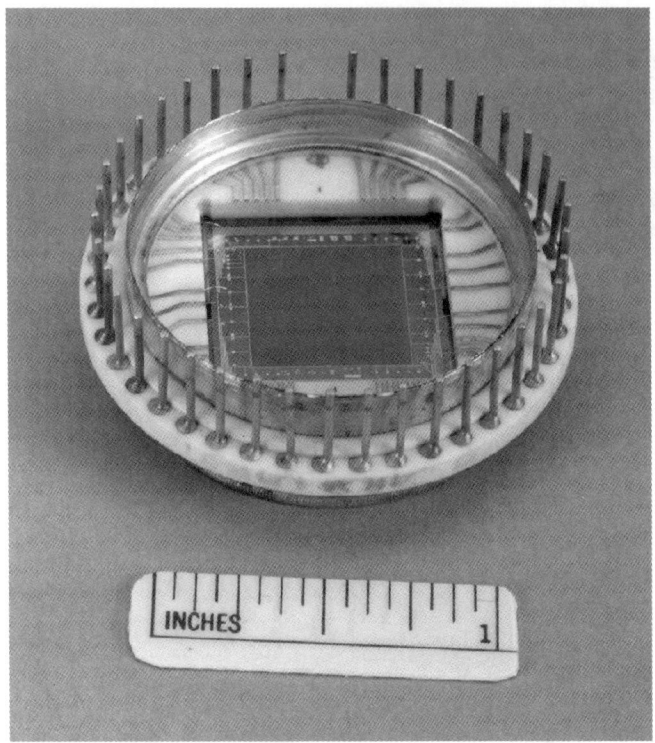

Figure 3.31(c) Bottom view of a WF/PC I CCD.

quency was in the band-pass of the vibrations generated by the Space Shuttle. Therefore, a concern existed that the CCDs would self-destruct when the shuttle was launched. The problem was warranted, because when a membrane broke, it rolled up instantly on side on the array (called the "venetian blind effect"). The sensor totally self-destructs. Nevertheless, ground-based reliability vibration tests were performed to show that the membrane could handle shuttle vibrations, and therefore, the CCDs were launched with less apprehension.

Today, window thinning is not employed. CCD wafers or die are usually fully supported before thinning is performed.[29–31] For example, support can be accomplished by taking another blank silicon wafer of the same size and epoxying it to the wafer to be thinned. Getting to the bond pads after this support structure is mounted has been the main challenge to thinning efforts. Several different approaches are taken to solve this problem, but these process secrets are well guarded by the manufacturer and cannot be revealed here.

The window-thinning technique introduced variations in membrane thickness on a single device large enough to allow the QE to be studied as a function of membrane thickness and p^+ concentration. For example, Fig. 3.32(a) presents a 800×800 flat-field image taken with 10,000-Å narrowband (5-nm) light showing spectral interference patterns. Interference is generated by light being reflected

CHARGE GENERATION

back and forth between the front and back surfaces of the CCD. Note that active fringing is greater in the corners of the CCD because of acid eddy currents that developed there (i.e., the devices is thinned more rapidly in the corners than the center of the chip). The fringing pattern seen is used to determine the membrane thickness variation through the equation

$$\Delta x = \frac{\delta \lambda \cos \theta}{4 \pi n_{SI}}, \qquad (3.5)$$

where Δx is the membrane thickness variation (µm), δ is the phase difference (2π from cycle to cycle), λ is the wavelength of stimulation (µm), θ is the angle from the normal in the silicon, and n_{SI} is the index of refraction. Note that when the total round-trip phase shift is an integral multiple of 2π, standing waves will show the greatest intensity and generate the most charge in the thinned membrane.

For thin (15-µm) CCDs, the order of the fringes at 9200 Å is about 120, so the fringe phase is very sensitive to small changes in thickness and wavelength. The wavelength spacing of the fringes is about 60 Å, the thickness spacing just over 0.1 µm. Also, the index of refraction is not constant with wavelength (between 3–4 over 7,000–10,000 Å). As the device becomes thicker, absorption characteristics move the fringing effect further into the red. Also, the fringes become much tighter due to the range of incident angles from the incoming beam reducing their contrast. This effect is demonstrated in Fig. 3.32(b), where predicted QE performance is shown for a 15-µm backside-illuminated CCD stimulated with normal incident narrowband light.

Example 3.5

What is the membrane thickness variation required to see one fringe in Fig. 3.32(a)? Assume $\lambda = 1$-µm normal incidence illumination and an index of refraction of 3.6. Also, determine the thickness variation in the corner of the image.
Solution:
From Eq. (3.5),

$$\Delta x = 2 \times 3.14 \times 1/(4 \times 3.14 \times 3.6),$$

$$\Delta x = 0.14 \, \mu m.$$

For the image shown, each fringe cycle represents a thickness change of approximately $1/7$ µm. Therefore, thinning nonuniformity is approximately 1 µm for the seven fringes counted from the corner of the device to its center. Figure 3.32(c) shows a line trace within the corner region used to count fringes.

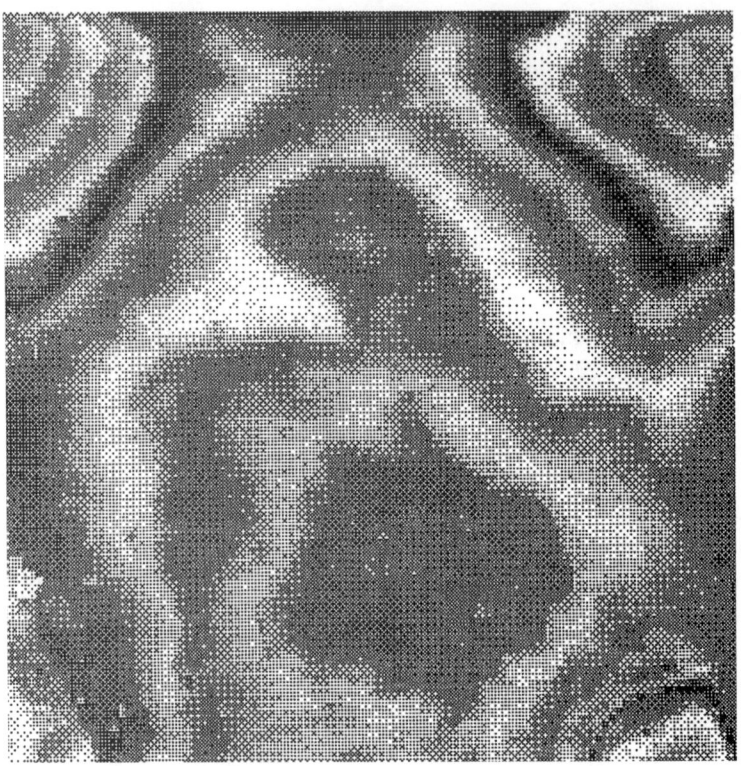

Figure 3.32(a) Narrowband near-IR image showing interference fringing due to thinning nonuniformities. [This image, along with others found in this book, were created with primitive pre-PC photomechanical transfer technology and are therefore low resolution.]

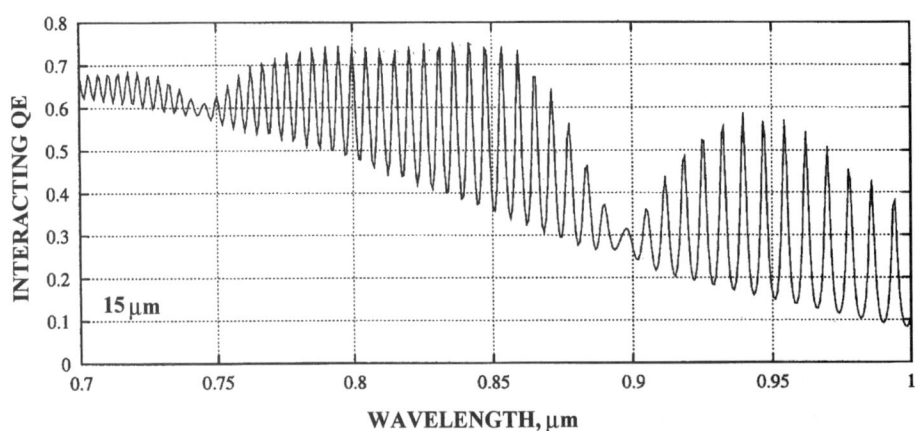

Figure 3.32(b) Theoretical near-IR QE showing interference fringing for 15-μm-thick CCD.

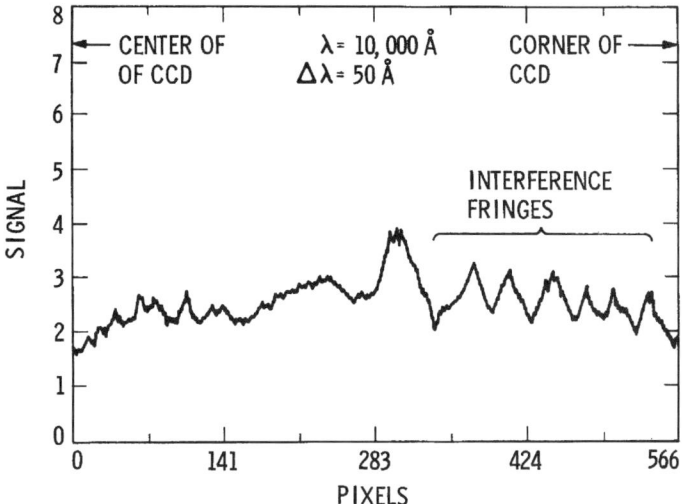

Figure 3.32(c) Line trace showing seven fringes.

Sensors with thinning nonuniformities >1 μm provide an excellent opportunity to study surface conditions and QE performance as a function of distance from the substrate to the front depletion edge. For example, Fig. 3.33(a) shows a narrowband flat-field 4000-Å response taken from a WF/PC I CCD that exhibits nonuniformity thinning problems. The corners and edges are thinnest and exhibit very poor QE performance for reasons that will be explained below. Figure 3.33(b) plots 4000-Å sensitivity as a function of membrane thickness for a small region on the CCD. This curve is labeled "QE uncharged" (the curves labeled "dark current" and the "QE charged" curves will be discussed later). The p^+ diffusion region is the same region that is measured in Fig. 3.30(b). The range of thicknesses characterized can be divided into five regions of interest [labeled in Fig. 3.33(b) as 1 through 5]. Figure 3.33(c) shows energy band diagrams for each region and the direction in which photo-generated charge moves.

Region 1 is called the "thick region" and is where a substantial amount of substrate material remains at the backside after thinning. Two factors are responsible for low QE observed in this region. First, a thick layer of p^+ material causes recombination of signal carriers due to the low minority-carrier lifetime in this region. Second, fixed positive charge at the Si–SiO$_2$ interface generates a small depletion region and backside well. Signal carriers generated near this well will be swept to the Si–SiO$_2$ interface and lost through recombination.

Region 2 is called the "thin region." Compared to region 1, the sensitivity is higher because the carrier lifetime increases. The local electric field also increases because of the p^+ boron doping gradient. The built-in field is generated by the diffusion of holes from the regions of high concentration to regions of low concentration. Therefore, a signal electron would "feel" an electric field that would direct it out of the region toward the front of the device. The region of greatest doping

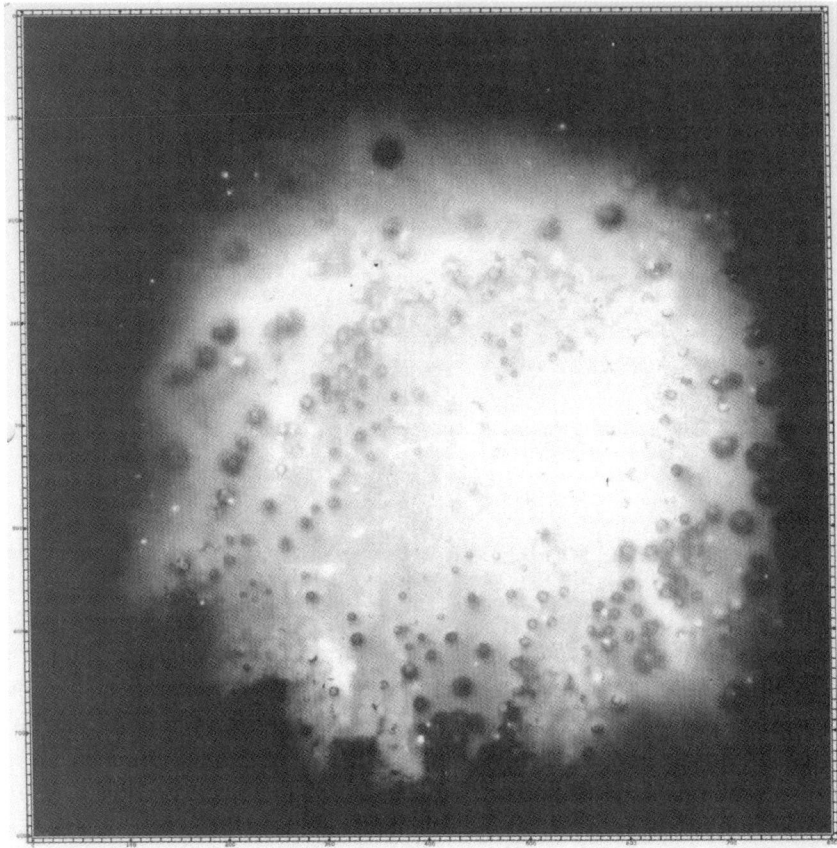

Figure 3.33(a) Flat-field 4000-Å image showing QE sensitivity nonuniformity.

change will generate the greatest electric field. The local electric field is

$$E_A = -\frac{kT}{qN_A(x)}\frac{dN_A(x)}{dx}, \qquad (3.6)$$

where $N_A(x)$ is p-acceptor doping concentration distribution (atoms/cm^3), k is Boltzmann's constant (8.62×10^{-5} eV/K) and T is temperature (K).

Equation (3.6) can be solved numerically with the equation

$$E_A = -\frac{kT}{qN_A(x)}\frac{N_A(x) - N_A(x + \Delta x)}{\Delta x}, \qquad (3.7)$$

where Δx is the increment used for numerical integration (typically 10^{-8} cm).

The electric field intensity from Eq. (3.7) is shown in Fig. 3.34, where impurity concentration, N_A, is also plotted in the p$^+$ diffusion region as a function of device thickness. The plot shows that a maximum field strength of 1.5×10^3 V/cm occurs

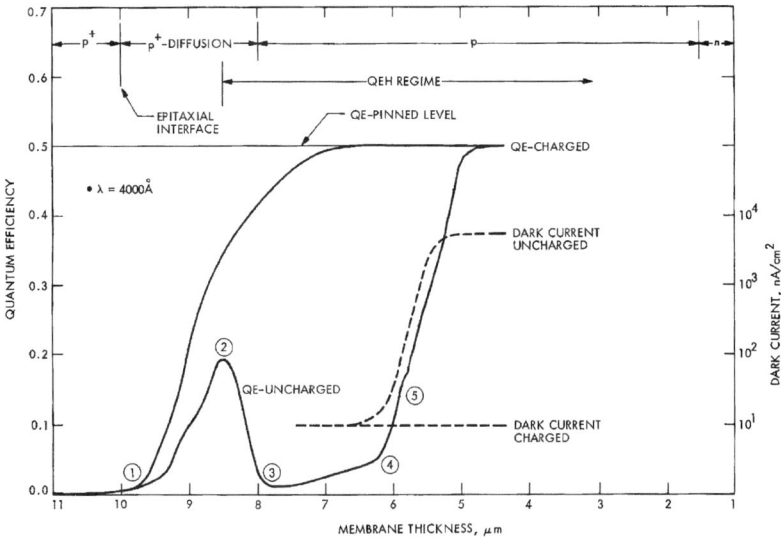

Figure 3.33(b) QE (4000 Å) and dark current (300 K) responses as a function of membrane thickness.

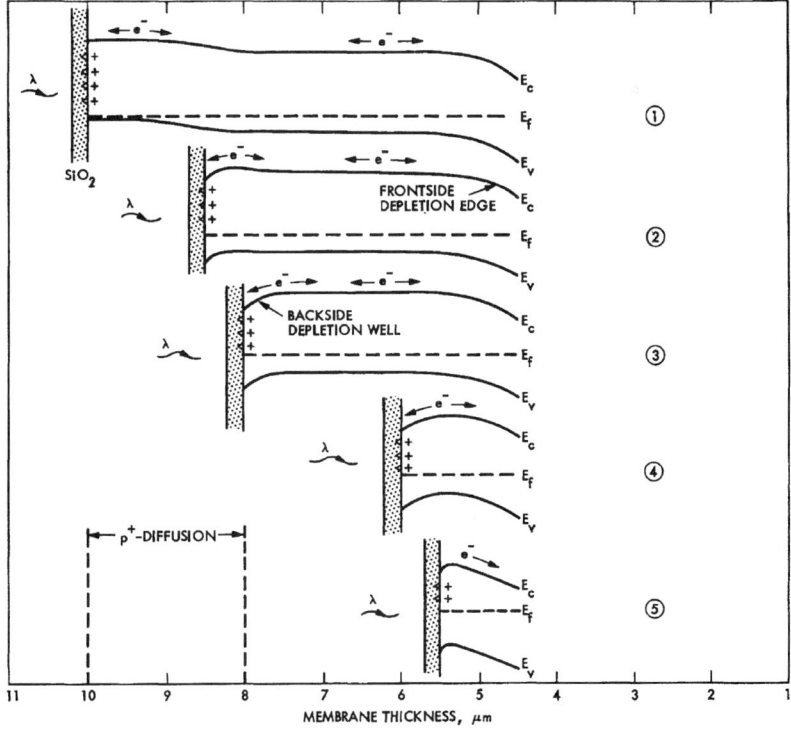

Figure 3.33(c) Energy band diagrams for the five regions labeled in Fig. 3.33(b).

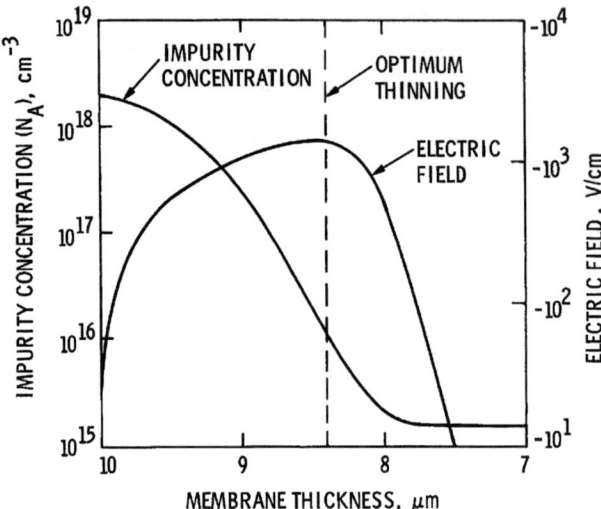

Figure 3.34 Electric field as a function of membrane thickness for self-accumulation.

when the device thickness is 8.4 μm (shown as a dashed line). Discussions below will show that this field strength is insufficient to direct photo-generated charge to the frontside for photon absorption distances of less than 2000 Å. Carriers at depths less than this distance have a finite probability of thermally diffusing to the backside even though an electric field toward the front exists. This is the main problem for self-accumulation: the internal fields generated are not strong enough to avoid surface recombination.

As more p$^+$ material is thinned away, the QE will decrease as shown in Fig. 3.33(b). This is because the backside electric field decreases and the backside potential well grows in size. The depth of the backside well, x_{BW}, increases as the impurity concentration decreases. The depth can be estimated by solving Poisson's equation [i.e., Eq. (1.8) with $p = n = 0$ for the depletion approximation],

$$\frac{d^2 V}{dx^2} = -\frac{q N_A(x)}{\varepsilon_{SI}}, \qquad (3.8)$$

where the boundary conditions are set at $V = 0$ and $dV/dx = 0$ at $x = x_{BW}$.

The numerical solution of Eq. (3.8) is an equation for the surface potential, V_S, at the Si–SiO$_2$ interface created by the positive charge in terms of the impurity concentration that must be self consistently solved for the integer N,

$$V_S = \frac{q}{\varepsilon_{SI}} \sum_{n=1}^{N} n N_A(x + n\Delta x) \Delta x^2. \qquad (3.9)$$

Once N is determined, the backside well depth, x_{BW}, can be found from the equa-

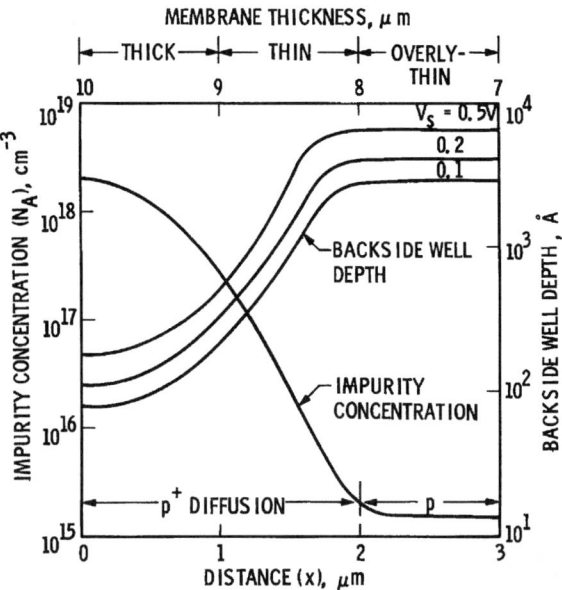

Figure 3.35 Backside well depth as a function of membrane thickness.

tion

$$x_{BW} = N \Delta x. \tag{3.10}$$

Figure 3.35 plots backside depletion depth for $V_S = 0.1, 0.2$, and 0.5 V as a function of device thickness. The actual surface potential that develops again depends on the positive charge generated during native oxide growth. Note that for $V_S = 0.2$ V, the backside well depth increases rapidly with thinning, starting at a depth of only 250 Å in the substrate and growing to 6000 Å in the p-epitaxial layer.

The sensitivity continues to drop as the device is thinned from the optimum QE point into region 3, called the "overly thinned" region. As with other regions, carriers are also lost by diffusion from field free material situated between the frontside depletion edge and backside well. Approximately half the carriers generated in the field free material are lost to the backside well. The sensitivity for the CCD is very low at this point, less than 5% at 4000 Å. The corners and sides of the CCD shown in Fig. 3.33(a) are overly thinned and exhibit a low QE for this reason. In contrast, the center of the CCD is thinned near the optimum thinning point, where the QE is higher.

Figure 3.36(a) shows a small backside region where QE deviates significantly because of a local thinning nonuniformity [taken from the device shown in Fig. 3.33(a)]. Figure 3.36(b) shows a three-dimensional image of the region, plotting absolute QE in two dimensions. The 100×100 area displayed is a thick spot on the CCD that was not thinned completely. Such regions have been nicknamed "p^+ donuts." The bright ring shows where optimum thickness for highest

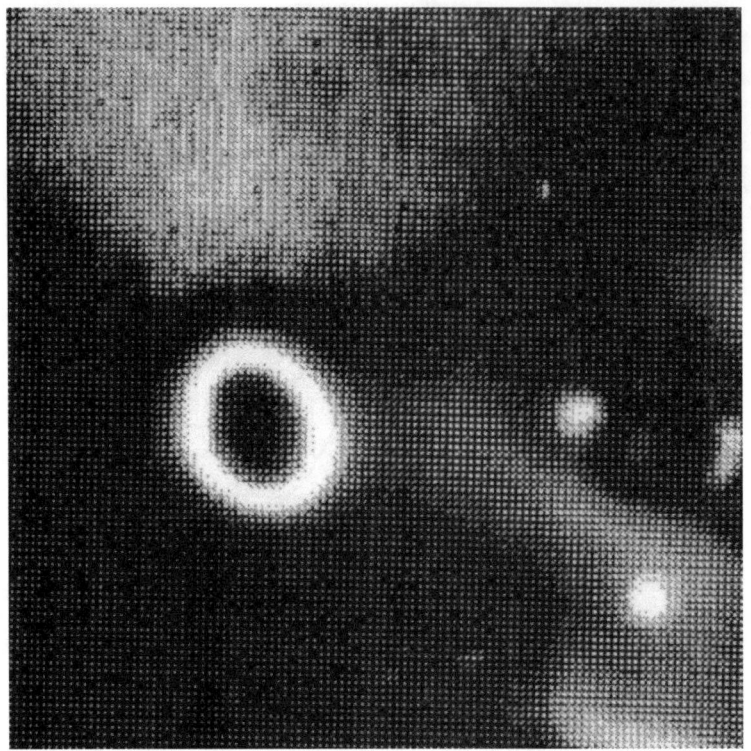

Figure 3.36(a) A 100 × 100 pixel region showing a p$^+$ donut in response to 4000-Å light.

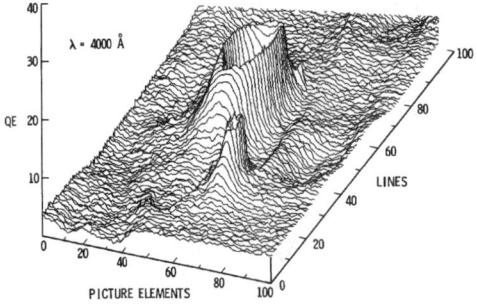

Figure 3.36(b) QE response at 4000 Å for a p$^+$ donut.

sensitivity has been achieved. The region in the middle of the donut is thicker than optimum, whereas the region outside the ring is overly thinned. A 4000-Å line trace through another p$^+$ donut is shown in Fig. 3.36(c) and again reveals locations of optimum thickness and response. Figure 3.36(d) presents the same region when stimulated with 10,000-Å light. Note that maximum response at this wavelength is where a minimum response is seen with 4000-Å light. The near-IR response

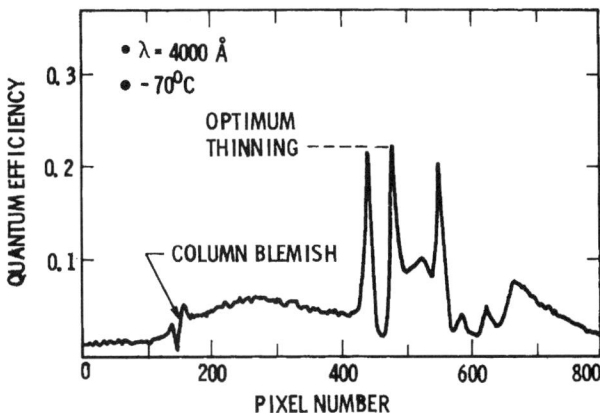

Figure 3.36(c) Line trace through a p$^+$ donut at 4000 Å.

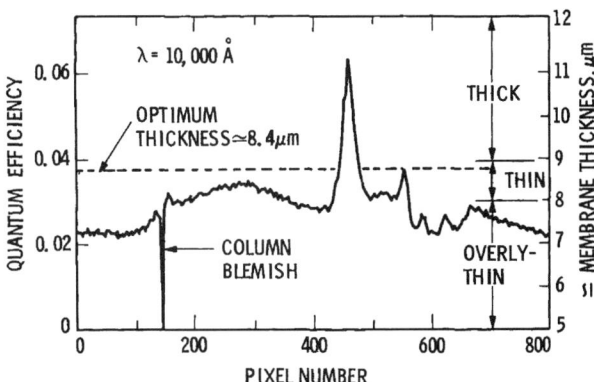

Figure 3.36(d) Line trace through a p$^+$ donut at 10,000 Å.

is directly proportional to device thickness because of the long absorption length of 10,000-Å light (> 200 μm). Figure 3.36(e) shows the response of another p$^+$ donut as a function of wavelength. Note how the p$^+$ donut turns itself out for longer wavelengths.

As the device is thinned further from region 4 to 5, the QE sensitivity increases dramatically. As the CCD is made thinner, the backside well will eventually meet the frontside depletion edge. The front depletion edge will overpower the back depletion region, resulting in near-perfect QE performance. Figure 3.37 shows line traces taken across a small region of a very thin CCD, where the depletion edge "hits" the backside as a function of substrate voltage. Elevating the substrate potential is equivalent to reducing the positive clock level to the CCD. This in turn reduces the frontside depletion depth, causing the backside well to reform and QE to decrease. Note that as the substrate voltage increases, the QE-pinned edge moves from the center of the device toward its edge where it is thinner.

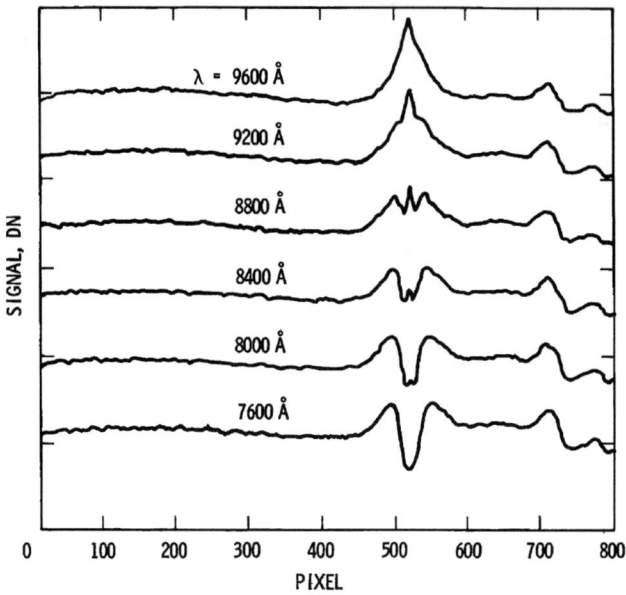

Figure 3.36(e) Line traces through a p$^+$ donut at different wavelengths.

Depleting to the backside Si–SiO$_2$ interface also generates a very high dark current signal as indicated in Fig. 3.33(b) (i.e., the curve labeled "dark current uncharged"). In comparison, the dark current rate is much greater than the frontside dark current level, typically by several orders of magnitude. This is because the interface state density associated with a native oxide is much greater than the front thermally grown gate oxide. Figure 3.38(a) shows a dark current image for a CCD that has been thinned very near the front depletion edge. The high backside dark current that abruptly appears in the corners is where the membrane thickness is less than 6 μm, the depth of the frontside depletion edge. Figure 3.38(b) plots backside dark current as a function of substrate voltage. Increasing this voltage will force the depletion edge from the backside, causing the backside well to reform and backside dark current to decrease. Backside dark current is lowered from 10,000 nA/cm^2 to less than 10 nA/cm^2, the dark current rate equal to the frontside.

3.4.5 Accumulation Theory

The key to achieving maximum QE, (i.e., QE-pinned) and stability lies in preventing electrons from reaching the backside Si–SiO$_2$ interface. In this section we present a simple theoretical model to show how an accumulation layer will accomplish this goal. The model is based on analytical relationships for the charge density, electric field, and electrostatic potential that exist at the immediate backside surface of the CCD when negative charge is provided. This study is similar to analysis done on a MOS capacitor in the accumulation state. An excellent study

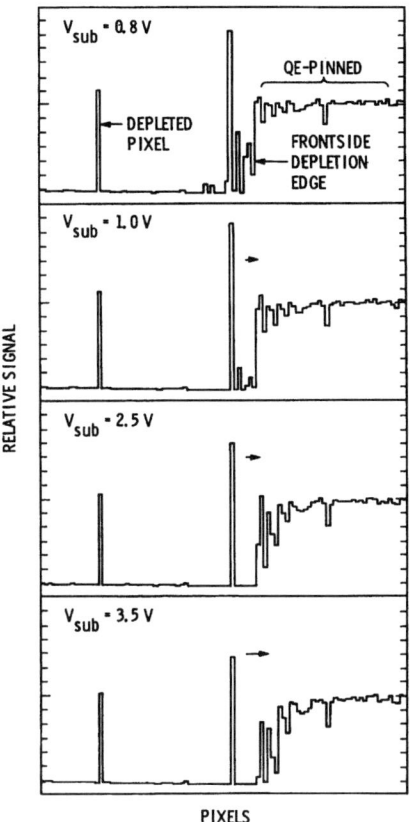

Figure 3.37 Line traces through a region where the depletion region has reached the backside.

on this problem is given by Robert Pierret in his book on MOSFETs.[32] We use his equations almost directly to explain the effects of an accumulation layer on QE performance for the backside-illuminated CCD. His theory is the solution of the full Poisson equation assuming uniform doping, whereas the theory presented in Chapter 1 used the depletion approximation.

Figure 3.39 shows an energy band diagram with the surface of the CCD in the accumulation state induced by negative charge. We will later show how this negative charge can be provided. The concentration of holes is greatest at the surface, producing an electric field that sends signal electrons to the frontside. This is similar to self-accumulation where the hole gradient and electric field were also greater at the surface because of a nonuniform doping concentration. The plot shows the conduction band E_c, the intrinsic level, E_i, the Fermi level, E_f, and the valence band, E_v. The potential voltage, V, in the silicon is defined as

$$V = \frac{E_i(\text{bulk}) - E_i(x)}{q}, \tag{3.11}$$

Figure 3.38(a) Dark current image showing high backside dark current in the corner regions of the CCD.

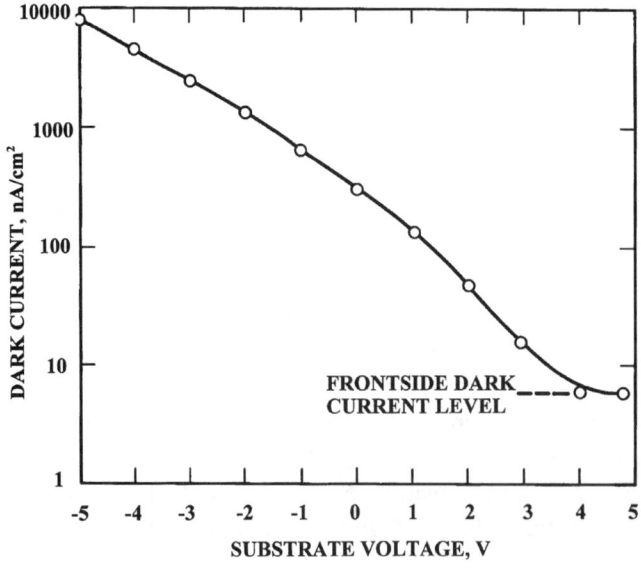

Figure 3.38(b) Backside dark current as a function of substrate voltage.

CHARGE GENERATION

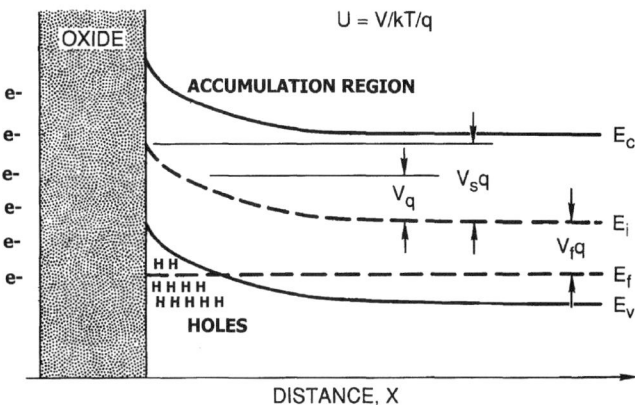

Figure 3.39 Energy band diagram of a backside CCD in accumulation.

where E_i(bulk) is the intrinsic energy level outside the accumulation region and $E_i(x)$ is the intrinsic energy level within the accumulation layer. Note that to convert energy measured in electron volts into potential, we divide by q.

The surface potential at the Si–SiO$_2$ interface, V_S, is the variable that will define the strength of the accumulation layer. Note that the Fermi level is constant throughout the semiconductor because of the insulating properties of the oxide layer. The variable $V_f q$ is the energy difference between the Fermi, E_f, and intrinsic bulk, E_i, energy levels. The Fermi potential is related to backside doping through

$$V_f = \frac{kT}{q} \ln\left(\frac{N_A}{n_i}\right), \qquad (3.12)$$

where n_i is the intrinsic carrier concentration (1.45×10^{10} carriers/cm^3 at 300 K).

Example 3.6

Find V_f assuming $T = 300$, $N_A = 10^{15}$ cm^{-3}, and $kT/q = 0.0259$ V.
Solution:
From Eq. (3.12),

$$V_f = 0.0259 \times \ln\left[10^{15}/(1.45 \times 10^{10})\right],$$

$$V_f = 0.288 \text{ V}.$$

Following Pierret's analysis we normalize V, V_f and V_S to kT/q, producing dimensionless qualities U, U_f, and U_S. For example, the potential V within the accumulation layer when normalized is given by $U = V/(kT/q)$.

Example 3.7

Find the normalized potential U_f for Example 3.6. Assume $kT/q = 0.0259$ V at room temperature.
Solution:

$$U_f = 0.288/0.0259,$$

$$U_f = 11.1.$$

The electric field, $E_A(x)$, as a function of distance from the back surface for a CCD in the accumulation state is

$$E_A(x) = -\frac{kT\,F(U, U_f)}{qL_{ID}}, \qquad (3.13)$$

where

$$F(U, U_f) = \left[e^{U_f}(e^{-U} + U - 1) + e^{-U_f}(e^U - U - 1)\right]^{1/2}, \qquad (3.14)$$

and

$$x = -L_{ID} \int_U^{U_S} \frac{dU}{F(U, U_f)}, \qquad (3.15)$$

where L_{ID} (cm) is the intrinsic Debye length given as

$$L_{ID} = \left(\frac{\varepsilon_{SI}kT}{2q^2 n_i}\right)^{1/2}. \qquad (3.16a)$$

The Debye length, L_D, for the accumulation layer estimates that length over which the electric field generated is effective and varies with doping concentration as

$$L_D = \left(\frac{\varepsilon_{SI}kT}{2q^2 N_A}\right)^{1/2}. \qquad (3.16b)$$

Example 3.8

Find the Debye length for the CCD specified in Example 3.6. Assume $kT/q = 0.0259$ V.

Solution:

$$L_D = \{1.044 \times 10^{-12} \times 0.0259/[2 \times (1.6 \times 10^{-19}) \times 10^{15}]\}^{1/2},$$

$$L_D = 9.192 \times 10^{-6} \text{ cm or } 919 \text{ Å}.$$

The estimate is good for small V_s where band bending is not significant.

Numerical techniques are used to calculate the electric field with x once U as a function of x is determined. This information is also used to calculate the accumulation charge density, ρ, in the silicon,

$$\rho = qn_i\left(e^{U_f - U} - e^{U - U_f} + e^{-U_f} - e^{U_f}\right). \tag{3.17}$$

Figure 3.40 presents calculations for charge density, potential, and electric field using the equations above for different V_S. These calculations are based on an epitaxial resistivity of 10 Ω-cm (1.3×10^{15} cm^{-3}). Note that the accumulation layer occupies a very small but concentrated region near the surface of the CCD.

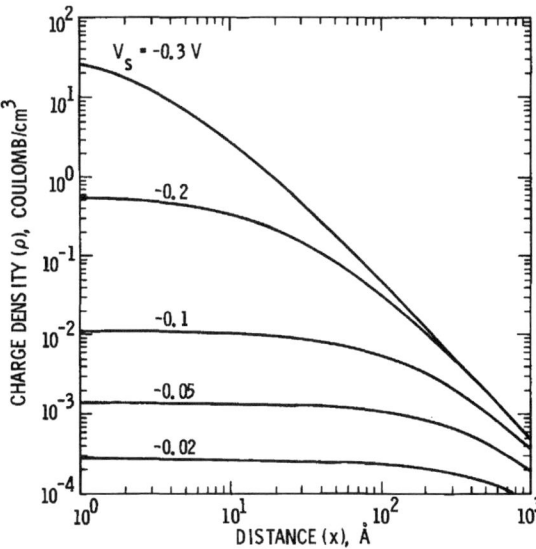

Figure 3.40(a) Charge density as a function of distance for different surface potentials.

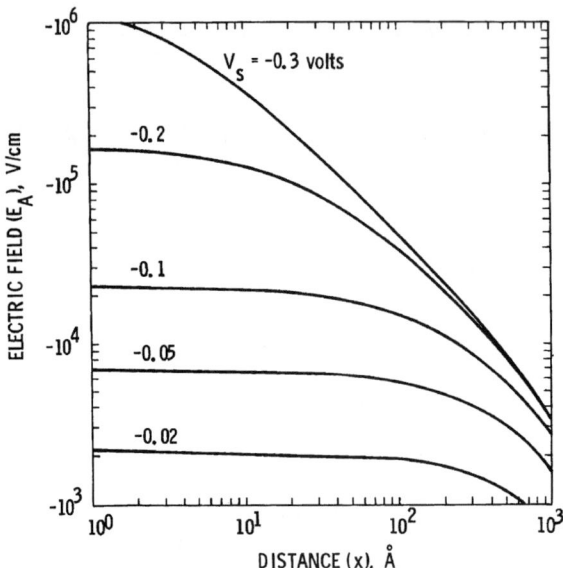

Figure 3.40(b) Electric field as a function of distance for different surface potentials.

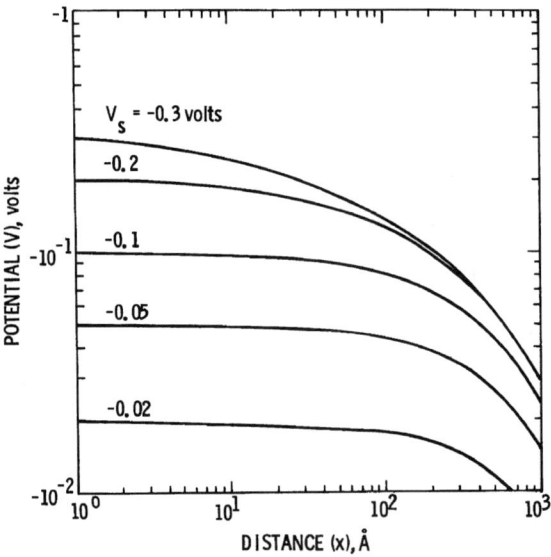

Figure 3.40(c) Potential as a function of distance for different surface potentials.

This leads to a high electric field and good QE performance. For a small V_S the extent of this region is equal to the Debye length.

The analysis above is based on the surface potential at the Si–SiO$_2$ interface. It is advantageous to specify the potential, V_G, at the outer surface of the native

Charge Generation

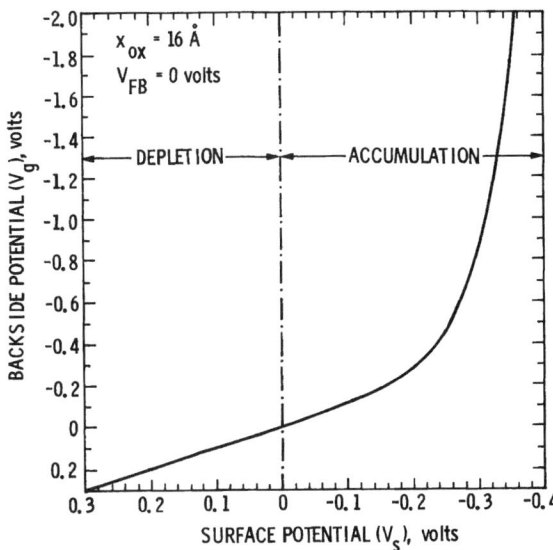

Figure 3.41 Backside potential as a function of surface potential at the Si–SiO$_2$ interface.

oxide layer. The relationship between the backside potential, V_G, and the surface potential, V_S, is,

$$V_G = \frac{kT}{q}\left[U_S + \frac{\varepsilon_{SI} x_{OX}}{\varepsilon_{OX} L_{ID}} F(U_S, U_f)\right] - V_{FB}, \quad (3.18)$$

where x_{OX} is the native oxide thickness (cm) and V_{FB} is the flat-band voltage created by the fixed positive charges at the Si–SiO$_2$ interface (V).

Equation (3.18) is plotted in Fig. 3.41 assuming a flatband voltage of 0 V, 10 Ω-cm silicon, and 16 Å of oxide. Note that the surface potential is linearly dependent on the backside voltage from depletion to approximately −0.1 V into accumulation. Above −0.1 V a large change in backside voltage produces a small change in surface potential. In this range, the voltage drop is primarily across the oxide layer because of the shielding effects of the accumulation layer at the surface.

3.4.5.1 Critical Electric Field and Absorption Distance

We can now turn to the issue of determining the required field strength to prevent photo-generated electrons from reaching the Si–SiO$_2$ interface. It is important to note that not just any field directed away from the backside will be satisfactory, since a signal electron can travel by thermal diffusion away from the point at which the photon is absorbed (refer to Fig. 3.42). The accumulation layer must create a

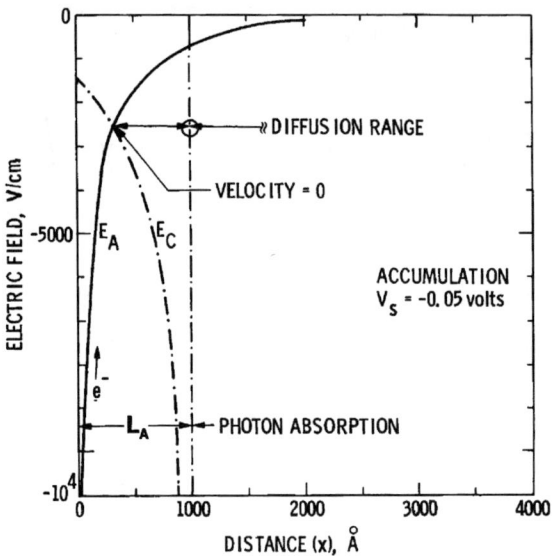

Figure 3.42 Accumulation model used to predict electric field requirements for the QE-pinned condition.

field strong enough to stop the electron from diffusing too far if recombination losses at the surface are to be avoided. We seek to define a "critical electric field" in such a way that if the electric field generated by the accumulation layer is anywhere greater than the critical field, the electron will be prevented from reaching the Si–SiO$_2$ interface. A simple definition of this critical electric field can be defined as follows.

The drift velocity of the electron due to the local electric field at the backside of the CCD is

$$v_E = \mu_{SI} E_A(x), \quad (3.19)$$

where μ_{SI} is the electron mobility (1350 cm^2/V-sec) and the coordinate x is the distance measured from the back surface of the CCD. We have adopted the convention that the electric field is negative when electrons are directed away from the backside so that accumulation causes a positive drift velocity, while a depletion region would cause electrons to move in the $-x$ direction.

Denoting L_A as the distance into the CCD at which the photon is absorbed and the photoelectron is created, the electron's average velocity due to diffusion will vary inversely with the distance $(x - L_A)$ as

$$v_{th} = \frac{D_n}{x - L_A}, \quad (3.20)$$

where D_n is the electron diffusion constant (35 cm^2/sec).

The net velocity of the electron under the combined influence of an electric field and diffusion is found by combining Eqs. (3.19) and (3.20), yielding the total velocity

$$v_T = \frac{D_n}{x - L_A} - \mu_{SI} E_A(x). \tag{3.21}$$

The electron can diffuse against the direction of the field only when the first term on the right-hand side of Eq. (3.21) dominates over the second. As $(x - L_A)$ increases, the diffusion term will weaken and the electron's velocity may become zero, depending upon the electric field. Setting $v_T = 0$ in Eq. (3.21), we define the critical electric field as the field strength required at any point x to cancel the diffusion velocity and stop the electron's progress in the direction opposite to that of the field. In equation form this is

$$E_C = -\frac{D_n}{\mu_{SI}(x - L_A)}. \tag{3.22a}$$

It is interesting to note that Eq. (3.22a) can also be expressed as

$$E_C = -\frac{kT}{q(x - L_A)}, \tag{3.22b}$$

using the Einstein relation $D_n/\mu_{SI} = kT/q$. This equation implies that the backside field requirement equals the thermal voltage divided by the absorption distance.

Using this simple model it is possible to determine whether or not the electron is able to diffuse far enough to reach the backside and recombine at the Si–SiO$_2$ interface, as shown in Fig. 3.43. The solid lines represent values of the local electric field found within the CCD generated by the accumulation layer. The critical electric field lines, as described by Eq. (3.22), are shown as dashed curves. For a photon absorbed at a depth of 1000 Å and a surface potential of -0.05 V, the local field is stronger than the critical field to the left of their intersection at $x = 400$ Å. We conclude that the electron cannot reach the backside under these conditions because its leftward velocity decreases to zero. Only if the local field were everywhere less than the critical field would the electron be capable of diffusing to the backside (e.g., -0.02 V). Note that the electric field strength increases toward the backside while the critical field is decreasing. Thus, if the absorption length is reduced from the value of 1000 Å, the point of intersection of the two curves will eventually reach the back surface (i.e., $x = 0$). Therefore, the photoelectron can reach the Si–SiO$_2$ interface if absorbed at any absorption distance less than the "critical absorption distance," given as

$$X_C = \frac{D_n}{\mu_{SI} |E_S|}, \tag{3.23a}$$

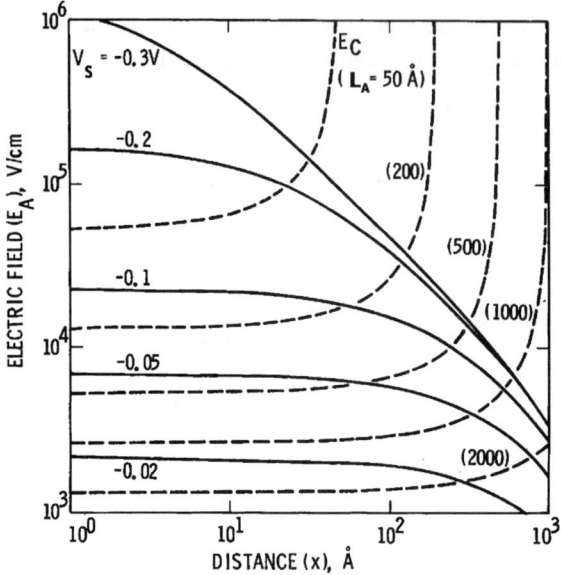

Figure 3.43 Critical electric field required for different photon absorption lengths.

where E_S is the electric field at the surface. Or, equivalently,

$$X_C = \frac{kT}{q|E_S|}. \tag{3.23b}$$

Example 3.9

Determine the critical absorption distance for a backside field of $E_S = -1.2 \times 10^4$ V/cm.

Solution:
From Eq. (3.23),

$$X_C = 35/\left[1350 \times \left(1.2 \times 10^4\right)\right],$$

$$X_C = 216 \, \text{Å}.$$

Interpreted another way, the incident photon must be absorbed at a distance $L_A > X_C$ to prevent the electron from reaching the back surface.

Example 3.10

Determine X_C, assuming a surface potential of 0.22 V, 10 Ω-cm silicon and 16 Å of oxide.

Solution:
Figure 3.44(a) plots X_C as a function of surface potential. The critical distance for a surface potential of 0.22 V is 100 Å. Figure 3.44(b) plots X_C as a function of surface electric field.

Unless the critical distance is much less that the absorption length, QE loss at the backside will be experienced. This constraint is especially important at wavelengths whose absorption length is small. Recall from Fig. 3.3(b) that the shortest absorption length occurs at a wavelength of 2700 Å ($L_A = 50$ Å). The surface and electric field required are greater than -0.2 V and 10^5 V/cm, respectively, to shield photo-generated carriers from the Si–SiO$_2$ interface at this wavelength (refer to Fig. 3.44).

3.4.5.2 Self-accumulation

We now go back to the electric field generated by self-accumulation to show how weak it is. In Fig. 3.45, the electric field is calculated about the optimum thinning point along with different critical electric fields and photon depths. As the plot shows, the optimum thinning point is far from delivering high QE for photon

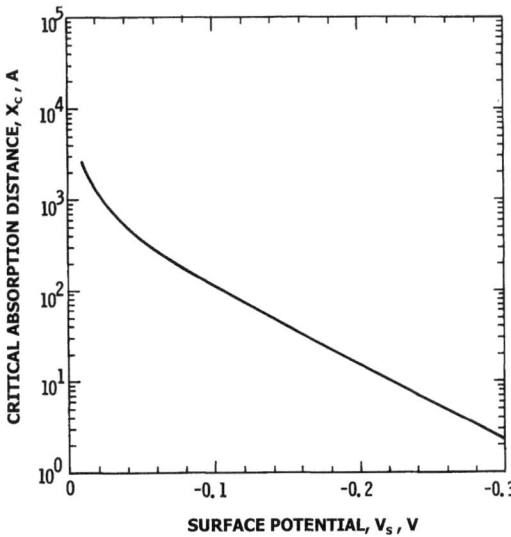

Figure 3.44(a) Critical distance as a function of surface potential.

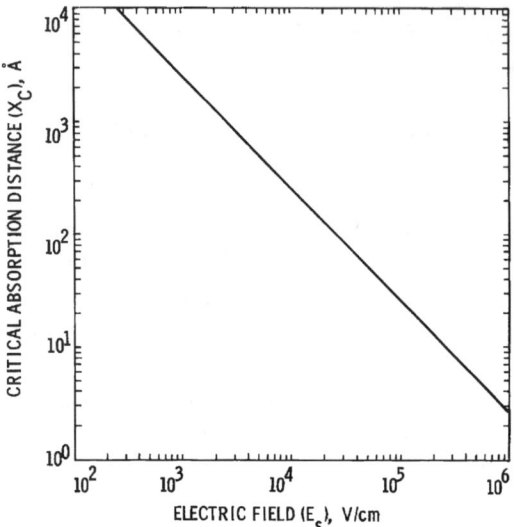

Figure 3.44(b) Critical distance as a function of electric field.

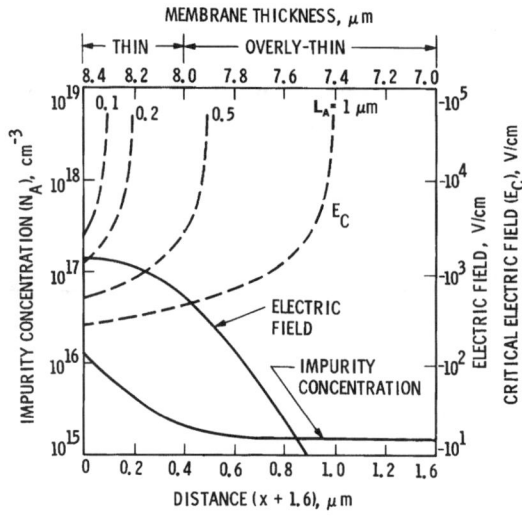

Figure 3.45 Critical electric fields for self-accumulation.

depths less than 2000 Å ($\lambda = 4000$ Å). Self-accumulation provides some improvement in short-wavelength response, but falls short of yielding the sensitivity and stability for three reasons. First, optimum thinning is extremely difficult to realize in practice. This is because the array must be etched just slightly past the epitaxial interface within an accuracy of approximately 0.2 µm, if the maximum field condition is to be achieved (this is a hit or miss game for large CCD sensors). Achieving this is further complicated by the fact that both the thickness of the epitaxial layer and the depth of the p$^+$ diffusion region vary significantly from wafer to wafer, depending on the history of epitaxial growth and subsequent high-temperature processing used for fabrication. Second, the maximum electric field produced at the backside, at the optimum thinning point, is only slightly greater than 10^3 V/cm. Estimates above show that fields $> 10^5$ V/cm are required to prevent diffusion to the backside of signal electrons that are generated by UV photons. Third, a small backside potential well will always be present because the source of the problem—positively charged states in the oxide—remains. It is for these reasons that the technique of self-accumulation was abandoned in the early development of the backside-illuminated CCD.

3.4.6 Passive Accumulation

There are five passive accumulation techniques to QE-pin the CCD: (1) backside charging, (2) flash gate, (3) chemisorption charging, (4) biased flash gate, and (5) biased ohmic contact on high-resistivity substrate.

3.4.6.1 Backside Charging

A direct method of obtaining QE pinning is to provide a surplus of negative charge on the back surface on the CCD.[33] The negative charge will overcome the positive oxide charge and the downward band bending shown in Fig. 3.27, and induce an accumulation layer. Backside charging, as it is called, can be readily accomplished by flooding the CCD with 2500-Å (5-eV) light, commonly provided by a mercury lamp.[34,35] The phenomenon of backside charging was found by accident while looking for a solution to the QEH problem with the WF/PC I CCDs. Some of this story will be related below.

In flooding the CCD, the UV photons pass through the native oxide and immediately stop in the silicon (recall the absorption length for a 2500-Å photon is less than 50 Å). A photoelectron is generated and photoemitted from the valence band, as shown in Fig. 3.46.[36–38] The electron acquires sufficient energy to overcome the conduction band of the oxide layer, an energy difference between the silicon valence band and oxide conduction band of about 4.25 eV. Some electrons collect on the surface of the CCD, charging it negatively.

Photons with energy < 4.25 eV do not efficiently charge the CCD. This fact has been confirmed experimentally, for the charging process nearly stops for wavelengths > 2900 Å (4.25 eV). UV flood is also limited to wavelengths < 1549 Å (8 eV). This energy level corresponds to the bandgap of SiO_2 and is the wavelength where photons begin to be absorbed by the oxide layer. When a photon is absorbed in the oxide layer, an electron-hole pair is generated. The electron is swept to the Si–SiO_2 interface because of the downward bending in the oxide if the device is charged, and the hole migrates to the backside where it recombines with backside charge. The process discharges the CCD to a flatband condition. An efficient UV

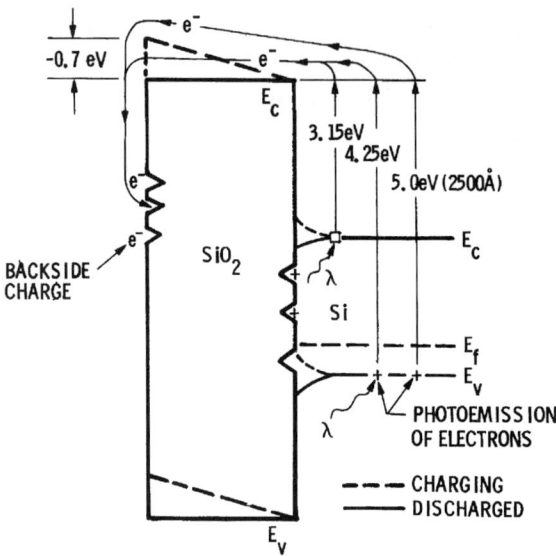

Figure 3.46 Photoemission of electrons induced by UV light.

flood wavelength uses a zinc lamp (2100 Å) to provide maximum energy to photoelectrons without oxide absorption. WF/PC I flooding experiments have shown that at this wavelength, approximately 10^9 photons/pixel are required in order to reach the QE-pinned state under ambient conditions.

For electrons to remain at the backside, a trapping mechanism must be provided. Experimentally good charging characteristics result under ambient conditions. The effects of charging are reduced considerably if the sensor is flooded under vacuum or inert gas environments. It is known that oxygen is adsorbed on a silicon surface when free electrons are present, producing acceptor states. For this reason, the CCD is normally charged in an oxygen-rich environment.

Figure 3.47 shows QE sensitivity at 2500 Å as an overly thinned CCD (i.e., without p^+ material) is backside charged with UV flood. The device exhibits low QE before charging, indicating that the surface of the CCD is depleted (i.e., a backside well exists). The sensitivity increases as an accumulation layer forms, induced by the negative charge created by the flood. A final QE value of 65% is achieved, a factor of 2.6 times greater than the theoretical QE shown in Fig. 3.4. The higher QE is due to multiple electron generation promoted by the high surface electric fields generated by backside charge (i.e., field-assisted emission). In addition, 2500-Å photons have sufficient energy to create multiple e-h pairs directly, on the average 1.1 e-h pairs within field free material [i.e., Fig. 2.7(a)]. Photon transfer was used to verify that impact ionization was truly taking place by measuring the camera gain constant (i.e., e^-/DN) throughout the charging process. A decrease in e^-/DN of 2.6 times was noted when fully charged, compared with the uncharged state.

Although initial QE will be higher and more stable, thick CCDs with p^+ material remaining on the surface are difficult to charge because the energy bands

CHARGE GENERATION

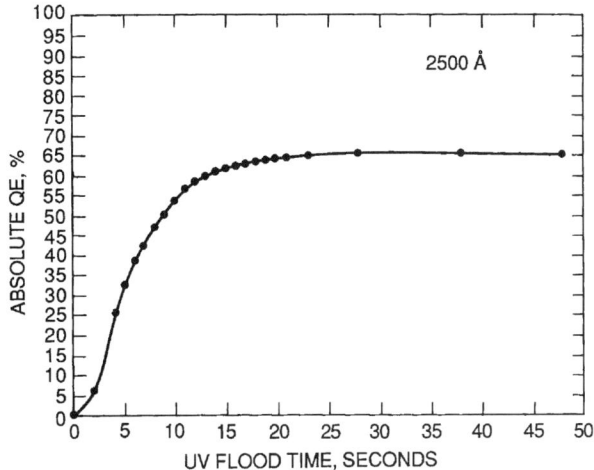

Figure 3.47 QE at 2500 Å as a function of UV flood time.

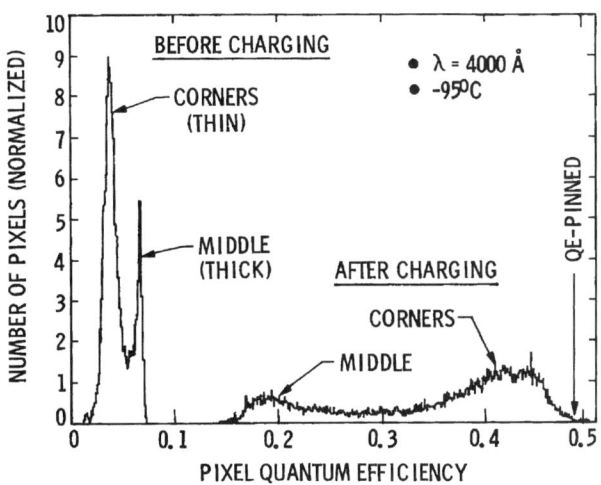

Figure 3.48 Charging characteristics for p and p^+ surfaces.

are more fixed. Figure 3.33(b) shows the final charged state as a function of thinning and removal of p^+ material. The QE-pinned state is achieved when all p^+ is removed. Figure 3.48 also demonstrates this characteristic by showing QE histograms taken from two regions, labeled thin and thick, before and after UV flood. Before UV flood is applied, the QE in the thick region is higher because the concentration of p^+ dopant and electric field is greater (i.e., self-accumulation). After partial charging, the thin region exhibits a higher QE because p-material charges more readily. Note that corner pixels come close to the QE-pinned state. Additional UV flood results in the QE-pinned condition for the p-regions, whereas the center of the CCD never quites reaches the ideal state.

3.4.6.1.1 QEH. QEH is suppressed after the CCD is backside charged. Figure 3.49(a) shows 1% QEH contours generated by four of the Hubble Planetary flight CCDs (1% contours are used because Hubble's long-term photometry accuracy is 1%). The first column of UV images represents the discharged state where the backside well is largest and QEH is worse. Some regions on the PC-8 CCD exhibit QEH > 200%. The second column of images was taken after a short solar simulation flood was attempted. Some sensors charge sufficiently to eliminate QEH to requirement. The third column shows responses for a full UV flood. Except for a small region for PC-5, QEH is eliminated.

It is interesting to note that the charged state for the WF/PC I CCDs remains for several weeks at room temperature. At cold operating temperatures the charging condition lasts much longer. Figure 3.49(b) shows discharge characteristics after a CCD was fully flooded under ambient conditions and then measured under vacuum conditions. The QE is measured at 4000 Å over the time periods shown. Note that QE is stable at $-100°C$. In fact, it was shown by experiment that WF/PC I CCDs could remain charged for several years when stored at $-90°C$, WF/PC's nominal operating temperature.

Unfortunately, the long time constants related to backside discharging caused Hubble's engineers to miss the QEH mechanism entirely. It was only after the

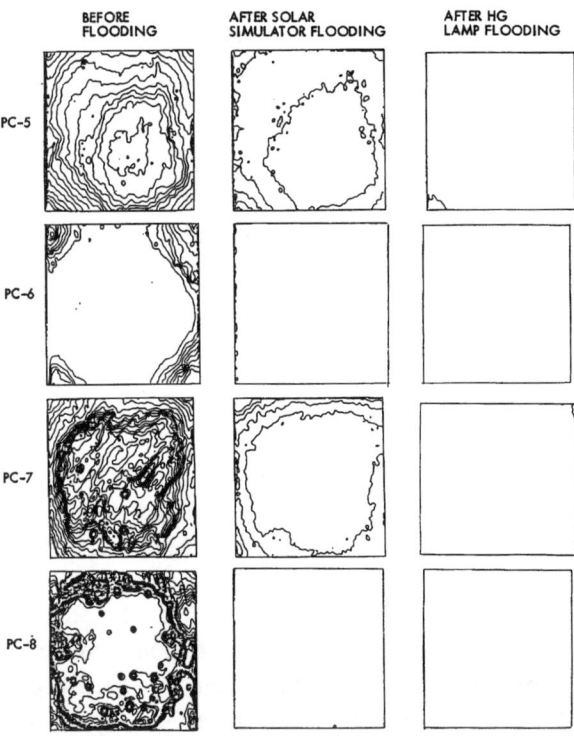

Figure 3.49(a) Hubble QEH flat-field 4000-Å contour images in response to UV flood.

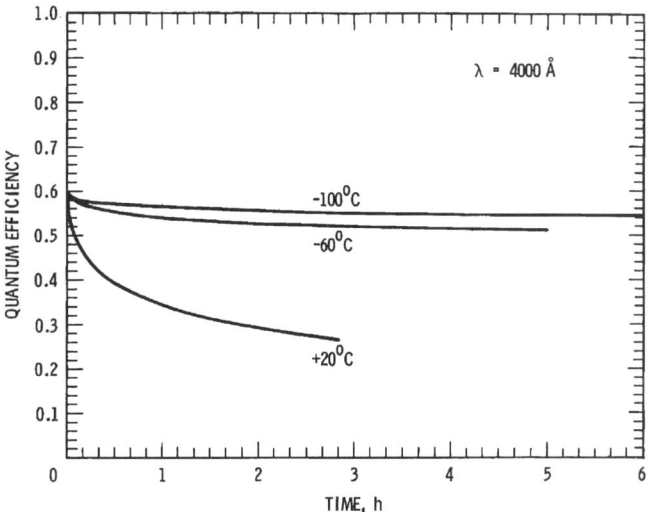

Figure 3.49(b) Discharge rates at different operating temperatures.

flight camera system was built and tested that QEH reared its head. Component screening before that time involved an intense UV flood to inspect a UV phosphor (coronene) deposited on the backside of the WF/PC I sensors. The inspection unknowingly charged the sensors, suppressing the QEH problem as detailed screening tests were performed. Only after the sensors were installed into the flight instrument and allowed to discharge was the QEH problem noticed (this was over approximately a three-month period). To circumvent the QEH problem in flight, a very expensive "light pipe" was flown to provide filtered sunlight to charge the CCDs. The next-generation camera system, WF/PC II, did not elect to use backside-illuminated CCDs, primarily because of the QEH problem. Instead, lumigen-coated frontside-illuminated CCDs were employed (the WF/PC II QE response is shown in Fig. 3.12).

3.4.6.1.2 NO, chlorine, corona and solar charging. Backside charging and the QE-pinned condition can be achieved by other charging techniques. For example, certain gases adsorbed on the surface, such as nitrogen monoxide and nitric oxide, can charge the CCD. The affinity of NO as a strong oxidizer promotes quantum mechanical tunneling of electrons through the native oxide to its surface. The adsorption of NO and high QE performance is long-lived when the CCD is kept cold and dry. Although NO gas is very effective in charging the CCD, there are a number of safety precautions to note. For example, inhalation of even a small dose (> 300 ppm) is considered lethal. Therefore, extreme care must be employed when charging the CCD with NO (e.g., gas masks are recommended). Also, in that the gas is a very strong oxidizing agent, many chemical reactions with other materials are possible. The main reaction observed is with copper, and therefore injection of NO gas with copper lines is not advisable. Also, immediate reaction with the atmosphere occurs when NO reacts with O_2, producing NO_2 (red in color), which is also deadly if inhaled.

Chlorine gas is also effective in charging the CCD. This charging mechanism was originally discovered by mixing a "Spa-tub" chlorine tablet in a squeeze bottle with some water. The chlorine gas liberated from the mixture was then blown over the CCD, and in a matter of minutes the CCD was charged into the QE-pinned state. However, there is one major drawback to this charging approach. Chlorine is a strong oxidizer and excessive use has detrimental effects on the aluminum bond pads of the CCD. The reaction also produces aluminum chloride, which is not conductive.

Corona charging is another technique used to charge the CCD.[39] Here, a corona discharge from a cold filament at -2000 V, placed approximately six inches from the CCD, at a pressure of 0.1 torr deposits electrons directly on the surface of the sensor. However, care must be exercised when employing this approach, for it is possible to overcharge the device to the point where the native oxide breaks down. Also, it is possible to induce radiation damage to the Si–SiO_2 interface because of electron bombardment.

Although inefficient, charging will also take place if the CCD is exposed to direct sunlight. The Earth's atmosphere is semitransparent down to approximately 3200 Å, allowing some UV light to charge the CCD. Fluorescent lights also induce the charging effect.

3.4.6.1.3 Backside dark current.
As shown in Figs 3.33(b) and 3.38, dark current increases dramatically when the frontside depletion edge reaches the backside of the CCD. However, an accumulation layer created by charging passivates the interface with holes, inhibiting this dark current source. Figure 3.33(b) shows that low dark current can be achieved after charging even when the depletion and accumulation regions meet.

Very low dark current generation is seen when the backside surface is accumulated by flood and when the front surface is totally inverted (i.e., MPP operation discussed in Chapter 4 and 7). For these very special conditions, both Si–SiO_2 interfaces (back and front) are populated with holes. Figure 3.50(a) presents dark current measurements taken from a backside-illuminated Reticon MPP HIRIS CCD with thick-p^+ and thin-p regions as a function of UV flood time. MPP and noninverted operating modes are measured where noninverted dark current is dominated by frontside dark current. Under inverted conditions, the thin region generates more dark current than the thick region before charging because of backside dark current. After charging, backside dark current is eliminated for both regions. The thick region exhibits a higher level because of more bulk current. Dark current levels < 10 pA/cm^2 were measured within the thin region after QE pinning was established.

Figure 3.50(b) shows backside dark current as a function of UV flood time for a CCD in which the backside is initially in depletion (i.e., uncharged). The dark current initially rises with charging as the backside well decreases in size. It

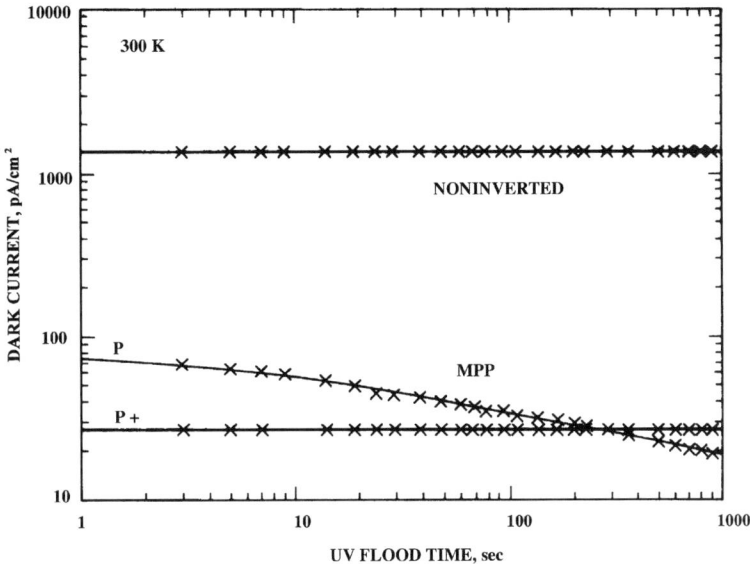

Figure 3.50(a) Dark current as a function of UV flood time.

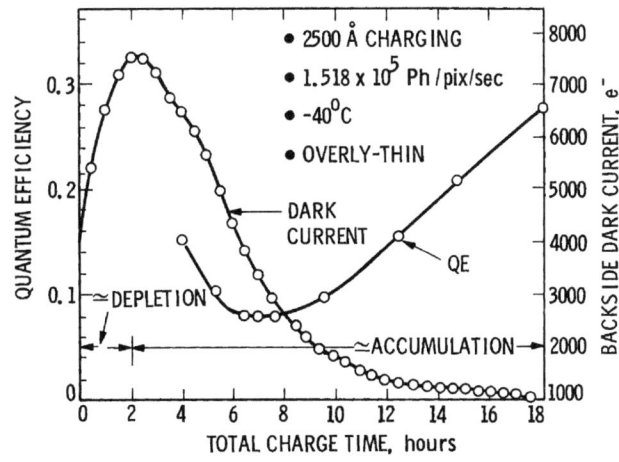

Figure 3.50(b) Dark current and QE responses when the device is taken from the depletion state to the accumulated state with UV flood.

then rapidly decreases in level as the Si–SiO$_2$ interface begins to accumulate with holes. Finally, when the sensor reaches the QE-pinned condition, the backside dark current is eliminated, leaving only frontside dark current as the source. The QE response at 4000 Å is also shown.

3.4.6.1.4 Antireflection coating charging. For some CCDs, UV flooding may cause unexpected charging characteristics. For example, Fig. 3.51 shows a 3500-Å

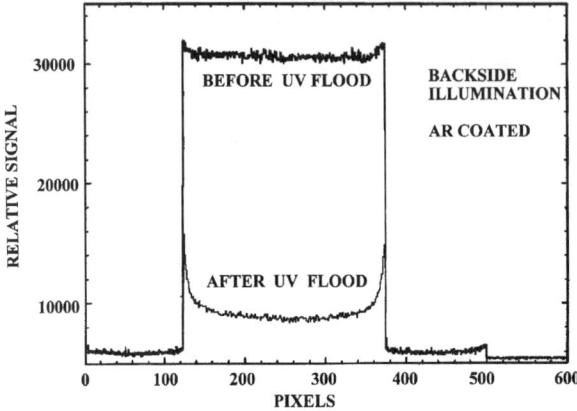

Figure 3.51 CCD with an antireflection coating that exhibits positive charging to UV flood.

response taken from a backside-illuminated CCD with an antireflection coating. For this sensor, the UV flood causes the sensitivity to decrease, indicating that positive charging is occurring. We attribute this effect to photoemission of electrons from the antireflection coating, resulting in a positive charge on its surface. Antireflection coatings that are semiconductive short out this charging effect, and therefore do not exhibit this characteristic.

3.4.6.1.5 Flash oxide. Backside charging is ineffective when attempted immediately after the CCD is thinned. However, sensors that are "aged" for 2–5 years exhibit good charging characteristics. This implies that a native oxide layer is required for the charging process, an observation verified experimentally in Fig. 3.52. Here, a five-year-old native oxide is intentionally removed from the left-hand side of a backside-illuminated CCD with hydrofluoric (HF) acid. The sensor is then charged with UV light and its QE measured. The two line traces shown before and after light flood show that the right-hand side of the CCD accepts backside charge, whereas the left-hand side, which is without an oxide, does not. The results of Fig. 3.52 imply that the maturity of the oxide layer is critical to establishing the desired surface potential and corresponding electric field at the surface of the CCD (i.e., specifically the number of dangling bonds and positive charge centers present at the Si–SiO$_2$ interface). This requirement is important for any backside accumulation method employed but is particularly important to passive methods.

The backside voltage induced by UV flood drops across both the oxide and the silicon according to

$$V_G = V_{OX} + V_S + V_{oq}, \quad (3.24)$$

where V_G is the voltage produced by the backside charge, V_{OX} is the voltage drop across the oxide, V_S is the surface potential (i.e., voltage drop across silicon) and

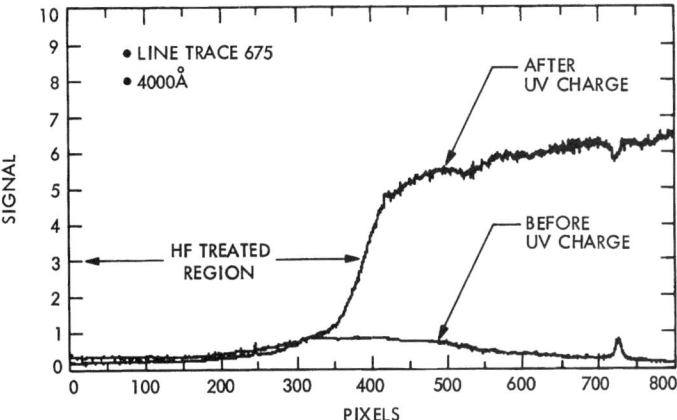

Figure 3.52 Charging characteristics after native oxide is removed by HF etch.

V_{oq} represents the negative voltage required to counterbalance the voltage induced by positive oxide charge.

Equation (3.24) describes the effect of positive oxide charges that "tie up" negative surface charges that would otherwise be coupled to the free accumulated holes in the silicon space charge layer. If the total oxide charge is large enough, the oxide charge voltage V_{oq} will compensate for the effect of the surface charge and will have almost no change in the surface band structure (i.e., the backside well of the CCD will remain and the device will exhibit low QE performance). This situation occurs for recently thinned CCDs that contain a high concentration of positively charged interface states due to the poorly formed native oxide.

The oxide charge voltage, V_{oq}, is not a constant, but is dependent on surface potential, V_S, and Fermi level, E_f. Interface charge results from traps with energy levels that lie within the bandgap of silicon, typically in a distribution centered at about one-third of the bandgap above the valence band. Under equilibrium conditions (i.e., when the CCD is not exposed to light) the interface traps will be empty for all energy levels above E_f and full for energies below E_f. Since the interface levels above E_f remain fixed in energy about E_c and E_v at the surface, a change in surface potential toward accumulation uncovers more positively charged interface states, thereby increasing V_{oq}. Also, increasing the dopant concentration in the silicon moves the Fermi level closer to the valence band edge, causing the trap population to increase. This effect is experienced with thicker CCDs with greater p$^+$ doping and is one reason that charging a thick CCD is more difficult than an overly thinned one. In addition, it is also known that oxides grown on p$^+$ material contain more interface states than oxides grown on p-material, making it more difficult to charge the CCD. Therefore, for successful charging, CCDs should be thinned past the epitaxial interface into the high-resistivity epitaxial region to reduce V_{oq}.

It is interesting to calculate the total number of negative charges required on the backside to QE pin the device. The number of charges is found by adding the number of positive charges associated with the oxide to the number of accumulated holes. That is,

$$N_g = N_{IT} + N_H, \qquad (3.25)$$

where N_g is the charge density on the backside (cm^{-2}), N_{IT} is the effective number of charges at the Si–SiO$_2$ interface (cm^{-2}), and N_H is the number of accumulated holes in the CCD (cm^{-2}).

Equation (3.25) is equivalent to

$$N_g = \frac{C_{OX}}{q}(V_G - V_S), \qquad (3.26)$$

where C_{OX} is the capacitance of the native oxide layer (F/cm^2).

Example 3.11

Referring to Fig. 3.53, estimate the number of electrons at the backside of the CCD required to achieve the QE-pinned condition. Assume a native oxide thickness of 16 Å and $V_{oq} = 0.1$ V due to positive charge at the interface.

Solution:
From Eq. (1.4), the capacitance associated with a native oxide thickness of 16 Å is

$$C_{OX} = 3.45 \times 10^{-13}/1.6 \times 10^{-7},$$

$$C_{OX} = 2.1 \times 10^{-6} \, \text{F/cm}^2.$$

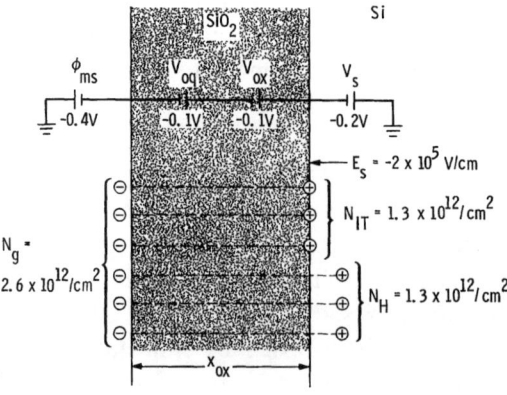

Figure 3.53 Backside charge required to QE pin a CCD.

An electric field greater than $E_S = -2 \times 10^5$ V/cm is required to achieve the desired accumulation state without oxide charge. This condition corresponds to a surface potential of $V_s = -0.2$ with an oxide drop of $V_{OX} = -0.1$ V (see Fig. 3.41). The required voltage on the backside is therefore 0.4 V, because an additional 0.1 V is required to neutralize positive oxide charges.

From Eq. (3.26),

$$N_g = 2.1 \times 10^{-6}/1.6 \times 10^{-19}(0.4 - 0.2),$$

$$N_g = 2.6 \times 10^{12} \text{ charges/cm}^2.$$

This is equivalent to approximately 6×10^6 charges/pixel for a 15-μm pixel. This number is considerably less than one monolayer of charge, assuming that the total number of possible states for the electrons that occupy the surface is 10^{15} sites/cm² or approximately 2×10^9 sites/(15-μm pixel). This example shows the extreme sensitivity of surface charging conditions. The CCD can switch from the deep depletion state (very low QE) to the accumulated state (QE-pinned) when charged with only 6×10^6 electrons/pixel.

For good charging characteristics it is important to help Mother Nature grow an oxide layer. Several attempts were initially made to develop a growth technique that proceeded more rapidly than native oxide growth. As mentioned above, mature native oxides can take several years before the oxide charge voltage, V_{oq}, is sufficiently small compared to V_G. Many different schemes for oxide growth have been investigated, including wet processes (such as sulfuric acid, nitric acid, and peroxide), numerous dry ambients (such as O_3, high-pressure H_2, and O_2), ionized H and O ambients, electrochemical processes, and laser annealing, all of which failed to produce the desired oxide layer (i.e., Mother Nature won out over these methods). Ironically, one of the last procedures tried turned out to be the simplest. This method is referred to as "flash oxide" growth.

The growth rate and quality of an oxide are determined primarily by the concentration of the oxidizing species and growth temperature. For native oxide growth in ambient conditions, where oxygen is the reactive species, the rate is extremely slow after the initial growth phase of about 10 Å. Thicker and higher-quality oxides are commonly grown by oxidation at very high temperatures. Unfortunately, high-temperature processes cannot usually be applied after the CCD is fabricated and thinned (and sometimes even packaged). In the growth of flash oxide, water, which oxidizes silicon much faster than oxygen, is the reactive species, with the following reaction taking place:

$$Si + 2H_2O \rightarrow SiO_2 + 2H_2. \qquad (3.27)$$

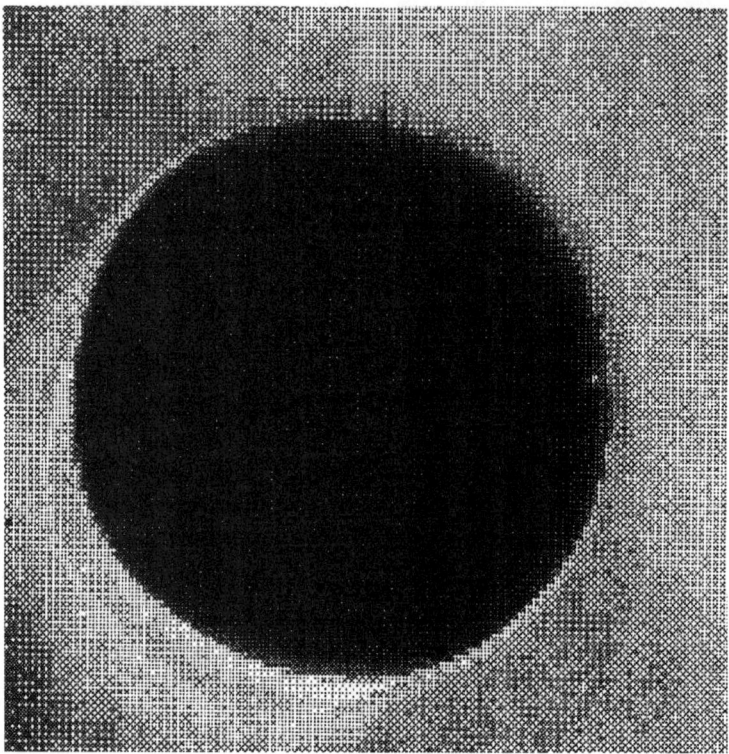

Figure 3.54(a) UV flood response after spot thinning a small region on a CCD.

The actual mechanism of this reaction is complex and involves many intermediate steps in which bonds in both the oxide and silicon are broken, rather than just those in the silicon lattice as occurs in the dry oxygen reaction.

One of the first successful flash oxide experiments performed is shown in Fig. 3.54. Figure 3.54(a) shows a 4000-Å flat-field image taken from a UV-charged CCD. Before charging, a small region on the array was spot thinned with a mixture of nitric, hydrofluoric, and acetic acids [seen as the black area in Fig. 3.54(a)]. Figure 3.54(b) is a near-IR line trace response at 1 μm, showing that approximately 1 μm of silicon material was removed in the process (i.e., approximately 7 fringes are counted). Figure 3.54(c) shows a UV flood response after a flash oxide was grown by applying one drop of deionized water into the "cupped" region (the water was allowed to evaporate under ambient conditions). Note that enough oxide has grown to support backside charge better than surrounding regions whose oxide was damaged by HF fumes. Figure 3.54(d) compares discharge qualities for the flash oxide grown. The discharge rate is a measurement of oxide quality. The more rapidly a CCD discharges after charging, the less mature is the oxide. Although the flash oxide region exhibits some charging characteristics, the five-year-old oxide is still superior in this case.

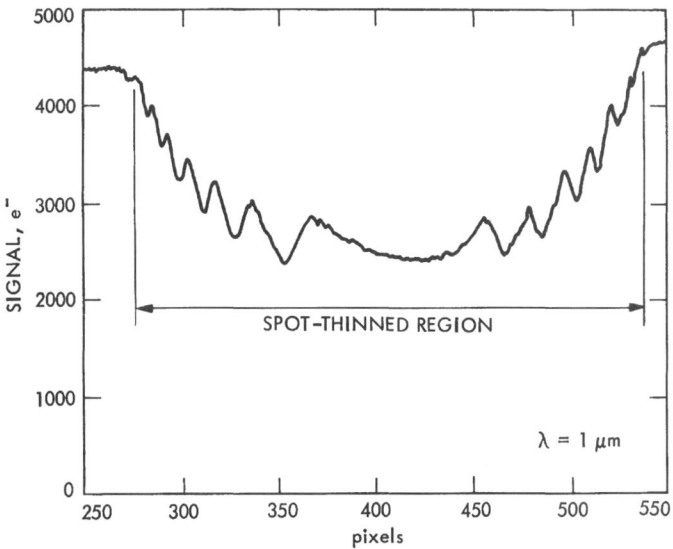

Figure 3.54(b) Interference pattern within the spot-thinned region showing approximately 7 fringes.

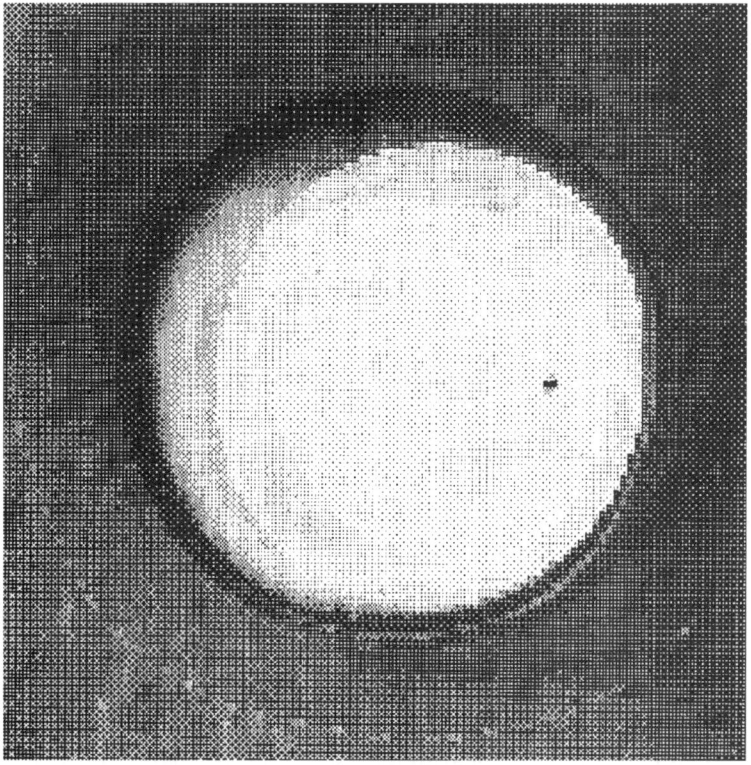

Figure 3.54(c) UV flood response after flash oxide is grown.

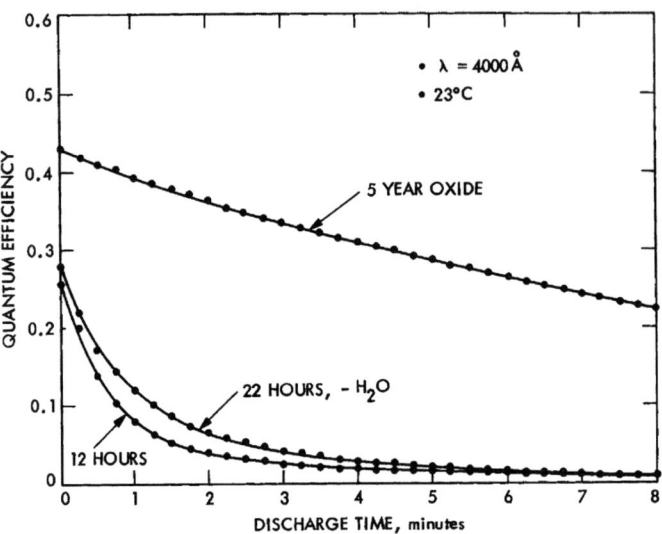

Figure 3.54(d) Discharge characteristics of flash oxide compared to native oxide.

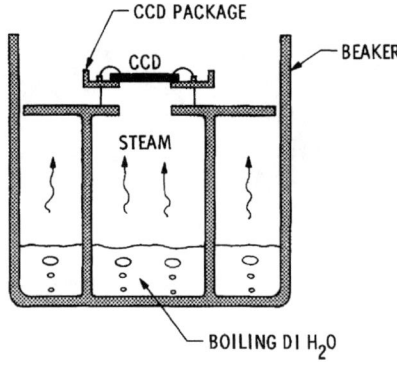

Figure 3.55 Rudimentary setup to grow flash oxide at higher temperatures.

The flash oxide grows much more quickly at elevated temperatures. For example, Fig. 3.55 shows a crude but effective way to grow flash oxide at 100°C. The steam method produces an oxide that is usually superior to the five-year-old native oxides within a few minutes of growth. For example, Fig. 3.56 shows a family of 4000-Å QE curves taken from a recently thinned CCD with a flash oxide grown with steam. The ability of an oxide to hold a UV charge at room temperature increases with oxidation time, until after 62 min when no apparent discharge occurs.

Figure 3.57(a) plots discharge characteristics as a function of flash oxide thickness taken with ellipsometry measurements. As before, these represent room temperature discharge curves, taken immediately after 30 sec of UV flood. The CCD tested was overly thinned into the epitaxial region without p^+ material. For a min-

Charge generation

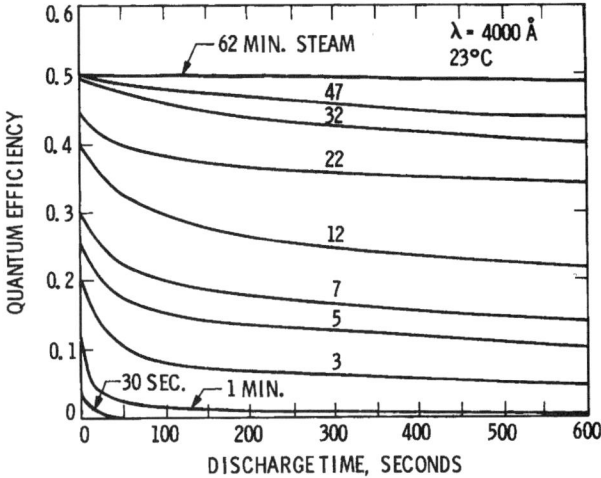

Figure 3.56 Discharge characteristics for a flash oxide grown with steam.

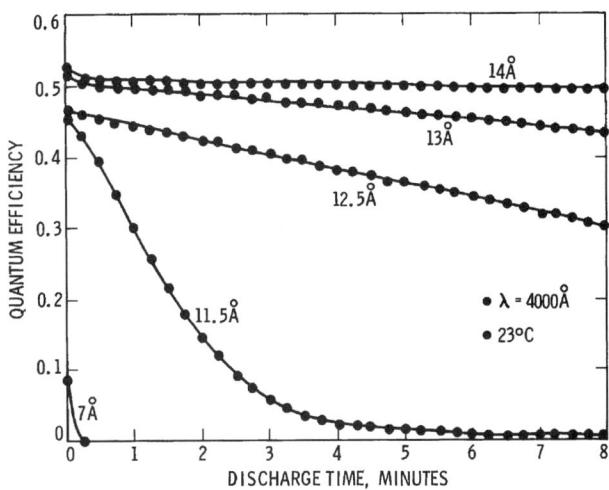

Figure 3.57(a) Discharge characteristics for different flash oxide thicknesses.

imal oxide thickness, the discharge rate decreases dramatically, perhaps due to a decrease in tunneling probability through the oxide layer. Full UV flood QE transfer curves are presented in Fig. 3.57(b) for a different backside-illuminated CCD as a function of oxide growth.

Today, environmental chambers are used to grow flash oxides where the temperature and humidity are much better controlled. Process furnaces have also been used to grow high-temperature (400°C) flash oxides. The quality of these oxides can QE pin a CCD for several months under ambient conditions after charging.

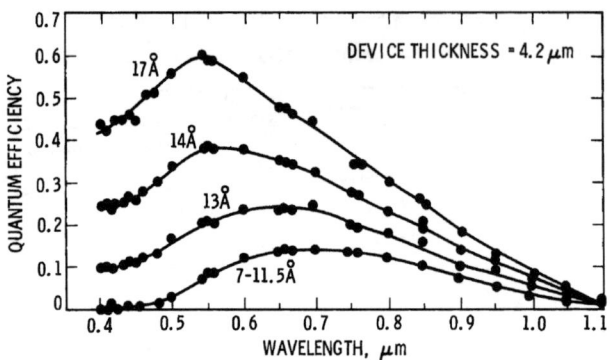

Figure 3.57(b) QE transfer curves for different flash oxide thicknesses.

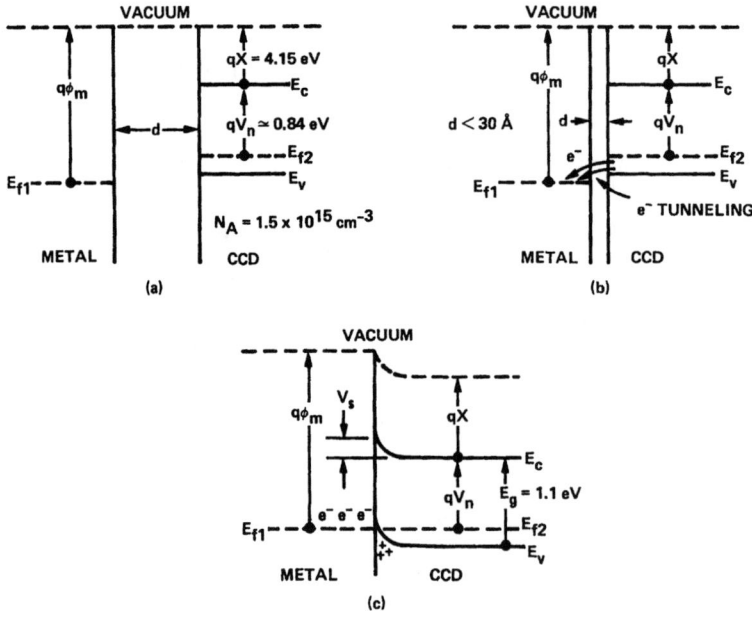

Figure 3.58 Energy band diagrams for a flash gate that promotes negative charge on the surface of the CCD.

3.4.6.2 Flash Gate

When a metal with a high work function (higher than silicon) makes intimate contact with the backside of the CCD, a contact potential develops which can generate an accumulation layer.[40–42] The work function measures the tenacity with which metals and other materials confine their electrons to themselves, i.e., the energy required to remove an electron from a material to vacuum. A high-work-function implies that electrons are kept close to the parent material. Figure 3.58 shows how this negative potential is generated. In Fig. 3.58(a), the CCD (less the oxide) and high-

CHARGE GENERATION

work-function metals are separated by distance d. Under these circumstances, the Fermi levels of the two materials do not coincide, and the system is not in equilibrium. When the metal and CCD are moved closer together as shown in Fig. 3.58(b), electrons will tunnel from the CCD to the metal when their physical separation is less than about 30 Å or about 6 interatomic distances. The flow of electrons creates a negative charge in the metal. This in turn induces an accumulation layer of holes in the CCD, as shown in Fig. 3.58(c). The accumulation layer generates an electric field within the CCD and raises the potential energy of the electrons with respect to those on the metal until the two Fermi levels coincide, at which point the tunneling process stops. The contact potential that develops after this current flow is given by the work function difference between the CCD and metal gate,

$$qV_S = q(\chi + V_n - \phi_m), \quad (3.28)$$

where $q\phi_m$ is the work function of the metal (eV), $q(\chi + V_n)$ is the work function of the CCD (eV), χ is the energy difference between the conduction band and vacuum level (4.15 eV), and V_n is the voltage difference between E_f and E_c given by

$$V_n = \frac{E_g}{2} + V_f, \quad (3.29)$$

where E_g is the energy bandgap of silicon (eV) and the Fermi voltage is given by Eq. (3.12).

The electric field, $E_A(x)$, generated by the flash gate is found by interchanging the backside charging voltage, V_G, with the work function difference of the CCD and metal.

Example 3.12

Calculate the surface potential generated by a platinum flash gate, assuming the following parameters: $q\phi_m = 5.6$ eV and $q(\chi + V_n) = 4.99$ eV. Also determine the electric field generated in the CCD using Fig. 3.40(b).
Solution:
From Eq. (3.28),

$$V_S = 5.6 - 4.99 = 0.61 \text{ V}.$$

A surface potential of 0.61 V is greater than any field condition shown in Fig. 3.40(b), and, therefore, would QE pin the CCD. However, oxide charge is not included in the calculation, which can reduce the field strength significantly.

In practice, the full contact potential developed by the flash gate does not drop entirely across the CCD, but in part develops across the native oxide layer. To understand the effects of the native oxide layer on the flash gate, the energy band diagrams shown in Fig. 3.59 are used. Figure 3.59(a) shows interface states at the Si–SiO$_2$ interface that are positively charged above the Fermi level (or when empty of electrons) and neutral when below the Fermi level (or filled with electrons). The presence of positively charged interface states will result in the formation of a surface depletion layer as already described (i.e., a backside well). As the metal approaches the CCD at a distance where tunneling probability becomes high, electrons will begin to tunnel from the interface states to the metal instead of from the silicon. If the density is high enough, the interface states will supply the necessary charge to establish the contact potential required to align the Fermi levels between the CCD and the metal. The condition is shown in Fig. 3.59(b). The resulting band structure within the silicon is essentially the same as before the metal contact, except that the bands may straighten out slightly by ΔV_S if some electrons are supplied by the silicon. Here, the contact potential that develops between the metal and CCD is dropped almost entirely across the native oxide instead of within the CCD. The backside remains in depletion as opposed to accumulation. As with backside charging, the success of a flash gate has inducing an accumulation layer is dependent on the quality of the flash oxide grown. It is standard practice to QE pin the CCD with UV flood before a flash gate is deposited. If the CCD can be QE pinned with UV flood, the flash gate will generally have no difficulty in accumulating the surface of the CCD.

A variety of gate materials and deposition techniques were originally used for the flash gate. Three vapor techniques were employed: sputtering, electron beam evaporation and thermal evaporation. Sputtering involves the creation of an ion-

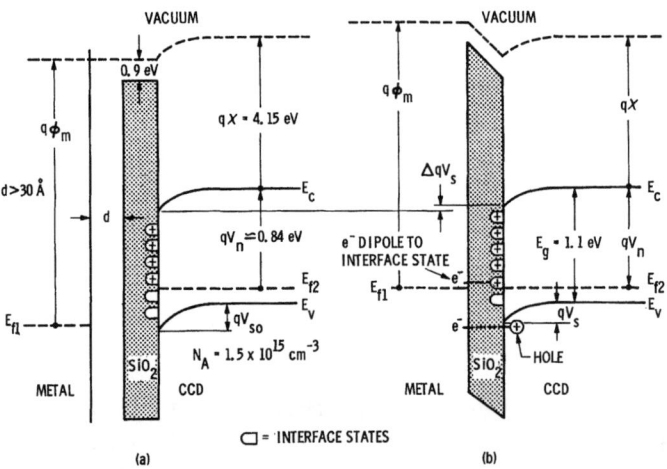

Figure 3.59 Energy band diagrams that illustrate how surface states negate the effect of the flash gate.

plasma that releases metal atoms from a target by collision. This method is useful for coating delicate specimens like the CCD, due to the omnidirectional scattering of material, that results in uniform films with a stationary sample. With this technique, a large negative potential (-3000 V is typical) is applied to the target, and an inert gas is leaked into the evacuated chamber, thereby producing a plasma contained by a magnetron system. The ionized gas molecules collide with metal target atoms, ejecting them out of the target. These metal atoms then coat the CCD back surface (through a suitable mask) which acts as an anode at ground potential. Another technique, electron beam evaporation, is more versatile and allows faster deposition rates, but requires rotation of the CCD to assure uniform coverage due to the smaller sources generally used. In this method, a boule of target material is heated by an electron beam which then releases atoms by evaporation. The result is deposition of low-energy atoms on the CCD surface.

Both sputtering and electron beam evaporation have been shown to induce radiation damage to the CCD; that is, an increase in the Si–SiO$_2$ interface density at the front and backside due to the generation of high-energy x rays (some of this damage will slowly anneal with time). Because of this problem, a third technique was used with simple resistance heating of the metal. In this method, a tungsten filament and metal are heated by applying a high dc current at a low voltage, which heats and evaporates the metal onto the CCD surface. While the thickness control is not as precise as sputtering or electron beam evaporation, tungsten evaporation by comparison is a benign process because no high-energy photons are involved. The flash gate films deposited are very thin (about a monolayer), which requires a brief surge of current through the filament to deposit the metal (hence the name, flash gate).

Three metals have been employed successfully for flash gates: platinum, gold and nickel, all of which have higher work functions than that of p-silicon. Of these candidates, experiments have shown that platinum is the best, because of its stability characteristics on silicon. Gold, for example, rapidly diffuses in silicon at room temperature, yielding a short-term QE-pinned response. It should be mentioned that the flash gate is very sensitive to contamination problems. For example, a Pt flash gate can be easily contaminated by metals that exhibit a work function less than silicon. For example, aluminum flash gates will deplete the CCD (i.e., promoting a backside well), giving free electrons to silicon and charging the gate positive.

Figure 3.60(a) shows a 4000-Å flat-field response taken from a Hubble WF/PC I CCD before thermally depositing 10 Å of platinum. The variations in QE seen in the image are due to irregularities in thinning and p$^+$ material left on the backside as explained above. The QE in the center region is only 5% because of a backside well (i.e., it is overly thinned). The corners of the device are thinned approximately to the frontside depletion edge. With the flash gate incorporated, as shown in Fig. 3.60(b), the CCD is nearly QE pinned, resulting in a high and uniform response.

The QE sensitivity of the flash gate is improved by depositing an antireflection coating to the backside after flash gate deposition. Figure 3.61 compares the

Figure 3.60(a) 4000-Å flat-field image before a 10-Å Pt flash gate is deposited.

performance of three backside structures employed on the same CCD: untreated, flash gate, and flash gate/antireflection coating. The antireflection coating used in this experiment (1135 Å of MgF_2 and 535 Å of ZnS) was tuned to approximately 5500 Å. A sensitivity of nearly 90% is achieved at this wavelength.

As with backside charging, the flash gate eliminates the QEH problem. Figure 3.62 shows a 4000-Å QEH image when a partial flash Pt gate is deposited on the CCD shown in Fig. 3.28(a). The lower right-hand corner is free of QEH where the 10-Å flash gate is applied.

3.4.6.3 Chemisorption Charging

Chemisorption charging is a popular technique that was developed by Michael Lesser of the University of Arizona.[43–47] For this method, backside charging occurs through the disassociation of oxygen or other gases promoted by the catalyst action of the metal film. In the case of oxygen, the disassociated molecules form O^- ions that are chemisorbed to the film, creating the required negative charge layer. The lifetime of ionic species such as O^- is many years at room temperature, producing a permanent backside charging mechanism. The strong chemical bonds of the ions to the metal film account for the mechanism's immunity to environmen-

Figure 3.60(b) 4000-Å flat-field image after a 10-Å Pt flash gate is deposited.

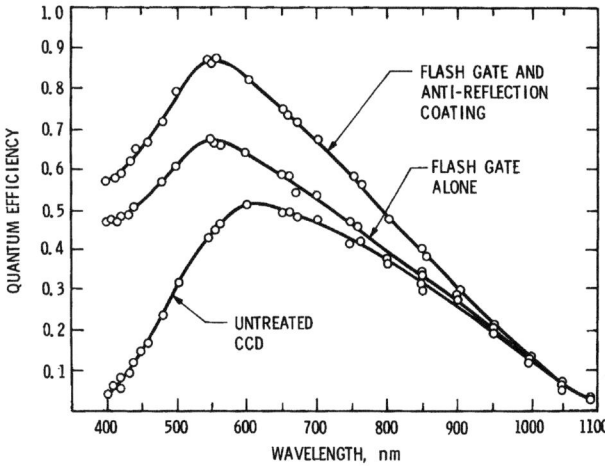

Figure 3.61 QE transfer responses before and after flash-gate deposition.

tal effects. QE pinning is achieved. Figure 3.63 shows the high QE obtained from this technique with a UV antireflection coating.

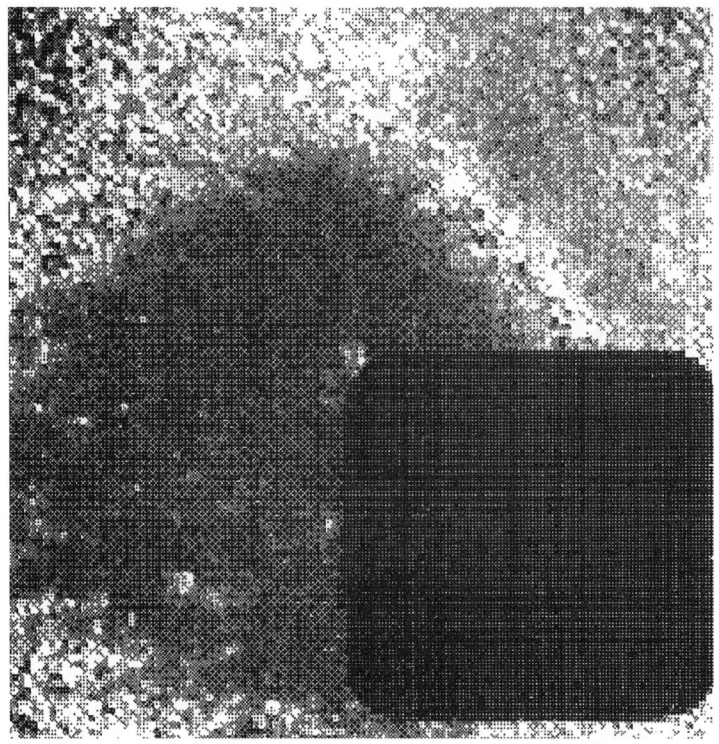

Figure 3.62 QEH behavior after a partial flash gate is deposited.

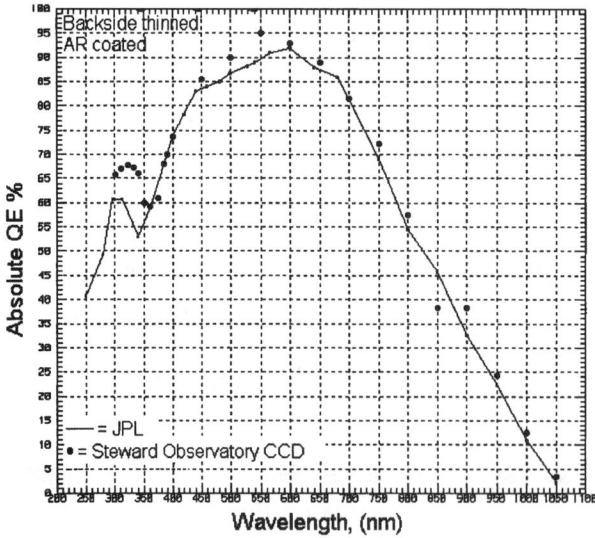

Figure 3.63 QE-pinned CCD that has been accumulated with the chemisorption charging process.

3.4.6.4 Biased Flash Gate

QE pinning cannot always reliably be achieved when a flash-gate CCD is subjected to high vacuum or dry inert gases.[48] Also, the platinum system is easily poisoned in the presence of hydrogen gas.[49] The reason for the high vacuum problem is not fully understood, but some theories offered point toward the oxide layer on which the gate resides. It is believed that oxygen and water vapor found in ambient environments play an important role in the structure of the oxide layer by participating in tying bonds at the Si–SiO$_2$ interface. Under high-vacuum conditions these constituents may be removed, which upsets the bonding equilibrium at the interface, resulting in more dangling bonds and positively charged interface states. An increase in surface state density effectively lowers the work function difference between the flash gate and CCD, thus reducing the strength of the accumulation layer and lowering QE performance.[50]

The biased flash gate was invented to compensate for the oxide voltage shift that occurs for the flash gate system. The biased flash gate is similar in structure to the flash gate, except that the platinum gate is made slightly thicker (approximately 20 Å as opposed to 5 Å) to assure that it is electrically conductive over the thinned membrane. The gate is then connected to an external bias voltage to control the surface potential. QE characteristics at 4000 Å for a platinum-biased flash gate are shown in Fig. 3.64(a). The test CCD is half-coated with lumigen. Note that QE pinning is fully achieved when a slight negative bias is applied to the gate, which promotes accumulation. A positive bias forces the device into depletion. The QE does not obtain a theoretical level of 50% (at 4000 Å) because some photons are absorbed by the biased flash-gate layer. Figure 3.64(b) shows line traces through the middle of a biased flash-gate CCD as a function of gate bias. For this CCD the QE-pinned condition is achieved when the gate is biased to approximately −0.2 V.

Figure 3.65(a) plots QE characteristics in the far UV for a QE-pinned Hubble WF/PC-I CCD that has been processed four different ways: bare region, a lumigen-coated bare region, 40-Å biased flash gate, and a region with the biased flash gate and lumigen. Measurements were performed at an operating temperature of −95°C. The biased flash gate alone achieves the highest QE compared to the other three regions. The bare silicon lacks sensitivity because of a backside well. The lumigen region shows increased sensitivity as some 5600-Å photons make it beyond the backside well. For the lumigen/biased-flash-gate region, most photo-generated electrons are collected; however, half the down-converted photons are emitted away from the CCD. Therefore, QE achieved is about 50% below what is theoretically possible (30% at 2500 Å without multiple e-h generation). In theory, the QE of the biased flash gate region should be higher than measured. The biased flash gate deposited on this test sensor was excessively thick (> 40 Å), and therefore shows some absorption loss. The sudden dip in QE at 1700 Å is due to a small amount of oxygen (< 1%) contained in the package, which is normally filled with pure argon gas (1/2 atmosphere). Also, some water ice in the package plays a role in lowering QE performance shortward of 1700 Å (e.g., the absorption length

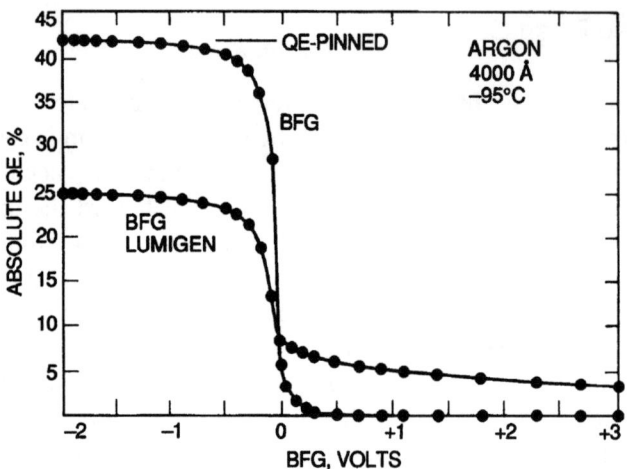

Figure 3.64(a) Biased flash-gate response at 4000 Å with and without lumigen.

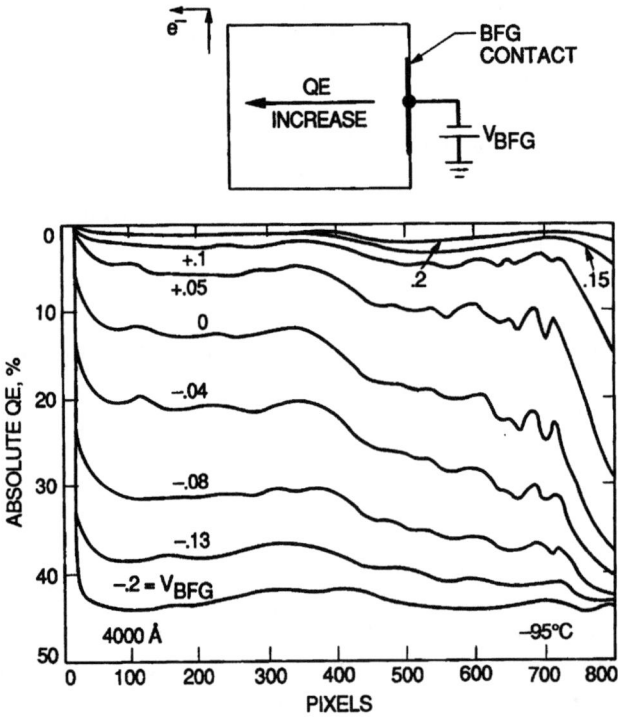

Figure 3.64(b) Biased flash-gate line trace responses at different bias voltages.

of water ice at 1400 Å is only 500 Å). Figure 3.65(b) shows an improved QE response when both problems are rectified. Degradation below 1700 Å is caused by absorption loss of the MgF_2 window above the CCD.

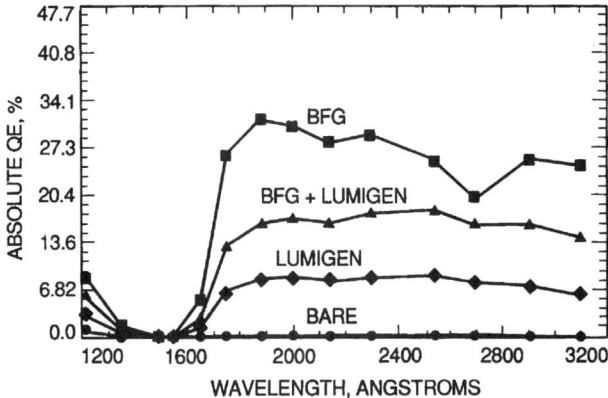

Figure 3.65(a) Extreme UV responses for four different processed regions on a single CCD.

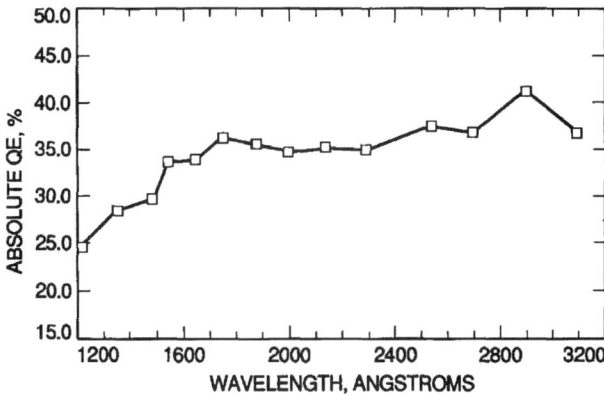

Figure 3.65(b) Extreme UV response for a biased flash-gate CCD.

It is interesting to note that the biased flash gate can be used as an electronic shutter and control QE performance. Figure 3.66(a) shows photon transfer data generated without a mechanical shutter in front of the CCD. The device was continuously exposed with 4000-Å light and electronically shuttered with the biased flash gate. The linearity data in Fig. 3.66(b) show $< 1\%$ nonlinearity for exposure periods < 1 sec. Electronic exposure control is only possible for wavelengths < 4500 Å. Long-wavelength photons penetrate beyond the backside depletion region that the biased flash gate provides.

3.4.6.5 Biased Ohmic Contact on High-Resistivity Substrate

The QE curves presented in this chapter all show a sharp decline in sensitivity as we pass from the visible into the near IR or from the soft x-ray into the hard x-ray. This

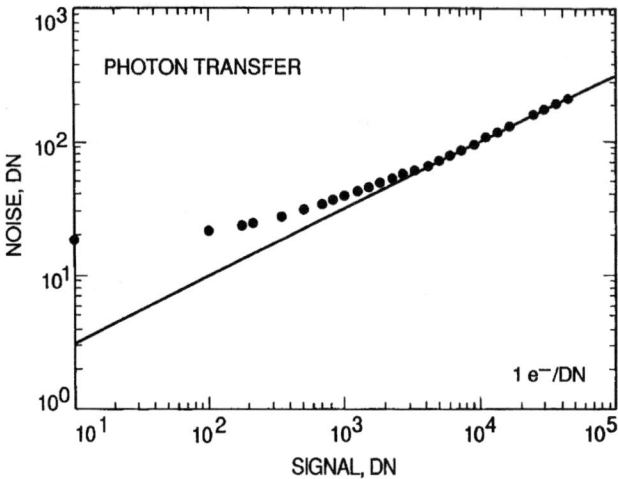

Figure 3.66(a) Photon transfer using the biased flash gate as an electronic shutter.

behavior, of course, is because these photons do not interact but instead penetrate the epitaxial layer, or the thinned membrane in the case of backside-illuminated CCDs. The need for improved near-IR and high-energy x-ray sensitivity has led to a class of sensors called deep depletion CCDs.[51–54] For this technology, the photosensitive thickness is considerably greater than conventional CCDs, allowing long- and short-wavelength photons to interact. Recall from Chapter 1 that depletion depth increases by the square-root of silicon resistivity. Therefore, to fabricate very deep potential wells, high-resistivity silicon must be used.

Deep depletion CCDs have been produced with a depletion thickness as great as 300 μm.[55,56] In order to fully deplete such a large thickness, very high resistivity ($> 10,000$ Ω-cm), n-type silicon is used. This type of silicon is commonly used in detectors for high-energy physics. For that application, the basic detector unit is a fully depleted p-i-n diode. The p-i-n photodiode is advantageous because the depletion-region thickness (i.e., the intrinsic layer) can be designed to optimize the QE especially in the near-IR. Figure 3.67(a) shows a cross-sectional view of this type of CCD. A unique feature of this technology is the application of a bias voltage to the ohmic contact on the backside of the wafer, which corresponds to the n region of the p-i-n structure. This bias voltage can fully deplete the entire 300-μm thickness. As a result, the CCD has good QE in the near IR and interference fringing associated with thin-membrane CCDs is eliminated. QE performance is shown in Fig. 3.67(b). Since the wafer is relatively thick and self-supporting, the backside ohmic contact layer can be put on the wafer before metallization. Therefore, concerns about process compatibility with the aluminum metallization are eliminated. The backside ohmic contact used in the CCD of Fig. 3.67(a) is formed by the LPCVD deposition of a thin layer of in situ doped (phosphorus) polysilicon at 650°C. Because the layer is thin (about 20 nm), the blue QE response is also high.

CHARGE GENERATION 253

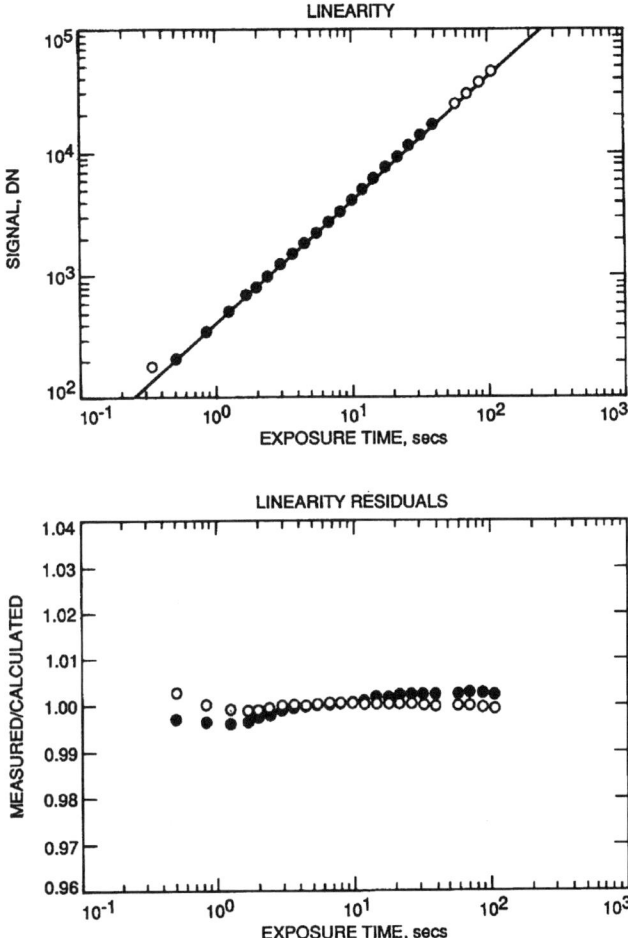

Figure 3.66(b) Linearity curves using the biased flash gate as an electronic shutter.

An indium tin oxide (ITO) layer can also be deposited to the n^+ polysilicon layer to reduce the sheet resistance of the backside layer, and to act as the first layer of the AR coating.

Since the substrate thickness is large compared to the thickness of the buried channel/gate dielectric, the capacitive coupling from the substrate bias voltage to the CCD channel is small, due to the capacitor divider effect. As a result, the CCD clock voltages can be set independently of the substrate bias voltage, which simply provides full depletion of the substrate. Conventional deep depletion CCDs, as used for x-ray astronomy, rely on gate bias alone to determine the depletion depth.

The main concern for very deep depletion CCDs is MTF degradation. However, by selecting a proper thickness and bias voltage, high MTF can be maintained.

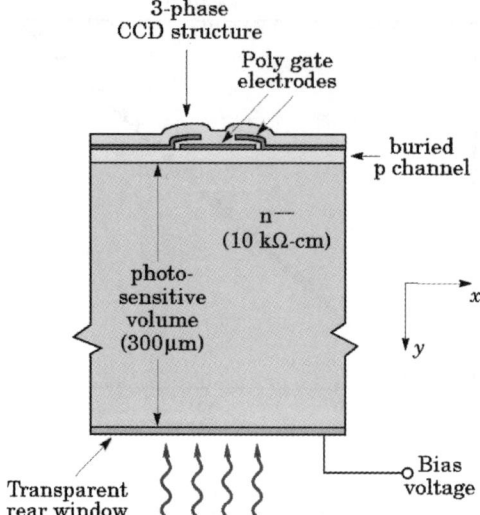

Figure 3.67(a) Cross-section of a deep-depletion CCD with external substrate bias.

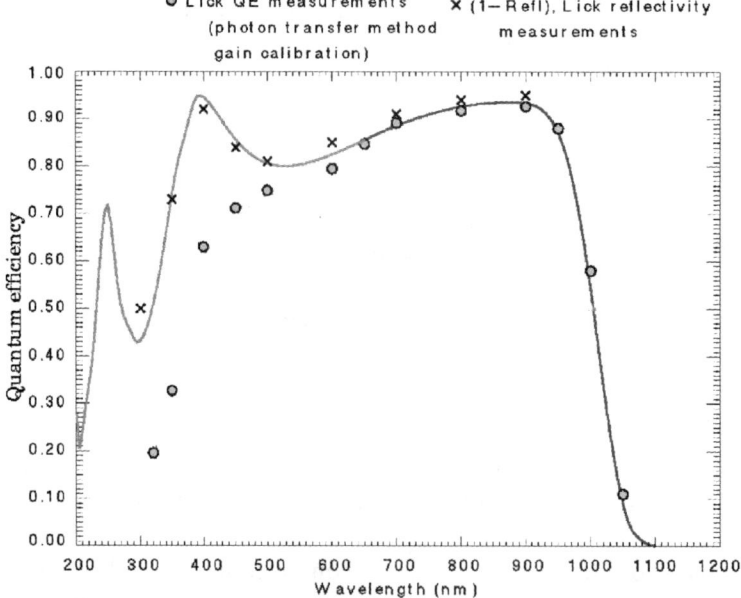

Figure 3.67(b) QE transfer for a deep depletion CCD exhibiting high near-IR sensitivity.[57]

The problem is caused by charge diffusion during transit of the photo-generated holes from the backside to the CCD potential wells. The effect is minimized by the drift electric field which can be adjusted by the bias applied to the CCD. The rms

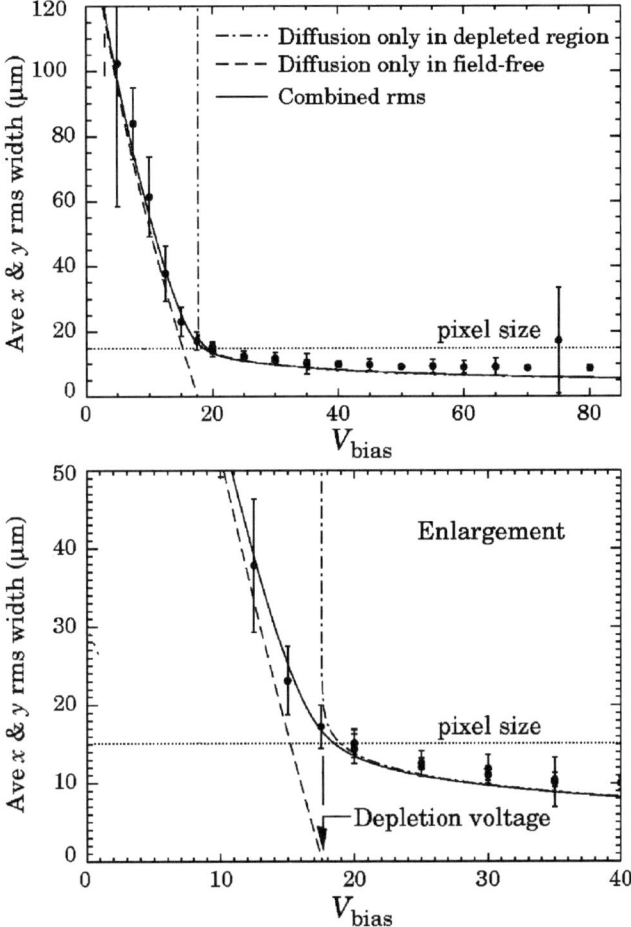

Figure 3.67(c) Point spread as a function of substrate bias.[58]

standard deviation of charge spreading, σ_D, is given by

$$\sigma_D = (2D_P t_R)^{1/2}, \qquad (3.30)$$

where D_P is the diffusion constant (cm²/sec) for holes and t_R is the transit time (sec) given by

$$t_R = \frac{x_{DEP}^2}{\mu_P V_{SUB}}, \qquad (3.31)$$

where x_{DEP} is the thickness of the depletion region, μ_P is the mobility for holes, and V_{SUB} is the substrate voltage across the device.

Making use of the Einstein relation $D_P/\mu_P = kT/q$ and substituting Eq. (3.31) into Eq. (3.30) yields

$$\sigma_D = \left(\frac{2x_{DEP}^2 kT}{q V_{SU}}\right)^{1/2}. \tag{3.32}$$

Example 3.13

Estimate the charge spreading that would occur for a 300-μm depletion region biased at 100 V. Assume 300 K operation.
Solution:
From Eq. (3.32),

$$\sigma_D = \left[2 \times (3 \times 10^{-2})^2 \times 0.025/100\right]^{1/2},$$

$$\sigma_D = 6.71 \times 10^{-4} \ (6.7 \, \mu m).$$

Figure 3.67(c) shows theoretical and experimental data of point spread versus bias voltage. For this nominal 280-μm-thick CCD operated at −120°C, the sigma due to charge diffusion during drift is about 10 μm. The 4000-Å light was illuminated through a pinhole mask placed directly on the backside of the CCD (which is the worst case with absorption near the backside contact). The plots show separate curves for diffusion in the depleted and field-free regions. Note that a bias voltage of approximately 19 V is where full depletion occurs (i.e., no field-free material).

3.4.7 ACTIVE ACCUMULATION

Active accumulation methods discussed below include (1) ion-implantation, (2) molecular beam epitaxial (MBE) and (3) chemical vapor deposition (CVD) accumulation. Self-accumulation, already discussed earlier, is also considered an active accumulation technique but, as mentioned above, is not utilized today.

3.4.7.1 Ion Implantation

It was shown earlier that the doping concentration for self-accumulation was insufficient in providing a strong electric field required to direct signal charge away from the backside. To generate stronger fields, groups accumulate the surface of the CCD by ion implantation.[59–63] In theory, ion implantation can QE pin the CCD. However, the primary difficulty with this method is that the energetic ions that are implanted induce lattice damage. Such damage induces charge traps, which lowers

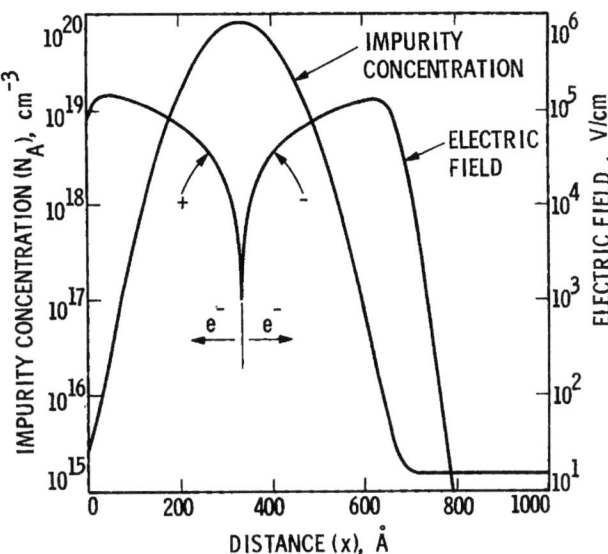

Figure 3.68(a) Ion-implantation doping profile as a function of distance and the electric field induced.

QE performance and generates high levels of backside dark current. Therefore, after implantation, the implant must be "activated" or annealed to repair the silicon lattice structure to its original state. Ideally, an activation temperature of 1000°C is required for full annealing. However, such a high processing temperature will melt the metal bus lines and pads located on the front of the CCD. To avoid this problem, fast annealing techniques such as rapid thermal, laser, and electron beam have been used. Other groups thin, implant, and anneal the CCD wafer before metal is deposited. Still other groups rely on a partial anneal, producing a sensor that is not fully QE-pinned and a backside dark current that usually dominates frontside dark current (especially MPP CCDs).

Figure 3.68(a) shows a 10-keV boron implant into single-crystal silicon before activation as a function of distance. The impurity distribution is given by

$$N_A(x) = \frac{Q_o}{(2\pi)^{1/2} \Delta R_P} e^{-0.5[(x-R_P)/\Delta R_P]^2} + 1.5 \times 10^{15}, \qquad (3.33)$$

where Q_o is the dose (1.5×10^{14} atom-cm^{-3}), ΔR_P is the straggling coefficient (7×10^{-7} cm), and R_P is the projected range of the ions (3.33×10^{-6} cm), the latter two terms being implant energy dependent. The last term represents the epitaxial layer doping (1.5×10^{15} in this case).

The implant distribution shown has a maximum range at R_P and falls off rapidly on either side of this value. Also shown in Fig. 3.68(a) is the electric field generated by the implant found by Eq. (3.6). Note that only a portion of the implant profile ($x > 330$ Å) is useful because the field provided by the first half ($x < 330$ Å)

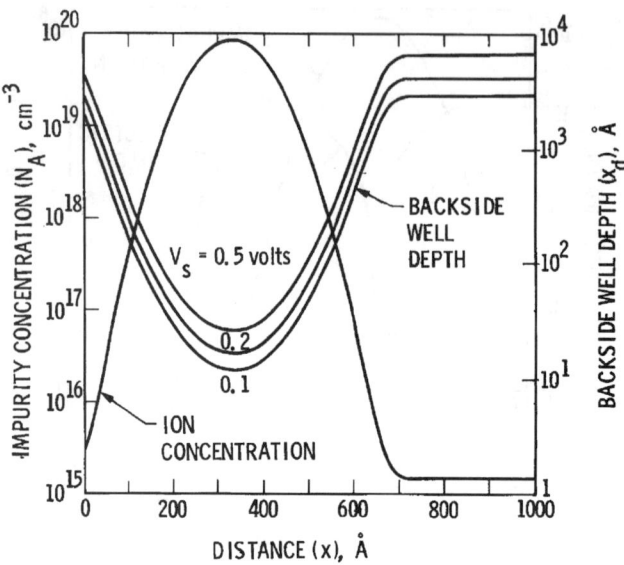

Figure 3.68(b) Backside well depth as a function of distance for an ion-implanted CCD.

directs signal electrons in the wrong direction (i.e., toward the backside). Also, the region between $330 < x < 500$ Å does not generate a field sufficient to stop carriers from diffusing to the backside. To the extent that it is impractical to thin away these unwanted regions of the implant profile, the only alternative is to reduce the projected range, R_P, to the point where the first half of the profile is negligibly small compared with the photon absorption length or thickness of the native oxide layer. However, this represents a very shallow implant if the CCD is to be sensitive in the UV regime. Implants at these depths are difficult to achieve in practice because of the relationship between penetration depth R_P and the anneal process, which modifies the implant profile and lowers the corresponding electric fields. Nevertheless, some backside manufactures are based on ion implantation to accumulate their CCDs. Unfortunately, the QE-pinned condition is rarely achieved.

Figure 3.68(b) plots backside well depth as a function of distance through the implant gradient for various surface potentials. Note that as the doping concentration increases, a smaller backside well is produced. For example, a backside well depth of approximately 20 Å is calculated at the peak doping concentration (10^{20} cm^{-3}) for a 0.2-V surface potential. Although this well depth may seem small, it is still large enough to affect QE performance in the UV.

3.4.7.2 Delta Doping

Delta-doped CCDs were invented at JPL in 1992[59] for producing backside-illuminated CCD with an ideal QE response, (i.e., QE-pinned). The term delta doping refers to modification of the backside band structure of a CCD by epitaxial growth of a highly doped layer of silicon on the back surface of a thinned CCD. Delta-doping provides an ideal electronic potential near the CCD back surface. Despite the fact that only a few atomic layers are involved, the delta-doped layer has a dramatic effect on the quantum efficiency because of its effect on the electronic band structure near the surface. In contrast to ion implantation, the MBE growth process places an extremely high density of dopant atoms ($> 10^{14}$ boron/cm^2) within a few atomic layers of the surface with no observable crystal defects and no requirement for postgrowth annealing. Because this delta-doped layer is a permanent part of the silicon crystal structure, the resulting improvement in CCD performance is stable and insensitive to environmental conditions. Delta-doping is versatile and compatible with most CCD fabrication and metallization processes, as demonstrated by successful adaptation of the delta-doping process to modify CCDs obtained from several CCD manufacturers.

Figure 3.69 shows a schematic of a delta-doped epitaxial layer grown on the back surface of a thinned CCD using molecular beam epitaxy (MBE). In preparation for MBE growth, the backside surface of the CCD is carefully cleaned using a multistep procedure.[60,61] The epitaxial layer is then grown in the MBE system under ultrahigh-vacuum conditions (10^{-10} torr) using electron-beam evaporation of elemental silicon and thermal evaporation of elemental boron. The CCD temperature is held below 450°C during the procedure. A 1-nm-thick p$^+$ silicon layer is grown first, followed by depositing $\sim$30% of a monolayer of boron atoms. A 1.5-nm-thick capping layer of epitaxial silicon is then grown. After removal from the MBE system, oxidation of the silicon capping layer protects the buried delta-doped layer. Characterization of the structure with high-resolution transmission electron microscopy has not shown any structural defects in the MBE-grown epitaxial layer.[62]

Delta-doped CCDs have been characterized in the visible and near-UV (250–750 nm) region of the spectrum in several different measurement setups. Because the photon absorption depth is shortest at 270 nm (only $\sim$ 4 nm), the quantum efficiency measured at this wavelength offers the most stringent test of the backside treatment. It is expected, therefore, that the delta-doped CCD should perform at the theoretical detection limit below 250 nm because of the longer photon absorption depth for these wavelength. QE measurements of a delta-doped CCD have been made into the far-UV region (121.6–310 nm).[62,63] Figure 3.70 displays the results of these measurements, along with measurement in the 250- to 700-nm waveband. The quantum efficiency measurements in the 250–700-nm waveband and the 121.6–310-nm waveband were performed in two different experimental setups. Figure 3.70 also shows the extension of characterization of delta-doped CCDs into the extreme ultraviolet region of the spectrum using the synchrotron source at the Stanford Synchrotron Radiation Laboratory (in collaboration with Professor Art Walker). The solid line in the figure is the transmittance of silicon with a 20-Å layer of oxide and represents the maximum quantum efficiency that can be achieved without antireflection coatings. As can be seen, the delta-doped CCD shows 100% internal quantum efficiency in the entire EUV through the visible region of the spectrum.

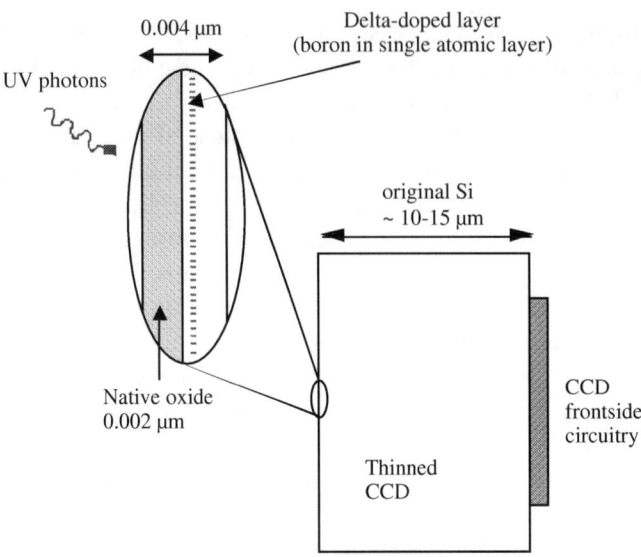

Figure 3.69 Cross section schematic of a delta-doped CCD. The epitaxially grown delta-doped layer on the back surface of a thinned CCD places a high density of boron atoms approximately 0.5 nm below the silicon epilayer surface and protected by an oxide overlayer.

Because the only measurable loss in delta-doped CCDs is due to reflection of photons from the back surface, reducing the reflection loss can produce devices with overall quantum efficiency near 100% over various regions of the spectrum. The feasibility of direct deposition of antireflection (AR) layers on delta-doped CCDs was demonstrated by depositing HfO_2 films to enhance the quantum efficiency in two different regions of the spectrum (the regions around 270 nm and 300–400 nm).[62] HfO_2 layers were deposited in the CCD laboratory at the University of Arizona in collaboration with Dr. M. Lesser, using resistive heating to evaporate the HfO_2 (resistive heating avoids exposure to damaging x rays encountered in e-beam evaporation). Using a shadow-masking technique, two separate 3×5 mm^2 areas were selected for the deposition of the two films on the same CCD. Figure 3.71 shows the response of a delta-doped CCD in the AR-coated and uncoated areas, together with the theoretical uncoated silicon transmittance curve. Because these are single-layer coatings, the AR-coated response will qualitatively follow the characteristic peaks in the Si reflection response. The response of the AR-coated regions shows the expected enhancement in the quantum efficiency, indicating that the AR coating process did not disrupt the δ-doped layer. Facilities at JPL's Microdevices Laboratory allow for in-situ application of AR coatings on CCDs immediately after growth of the delta-doped layer, without disruption of vacuum.[64] This capability makes it possible to deposit coatings on the back surface of the delta-doped CCD before the absorbing native oxide is formed, which is significant in the far-UV region of the spectrum.

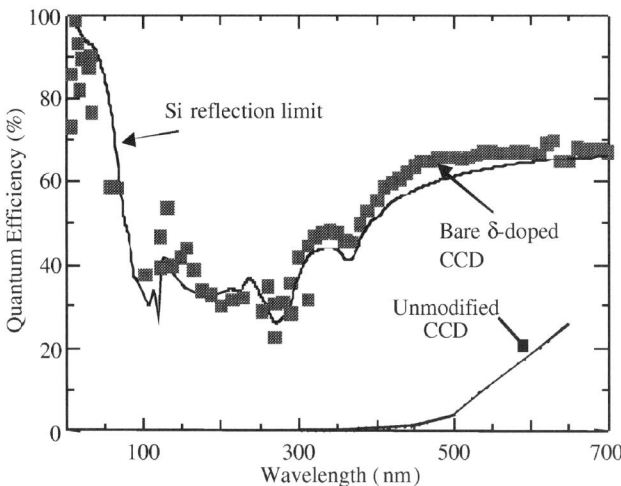

Figure 3.70 Measured quantum efficiency of a delta-doped Loral 1024 × 1024 CCD. The QE measurements in the 250–700-nm region and 120–310-nm region were taken in two different systems. The extreme-UV measurements (down to 6 nm) were performed at Stanford Synchrotron Radiation Laboratory. The solid line is the transmittance of silicon with a 20-Å layer of oxide, and corresponds to the maximum quantum efficiency possible without the addition of an antireflection coating. For ease of interpretation, the data below 200 nm has been corrected by dividing the measured response by the quantum yield, which we have approximated with the ratio $E(photon)/3.65$ eV, where $E(photon)$ is the energy of the incident photons measured in electron volts. With this convention, the maximum quantum efficiency is 100% at all incident energies.

Figure 3.72 shows the QE stability of a delta-doped CCD over the time period of years. QE measurements were performed on a delta-doped CCE at approximately one-year intervals. During the interval between QE measurements, the device was stored in air in an antistatic box with no further protection. The device has been subjected to repeated temperature cycling and exposed to different environments. No change from the ideal UV response has been observed within the accuracy of these measurements. In addition to long-term stability, QE hysteresis has also been examined by performing the standard test of measuring the UV response of the CCD before and after long exposure to ultraviolet light. No measurable hysteresis was observed.

Figure 3.73 shows the results of QE measurements on four different delta-doped CCDs (SITe 1100 × 330, SITe 512 × 512, Reticon 1024 × 1024, and Loral 1024 × 1024-pixel devices). Each device was observed to respond at the reflection-limit level and agree with silicon transmittance calculations (silicon transmittance is shown in the figure as a solid line).

Figure 3.74 shows a horizontal x-ray transfer response of a delta-doped CCD. The strong signal levels indicate good CTE performance. It can be seen that very

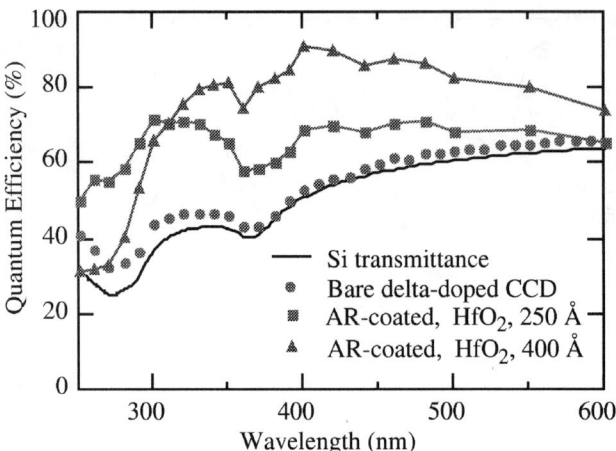

Figure 3.71 Enhancement of QE by the addition of antireflection coatings optimized for the 270- to 400-nm wavelength regions.

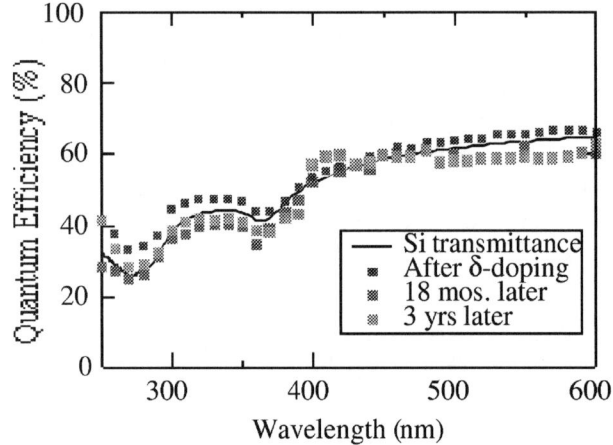

Figure 3.72 QE measured over a three-year period on the same delta-doped 512 × 512-pixel CCD. The CCD was stored unprotected and at room temperature. The CCD responds at the reflection limit in all three measurements.

low energy x rays such as Al K_α are detected. This is attributed to the fact that the backside well has been eliminated and the surface is QE-pinned. Additional tests were performed with carbon and fluorine x rays. The fluorine and carbon x rays exhibit very short absorption lengths in silicon, and, therefore, produce photoelectrons that are very vulnerable to backside recombination effects. The tests show that no significant surface recombination was occurring.

The test results in Fig. 3.74 are particularly significant for another reason. The CCD is exposed to Si K_α x rays during the MBE process by electron-beam evap-

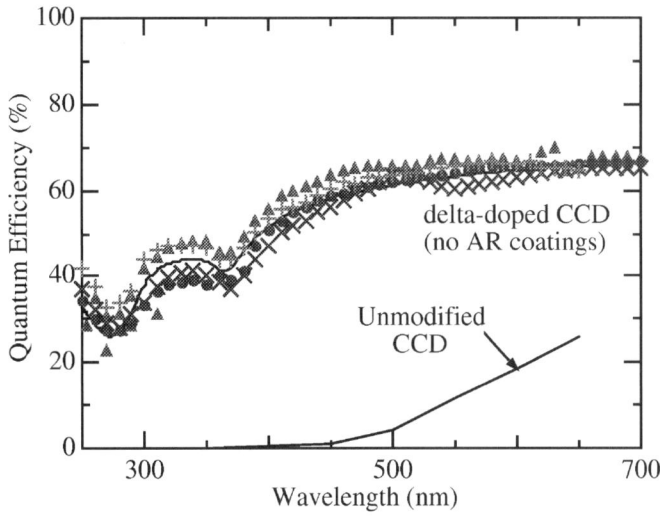

Figure 3.73 Quantum efficiency measurements of four different delta-doped CCDs (SITe 001, SITe 502, Reticon 1024 × 1024, Loral 1024 × 1024) modified at different times. The solid line is the bare silicon reflection limit.

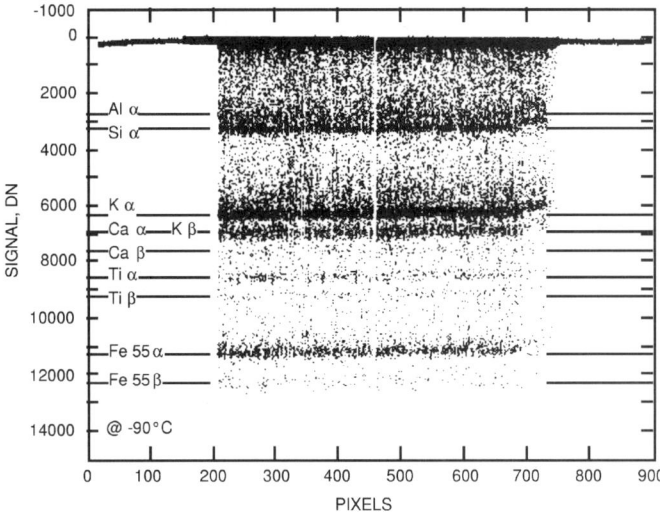

Figure 3.74 X-ray line-trace response generated by a delta-doped 1024 × 1024 pixel Loral CCD.

oration of silicon. The frontside gates of a thinned CCD are exposed to approximately 5 krads of Si K_α x rays during this process. Excessive exposure to Si K_α x rays produces radiation damage which could result in degraded CTE and higher dark current levels. These tests verify that this exposure does not degrade device performance.

The uniformity of a delta-doped 512 × 512 pixel Reticon CCD is demonstrated by the flat-field image in Fig. 3.75. The flat-field response of this delta-doped CCD shows good QE uniformity with only a few blemishes, well within the normal range for high-grade CCDs.

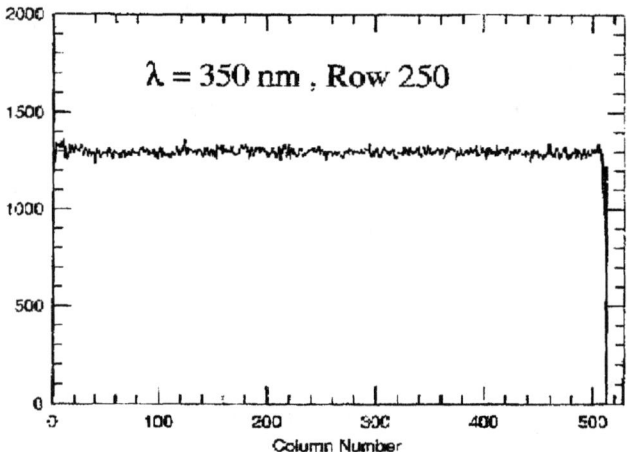

Figure 3.75 Flat-field image taken at 350 nm. The blemished area around the device is possibly indicative of the presence of fixtures during the process. Data taken at Reticon.

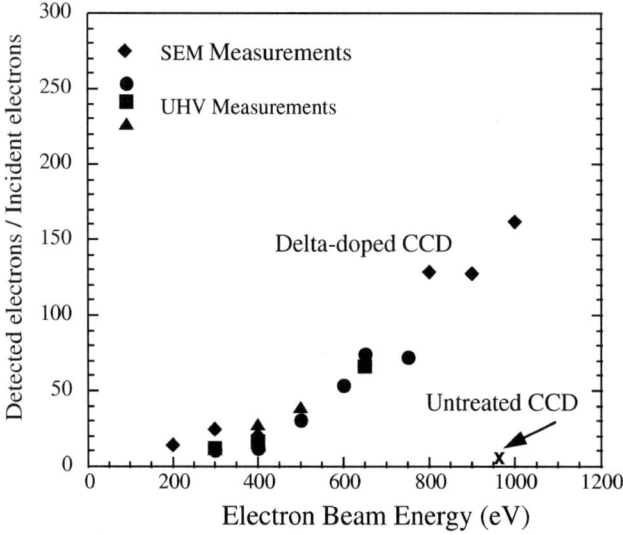

Figure 3.76 Quantum efficiency of delta-doped CCD with low-energy electrons using an SEM and an indirectly heated cathode gun. The increase in the QE with energy is the result of multiple electron-hole pair generation. The response of the untreated CCD has been plotted for comparison.

The UV response of delta-doped CCDs indicates that essentially all photoelectrons generated very near the back surface of these devices are detected. Because low-energy particles impart most of their energy near the surface, it was also expected that delta-doped CCDs would be good low-energy particle detectors. Experiments have shown that delta-doped CCDs exhibit an order of magnitude improvement over conventional solid state detectors in the low-energy detection cutoff for protons and electrons. Figure 3.76 shows the measured quantum efficiency of a delta-doped CCD in response to electrons in the range of 200–1000 eV.[70–72] These measurements were performed using an SEM and a custom UHV chamber equipped with an indirectly heated cathode electron source. Data in the electron energy range between 50–200 eV were also obtained with a thermal electron source. Long exposure of delta-doped CCDs to electrons was also examined. To determine the effect of this exposure, the quantum efficiency stability of the delta-doped CCD in the UV was examined before and after exposure to electrons in the energy range 50–1500-eV at fluxes of approximately 1 nA/cm^2. Measurements before and after the exposure showed no degradation in the UV quantum efficiency, indicating no change in the delta-doped layer. These results show that delta-doped CCDs are good candidates for use in electron-bombarded CCD (EBCCD) applications.

Single-event proton (1.2–12 keV) testing was also performed using a delta-doped CCD.[73] The results of these measurements are shown in Fig. 3.77. The delta-doped CCD detects protons in an energy range over an order of magnitude below the current detection threshold of solid state detectors. It should be pointed out that the detection energy threshold of delta-doped CCDs is not yet clear because the proton source was limited to energies greater than 1 keV. However, these results

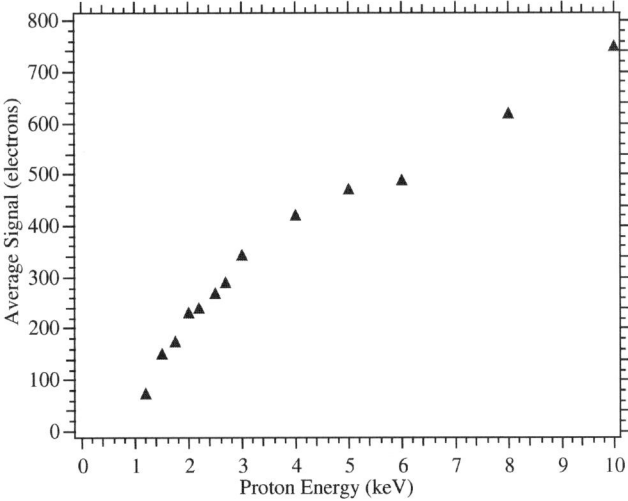

Figure 3.77 Low-energy proton detection with a p-type low-resistivity delta-doped CCD. The low-energy cutoff of 1.2 keV is approximately an order of magnitude improvement over the current solid state technology.

show that surface modification of the surface by delta doping greatly improves sensitivity to low-energy particles in terms of both the low-energy cutoff and the gain (quantum yield) of the detector. More work is underway for development of low-energy particle detectors using delta doping technology.

3.4.7.3 Chemical Vapor Deposition Accumulation

Chemical vapor deposition (CVD) techniques can be used to deposit a thin layer of oxide and accumulate the CCD at the same time. The deposition technique deposits oxide by the thermal decomposition and reaction of a gaseous compound. For example,

$$SiH_4 + 2O_2 \rightarrow SiO_2 + 2H_2O, \quad t = 400 - 500\,C, N_2 \text{ (carrier gas)}. \quad (3.34)$$

The inclusion of a gas compound containing a dopant permits the incorporation of that impurity in the deposition of an oxide layer. For example, diborane (B_2H_6) will dope the oxide with boron (i.e., B_2O_3/SiO_2). The CCD is then annealed at 750°C, which drives the boron into the silicon. This process can potentially yield a very shallow p^+ ($< 30\,\text{Å}$) accumulation layer with concentration of 10^{20} cm^3 (i.e., degenerate doping). The oxide layer that remains (approximately 100 Å) can be removed if desired for UV and x-ray applications. The main drawback to the technique is the high-temperature drive required. A similar technique, referred to as gas immersion laser doping (GILD), also produces a very shallow accumulation doping profile. Here the CCD is placed in an atmosphere containing diborane gas (B_2H_6). A pulsed laser light from a KrF excimer laser ($\lambda = 248$ nm) is then scanned over the rear surface of the CCD. The high-intensity light dissociates the diborane gas and melts the surface at a very small depth, depending on the energy of the laser and scanning time. The boron penetrates into the molten silicon to form the accumulation layer. The boron is fully activated and incorporated in the silicon lattice structure without further annealing. The frontside of the CCD remains relatively cool during the process, maintaining metal bus integrity. A peak boron concentration of 10^{20} cm^{-3} in a depth of 100 Å has been achieved.

3.4.8 ANTIREFLECTION COATINGS

The most significant QE loss for backside-illuminated CCDs is reflection loss. The thinning process creates a mirror like finish with an extremely high specular reflectivity. The reflection coefficient for normal incident light is

$$R_{REF} = \frac{(1 - n_{SI})^2}{(1 + n_{SI})^2}. \quad (3.35)$$

Antireflection coatings are employed to suppress reflection loss and are usually tailored for particular wavelength coverage, depending on the application.[73] Some

CHARGE GENERATION

candidate antireflection coatings include HfO_2, ZnS, SiO, MgF_2, and Si_3N_4. When an antireflection coating of proper index and 1/4 wave film is deposited onto a silicon surface, the reflectivity can be reduced to nearly zero. The film thickness at which this happens is

$$d_{AR} = \frac{\lambda}{4n_{AR}}, \qquad (3.36)$$

where λ is working wavelength (Å) and n_{AR} is the index of refraction for the coating.

The reflectance for a quarter wave film thickness is

$$R_{AR} = \frac{(n_{SI} - n_{AR}^2)^2}{(n_{SI} + n_{AR}^2)^2}. \qquad (3.37)$$

Note the reflectance is zero when $n_{AR} = (n_{SI})^{1/2}$.

Example 3.14

Determine the thickness of a MgF_2 coating to produce minimum reflection at 5500 Å. Also calculate the reflectance with and without coating. Assume $n_{SI} = 3.5$, and $n_{AR} = 1.38$ for MgF_2.

Solution:
From Eq. (3.36) the thickness is

$$d_{AR} = 5500/(4 \times 3.5),$$

$$d_{AR} = 392 \text{ Å}.$$

From Eq. (3.37) the reflectivity is

$$R_{AR} = (3.5 - 1.38^2)^2 / (3.5 + 1.38^2)^2,$$

$$R_{AR} = 0.087 \ (8.7\%).$$

From Eq. (3.35) the reflection coefficient without the antireflection coating is

$$R_{REF} = (1 - 3.5)^2 / (1 + 3.5)^2,$$

$$R_{REF} = 0.31 \ (31\%).$$

REFERENCES

1. E. A. Palik, *Handbook of Optical Constants of Solids,* Academic Press, London (1985).
2. K. Rajkanan, R. Singh and J. Shewchun, "Absorption coefficient of silicon for solar cell calculations," *Solid-State Elec.* Vol. 22, pp. 793–795 (1979).
3. G. Macfarlane, T. McLean, J. Quarrington and V. Roberts, "Fine structure in the absorption-edge spectrum of Si," *Phys. Rev.* Vol. 111, No. 5, pp. 1245–1254 (1958).
4. R. Braunstein, A. Moore and F. Herman, "Intrinsic optical absorption in germanium-silicon alloys," *Phys. Rev.* Vol. 109, No. 3, pp. 695–710 (1958).
5. G. Mcfarlane and V. Roberts, "Infrared absorption of silicon near the lattice edge," *Phys. Rev.* Vol. 98, pp. 1865–1866 (1955).
6. N. Kristianpoller and D. Dutton, "Optical properties of 'liumogen': A phosphor for wavelength conversion," *Applied Optics* Vol. 3, No. 2 (1964).
7. M. Cowens, M. Blouke, T. Fairchild and J. Westphal, "Coronene and liumogen as VUV sensitive coatings for Si CCD imagers: a comparison," *Applied Optics* Vol. 19, No. 22, p. 3727 (1980).
8. M. Blouke, M. Cowens, J. Hall, J. Westphal and A. Christensen, "Ultraviolet downconverting phosphor for use with silicon CCD imagers," *Applied Optics* Vol. 19, No. 19 (1980).
9. J. Hynecek, "Virtual phase charge transfer device," U.S. Patent No. 4,229,752.
10. J. Hynecek, "Virtual phase technology: A new approach to fabrication of large-area CCDs," *IEEE Trans. Electron Devices* Vol. ED-28, No. 5 (1981).
11. J. Janesick, J. Hynecke and M. Blouke, "A virtual phase imager for Galileo," *Proc. SPIE* Vol. 290 (1981).
12. J. Janesick, "Open-pinned phase charge coupled device," *NASA Tech Briefs* Vol. 16, No. 1, p. 16 (1992).
13. J. Janesick, "CCD with a 'thin gate'," *Laser Tech. Briefs* Vol. 2, No. 2, p. 18 (1994).
14. J. Janesick, "Frontside illuminated charge-coupled device with high sensitivity to the blue ultraviolet and soft x-ray spectral range," U.S. Patent No. 5,365,092 (Nov. 1994).
15. R. Kraft, D. Burrows, G. Garmire, J. Nousek, J. Janesick and P. Vu, "Soft x-ray spectroscopy with subelectron read noise charge coupled devices," *Nuclear Instruments and Methods in Physics Research A* Vol. 361, pp. 372–383 (1995).
16. R. Kraft, D. Burrows, G. Garmire, D. Lumb, J. Nousek and M. Skinner, "Soft x-ray spectroscopy using charge-coupled devices with thin poly gates and floating gate output amplifiers," *Proc. SPIE* Vol. 2280, pp. 320–332 (1994).
17. W. Keenan and D. Harrison, "A tin oxide transparent-gate buried channel virtual phase CCD imager," *IEEE Trans. Electron Devices* Vol. ED-32, No. 8 (1985).

18. A. Turley, B. Frias, A. Santos, R. Tacka, L. Colquitt, M. Polinsky, J. Ebner, E. Juergensen, R. Bishop and P. Difonzo, "Design and fabrication of tin oxide gate CCD visible imaging arrays," *Proc. SPIE* Vol. 1693, pp. 113–121 (1992).
19. E. Meisenzahl, W. Chang, W. DesJardin, S. Kosman, J. Shepherd, E. Stevens and K. Wong, "A 3.2 million pixel full-frame true 2-phase CCD image sensor incorporating transparent gate technology," *Proc. SPIE* Vol. 3965, pp. 92–100 (2000).
20. S. Shortes, W. Chan, W. Rhines, J. Barton and D. Collins, "Characteristics of thinned backside illuminated charge coupled device imagers," *Applied Physics Letters* Vol. 24, No. 11 (1974).
21. A. Stoller, R. Speers and S. Opresko, "A new technique for etch thinning silicon wafers," *RCA Review* (1970).
22. W. Kern, "Chemical etching of silicon, germanium, gallium arsenide, and gallium phosphide," *RCA Review* Vol. 39, pp. 278–308 (1978).
23. E. Savoye, D. Battson, T. Edwards, W. Henry, D. Tshudy, L. Wallace, G. Hughes, W. Kosonocky, P. Levine and F. Shallcross, "High sensitivity charge-coupled device (CCD) imagers for television," *Proc. SPIE* Vol. 501 (1984).
24. H. Robbins and B. Schwartz, "Chemical etching of silicon II. The system HF, HNO_3, H_2O, and $HC_2H_3O_2$," *J. Electrochem. Soc.* Vol. 107, pp. 108–111 (1960).
25. B. Schwartz and H. Robbins, "Chemical etching of silicon III. A temperature study in the acid system," *J. Electrochem. Soc.* Vol. 108, pp. 365–372 (1961).
26. D. Klein and D. D'Stefan, "Controlled etching of silicon in the HF-HNO_3 system," *J. Electrochem. Soc.* Vol. 109, pp. 37–42 (1961).
27. H. Robbins and B. Schwartz, "Chemical etching of silicon I. The system HF, HNO_3, H_2O, and $HC_2H_3O_2$," *J. Electrochem. Soc.* pp. 505–508 (1959).
28. H. Muraoka, T. Ohhashi and Y. Sumitomo, "Controlled preferential etching technology," H. Huff and R. Burgess, eds., Electrochem. Soc., Princeton, New Jersey, pp. 327–338 (1973).
29. B. Burke, J. Gregory, R. Mountain, J. Huang, M. Cooper and V. Dolat, "High-performance visible/UV CCD imagers for space-based applications," *Proc. SPIE* Vol. 1693, pp. 86–100 (1992).
30. R. Winzenread, "Flat, thinned scientific CCDs," *Proc. SPIE* Vol. 2198, pp. 886–894 (1994).
31. R. Schaefer, R. Varian, J. Cover and R. Larsen, "Megapixel CCD thinning/backside progress at SAIC," *Proc. SPIE* Vol. 1447, pp. 165–176 (1991).
32. R. Pierret, *Modular Series on Solid State Devices*, Addison-Wesley Publishing Co. (1983).
33. J. Janesick and T. Elliott, "Increased spectral response for charge-coupled devices," *NASA Tech. Briefs* Vol. 10, No. 1, p. 38 (1986).
34. P. Caplan and E. Poindexter, "Ultraviolet bleaching and regeneration of Si-Si_3 centers at the Si/SiO_2 interface of thinly oxidized silicon wafers," *J. Appl. Phys.* Vol. 53, No. 1, pp. 541–545 (1982).

35. J. Janesick, T. Elliott, T. Daud and J. McCarthy, "Backside charging of the CCD," *Proc. SPIE* Vol. 570, pp. 46–79 (1985).
36. R. Williams, "Photoemission of electrons from silicon into silicon dioxide," *Physical Review* Vol. 140, No. 2A (1965).
37. R. Williams, "Properties of the silicon-SiO_2 interface," *J. Vac. Sci. Technol.* Vol. 14, No. 5 (1977).
38. T. DiStefano, "Field dependent internal photoemission probe of the electronic structure of the Si-SiO_2 interface," *J. Vac. Sci. Technol.* Vol. 13, No. 4 (1976).
39. R. Williams and M. Woods, "High electric fields in silicon dioxide produced by corona charging," *J. Appl. Phys.* Vol. 44, No. 3 (1973).
40. J. Janesick, "Metal film increases CCD quantum efficiency," *NASA Tech. Briefs* Vol. 13, No. 4, pp. 24–25 (1989).
41. J. Janesick, "Metal coat increases output sensitivity," *NASA Tech. Briefs* Vol. 13, No. 5, pp. 26–28 (1989).
42. J. Janesick, T. Elliott, T. Daud and D. Campbell, "The CCD flash gate," *Proc. SPIE* Vol. 627, pp. 543–582 (1986).
43. M. Lesser, "Back illuminated large format Loral CCDs," *Proc. SPIE* Vol. 1447, pp. 177–183 (1991).
44. M. Lesser, L. Bauer, L. Ulrickson and D. Ouelliette, "Bump bonded back illuminated CCDs," *Proc. SPIE* Vol. 1656 (1992).
45. M. Lesser, "Backside coatings for back illuminated CCDs," *Proc. SPIE* Vol. 1900, pp. 219–227 (1993).
46. M. Lesser, "Improving CCD quantum efficiency," *Proc. SPIE* Vol. 2198, pp. 782–791 (1994).
47. M. Lesser and V. Iyer, "Enhancing back illuminated performance of astronomical CCDs," *Proc. SPIE* Vol. 3355, pp. 446–456 (1998).
48. J. Oke, F. Harris, D. Oke and D. Wang, "CCD Testing at Palomar Observatory," *Publications of the Astronomical Society of the Pacific* Vol. 100, No. 623 (1988).
49. T. Daud and J. Janesick, "Photovoltaic hydrogen sensor," *NASA Tech. Briefs* Vol. 13, No. 1, p. 36 (1989).
50. L. Robinson, W. Brown, K. Gilmore, R. Stover and M. Wei, "Performance tests of large CCDs," *Proc. SPIE* Vol. 1447 (1991).
51. B. Burke, J. Gregory, M. Bautz, G. Prigoshin, S. Kissel, B. Kosicki, A. Loomis and D. Young, "Soft-x-ray CCD imagers for AXAF," *IEEE Trans. Electron Devices* Vol. 44, No. 10 (1997).
52. Tsoi, J. Ellul, M. King, J. White and W. Bradley, "A deep-depletion CCD imager for soft x-ray, visible and infrared sensing," *IEEE Trans. Electron Devices* Vol. ED-32, No. 8 (1985).
53. A. Owens and K. McCarthy, "Energy deposition in x-ray CCDs and charge particle discrimination," *Nucl. Instrum. and Methods in Phys. Res. A* Vol. 366, pp. 148–154 (1995).
54. P. Bailey, C. Castelli, M. Cross, P. van Essen, A. Holland, F. Jansen, P. de Korte, D. Lumb, K. McCarthy, P. Pool and P. Verhoeve, "Soft x-ray performance of backside illuminated EEV CCDs," *Proc. SPIE* Vol. 1344, pp. 356–371 (1990).

55. S. Holland et al., "A 200 × 200 CCD image sensor fabricated on high-resistivity silicon," *IEDM Technical Digest* pp. 911–914 (1996).
56. R. Stover et al., "Characterization of a fully-depleted CCD on high-resistivity silicon," *Proc. SPIE* Vol. 3019, pp. 183–188 (1997).
57. R. Stover et al., "A $2k \times 2k$ high resistivity CCD," to appear in *Proc. 4th ESO Workshop on Optical Detectors for Astronomy*, Garching, Germany (1999).
58. D. Groom et al., "Point-spread function in depleted and partially depleted CCDs," to appear in *Proc. 4th ESO Workshop on Optical Detectors for Astronomy*, Garching, Germany (1999).
59. M. Blouke, "A simplified model of the back surface of a charge-coupled device," *Proc. SPIE* Vol. 1147 (1991).
60. C. Huang, B. Kosicki, J. Theriault, J. Gregory, B. Burke, B. Johnsom and E. Hurley, "Quantum efficiency model for p^+ doped backside illuminated CCD," *Proc. SPIE* Vol. 1447 (1991).
61. R. Stern, R. Catura, R. Kimble, A. Davidson, R. Winsenread, M. Blouke, R. Hayes, D. Walton and J. Culhane, "Ultraviolet and extreme ultraviolet response of charge coupled detectors," *Optical Engineering* Vol. 26, p. 875 (1987).
62. C. Huang, B. Burke, B. Kosicki, R. Mountain, P. Daniels, D. Harrison, G. Lincoln, N. Usiak, M. Kaplin and A. Forte, "A new process for thinned, back-illuminated CCD imager devices," *1989 International Symposium on VLSI Technology, Systems and Applications*, Taipei, Taiwan, ROC (1989).
63. C. Tassin, Y. Thenoz and J. Chabbal, "Thinned backside illuminated CCD for ultraviolet imaging," *Proc. SPIE* Vol. 1140, 1988 Technical Symposium Southeast on Optics, Electro-optics and Sensors, pp. 139–148 (1988).
64. M. E. Hoenk, P. J. Grunthaner, F. J. Grunthaner, M. Fattahi, H.-F. Tseng and R. W. Terhune, "Growth of a delta-doped silicon layer by molecular-beam epitaxy on a charge-coupled device for reflection-limited ultraviolet quantum efficiency," *Appl. Phys. Lett.* Vol. 61, p. 1084 (1992).
65. P. J. Grunthaner, F. J. Grunthaner, R. W. Fathauer, T.-L. Lin, M. H. Hecht, L. D. Bell, W. J. Kaiser, F. D. Schowendgerdt and J. H. Mazur, "Hydrogen-terminated silicon substrates for low-temperature molecular-beam epitaxy," *Thin Solid Films* Vol. 183, p. 197 (1989).
66. M. E. Hoenk, P. J. Grunthaner, F. J. Grunthaner, R. W. Terhune and M. Fattahi, "Epitaxial growth of p+ silicon on a backside-thinned ccd for enhanced UV response," *Proc. SPIE* Vol. 1656, p. 488 (1992).
67. S. Nikzad, M. E. Hoenk, P. J. Grunthaner, R. W. Terhune, F. J. Grunthaner, R. Wizenread, M. Fattahi and H. F. Tseng, "Delta-doped CCDs: High QE with longterm stability at UV and visible wavelengths," *Proc. SPIE* Vol. 2198, p. 907 (1994).
68. S. Nikzad, M. E. Hoenk, P. J. Grunthaner, R. W. Terhune, R. Wizenread, M. Fattahi, H. F. Tseng and F. J. Grunthaner, "Delta-doped CCDs as stable, high sensitivity, high resolution UV imaging arrays," *Proc. SPIE* Vol. 2217, p. 355 (1994).

69. P. Deelman, S. Nikzad and M. Blouke, "Delta doped CCDs with integrated UV coatings," *Proc. SPIE* Vol. 3965 (2000).
70. A. L. Smith, Q. Yu, S. T. Elliott, T. A. Tombrello and S. Nikzad, "Modification of the surface band bending of a silicon CCD for low-energy electron detection," *Proc. of the MRS* Vol. 448, pp. 177–186 (1996).
71. S. Nikzad, A. L. Smith, S. T. Elliott, T. J. Jones, T. A. Tombrello and Q. Yu, "Low-energy electron detection with delta-doped CCDs," *Proc. SPIE* Vol. 3019, p. 241 (1997).
72. S. Nikzad, Q. Yu, A. L. Smith, T. J. Jones, T. A. Tombrello and S. T. Elliott, "Direct detection and imaging of low-energy electrons with delta-doped charge-coupled devices," *Appl. Phys. Lett.* Vol. 73, p. 3417 (1998).
73. S. Nikzad, D. Croley, S. T. Elliott, T. J. Cunningham, W. P. Proniewicz, G. Murphy and T. J. Jones, "Single-event keV proton detection using a delta-doped charge-coupled device," *Appl. Phys. Lett.* Vol. 75, p. 2686 (1999).

CHAPTER 4

CHARGE COLLECTION

4.1 CHARGE COLLECTION

After signal charge is generated it is immediately collected in individual pixel cells. This chapter will discuss three parameters related to this process: (1) well capacity, (2) pixel nonuniformity and (3) charge diffusion. Well capacity or full well is a measure of the maximum amount of charge that a pixel can hold. Some CCDs accommodate more charge than others, depending on the size of pixel and the process technology employed. Charge capacity has dramatically increased over the years. For example, full well performance for the Wide Field/Planetary Camera I CCD was limited to 60,000 e$^-$ (refer to Fig. 2.3). Charge capacities of 500,000 e$^-$ are being achieved today for the same pixel size (i.e., 15 μm). The improvement has come from process advancements and inverted clocking. Charge capacity for current three-phase technology is averaging ~ 8000 e$^-$/μm^2.

Pixel-to-pixel nonuniformity is associated with how signal charge divides up among pixels. When the CCD is uniformly illuminated, one finds an average charge level collected by the pixels and a variance about this mean. This departure from the mean is physically related to lithography and process variations in defining pixel boundaries. Clock bias can also influence the amount of pixel nonuniformity seen. Pixel nonuniformity noise for high-performance CCDs can be less than 1% as measured by photon transfer. This noise level may seem small, but as discussed in Chapter 2, it will dominate S/N performance over most of the sensor's dynamic range. However, pixel nonuniformity is usually not a concern to the scientific user because it can be eliminated by a technique called flat fielding, a subject discussed in this chapter.

Charge diffusion is the third charge collection efficiency parameter discussed in this chapter. The ability of the CCD to record and reproduce the spatial information in a scene is an important measure of the utility of the sensor. For the CCD, this means that all charge generated by photons incident on a target pixel should be collected by that pixel. Charge that escapes the target pixel indicates a charge collection problem. Charge diffusion and crosstalk among pixels is primarily generated in regions of the CCD that lie outside the depletion edge where field-free material exists. Field-free material allows charge to diffuse and randomly wander away from the target pixel into neighboring pixels. The modulation transfer function (MTF) is the primary transfer curve used to quantify charge diffusion problems. As we will show, CCDs can approach theoretical MTF levels when properly designed and processed.

It should be kept in mind when reading through this chapter that high charge collection efficiency (i.e., high full well, low pixel nonuniformity, high MTF) implies high S/N performance. For example, when shot noise limited, S/N of an image increases by the square root of full well [Eq. (2.26) and Eq. (2.35)]. Therefore, the greater the well capacity, the better the image quality especially for low-contrast scenes. The S/N is fixed with signal when pixel nonuniformity noise dominates. This noise source can seriously limit S/N and is why scientific groups correct pixel nonuniformity to make the images shot noise limited. High MTF is also very influential on low-contrast images, especially high-spatial-frequency scenes. Image modulation and S/N are directly related to MTF response.

4.2 Well Capacity

Well capacity or full well is a complex performance parameter that requires in-depth discussions involving CCD potential well physics. In theory, maximum storage occurs when signal charge compensates all buried channel phosphorus atoms located between the surface of the CCD and the potential maximum, V_{max}. That is,

$$Q_{FW} = N_D(t - x_n), \tag{4.1}$$

where Q_{FW} is the full well charge density (e$^-$/cm^2), N_D is the buried channel doping (cm^{-3}) and $t - x_n$ is the distance from the surface of the CCD to the potential maximum V_{max} when the well is empty (refer to Fig. 1.25). Note from Fig. 1.27 that the potential maximum moves toward the surface as charge fills a well, explaining why only those phosphorus atoms within the $t - x_n$ region are involved.

Example 4.1

Determine Q_{FW} assuming $N_D = 2 \times 10^{16}$ cm^{-3} and $(t - x_n) = 5 \times 10^{-5}$ cm.
Solution:
From Eq. (4.1),

$$Q_{FW} = (2 \times 10^{16})(5 \times 10^{-5}),$$

$$Q_{FW} = 10^{12} \text{ e}^-/\text{cm}^2.$$

Note that the charge capacity for a 4 × 8-µm collecting area would be 320,000 e$^-$ (10,000 e$^-$/µm^2).

There is a maximum limit to the buried channel doping concentration. For example, the greater the doping, the greater the clock swing required as discussed below. If the clock voltage becomes too high, some pixels on an array will show signs of leakage through the gate dielectric with the possibility of turning into a column blemish. The pixel may even short out, creating a blemish referred to as "catastrophic blem" (refer to Chapter 7).

Silicon breakdown represents another limit. Electric fields associated with the potential well may approach the critical avalanche field if the channel doping is too high. This high-field condition occurs at approximately $3\text{--}5 \times 10^5$ V/cm (30–50 V/µm) for silicon material. Avalanche is initiated when an electron-hole pair is generated by thermal means through an impurity or lattice detect. If the electric field is sufficiently high, the electron can gain enough kinetic energy to slam into the lattice and break silicon-silicon covalent bonds, a process called impact ionization. This will lead to the formation of other electron-hole pairs. These secondary carriers can in turn accelerate and ionize and create even more pairs. The first indicator of avalanche breakdown for the CCD is the sign of hot pixels or dark spikes (a bulk dark current problem discussed in Chapter 7). It has been shown experimentally that dark spike generation is manageable for channel doping levels as high as 3.0×10^{16} cm^{-3}. Doping concentrations at this level have produced charge capacities of more than 12,000 e$^-$/µm^2 for multiphase CCDs.

Although charge capacity can be a large fraction of Q_{FW}, many practical restrictions do not allow Eq. (4.1) to be completely realized. The next sections introduce two very important full well limits.

4.2.1 BLOOMED FULL WELL

For a multiphase CCD, signal charge will spill into neighboring pixels when the channel potential maximum of the collecting phase equals that of the adjacent barrier phase (refer to Fig. 1.19). This example of charge spreading among pixels is called blooming. When a CCD blooms, we say that the sensor has reached the "bloomed full well (BFW) state." Figure 4.1 shows the blooming effect for three bright regions imaged on a CCD array when BFW is exceeded. Note that blooming only occurs in the vertical direction. This is because the channel stop region is at a much lower potential (i.e., 0 V) than the barrier phases. It is for this reason that charge will never bloom over a channel stop. Only when the full well is exceeded in the horizontal register do we see horizontal blooming. This blooming effect usually takes place when pixel binning is performed in the horizontal register.

Charge capacity under BFW conditions is estimated by

$$Q_{BFW} = \frac{C_{EFF} V_{CB}}{q}, \qquad (4.2)$$

where Q_{BFW} is BFW capacity (e$^-$/cm^2), C_{EFF} is the effective capacitance defined in Eq. (1.53), and V_{CB} is the gate potential difference between the collecting and barrier phases,

$$V_{CB} = V_C - V_B, \qquad (4.3)$$

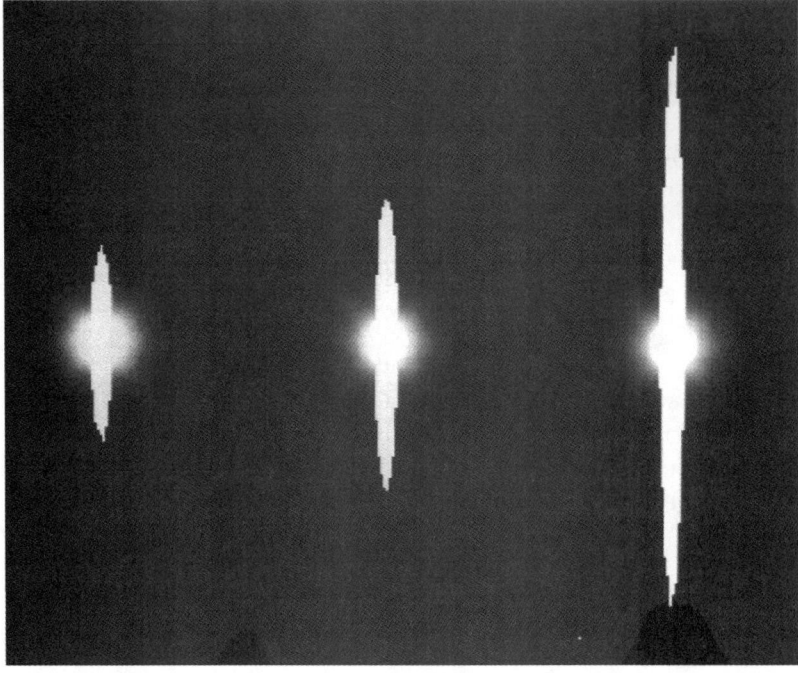

Figure 4.1 Blooming signature for three point images.

where V_C is the potential maximum of the collecting phase and V_B is the potential maximum of the barrier phase as shown in Fig. 1.19.

We can think of Eq. (4.3) as that amount of charge stored on the effective capacitance to produce a voltage change of V_{CB} volts. At that point the barrier and collecting potentials are equivalent, the point where blooming will occur.

Example 4.2

Determine BFW assuming $C_{EFF} = 2.53 \times 10^{-8}$ F/cm^2 and $V_{CB} = 5$ V.
Solution:
From Eq. (4.2),

$$Q_{BFW} = (2.53 \times 10^{-8})5/1.6 \times 10^{-19},$$

$$Q_{BFW} = 7.91 \times 10^{11} \text{ e}^-/\text{cm}^2.$$

Full well for a 4×8-µm area would be 253,000 e$^-$ (7906 e$^-$/µm^2).

Note that BFW is actually greater than calculated here because the value of C_{EFF} assumed is without the presence of a signal. C_{EFF} increases slightly because $t - x_n$ width decreases with signal.

CHARGE COLLECTION

4.2.2 SURFACE FULL WELL

From Eq. (4.2) it would appear that full well is limitless because V_{CB} can be as large as desired by increasing the clock voltage on the collecting phase. Unfortunately, increasing the gate voltage to will cause the potential maximum to move toward the surface as more channel depletion comes from the junction side. In turn, the difference in surface potential, V_S, and the maximum potential, V_{max}, is reduced. As charge fills the potential well, it now has the opportunity to interact with the Si–SiO$_2$ interface states before blooming occurs. Full well by definition is exceeded when this effect takes place, because CTE degrades significantly. The full well condition is called surface full well (SFW). From Eq. (4.1), SFW is

$$Q_{SFW} = N_D(t - x_n). \tag{4.4}$$

Note that SFW decreases with increasing x_n and equals zero when $x_n = t$. At that point the buried channel CCD operates as a surface channel device (i.e., $V_{max} = V_S$).

Example 4.3

Calculate SFW for $x_n = 0.5$, 0.7, and 0.9 μm as the gate voltage increases. Also assume, $t = 1 \times 10^{-4}$ cm and $N_D = 2 \times 10^{16}$ cm^{-3}.

Solution:
From Eq. (4.4),

$$Q_{SFW} = (2 \times 10^{16})(1 \times 10^{-4} - 0.5 \times 10^{-4}),$$

$$Q_{SFW} = 10^{12} \text{ e}^-/\text{cm}^2 \ (x_n = 0.5 \text{ μm}),$$

$$Q_{SFW} = 6 \times 10^{11} \text{ e}^-/\text{cm}^2 \ (x_n = 0.7 \text{ μm}),$$

$$Q_{SFW} = 2 \times 10^{11} \text{ e}^-/\text{cm}^2 \ (x_n = 0.9 \text{ μm}).$$

4.2.3 OPTIMUM FULL WELL

Which happens first, BFW or SFW, depends on the positive clock level of the collecting phases set by the user. Optimum well capacity occurs when BFW = SFW, a very important operating condition for multiphase CCDs. Figure 4.2 is an illustration showing potential well plots for the BFW and SFW conditions and the optimum full well bias point. Optimum full well occurs when the barrier phase is inverted and pinned as shown.

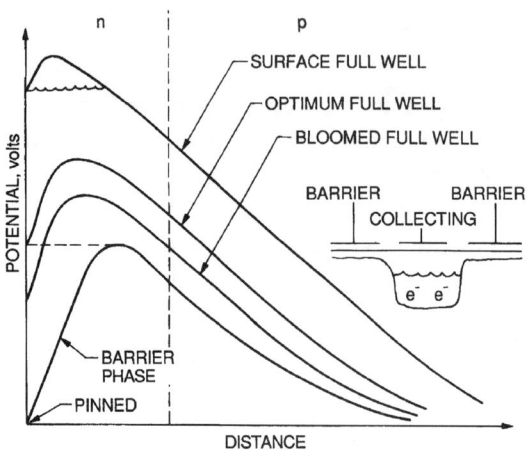

Figure 4.2 Illustration showing BFW, SFW and optimum full well bias conditions for the collecting phase.

It should be mentioned that full well discussions here apply to multi-phase CCDs where the gate potentials under the collecting and barrier phases can be independently adjusted (e.g., three-phase, four-phase, etc.). Virtual and two-phase CCDs are fabricated with implanted barriers, and, therefore, the barrier height (V_{CB}) is built into the sensor. These CCD technologies are processed to bloom before SFW is encountered. However, SFW characteristics can still be seen if these CCDs are overdriven.

Figure 4.3 shows channel potential plots as a function of gate voltage for a multiphase CCD. Three special gate voltages are indicated: (1) the gate voltage where inversion occurs ($V_G = V_{INV}$), (2) the gate voltage where the effective threshold, V_{EFF}, is defined ($V_G = 0$) and (3) the gate voltage where optimum full well occurs ($V_G = V_{FULL}$). Figure 4.3(a) shows the channel potential for two different channel doping levels. Note that V_{INV}, V_{EFF} and V_{FULL} all increase with doping concentration. As mentioned above, the main advantage for higher doping is greater full well performance. Figure 4.3(b) shows how channel potential varies with channel drive, effectively the processing temperature and time that defines the depth of the channel. Note that V_{INV} remains fixed, whereas V_{EFF} and V_{FULL} both increase. The main advantage for driving the channel deeper is for greater speed performance and larger fringing fields between phases, an important subject discussed in Chapter 5.

It can been shown through device simulation that V_{INV}, V_{EFF} and V_{FULL} are approximately related through the equation

$$V_{EFF} = V_{FULL} - V_{INV}. \tag{4.5}$$

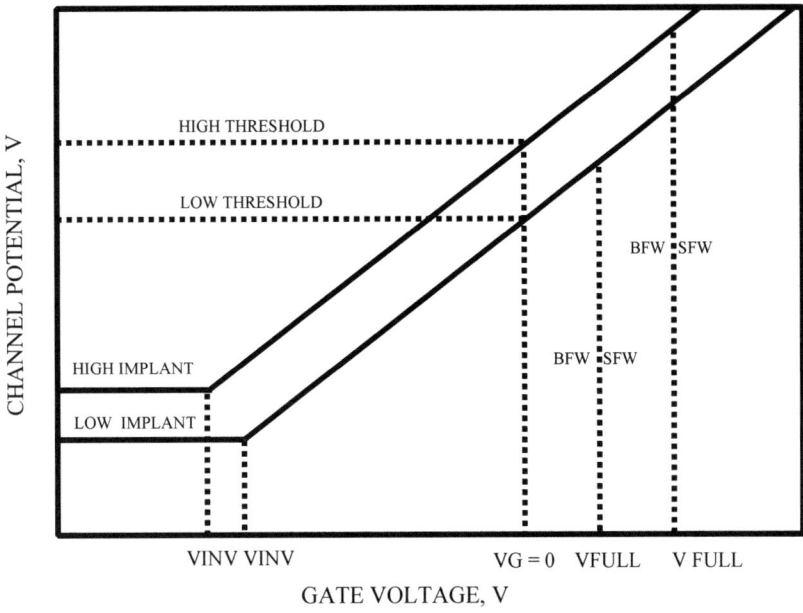

Figure 4.3(a) Channel potential (V_{max}) plots showing gate bias levels for inversion, threshold and optimum full well for two different channel dopings.

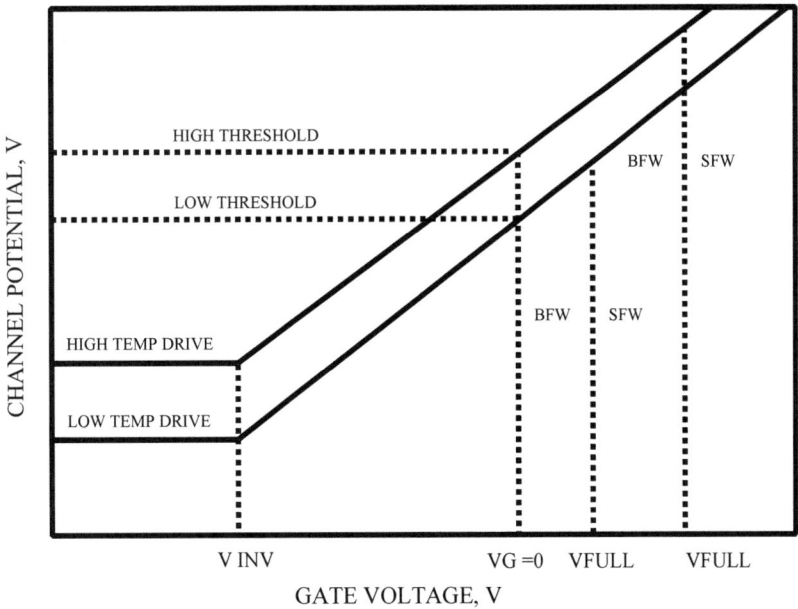

Figure 4.3(b) Channel potential plots showing bias levels for inversion, threshold and optimum full well for two different channel drives.

Example 4.4

Estimate the positive clock voltage required for optimum full well. Assume the following process parameters: $V_{EFF} = 12$ V and $V_{INV} = -9$ V.
Solution:
From Eq. (4.5),

$$V_{FULL} = 12 - 9 = 3 \text{ V}.$$

Therefore, the vertical clocks must swing from -9 to 3 V for optimum charge capacity.

4.2.4 Clocking Modes

Three clocking options are used to clock CCDs. These modes are called "noninverted," "partially inverted" and "inverted" clocking. Noninverted clocking implies that the barrier phases are not inverted or pinned (i.e., $V_B > V_{INV}$). Partially inverted clocking is when the barrier phases are taken into inversion (i.e., $V_B < V_{INV}$). Inverted clocking is when all phases, including the collecting phase, are inverted ($V_B = V_C < V_{INV}$). This last mode of operation applies to a special class of CCDs called multipinned-phase (MPP) CCDs, to be discussed below. Many CCD performance parameters are dependent on which clocking mode is used. For example, dark current generation, spurious charge and residual image are highly dependent on inverted clocking, as discussed in Chapter 7. Full well, antiblooming (AB) and pixel nonuniformity are also dependent on inverted clocks, as discussed in this chapter. It should also be mentioned that two-phase and virtual-phase CCDs with implanted barriers can also be clocked in these modes.

4.2.5 Full Well Transfer

Figure 4.4(a) presents a mainstream transfer curve known as full well transfer. This very important transfer curve will be seen often in this book. The transfer curve plots charge capacity for a three-phase 18-μm-pixel CCD as a function of V_C with V_B pinned at -8 V. Optimum full well occurs at $V_C = V_{FULL} = 2.5$ V yielding 350,000 e$^-$. SFW and BFW regimes are indicated above and below V_{FULL}. Note that at some high clock voltage, $S_{SFW} = 0$ e$^-$ when $V_{max} = V_S$. At some low clock voltage $V_C = V_B$ and $S_{BFW} = 0$ e$^-$.

The full well transfer curve can be generated through photon transfer for different collecting voltages. The curve can also be generated by a simple scope measurement (a computer-derived method will also be discussed below). We begin

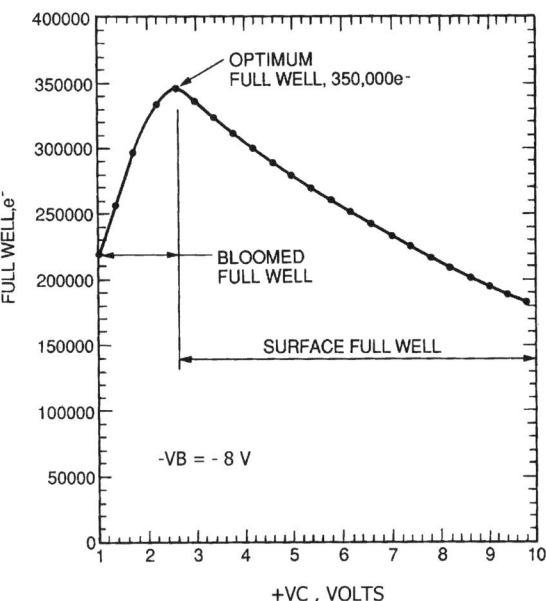

Figure 4.4(a) Full well transfer curve showing BFW and SFW regimes and the optimum full well bias point.

this method by measuring BFW characteristics. The device is exposed to a semi-uniform light source while the sensor is constantly reading out (an LED above the CCD connected to a variable power supply is a good light source for this test). While monitoring the output signal, V_C is manually dithered up and down by a few tenths of a volt. A signal movement on the oscilloscope indicates that the device is beyond full well and is blooming. No movement implies that the device is transferring charge properly and is not above full well. In the latter case, the light is increased or clock voltage decreased until a slight signal movement is noticed, indicating the onset of blooming. The signal and corresponding V_C at this light level represent the first data point on the full well transfer curve. The signal can be converted to electrons if the sensitivity (V/e$^-$) is known at the point of measurement (this measurement is usually performed at the input of the ADC). More points on the curve are found in the same manner, using different illumination levels. These measurements form the left-hand side of the full well transfer curve (i.e., BFW).

The SFW regime or the right-hand side of the curve is produced using the same illumination levels made for the BFW points (i.e., these settings were recorded earlier). However, instead of dithering V_C we increase the voltage until the signal drops in amplitude (one might need to magnify the trace on the oscilloscope to clearly see the effect). Increasing the collecting voltage will force the device deeper into the SFW regime. At some point a decrease in signal will be seen, because recombination at the surface will occur as charge interacts at the Si–SiO$_2$ interface. It is important that the barrier phases be placed into inversion to promote hole recombination (without inversion it is impossible to see when the CCD actually goes into a SFW). Different illumination levels will produce the SFW portion of the full well transfer curve as shown in Fig. 4.4(a).

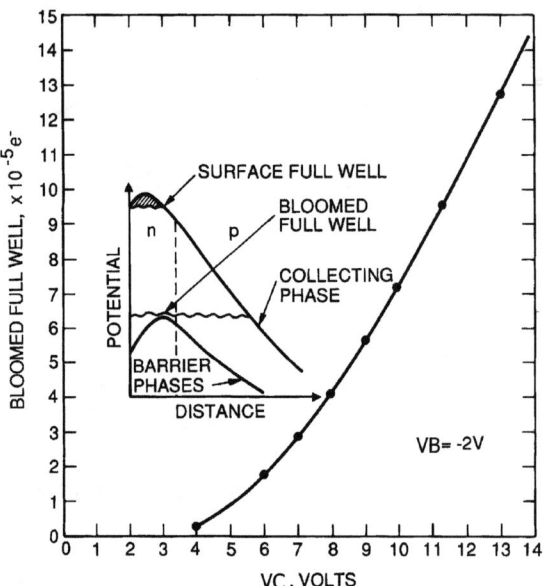

Figure 4.4(b) BFW as a function of gate bias with SFW exceeded.

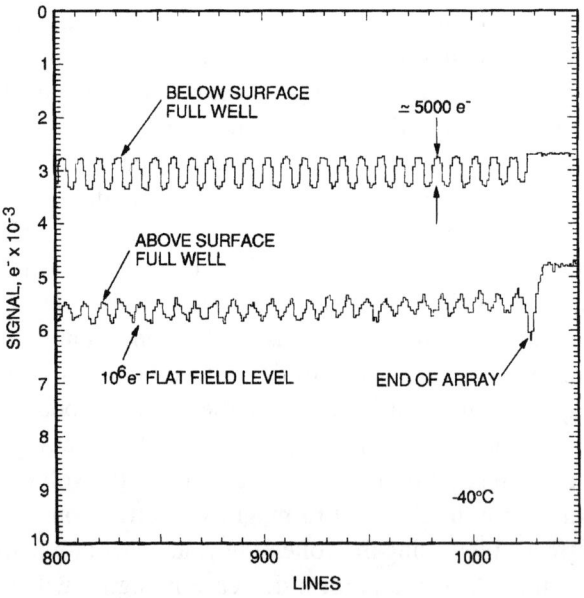

Figure 4.4(c) Modulation loss when the CCD is driven into the SFW regime.

4.2.6 FULL WELL DATA

Figure 4.4(b) plots BFW characteristics for the same CCD shown in Fig. 4.4(a). Clock bias conditions are such that SFW occurs when $V_C > 6$ V (the barrier phase is set at -2 V, out of inversion, to avoid recombination when SFW is exceeded). Note that the full well signal level $S_{BFW} = 1.5 \times 10^6$ e$^-$ at $V_C = 14$ V, which is significantly greater than the optimum full well level of 350,000 e$^-$ shown in Fig. 4.4(a). However, the penalty paid for operating above SFW is poor CTE, because of signal interaction at the Si–SiO$_2$ interface. For example, Fig. 4.4(c) shows the modulation of a 5000 e$^-$ square-wave signal above and below SFW. In the upper trace, which is below SFW, CTE is well behaved. The lower trace was generated by adding a flat-field level of 10^6 e$^-$ to the square-wave image, forcing the CCD deep into SFW. Note that the modulation decreases because of surface interaction.

Optimum full well for a multiphase CCD may occur at a drive voltage less than substrate potential (i.e., 0 V). For example, Fig. 4.5 shows a full well transfer curve generated by a 1024 × 2048 × 12-μm-pixel CCD. The channel implant (1.6×10^{16} cm^{-3}) and low-temperature channel drive produce an optimum drive level of only $V_{FULL} = -0.5$ V ($V_B = -6.5$ V $= V_{INV}$). An effective threshold of only 6 V is measured for the CCD. This full well characteristic may be difficult to measure if the CCD clock drivers cannot drive V_C below ground level.

Figure 4.6 shows several full well transfer curves as a function of channel stop width. Measurements were taken from an experimental 1024 × 1024 × 18-μm CCD implanted at 1.6×10^{16} cm^{-3}. Note that V_{FULL} decreases with channel stop width.

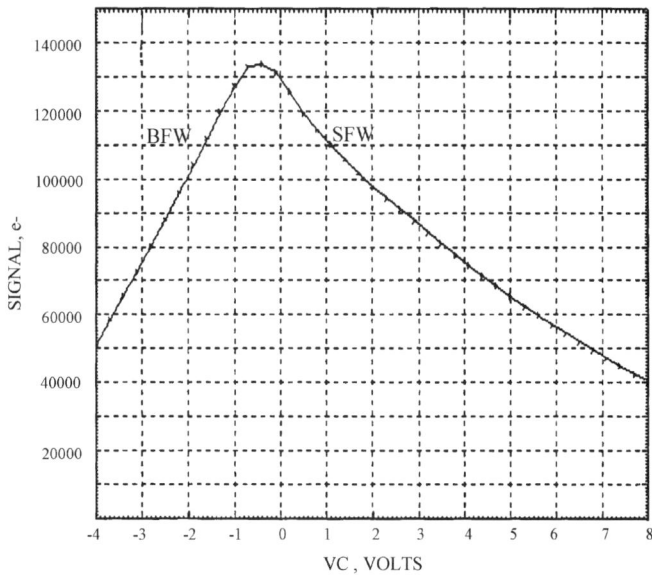

Figure 4.5 Full well transfer curve showing an optimum full well at $V_{FULL} = -0.5$ V.

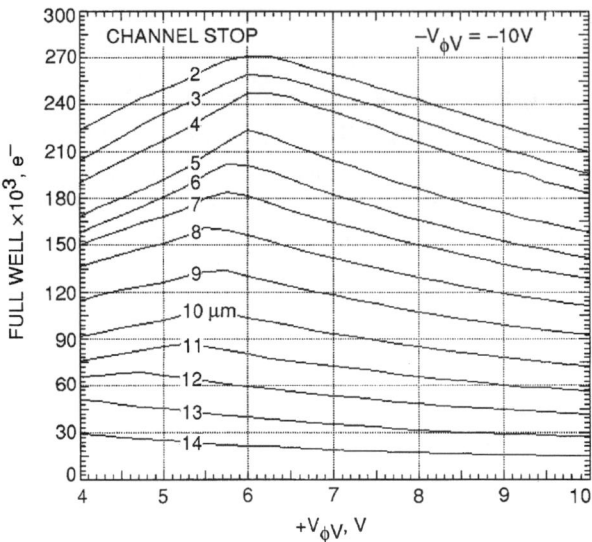

Figure 4.6 Full well transfer curves for a test CCD with different channel stop widths.

This is because the effective threshold decreases as the boron implant in the channel stop region encroaches into the main channel that compensates phosphorus atoms.

Figure 4.7 shows two saturated (i.e., above full well) spot images taken from an 800 × 800 CCD that is clocked noninverted ($V_B = -6$ V, $V_{INV} = -8$ V). The light level to the CCD is identical in both images. Figure 4.7(a) was taken at $V_C = +3$ V, which places the CCD in the BFW regime slightly below optimum full well. For this bias setting, charge will bloom before SFW can occur. Figure 4.7(b) is taken at $V_C = 7$ V, forcing charge to interact with the surface at full well before blooming occurs. Although BFW characteristics are clearly seen, SFW traits are much more subtle. On closer inspection, a deferred charge tail is seen following the image as charge slowly escapes the Si–SiO$_2$ interface (charge is transferred downward in both images).

The presence of trapped charge at the surface becomes even more difficult to detect when the CCD is clocked partially inverted ($V_B = -8$ V) because this charge recombines with the inversion layer of holes. Figure 4.8 shows alpha particle events (Am^{95}) that saturate the target pixel above SFW with approximately 250,000 e$^-$/event. Charge is transferred toward the bottom of the CCD in both images. Figure 4.8(a) was taken under partial inversion conditions ($V_B = -8$ V), whereas Fig. 4.8(b) under the noninverted state ($V_B = -5$ V). In both cases the device was biased deep into the SFW regime ($V_C = +8$ V), forcing surface interaction. For the noninverted case [Fig. 4.8(b)], deferred tails follow each event as charge escapes the Si–SiO$_2$ interface (i.e., residual image during readout). Partial inversion allows trapped charge to recombine with holes during readout, thus eliminating the tails completely [Fig. 4.8(a)].

CHARGE COLLECTION

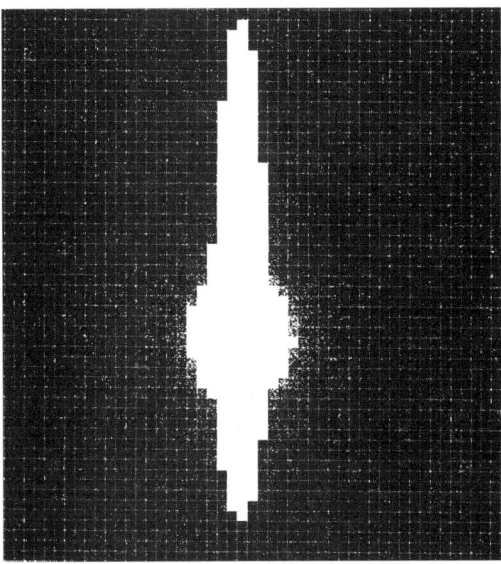

Figure 4.7(a) A saturated image with collecting phase biased into the BFW regime.

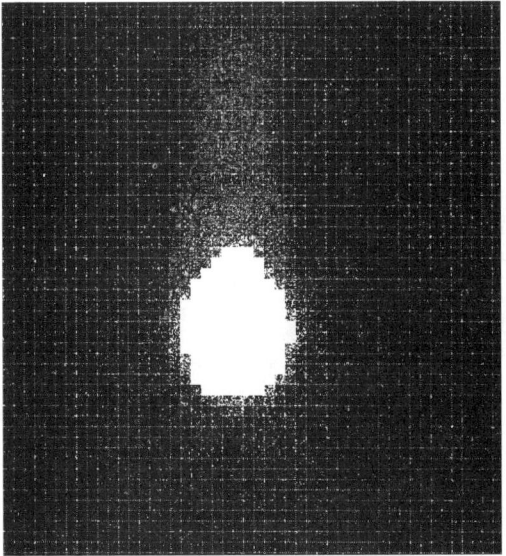

Figure 4.7(b) A saturated image with the collecting phase biased into the SFW regime.

Figure 4.9 shows a photon transfer curve produced by a CCD clocked under partial inverted conditions. The sensor reaches SFW at approximately 240,000 e^- and BFW at 600,000 e^-. Note that BFW breakpoint is clearly seen, whereas SFW only shows a slight deviation in the shot noise measured. This is caused by a slight

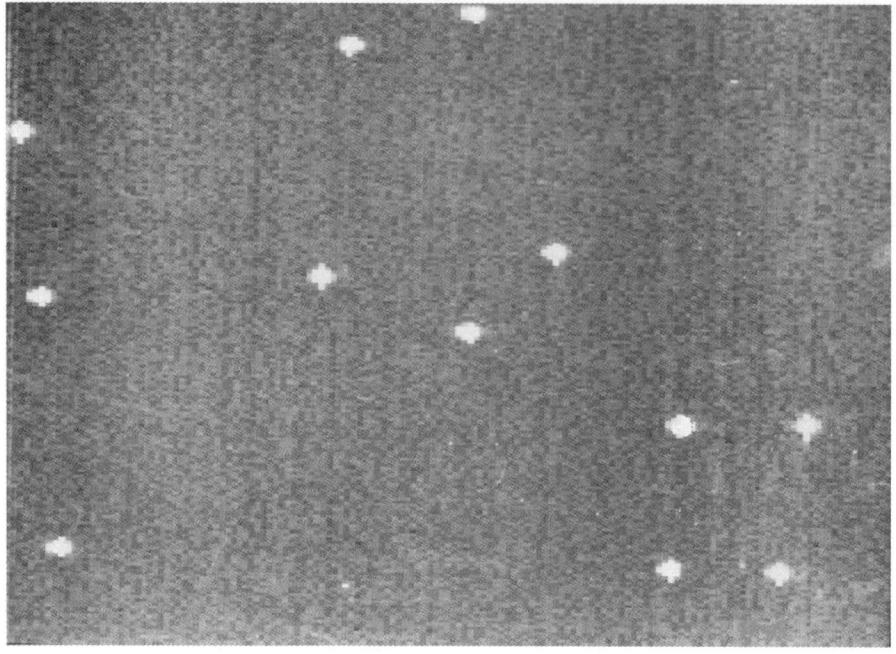

Figure 4.8(a) An alpha particle image in response to partial inverted clocked conditions.

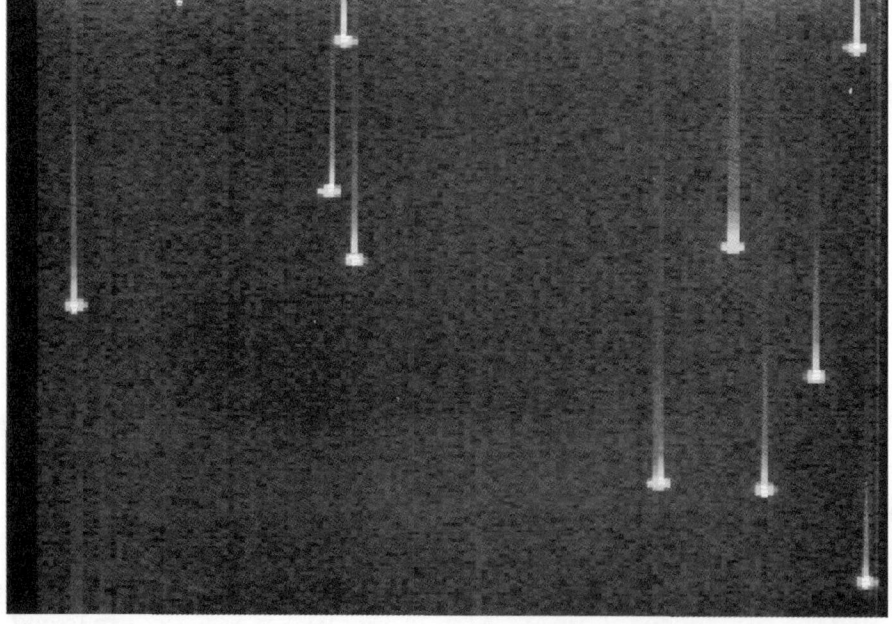

Figure 4.8(b) Alpha particle image when clocked noninverted, showing deferred charge tails indicative of SFW.

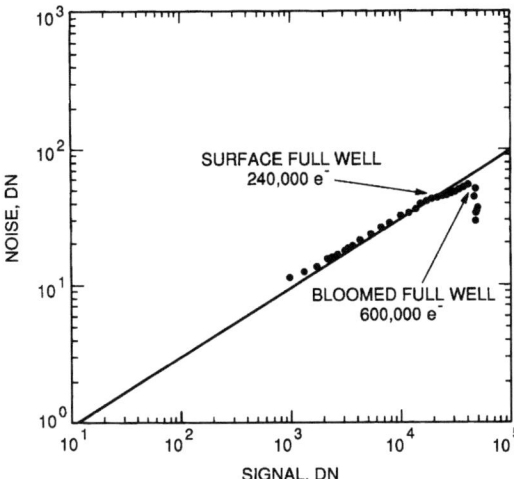

Figure 4.9 Photon transfer curve showing BFW and SFW breakpoints.

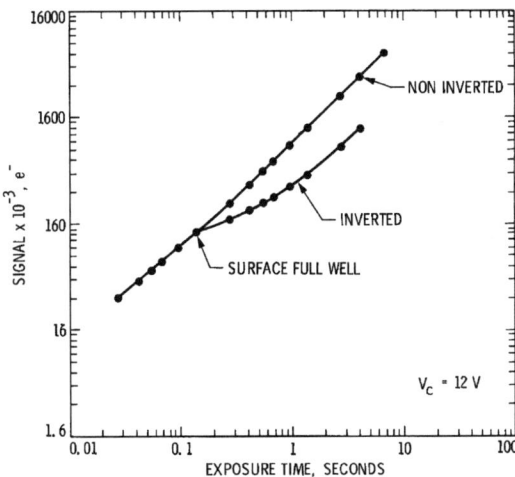

Figure 4.10(a) Linearity characteristics under partially inverted and noninverted conditions.

CTE reduction when signal charge interacts with the Si–SiO$_2$ interface, which produces lower noise values.

Figure 4.10(a) is a linearity plot taken under partially inverted ($V_B = -8$ V and $V_C = 12$ V) and noninverted ($V_B = -4$ V and $V_C = 12$ V) clocked conditions. The collecting phase is set at a high level to force SFW operation. For the inverted case, the linearity abruptly changes at SFW because of recombination. However, for the noninverted case, surface recombination does not take place, because there are no holes. The CCD continues to exhibit a linear response until at some high signal level BFW occurs (not shown in the plot). Figure 4.10(b) shows linearity residuals

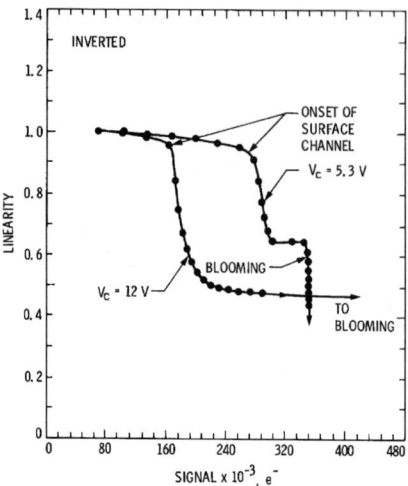

Figure 4.10(b) Linearity residuals under partially inverted clocking conditions showing BFW and SFW breakpoints.

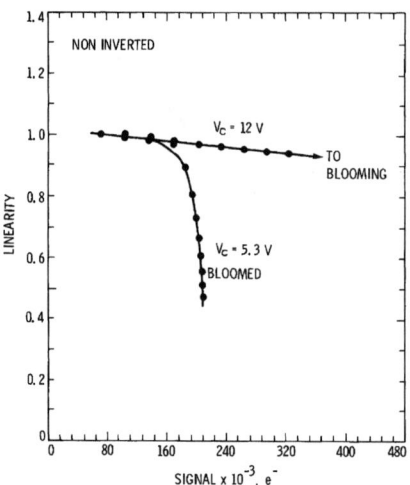

Figure 4.10(c) Linearity residuals under noninverted clocking conditions showing BFW and SFW breakpoints.

for the same CCD operating partially inverted ($V_B = -8$ V) for two different V_C. As explained above, as V_C increases, BFW increases and SFW decreases. When $V_C = 5.3$ V, $S_{SFW} = 240{,}000$ e$^-$ and $S_{BFW} = 340{,}000$ e$^-$. At $V_C = 12$ V, $S_{SFW} = 160{,}000$ e$^-$ and $S_{BFW} > 480{,}000$ e$^-$. Figure 4.10(c) is a similar plot when the device is clocked noninverted ($V_B = -5$ V). Only the BFW break point can be detected in the data. The data in Fig. 4.10 again show that SFW is difficult to detect when the CCD is clocked noninverted.

CHARGE COLLECTION

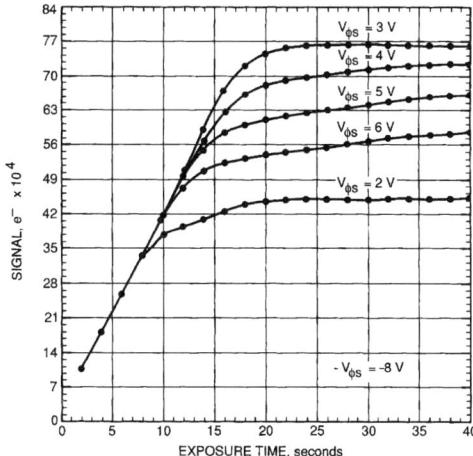

Figure 4.11 Horizontal register linearity characteristics under partially inverted clocking conditions showing BFW and SFW breakpoints.

Figure 4.11 measures full well performance for the horizontal register that is clocked partially inverted ($V_B = -8$ V). Vertical pixel binning is used to fill the horizontal register with charge beyond vertical full well. The array is illuminated with a spot of light to allow blooming within the horizontal register. Note that $V_{FULL} = 3$ V, yielding 770,000 e^- for this CCD.

4.2.7 SELF-INDUCED EMISSION AND THERMIONIC EMISSION

Charge naturally wants to escape a potential well because of two physical mechanisms. First, a self-induced electric field between electrons will try to force carriers out of the well, lowering full well performance. This field will produce a "cusp" of charge that climbs the edges of the potential well. Second, electrons will acquire thermal energy caused by random lattice vibrations (i.e., by phonons) which will also try to make the electrons escape the well. This effect is called thermionic emission. The thermal energy distribution is described by Boltzmann's statistics with a probability proportional to $e^{-E/kT}$. The average thermal energy of an electron is equal to kT which at room temperature is about 25 mV. If the energy of an electron is sufficiently large, it can jump from the collecting phase over the barrier phase, lowering BFW. The carrier can also jump into the Si–SiO$_2$ interface region, lowering SFW. This kT effect is illustrated in Fig. 4.12, which shows the surface full well charge level at two operating temperatures (300 and 178 K). Note that a higher barrier potential from the surface is required at room temperature. From this figure we would expect that the charge capacity will depend on operating temperature. Figure 4.13 presents two full well transfer curves, taken at 300 K and 178 K. Both S_{SFW} and S_{BFW} increase with decreasing operating temperature because of the kT effect.

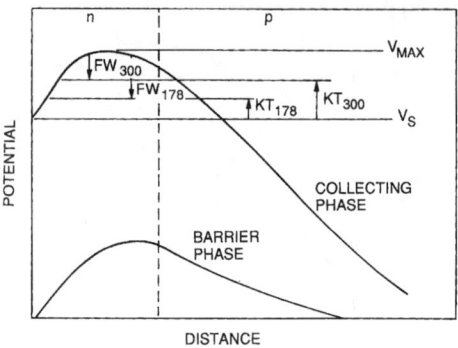

Figure 4.12 Potential well showing 178 and 300 kT barriers required to prevent charge interaction at the Si–SiO$_2$ interface.

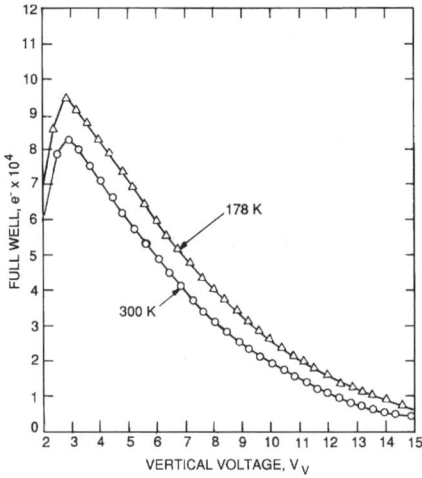

Figure 4.13 Full well transfer curves taken at 300 and 178 K.

Self-induced and thermionic emission can be significant and must be considered when CCDs are designed and processed.[1–3] For example, these effects are important to MPP, notch channel and antiblooming CCD designs, where small built-in potentials exist. We can estimate the thermal barrier height required to prevent charge from escaping the collecting phase through the equation

$$N_e(t) = \frac{2S_{FW}v_{th}t}{L}e^{-\frac{V_{TB}}{kT}}, \qquad (4.6)$$

where $N_e(t)$ is the number of electrons that escape the potential well in time t, S_{FW} is the full well level (e$^-$), L is the gate length (cm), V_{TB} is the thermal barrier (V) and v_{th} is the thermal velocity of an electron,

CHARGE COLLECTION

$$v_{th} = \left(\frac{kT}{2\pi m_e}\right)^{1/2}, \tag{4.7}$$

where m_e is the effective mass of an electron (0.9×10^{-30} kg).

Solving for the thermal barrier height in Eq. (4.6) yields

$$V_{TB} = -kT \ln\left(\frac{N_e L}{2S_{FW} t v_{th}}\right). \tag{4.8}$$

Example 4.5

Determine the barrier height required to prevent an electron from escaping a potential well in a 10-msec time period, given $L = 7\,\mu$m, $S_{FW} = 150{,}000$ e$^-$ and $T = 300$ K. Also, find the barrier height at 178 K.

Solution:
300 K:

$$kT = 0.0259\,\text{eV}.$$

From Eq. (4.7),

$$v_{th} = \{(1.3 \times 10^{-23}) \times 300/[2 \times 3.14 \times (0.9 \times 10^{-30})]\}^{1/2},$$

$$v_{th} = 2.5 \times 10^4 \text{ m/sec } (2.5 \times 10^6 \text{ cm/sec}).$$

From Eq. (4.8), the thermal barrier height is

$$V_{TB} = -0.0259 \times \ln\{1 \times (7 \times 10^{-4})/[2 \times (1.5 \times 10^5) \times 10^{-2} \times (2.5 \times 10^6)]\},$$

$$V_{TB} = 0.777\,\text{V}.$$

178 K:

$$kT = 0.0153\,\text{eV}.$$

From Eq. (4.7),

$$v_{th} = 2.02 \times 10^4 \text{ m/sec } (2.02 \times 10^6 \text{ cm/sec}).$$

From Eq. (4.8), the thermal barrier height is

$$V_{TB} = 0.456\,\text{V}.$$

Note that the thermal barriers calculated are significantly greater than the average kT at room temperature (0.0259 V).

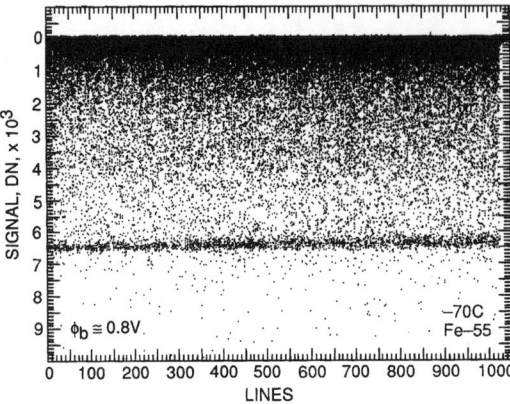

Figure 4.14(a) X-ray stacking plot where $V_C - V_B = 0.8$ V.

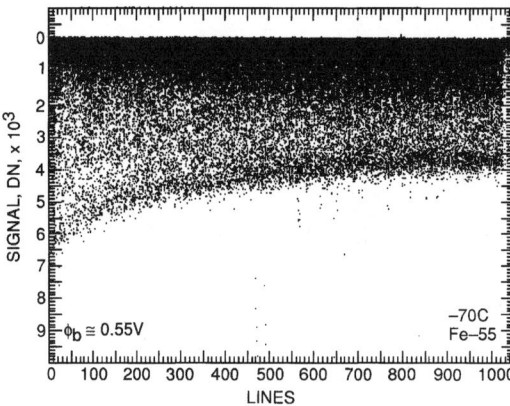

Figure 4.14(b) X-ray stacking plot where $V_C - V_B = 0.55$ V.

Figure 4.14 shows two vertical x-ray transfer responses taken from a 1024 × 1024 × 12-μm three-phase CCD operating at −70°C. The CTE data were generated by setting a small potential difference between the collecting and barrier phases to measure the thermal barrier involved (i.e., $V_C - V_B = V_{TB}$). Figure 4.14(a) shows that charge transfer is near ideal when $V_{TB} = 0.8$ V. However, in Fig. 4.14(b) at $V_{TB} = 0.55$ V, charge blooms from the collecting well because of thermionic emission. Note that the charge level does eventually reach an equilibrium level at the end of the horizontal readout.

Figure 4.15 shows how charge spreads vertically (i.e., blooms) as a function of time immediately after the CCD is quickly overexposed to a spot of light (several times greater than full well). After a 20-ms exposure period the device is unclocked for the specified time shown before fast readout commences. During this time, charge is allowed to bloom due to the kT effect. The upper traces were taken at −70°C, whereas the lower traces were taken at −30°C. Note that more blooming

CHARGE COLLECTION

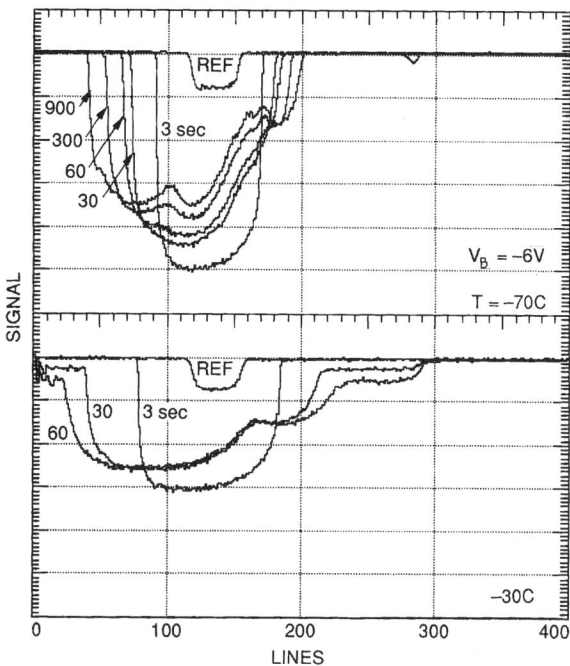

Figure 4.15 Temporal blooming characteristics at $-70°C$ and $-30°C$.

occurs for $-30°C$ operation because electrons are more energetic. The low light level trace labeled "REF" shows where the spot of light was projected. The test result indicates that full well is also a dynamic performance parameter dependent on collection and readout time.

4.2.8 CLOCKED ANTIBLOOMING

This section discusses a special clocking technique used to eliminate blooming in multi-phase CCDs that do not have built-in antiblooming (AB) structures (built-in AB pixel designs will be discussed in the next section). The method is referred to as "clocked antiblooming" or CAB for short.[4,5] Figure 4.16 illustrates how CAB works during an exposure. The method starts out by first taking phase-3 into inversion. This phase will not take part in the CAB process but will remain pinned and act only as a barrier. Phase-1 is clocked into the SFW regime at a gate voltage slightly above optimum full well. Phase-2 is clocked into inversion. We then open a shutter and expose the CCD to light. At some point in the exposure, the charge level will reach the SFW state under phase-1, trapping some electrons at the Si–SiO$_2$ interface. Before blooming occurs, charge is transferred from phase-1 to phase-2. As phase-1 inverts, trapped electrons will recombine with holes. At the same time, charge begins to collect under phase-2, which begins at SFW, forcing charge into the Si–SiO$_2$ interface. Again, before blooming occurs under phase-2,

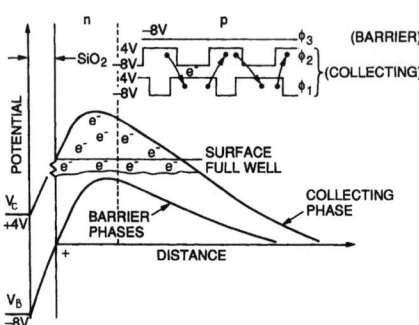

Figure 4.16 Clocked antiblooming (CAB) potential well diagram and clocking set up.

the clocks are returned to their initial states (i.e., phase-1 high and phase-2 low). Trapped charge during this time recombines under phase-2. This clocking process is repeated between phases-1 and -2 at a sufficient rate to maintain charge level less than blooming. In other words, charge recombines at the Si–SiO$_2$ interface at the same rate as charge is being generated, preventing blooming. When integration ends and the shutter is closed, the CCD is read out in a normal fashion.

Note that CAB relies on two important operating conditions. First, the CCD must be clocked into the SFW regime to allow signal charge to enter the Si–SiO$_2$ interface. This operating point is found through full well transfer. If the CCD is biased in the BFW regime, no interaction with the surface can take place, resulting in blooming. Second, the CCD must be inverted to allow holes to recombine with trapped signal electrons when reaching SFW.

Figure 4.17 shows an image of a golf ball with a bright LED above it. The setup is used to simulate the backside of planet Saturn, with the LED acting as the Sun (Cassini incorporates the CAB technique). Note that blooming for the image on the left extends across the entire length of the array. This image was taken without CAB protection. The image on the right shows no blooming when CAB is applied.

Figure 4.18 shows flat-field column trace responses generated by a frame transfer CCD. Figure 4.18(a) shows a response near full well. The left-hand side of the trace is the storage region and the right side the imaging region. Figure 4.18(b) is a similar trace after the CCD is exposed to three times full well without CAB applied. Blooming from the image region into the storage region is apparent. Figure 4.18(c) shows the response when the light level is increased by a hundred fold with CAB applied. No blooming into the storage region is observed. Note also that the metal shield above the storage region leaks light through very small pinholes in the aluminum structure (this is a common characteristic of light shields). Figure 4.18(d) shows a thousand times full well response again with CAB applied. CAB now prevents the pinholes from blooming. This imaging feature would not have been seen without CAB.

CHARGE COLLECTION

Figure 4.17 Blooming characteristics for a three-phase CCD with and without CAB applied.

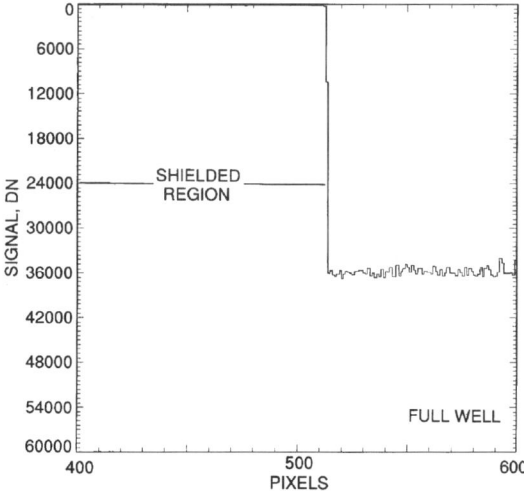

Figure 4.18(a) Full well column trace generated by a frame transfer CCD.

It is important to note in Fig. 4.18(c) and 4.18(d) that the pixel-to-pixel full well signature within the imaging region remains fixed. This is because the charge level for each pixel is forced to SFW when CAB is applied. This represents a very

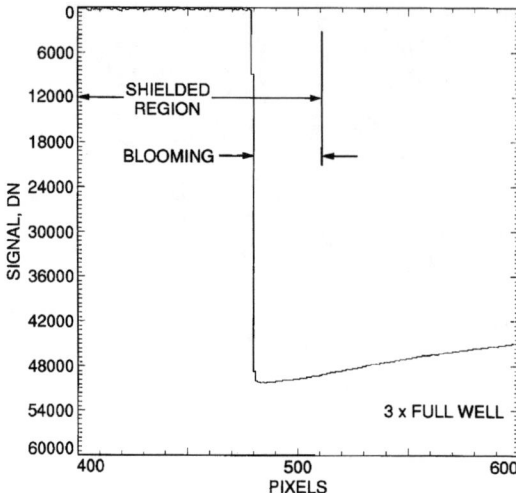

Figure 4.18(b) Blooming within the storage region for a 3× full well exposure without CAB applied.

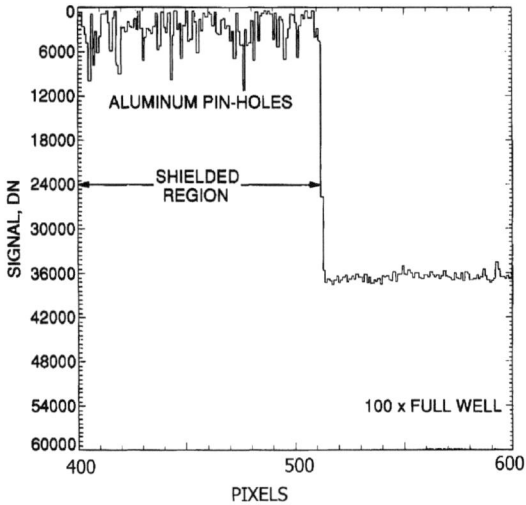

Figure 4.18(c) CAB control for a 100× full well exposure showing aluminum pin-holes in light shield.

sensitive technique in measuring SFW characteristics and is used to generate the full well transfer curves presented above.

Figure 4.19 shows two linearity plots taken of a point image with CAB applied. In the lower curve, phases-1 and 2 are clocked from $V_B = -8$ V to $V_C = 4$ V. This bias condition allows SFW interaction and CAB to function. The signal recombines at the onset of blooming, producing a horizontal response. The upper curve ($V_C =$

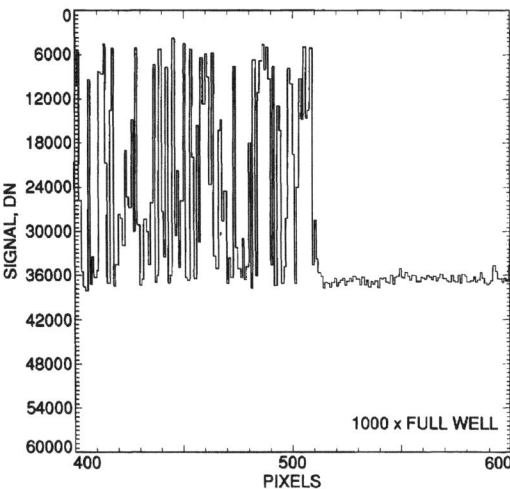

Figure 4.18(d) CAB control for a 1000× full well exposure.

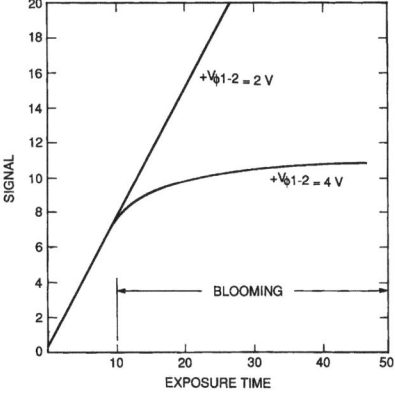

Figure 4.19 Linearity characteristics operating within the SFW ($V_C = 4$ V) and BFW ($V_C = 2$ V) regimes with CAB applied.

2 V) does not allow surface recombination, because blooming occurs instead. Here the signal continues to increase beyond the bloomed level (i.e., charge is conserved around the point source).

Discussions above indicate that the amount of blooming control that CAB provides depends on the rate of charge generated. For some CCD applications, the charge generation rate may be faster than the recombination allows. The efficiency of CAB depends on the clock dither rate and how far the CCD is driven into SFW. Figure 4.20(a) presents a calibration technique that determines the dither rate required for a given charge generation rate. Calibration begins by partially illuminating a CCD with a flat field. A frame transfer CCD will be employed in our discussions here as in Fig. 4.18. The shielded region is used to decide whether or not the CCD blooms. Six operations labeled in the plot are used to calibrate CAB. At point 1 the CCD is cleared of charge. At point 2 the shutter is opened while the CCD is continuously read out. Charge increases linearly as lines are read out until saturation occurs (normally set by the ADC). The slope of the curve signifies the charge generation rate (i.e., electrons/sec), thus providing an absolute calibration of the light source. At point 3 the shutter is closed and the CCD is erased again. At point 4 the shutter is opened with CAB applied. The shutter is then closed and CAB inhibited. At point 5 the CCD is read out under normal read conditions.

The readout portion of the curve (point 6) determines if antiblooming was effective for the rate of charge generated. One examines the shield edge to decide if blooming is controlled as shown in the figure. For example, if the device blooms under the light shield, then the clocking rate was not fast enough or the device was not driven far enough into the SFW regime. It is also possible that the charge level readout is still greater than SFW but less than BFW, indicating that the clock rate is slightly below optimum. Here, an exponential decrease in signal is seen, as shown in the figure. The exponential decrease in signal is because the phases are clocked in and out of inversion during readout, which, in turn, slowly consumes charge above SFW. Ideally, CAB should be applied for a brief amount of time after the shutter is closed to eliminate this effect. This will guarantee that all pixels are at SFW before readout commences.

Figure 4.20(b) shows actual data generated by a test 1024×1024 CCD to demonstrate the calibration procedure discussed. The light level is adjusted for a charge generation rate of 250,000 e^-/sec as determined by the slope of the curve (point 2). This value corresponds to the maximum charge rate that will be seen by the CCD for the application at hand. CAB is applied at three different rates, labeled in the figure as 2, 4 and 8 cycles per line time (the line time for the test camera is 20×10^{-3} sec). Note that 2 cycles/line are not sufficient, since blooming is observed under the light shield region. However, 4 and 8 cycles prevent blooming. Note the exponential decay in signal level during readout for the 4-cycle curve (point 6). As mentioned previously, this portion of the curve can be eliminated if CAB were applied momentarily after the shutter was closed.

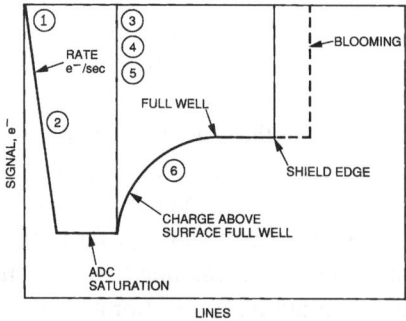

Figure 4.20(a) Illustration for CAB calibration procedure described in text.

CHARGE COLLECTION

Figure 4.20(c) plots the number of bloomed pixels under the light shield as a function of the number of seconds CAB is applied at 2500 Hz within a 10-sec exposure period. The light source was adjusted for a 10× full well level during this time. A family of curves are shown for different clock bias, showing that CAB is more efficient as the device is pushed further into SFW. However, the penalty paid for operating the device deeper into SFW is a full well loss as governed by the full well transfer curve.

The CAB technique was intended for slow-scan scientific applications. The technique is not applicable for fast framing or pulsed light source applications (e.g., pulsed lasers). The technique has been used to prevent blooming in three-phase CCDs at charge generation rates up to approximately 10^9 electron/sec. The CAB technique is not applicable for two-phase CCDs because they are difficult to drive into SFW. However, MPP CCDs can employ the technique.

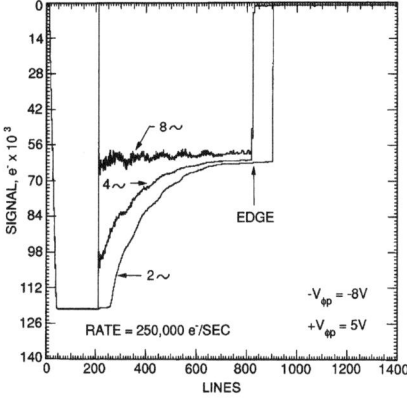

Figure 4.20(b) CAB calibration data showing that 8 cycles/line time is required to control blooming for 250,000 e⁻/sec rate.

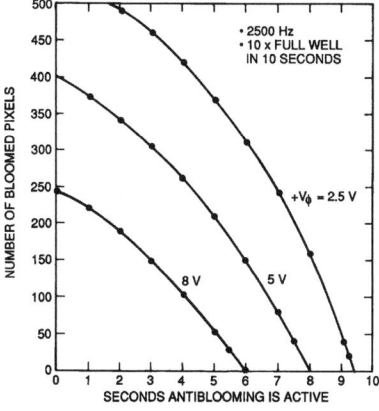

Figure 4.20(c) CAB efficiency data for different V_C.

As mentioned above, CAB is used to generate full well transfer curves by the following procedure. First, the negative vertical clock rail is adjusted deep into the pinned state (e.g., $V_B = -8$ V). The positive clock level is varied to cover the BFW and SFW regimes. A small spot of light is projected onto the center of the CCD to allow blooming (a flat-field source will result in charge backup problems near the horizontal register, preventing the CCD from blooming properly). The CCD is then overexposed to light, approximately two times greater than optimum full well. Then, with the shutter closed, CAB is applied until blooming and surface interaction have reached an equilibrium level on the array. Then, CAB is inhibited and the device is read out normally. The signal is then measured within the exposed region and plotted against gate voltage (i.e., V_C). The sequence is repeated for different V_C, thus covering the full well range specified.

4.2.9 ANTIBLOOMING STRUCTURES

This section discusses three antiblooming structures built into the pixel to prevent blooming.[6,7] As we will discuss below, most scientific CCDs do not incorporate AB protection because there are significant QE and full well penalties paid when this option is incorporated. However, for some imaging applications AB is essential. Popular AB designs implemented in CCDs include (1) lateral antiblooming gate (LABG), (2) lateral antiblooming implant (LABI) and (3) vertical antiblooming (VAB).

4.2.9.1 Lateral Antiblooming Gate

The LABG structure is shown in Fig. 4.21. The AB gate is found between the signal channel and a reverse biased n$^+$ region where bloomed charge drains. For example, when the barrier phases are pinned (e.g., at $V_B = -8$ V), the AB gate can be biased out of inversion (e.g., $V_{ABG} = -4$ V), allowing charge into the drain region before blooming occurs. In addition, the AB gate can be used to rapidly erase the CCD by biasing the AB gate to a potential greater than the collecting phase. In this way the AB gate serves as an electronic exposure control. For a three-phase CCD, the LABG structure is based on a four-level poly process, whereas a four-phase CCD requires three levels of poly. Here, poly 1 and 2 are used for the clocked phases and a poly 3 layer for the AB gate. The AB gate length and drain region are designed as small as possible to maintain high charge capacity (typically 2.5–3.0 μm each). It should also be noted that LAB structure results in QE degradation because photogenerated charge made in the AB region will be lost to the drain. Also, the LABG overlap capacitance with clocked gates is not desirable for high-speed readout.

4.2.9.2 Lateral Antiblooming Implant

An LABI design replaces the AB gate with a p$^-$ boron implant as shown in Fig. 4.21. The barrier potential, V_{AB}, in the region is controlled by the implant

CHARGE COLLECTION

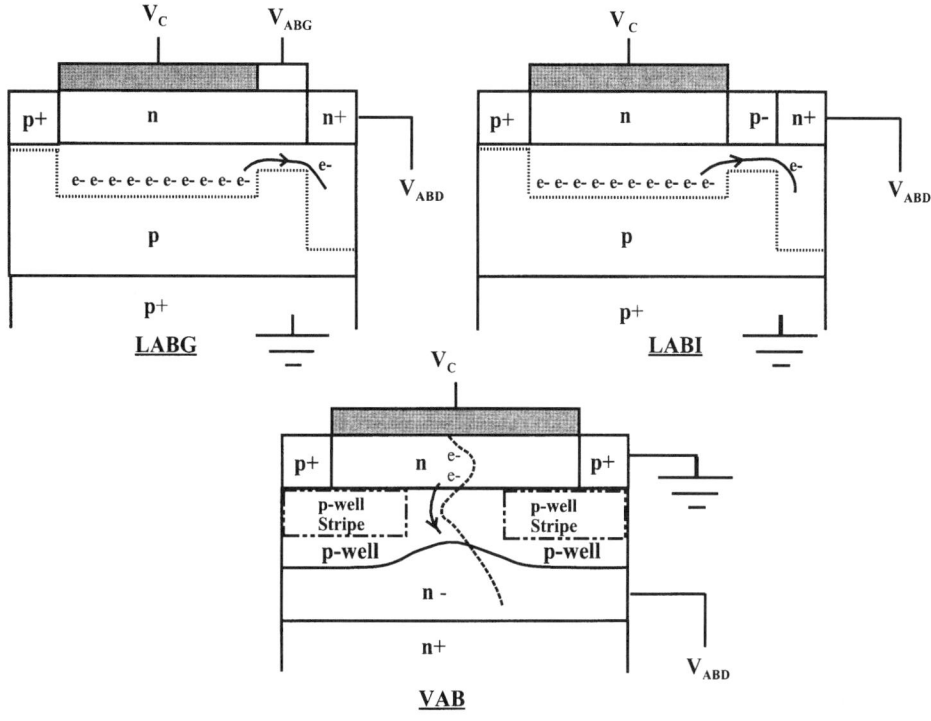

Figure 4.21 Popular built-in antiblooming structures.

concentration, the bias on the n^+ AB drain (V_{ABD}) and the collecting phase potential (V_C). Figure 4.22(a) plots V_{AB} for a test device as a function of V_{ABD} for different V_C. For this CCD, the main signal channel is doped to produce an inversion level of $V_{INV} = -9$ V and $V_{EFF} = 15$. Therefore, to prevent blooming and allow charge to spill in the drain, V_{AB} must be greater than $V_{EFF} + V_{INV}$, assuming that the barrier phases are inverted (i.e., $V_B = 15 - 9 = 6$ V). Figure 4.22(b) shows the potentials within each region for this setup condition.

Figure 4.22(c) plots charge capacity for a LABI CCD as a function of light intensity and V_{ABD}. For this sensor, full well without AB applied is approximately 162,000 e^- (optimized at $V_{FULL} = V_C = 6$ V and $V_B = -9$ V). The graph shows three different plots at $V_{ABD} = 20$, 15 and 11 V for a fixed V_C and exposure time. These bias levels produce corresponding full well levels of 70,000 e^-, 103,000 e^- and 120,000 e^-, respectively. A higher V_{ABD} voltage offers more AB protection at the expense of full well performance. For example, when $V_{ABD} = 20$ V, the sensor can accept a 1000× increase in light intensity before blooming occurs. For comparison, a $V_{ABD} = 11$ V setting will result in blooming at a lower light level, approximately a 30× increase. The difference in AB protection is because a finite amount of time is required for signal charge to leave the collecting phase and diffuse into the AB drain. If charge is generated too quickly, the potential under the collecting phase may become equal to the barrier phase. This results in bloom-

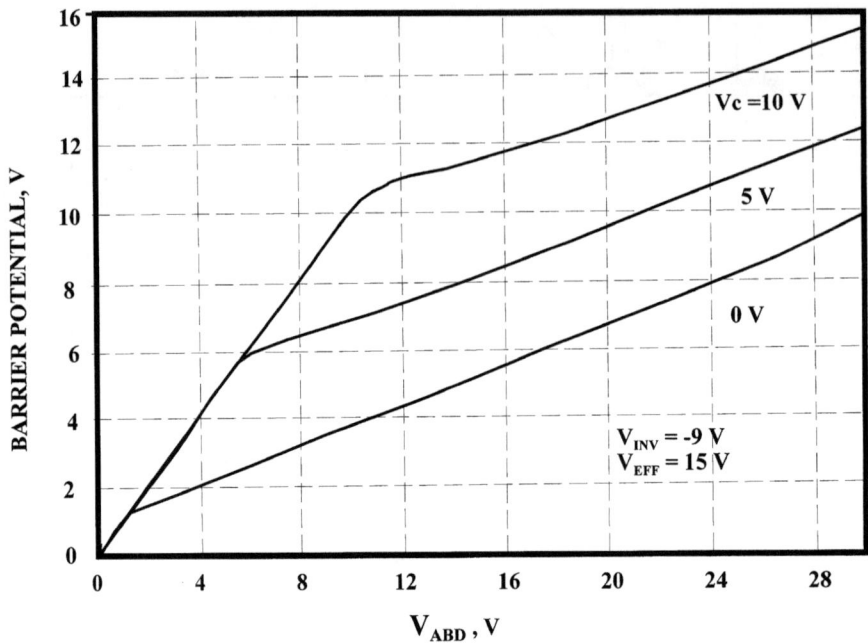

Figure 4.22(a) LABI barrier potential as a function of drain voltage for different V_C.

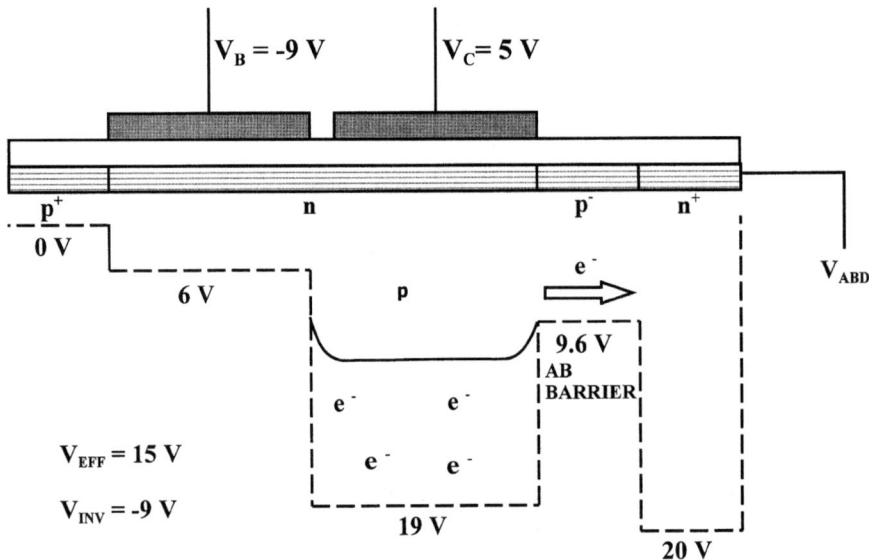

Figure 4.22(b) LABI setup condition to prevent blooming.

ing, even though the AB barrier potential is greater than the barrier phase. These characteristics give rise to the slopes seen in the full well curves of Fig. 4.22(c). The intersection of the curves with the non-AB full well level is where the gener-

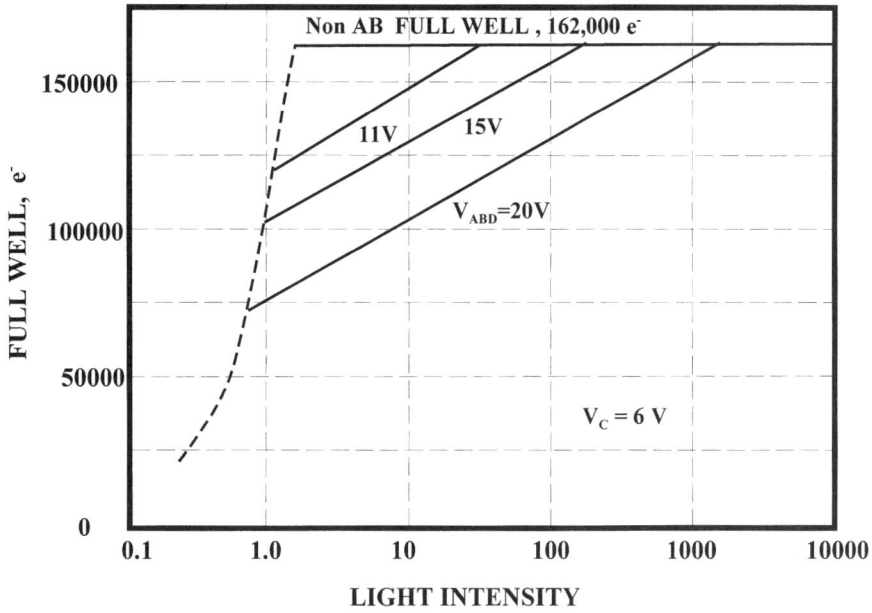

Figure 4.22(c) LABI full well transfer curve as a function of light intensity.

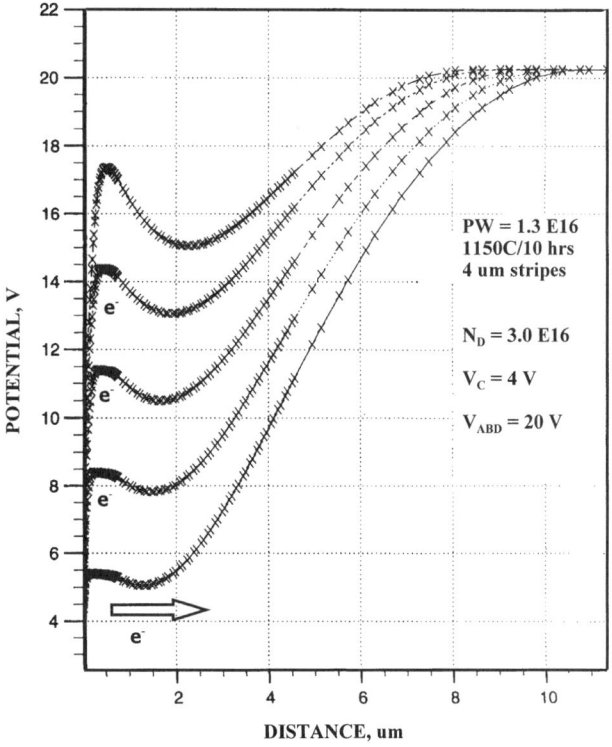

Figure 4.22(d) VAB potential wells for different signal levels.

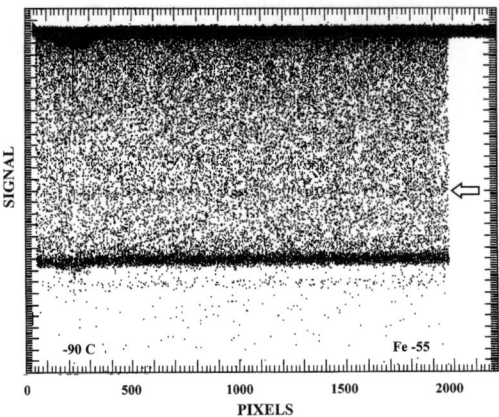

Figure 4.22(e) Horizontal x-ray transfer for a VAB CCD showing relatively few partial events.

ation rate becomes too fast for AB drain to function properly. For these reasons, the amount of antiblooming protection is specified by the maximum rate of charge generated (i.e., e⁻/sec).

Compared to LABG, the built-in barrier potential exhibits a limited bias range to work within. The boron implant dose is very critical in establishing proper blooming control. Too much of an implant will result in a large barrier potential with no blooming control. Too little implant will result in a small barrier and premature blooming into the drain region, producing low full well. Also, the drain voltage is limited because of breakdown characteristics. For example, the drain voltage for the device tested in Fig. 4.22(c) is limited to approximately 20 V. The main breakdown problem is associated with the gate oxide that separates the poly gates and AB drain region. The voltage across the oxide cannot be too high, or a leakage problem will develop (see Chapter 7 on excess charge). In addition, V_{ABD} cannot be too low, or charge injection will occur between the AB drain and collecting phase (i.e., when $V_{ABD} = V_{AB}$). For example, this occurs when $V_{ABD} = 10$ V and $V_C = 10$ V, as seen in Fig. 4.22(a). The onset of injection increases proportionally with V_C.

4.2.9.3 Vertical Antiblooming

The third AB protection method, VAB, is considerably different than LAB.[2,8,9] Bloomed charge for this technology leaves vertically through the substrate, as opposed to laterally. The VAB CCD is built on n-epitaxial material as shown in Fig. 4.21. The CCD array itself is built in a p-well formed by a high-energy boron implant (approximately 1 MeV) and extended high-temperature drive. Phosphorous is implanted within the p-well to form the n-buried channel. The potential in the center of the channel is made locally smaller by implanting the p-well in stripes centered about the channel stops, as shown. During the high-temperature drive the p-well stripes diffuse and merge, leaving slightly less concentration directly below the center of the channel. This is the location where charge leaves the channel to provide a blooming path into the substrate.

Figure 4.22(d) shows potential profiles for a VAB structure within the center of the channel as a function of signal level. The top potential curve is when charge is absent, whereas the bottom potential well has been filled to the point where charge blooms into the substrate. For this model simulation the p-well is doped to 1.3×10^{12} ions/cm^2 followed by a 1150°C, 10-hour temperature drive.

A VAB device has a very shallow photosensitive region. For example, as shown in Fig. 4.22(d), charge that is generated below approximately 2 μm will be lost to the substrate. Therefore, near-IR sensitivity for VAB CCDs is typically low (e.g., a few percent at 9000 Å). This characteristic also produces an overall low QE (e.g., the peak QE is about 25%). On the other hand, charge collection efficiency for the technology is exceptional because there is very little field-free material for charge to diffuse. For example, Fig. 4.22(e) shows a horizontal Fe^{55} x-ray response taken from a Philips 1920(H) × 1152(V) VAB CCD. The response shows a strong single-pixel-event line with relatively few split events compared to a conventional CCD.

VAB is typically less efficient in preventing blooming than LAB. This is because when the well fills with charge, the barrier moves further in the VAB case. This causes the slope in Fig. 4.22(c) to increase compared to the LAB case. The effect is quantified in a parameter called the "nonideality factor." For LAB devices, the factor is usually between 1 and 2. For VAB, it can be as high as 10. Therefore, compared to LAB devices, additional full well must be sacrificed for a specified amount of AB protection.

VAB allows for electronic exposure control by driving V_{ABD} high and clocking all phases into inversion. VAB devices cannot be thinned and backside-illuminated because the signal charge generated would not go to the front of the CCD. Also, the technology is sensitive to substrate bounce problems as discussed in Chapter 5.

4.2.10 HIGH-SPEED ERASURE

Rapidly erasing a CCD is often necessary before images and test data can be taken. For example, when power is first applied to the sensor, it is saturated. For large slow-scan CCD arrays it may take several minutes to read and clear the sensor for image taking. Also, there are instances where light flooding the CCD is necessary, which requires high-speed erasure (e.g., refer to Chapter 7 on residual bulk image).

One erasure method often suggested is to simply collapse the collecting phases into barrier phases. This bias state would force charge to bloom from the array into the horizontal register. From there it would very slowly diffuse out through the reset switch to V_{REF} (i.e., the horizontal register and reset switch act as drains for charge to escape). However, even if sufficient diffusion time is given, this scheme does not completely erase the CCD because there are always small potential differences between phases that hold some charge back.

The only effective way to ensure complete erasure is to read the CCD by clocking the vertical registers very rapidly. However, high-speed vertical clocking can lead to "charge backup" problems in the array. Charge backup occurs when the horizontal register saturates because it cannot keep up with the charge delivered by the vertical registers. The horizontal clock rate required to efficiently accept charge and prevent charge backup is

$$f_H = \frac{N_{PIXH} S_{AVG}(e^-)}{t_{LT} S_{FWH}}, \quad (4.9)$$

where f_H is the horizontal clock frequency (pixels/sec), N_{PIXH} is the number of horizontal pixels, $S_{AVG}(e^-)$ is the average signal level to be erased from the vertical registers, t_{LT} is the vertical line time when high-speed erasure is applied (sec) and S_{FWH} is the charge capacity of the horizontal registers (e^-).

Example 4.6

Determine the horizontal clock rate for a 1024 × 1024 CCD erased at a vertical rate of 0.0001 sec/line (0.1-msec frame erase time). Assume $S_{FWH} = 200{,}000\ e^-$ and a flat-field signal level in the vertical registers of $S_{AVG}(e^-) = 100{,}000\ e^-$. Also, calculate the horizontal clock rate to erase a 4096 × 4096 CCD (0.41-msec frame erase time).
Solution:
From Eq. (4.9),

$$f_H = 1024 \times (1 \times 10^5)/[10^{-4} \times (2 \times 10^5)],$$

$$f_H = 5.12 \times 10^6 \text{ pixels/sec } (1024 \times 1024),$$

$$f_H = 2 \times 10^7 \text{ pixels/sec } (4096 \times 4096).$$

This example shows that it may become impossible to clock the horizontal register at high speeds to keep up with the vertical register line rate. Also, clock driver speed and power consumption may limit how fast charge can be moved.

Erasure speed can be significantly reduced over Eq. (4.9) by offsetting the barrier potentials between the vertical and horizontal registers as shown in Fig. 4.23. Here the vertical registers are clocked partially inverted (e.g., $V_B = -8$ to $V_C = 4$ V), whereas the horizontal register is clocked noninverted (e.g., $V_B = -3$ to $V_C = 7$ V). This bias condition makes it much more difficult for charge to backup

CHARGE COLLECTION

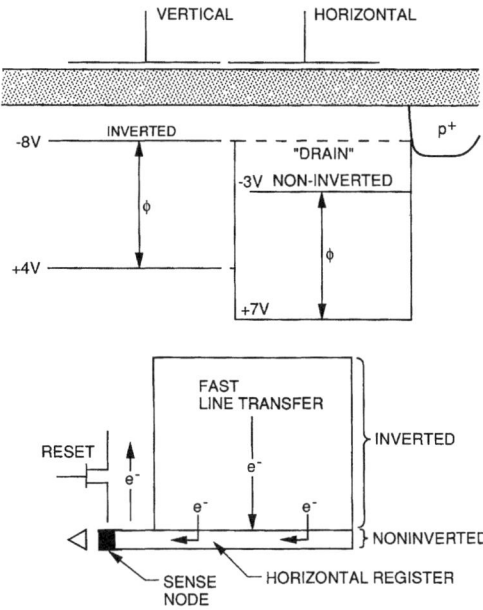

Figure 4.23 High-speed erasure bias arrangement to avoid horizontal charge backup problems.

when the horizontal register saturates because of the large barrier between the registers (in this case a 5-V barrier exists between the horizontal and vertical register). However, the erasure technique becomes limited when the horizontal blooming rate is less than the charge transferred from the vertical registers. When this happens, charge does not diffuse fast enough, causing the potential difference between registers to decrease. At some high speed, the potentials become equal and charge backup occurs. The efficiency can be improved by increasing the barrier height between the horizontal and vertical registers. Also, leaving the reset gate high during erasure allows charge to leave the device more efficiently.

The high-speed erasure technique described is intended for multi-phase CCDs where the barrier levels can be independently adjusted. Fortunately, the offset between registers is usually required for optimal performance (i.e., the vertical registers are normally operated inverted whereas the horizontal registers are not). The high-speed clocking method may not work for some CCD technologies. For example, the horizontal and vertical registers for a virtual-phase CCD must be both clocked inverted to transfer charge. Therefore, one cannot have a barrier height difference between registers. The horizontal register cannot saturate and must keep up with the vertical rate in dumping charge quickly (i.e., Eq. (4.9) applies).

Figure 4.24 shows the charge backup problem for a 1024 × 1024 CCD when the barrier phases of the vertical and horizontal registers are clocked at the same

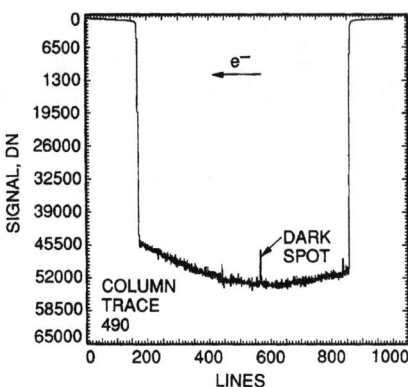

Figure 4.24(a) Reference column trace at full well level.

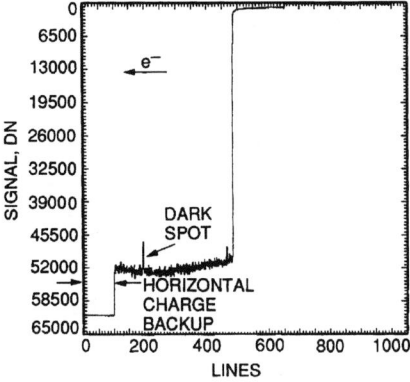

Figure 4.24(b) Readout showing the horizontal charge backup problem.

level. Figure 4.24(a) shows a full well column trace in response to a circular light stimulus. Figure 4.24(b) shows the response after the vertical registers are clocked 380 times at a rate of 50 microsec/line. Note that charge backup into the array has occurred. The dark spot (a dust particle on the surface of the CCD) labeled is used for reference purposes.

Figure 4.25 performs the same experiment where the barrier phases between the vertical and horizontal registers are offset by 5 V. Figure 4.25(a) is a baseline column trace through the circular image before rapid readout is initiated. Figure 4.25(b) is an expanded view of the dark spot region again used for reference purposes. Figure 4.25(c) shows the resultant image after 635 lines are clocked rapidly at 50 microsec/line. Note for this bias condition that charge backup does not occur, allowing the dark spot to be seen.

The 1024×1024 Cassini CCD is erased from a full well state by using a vertical line rate of 10 μsec/line and a 4-V barrier between the vertical and horizontal

CHARGE COLLECTION

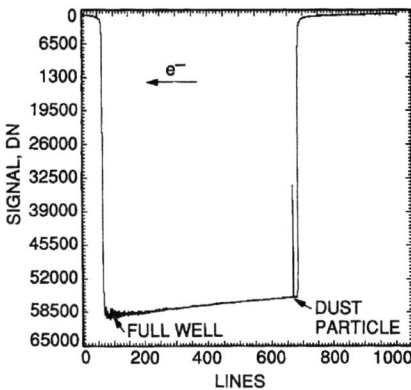

Figure 4.25(a) Reference column trace above full well level.

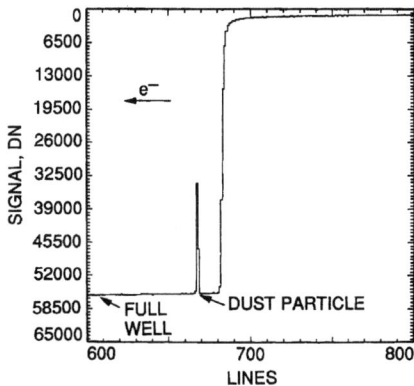

Figure 4.25(b) Magnified view of Fig. 4.25(a) showing dust particle for reference.

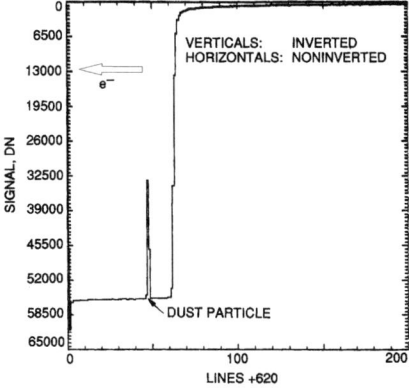

Figure 4.25(c) Readout showing the absence of the horizontal charge backup problem when the barrier phases of the horizontal and vertical registers are offset.

registers. The sensor can be erased in 10 msec. The horizontal and reset clock rates are held at the normal 100,000 pixels/sec readout rate during erasure.

It should be mentioned that AB structures discussed above can perform very high speed erasure. For example, in the case of the LABG, it is pulsed high enough to allow charge in the collecting phases to transfer to the ABD. It is also possible to include an LABG next to the horizontal register to eliminate charge backup problems.

4.2.11 Window Clocking

Window clocking is an important clocking mode related to high-speed erasure. Window clocking allows a subarray of pixels on the array to be interrogated very rapidly, for regions far from the on-chip amplifier (i.e., similar to random access). For example, a 20 × 20-pixel region at the very top of a 4096 × 4096 CCD can be acquired in a very short time compared with normal readout. This is accomplished by clocking approximately 4076 lines (i.e., 4096 − 20) very quickly at the same rate as high-speed erasure, thus guaranteeing that no charge backup will occur. Once the region of interest is reached, normal clocking and readout resumes. Windowing can be also be applied in the horizontal direction for even faster acquisition.

Window clocking is used in CCD star-tracking applications where a guide star is first found by acquiring a full-frame image. Subsequent images are then taken in rapid succession using window clocking to the region that contains the star. These data are analyzed in real time, finding the star's centroid with high precision (e.g., < 1/100 of a pixel is achieved). Multiple windows can also be designated to track several stars simultaneously.

Window clocking is also very important to testing and characterizing CCDs. For example, generating a photon transfer curve for a 4096 × 4096 CCD can be a lengthy process if normal full-frame readout is employed. Window clocking allows rapid clocking to a small array of pixels (say, a 20 × 20-pixel region) used to generate signal and noise data. The CCD is quickly erased after these pixel values are stored into the computer. Photon transfer curves can be generated in a matter of minutes using window clocking for any size CCD. Photon transfer data can also be collected from several different regions on the CCD simultaneously if desired.

4.2.12 Multipinned Phase (MPP)

Multipinned phase (MPP) technology was invented to permit multiphase CCDs to operate fully inverted.[10–12] We will discuss the effect of full inversion on dark current and residual image in Chapter 7. MPP history is worth a brief review here because of its tie to full well performance. MPP came about by accident in the late '70s when the CCD was experiencing a serious residual image problem. That is, when the CCD was exposed to a bright light source, an "after" image was seen in subsequent images (refer to Chapter 7 for example residual images). Some residual images were seen several weeks after the original exposure was taken! In fact, astronomers considered residual image to be the most serious deficiency for the CCD at that time. During the same time period it was discovered that residual image could be completely eliminated by very briefly pulsing the substrate potential positively. Although there was no solid state explanation for why residual image disappeared, the solution was so effective that it was flown on WF/PC I as a means to control residual image problems. Today we know that pulsating the substrate positively or driving the gates negatively inverts the channel which eliminates residual image by hole recombination. The WF/PC II camera elected to keep the substrate at 0 V and use bipolar drivers to invert the CCD to eliminate residual image.

In the course of residual image studies, the WF/PC I CCD channel potential plots shown in Fig. 4.26 were generated. The plots show clocking requirements to obtain inverted mode operation for each phase. Similar to the output transfer gate transfer curve shown in Fig. 1.23, inversion takes place when the channel potential becomes constant at some negative clock level. It is at the inversion point that residual image magically disappears.

Note from Fig. 4.26 that the inversion break points and channel potentials are different for each phase. After some investigation, the variation was found to be related to a very subtle process problem. Before each poly phase was deposited and patterned, the gate oxide was removed. In this step, some silicon was also unintentionally etched, removing some phosphorus from the channel. For example, the region under poly-3 experienced two etch steps, and, therefore, phase-3 had less phosphorus than phases-1 and -2. Similarly, phase-2 was doped less than phase-1 because of an additional etching step. From Chapter 1 and Fig. 4.3(a)

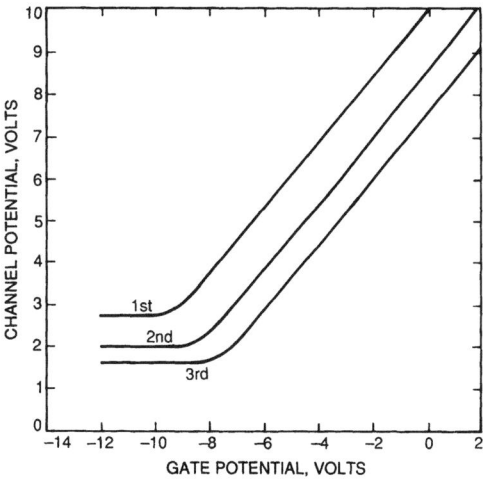

Figure 4.26 Early Hubble MPP channel potential plots.

we know that a reduction in reducing phosphorus concentration reduces the channel potential, explaining the small potential differences seen. Note that a 1-V potential difference exists between phases-1 and -3 when these phases are inverted. This observation implied that the CCD would collect some charge under totally inverted conditions. An experiment was performed which showed that charge capacity under total inversion was approximately 15,000 e^-. Although small, well capacity was large enough to measure dark current characteristics when the CCD was driven from noninverted to fully inverted conditions. Amazingly, the dark current dramatically dropped from 15 nA/cm^2 to less than 10 pA/cm^2 at room temperature!

Three-phase CCDs today do not normally exhibit potential characteristics as in Fig. 4.26. This is because most CCD manufacturers leave the gate oxide in place throughout the entire process, thus maintaining the same channel doping under each phase. For example, a standard gate insulator process is to first grow a thermal oxide (500 Å) followed by depositing a layer of silicon nitride (500 Å). However, during poly etching a fraction of the nitride layer is etched away. The nitride loss causes the channel potential to increase, but only under noninverted conditions. The channel potential under inverted conditions remains fixed (recall from Fig. 4.3 that the inversion point can only be changed through doping or channel drive). Figure 4.27 measures well capacity for a three-phase CCD with an oxide/nitride insulator when all phases are driven into inversion. A well capacity of 12,000 e^- is measured within the noninversion regime. When inverted, full well drops to zero because the channel potentials for each phase are equivalent. Therefore, CCDs with oxide/nitride cannot be driven fully inverted, as was accidently done for the early Hubble CCD.

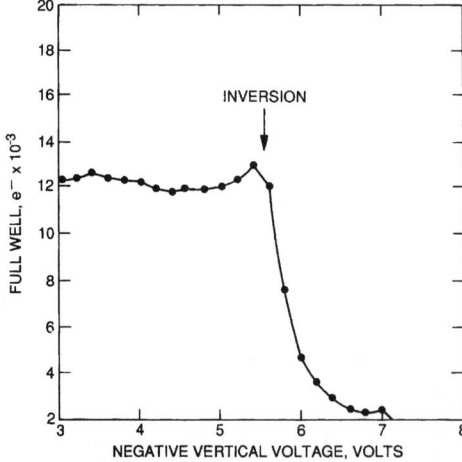

Figure 4.27 Full well characteristics for an oxide/nitride CCD when all phases are driven into inversion.

The test results of Fig. 4.26 opened the door for the MPP CCD. For this technology, an additional implant beneath one or more of the phases is included (we will be specific later). The implant increases the potential difference between the collecting and barrier phases, similar to the unintentional doping difference that occurred for the Hubble CCD. A barrier height of a few volts is typically employed, which yields respectable full well and ultra-low dark current generation when fully inverted.

The MPP implant is only applied to the vertical registers where full inversion is important to performance (i.e., dark current, residual image, etc.). The MPP implant is masked off to the horizontal register because this register is always clocked noninverted to suppress spurious charge generation, an important source of unwanted charge discussed in Chapter 7. The MPP implant is applicable to all multiphase CCDs (e.g., three-phase, four-phase, etc.). Two-phase CCDs can be driven inverted and operate like MPP because the barrier implant is already built-in.

An MPP CCD can be processed many ways. The MPP implant, for example, can be incorporated by implanting additional phosphorus under phases-2 and -3 after poly-1 has been patterned (called an n-MPP CCD). Here, charge would collect under phases-2 and -3 with phase-1 as the barrier. A four-phase MPP CCD can be fabricated by doping the poly-2 region with boron after poly-1 is patterned. Charge for this arrangement would collect under phases-1 and -2. Most three-phase MPP CCDs fabricated today dope phase-3 with boron after phases-1 and -2 are patterned (called a p-MPP CCD). Figure 4.28 shows this arrangement and the potential wells produced. Note that the MPP implant is self-aligned to poly gates. To simplify discussions below we will assume a p-MPP three-phase CCD.

Early three-phase MPP CCDs exhibited low charge capacity because the MPP implant was far from optimum. For maximum well capacity the boron concentration under the MPP phase must be carefully defined. As it turns out there is only

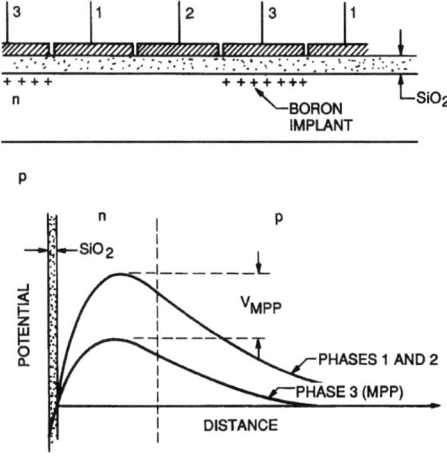

Figure 4.28 A three-phase p-MPP pixel employing boron in phase-3 CCD.

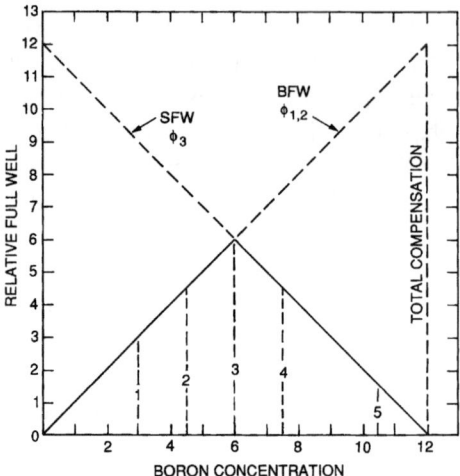

Figure 4.29 SFW and BFW characteristics as a function of MPP boron concentrations.

one optimum implant dose. As boron is implanted into phase-3, well capacity increases as the potential difference between phase-3 and phases-1 and -2 increase. However, the boron atoms also compensate phosphorus atoms, leading to a reduction of SFW under the MPP phase (this loss is experienced when reading the CCD). Optimum full well occurs when the charge capacity under phases-1 and -2 (during integration) equals the charge capacity of phase-3 (during readout). Increasing the implant beyond this point lowers SFW further until all phosphorus atoms are compensated, resulting in SFW = 0. Figure 4.29 illustrates, with two lines, the characteristics just described by plotting full well as a function of boron concentration. The first line plots BFW characteristics for phases-1 and -2. The second line is for SFW characteristics for phase-3. Charge capacity for the CCD is determined by the lowest full well curve shown. Optimum full well occurs at the point where the two curves intersect, a condition that manufacturers want to achieve. Channel doping and drive are secondary factors that also influence the optimum MPP implant.

Figure 4.30 illustrates how full well varies as a function of gate voltage for five different hypothetical p-MPP CCDs. Consider sensor #1 first, which is doped below the optimum level. At $V_C = -2$ V full well is zero because the channel potentials for all phases are equivalent. Increasing the gate voltage increases the charge capacity under phase-3 during readout. At $V_C = 0$ V phase-3 holds more charge during readout than phases-1 and -2 during integration. Full well becomes limited by these phases, producing a flat response with increasing V_C as shown. SFW for phase-3 will eventually occur at $V_C = 8$ V. At this point full well decreases as phase-3 is driven deeper into the SFW regime (again limited during readout). Device #2 is fabricated with a higher boron dose than the first CCD. Full well = 0 when $V_C = -1$ V, a shift of 1 V compared to device #1 because of the ad-

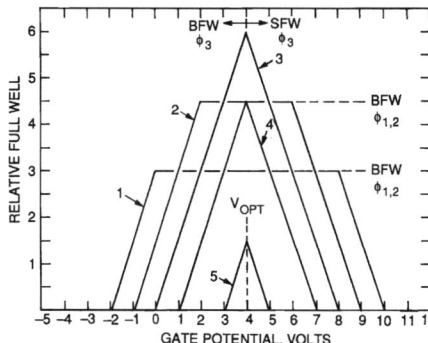

Figure 4.30 Full well transfer illustrations for different MPP boron doping concentrations.

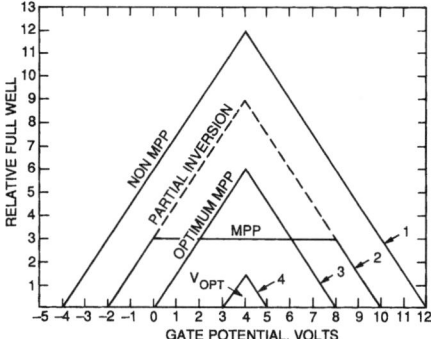

Figure 4.31 Full well transfer curves for non-MPP, optimum MPP and over optimum MPP CCDs.

ditional boron implant. Increasing the doping increases full well for phases-1 and -2 because the barrier height increases under integration. SFW for phase-3 is also offset by 1 V since more boron reduces SFW characteristics. Therefore, both BFW and SFW shift toward V_{FULL} by the same amount. The implant used for device #3 is optimum because BFW = SFW for phase-3. Devices #4 and #5 employ an implant dose greater than optimum (limited by SFW for phase-3).

Figure 4.31 illustrates full well characteristics for the four types of CCDs just described: (1) non-MPP clocked partially inverted, (2) MPP with boron implant below optimum full well, (3) MPP with optimum boron implant and (4) MPP with boron implant above optimum. Sensor #2 can be clocked either partially inverted (shown dashed) or MPP (solid). As discussed above, when clocked MPP full well is limited by the charge that can be stored under phases-1 and -2 during integration. However, when clocked partially inverted these phases are biased high, allowing for more charge capacity. Under these conditions full well is limited by SFW and BFW characteristics for phase-3. Charge capacity can be significantly different between the two clocking modes. Partial inversion and MPP clocking yield equivalent full well characteristics for sensors #3 and #4 and are dependent on SFW characteristics for phase-3. There is no advantage to implanting a CCD beyond the optimum MPP level (sensor #4).

Figure 4.32 shows an experimental full well transfer curve for an MPP CCD clocked MPP. The curve was generated using the CAB measuring technique discussed above. It is important that the MPP phase be one of the CAB clocked phases because the MPP phase will always limit SFW. The plateau response observed shows that the MPP implant is not quite at the optimum level.

Determining the optimum MPP implant through process simulation is not possible, due to the process variations involved. Therefore, the implant is normally found through trial and error procedures. For some applications it may be desirable to implant an MPP CCD below optimum, which would allow for both high charge capacity when clocked partially inverted (high-signal applications) and low dark current when clocked MPP (low-signal applications). The CCD camera can be commanded to clock either way.

Discussions above have assumed that all phases are clocked to the same positive level, including the MPP phase. Additional charge capacity can usually be obtained if phase-3 is clocked higher relative to phases-1 and -2. The improvement is related to the charge transfer speed, which is slower through a MPP phase than a normal phase. The problem has been traced to a processing detail related to the MPP implant. During high-temperature processing the MPP implant encroaches under adjacent phases (typically phases-1 and -2 for a three-phase p-MPP CCD). This effect decreases the potential at the edges of these phases, generating a small but influential potential barrier to electrons. When phase-3 is empty, these barriers are overwhelmed by fringing fields and charge transfer is well behaved. However, as the signal level increases, the fringing fields collapse and the barriers become ap-

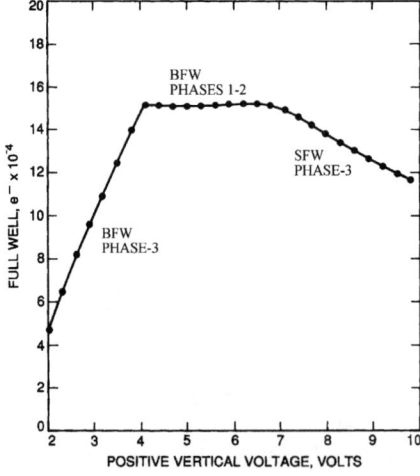

Figure 4.32 Experimental data generated by a MPP CCD with nonoptimum MPP implant.

CHARGE COLLECTION

parent, making it more difficult to transfer charge. Therefore, the full effect of this problem becomes present at full well conditions where fringing fields are smallest. Charge can only escape over the barriers by thermal diffusion. Since the potentials associated with the barriers are many times greater than kT/q, more transfer time is required compared with CCDs without the MPP implant. To overcome the built-in barrier, phase-3 can be driven harder. Although SFW decreases, the overall charge capacity increases. Improvement increases with clocking rate. Optimum clock voltages are determined by plotting full well as a function of clock drive for phases-1 and -2 at various clock levels for phase-3. It should be noted that additional clock drive on phase-3 is only beneficial when the MPP implant is above optimum. Below optimum, BFW for phases-1 and -2 will always set charge capacity (not phase-3).

Figure 4.33 shows MPP and partial inverted timing diagrams for comparison. For the MPP case the phases assume a fully inverted state during integration. After integration the CCD is readout as follows. Phase-1 goes high, forcing charge from phase-2 into phase-1. Next, charge is transferred to phase-3 by elevating phase-3 and lowering phase-1. Then phase-2 goes high and phase-3 low, forcing charge into phase-2. Then phase-2 goes low, completing one line cycle. This sequence is repeated until the CCD is read out. Note that during horizontal transfer all phases assume the inverted state. For a non-MPP CCD there is always one phase that remains high, which leads to higher dark current.

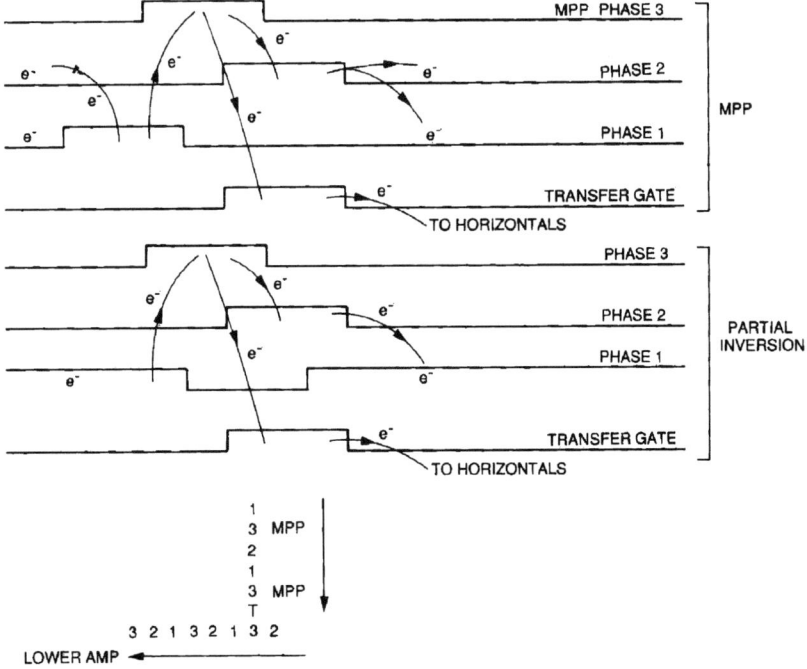

Figure 4.33 MPP and conventional three-phase timing diagrams.

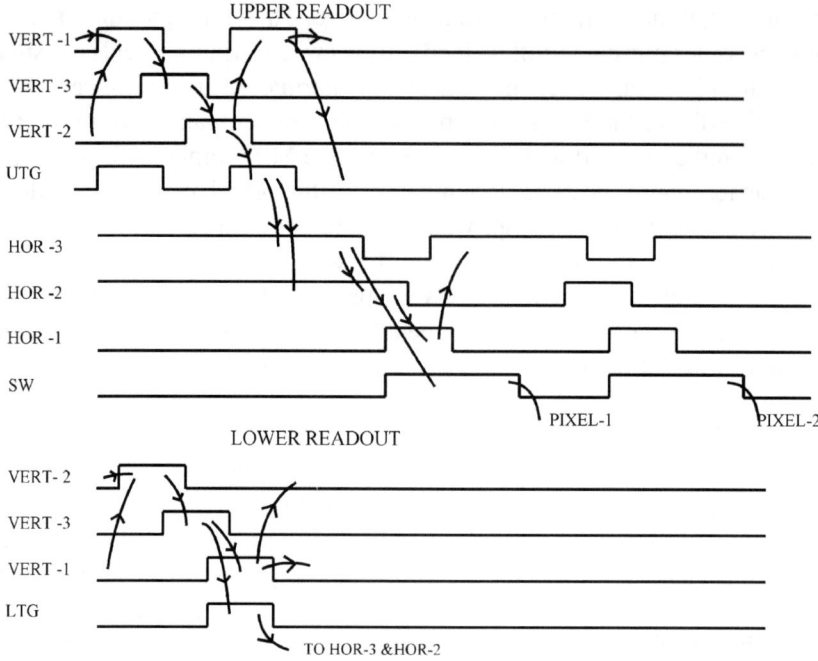

Figure 4.34 MPP timing diagrams for the functional diagram shown in Fig. 1.11(b).

It should be mentioned that a vertical clocking conflict can occur for some MPP CCDs. The situation arises when two non-MPP phases are situated between the vertical and horizontal registers, as opposed to a single phase. For example, the vertical/horizontal interface at the top and bottom of the CCD layout shown in Fig. 1.11(b) exhibits this problem, assuming the MPP implant is in all phase-3 regions. Figure 4.34 shows the required MPP clocking to transfer charge properly for each region. Note for the upper horizontal register that a double pulse clocking scheme is required.

4.3 FIXED-PATTERN NOISE

4.3.1 PIXEL NONUNIFORMITY

Pixel nonuniformity, or fixed pattern noise, is the variation in sensitivity from pixel to pixel. Photon transfer quantifies pixel-to-pixel nonuniformity. For example, Fig. 4.35 presents three photon transfer plots that separate random and fixed-pattern noise sources. The first curve (triangles) plots only shot noise. The second plot (circles) contains all three noise sources, including fixed-pattern noise. The third curve (squares) subtracts shot noise from the total noise curve leaving only fixed-pattern noise. Any point on this curve defines pixel nonuniformity through

CHARGE COLLECTION

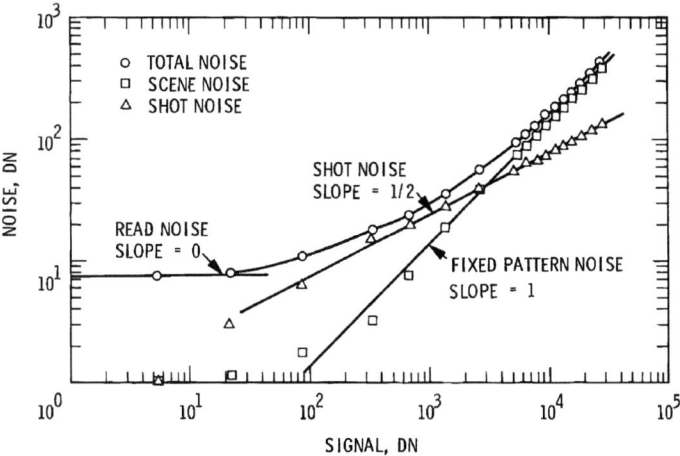

Figure 4.35 Photon transfer curves showing shot and fixed-pattern noise components.

the equation

$$P_N = \frac{\sigma_S(\text{DN})}{S(\text{DN})}. \tag{4.10}$$

Example 4.7

Determine the pixel nonuniformity for the CCD in Fig. 4.35.
Solution:
From the fixed-pattern noise curve $\sigma_S = 100$ DN at $S(\text{DN}) = 8000$ DN.
From Eq. (4.10),

$$P_N = 100/8000,$$

$$P_N = 0.0125 \ (1.25\% \ \text{nonuniformity}).$$

As seen in Fig. 4.35, pixel nonuniformity noise dominates most of the sensor's dynamic range and, therefore, will significantly influence signal-to-noise performance. S/N that includes pixel nonuniformity for a low-contrast scene is

$$(\text{S/N})_{CON} = \frac{C_{CON} S_{AVG}(\text{e}^-)}{\{S_{AVG}(\text{e}^-) + [P_N S_{AVG}(\text{e}^-)]^2 + [\sigma_R(\text{e}^-)]^2\}^{1/2}}, \tag{4.11}$$

where C_{CON} is related to the signal contrast defined in Eq. (2.33).

When S/N is limited only by pixel nonuniformity this equation reduces to

$$(S/N)_{CON} = \frac{C_{CON}}{P_N}. \qquad (4.12)$$

Note that the S/N is constant and independent of signal level. This is a different result compared with the shot-noise-limited case where S/N increases by the square root of signal [refer to Eq. (2.35)].

Example 4.8

Determine the S/N where the contrast is 0.01 (1%) and the pixel nonuniformity is 0.01 (1%).
Solution:
From Eq. (4.12),

$$(S/N)_{CON} = 0.01/0.01,$$

$$(S/N)_{CON} = 1.$$

For this example, the modulation would be nearly lost in fixed pattern noise.

The signal level where pixel nonuniformity begins to dominate shot noise (i.e., $S_{AVG}^{1/2} = P_N S_{AVG}$) and where Eq. (4.12) applies is

$$S_{OPN}(e^-) = \frac{1}{P_N^2}, \qquad (4.13)$$

where S_{OPN} is the onset of pixel nonuniformity (e^-).

Example 4.9

Determine the signal level where the S/N becomes constant when limited by pixel nonuniformity of 0.01 (1%).
Solution:
From Eq. (4.13),

$$S_{OPN}(e^-) = 1/(0.01)^2,$$

$$S_{OPN}(e^-) = 10,000 \ e^-.$$

Note that this signal level represents a very small portion of the CCD's dynamic range. Above 10,000 e^- the S/N is constant. To achieve higher S/N, pixel nonuniformity must be removed. In a sense, dynamic range for the CCD becomes limited at this point.

4.3.2 FLAT FIELDING

The examples above show that S/N for low-contrast scenes is limited by pixel nonuniformity over most of the dynamic range. Fortunately, the excellent linear characteristics exhibited by the CCD permit simple computer algorithms to eliminate pixel nonuniformity easily and accurately. One technique used extensively in scientific CCD applications is called "flat fielding," a computer process that adjusts the sensitivity of each pixel to the same level. The computer algorithm used is based on the following linear equation:

$$S_i = \mu_C \frac{b_i}{a_i}, \qquad (4.14)$$

where μ_C is the average signal level of the calibration frame (e$^-$), S_i is the value of the corrected ith pixel (e$^-$), b_i is the value of the data ith pixel (e$^-$) and a_i is the value of the ith flat-field calibration pixel (e$^-$).

The example below will show how this flattening process is performed.

Example 4.10

Assume the following signal levels for a subarray of data pixels (b_i):

25_{01} 38_{02} 39_{03} 45_{04}
34_{05} 33_{06} 35_{07} 44_{08}
26_{09} 31_{10} 36_{11} 27_{12}
23_{13} 28_{14} 29_{15} 32_{16}

and a subarray of calibration pixels (a_i):

225_{01} 342_{02} 351_{03} 405_{04}
306_{05} 297_{06} 315_{07} 396_{08}
234_{09} 279_{10} 234_{11} 243_{12}
207_{13} 252_{14} 261_{15} 288_{16}

Determine the resultant flattened subarray by applying Eq. (4.14).
Solution:
The average level for the calibration array of pixels is $\mu_C = 288$ e$^-$. From

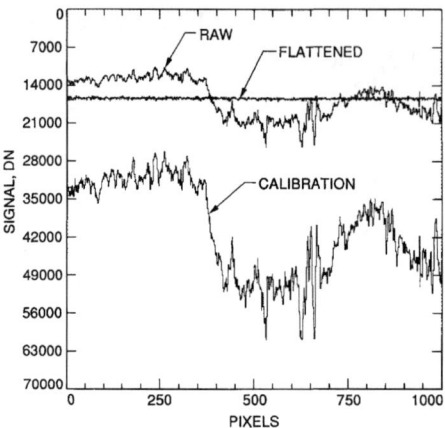

Figure 4.36 Line plots demonstrating the flat-fielding process.

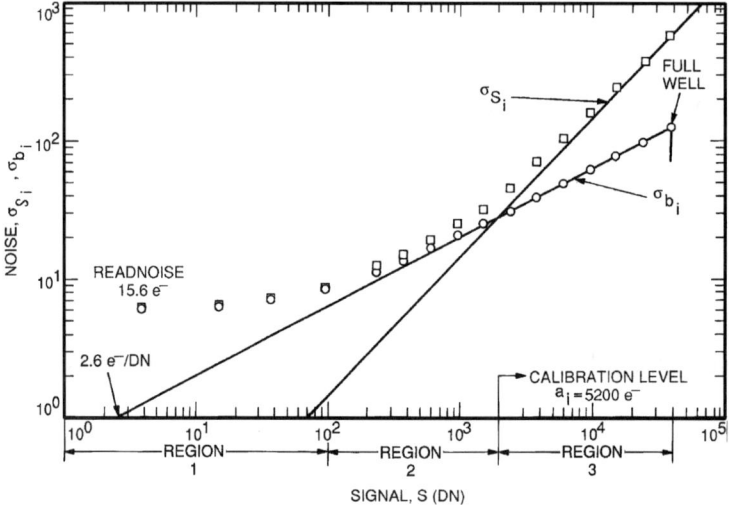

Figure 4.37 Photon transfer curves showing S/N performance before and after flat fielding.

Eq. (4.14) the resultant flattened array, S_i, becomes

$$32_{01}\ 32_{02}\ 32_{03}\ 32_{04}$$
$$32_{05}\ 32_{06}\ 32_{07}\ 32_{08}$$
$$32_{09}\ 32_{10}\ 32_{11}\ 32_{12}$$
$$32_{13}\ 32_{14}\ 32_{15}\ 32_{16}$$

CHARGE COLLECTION

Figure 4.36 presents experimental data generated by a 1024×1024 CCD showing the flattening process. Shown at the top is a raw line trace of pixels (b_i) and below it a calibration line trace (a_i) for the same pixels taken at a higher signal level (both are generated from a flat field light source). The calibrated flattened line trace is also shown (S_i).

A flattened image will contain additional shot noise from the calibration frame, thereby degrading S/N performance. The amount of degradation is found through the propagation of errors formula when applied to Eq. (4.14), which yields

$$\sigma_{S_i}^2 = \sigma_{bi}^2 + \left(\frac{b_i}{a_i}\sigma_{ai}\right)^2 + \sigma_R^2, \qquad (4.15)$$

where σ_{S_i} is the resultant noise of the corrected pixel, σ_{bi} and σ_{ai} are the shot noise components of the raw and calibration pixels, respectively, and σ_R is the read noise.

Equation (4.15) shows that optimum S/N is achieved when the signal level of the calibration frame is greater than the data frame. If the level of the calibration frame is less than the raw frame, S/N will degrade by the factor $\sigma_{ai}(b_i/a_i)$ from ideal. Figure 4.37 shows S/N performance where σ_{S_i} (squares) and σ_{bi} (circles) are plotted as a function of signal. The raw pixels are corrected by Eq. (4.14) assuming an average calibration level of $\mu_C = 5200$ e$^-$ (or 2000 DN assuming 2.6 e$^-$/DN). There are three regions of interest:

> *Region 1* $\sigma_R > \sigma_{ai} > \sigma_{bi}$, $b_i/a_i \ll 1$.
> In this region, read noise dominates (i.e., $\sigma_{S_i} = \sigma_R$). A characteristic slope of zero is measured for σ_{S_i}. In this region there is no loss of information due to flat fielding.
> *Region 2* $\sigma_{bi} > \sigma_R$, $\sigma_{bi} < \sigma_{ai}$, $b_i/a_i < 1$.
> In this region, shot noise dominates such that $\sigma_{S_i} = \sigma_{bi}$. A slope of half is measured. In this region there is no loss of information.
> *Region 3* $\sigma_{bi} \gg \sigma_R$, $\sigma_{ai} > \sigma_{bi}$, $b_i/a_i > 1$.
> In this region, the shot noise of the calibration frame dominates so that $\sigma_{S_i} = (b_i \sigma_{ai})/a_i$. A slope of unity results. In this region there is a loss of information (i.t., S/N).

Note that when the raw and calibration signal levels are equivalent, the noise of the corrected image increases by the square root of 2 over σ_{bi}.

Example 4.11

> Determine the S/N for a corrected image when $b_i = 10{,}400$ e$^-$ and $a_i = 5200$ e$^-$ for the data presented in Fig. 4.37.
> *Solution*:
> From Figure 4.37,
>
> $$b_i = 10{,}400 \text{ e}^-, \quad \sigma_{bi} = 102 \text{ e}^-,$$

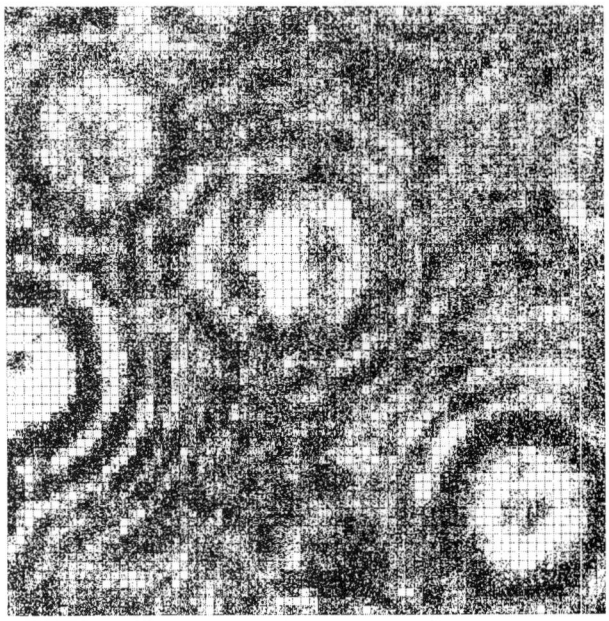

Figure 4.38(a) A flat-field, narrowband, colliminated image showing fixed-pattern noise caused by dust on a CCD window.

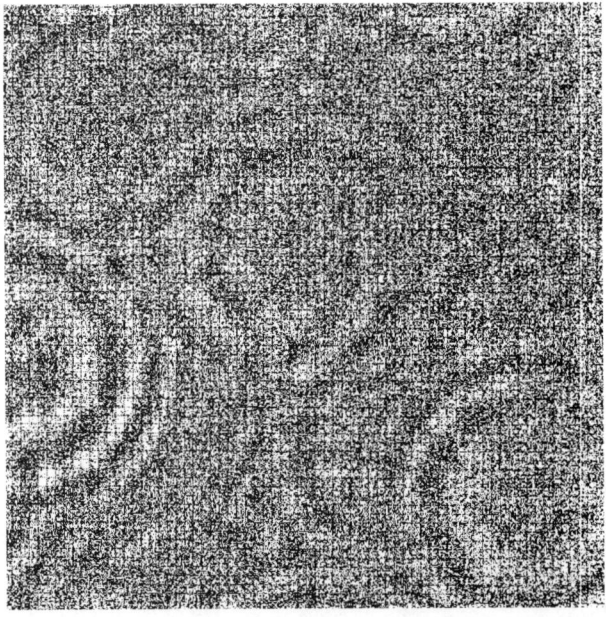

Figure 4.38(b) The same image flattened showing remnants of donut patterns caused by a slight spatial shift in the raw and calibration frames.

$a_i = 5200 \text{ e}^-$, $\sigma_{ai} = 72 \text{ e}^-$,

$\sigma_R = 15.6 \text{ e}^-$.

Using Equation (4.15),

$\sigma_{S_i} = \{102^2 + [(1.04 \times 10^4) \times 72/(5.2 \times 10^3)]^2 + (15.6)^2\}^{1/2}$,

$\sigma_{S_i} = 177 \text{ e}^- \text{ rms}$.

The resultant S/N is

S/N $= (1.04 \times 10^4)/177$,

S/N $= 58$.

The ideal S/N is

S/N $= (1.04 \times 10^4)/102$,

S/N $= 102$.

Flat fields are taken immediately after raw CCD images have been collected. It is important that these frames be collected under identical conditions or reminant fixed pattern noise will be in the processed image. This includes using the same camera gain, optical setup (e.g., filters), CCD operating temperature and clock voltages. For example, pixel nonuniformity is sensitive to clocking conditions as discussed below. Vignetting, dust particles above the CCD and QEH also contribute to fixed-pattern noise which can change with time.

Figure 4.38(a) shows a flat-field response generated by a CCD exposed to collimated narrowband (4000 Å) light. The round circular patterns seen in the image are diffraction patterns generated by dust particles that are on the window of the CCD. These patterns are called "donuts" and are commonly seen in raw unprocessed CCD images. Figure 4.38(b) shows a processed image after flat fielding is applied. Although fixed-pattern noise is reduced, some interference remains, indicating that movement between the calibration and raw frames occurred (the actual shift here being less than a pixel). It is usually difficult to flatten narrowband images completely.

4.3.3 FIXED-PATTERN SOURCES

The main source of pixel nonuniformity is related to process variations and photomask alignment errors.[13] Curiously, not much attention has been given by the CCD manufacturer to reducing pixel nonuniformity below 1% even though the noise

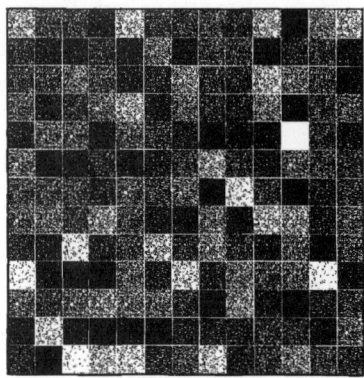

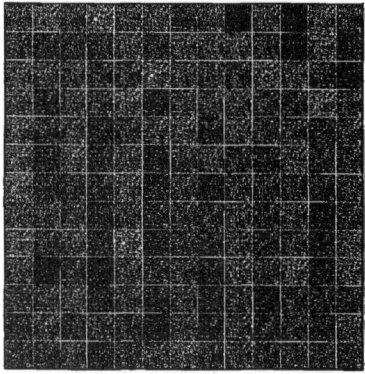

Figure 4.39 Partial flat-field images showing pixel nonuniformity for noninverted and inverted conditions.

source dominates most of the CCD's dynamic range. The main reason for the lack of development in this area is that flat fielding takes care of the problem.

Pixel nonuniformity is also sensitive to clock drive. How charge collects within a pixel depends on the distribution of small potential bumps under the barrier phase. The potential bumps will preferentially direct charge to the collecting phases. The potential bumps can be reduced by pinning the barrier phases. For example, potential bumps created by local thickness variations of the gate insulator will be flattened when inverted (recall from discussions above that the thickness of the gate oxide layer has no effect on channel potential when inverted). Figure 4.39 shows two magnified flat-field images (169 pixels) taken with a three-phase CCD under noninverted ($V_B = -2$ V) and partially inverted ($V_B = -8$ V) clocking conditions. Phases-2 and -3 act as barriers with phase-1 as the collecting phase. Fixed-pattern noise is evident in both images. However, nonuniformity is less for the inverted case (1% compared with 5%).

Figure 4.40(a) plots pixel nonuniformity for a three-phase CCD as a function of V_B on phase-2 for various phase-3 levels with phase-1 fixed at $V_C = 7$ V dur-

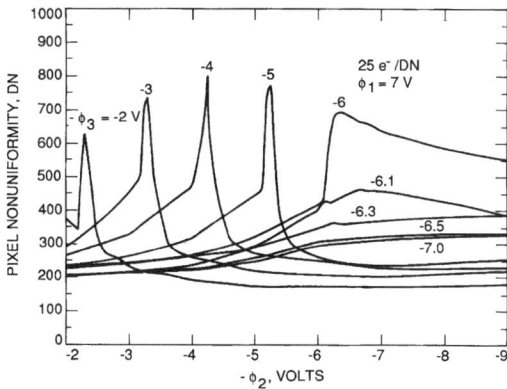

Figure 4.40(a) Pixel nonuniformity as a function of clock bias.

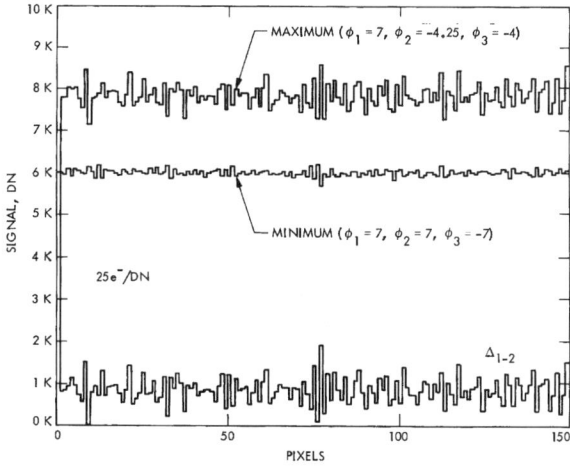

Figure 4.40(b) Line traces showing pixel nonuniformity at two different bias conditions.

ing integration. Note that pixel nonuniformity peaks when the barrier phases are at approximately the same voltage. Under this clocking condition the channel potential is flat between the barrier phases, accentuating the effects of small potential bumps. Low pixel nonuniformity results when the barrier phases are offset from each other, directing charge more efficiently to the collecting phase. Also note that fixed-pattern noise is reduced when both barrier phases are driven into inversion (-7 V) for reasons explained above. Figure 4.40(b) shows line traces for two bias conditions showing maximum and minimum fixed pattern noise. Note that nonuniformity exhibits an even-odd structure as charge splits between neighboring pixels. Figure 4.40(c) measures pixel nonuniformity as a function of signal for maximum pixel nonuniformity (5%) and minimum nonuniformity (1%) under the bias conditions indicated. Again, maximum pixel nonuniformity oc-

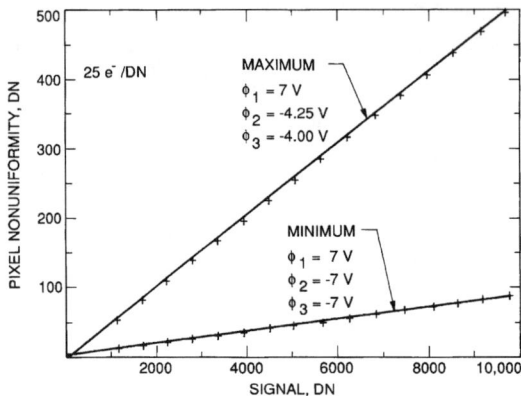

Figure 4.40(c) Maximum and minimum pixel nonuniformity as a function of signal for two different bias conditions.

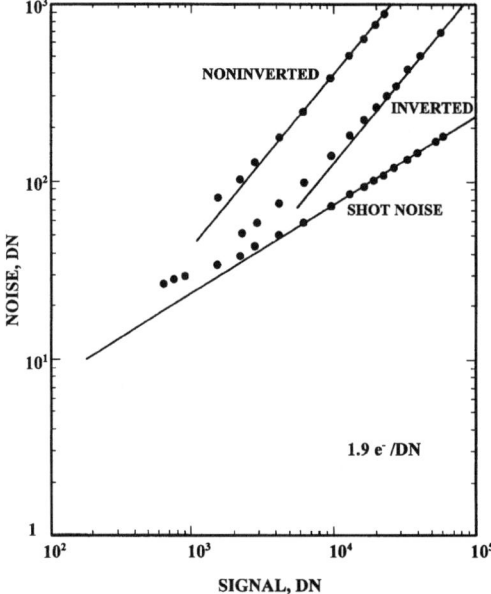

Figure 4.40(d) Photon transfer curves that measure pixel nonuniformity under partially inverted and noninverted clocking conditions.

curs when the barrier phases are driven noninverted with near-equal drive. Minimum fixed-pattern noise is achieved when barrier phases are driven into inversion. Figure 4.40(d) shows two photon transfer curves taken under noninverted and partially inverted conditions, showing pixel nonuniformities of 4.5 to 1.25%, respectively.

Three-phase data presented above show that minimum pixel nonuniformity occurs when charge is collected under two phases and with the barrier phase pinned.

CHARGE COLLECTION

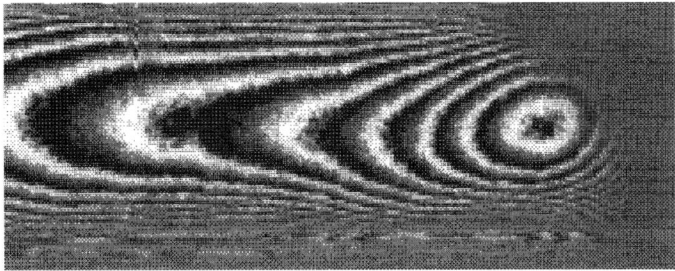

Figure 4.41 Narrowband 10,000-Å interference pattern generated by a backside CCD.

Similarly, three collecting phases should be employed for a four-phase CCD for minimum pixel nonuniformity. The disadvantage of collecting charge under one phase is that dark current will be greater because fewer barrier phases are pinned. Also, full well will be lower in the case of four-phase.

Narrowband spectral interference is another form of fixed-pattern noise. As mentioned in Chapter 3, fringing patterns are caused by multiple reflections internal to the CCD. Collimated (or high $f/\#$), narrowband (< 10 nm) light sources typically show the fringing effect. Fringing is especially important for thinned CCDs at wavelengths greater than 8000 Å. For example, Fig. 4.41 shows a narrowband 10,000-Å flat-field response generated by a rear-illuminated 404×64 HRIS CCD, showing a fringing pattern.

Fringing is also a problem for frontside-illuminated CCDs due to multiple reflections within the gates and other topside layers. For example, Fig. 4.42 shows fringing for an early front-illuminated 1×500 Fairchild linear CCD. Data were taken from a single pixel using a monochrometer source set at a bandwidth of 5 nm.

Fringing can also occur between the CCD and optical elements above it. Figure 4.43 shows 5461-Å flat-field image with deep fringes for a Galileo CCD. The interference seen is caused by reflections between the CCD and an optical window that is nearly perfectly flat to the sensor's surface.

Figure 4.44 shows a narrowband interference pattern generated by a backside-illuminated RCA 512(V) $\times$ 320(H) CCD. The fringing is caused by light being reflected from the back surface of the CCD and the glass support structure epoxied to its surface. The pattern is called the "zebra effect" for obvious reasons.

A "step-and-repeat pattern" associated with the reticles used in fabricating the CCD is another source of pixel nonuniformity. CCD reticles are manufactured using an e-beam process that patterns chrome on glass. The e-beam has limited field and can only work on a small region at a time, approximately 32×32 pixels in the case of the WF/PC II CCD (i.e., step and repeat). The process of transferring the entire design onto a reticle was not perfect for Hubble and left "seams" between regions on the order of a tenth of a micron. The seams in turn result in pixel nonuniformity that show up as a low-contrast checkerboard pattern. For example,

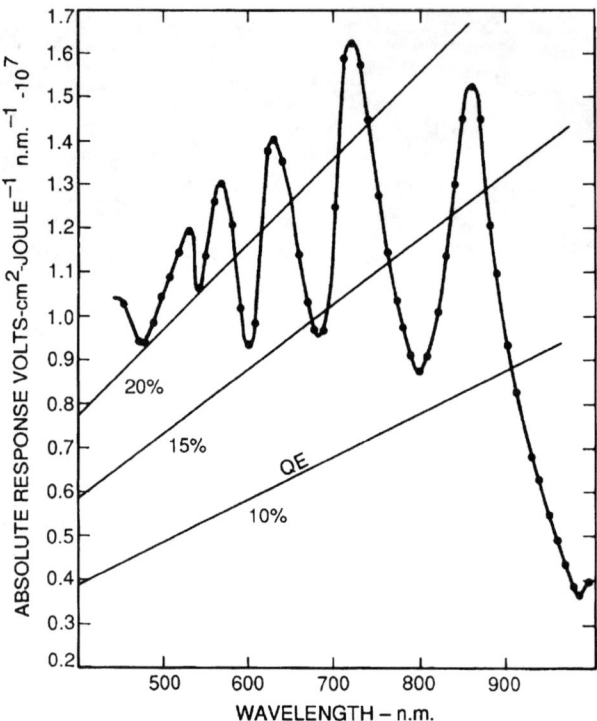

Figure 4.42 Narrowband QE response for a single pixel as a function of wavelength.

Fig. 4.45(a) is a highly "stretched" 9500-Å flat-field image showing the pattern. Figure 4.45(b) is a line trace through the image showing the pattern in more detail. Here, several frames of data were taken and averaged together to reduce the shot noise and increase the S/N of the pattern.

Another nonuniformity problem is related to contamination of the surface of the CCD or the window above it. The problem is sometimes associated with vacuum pumps, which can leave oil droplets that act as microlenses on the surface of the CCD (the small bright circular spots in Fig. 4.43 are caused by this problem). Groups have been successful in cleaning the CCD with acetone and afterwards flushing with isopropyl alcohol to eliminate the surface residue left on by acetone. The cleaning process is finished by carefully blowing the CCD with dry air.

Figure 4.46 shows an interesting nonuniformity pattern related to the first column of pixels taken from a virtual-phase SXT CCD. The fixed pattern observed is related to the bond pads that are around the CCD. When the CCD is exposed to light in the bond pad regions, charge diffuses into the first and last columns of the array. The bond pads shield the CCD from incoming light, resulting in the dark patterns seen. This problem can be eliminated by using n^+ guard columns on either side of the array to drain charge away.

CHARGE COLLECTION

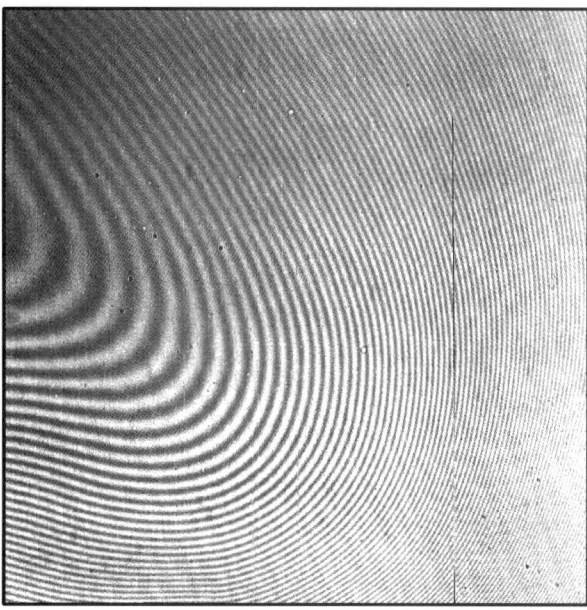

Figure 4.43 Interference pattern generated between the surface of the CCD and an optical window above the sensor.

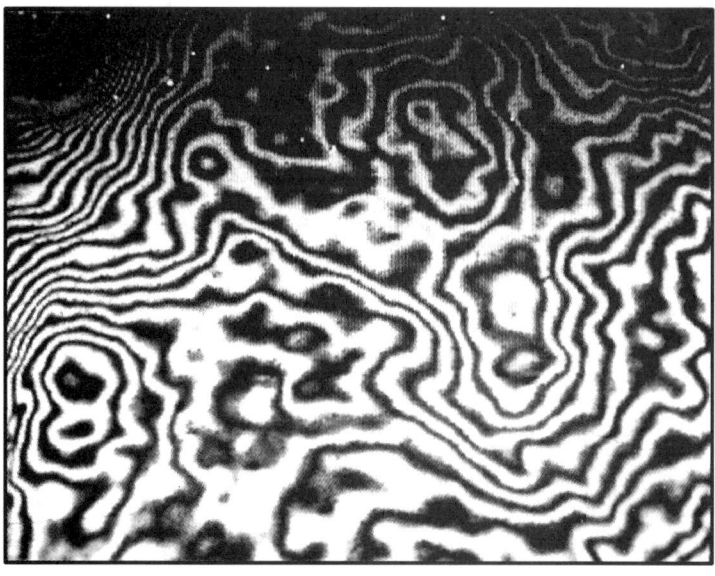

Figure 4.44 Interference pattern generated by a backside-illuminated CCD that is supported by a glass substrate mounted to the rear surface.

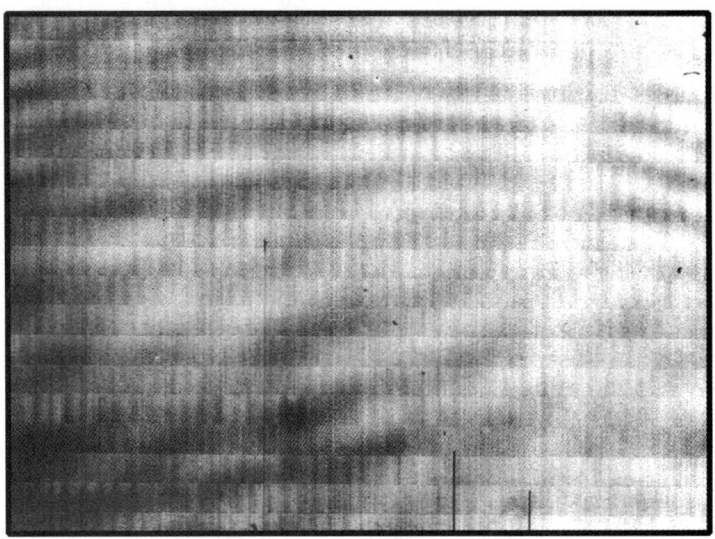

Figure 4.45(a) Step-and-repeat pixel nonuniformity.

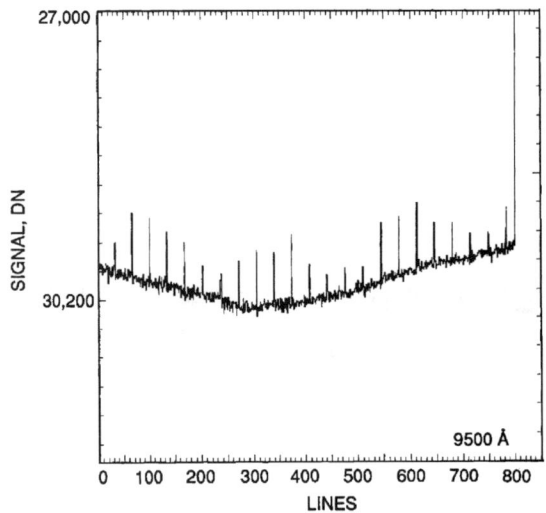

Figure 4.45(b) Line trace through step-and-repeat pattern.

4.4 CHARGE DIFFUSION

4.4.1 CHARGE DIFFUSION

Charge diffusion primarily takes place in field-free material below the potential wells and within the substrate layer. Figure 4.47(a) presents a cross-sectional view of the epitaxial and substrate regions showing where field-free regions exist. The

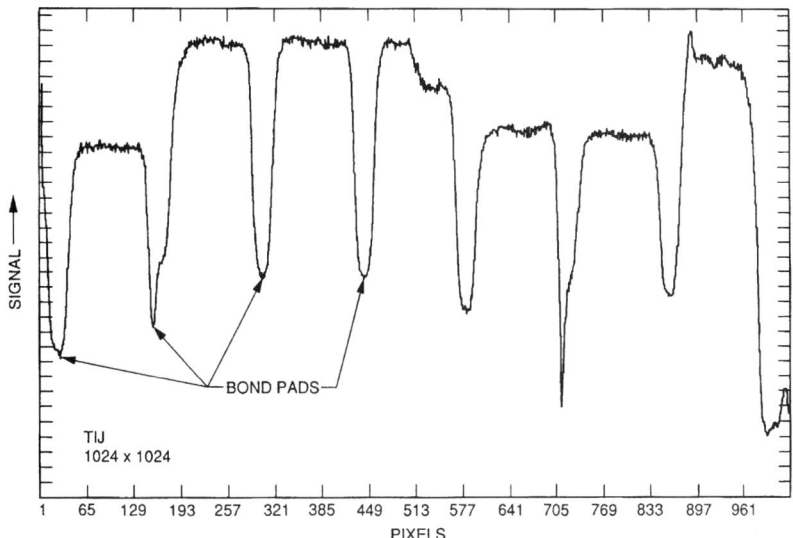

Figure 4.46 Pixel nonuniformity signature generated by bond pads at the edge of a SXT CCD.

potential distribution and electric field strength throughout the silicon layer are also shown. The primary diffusion problem is related to the epitaxial layer because the substrate layer for the most part is optically dead. This is not the case for CCDs built on bulk material (i.e., nonepitaxial), which exhibit very poor charge collection efficiency (CCE) in the near IR. For the discussion below, we will assume the CCD is fabricated on epitaxial material.

Note in Fig. 4.47(a) that there is a small field region generated at the epitaxial interface because of the doping gradient found there. This built-in field will reflect carriers that try to diffuse toward the substrate region, thus improving CCE. The field will also attract electrons generated in the substrate region that do not recombine, degrading CCE. For backside-illuminated CCDs the epitaxial interface is thinned away, eliminating this built-in field. In its place the accumulation region will generate a much stronger reflecting wall. Thinning, if done properly, will also remove most field-free material. It is for these two reasons that backside CCDs can offer superior CCE performance compared to front-illuminated CCDs.

Figure 4.47(b) shows a Monte Carlo response for a single electron that diffuses between the rear surface of a backside-illuminated CCD and front depletion edge. The reflecting wall at the backside is provided by the electric field produced by the accumulation layer. If the electron reaches the backside by the random walk, the backside field will reflect it toward the frontside to allow it to wander again. Diffusion may cause it to wander away from the target pixel, degrading CCE. By luck, this electron will probably find its way to the target pixel. This analysis shows the importance of thinning very near the depletion edge to prevent photo electrons from wandering too far from the target pixel.

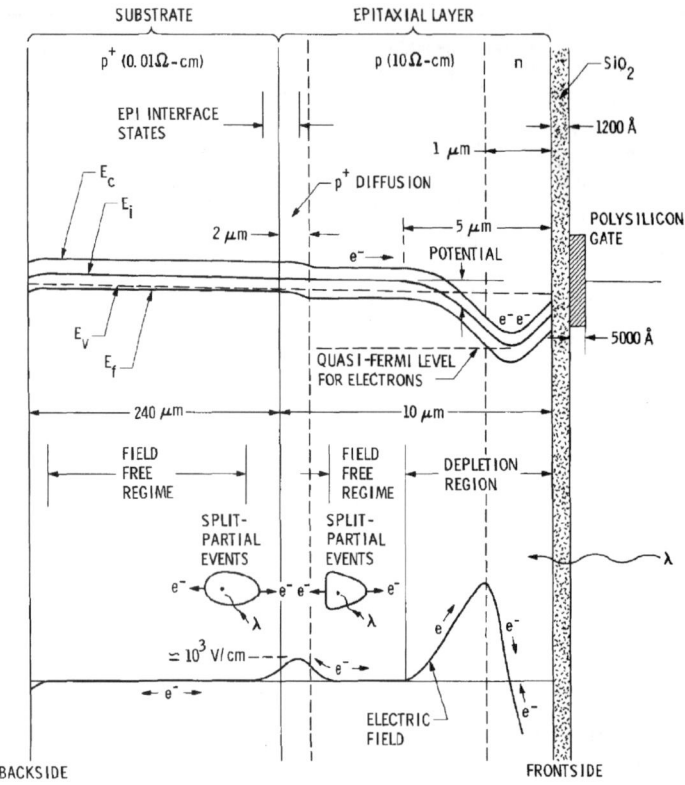

Figure 4.47(a) Substrate and epitaxial cross-section showing field and field-free regions.

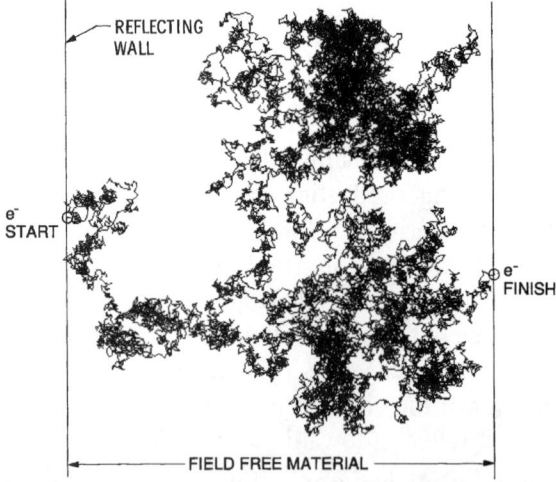

Figure 4.47(b) A Monte Carlo random walk of a photoelectron generated at the backside of a CCD within field-free region.

CHARGE COLLECTION 335

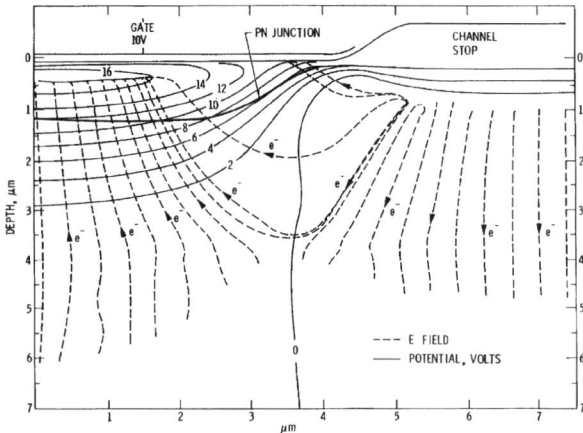

Figure 4.48 Cross-sectional view near the channel stop region showing equal potentials and electric field lines.

A measurable amount of charge diffusion also occurs under the channel stop region. Figure 4.48 shows a CCD cross-section with equal-potential contours and electric field lines in the signal channel and a channel stop region. Electric fields are present throughout the depletion region, the strongest found in the buried channel region located at the immediate surface. CCE is essentially perfect in this region. Beneath the channel stop region we find field-free material where charge diffusion takes place.

It should be mentioned that charge generated under a barrier phase must diffuse into collecting phases on either side, effecting CCE. The amount of diffusion that takes place can be reduced by making the collecting region of the target pixel as large as possible (e.g., two phases of a three-phase CCD would act as collecting phases, leaving only one phase as the barrier). The fringing field between the collecting and barrier phases will also reduce this diffusion problem. We will discuss the extent of these fringing fields below. However, for large pixels the fringing fields may be limited and not reach the center of the barrier phase (this becomes a problem for gate lengths approximately 5 μm and greater). Also, as charge fills the collecting potential wells the fringing fields disappear, increasing charge diffusion under the barrier phases.

Monte Carlo experiments as demonstrated in Fig. 4.47(b) allow one to quantify the diffusion process.[14] It is found that the average lateral dimension from the point of generation gives a two-sigma event diffusion diameter equal to

$$C_{ff} = 2x_{ff}\left(1 - \frac{L_A}{x_{ff}}\right)^{1/2}, \tag{4.16}$$

where C_{ff} is the lateral diameter of diffusion (cm), x_{ff} is the amount of field-free material (cm) and L_A is the distance from the back surface at which the photon interacts (cm).

Example 4.12

Determine the amount of lateral diffusion for a backside CCD with a field-free region of 5 μm. Assume the photon interacts at the immediate back surface.

Solution:
From Eq. (4.16),

$$C_{ff} = 2 \times (5 \times 10^{-4}),$$

$$C_{ff} = 10 \text{ μm}.$$

This example shows that charge can readily escape a small target pixel when this amount of field-free material exists.

The charge diffusion problem can be quite striking for small-pixel CCDs. Figure 4.49 shows a collection of images illustrating the field-free epitaxial charge diffusion problem for a frontside-illuminated 800 × 800 × 7.5-μm-pixel CCD fabricated on 20-μm epitaxial material (40–50 ohm-cm). The depletion depth for the

Figure 4.49(a) Dollar bill image taken at 4000 Å exhibiting good CCE characteristics.

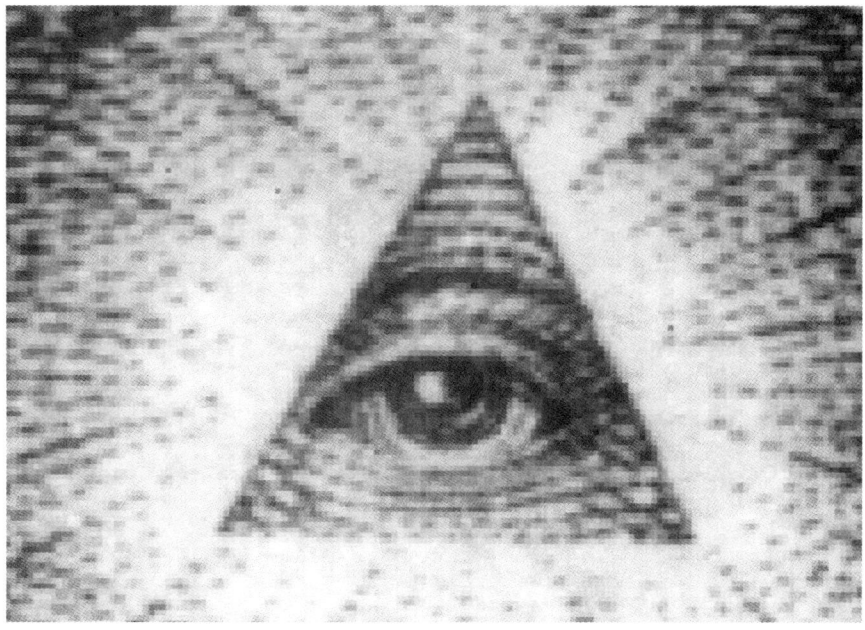

Figure 4.49(b) Magnified view of Fig. 4.49(a).

Figure 4.49(c) Dollar bill image taken at 9000 Å showing poor CCE characteristics.

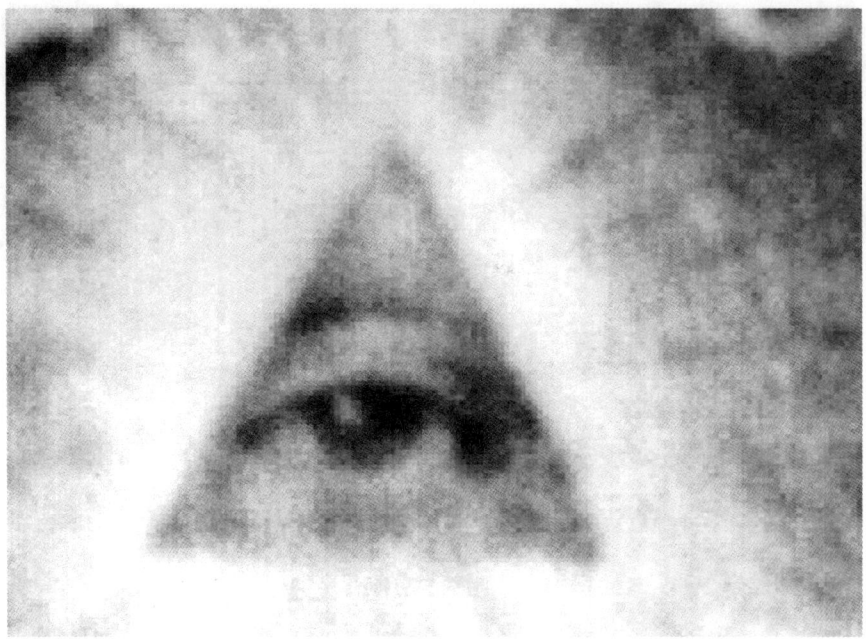

Figure 4.49(d) Magnified view of Fig. 4.49(c).

sensor is only 7 μm, leaving 13 μm of field-free material between the depletion edge and epitaxial interface. Figure 4.49(a) shows an image of a dollar bill taken at a wavelength of 4000 Å. Recall from Chapter 3 that the absorption length for a 4000-Å photon is approximately 2000 Å. Therefore, most incoming photons are absorbed within the depletion region, yielding excellent CCE characteristics. For example, Fig. 4.49(b) is a magnified view of the eyeball region showing good resolution performance. Figure 4.49(c) is an image taken at a wavelength of 9000 Å. Recall that the absorption depth for a 9000-Å photon is much longer. Over half the 9000-Å photons interact with the field-free material, degrading resolution through charge diffusion. Figure 4.49(d) shows a magnified view of the resolution problem. The difference in resolution characteristics is striking between Figs. 4.49(b) and 4.49(d). The CCD tested here was fabricated for visible applications. Therefore, a near-IR blocking filter was employed on the CCD package to reject photons greater than 7000 Å to maintain good resolution. Resolution performance seen in Fig. 4.49 shows the importance of selecting a proper epitaxial thickness and resistivity to reduce the extent of the field-free region. A better match would be an epitaxial thickness of 10 μm for the resistivity and pixel size selected here.

4.4.2 Measurement and Modeling Techniques

This section presents five measurement techniques to quantify CCE performance. They are referred to as (1) modulation transfer function (MTF), (2) contrast transfer

CHARGE COLLECTION

function (CTF), (3) knife edge, (4) point spread and (5) photon transfer. MTF, CTF and knife edge are relative CCE test measurements that only characterize charge diffusion effects (i.e., recombination loss is not measured). Point spread is based on x-ray stimuli which can measure both charge diffusion and recombination, providing an absolute measurement of CCE. Photon transfer characterizes quantum yield performance and also represents an absolute CCE measurement.

4.4.2.1 Modulation Transfer Function

An important measure of the resolution capabilities of an imager is the sine wave MTF response. MTF describes the ability of the CCD to reproduce the contrast or modulation present in the scene at any given spatial frequency. The overall MTF of the chip is made of three components: (1) integration MTF (MTF_I), (2) diffusion MTF (MTF_D) and (3) CTE MTF (MTF_{CTE}). MTF_I is a fixed modulation loss that is specific to the geometry of the pixel. MTF_D is dependent on the amount of field-free material present and occurs when charge is collected by the pixel. MTF_{CTE} loss occurs when charge is transferred between pixels.

4.4.2.1.1 Integration MTF.
Since the CCD is a discrete sampling device, the ideal integration MTF will exhibit a sampling response given by

$$MTF_I = \frac{\sin\left(\frac{\pi f \Delta_P}{2 f_N p}\right)}{\frac{\pi f \Delta_P}{2 f_N p}}, \qquad (4.17)$$

where the pixel has an open aperture of length Δ_P that is repeated with periodicity or pixel pitch p (cm), f is the scene spatial frequency (cycles/cm) and f_N is the Nyquist frequency given by

$$f_N = \frac{1}{2p}. \qquad (4.18)$$

Equation (4.17) will be derived below. Nyquist frequency is defined when a complete sinusoidal cycle occurs in 2 pixels. The geometry of the variables involved is shown in the top illustration of Fig. 4.50.

Example 4.13

Calculate the theoretical MTF_I at Nyquist for a 15-μm pixel. Assume $\Delta_P = p$ (i.e., 100% fill factor).
Solution:
From Eq. (4.18),

$$f_N = 1/\left[2 \times (15 \times 10^{-4})\right],$$

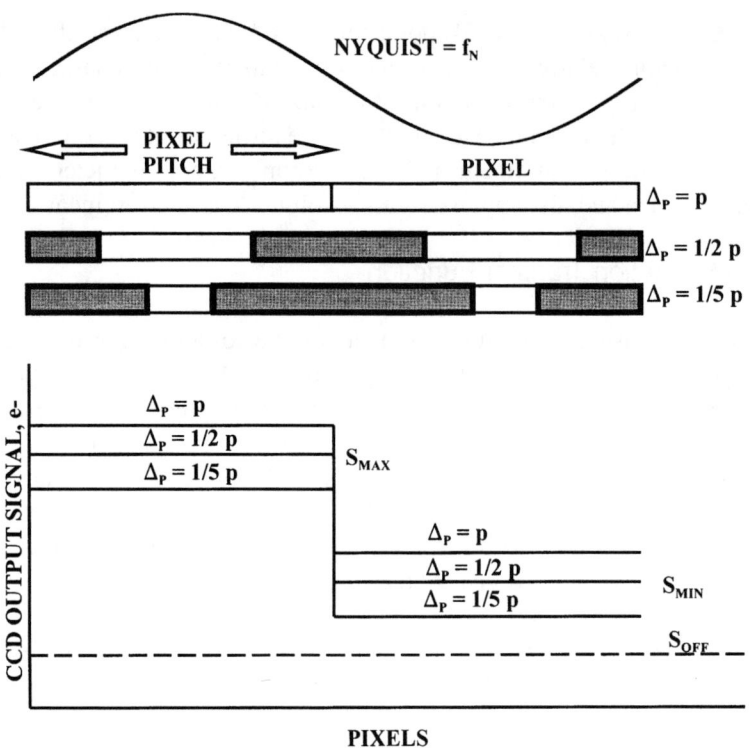

Figure 4.50 Illustration showing sinusoidal stimulus above two pixels at Nyquist frequency with different aperture openings.

$f_N = 333$ cycles/cm.

The MTF_I response at Nyquist from Eq. (4.17) is

$\text{MTF}_I = \text{Sinc}(\pi/2)$,

$\text{MTF}_I = 0.63\ (63\%)$.

Figure 4.51 plots MTF_I as a function of normalized spatial frequency (f/f_N) for different Δ_P ($\Delta_P = p, 1/2p$ and $1/5p$) with corresponding zero crossovers at $2f_N$ (e.g., full-frame imager with 100% pixel fill factor), $4f_N$ (e.g., interline CCD with 50% fill factor in the horizontal direction) and $10\ f_N$.

MTF is measured by comparing the maximum modulation level, S_{MAX}, to the minimum level, S_{MIN}, of a sinusoidal light input using the formula

$$\text{MTF} = \frac{S_{MAX} - S_{MIN}}{S_{MAX} + S_{MIN}}. \qquad (4.19)$$

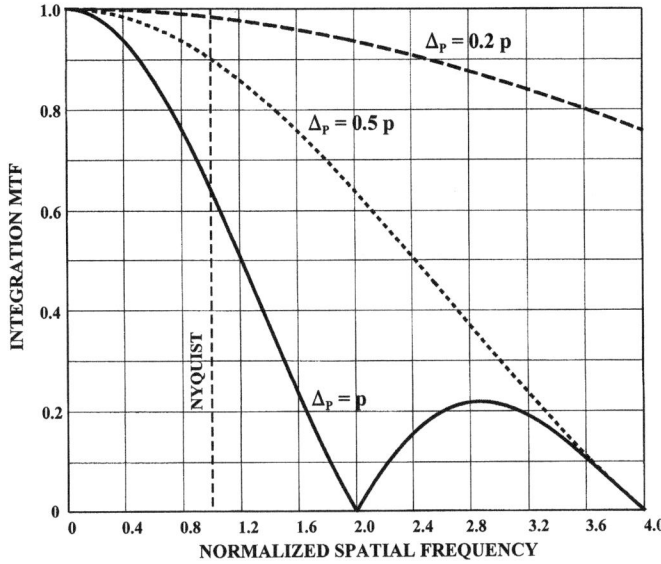

Figure 4.51 Integration MTF as a function of normalized spatial frequency.

The bottom illustration of Fig. 4.50 shows S_{MAX} and S_{MIN} produced at Nyquist for three pixel window openings. It is important to remove the signal offset, S_{OFF}, from each value when calculations are performed. The next example shows how Eq. (4.19) is applied to CCD data.

Example 4.14

Calculate the horizontal MTF for the line trace shown in Fig. 4.52.
Solution:
From Eq. (4.19),

$$\text{MTF} = (14.7 - 2.3)/(14.7 + 2.3),$$

$$\text{MTF} = 0.73.$$

Note the spatial frequency measured here is approximately $(1/5)f_N$.

Equation (4.17) can be derived through simple integration, as opposed to Fourier analysis, which is commonly used. That is, we can integrate the charge that would be generated in pixels when stimulated with a sinusoidal spatial frequency. The sinusoidal input is aligned on two pixels (not necessarily neighbors to each other) to produce a maximum and minimum response. For example, Figure 4.53 shows integration of a $(1/2)(1 + \cos\theta)$ input wave at Nyquist and 3/2 of Nyquist aligned over two pixels. The areas under each curve correspond to S_{MAX} and S_{MIN}

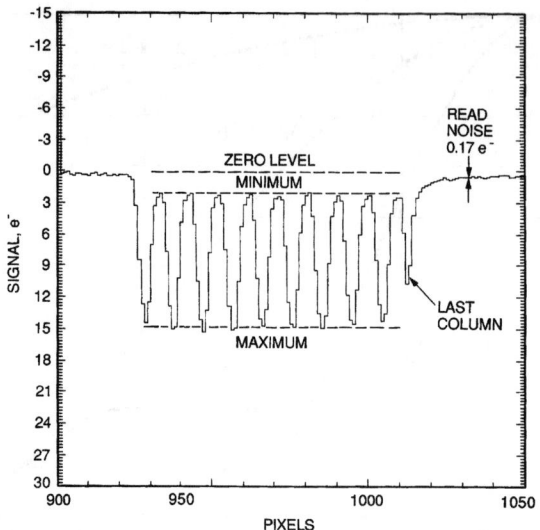

Figure 4.52 Low-level modulation response used to make MTF calculation in Example 4.14.

and are used to determine MTF$_I$ using Eq. (4.19). The next example shows how this is accomplished.

Example 4.15

Calculate the horizontal MTF$_I$ for the integrated areas indicated in Fig. 4.53.
Solution:
From Eq. (4.19) at Nyquist,

$$\text{MTF}_I = (2.57 - 0.57)/(2.57 + 0.57),$$

$$\text{MTF}_I = 0.64.$$

From Eq. (4.19) at 3/2 Nyquist,

$$\text{MTF}_I = (3.07 - 1.65)/(3.07 + 1.65),$$

$$\text{MTF}_I = 0.30.$$

Note that these results agree with the $\Delta_P = p$ curve in Fig. 4.51 generated with Eq. (4.17).

CHARGE COLLECTION

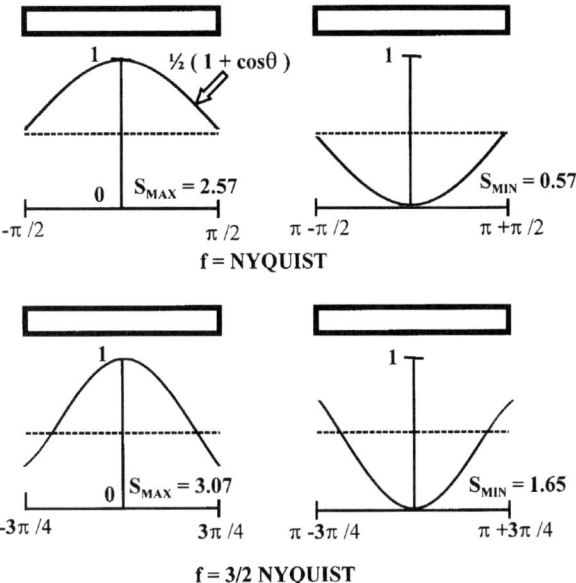

Figure 4.53 Sinusoidal integration required to find S_{MAX}, S_{MIN} and MTF.

In general, MTF_I is found by integration through

$$S_{MAX} = \int_{-\pi/\lambda}^{\pi/\lambda} \frac{1}{2}[1 + \cos(\theta)]d\theta, \qquad (4.20)$$

$$S_{MIN} = \int_{-\pi/\lambda+\pi}^{\pi/\lambda+\pi} \frac{1}{2}[1 + \cos(\theta)]d\theta, \qquad (4.21)$$

where

$$\lambda = \frac{2f_N}{f}. \qquad (4.22)$$

Note that at Nyquist, f_N (0.5 cycles/pixel) $\lambda = 2$. Substituting Eq. (4.22) into Eqs. (4.20) and (4.21) and integrating yields

$$S_{MAX} = \frac{\pi}{2}\frac{f}{f_N} + \sin\left(\frac{\pi}{2}\frac{f}{f_N}\right), \qquad (4.23)$$

and

$$S_{MIN} = \frac{\pi}{2}\frac{f}{f_N} - \sin\left(\frac{\pi}{2}\frac{f}{f_N}\right). \qquad (4.24)$$

Taking these equations through Eq. (4.19) yields

$$\text{MTF}_I = \text{Sinc}\left(\frac{\pi}{2}\frac{f}{f_N}\right), \qquad (4.25)$$

the same as Eq. (4.17) assuming a 100% fill factor. The integrals defined by Eqs. (4.20) and (4.21) are powerful and can be used to find the MTF response for any shaped pixel by setting the limits of integration accordingly.

4.4.2.1.2 Diffusion MTF.

Integration MTF sets the theoretical limit for MTF performance. High-performance CCDs can be designed and fabricated to perform very near this optimum level; however, there will always be a finite MTF loss because of the presence of field-free material and related diffusion effects. The following simple equation illustrates the important variables that influence diffusion MTF for a frontside-illuminated CCD :[5,6]

$$\text{MTF}_D = \frac{e^{d_{ff}/L_n} + e^{-d_{ff}/L_n}}{e^{d_{ff}/L} + e^{-d_{ff}/L}}, \qquad (4.26)$$

where d_{ff} is the distance the photon penetrates into field-free material beyond the depletion region (cm) and L is the spatial-frequency-dependent effective diffusion length

$$L = \left\{\left[\frac{1}{L_n^2} + (2\pi f)^2\right]^{-1}\right\}^{1/2}, \qquad (4.27)$$

where L_n is the diffusion length (cm) and f is the spatial frequency (cycles/cm).

This MTF_D model assumes that all photons have the same penetrating depth. In real life, photons are absorbed exponentially with distance (this characteristic is accounted for in Ref. 17). Also, signal charge is generated in the substrate region of the CCD, not included in the model. Carriers generated in this region exhibit a diffusion length of about 5 μm (0.01-ohm-cm material) and, therefore, can diffuse and reduce MTF_D performance. Also, the built-in field at the epitaxial interface contributes to MTF_D not included in Eq. (4.26).

Example 4.16

Find the diffusion MTF for a frontside-illuminated CCD at Nyquist with the following parameters:

$$d_{ff} = 5 \times 10^{-4} \text{ cm},$$

$$L_n = 0.06 \text{ cm},$$

$$p = 12 \times 10^{-4} \text{ cm (100\% fill factor)},$$

$$f_N = 1/[2 \times (12 \times 10^{-4})] = 416 \text{ cycles/cm}.$$

Solution:
From Eq. (4.27),

$$L = \left\{ \left[(1/0.06)^2 + (2 \times 3.14 \times 416)^2 \right]^{-1} \right\}^{1/2},$$

$$L = 3.82 \times 10^{-4} \text{ cm}.$$

From Eq. (4.26),

$$\text{MTF}_D = \frac{\exp\left[(5 \times 10^{-4})/0.06\right] + \exp\left[-(5 \times 10^{-4})/0.06\right]}{\exp\left[(5 \times 10^{-4})/(3.82 \times 10^{-4})\right] + \exp\left[-(5 \times 10^{-4})/(3.82 \times 10^{-4})\right]},$$

$$\text{MTF}_D = 0.501.$$

Note that the diffusion length is normally much greater than the epitaxial thickness. Therefore, it will not affect MTF calculations unless it is comparable to the field-free thickness.

Figure 4.54(a) plots MTF_D as a function of normalized spatial frequency for different d_{ff} using Eq. (4.26). We assume a pixel size of 10 µm and a diffusion length of $L_n = 600$ µm. MTF_D rapidly decreases with d_{ff} as more charge diffuses from the target pixel. Figure 4.54(b) is a similar plot for a 20-µm pixel showing an improved response because less charge diffusion from the target pixel occurs. The above results show that the field-free region must be a small fraction of pixel size to maintain high MTF.

Figure 4.55 plots results taken from a more complex frontside-illuminated CCD model in predicting diffusion MTF.[18] The plots include MTF_I for a 100% fill factor pixel (i.e., $\text{MTF} = \text{MTF}_I \times \text{MTF}_D$). The MTF response is evaluated at Nyquist as a function of undepleted bulk. The model assumes that the edge of the depletion region is at 6.0 µm (the dotted line labeled "TI"). Note that MTF is optimum and independent of thickness for wavelengths shorter than 5000 Å (i.e., $\text{MTF} = \text{MTF}_I = 63\%$). For these wavelengths, charge is generated directly in the potential well. As the wavelength increases, photons penetrate beyond the depletion edge, resulting in MTF loss.

The following detailed model predicts MTF_D performance for a backside-illuminated CCD.[15,18,19] Here we specify the depletion depth, the field-free region thickness, wavelength of light and the pixel pitch. MTF_D for this model begins with the equation,

$$\text{MTF}_D = \frac{N_K}{N_O}. \tag{4.28}$$

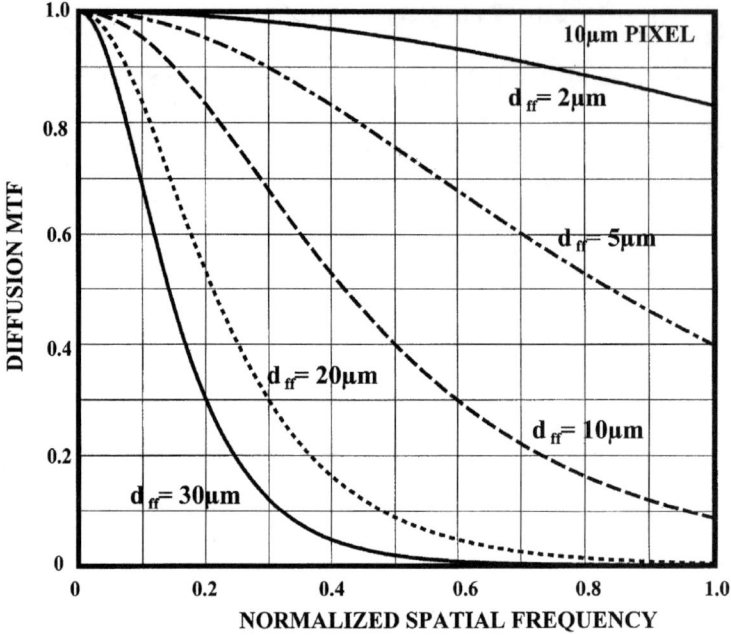

Figure 4.54(a) Diffusion MTF as a function of normalized spatial frequency for different photon absorption depths.

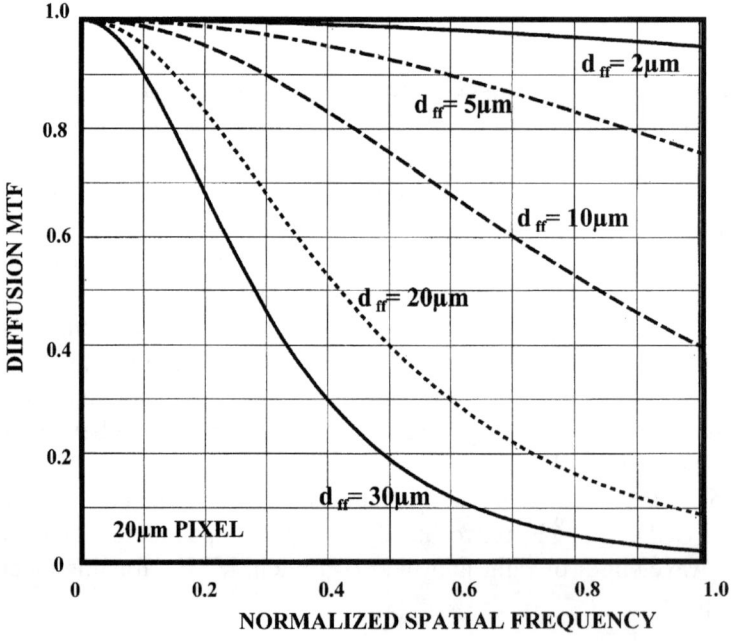

Figure 4.54(b) Similar to Fig. 4.54(a) for a larger 20-μm pixel.

CHARGE COLLECTION

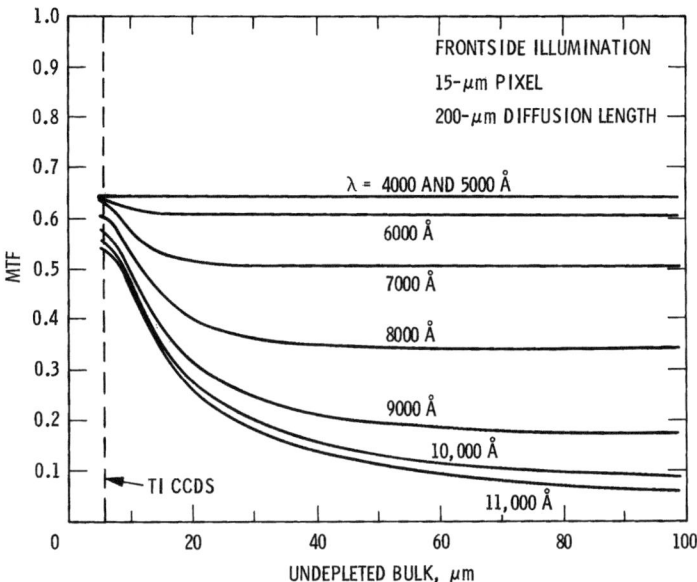

Figure 4.55 MTF as a function of field-free thickness for a frontside-illuminated CCD.

The constant N_O is given by

$$N_O = \frac{\alpha L_n}{\alpha^2 L_n^2 - 1} \left[\frac{2\left(\alpha L_n + \frac{V_S L_n}{D_n}\right) - (\beta_+ - \beta_-)e^{-\alpha x_{ff}}}{\beta_+ + \beta_-} - \frac{e^{-\alpha x_{ff}}}{\alpha L_n} \right] - e^{[-\alpha(x_{ff}+x_p)]},$$

(4.29)

where

$$\beta^+ = \left(1 + \frac{V_S L_n}{D_n}\right) e^{\frac{x_{ff}}{L_n}}, \qquad (4.30)$$

and

$$\beta_- = \left(1 - \frac{V_S L_n}{D_n}\right) e^{-\frac{x_{ff}}{L_n}}. \qquad (4.31)$$

In these equations x_{ff} is the field-free thickness (cm), V_S is the surface recombination velocity (cm/sec) at the backside, D_n is the diffusion coefficient (39 cm^2/sec), L_n is the diffusion length in the epitaxial silicon (cm), α is the optical absorption coefficient (cm^{-1}) (see Table 3.1), and x_p is the depletion depth of the p-region as defined in Eq. (1.21).

N_K is found by replacing L_n by L into Eqs. (4.30), (4.31) and (4.32), where

$$L = \left(\frac{1}{L_n^2} + K^2\right)^{-1/2}, \qquad (4.32)$$

and

$$K = \frac{\pi}{p}\frac{f}{f_N}, \qquad (4.33)$$

where p is the pixel pitch (cm).

Example 4.17

Calculate the MTF at Nyquist for a backside-illuminated CCD at $\lambda = 4000$ Å for the following parameters:

$$x_p = 6 \times 10^{-4} \text{ cm},$$

$$x_{ff} = 5 \times 10^{-4} \text{ cm},$$

$$p = 12 \times 10^{-4} \text{ cm},$$

$$V_S = 1000 \text{ cm/sec},$$

$$D_n = 39 \text{ cm}^2/\text{sec},$$

$$L_n = 0.0600 \text{ cm},$$

$$\alpha = 50,000 \text{ cm}^{-1} \ (L_A^{-1} \text{ from Table 3.1}).$$

Solution:
From the equations above,

$$\beta_+ = (1 + 10^3 \times 0.06/39)e^{+(0.0005/0.06)},$$

$$\beta_+ = 2.56,$$

$$\beta_- = (1 - 10^3 \times 0.06/39)e^{-(0.0005/0.06)},$$

$$\beta_- = -0.534.$$

Breaking N_O into several terms,

$$e^{-\alpha x_{ff}} = e^{-(50,000 \times 0.0005)} = 1.38 \times 10^{-11},$$

CHARGE COLLECTION

$$e^{-\alpha(xff+xp)} = e^{-50,000(0.0005+0.0006)} = 1.29 \times 10^{-24},$$

$$\alpha L_n/(\alpha^2 L_n^2 - 1) = 50,000 \times 0.06/[(50,000 \times 0.06)^2 - 1]$$
$$= 3.3 \times 10^{-4},$$

$$2(\alpha L_n + SL_n/D_n) = 2(50,000 \times 0.06 + 1000 \times 0.06/39)$$
$$= 6 \times 10^3,$$

$$(\beta_+ - \beta_-)e^{-\alpha xff} = (2.56 + 0.534)1.38 \times 10^{-11} = 4.26 \times 10^{-11},$$

$$(\beta_+ + \beta_-) = 2.56 - 0.534 = 2.026,$$

$$e^{-\alpha xff}/\alpha L_n = 1.38 \times 10^{-11}/(50,000 \times 0.06) = 4.6 \times 10^{-15}.$$

Substituting into N_O

$$N_O = 3.33 \times 10^{-4}[(6 \times 10^3 - 4.26 \times 10^{-11})/$$
$$2.026 - 4.26 \times 10^{-11}] - 1.29 \times 10^{-24},$$

$$N_O = 0.988.$$

Calculating K at Nyquist,

$$K = \pi/(12 \times 10^{-4}),$$

$$K = 2618.$$

L for N_K is

$$L = (1/0.06^2 + 2618^2)^{-1/2},$$

$$L = 3.819718 \times 10^{-4}.$$

Performing similar calculations for N_K using the new L,

$$\beta_+ = [1 + 10^3 \times (3.819718 \times 10^{-4})/39]e^{+(0.0005/3.819718 \times 10^{-4})},$$

$$\beta_+ = 3.702,$$

$$\beta_- = (1 - 10^3 \times 3.819718 \times 10^{-4}/39)e^{-(0.0005/3.819718 \times 10^{-4})},$$

$$\beta_- = 0.2700,$$

$$\alpha L_n/(\alpha^2 L_n^2 - 1) = 50{,}000 \times 3.819718 \times 10^{-4}/$$
$$\left[(50{,}000 \times 3.819718 \times 10^{-4})^2 - 1\right]$$
$$= 5.25 \times 10^{-2},$$

$$2(\alpha L_n + SL_n/D_n) = 2(50{,}000 \times 3.819718 \times 10^{-4}$$
$$+ 1000 \times 3.819718 \times 10^{-4}/39)$$
$$= 38.197,$$

$$(\beta_+ - \beta_-)e^{-\alpha x_{ff}} = (3.702 - 0.2700)1.38 \times 10^{-11}$$
$$= 4.736 \times 10^{-11},$$

$$(\beta_+ + \beta_-) = 3.702 + 0.2700 = 3.97,$$

$$e^{-\alpha x_{ff}}/\alpha L_n = 1.38 \times 10^{-11}/(50{,}000 \times 3.819718 \times 10^{-4})$$
$$= 7.2256 \times 10^{-13},$$

$$N_K = 5.25 \times 10^{-2} \left[\frac{38.197 - 4.736 \times 10^{-11}}{3.97 - 7.2256 \times 10^{-13}}\right] - 1.29 \times 10^{-24},$$

$$N_K = 0.505.$$

The diffusion modulation is

$$\text{MTF}_D = 0.505/0.988,$$

$$\text{MTF}_D = 0.511.$$

Figure 4.56 presents diffusion MTF curves using the equations above. The plots are at Nyquist for a backside-illuminated 12-μm-pixel CCD as a function of field-free material for different wavelengths (a depletion depth of 6 μm is assumed). MTF_D is worse at 4000 Å because all photons interact in the field-free regime at the immediate surface of the CCD (recall the absorption length for a 4000-Å photon is only 2000 Å). At longer wavelengths, MTF_D improves as photons penetrate past the field-free region. Note that MTF_D is very sensitive to field-free material at the backside. Ideally no more than a couple microns can be present for a small-pixel device before MTF_D decreases significantly.

CHARGE COLLECTION

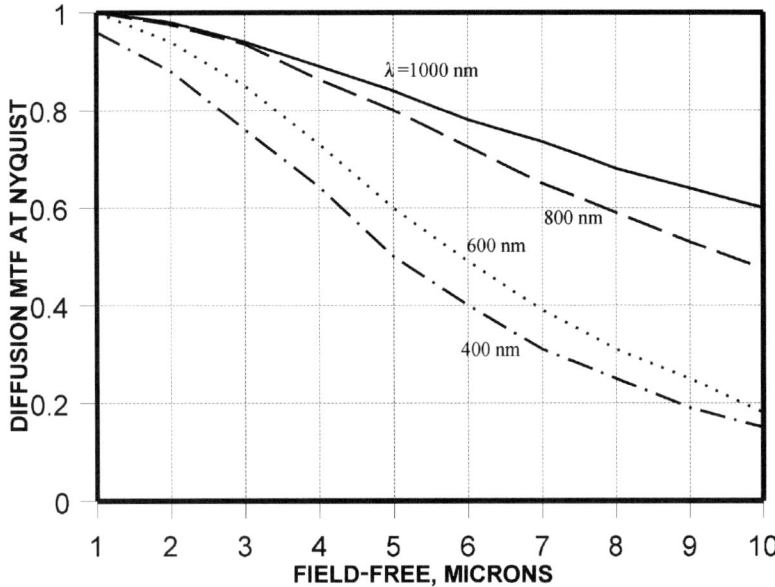

Figure 4.56 Diffusion MTF as a function of field-free thickness for a backside-illuminated CCD.

4.4.2.1.3 CTE MTF. It is unlikely that CTE will influence MTF significantly; however, as CCD arrays become larger the effect may become noticeable in regions furthest from the on-chip amplifier. The expected MTF loss due to CTE can be estimated by[20]

$$\text{MTF}_{CTE} = \exp\left\{-N_P(1-\text{CTE})\left[1-\cos\left(\frac{\pi f}{f_N}\right)\right]\right\}, \quad (4.34)$$

where N_P is the number of pixel transfers and CTE is the charge transfer efficiency per pixel transfer. MTF_{CTE} is calculated separately for the horizontal and vertical registers.

Example 4.18

Find the horizontal MTF_{CTE} at Nyquist for a 4096 × 4096 CCD at the furthest corner from the readout amplifier. Assume a $\text{CTE}_H = 0.9999$.
Solution:
From Eq. (4.34),

$$\text{MTF}_{CTE} = \exp\{(-4096 \times (1-0.9999) \times [1-\cos(3.14)]\},$$

$$\text{MTF}_{CTE} = 0.44078.$$

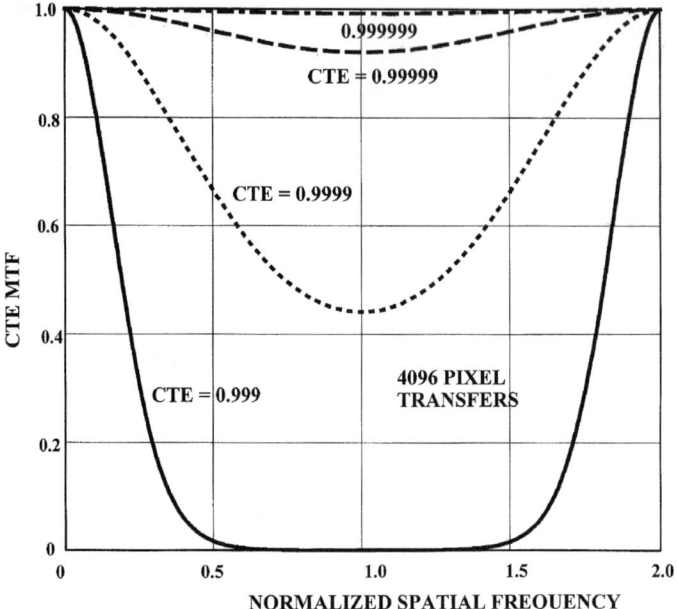

Figure 4.57 CTE MTF as a function of normalized spatial frequency for a 4096 × 4096 CCD.

Figure 4.57 plots MTF$_{CTE}$ as a function normalized spatial frequency for a 4096-element CCD and different CTEs. MTF loss is significant for a CTE of only 0.9999. There is an interesting reason why CTE MTF increases for spatial frequencies above Nyquist, an effect analyzed in Section 4.4.3 on aliasing.

4.4.2.1.4 System MTF. The overall MTF for the CCD when MTF$_I$, MTF$_D$ and MTF$_{CTE}$ are combined is

$$\text{MTF} = \text{MTF}_I \, \text{MTF}_D \, \text{MTF}_{CTE}. \quad (4.35)$$

Example 4.19

Determine the combined MTF at Nyquist for Example 4.17. Assume that MTF$_{CTE}$ and MTF$_I$ are ideal.
Solution:
The diffusion MTF was calculated to be 0.511 at Nyquist. From Fig. 4.56, integration MTF at Nyquist is 0.63.
From Eq. (4.35),

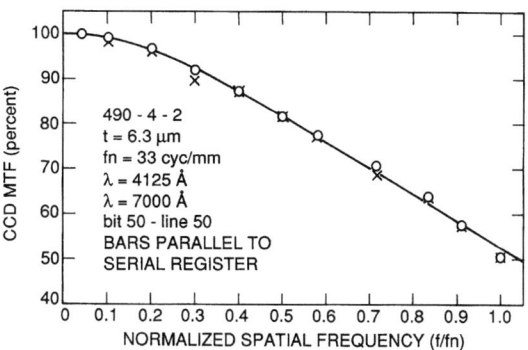

Figure 4.58(a) Experimental MTF curves for a backside-illuminated CCD at two different wavelengths.

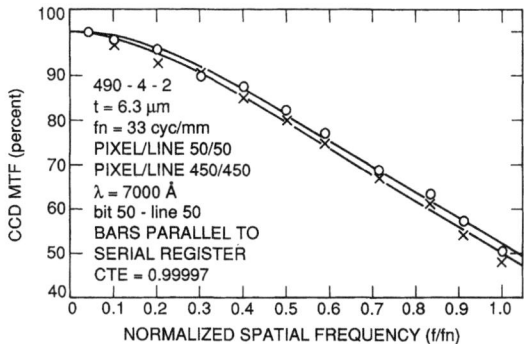

Figure 4.58(b) Experimental MTF curves for a backside-illuminated CCD at two different locations on the array.

$$\text{MTF} = 0.63 \times 0.511 \times 1,$$

$$\text{MTF} = 0.322.$$

4.4.2.1.5 MTF data. Figure 4.58 presents experimental vertical sinusoidal MTF responses taken from a 500 × 500 × 15-μm-pixel backside-illuminated CCD. The MTF response shown in Fig. 4.58(a) was taken at two different wavelengths, 4125 Å and 7000 Å, respectively. The shorter wavelength represents a worst-case MTF measurement because this wavelength will exaggerate the charge diffusion problem. The penetrating depth of 7000-Å light is much longer, creating most carriers in the depletion region. Because the two responses are nearly identical, charge diffusion and field-free material for the CCD must be minimal. The test CCD was thinned to 6.3 μm leaving the frontside depletion edge almost to the back surface. The only significant region where field-free material exists is under the channel stop. This region could be responsible for the 50% MTF observed, which is just shy of the theoretical level of 63%. Figure 4.58(b) is an MTF plot for the same CCD taken at 7000 Å at two different locations on the array (the bottom and top of the array). The slight difference in MTF measured is caused by a CTE problem, which degrades modulation from the theoretical level (refer to Section 5.4.1 for a discussion on this problem).

4.4.2.2 Contrast Transfer Function

Quite often a square-wave target is used as a substitute for a sinusoidal stimulus. The frequency response from a square-wave target is referred to as contrast transfer function (CTF). Figure 4.59 shows a "slant bar" target with several square-wave frequencies above and below Nyquist. For example, the region labeled 1.0 represents 400 cycles per picture height or the Nyquist frequency for a 800 × 800-pixel CCD (this particular target was custom designed for the Hubble 800 × 800 CCD). The region labeled 0.4 is a frequency at $0.4 f_N$. The bar targets are slanted to avoid alignment problems which could give erroneous results for S_{MAX} and S_{MIN}. For example, Fig. 4.60 illustrates an alignment problem. On the left side of the figure we show a Nyquist square-wave target that is perfectly aligned to the array. The modulation response is zero both horizontally and vertically because all pixels are half illuminated. On the other hand, the slanted arrangement shown on the right-hand side exercises some pixels to their maximum and minimum levels required for MTF measurement.

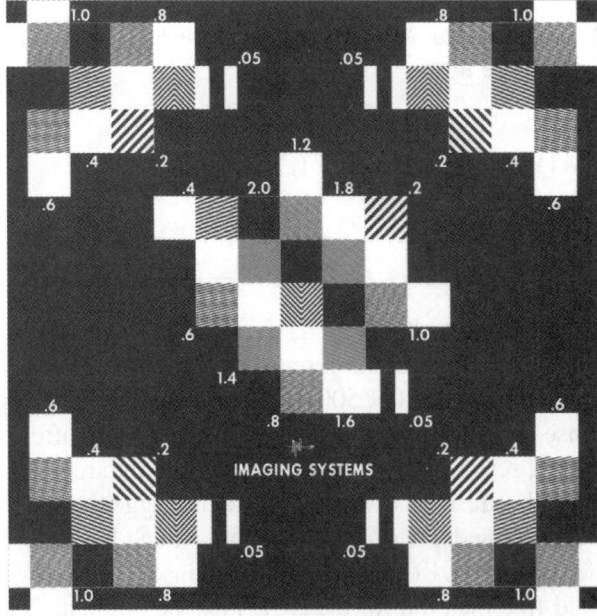

Figure 4.59 Slant bar target used in taking CTF measurements.

CHARGE COLLECTION

Theoretical CTF_I is derived by integrating over a pixel to find S_{MAX} and S_{MIN} in the same fashion as with a sinusoidal waveform [i.e., Eqs. (4.20) and (4.21)]. Doing the required integration produces a CTF_I for 100% fill factor of

$$\text{CTF}_I = 1, \quad (0 < f < f_N), \tag{4.36}$$

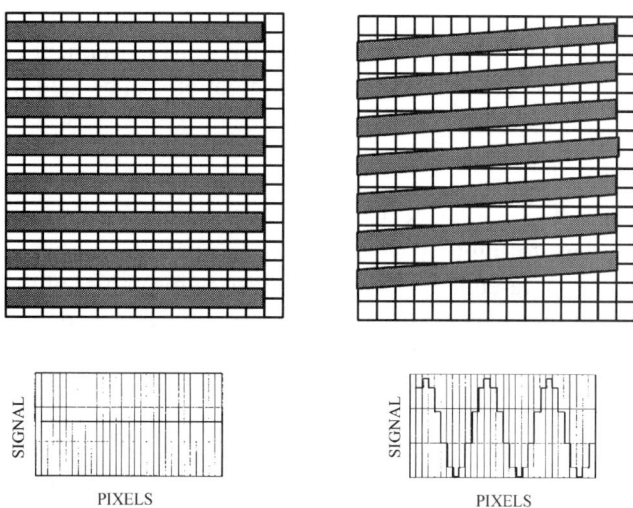

Figure 4.60 Alignment problem that is solved by a slant bar target.

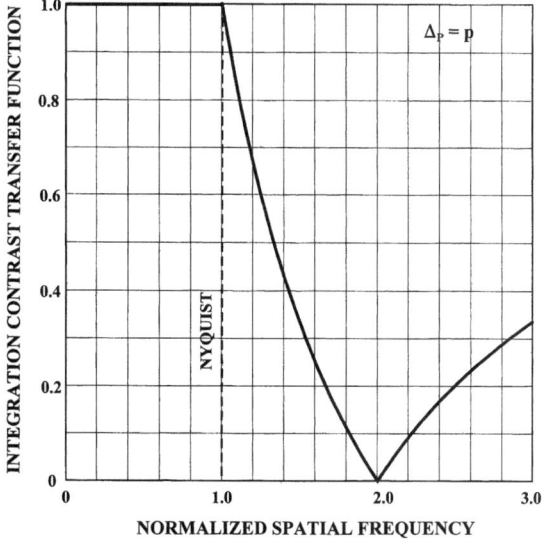

Figure 4.61 Integration CTF as a function of normalized spatial frequency.

and

$$\text{CTF}_I = \frac{2}{(f/f_N)} - 1, \quad (f_N < f < 2f_N), \tag{4.37}$$

and

$$\text{CTF}_I = 1 - \frac{2}{(f/f_N)}, \quad (2f_N < f < 3f_N). \tag{4.38}$$

Figure 4.61 plots CTF_I as a function of normalized spatial frequency for spatial frequencies out to $f = 3f_N$. Note that 100% modulation can be achieved at Nyquist.

4.4.2.3 Edge Response

A knife-edge response measures the number of pixels required for the CCD to transition from dark to light. A vertical edge response for a three-phase CCD is obtained when the knife-edge is optically imaged in the center of the barrier phase or in the middle of the channel stop for the horizontal edge response. This lighting condition is illustrated in Fig. 4.62.

In Section 4.4.2.4 we will also discuss the point spread response. Point spread is similar to the edge response and is measured by only illuminating the target pixel, also illustrated in the figure. Note again that the illumination boundary for point spread is defined by the middle of the barrier phase and the channel stop region. We will show below why the pixel boundaries are defined in this manner for optimum MTF, CTF, edge and point spread responses.

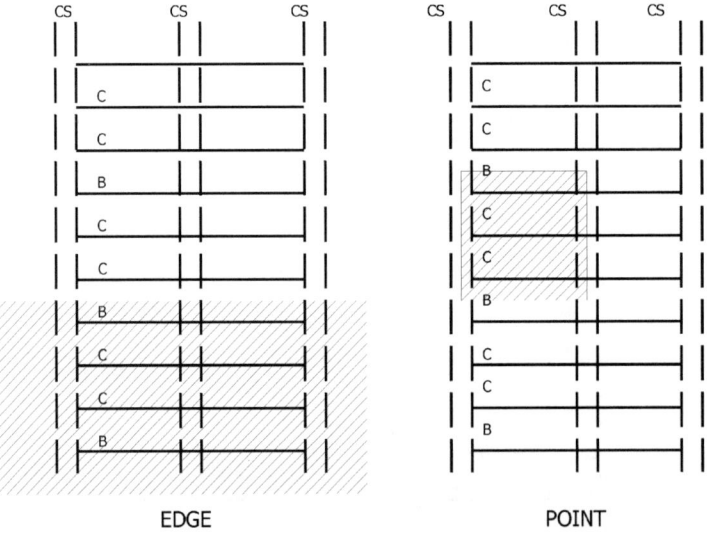

Figure 4.62 Pixel boundaries for a three-phase CCD to obtain maximum edge and point spread responses.

A knife-edge response is best obtained by placing the edge directly on the surface of the CCD and stimulating the arrangement with colliminated light (as opposed to optical projection, which requires precise focus and alignment). This has been accomplished by vacuum depositing a thin metal layer onto the CCD to act as the knife-edge. The metal edge is typically slanted, as shown in Fig. 4.63, to reduce the required precision in aligning the metal layer to the pixel boundaries. When slanted, a computer can scan the edge and find where perfect alignment (i.e., highest modulation) takes place. A knife-edge can also be mounted onto the CCD.

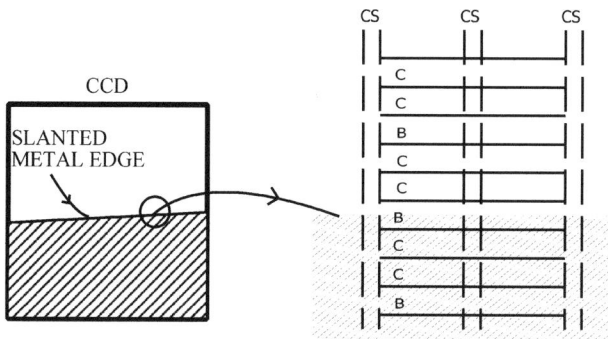

Figure 4.63 Optical arrangement that slants a knife-edge for maximum modulation.

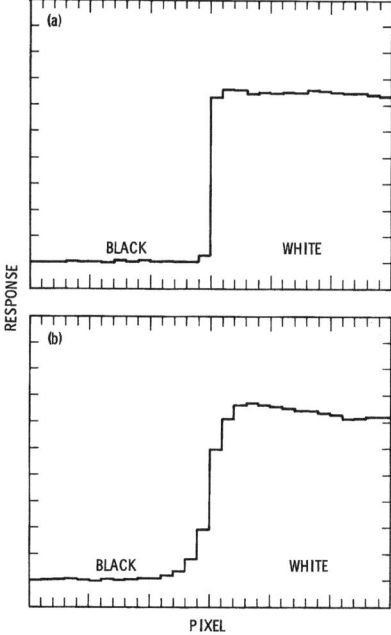

Figure 4.64 Experimental knife-edge response at two different wavelengths.

A thin metal shim with a very sharp edge can temporarily be "glued" onto a test CCD, again with a slight slant.

When a knife-edge is properly projected, a 100% contrast change is possible in a single pixel transition. This result implies perfect CTF_D and MTF_D (i.e., 100% and 63% respectively at Nyquist). Figure 4.64 shows a knife-edge response taken from a frontside-illuminated three-phase CCD stimulated with 4000-Å and 10,000-Å collimated light. For this setup the edge is directly glued onto the CCD with a slight tilt. Note that while the transition from dark to light is made essentially in one pixel for 4000-Å light, the same transition requires more than nine pixels for 10,000-Å illumination. The lack of a sharp transition for near-IR light is associated with field-free material below the depletion region. This CCD was fabricated on 40-μm-thick, 10-ohm-cm resistivity epitaxial silicon which leaves about 35 μm of field-free material for charge to diffuse.

Equation (4.36) shows that 100% CTF is possible at Nyquist. However, even if charge diffusion effects are neglected outside the depletion region, perfect CTF_D (or MTF_D) may not be totally achieved because of a charge diffusion problem related to the barrier phase. Table 4.1 illustrates the problem for a three-phase CCD. The second column shows the phase number assuming phases-1 and -2 are the collecting phases (each labeled C) and phase-3 the barrier phase (labeled B). The next column, labeled S_P, shows the charge generated in each phase. We assume Nyquist frequency and 100 e^-/phase. The next column, labeled S_C, is the charge collected by the collecting phases. Note that charge is absent from the barrier phase because it will equally divide between the target pixel and a neighbor. The column labeled S_T shows the total charge contained in the target pixel after integration. The last line of the table calculates CTF_D at Nyquist frequency. Ideally, for a perfect CTF_D

Table 4.1 Table showing how target alignment influences MTF and CTF measurements.

Pixel	Phase	S_P	S_C	S_T	S_P	S_C	S_T	S_P	S_C	S_T
	1 (C)	100 e^-	100 e^-		0 e^-			0 e^-	0 e^-	
1	2 (C)	100 e^-	150 e^-	250 e^-	100 e^-	150 e^-	150 e^-	0 e^-	50 e^-	50 e^-
	3 (B)	100 e^-	0 e^-		100 e^-	0 e^-		100 e^-	0 e^-	
	1 (C)	0 e^-	50 e^-		100 e^-	150 e^-		100 e^-	150 e^-	
2	2 (C)	0 e^-	0 e^-	50 e^-	0 e^-	0 e^-	150 e^-	100 e^-	100 e^-	250 e^-
	3 (B)	0 e^-	0 e^-		0 e^-	0 e^-		0 e^-	0 e^-	
	1 (C)	100 e^-	100		0 e^-	0 e^-		0 e^-	0 e^-	
3	2 (C)	100 e^-	150 e^-	250 e^-	100 e^-	150 e^-	150 e^-	0 e^-	50 e^-	50 e^-
	3 (B)	100 e^-	0 e^-		100 e^-	0 e^-		100 e^-	0 e^-	
	1 (C)	0 e^-	50 e^-		100 e^-	150 e^-		100 e^-	150 e^-	
4	2 (C)	0 e^-	0 e^-	50 e^-	0 e^-	0 e^-	150 e^-	100 e^-	100 e^-	250 e^-
	3 (B)	0 e^-	0 e^-		0 e^-	0 e^-		0 e^-	0 e^-	
	CTF			0.67			0			0.67

100 e^-/phase square wave at Nyquist frequency.
S_P = charge generated in each phase.
S_C = charge collected by each phase.
S_T = total charge collected by target pixel.

response $S_{MAX} = 300$ e$^-$ and $S_{MIN} = 0$ e$^-$. However, because of barrier diffusion, CTF$_D$ is limited to 0.67 for the three alignment conditions shown in Table 4.1. The highest modulation that can be achieved occurs when the edges of the square wave

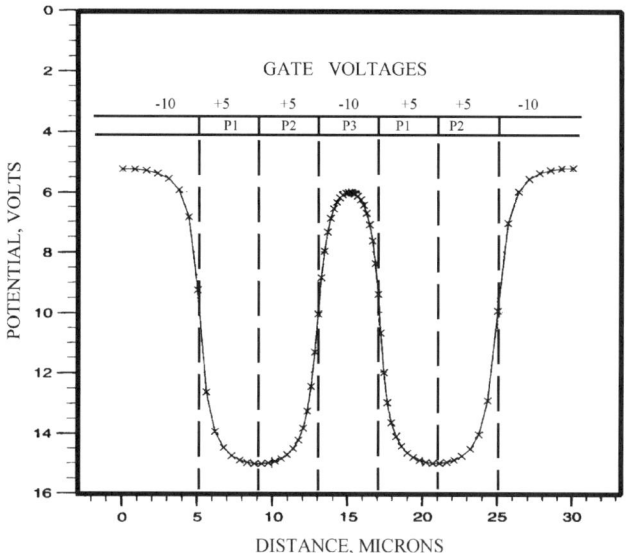

Figure 4.65(a) Potential well plots for collecting and barrier phases in the absence of signal charge.

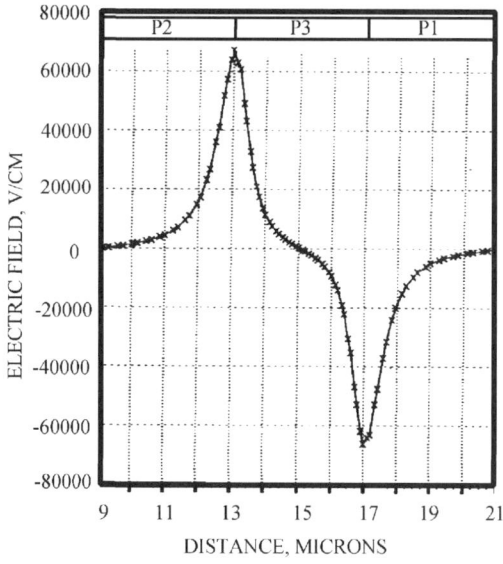

Figure 4.65(b) Fringing electric fields between phases for the potential wells shown in Fig. 4.65(a).

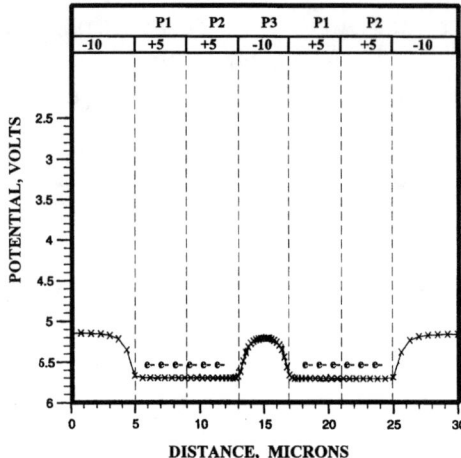

Figure 4.66(a) Potential well plots for collecting and barrier phases near full well.

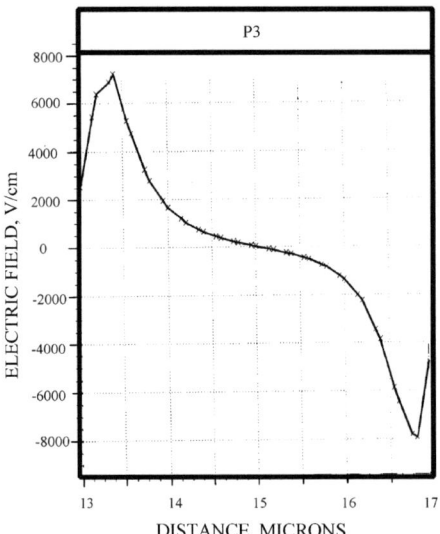

Figure 4.66(b) Fringing electric fields between phases for the potential wells shown in Fig. 4.66(a).

are projected in the middle of the barrier phase as shown in Fig. 4.62 (this geometry is not analyzed in Table 4.1). By definition, pixel boundaries are normally defined where maximum MTF occurs.

Results in Table 4.1 indicate that a three-phase pixel does not completely confine charge to the target pixel, because the potential under the barrier phase was assumed to be flat and field-free. However, in reality, fringing fields from the collecting phases will extend under the barrier phase to aid the collection process. In fact, if the gate length of the barrier phase is made sufficiently small, fringing fields will extend throughout the region, producing an ideal CTF_D response (i.e., 100% at Nyquist). Figure 4.65(a) shows the channel potential for a three-phase 12-μm pixel with $V_C = 5$ V and $V_B = -10$ V (i.e., inverted and pinned). Note that the potential distribution in the barrier region is not flat, indicative of fringing fields. Charge generated within the barrier phase for the most part will divide equally between the two collecting pixels. Figure 4.65(b) plots the electric field strength for the same region. Note that maximum field strength occurs at the interface between the barrier and collecting phases. Electric fields < 5000 V/cm at the immediate center of the barrier phase may not be sufficiently strong to override thermal diffusion effects. This could result in a small amount of barrier crosstalk producing a $\text{CTF}_D < 100\%$. As charge collects in a potential well, the fringing fields become weaker and eventually collapse at full well. Figure 4.66(a) shows collapsed potential wells for the same pixel at full well. A magnified view of the electric field is shown in Fig. 4.66(b). The fringing fields are reduced significantly compared with an empty well. The effect increases the crosstalk between pixels as the signal level increases. This MTF effect has been seen experimentally.

The discussions above have assumed that the bar stripes run horizontally to measure vertical MTF_D (with a slight slant to avoid alignment problems). CTF_D will be slightly different if the square-wave pattern runs vertically. In this direction, the channel stop becomes a source of charge diffusion. Fringing fields do not extend very far under the channel stop region, as they do under a barrier phase. Therefore, horizontal CTF_D is usually worse than vertical CTF_D. However, CTF_D is also dependent on wavelength, pixel size, clock bias, CTE and CCD technology employed (1-phase, 2-phase, 3-phase, etc.), which may produce the opposite result.

4.4.2.4 Point Spread Response

X rays can be used to generate a point spread response and to estimate the MTF response. Fe^{55} x rays are advantageous because they interact uniformly and stimulate all field-free regions that contribute to charge diffusion losses. The response represents a worst-case MTF measurement for frontside-illuminated CCDs. For backside-illuminated CCDs, lower-energy x rays (< 1 keV) are used to generate charge packets at the immediate surface of the CCD, thus representing a worst-case point spread response for this family of sensors.

X rays generate a near-perfect point source of charge. When interacting, the x-ray photon imparts its energy to a K-shell electron which in turn excites outer shell electrons in nearby atoms through a series of quasi-elastic Rutherford scattering collisions. The electron-hole pair creation zone can be approximated as a sphere, centered at the absorption depth of the impinging photon, with X_D as its diameter (μm) of[21–24]

$$X_D = 0.0171[E(\text{eV})]^{1.75}, \qquad (4.39)$$

where $E(\text{eV})$ is the energy of the photon (keV).

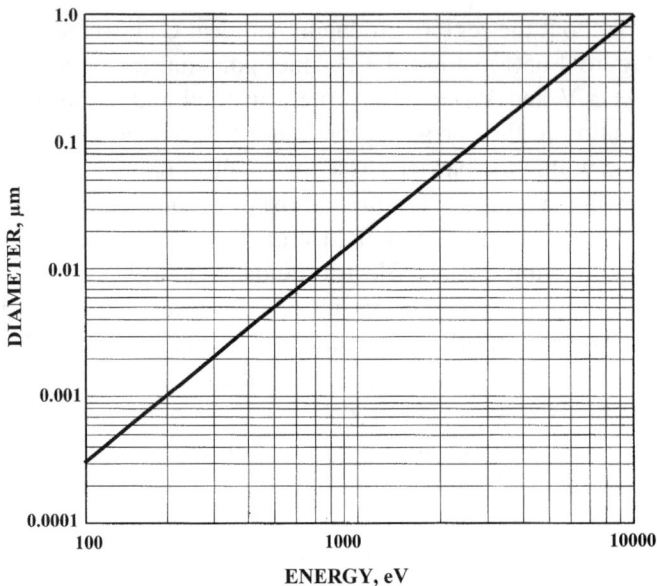

Figure 4.67 X-ray event diameter as a function of energy.

Example 4.20

Determine the average cloud size of a Fe^{55} x-ray event.
Solution:
From Eq. (4.39),

$$X_D = 0.0171 \times (5.9)^{1.75},$$

$$X_D = 0.38 \text{ μm}.$$

Figure 4.67 plots X_D as a function of x-ray energy.

Charge collection performance is perfect when all x-ray charge is collected by the target pixel without recombination loss or diffusion. CCE loss depends upon where in the pixel the x rays are absorbed. As discussed in Chapter 2, there are three types of x-ray events generated in the CCD: (1) single-pixel events, (2) split events and (3) partial events. Single-pixel events interact within the depletion region and do not contribute to CCE loss. Split events occur when the interaction occurs within field-free material. Partial events mainly originate in the substrate and in the channel stop regions where recombination with holes takes place. Also, backside CCDs can show recombination loss when x-ray events interact near the back surface due to a dead layer associated with this region (e.g., backside well).

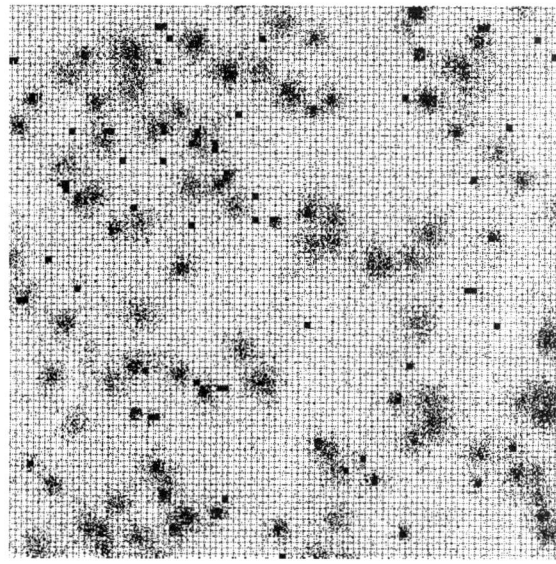

Figure 4.68(a) X-ray events for a frontside CCD that exhibits a large field-free region.

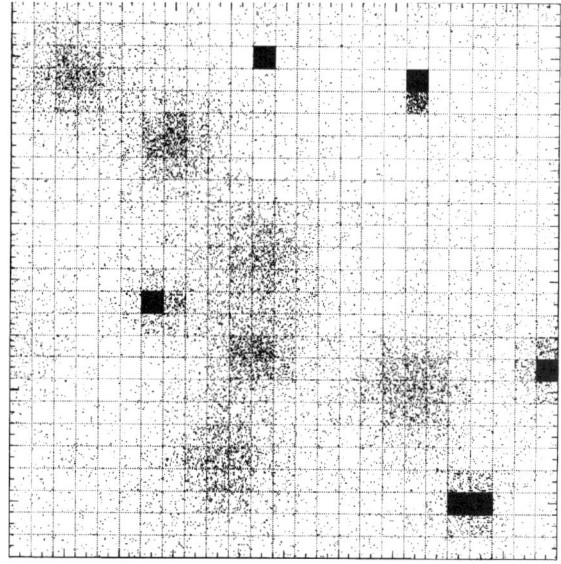

Figure 4.68(b) Magnified view of Fig. 4.68(a) showing severe event diffusion.

Figure 4.68(a) shows Fe^{55} x-ray events produced by a frontside-illuminated $800 \times 800 \times 7.5$-μm-pixel CCD built on 20-μm, 30–40-Ω-cm epitaxial material. Single, split and partial events are seen. The depletion depth for the sensor is approximately 7 μm, which leaves 13 μm of field-free material between the epitaxial

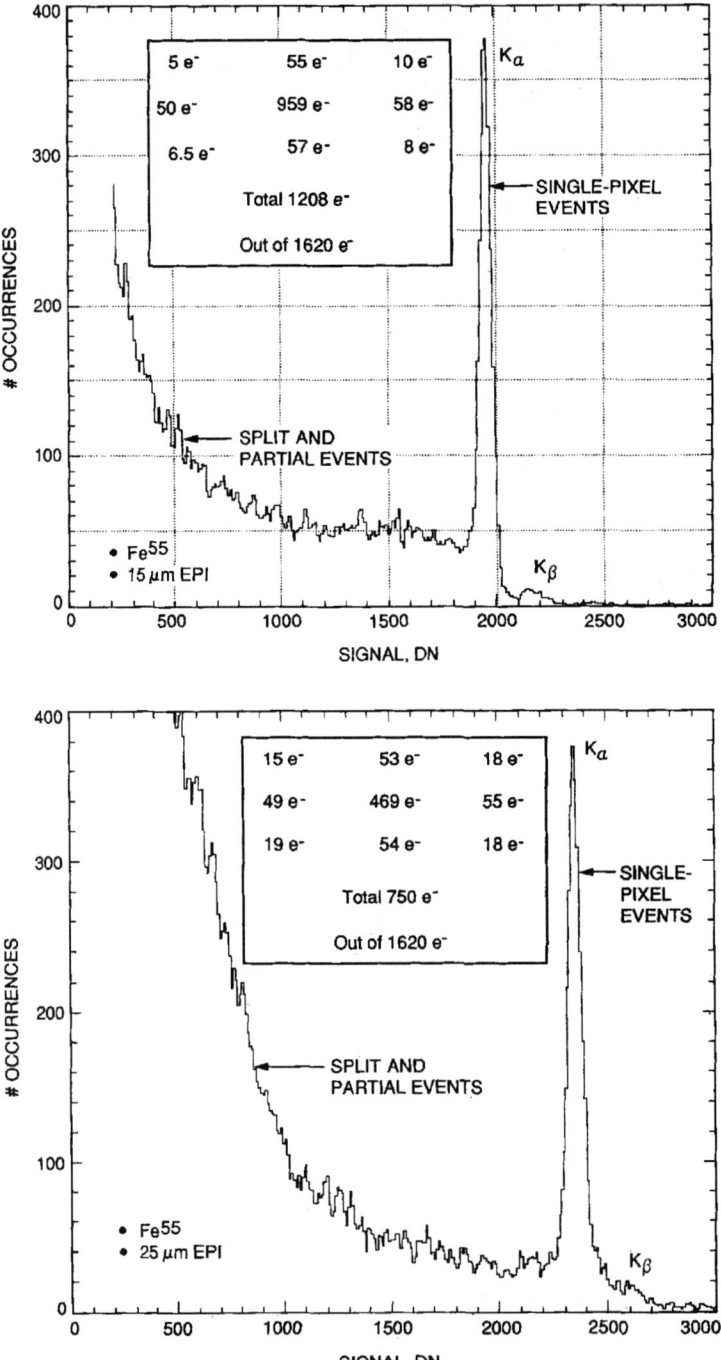

Figure 4.69 Fe^{55} x-ray histograms for 15-μm and 25-μm frontside-illuminated CCDs.

interface and the front depletion edge. Figure 4.68(b) is a magnified view showing each event type. X-ray photons that interact within the field-free region occupy up to 16 pixels, indicative of very poor CCE. This image should be compared with the x-ray image in Fig. 2.22 generated by a 10-μm epitaxial CCD with the same depletion depth. X-ray events for this CCD are contained in 4 pixels or less, indicative of better CCE performance.

Figure 4.69 shows Fe^{55} x-ray histograms for two frontside-illuminated CCDs fabricated on 15-μm and 25-μm epitaxial silicon. Point spread signatures also shown in each figure were generated by stacking 3000 target pixels together (i.e., the pixel with the largest x-ray signal is defined as the target pixel). Ideally 1620 e^- should be collected; however, both devices fall short of this response because of split and partial events generated. The 25-μm epitaxial CCD exhibits inferior CCE performance because it has 10 μm more field-free material than the 15-μm epitaxial CCD. The average charge collected by the target pixel for the 15-μm epitaxial device is 959 e^- compared with only 469 e^- for the 25-μm epitaxial CCD.

Figure 4.70 compares horizontal knife-edge and bar target responses with a point spread response. The responses are closely related and can all estimate MTF_D performance. For high-performance CCDs, diffusion is usually only associated with pixels adjacent to the target pixel as shown. For illustration purposes, the figure shows a charge level of 100 e^- generated within the target pixel with a specific amount of charge diffusion going to adjacent pixels. For example, 10 e^- from the target pixel diffuses into the neighboring pixel for the edge response. For the point spread response, 40 e^- total diffuse in all directions from the target pixel. For the bar target 20 e^- diffuse from adjacent pixels into the target pixel, 10 e^- from each side. All of these diagrams represent the same amount of charge diffusion. They are only different because of the differences in the way the CCD is stimulated. It

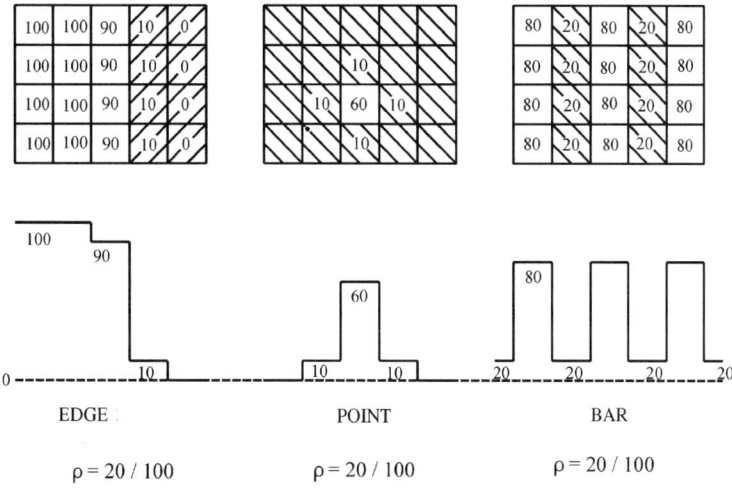

Figure 4.70 Edge and bar target responses compared to point spread.

should be pointed out that x rays stimulate the entire boundary of the pixel shown in Fig. 4.62. We can think of an x ray point spread as the average impulse response for the pixel, including the depth dimension.

To convert the responses shown in Fig. 4.70 to MTF, it is necessary to ratio the diffusion signal component to the ideal target pixel signal level as

$$\rho = \frac{S_{DIF}}{S_{TAR}}, \qquad (4.40)$$

where S_{TAR} is the ideal target pixel signal level and S_{DIF} is the diffusion level defined in Fig. 4.71(a). Note that ρ is that portion of the total target pixel charge that diffuses in the direction of interest (i.e., vertically or horizontally). Note in Fig. 4.70 that ρ for the edge and bar responses are made equivalent to point spread because charge diffusion is the same for a given CCD.

MTF(f) is estimated from point spread by weighting Eqs. (4.20) and (4.21) with ρ as

$$S_{MAX}^* = S_{MAX}(1 - \rho), \qquad (4.41)$$

and

$$S_{MIN}^* = S_{MIN} + \rho \int_{-3\pi/\lambda+\pi}^{-\pi/\lambda+\pi} \frac{1}{2}[1 + \cos(\theta)]d\theta. \qquad (4.42)$$

From Fig. 4.71(a), S_{MAX} loses charge because of diffusion by ρ giving S_{MAX}^*. S_{MIN} gains charge through diffusion from the neighboring pixel, which contains an amount of charge given by the integral as in Eq. (4.42).

Integrating Eq. (4.42) and substituting into Eq. (4.19) along with Eq. (4.41) yields the MTF response for the edge, point spread and bar responses shown in Fig. 4.70 of

$$\text{MTF} = \frac{-\rho \frac{\pi}{\lambda} + \sin\frac{\pi}{\lambda} - \rho \sin\left(\frac{\pi}{\lambda}\right)^3}{\frac{\pi}{\lambda} - \rho \sin\frac{\pi}{\lambda} + \rho \sin\left(\frac{\pi}{\lambda}\right)^3}. \qquad (4.43)$$

Note that the MTF response calculated here is the combination of integration MTF and diffusion MTF. If desired, MTF_D can be found through Eqs. (4.35) and (4.17) assuming perfect MTF_{CTE}. Figure 4.71(b) plots MTF as a function of normalized frequency for $\rho = 0$, 0.09 and 0.2 using Eq. (4.43). This formula is only valid if charge diffusion is confined to adjacent pixels, which for high-performance scientific CCDs is normally the case.

Charge Collection

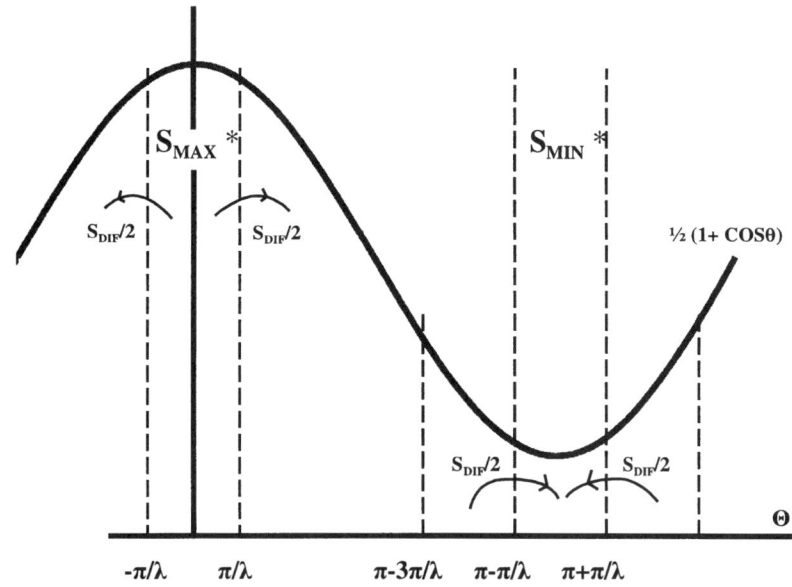

Figure 4.71(a) Illustration showing integration limits used to convert a point spread response to MTF.

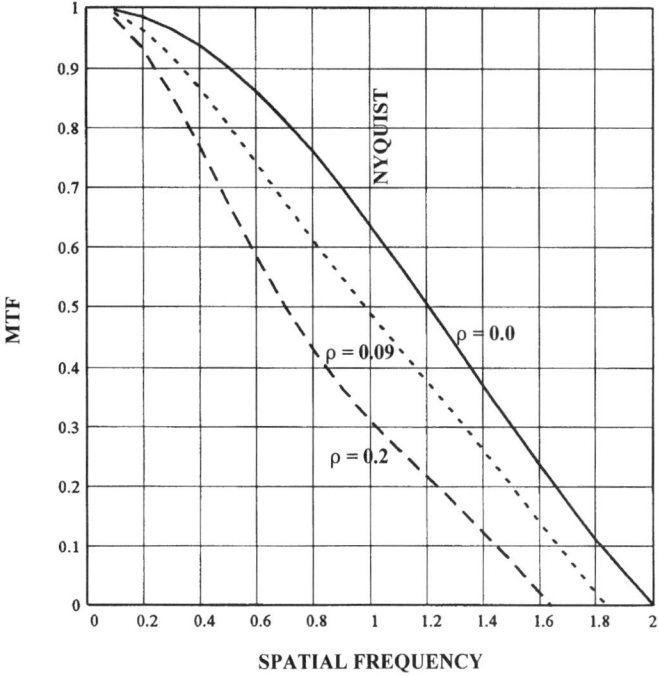

Figure 4.71(b) MTF as a function of normalized spatial frequency for different ρ.

Example 4.21

Table 4.2 shows Fe^{55} x-ray point spread data generated by a backside-illuminated 12-μm-pixel three-phase CCD. The charge distribution shown within and about the target pixel is given in percent. Estimate the horizontal and vertical MTF at Nyquist from Eq. (4.43). Also, find MTF$_D$ in each case.

Solution:
From Eq. (4.22),

$$\lambda = 2 \text{ at Nyquist,}$$

$$\pi/\lambda = 1.57.$$

The total charge generated in the target pixel is 100%. The amount of charge that diffuses in the horizontal direction is $10.5 + 9.5 + 0.7 + 0.8 = 21.5\%$. From Eq. (4.40),

$$\rho = 21.5/100,$$

$$\rho = 0.215.$$

From Eq. (4.43) the horizontal MTF(f) is,

$$\text{MTF} = \frac{-0.215 \times 1.57 + \sin(1.57) - 0.215 \sin^3(1.57)}{1.57 - 0.215 \sin(1.57) + 0.215 \sin^3(1.57)},$$

$$\text{MTF} = (-0.3375 + 1 - 0.215)/(1.57 - 0.215 + 0.215),$$

$$\text{MTF} = 0.285,$$

$$\text{MTF}_D = 0.285/0.63,$$

Table 4.2 Fe^{55} x-ray point spread response for a backside-illuminated CCD.

Column	1	2	3	4	5
Line					
1	0.1%	0.0%	0.3%	0.1%	0.0%
2	0.1%	1.4%	5.3%	1.5%	0.0%
3	0.7%	10.5%	60.2%	9.5%	0.8%
4	0.0%	1.3%	4.2%	1.4%	0.1%
5	0.0%	0.1%	0.2%	0.1%	0.0%

$MTF_D = 0.45.$

The amount of charge that diffuses in the vertical direction is $5.3 + 4.2 + 0.2 + 0.3 = 10\%$. From Eq. (4.41),

$$\rho = 10/100,$$

$$\rho = 0.10.$$

From Eq. (4.43) the vertical MTF is

$$\text{MTF} = \frac{-0.1 \times 1.57 + \sin(1.57) - 0.1 \sin^3(1.57)}{1.57 - 0.1\sin(1.57) + 0.1\sin^3(1.57)},$$

$$\text{MTF} = (-0.157 + 1 - 0.1)/(1.57 - 0.1 + 0.1),$$

$$\text{MTF} = 0.47,$$

$$\text{MTF}_D = 0.47/0.63,$$

$$\text{MTF}_D = 0.75.$$

Note that the horizontal MTF(f) is worse compared to the vertical direction. The difference is caused by channel stop diffusion. The channel stop region for this CCD is approximately 4 μm wide, a good portion of the pixel area.

4.4.2.5 Absolute CCE

MTF, CTF and knife-edge responses represent a family of relative CCE measurements that quantify average charge diffusion characteristics among pixels. X-ray point spread simultaneously characterizes charge diffusion as well as recombination loss. Sometimes it is desirable to isolate the two effects to specific regions in the CCD. For example, a CCD may exhibit significant recombination loss while achieving good MTF performance. Backside-illuminated CCDs exhibit this characteristic.

Absolute CCE based on x rays is defined by comparing the average charge collected by the target pixel to the ideal level.[25] That is,

$$\text{CCE} = \frac{\eta_E}{\eta_i}, \tag{4.44}$$

where η_E is called the effective quantum yield, the amount of charge collected by the target pixel and η_i is the ideal quantum yield that is defined by Eq. (1.2).

Example 4.22

Calculate the CCE performance for the sensors characterized in Fig. 4.69.
Solution:
From Eq. (4.44),

$$CCE = 959/1620,$$

$$CCE = 0.59 \text{ (15-}\mu\text{m epitaxial)},$$

$$CCE = 469/1620,$$

$$CCE = 0.29 \text{ (25-}\mu\text{m epitaxial)}.$$

CCE can also be calculated by summing split charge around the target pixel and adding it to the target pixel. For example, the average charge collected under summed conditions is 1208 e^- for the 15-μm epitaxial CCD and 750 e^- for the 25-μm epitaxial CCD. Recombination loss is, therefore, 25% and 54% respectively for the average event. This loss mainly occurs in the field free and p^+ substrate region of the device. CCE when summing is included is

$$CCE = 1208/1620,$$

$$CCE = 0.75 \text{ (15-}\mu\text{m epitaxial)},$$

$$CCE = 750/1620,$$

$$CCE = 0.46 \text{(25-}\mu\text{m epitaxial)}.$$

As mentioned above, stimulating a frontside-illuminated CCD with Fe^{55} x rays represents a CCE acid test because these photons interact uniformly throughout the device. Low-energy x rays, which exhibit a short absorption length, will produce a better response because interaction occurs in the depletion region. On the other hand, low-energy x rays represent a worst-case response for backside-illuminated CCDs because interaction occurs directly in the field-free/recombination regions. For these reasons, x-ray energy is always specified when quoting CCE performance. This specification is analogous to specifying the wavelength when optical MTF tests are performed.

4.4.2.6 Photon Transfer

Chapter 2 discussed how the photon transfer technique is used to measure quantum yield (refer to Section 2.2.4). Figure 4.72 shows a Fe^{55} quantum yield histogram taken from a frontside-illuminated CCD built on 15-μm 30–40-ohm-cm epitaxial material. The 15-μm-pixel CCD was uniformly stimulated with a Fe^{55} x-ray flux of approximately five x rays per pixel. The effective quantum yield was calculated for several different 40 × 40-pixel subarrays across the sensor. The data were then displayed in the histogram shown. The average quantum yield is approximately 900 e^-, significantly less than the ideal quantum yield of 1620 e^-. The less than optimum response is caused by field-free material and charge recombination in the substrate. The majority of frontside scientific CCDs fabricated exhibit this level of performance.

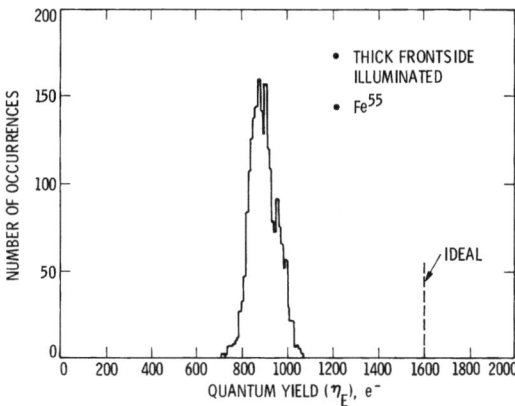

Figure 4.72 Fe^{55} quantum yield measurement for a frontside-illuminated CCD.

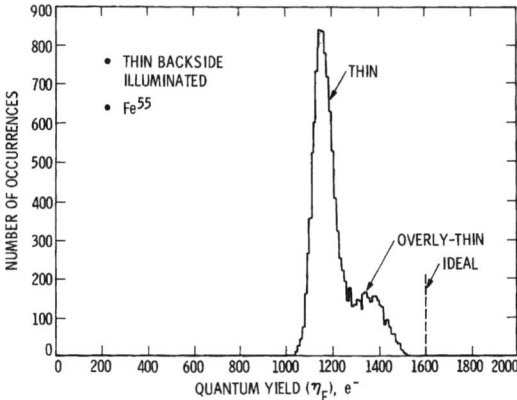

Figure 4.73(a) Fe^{55} quantum yield measurement for a backside-illuminated CCD.

Figure 4.73(b) Quantum yield image showing a higher response in the corners of the CCD where less field-free material exists.

```
1248 1251 1323 1346 1382 1369 1348 1440 1420 1478 1526 1434 1415
1273 1246 1261 1344 1354 1314 1339 1420 1420 1441 1471 1475 1437
1192 1157 1230 1248 1255 1321 1344 1303 1413 1375 1391 1489 1431
1058 1117 1149 1139 1232 1416 1374 1293 1337 1342 1462 1500 1428
1131 1118 1141 1129 1280 1270 1328 1339 1402 1413 1456 1391 1358
1099 1127 1183 1154 1241 1185 1270 1337 1399 1326 1332 1373 1411
1157 1154 1207 1142 1171 1116 1116 1186 1231 1280 1324 1461 1441
1199 1221 1217 1193 1151 1168 1171 1142 1180 1282 1379 1461 1369
1130 1130 1207 1172 1185 1191 1116 1152 1202 1146 1236 1318 1319
1166 1167 1198 1204 1220 1188 1121 1171 1117 1131 1157 1202 1325
1231 1203 1170 1135 1207 1207 1134 1121 1046 1216 1171 1159 1316
1202 1221 1279 1241 1264 1129 1094 1157 1204 1239 1175 1136 1187
1135 1063 1147 1189 1203 1082 1153 1175 1186 1146 1116 1157 1142
```

Figure 4.73(c) Quantum yield numbers in one corner of the CCD.

For backside-illuminated CCDs, field-free epitaxial material can be removed, which significantly reduces the number of split and partial events. As demonstrated in Chapter 3, properly accumulated CCDs can deliver 100% internal QE where very few partial events are observed. Figure 4.73(a) shows a quantum yield histogram taken from a backside-illuminated CCD thinned to approximately 6 μm. The corners of the array were overly thinned by approximately 1 μm because of thinning method employed (i.e., window thinning as discussed in Chapter 3). Therefore, the corners have less field-free material and exhibit a higher quantum yield compared to the center of the array (i.e., approximately 1400 e^-/photon compared with 1200 e^-/photon). Corresponding CCEs based on Eq. (4.44) are 0.86

CHARGE COLLECTION

and 0.74 respectively. The variability of quantum yield across the CCD is seen in Fig. 4.73(b). Here quantum yield is displayed in image format by calculating it for 160,000 subarrays across the CCD. Figure 4.73(c) shows quantum yield values in one corner of the array.

4.4.3 ALIASING AND BEATING

"Aliasing" and "beating" are two interesting imaging artifacts that are related to MTF measurements. Each effect can be analyzed by projecting a sinusoidal frequency over a line of pixels and integrating the resultant response. The formula for doing this is

$$P_i = \int_{-\pi i c}^{\pi(i+1)c} \frac{1}{2c}[1 + \cos(\theta)]d\theta, \tag{4.45}$$

where P_i is the ith pixel signal value and c is that fraction of Nyquist frequency (i.e., f/f_N) input over the pixels. The equation is a powerful formula used to understand unusual modulation characteristics such as beating and aliasing. The formula can also be slightly modified to show how the responses change with different pixel aperture openings. The equation above assumes a 100% fill factor pixel.

Figure 4.74 shows a response from Eq. (4.45) for three input spatial frequencies of $c = 0.2$, 1.8 and 4.2. Note that the frequencies 1.8 and 4.2 are both above Nyquist. According to Nyquist, to preserve image information the sampling frequency ($f_s = 1/p$) must be at least twice the highest spatial frequency contained in the image. Frequencies greater than Nyquist will appear in the image as false low-frequency components, a sampling effect called aliasing. Those frequencies that are undersampled will result in lower frequencies at

$$\frac{f_{ALIAS}}{f_N} = 2 - \frac{f}{f_N}, \quad f_N < \frac{f}{f_N} < 2f_N, \tag{4.46}$$

and

$$\frac{f_{ALIAS}}{f_N} = \frac{f}{f_N} - 2, \quad 2f_N < \frac{f}{f_N} < 3f_N, \tag{4.47}$$

and

$$\frac{f_{ALIAS}}{f_N} = 4 - \frac{f}{f_N}, \quad 3f_N < \frac{f}{f_N} < 4f_N, \tag{4.48}$$

and

$$\frac{f_{ALIAS}}{f_N} = \frac{f}{f_N} - 4, \quad 4f_N < \frac{f}{f_N} < 5f_N, \tag{4.49}$$

and so on.

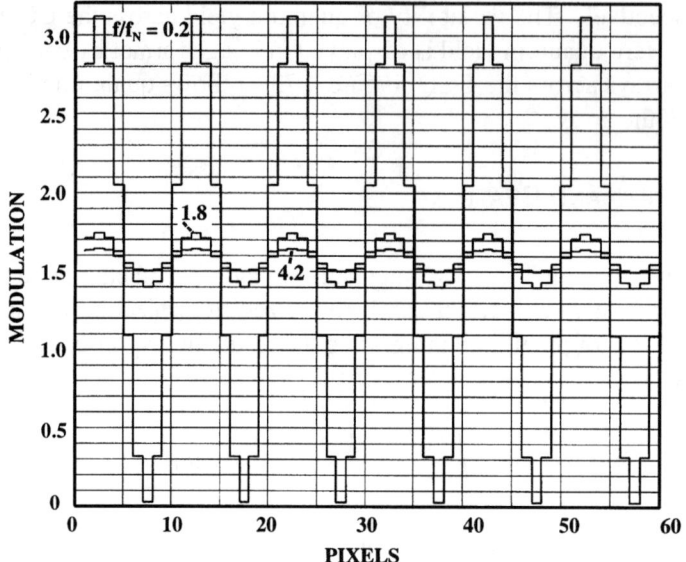

Figure 4.74 Pixel response from Eq. (4.45) showing aliased frequencies.

Example 4.23

Find the aliasing frequencies and the corresponding modulation MTF_I from the results shown in Fig. 4.74.

Solution:
From Eq. (4.46), the aliasing frequency at $f/f_N = 1.8$ is

$$f_{ALIAS}/f_N = 2 - 1.8,$$

$$f_{ALIAS}/f_N = 0.2.$$

From Eq. (4.49), the aliasing frequency at $f/f_N = 4.2$ is also

$$f_{ALIAS}/f_N = 4.2 - 4,$$

$$f_{ALIAS}/f_N = 0.2.$$

The number of pixels covered by each cycle is

$$p = 2/f,$$

$$p = 2/0.2,$$

$$p = 10 \text{ pixels}.$$

CHARGE COLLECTION 375

From Fig. 4.74 S_{MAX} and S_{MIN} are approximately $S_{MAX} = 1.8$ and $S_{MIN} = 1.74$ at $f/f_N = 1.4$.

$$\text{MTF}_I = 1.8 - 1.4/(1.8 + 1.4),$$

$$\text{MTF}_I = 0.125,$$

$S_{MAX} = 1.65$ and $S_{MIN} = 1.5$ for $f/f_N = 4.2$,

$$\text{MTF}_I = 1.65 - 1.5/(1.65 + 1.5),$$

$$\text{MTF}_I = 0.0476.$$

These MTF_I values agree with the theoretical modulation given by Eq. (4.17).

Aliasing can also be described in the spatial frequency domain through

$$S(f) = [\text{MTF}(f)R(f)] \otimes Q(f), \qquad (4.50)$$

where $S(f)$ is the output spatial frequency response (cycles/cm), $\text{MTF}(f)$ is the MTF of the CCD (cycles/cm), $R(f)$ is the input spatial frequency spectrum (cycles/cm) and $Q(f)$ is the sampling transfer function given by

$$Q(f) = f_s \sum_{i=-\infty}^{\infty} \delta(f - if_s), \qquad (4.51)$$

which is convolved with $\text{MTF}(f)R(f)$ to produce $S(f)$. The variable f_s is the sampling frequency, which is equal to $1/p$.

Example 4.24

Show with an illustration the transfer functions of Equation (4.50) in response to input spatial frequencies of $f/f_N = 0.2, 1.8$ and 4.6.
Solution:
Figure 4.75 shows the input spatial frequencies, $R(f)$, the $\text{MTF}(f)$ of the CCD, the sampling transfer function, $Q(f)$, and the output spectrum, $S(f)$. Note that the input frequencies above Nyquist are translated down into the passband at 0.5 ($f/f_N = 1.5$) and 0.6 ($f/f_N = 4.6$) in accordance to Eqs. (4.46) and (4.49).

From discussions above we can now explain why the CTE MTF shown in Fig. 4.57 increases above Nyquist frequency. Note from Eq. (4.46) that frequencies above Nyquist will always be translated to lower frequencies. In that it is easier for a CCD to transfer a lower spatial frequency than a higher one, we should expect the modulation to increase above f_N as shown. The curves are symmetric about Nyquist.

There has been considerable debate in the imaging community about how to handle the aliasing problem. Solutions vary from optimizing the aperture size of the pixel element for a given scene to prefilter the scene before sampling takes place. In all cases, any attempt to eliminate aliasing will degrade the frequency response in the passband (i.e., information is lost). Unfortunately, prefiltering frequencies above Nyquist without afflicting frequencies in the passband requires a filter that is not physically realizable.

For most scientific CCD applications, prefiltering the response above Nyquist probably does more harm than good. The concern for aliasing is usually related to taking "pretty" pictures rather than its effect on scientific results. Scientific information comes in the form of edges, lines and points, all of which have frequency components that lie outside the passband f_N. Although these features may not be reproduced by the CCD exactly (e.g., star images appear as abrupt jumps from pixel to pixel), in most situations prefiltering will take away from scientific return. For example, defocusing will blur star images and reduce aliasing. However, defocusing will also spread energy into adjacent pixels and, in turn, degrade S/N

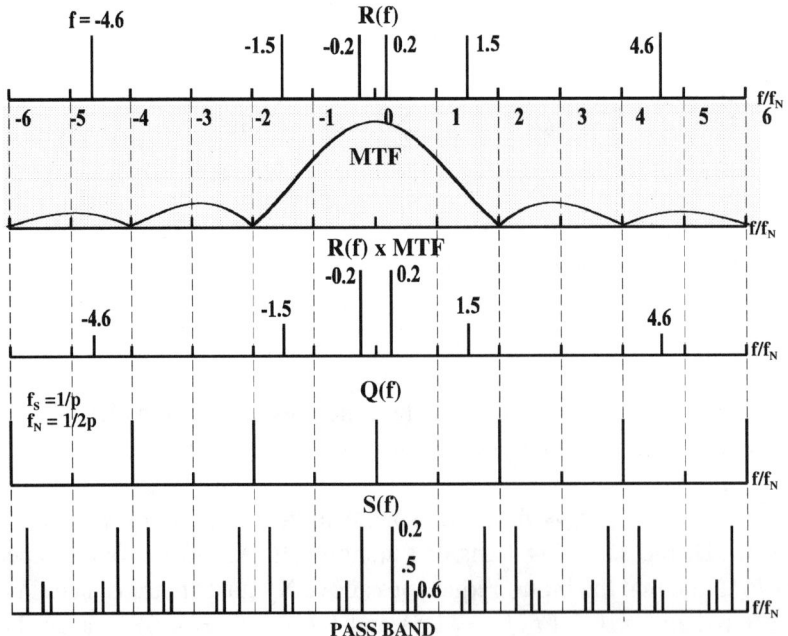

Figure 4.75 Spectral plots showing the aliasing effect.

performance. Line images generated by spectroscopy applications will suffer the same problem. It is important to realize that aliasing is spatially confined to the region with the problem. In other words, it does not affect neighboring pixels. On the other hand, defocusing spreads information outside the problem area. This is an important distinction. The process of prefiltering reduces information content where aliasing does not.

The other imaging artifact is called frequency "beating" or the moiré effect.[26] The beat effect is often confused with aliasing. Beating can be seen below and above Nyquist, whereas aliasing only occurs above Nyquist. As with other beat phenomena, beating is only observed when the two interacting frequencies are close to equal. In the case of imaging, beats are seen when the scene frequency is close to the imager's Nyquist frequency. For example, Fig. 4.76 shows the results of Eq. (4.45) at input frequencies of $f/f_N = 0.95$ and 1.05. The beat frequencies are clearly seen as a low-frequency envelope. The pattern is caused by the superposition of the pixel frequency $(1/2p)$ and the input frequency. The beat frequency is given by their difference,

$$f_{BEAT} = \left|\left(1 - \frac{f}{f_N}\right)\right|, \quad 0 < \frac{f}{f_N} < 2f_N. \quad (4.52)$$

Example 4.25

Find the beat frequencies seen in Fig. 4.76(a).

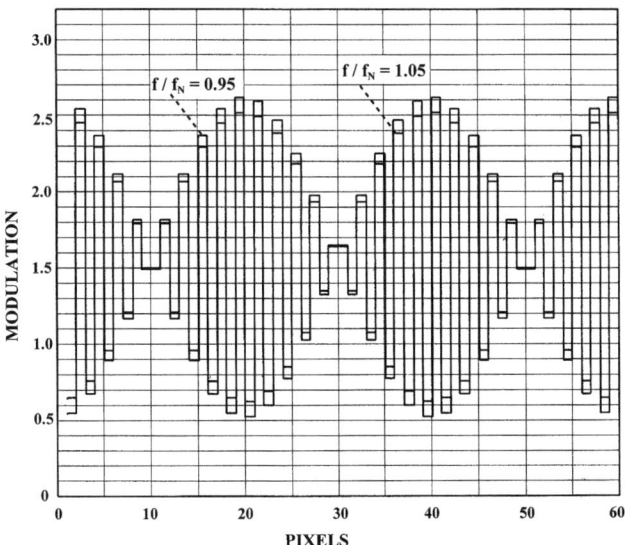

Figure 4.76(a) Low-frequency beat interference for input frequencies above and below Nyquist.

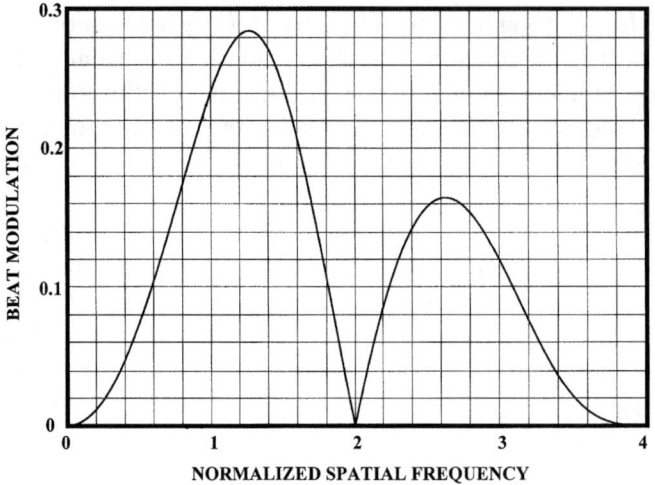

Figure 4.76(b) Predicted beat modulation as a function of normalized spatial frequency.

Solution:
The input frequencies 0.95 and 1.05 Nyquist will produce the same beat frequency. From Eq. (4.52),

$$f_{BEAT} = |1 - 0.95| = |1 - 1.05|,$$

$$f_{BEAT} = 0.05.$$

The number of pixels associated with this beat frequency is

$$p = 2/f,$$

$$p = 2/0.05,$$

$$p = 40 \text{ pixels}.$$

Note that the only way to tell the difference between the two input frequencies is by the modulation that is produced [i.e., $(S_{MAX} - S_{MIN})/(S_{MAX} - S_{MIN})$]. The lower input frequency will produce a slightly greater response as dictated by Eq. (4.17). This modulation difference is seen in Fig. 4.76(a).

The beat envelope modulation (from peak to valley) is found through

$$S_{MAX} = \int_{-\pi/\lambda}^{\pi/\lambda} \frac{1}{2}[1 + \cos(\theta)]d\theta, \qquad (4.53)$$

Charge collection

and

$$S_{MIN} = \int_0^{2\pi/\lambda} \frac{1}{2}[1+\cos(\theta)]d\theta. \qquad (4.54)$$

Integrating these equations and applying Eq. (4.19) yields

$$\text{MTF}_{BEAT} = \frac{\sin\left(\frac{\pi}{\lambda}\right) - \frac{1}{2}\sin\left(\frac{2\pi}{\lambda}\right)}{\frac{2\pi}{\lambda} + \sin\left(\frac{\pi}{\lambda}\right) + \frac{1}{2}\sin\left(\frac{2\pi}{\lambda}\right)}. \qquad (4.55)$$

Figure 4.76(b) plots Eq. (4.55) as a function of f/f_N.

Example 4.26

Find the envelope modulation of the beat frequency shown in Fig. 4.76(a). Compare with Fig. 4.76(b).
Solution:
From Eq. (4.19) and Fig. 4.76(a),

$$\text{MTF}_{BEAT} = 2.51 - 1.55/(2.51 + 1.55),$$

$$\text{MTF}_{BEAT} = 0.236.$$

The same result is found from Fig. 4.76(b).

Figure 4.77(a) presents three slant bar images taken by a 1024 × 1024 × 12-μm-pixel CCD at spatial frequencies of 0.286, 0.64 and 1.067. A strong "beat" signal occurs for the frequency above Nyquist. Figure 4.77(b) shows a line trace through the images.

Example 4.27

Determine the input frequency in Fig. 4.77(a) from the beat frequency assuming it is above Nyquist.
Solution:
The number of pixels associated with the beat frequency is 30 pixels. Therefore the frequency is

$$f/f_N = 2/30,$$

$$f/f_N = 0.067.$$

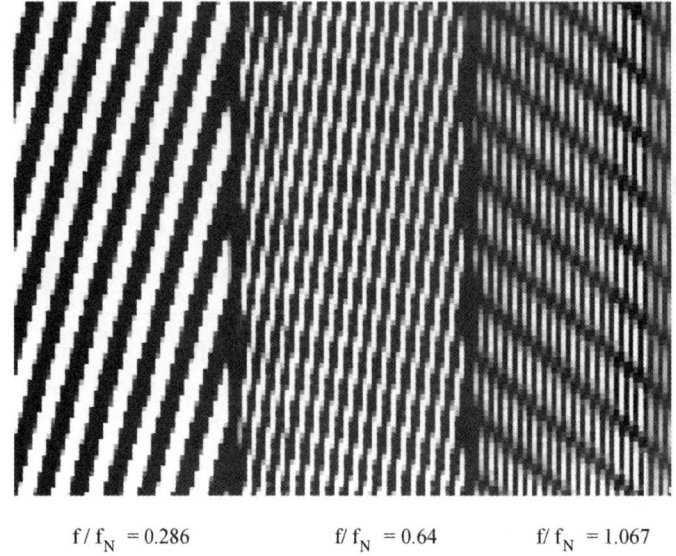

$f/f_N = 0.286$ $f/f_N = 0.64$ $f/f_N = 1.067$

Figure 4.77(a) Slant bar images showing interference (beating) at a frequency of $f/f_N = 1.067$.

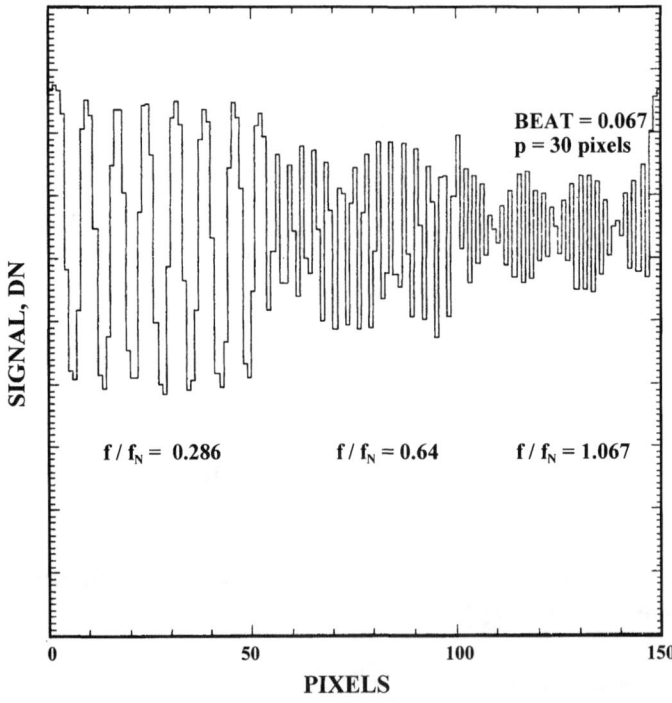

Figure 4.77(b) Line trace through the image shown in Fig. 4.77(a).

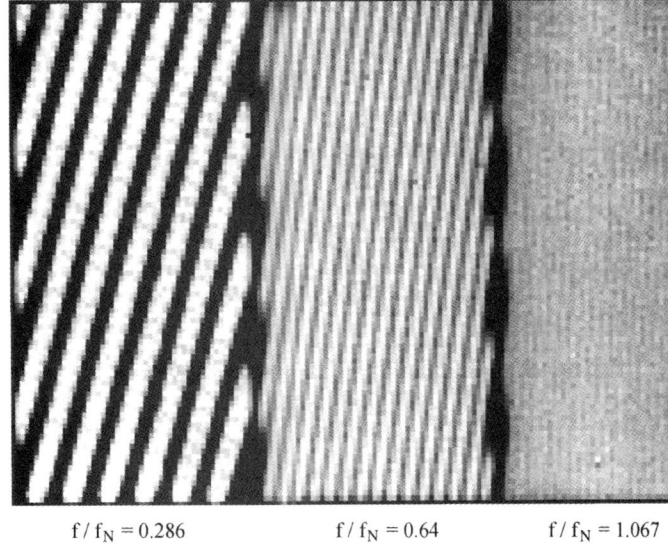

$f/f_N = 0.286$ $f/f_N = 0.64$ $f/f_N = 1.067$

Figure 4.77(c) Slant bar image showing how interference and aliasing can be reduced through defocusing.

The input frequency from Eq. (4.52) is

$$f/f_N = 1 + 0.067,$$

$$f/f_N = 1.067.$$

Figure 4.77(c) is the same image after the image is slightly defocused. Defocusing reduces the MTF response and, in turn, eliminates aliasing and beating caused by frequencies above Nyquist. This simple technique was used on the Galileo CCD camera, where the MTF was reduced by approximately a factor of two at Nyquist. Many other alternatives were evaluated and studied, but in the end a small metal shim behind the CCD focal plane was employed to suppress the aliasing and beating problem.

Figure 4.78 shows a computer imaging simulation generated by Jim McCarthy that also demonstrates the beating problem. The image in Fig. 4.78(a) shows a 1258 × 978-pixel region taken from a "sun burst" target with "rays" that are 5-deg wide and separated by 10 deg. Figure 4.78(b) is the same image magnified showing a 156 × 122-pixel region. Individual pixels are seen under this magnification. Figure 4.78(c) shows a 156 × 122-pixel region for a target with 0.5-deg-wide rays and 1-deg spacing. The high-spatial-frequency content in this image shows numerous regions where high-contrast beating takes place.

Figure 4.78(a)

Figure 4.78(b)

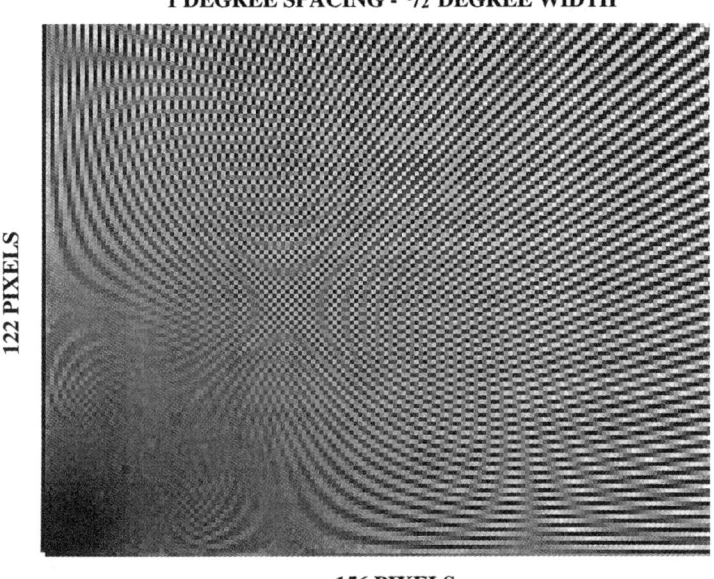

Figure 4.78(c)

Beating is particularly important to cinema movie applications, where such imaging artifacts are highly undesirable. The problem is usually unsetting to imaging groups who have used photographic film in the past. Film inherently does not have the beating problem because the light-sensitive silver grains are randomly spaced on the photographic emulsion. This arrangement destroys aliasing and beating patterns. To reduce aliasing and beating, the CCD pixels should be designed with random size and location. This requirement represents new design challenges.

REFERENCES

1. S. Kawai, N. Mutoh and N. Teranishi, "Thermionic-emission-based barrier height analysis for precise estimation of charge handling capacity in CCD registers," *IEEE Trans. Elec. Devices* Vol. 44, No. 10 (1997).
2. J. Bosiers, A. Kleimann, B. Dillen, H. Peek, A. Kokshoorn, N. Daemen, A. van der Sijde and L. Gaal, "A 2/3-in.1187(H) × 581(V) S-VHS-compatiable frame-transfer CCD for ESP and movie mode," *IEEE Trans. Elec. Devices* Vol. 38, No. 5 (1991).
3. P. Chatterjee and G. Taylor, "Optimum scaling of buried channel CCDs," *IEEE Trans. Elec. Devices* Vol. ED-27, No. 3 (1980).
4. J. Janesick, "Prevent blooming in CCD images," *NASA Tech. Briefs* Vol. 16, No. 7, p. 20 (1992).

5. J. Hynecek, "Electron-hole recombination antiblooming for virtual phase imager," *IEEE Trans. Elec. Devices* Vol. ED-30, No. 8 (1983).
6. W. Kosonocky, J. Carnes, M. Kovac, P. Levine and F. Shallcross, "Control of blooming in charge-coupled imagers," *RCA Rev.* Vol. 35, pp. 3–24 (1974).
7. Y. Ishihara, E. Oda, H. Tanigawa, N. Teranishi, E. Takeuchi, I. Akiyama, K. Arai, M. Nishimura and T. Kamata, "Interline CCD image sensor with an antiblooming structure," *ISSSCC Dig. Tech. Papers* pp. 168–169, 314 (1982).
8. M. Van de Steeg, H. Peek, J. Bakker, J. Pals, B. Dillen and J. Oppers, "A frame-transfer CCD color imager with vertical antiblooming," *IEEE Trans. Elec. Devices* Vol. ED-32, No. 8 (1985).
9. A. Theuwissen, H. Peek, M. van de Steeg, R. Boesten, P. Hartog, A. Kokshoorn, E. Koning, J. Oppers, F. Vledder, P. Centen, H. Blom and W. Haar, "A 2.2-m pixel FT-CCD imager according to the Eureka HDTV standard," *IEEE Trans. Elec. Devices* Vol. 40, No. 9 (1993).
10. J. Janesick, "Multi-pinned phase charge-coupled devices," *NASA Tech. Briefs* Vol. 14, No. 8, p. 22 (1990).
11. J. Janesick, "Multi-pinned-phase charge-coupled device," U.S. Patent 4,963,952, Oct. 16, 1990.
12. J. Janesick, T. Elliott, R. Bredthauer, C. Chandler and B. Burke, "Fano-noise-limited CCDs," *Proc. SPIE* Vol. 982, pp. 70–94 (1988).
13. K. Orihara and E. Oda, "Generation mechanism and elimination of fixed pattern noise in dual-channel horizontal-CCD register image sensor," *IEEE Trans. Elec. Devices* Vol. 35, No. 11 (1988).
14. J. Janesick, T. Eliott, S. Collins, T. Daud, D. Campbell and A. Dingzian, *Proc. SPIE* Vol. 597, p. 364 (1985).
15. M. Crowell and E. Labuda, "The silicon diode array camera diode," *Bell Syst. Tech. J.* Vol. 48, pp. 1481–1528 (1969).
16. D. Seib, "Carrier diffusion degradation of modulation transfer function in charge coupled imagers," *IEEE Trans. Elec. Devices* Vol. ED-21, pp. 210–217 (1974).
17. L. Schumann and T. Lomheim, "Modulation transfer function and quantum efficiency correlation at long wavelengths (greater than 800 nm) in linear charge coupled imagers," *J. Applied Optics* Vol. 28, No. 9 (1989).
18. M. Blouke, "A method for improving the spatial resolution of frontside illuminated CCDs," *IEEE Trans. Elec. Devices* Vol. ED-28, No. 3 (1981).
19. D. Seib, "Carrier diffusion degradation of modulation transfer function in charge coupled devices," *IEEE Trans. Elec. Devices* Vol. ED-21, No. 3 (1974).
20. D. Barbe, "Imaging devices using charge-coupled concept," *Proc. IEEE* Vol. 63, pp. 38–67 (1975).
21. H. Fitting, H. Glaefeke and W. Wild, *Phys. Status Solidi A.* Vol. 43, p. 185 (1977).
22. T. Everhart and P. Hoff, *J. Appl. Phys.* Vol. 42, p. 5837 (1971).
23. G. Hopkinson, *Optical Engineering* Vol. 26, p. 766 (1987).

24. G. Hopkinson, "Analytic modeling of charge diffusion in charge-coupled device imagers," *Optical Engineering* Vol. 26, No. 8 (1987).
25. J. Janesick, K. Klaasen and T. Eliott, "Charge-coupled-device charge-collection efficiency and the photon-transfer technique," *Optical Engineering* Vol. 26, No. 10 (1987).
26. T. Lomheim, J. Kwok, T. Dutton, R. Shima, J. Johnson, R. Boucher and C. Wrigley, "Imaging artifacts due to pixel spatial sampling smear and amplitude quantization in two-dimensional visible imaging arrays," *Proc. SPIE* Vol. 3701 (1999).

CHAPTER 5

CHARGE TRANSFER

5.1 CHARGE TRANSFER

Charge transfer efficiency (CTE) is a measure of the ability of the device to transfer charge from one potential well to the next. CTE is defined as the ratio of charge transferred from the target pixel to the initial charge stored in the target pixel. The transfer process is amazingly efficient. Typically, for well-made buried-channel scientific CCDs, the CTE will be between 0.99999 and 0.999999 for large charge packets (> 1000 e$^-$). Assuming a CTE of 0.999999, this means that for a CCD of 1000 pixels on the side, 99.8% of the charge will remain in the pixel farthest removed from the output after it has been transferred to the output (i.e., 2000 pixel transfers). The lost charge from the target pixel dribbles out as a deferred charge in trailing pixels. For computational purposes CTE is specified in terms of charge transfer inefficiency (CTI). CTI is the fraction of charge left behind in a single pixel transfer and is simply defined as

$$\text{CTI} = (1 - \text{CTE}). \tag{5.1}$$

Theoretically, the amount of charge that is lost for an N-stage CCD register is[1]

$$S_{N_P+n} = \frac{S_i N_P!}{(N_P - n)! n!} (1 - \text{CTI})^n \text{CTI}^{(N_P-n)}, \tag{5.2a}$$

where N_P is the number of pixel transfers (pixels), n is the trailing pixel number that follows the target pixel (i.e., $n = 0$ is the target pixel itself, $n = 1$ is the first trailer, etc.), S_i is the initial charge contained in the target pixel (e$^-$), S_{N_P+n} is the charge contained in the $N_P + n$ trailing pixel (e$^-$) and CTI is the charge transfer inefficiency.

Using Poisson's approximation to the binomial distribution above we find

$$S_{N_P+n} = \frac{S_i (N_P \text{CTI})^n}{n!} \exp(-N_P \text{CTI}). \tag{5.2b}$$

Example 5.1

Find the charge contained in the target pixel and the first trailing pixel after 1000 pixel transfers. Assume $S_i = 1620$ e$^-$ and CTI $= 0.0001$.

Solution:
From Eq. (5.2),
Target pixel: $n = 0$

$$S_{1000} = 1620 \times \exp(-10^3 \times 10^{-4}),$$

$$S_{1000} = 1465\,e^-.$$

First trailer: $n = 1$

$$S_{1001} = 1620 \times 10^3 \times 10^{-4} \exp(-10^3 \times 10^{-4}),$$

$$S_{1001} = 146.5\,e^-.$$

Figures 5.1(a) and 5.1(b) are plots of Eq. (5.2) as a function of N_P assuming a point sources $S_i = 10{,}000\,e^-$ for CTE $= 0.995$ and 0.9999 respectively. Note how the centroid of the point source shifts in the opposite direction to charge transfer as N_P increases. From the equation above it can be shown that the target pixel and first trailing pixel will have the same amount of charge when $N_P = 1/\text{CTI}$. For example, $N_P = 200$ transfers when CTE $= 0.995$ [i.e., $N_P = 1/(1 - 0.995)$]. In fact, for every $1/\text{CTI}$ transfers, the maximum level shifts by one pixel. This characteristic is seen in Fig. 5.1(a) for every 200 pixels. The effect results in a time delay such that the charge packet comes out of the CCD late by $N_P \times \text{CTI}/f$ where f is the clock frequency for either the horizontal or vertical registers. The centroiding problem is critical to CCD applications that require high geometric accuracy. For example, most star-tracking CCDs require centroiding precision on the order of $1/100$ pixel, and therefore, very high CTEs are required for these applications.

Table 5.1 tabulates deferred charge loss for a $1620\,e^-$ charge packet transferred through a CCD register for different pixel counts (i.e., $N_P = 64, 256, 512, 1024, 2048$ and 4096 pixel transfers) and CTEs.

Example 5.2

Using Table 5.1, find the deferred charge for a $1620\,e^-$ charge packet that is transferred through a 2048×2048 CCD beginning at the pixel furthest removed from the on-chip amplifier. Assume a CTE of 0.9999 for both the horizontal and vertical registers.

Solution:
There are a total of 4096 pixel transfers. From Table 5.1 we find a 34% deferred loss of $544\,e^-$. As shown below, horizontal and vertical CTE will usually be different and must be specified separately.

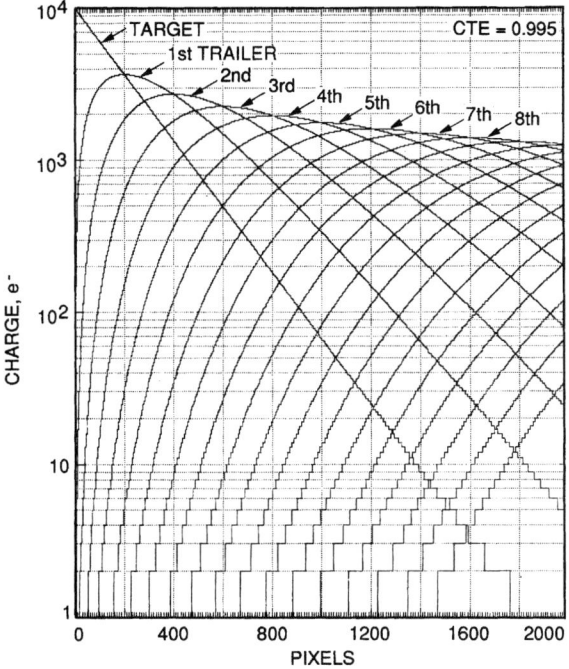

Figure 5.1(a) Theoretical 10,000-e^- point-source response assuming CTE = 0.995.

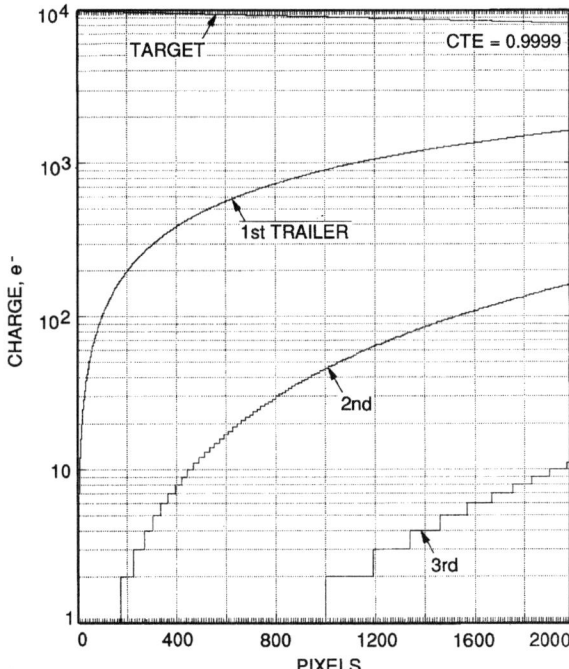

Figure 5.1(b) Theoretical 10,000-e^- point-source response assuming CTE = 0.9999.

Table 5.1 Deferred charge loss for a 1620-e⁻ charge packet as a function of CTE and register length.

CTE	64	256	512	1024	2048	4096
0.9999999	0.01	0.04	0.08	0.17	0.33	0.66
0.9999998	0.02	0.08	0.17	0.33	0.66	1.33
0.9999997	0.03	0.12	0.25	0.50	1.00	1.99
0.9999996	0.04	0.17	0.33	0.66	1.33	2.65
0.9999995	0.05	0.21	0.41	0.83	1.66	3.31
0.9999994	0.06	0.25	0.50	1.00	1.99	3.98
0.9999993	0.07	0.29	0.58	1.16	2.32	4.64
0.9999992	0.08	0.33	0.66	1.33	2.65	5.30
0.9999991	0.09	0.37	0.75	1.49	2.98	5.96
0.999999	0.10	0.41	0.83	1.66	3.31	6.62
0.999998	0.21	0.83	1.66	3.31	6.62	13.22
0.999997	0.31	1.24	2.49	4.97	9.92	19.78
0.999996	0.41	1.66	3.31	6.62	13.22	26.33
0.999995	0.52	2.07	4.14	8.27	16.50	32.84
0.999994	0.62	2.49	4.97	9.92	19.78	39.33
0.999993	0.73	2.90	5.8	11.57	23.06	45.79
0.999992	0.83	3.31	6.62	13.22	26.33	52.22
0.99999	1.04	4.14	8.27	16.5	32.84	65.01
0.99998	2.07	8.27	16.5	32.84	65.02	127.42
0.99997	3.11	12.39	24.69	49.01	96.54	187.32
0.99996	4.14	16.50	32.84	65.02	127.42	244.82
0.99995	5.18	20.60	40.95	80.86	157.68	300.01
0.99994	6.21	24.69	49.01	96.54	187.33	352.99
0.99993	7.24	28.77	57.03	112.06	216.37	403.84
0.99992	8.27	32.84	65.02	127.42	244.83	452.65
0.99991	9.30	36.90	72.96	142.63	272.71	499.51
0.9999	10.34	40.95	80.86	157.68	300.02	544.48
0.9998	20.61	80.86	157.69	300.03	544.5	905.99
0.9997	30.81	119.78	230.70	428.54	743.72	1146.00
0.9996	40.95	157.71	300.06	544.54	906.05	1305.35
0.9995	51.03	194.68	365.97	649.27	1038.32	1411.14
0.9994	61.05	230.73	428.59	743.80	1146.09	1481.37
0.9993	71.00	265.86	488.10	829.13	1233.91	1527.98
0.9992	80.89	300.12	544.63	906.16	1305.46	1558.93
0.9991	90.72	333.50	598.35	975.70	1363.75	1579.47

5.2 Transfer Mechanisms

Ideally, three mechanisms are responsible for charge transfer: thermal diffusion, self-induced drift and a fringing field effect.[2] The relative importance of each of these is primarily dependent on the charge packet size. Both thermal diffusion and fringing fields are important in transferring small charge packets, whereas self-induced drift, caused by mutual electrostatic repulsion of the carriers within a packet, dominates charge transfer for large packets.

Charge transfer for a multiphase CCD can be summarized as follows. We begin with a large charge packet in a collecting phase. Next, a neighboring barrier phase is switched high to also become a collecting phase. Charge will then equally divide between the two collecting phases. The main force that causes charge to

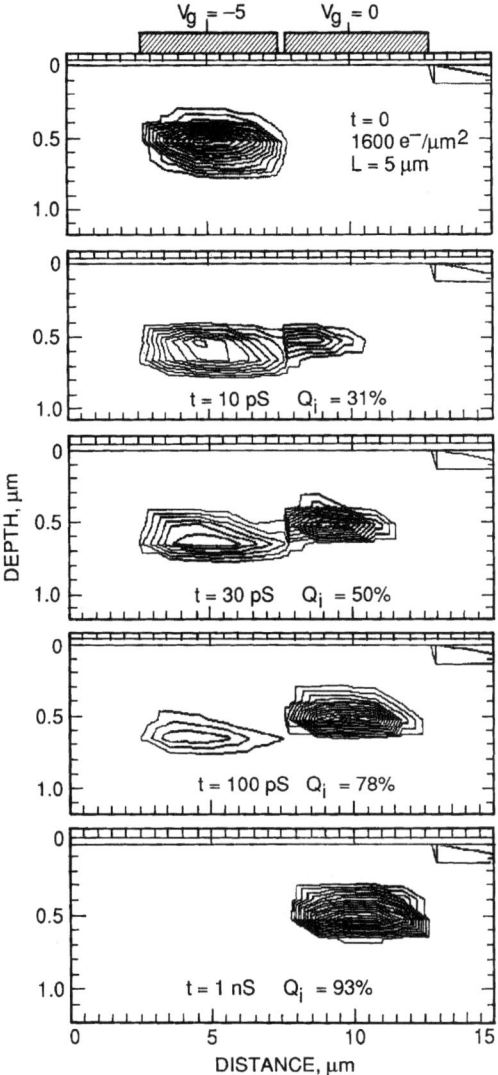

Figure 5.2 Charge transfer speed between two gates.

transfer between phases is electrostatic repulsion (i.e., self-induced drift). Shortly thereafter (i.e., one clock overlap period), the original collecting phase becomes a barrier phase. Again, electrostatic repulsion begins the transfer process. However, as charge density decreases, the self-induced drift field becomes less effective. The remaining charge is transferred either by fringing fields, which are strongest near the edges of the phases, or by thermal diffusion for charge that is near the center of the phase. Note that for charge packet amounts near full well, fringing fields even near the edges of the phases disappear as the collecting well is filled.

Figure 5.2 presents a two-dimensional simulation of transferring a charge packet between two gates. We assume a 5-V potential differential exists between the phases. At $t = 0$ the charge packet is under the left-hand gate. In 10 ps, 31% of the charge is transferred into the neighboring phase primarily by self-induced drift. In one nsec 93% of the charge is transferred. At this point, charge will be transferred by diffusion and fringing fields. The illustration shows that it takes a finite time for charge to transfer. For early scientific applications this time period was not fully appreciated because charge was usually transferred at a slow rate. Today there are many scientific applications that require the CCD to run at its theoretical speed limit.

Nearly all high-speed CTE problems are associated with the rise and fall time of the applied clocks. In fact, if the slew rate (i.e., V/sec) of the clock edges was infinite, the CCD would not function! For many reasons, the rise and fall time of the clocks should be made as slow as possible (i.e., the clocks need to be wave shaped). For example, if the falling edge of the clock is faster than the time it takes for charge to diffuse from phase to phase, blooming will result. Also, a high-speed problem referred to as "substrate bounce" is associated with the rising edge of the clock. Its speed must also be limited or blooming will result. We will discuss both problems in considerable depth below.

Unless otherwise indicated, data presented in this book are based on the following clock wave-shaping recipe for a three-phase CCD,

$$\tau_{WS} = \frac{t_T}{12}, \tag{5.3}$$

where τ_{WS} is the RC clock wave-shaping time constant (sec) and t_T is either the time to transfer one line vertically (t_{LT}) or the time to transfer one pixel horizontally (t_{PT}). As shown in Fig. 5.3(a), two time constants are given to each clock-overlap period. Twelve time constants are required for a complete line or pixel clock transfer cycle.

Example 5.3

Determine the vertical clock wave-shaping time constant for a line transfer period of $t_{LT} = 0.60$ µsec.
Solution:
From Eq. (5.3),

$$\tau_{WS} = (6 \times 10^{-7})/12,$$

$$\tau_{WS} = 5 \times 10^{-8} \text{ sec.}$$

Figure 5.3(b) shows the high-speed clock driver circuit used to generate CCD data products presented in this book. The driver circuit is TTL compatible and uses off-the-shelf transistors (pnp-2N2907 and npn-2N2369). The RC network at

CHARGE TRANSFER

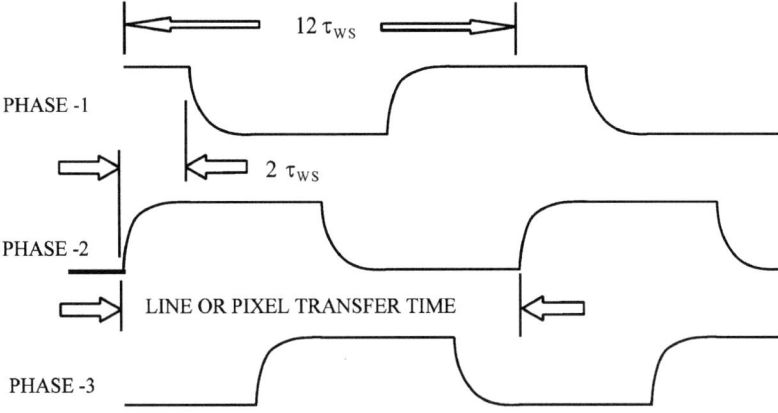

Figure 5.3(a) Three-phase timing diagram showing clock wave-shaping requirements.

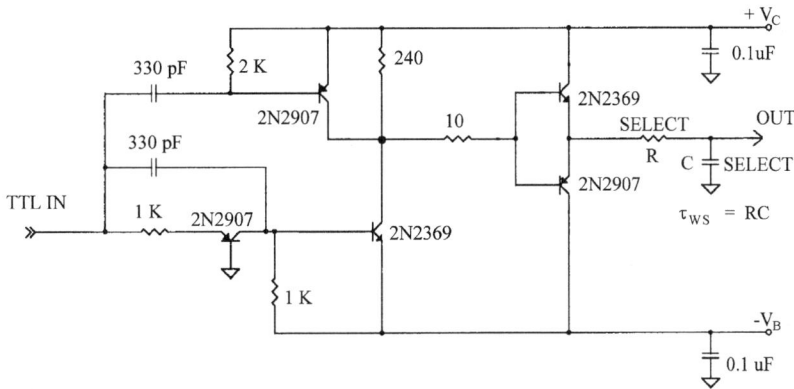

Figure 5.3(b) High-speed clock driver.

the output of the driver is where clock wave shaping is normally applied (i.e., $\tau_{WS} = \text{RC}$).

5.2.1 DIFFUSION DRIFT

Thermal diffusion defines the limiting transfer rate when self-induced and fringing field drift are not present. When only thermal diffusion applies, the fractional charge that remains under a phase after transfer time t is[3]

$$\text{CTI}_D = e^{-t/\tau_{th}}, \tag{5.4}$$

where the diffusion drift time constant, τ_{th}, is

$$\tau_{th} = \frac{L^2}{2.5 D_n}, \qquad (5.5)$$

where L is the gate length (cm) and the diffusion coefficient is

$$D_n = \frac{kT}{q}\mu_{SI}, \qquad (5.6)$$

where μ_{SI} is the electron mobility. Mobility will decrease with increasing operating temperature because lattice scattering of electrons. The effect is due to the thermal vibration of the atoms of the crystal lattice, which disrupts the periodicity of the lattice and impedes the motion of electrons. Decreasing operating temperature reduces lattice scattering and in turn increases the mobility. Although the mobility increases with decreasing temperature, the overall diffusion time will increase because of kT.

Example 5.4

Determine the diffusion time between two gates to achieve a CTE = 0.9999 at room temperature. Assume a gate length of $L = 10$ μm.

Solution:

From Eq. (5.6),

$$D_n = 0.0259 \times 1350,$$

$$D_n = 35 \text{ cm}^2/\text{sec}.$$

From Eq. (5.5),

$$\tau_{th} = (10 \times 10^{-4})^2 / (2.5 \times 35),$$

$$\tau_{th} = 1.14 \times 10^{-8} \text{ sec } (11.4 \text{ nsec}).$$

Ten time constants are approximately required to achieve a CTE = 0.9999 (CTI$_D$ = $10^{-4} \simeq e^{-10}$) or a time period of 114 nsec, which is equal to the clock overlap period.

5.2.2 SELF-INDUCED DRIFT

The fractional charge remaining after transfer time t caused when self-induced drift dominates is,[4]

$$\text{CTI}_{SID} = \left(1 + \frac{t}{\tau_{SID}}\right)^{-1}, \tag{5.7}$$

where τ_{SID} is

$$\tau_{SID} = \frac{2L^2 C_{EFF}}{\pi \mu_{SI} q Q}, \tag{5.8}$$

where C_{EFF} is the effective capacitance (F/cm^2) defined in Chapter 1 [Eq. (1.52)] and Q is the number of electrons per unit area (e$^-$/cm^2) for the charge packet at $t = 0$.

Note that CTI has a $1/x$ time dependence because the field strength decreases as charge is transferred (i.e., the process is not exponential as is charge diffusion and fringing field drift). The self-induced field decreases until thermal diffusion takes over. The time required to reach this condition is approximately equal to the diffusion time constant, τ_{th}.

Example 5.5

Calculate the time required to transfer 99% of the charge by self-induced drift. Assume $L = 10$ μm, $C_{EFF} = 2.05 \times 10^{-8}$ F/cm^2 and $Q = 8 \times 10^{11}$ e$^-$/cm^2 (8000 e$^-$/μm^2).

Solution:
From Eq. (5.8),

$$\tau_{SID} = 2 \times (10^{-3})^2 \times (2.05 \times 10^{-8})/[3.14 \times 1350 \times 1.6 \times 10^{-19} \times (8 \times 10^{11})],$$

$$\tau_{SID} = 7.55 \times 10^{-11} \text{ sec}.$$

Solving for t in Eq. (5.7),

$$t = \tau_{SID}(1/\text{CTI} - 1),$$

$$t = (7.55 \times 10^{-11}) \times (1/0.01 - 1),$$

$$t = 7.55 \times 10^{-9} \text{ sec (99\%)}.$$

The results of Example 5.4 and this example show that most (> 99%) charge is transferred by self-induced drift compared with thermal diffusion. Diffusion becomes dominant over self-induced drift in approximately 1.14×10^{-8} sec.

5.2.3 Fringing Field Drift

The fractional charge remaining after transfer time t due to fringing field drift is[5]

$$\text{CTI}_{FF} = e^{-t/\tau_{FF}}, \tag{5.9}$$

where τ_{FF} is the fringing field time constant,

$$\tau_{FF} = \frac{L}{2\mu_{SI}E_{\min}}, \tag{5.10}$$

where $E_{\min}$ is the minimum electric field strength under the gate (V/cm),

$$E_{\min} = \frac{2.09 \Delta V_G \varepsilon_{SI}}{L^2 C_{EFF}}, \tag{5.11}$$

where ΔV_G is the potential difference between gates (V), taking into account signal charge in the collecting phase.

Example 5.6

Calculate the fringing field time constant to yield CTE = 0.9999. Assume $L = 10$ μm, $C_{EFF} = 2.05 \times 10^{-8}$ F/cm^2 and $\Delta V_G = 5$ V.
Solution:
From Eq. (5.11) the minimum fringing field generated is

$$E_{\min} = 2.09 \times 5 \times 1.04 \times 10^{-12} / \left[(10^{-3})^2 \times 2.05 \times 10^{-8} \right],$$

$$E_{\min} = 530 \text{ V/cm}.$$

The corresponding time constant is found from Eq. (5.10),

$$\tau_{FF} = 10^{-3} / (2 \times 1350 \times 530),$$

$$\tau_{FF} = 6.98 \times 10^{-10} \text{ sec } (0.78 \text{ nsec}).$$

10 time constants yield CTE = 0.9999 or a total time of 6.98×10^{-9} sec.

Example 5.6 shows that fringing fields are dominant over diffusion fields for small gates (< 15 µm). However, fringing field strength is highly dependent on signal level and becomes nearly nonexistent at full well. This is because the potential difference between the collecting and barrier phases collapses to zero at full well (i.e., $\Delta V_G = 0$ V). Under full well conditions, charge diffusion is the only mechanism left to complete the transfer process. Therefore, transferring full well signals is usually a slow process.

Figure 5.4(a) illustrates the fringing field problem near full well. The top drawing shows that much stronger fringing fields exist when a potential well is empty compared to a full well level. The bottom drawing illustrates the diffusion blooming problem if the edge of a clock falls too quickly. That charge which has not diffused from the barrier phase has the opportunity to bloom backwards. Figure 5.4(b) is a computer simulation that shows the speed problem for a falling edge of only 0.25 nsec. The simulation uses two 4-µm gates that initially share a near full well amount of charge. Then phase-1 turns into a barrier phase, sending charge to its neighbor (i.e., phase-2) as indicated by the arrow in the illustration. The curve labeled "final" represents the final potential distribution after 0.25 nsec. Note before this time that the potential under the clocked phase is considerably less than in the barrier regions next to each gate, regions that are inverted and pinned. The only way the potential can be lower under phase-1 than the barrier phase is due to the presence of electrons. Charge that has not transferred before this time has the opportunity to bloom backwards. Figure 5.4(c) shows a similar simulation using a 1-nsec transition. The blooming situation is improved, but 26 e^- of charge remains

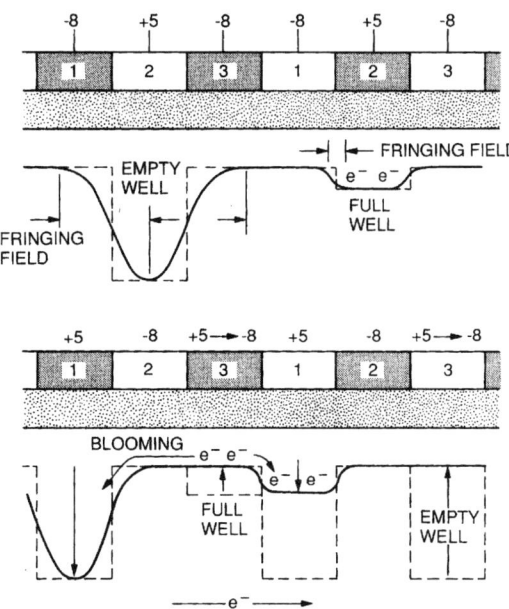

Figure 5.4(a) Fringing field comparison between an empty and full well.

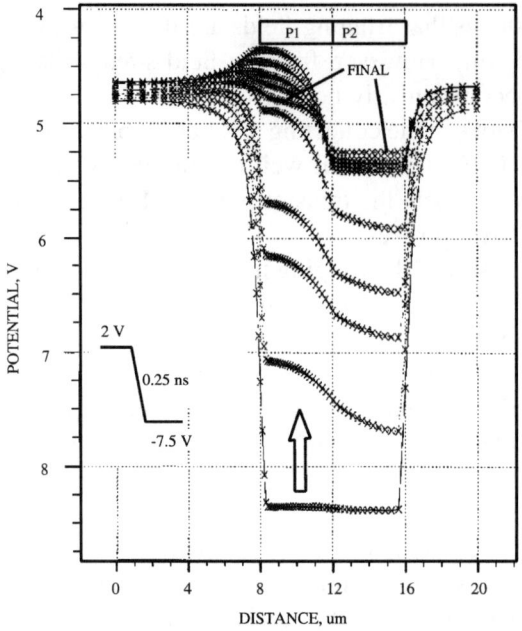

Figure 5.4(b) Charge backup problem when a collecting phase collapses into a barrier phase in 0.25 nsec.

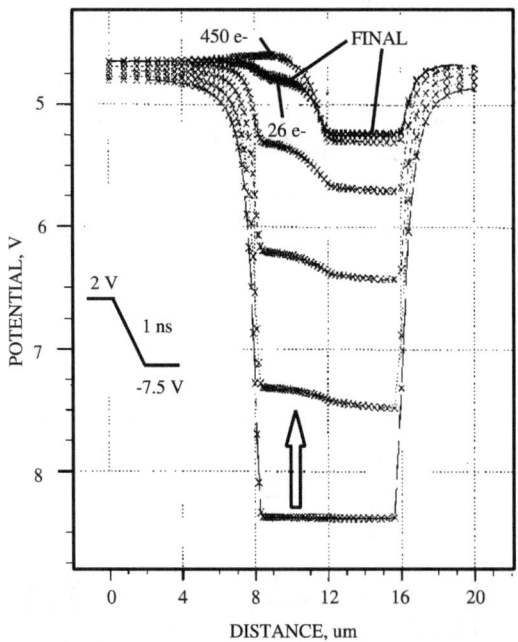

Figure 5.4(c) Charge backup problem when a collecting phase collapses into a barrier phase in 1 nsec.

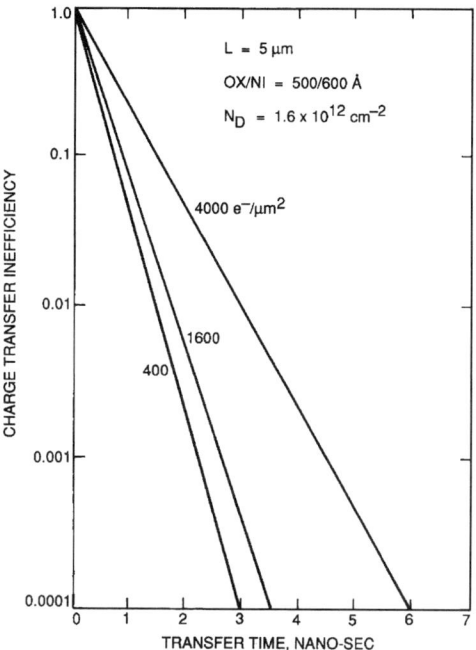

Figure 5.5(a) CTI as a function of transfer time and charge packet density.

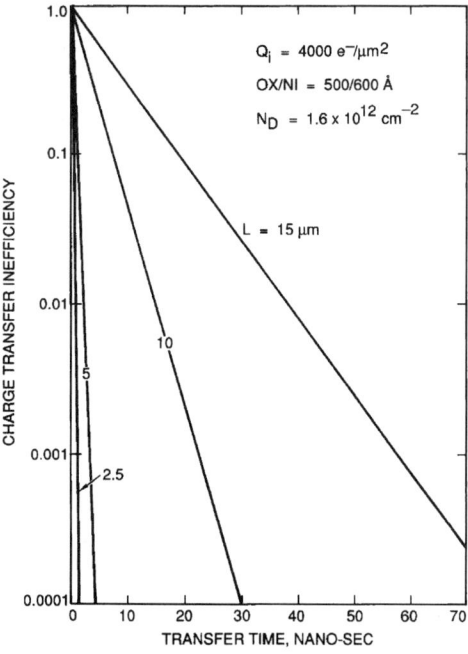

Figure 5.5(b) CTI as a function of transfer time and gate length.

in phase-1 after 1 nsec, indicating charge transfer for a single-phase transfer is far from complete.

CTI as a function of transfer time at different signal densities is modeled in Fig. 5.5(a). Note that transfer time increases as charge density increases because of the decreasing fringing field problem. The effect is enhanced when the gate length is made longer. Figure 5.5(b) plots CTI for various gate lengths assuming a charge density of 4000 $e^-/\mu m^2$. CTI increases by approximately the square of gate length [Eq. (5.5)].

5.2.4 HIGH-SPEED DATA

Figure 5.6(a) presents full well transfer data generated by a three-phase 1024 × 1024 CCD. The curves were taken at different line transfer times ($t_{LT} = 1.0, 0.86, \ldots$ µsec) and near the edge of the array. Wave shaping was employed using Eq. (5.3). For example, $t_{LT} = 1$ µsec corresponds to a wave-shaping time constant of $\tau_{WS} = 8 \times 10^{-8}$ sec. For line transfer periods < 1 µsec, full well degrades in part due to the charge diffusion problem on the falling clock edge being too fast. Figure 5.6(b) presents three column traces showing how charge blooms opposite to transfer. Charge moves from right to left in the figure.

Full well data shown in Fig. 5.6(a) were generated for the vertical registers. The horizontal register exhibits the same speed characteristics because their gate lengths are the same (i.e., the diffusion time constant is the same). However, the gates of the horizontal register are usually designed much wider and therefore exhibit a greater well capacity. Therefore, fringing fields exist when transferring a vertical full well signal, allowing much faster transfer speeds for this register.

Figure 5.7(a) plots full well as a function of wave shaping applied to an output summing well gate. Figure 5.7(b) illustrates the setup and voltages used in the experiment. For a fixed output summing well clock swing, the output transfer gate voltage is decreased until CTI = 0.01 is measured. Full well is greatest when the level of the output transfer gate is adjusted close to the low level of the output summing well. For this bias condition the fringing fields between the two gates are the smallest and wave shaping is greatest. If the falling clock edge of the output summing well is too fast, charge will diffuse backwards into the horizontal register. When the voltage difference between the two gates is increased, the fringing field will increase with sacrifice in well capacity. For example, full well drops from 680,000 e^- to 550,000 e^- when the potential difference changes by 3 V. However, the wave shaping can be reduced from 80 nsec to 4 nsec for much greater speed.

The results shown in Fig. 5.7(a) also apply to the horizontal register because the gate length and width are approximately the same for this test CCD. At full well (680,000 e^-), the maximum horizontal transfer time is approximately $t_{PT} = 0.96$ µsec (i.e., 12 × 80 nsec) for a CTI of 0.01 per transfer. Horizontal speed for a vertical full well signal (about 100,000 e^- for this CCD) is very fast. Theory predicts a falling edge of less than 5 nsec is satisfactory.

Charge transfer

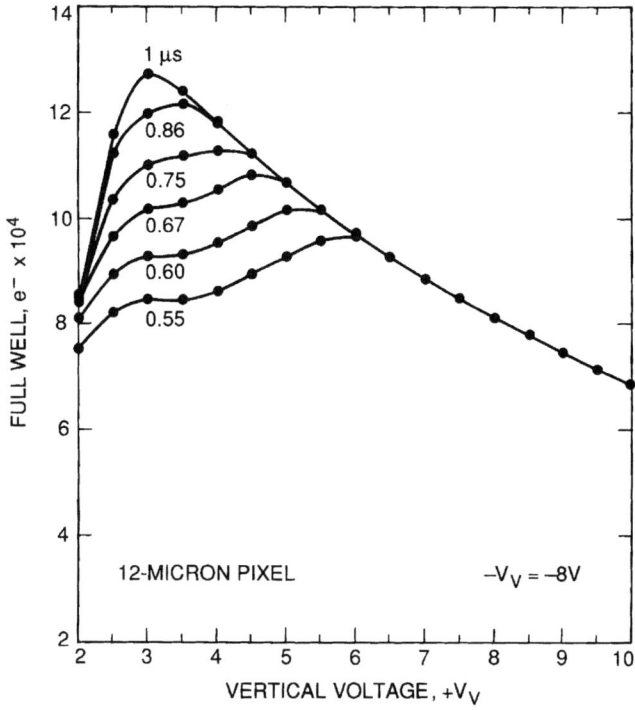

Figure 5.6(a) Full well transfer for different line transfer times.

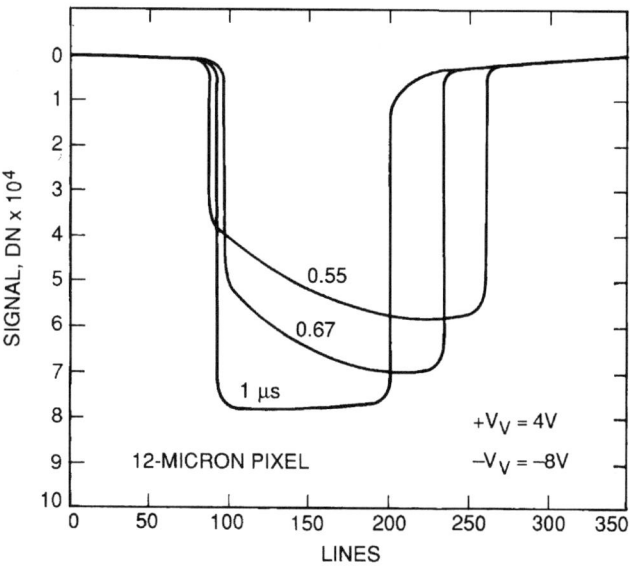

Figure 5.6(b) Column traces showing reverse blooming for different line transfer times.

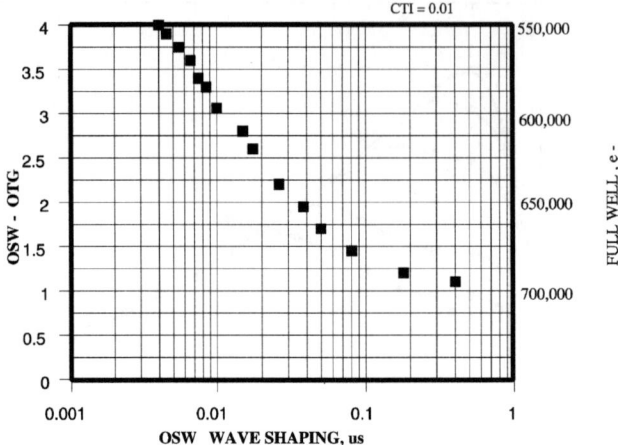

Figure 5.7(a) Full well as a function of wave shaping for the output summing well.

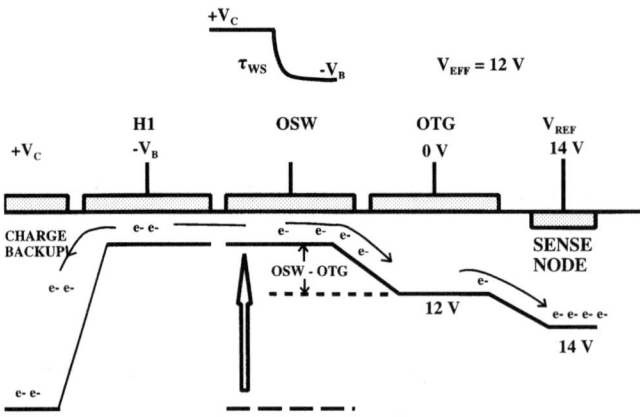

Figure 5.7(b) Potential diagram of the output summing well region.

The full well transfer curves above show that very high transfer rates are possible if the signal level is significantly below full well. For example, Fig. 5.8(a) shows an Fe^{55} x-ray transfer response taken at $t_{LT} = 200$ ns for a $1024 \times 1024 \times 12$-μm-pixel CCD. Fringing fields are primarily responsible for transferring charge at this signal level (i.e., $1620\,e^-$). Simulations predict that a $t_{LT} < 5$ ns is required to transfer charge packets this size. Unfortunately, generating three-phase vertical clocks this fast is difficult because of the high load capacitance associated with the vertical registers, thus presenting a challenge to verify this prediction experimentally.

Figure 5.8(b) shows an Fe^{55} x-ray transfer plot in which the horizontal register is only clocked from $V_B = -0.5$ V to $V_C = +0.5$ V. Fringing fields under these clocking conditions are very small, and therefore, charge is mainly transferred by diffusion. Nevertheless, CTE is well behaved for slow scan rates. The

Charge transfer

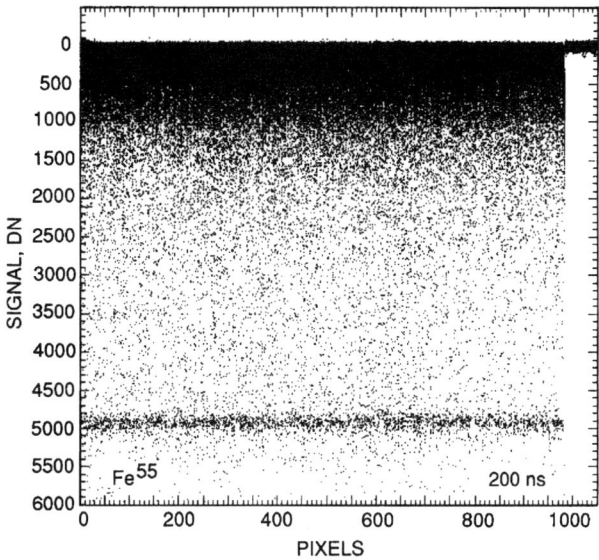

Figure 5.8(a) Fe^{55} x-ray transfer plot for a line transfer time of 200 ns.

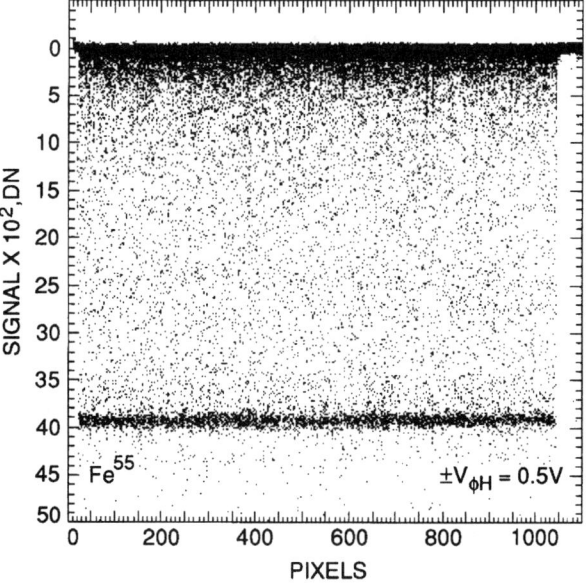

Figure 5.8(b) Fe^{55} x-ray transfer plot for a horizontal clock swing of only 1 V.

test also verifies the absence of small potential pockets or barriers within the signal channel.

Figure 5.9 shows full well characteristics taken from a 7.5-μm-pixel CCD. The speed limit for the CCD is $t_{LT} = 0.55$ μsec for a full well charge level of 26,000 e$^-$.

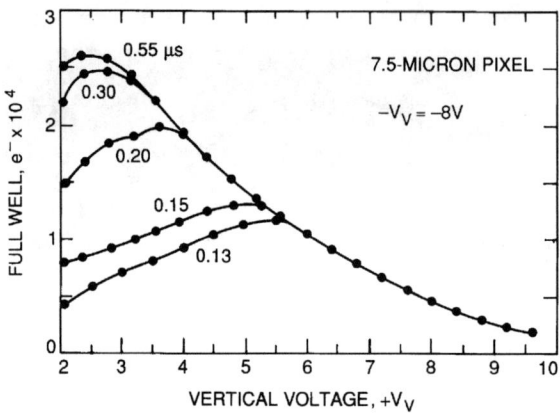

Figure 5.9 Full well transfer curves taken from a 7.5-μm-pixel CCD.

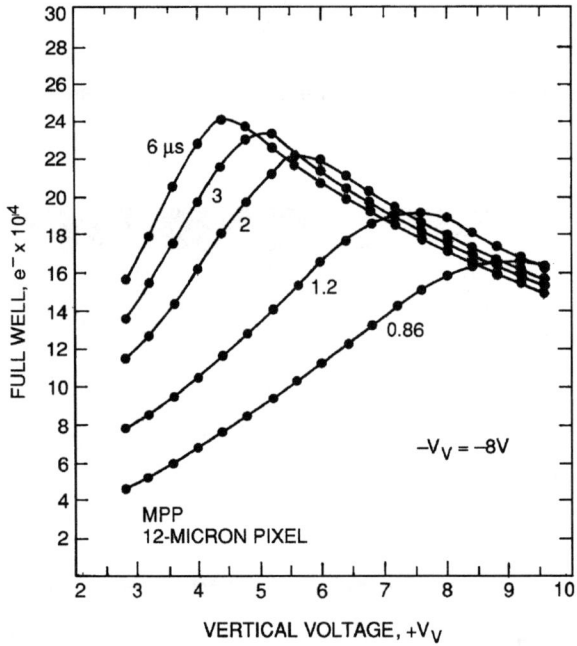

Figure 5.10 Full well transfer curves taken from a 12-μm-pixel MPP CCD.

Note from Fig. 5.6(a) that the full well for a 12-μm pixel is limited to 90,000 e^- at the same line transfer rate. This charge capacity level is still greater than the 7.5-μm-pixel device. In general, the larger pixel will always exhibit a greater charge capacity for any line transfer rate even though it suffers greater full well loss.

MPP CCDs are inherently slower than conventional multiphase CCDs. Recall from Chapter 4 that during high-temperature processing the MPP implant diffuses and encroaches under adjacent phases (phases-1 and -2 for a three-phase p-MPP CCD), generating a small potential barrier to electrons. When phase-3 is empty, these barriers are overwhelmed by the fringing fields and charge transfer is well behaved. However, as signal electrons are collected, the fringing fields collapse and the barriers become more important, making it more difficult to transfer charge. Therefore, the problem becomes present at full well conditions where fringing fields are smallest. Charge can only escape over the barriers by thermal diffusion, and therefore additional transfer time is required compared with CCDs without MPP. Figure 5.10 shows full well characteristics for a 12-μm-pixel CCD. Note that transfer speed is reduced significantly compared with the non-MPP CCD characterized in Fig. 5.6(a). Similar characteristics are observed for two-phase and virtual-phase CCDs because of potential barriers that form next to the implant regions (refer to Fig. 5.32).

5.2.5 CLOCK PROPAGATION

Discussions above show that charge motion is ultimately limited by thermal diffusion, self-induced drift and fringing field drift. Charge transfer can also be limited by a transfer problem called "clock propagation." The problem is associated with the poly gate resistance and pixel capacitance. The combined impedance will wave shape the clock as it works its way to the center of the array. Ideally the external wave-shaping time constant, τ_{WS}, should dominate the wave shaping that occurs internal to the CCD. However, at some high clock frequency, self-induced wave shaping will become more significant. At that point the clock amplitude will drop and, in turn, degrade charge capacity as dictated by full well transfer. The amount of wave shaping that takes place internal to the CCD is estimated by the formula

$$\tau_{POLY} = 1.5 R_{PIX} C_{PIX} N_{PIXC}^2, \quad (5.12)$$

where τ_{POLY} is the clock time constant at the center of the array (sec) and N_{PIXC} is the number of pixels to the center of the array. C_{PIX} is the drive capacitance associated with a single phase that drives a single pixel (F/pixel/phase). This includes the oxide, depletion, overlap, field and sidewall capacitances (refer to Section 5.5.3 for capacitance equations). The resistance associated with the poly gates is

$$R_{PIX} = \frac{R_S L}{W}, \quad (5.13)$$

where R_{PIX} is the pixel resistance (ohms/phase/pixel), R_S is the sheet resistance of the poly line (ohm/square), W is the width of the gate (cm) and L is the length of the gate (cm). Equation (5.12) assumes that the vertical registers are driven from both sides of the array.

Example 5.7

Estimate the poly wave shaping at the center of a three-phase 1024 × 1024 × 16-μm-pixel CCD. Assume poly resistance of $R_S = 40$ ohm/square and a pixel capacitance of $C_{PIX} = 1.83 \times 10^{-14}$ F/pixel/phase (see Example 5.24 for capacitance calculations). Also determine the minimum line transfer time, t_{LT}.

Solution:
From Eq. (5.13),

$$R_{PIX} = 40 \times (16 \times 10^{-4})/(5.33 \times 10^{-4}),$$

$$R_{PIX} = 120 \text{ ohm/pixel}.$$

From Eq. (5.12),

$$\tau_{POLY} = 1.5 \times 120 \times (1.83 \times 10^{-14}) \times 512^2,$$

$$\tau_{POLY} = 0.8635 \text{ μsec}.$$

From Eq. (5.3) and the timing diagram shown in Fig. 5.3(a),

$$t_{LT} = 12 \times (8.635 \times 10^{-7}),$$

$$t_{LT} = 10 \text{ μsec (12 wave-shaping time constants)}.$$

Figure 5.11 shows full well transfer plots taken from a 2048 × 2048, 15-μm-pixel MPP CCD clocked at $t_{LT} = 6$ and 30 μsec. The center and edges of the array are characterized (labeled as "C" and "E" respectively). Note that full well characteristics for the center of the CCD are worse than at the edges due to poly wave shaping. Attempting to increase the clock voltage to optimize the center region will only force the columns at the edge of the CCD into SFW. Therefore, each column of the CCD exhibits a different full well transfer response when poly wave shaping becomes present.

Frame transfer CCDs are especially prone to the poly wave-shaping problem because the vertical registers are clocked as fast as possible to avoid image smearing. Smearing occurs when light falls on the device while charge is being read into the storage region. The line transfer time required for frame transfer operation for a given smear specification is

$$t_{LT} = \frac{S_M t_I}{N_L}, \tag{5.14}$$

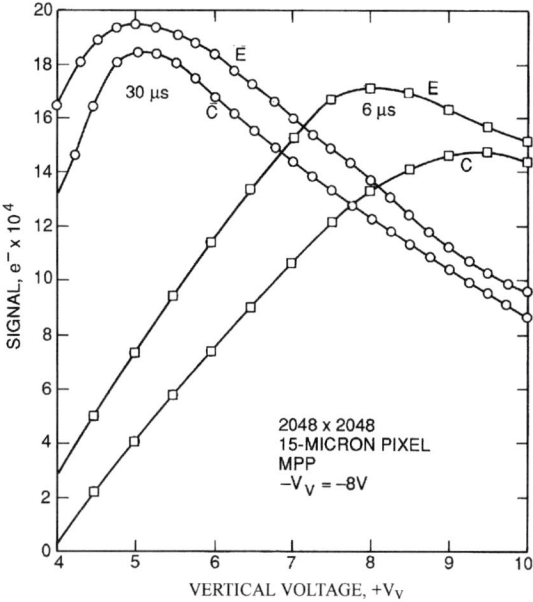

Figure 5.11 Full well transfer curves taken from a 2048 × 2048 ×15-μm MPP CCD.

where S_M is the smear allowed, t_I is the integration time (sec) and N_L is the number of lines transferred into the storage register.

Example 5.8

Determine t_{LT} for a frame transfer 1024(V) × 512(H) CCD. Assume that the integration time is 1/30 second with a smear of less than 1%. Also determine the wave shaping required.
Solution:
From Eq. (5.14),

$$t_{LT} = 0.01 \times 0.033/512,$$
$$t_{LT} = 0.644 \text{ μsec}.$$

From Eq. (5.3),

$$\tau_{WS} = 6.44 \times 10^{-7}/12,$$
$$\tau_{WS} = 5.3 \times 10^{-8} \text{ sec}.$$

The example above shows that the required line time may not be realizable because of clock propagation. Poly resistance can be essentially eliminated by

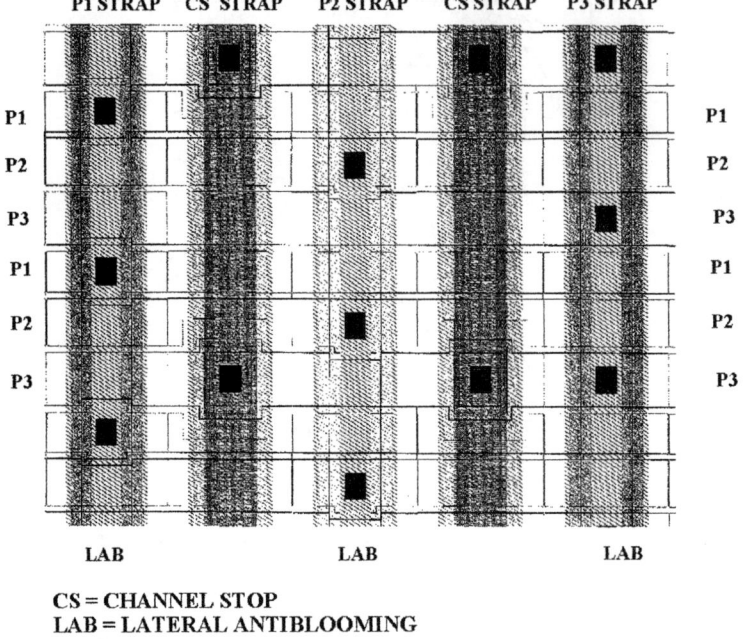

Figure 5.12 High-speed design showing poly and channel stop strapping.

strapping the poly gates with metal bus lines that run over the channel stops. Figure 5.12 shows a high-speed backside-illuminated CCD design that employs poly metal strapping. For this design the metal straps are connected to a poly phase over the channel stop region every other column (i.e., each poly phase gets strapped every six columns). Also shown is a metal strap made to the channel stop, a high-speed requirement for backside-illuminated CCDs discussed below.

It should be mentioned that metal buses used to strap the poly gates can also exhibit clock propagation. For example, some CCD manufacturers employ refractory metals that exhibit high impedance. Ti-tungsten metal buses exhibit a resistance of approximately 0.18 ohm/square compared with 0.04 ohm/square for aluminum. Even aluminum bus lines may limit speed for large CCD arrays depending on the bus width and drive capacitance. The required bus width is calculated for the maximum clock rate specified.

5.2.6 SUBSTRATE BOUNCE

Substrate bounce is an important mechanism that can affect high-speed operation. It is a major problem that limits speed performance for backside-illuminated CCDs. Bounce occurs when displacement currents flow in and out of the substrate as the CCD is clocked. For an n-channel CCD this current is in the form of holes that

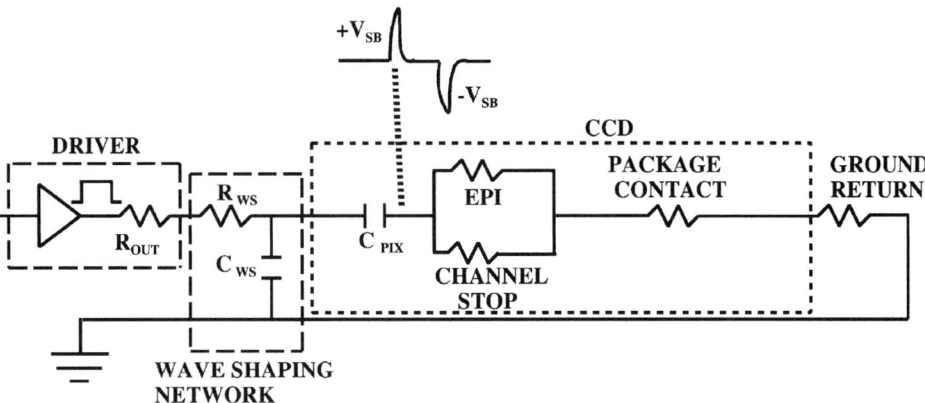

Figure 5.13 Impedances responsible for substrate bounce.

originate in the depletion and gate inversion regions. If a ground impedance exists, an IR drop will be generated, causing the substrate to "bounce" in potential. The rising edge of a clock causes a positive bounce because holes leave the CCD, whereas the falling edge produces a negative bounce as holes return to the sensor. Since the substrate voltage is the master reference level, substrate bounce can cause havoc on operating voltages and performance. For example, bounce can dramatically degrade charge capacity, as demonstrated below. Bounce problems are more prevalent for the vertical clocks because of high capacitance and related displacements currents. The discussions on substrate bounce below refer to a three-phase CCD; however, the problem is common to all CCD families.

Figure 5.13 shows important impedances related to the ground return. The connection made between the CCD and ground can generate bounce. An impedance of a few ohms in this path can cause significant bounce problems at high speed. Inductance associated with this return is normally the concern in this situation. For example, above 11 kHz, a 1-in. length of 22-gauge wire has more inductive reactance than its resistance (refer to Chapter 6 on grounding).

The epoxy layer between the CCD and metal package can also generate substrate bounce. This grounding problem has plagued the Cassini CCD. The wafers for these CCDs were backed with aluminum metal sintered into the silicon. The wafers were then diced and epoxied into a gold plated package. Over time, the silver epoxy chemically reacted with the aluminum, forming a thin oxide layer. The substrate impedance that developed produced a noticeable change in full well, a characteristic of substrate bounce. As shown below, clock wave shaping was the main solution to Cassini's bounce problem.

Backside-illuminated CCDs often use a frontside substrate contact to ground the CCD (i.e., a p^+ ohmic contact to the p^{++} field that surrounds the array). Single point substrate contacts can exhibit substrate bounce problems because of the finite contact resistance (tens of ohms typically). The resistance is reduced by encircling

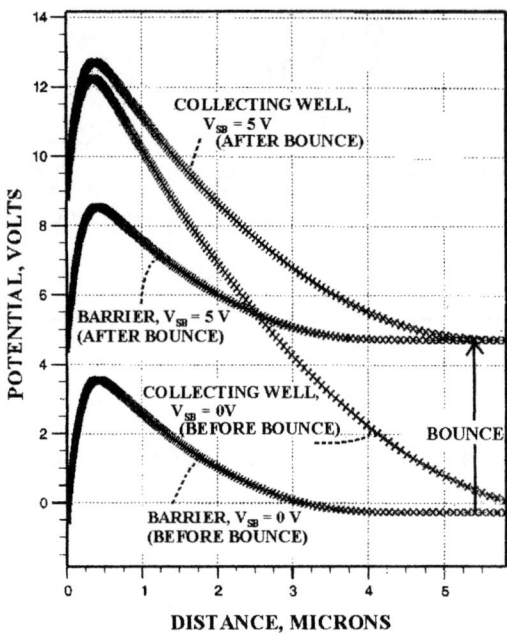

Figure 5.14 Potential wells in reaction to a 5-V substrate bounce.

the entire array with many substrate contacts for high-speed operation. A contact can also be made to the backside through the aluminum light shield. The aluminum must make intimate contact with the silicon to be effective.

The rising edge of the clock generates most substrate bounce problems. Figure 5.14 shows potential plots before and after a +5-V bounce on the ground return occurs. The substrate potential and the surface potential of the barrier phase before bounce assume a zero volt level (i.e., the barrier phase is inverted and pinned). When bounce occurs, the surface potential for the barrier phase follows the substrate (i.e., both move 5 V upward). On the other hand, the potential maximum for the collecting phase only shifts up slightly. Therefore, the potential difference between the collecting and barrier phases, $(V_C - V_B)$, decreases by approximately 5 V as bounce occurs. The problem therefore causes a full well reduction. The falling edge has negligable effect because the barrier phase comes out of inversion.

It is important to note that substrate bounce for a three-phase CCD will occur three times over a line transfer cycle period. Referring to Fig. 5.3, note that the rising edge of phase-1 will cause phase-2 to bounce, causing phase-3 to bloom. Likewise, the rising edge of phase-3 will cause phase-1 to bounce, causing phase-2 to bloom. Finally, the rising edge of phase-2 will cause phase-3 to bounce, causing phase-1 to bloom. Blooming will be more significant for that phase with the lowest charge capacity.

CHARGE TRANSFER 411

The amount of substrate bounce increases with the drive capacitance of the clock phase being driven and the substrate resistance. It also decreases for longer rise times of the applied clock and reduced amplitude. If the substrate resistance is external to the CCD, the substrate bounce as a function of time is given by

$$V_{SB}(t) = \frac{A_C}{1 - \tau_{WS}/\tau_{SB}}\left(e^{-t/\tau_{SB}} - e^{-t/\tau_{WS}}\right), \quad (5.15)$$

where A_C is clock amplitude with a wave-shaping time constant of τ_{WS} and τ_{SB} is the substrate bounce time constant given by

$$\tau_{SB} = R_{SB} C_G, \quad (5.16)$$

where R_{SB} is the external substrate impedance (ohms) and the gate capacitance is

$$C_G = C_T A_G, \quad (5.17)$$

where C_T is the gate and depletion capacitance [Eq. (1.7)] and A_G is the total array area of the clocked gate (cm^2).

Example 5.9

Using Eq. (5.15), plot substrate bounce as a function of $\tau_{WS}/\tau_{SB} = 0.1$, 1 and 10. Assume a three-phase 1024 × 1024 × 12-μm-pixel CCD, a gate capacitance of 1.43×10^{-9} F/cm^2 and a channel stop width of 4 μm. Also assume an external substrate resistance of 500 Ω and $A_C = 1$ V. Neglect poly gate impedance (i.e., strapped poly gates).

Solution:
From Eq. (5.17),

$$C_G = \left(1.43 \times 10^{-9}\right) \times \left(4 \times 10^{-4}\right) \times \left(8 \times 10^{-4}\right) \times 1024^2,$$

$$C_G = 4.8 \times 10^{-10} \text{ F}.$$

From Eq. (5.16),

$$\tau_{SB} = 500 \times \left(4.8 \times 10^{-10}\right),$$

$$\tau_{SB} = 2.4 \times 10^{-7} \text{ sec}.$$

Figure 5.15(a) plots substrate bounce for the wave-shaping time constants specified.

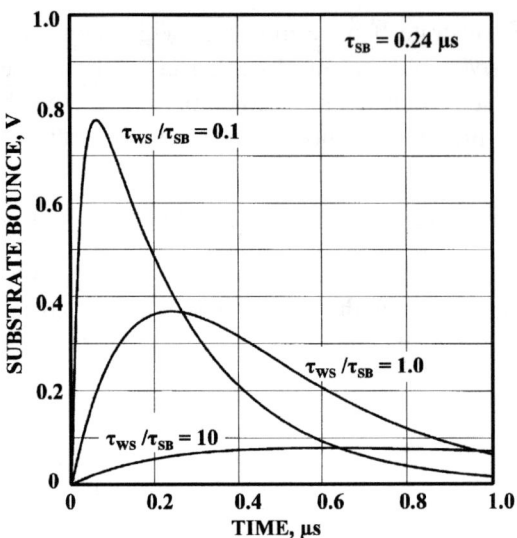

Figure 5.15(a) Substrate bounce as a function of time.

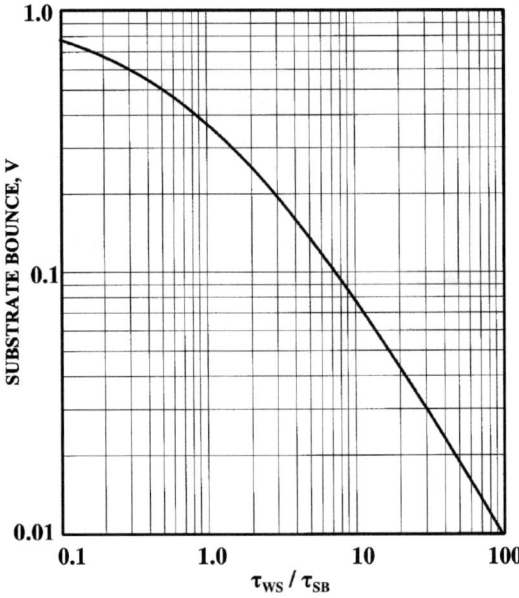

Figure 5.15(b) Maximum substrate bounce as a function of clock wave shaping.

The time at which maximum bounce occurs is

$$t_{MAX} = \frac{Ln(\tau_{WS}/\tau_{SB})}{1/\tau_{SB} - 1/\tau_{WS}}. \tag{5.18}$$

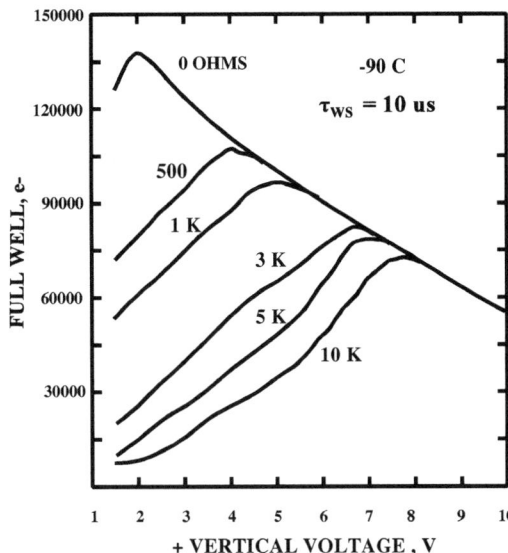

Figure 5.16 Full well transfer plots as a function of ground impedance.

Figure 5.15(b) plots maximum substrate bounce as a function of τ_{WS}/τ_{SB} assuming a 1-V clock transition. Note that τ_{WS} must be at least 10 times greater than τ_{SB} to suppress substrate bounce to a negligible level.

Figure 5.16 shows a full well transfer generated by a 1024 × 1024 × 12-μm-pixel Cassini CCD. A resistor was inserted between the CCD and ground to purposely generate the substrate bounce problem slow scan. Optimum full well shifts to a higher clock voltage as the substrate resistance increased. For $R_{SB} = 500$ Ω, a shift of 2 V in optimum full well was experienced for $\tau_{WS} = 10$ μsec.

As mentioned above, substrate bounce is an especially important problem for backside-illuminated CCDs. For these CCDs, holes must travel laterally through neutral epitaxial material below the depletion region or channel stops to reach ground. In comparison, holes for a frontside-illuminated CCD can come and go freely through the substrate region. The impedance seen by holes for backside devices can be particularly large, resulting in bounce problems when the clock edges become too fast.

Example 5.10

Calculate the resistance seen by holes for the channel stop and epitaxial regions shown in Fig. 5.17. Assume a 12-μm pixel, a 4-μm channel stop and a field-free region at the backside of the CCD of 7 μm. Also assume a channel stop resistivity = 1500 Ω/square and epitaxial resistivity = 40 Ω-cm. Also calculate the amount of channel stop bounce and required wave shaping

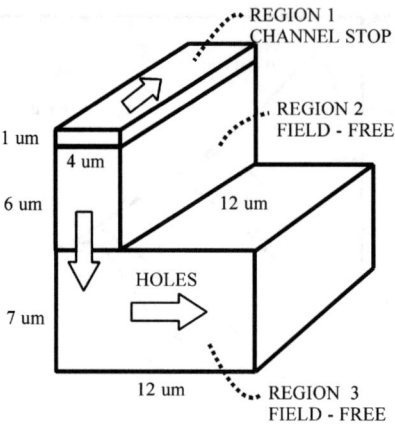

Figure 5.17 Regions where hole current flows in a backside-illuminated CCD.

for this structure for a thinned $1024 \times 1024 \times 12$-μm pixel CCD assuming $C_{pix} = 4.6 \times 10^{-16}$ F/pixel.

Region 1 = Channel stop resistance (per pixel),
Region 2 = Resistance under the channel stop (per pixel),
Region 3 = Lateral resistance of backside field-free material (per pixel).
Solution:
Region 1:

$$R_{CS} = 1500 \times (12 \times 10^{-4})/(4 \times 10^{-4}),$$

$$R_{CS} = 4.5 \times 10^3 \text{ ohms/pixel}.$$

Region 2:

$$R_{EPI} = 40 \times (6 \times 10^{-4})/[(4 \times 10^{-4})(12 \times 10^{-4})],$$

$$R_{EPI} = 1.38 \times 10^4 \text{ ohms/pixel}.$$

Region 3:

$$R_{FF} = 40 \times (12 \times 10^{-4})/[(7 \times 10^{-4})(12 \times 10^{-4})],$$

$$R_{FF} = 5.7 \times 10^4 \text{ ohms/pixel}.$$

This example shows that the majority of the hole current will flow through the channel stops.

We can assume that the net resistance is the parallel combination of the three regions. This impedance is 3.2×10^3 Ω/pixel.

One can also apply Eq. (5.12) to estimate the channel stop bounce time constant. That is,

$$\tau_{CB} = 1.5 R C_G N_{PIX}^2,$$

where R is the parallel resistance of the three regions shown in Fig. 5.17 (Ω/pixel), and C_G is the gate capacitance that stimulates the pixel into the bounce state (F/pixel/phase). Assuming $C_G = 4.6 \times 10^{-16}$ F/pixel/phase, $R = 3.2 \times 10^3$ Ω/pixel and $N_{PIX} = 512$ pixels to the center of the device, we find

$$\tau_{SB} = 1.5 \times (3.2 \times 10^3) \times (4.6 \times 10^{-16}) \times (512)^2,$$

$$\tau_{SB} = 0.58 \text{ }\mu\text{sec}.$$

To avoid channel stop bounce, the required wave shaping needs to be at least 5.8 μsec (i.e., $\tau_{WS} = 10\tau_{CB}$).

It is important to note that the gate capacitance here is only the oxide and depletion capacitance that involves hole movement. The poly overlap capacitance does not play a role in moving holes through the channel stops. This capacitance was important to the clock propagation problem where Eq. (5.12) was also applied. Refer to Example 5.24 where pixel capacitances are further discussed and defined.

It should be noted that CCDs with vertical antiblooming also exhibit bounce problems as backside-illuminated CCDs. The p-well associated with these sensors is depleted except for material beneath the channel stop regions (refer to Chapter 4). Holes must leave through the channel stops, resulting in substrate bounce for this family of sensors.

Substrate bounce increases when the phases are clocked into inversion because extra holes must move in and out of the pixels (i.e., C_G increases). It is for this reason that inversion may not be appropriate for very high speed applications. The full well transfer curves presented in Fig. 5.18 show the effect. Data are taken from a 656(H) × 1004(V) × 12-μm-pixel backside-illuminated split-frame transfer CCD using 200-nsec clock wave shaping. Inversion for the CCD occurs at −10 V. Note that full well degrades on either side of inversion. The charge capacity decreases when the barrier phases are set out of inversion because the potential between the collecting and barrier potential is reduced (i.e., $V_C - V_B$). Clocking the CCD deeper into inversion increases the substrate bounce problem, also degrading full well. Optimum wave shaping for this sensor is set at the onset of inversion.

Figure 5.19(a) shows a 7-sec dark current column trace for the backside-illuminated CCD characterized in Fig. 5.18. The sensor is clocked into deep inversion ($V_B = -15$ V) and wave shaped at 200 nsec, making the sensor vulnerable to substrate bounce. The first 256 lines read out are from the frame storage section.

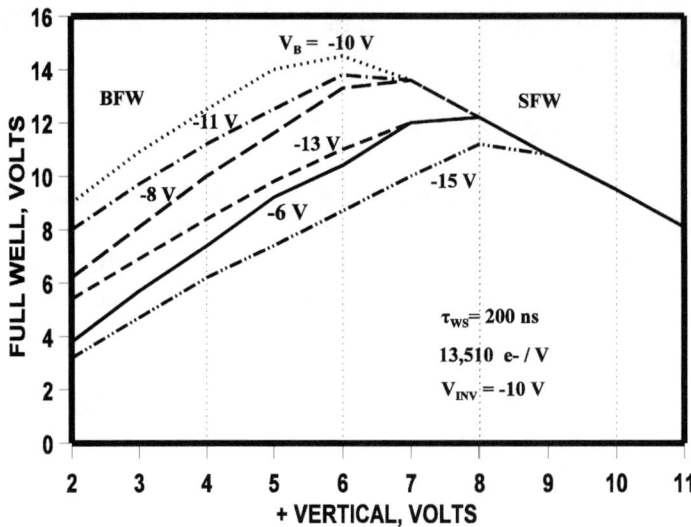

Figure 5.18 Backside-illuminated CCD full well transfer optimized for high-speed operation.

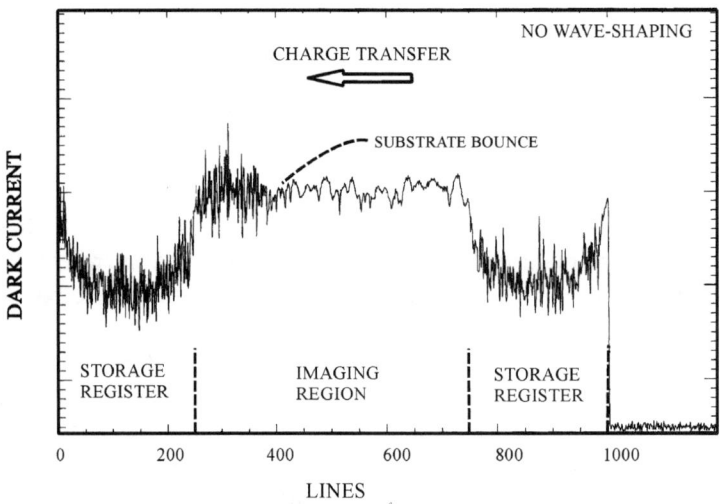

Figure 5.19(a) Column trace taken from a backside-illuminated split-frame transfer CCD showing substrate bounce within the thinned imaging region.

CTE is well behaved in this region because the light shield on the backside of the CCD acts as a ground plane. Holes in this region only need to travel down through field-free material to the light shield, a distance of approximately 14 μm for this CCD. This represents the same impedance as a frontside-illuminated CCD. However, within the imaging region (lines 257–768) the holes have no direct ground

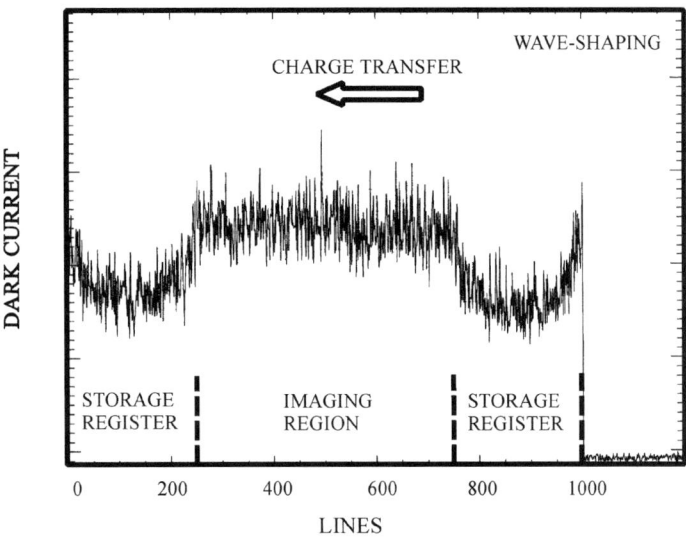

Figure 5.19(b) Same response as Fig. 5.19(a) with additional wave shaping.

path. In this region, holes must laterally flow down the channel stops toward the storage region. The center pixel of the imaging region sees the highest impedance and experiences the greatest "channel stop bounce." The column trace shown reflects this problem. Approximately 150 lines up from the image/storage interface the bounce causes the CCD to bloom, as seen by the dark current pixel nonuniformity signature (i.e. modulation is lost due to poor CTE). The second storage region (lines 769–1024) also has a light shield and ground return. However, charge in this region must transfer through the imaging region, and, therefore, its full well is limited to the same level as the imaging region. The backside dark current level within the storage region is lower than the imaging region caused by accumulation differences from the light shield. Figure 5.19(b) shows the same response with additional wave shaping. Optimum full well characteristics return (i.e., the -10 V curve of Fig. 5.18).

Two-phase CCDs are less sensitive to substrate bounce problems than multiphase. This is because the phases can be clocked 180 degrees out of phase with each another. Such clocking will transfer holes from phase to phase instead of leaving the substrate (holes will transfer in the opposite direction to the signal electrons). Therefore, substrate bounce is suppressed. Four-phase CCDs can also be clocked in a two-phase fashion, thus reducing substrate bounce. However, there is a full well penalty when clocking a four-phase in this manner. Virtual-phase CCDs are sensitive to substrate bounce in that this technology uses a single clock that must be inverted.

The ultimate solution to substrate bounce for backside-illuminated CCDs is to strap the channel stops with metal. Figure 5.12 shows, along with poly strapping already mentioned, channel stops that are strapped every other column. The wide straps shown do not interfere with QE performance because the sensor is illuminated from the back. Frontside strapping for CCDs with antiblooming is also possible; however, design and process rules must keep the straps and contacts as small as possible to maintain a high fill factor (contacts employed in Fig. 5.12 are 2.0 μm).

5.3 CTE Measurement Techniques

5.3.1 X-ray Transfer

X-ray transfer, as discussed in Chapter 2, is the standard method in measuring absolute CTE performance.[6] The technique is extremely valuable and is indispensable compared to other CTE measurement methods used which often give erroneous results. Figure 5.20 defines absolute CTE with the equation

$$\text{CTE}_X = 1 - \frac{S_D(\text{e}^-)}{X(\text{e}^-)N_P}, \quad (5.19)$$

where CTE_X is the CTE as measured by x rays, $X(\text{e}^-)$ is the x-ray signal (e.g., 1620 e$^-$ for an Fe^{55} x-ray source) and $S_D(\text{e}^-)$ is the average deferred charge after N_P pixel transfers (e$^-$).

Figure 5.21(a) shows a horizontal Fe^{55} x-ray transfer plot taken from a 520 × 520 CCD that exhibits a global CTE problem. The x-ray events interacting furthest from the on-chip amplifier experience the greatest charge loss. Note that the single-pixel-event line tilts "upward," whereas the deferred charge level for the first trailing pixel following the target pixel is seen as "downward" pattern

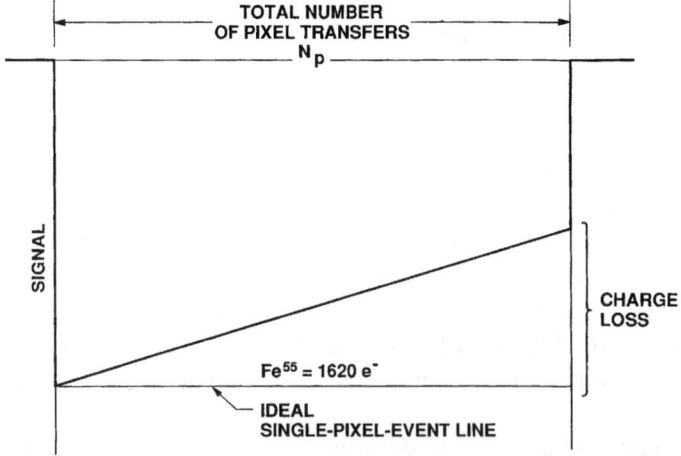

Figure 5.20 Absolute CTE definition using x-ray stimulation.

at a much lower level (seen slightly above the read noise floor near the end of the horizontal register). The total charge loss for the last pixel is approximately 261 e$^-$ assuming a gain of 0.52 e$^-$/DN. Figure 5.21(b) presents a square-wave response for a CCD with a global CTE problem, showing loss of modulation through the horizontal register.

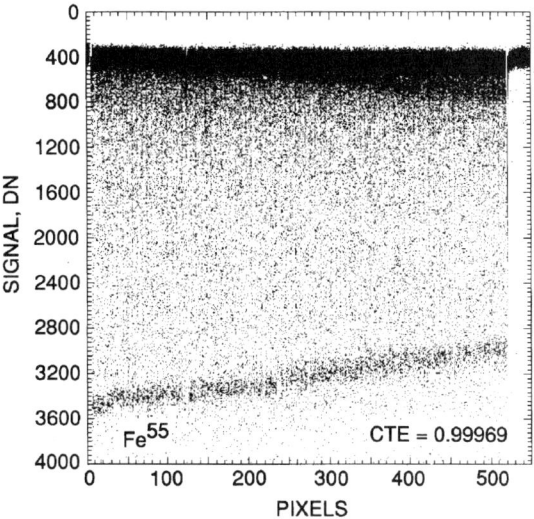

Figure 5.21(a) Horizontal x-ray transfer for a CCD that exhibits a global CTE problem.

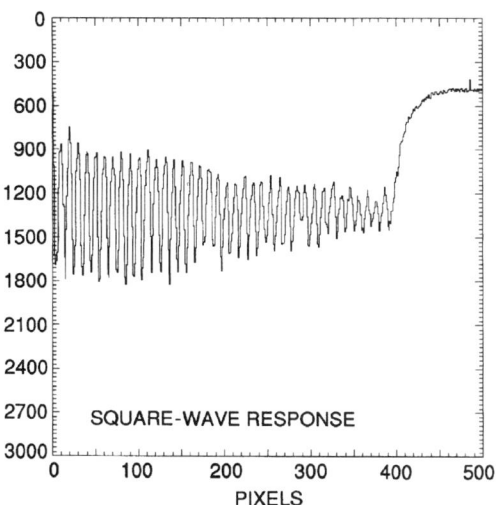

Figure 5.21(b) Bar target response for a CCD with a global horizontal CTE problem.

Example 5.11

Find the horizontal CTE for the characterized in Fig. 5.21(a). Assume $S_D = 261$ e$^-$, $X(e^-) = 1620$ e$^-$ and $N_p = 520$ pixel transfers.
Solution:
From Eq. (5.19),

$$\text{CTE} = 1 - [261/(1620 \times 520)],$$

$$\text{CTE} = 0.99969.$$

On the average, 0.5 e$^-$ are deferred for each horizontal pixel transfer.

CTI is specified for both the horizontal and vertical registers separately because performance is usually different. The deferred charge loss when vertical, CTI$_V$, and horizontal, CTI$_H$, are combined is

$$S_D(e^-) = X(e^-)(N_{VP}\text{CTI}_H + N_{HP}\text{CTI}_V), \tag{5.20}$$

where N_{HP} and N_{VP} are the number of horizontal and vertical pixel transfers respectively.

Example 5.12

Find the deferred charge loss when a 1620-e$^-$ charge packet is transferred 512 pixels vertically and horizontally. Assume CTI$_H = 10^{-4}$ and CTI$_V = 10^{-5}$.
Solution:
From Eq. (5.20),

$$S_D(e^-) = 1620\left(512 \times 10^{-4} + 512 \times 10^{-5}\right),$$

$$S_D(e^-) = 91 \text{ e}^-.$$

Note that 82 e$^-$ are deferred in the horizontal register alone.

CTE measurements using x-ray transfer become imprecise when CTE > 0.99998 and $N_P > 1024$ (i.e., the tilt in the x-ray line is difficult to measure). For example, Fig. 5.22 presents an x-ray transfer generated by a 1024 × 1024 CCD. The 34-e$^-$ CTE tilt measured yields a CTE = 0.999979. CTE for high-performance CCDs can be much better and, therefore, more sensitive CTE measurement techniques are required.

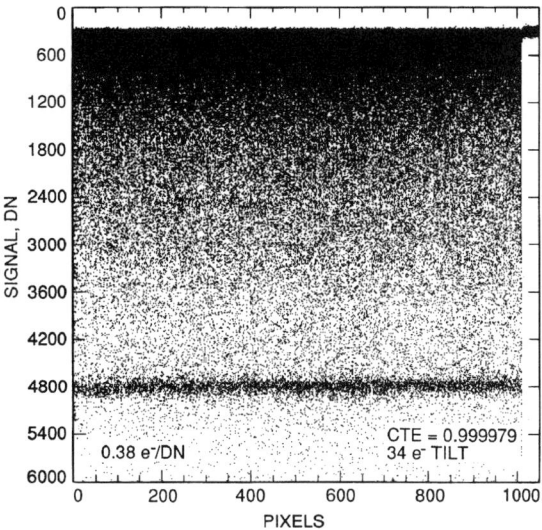

Figure 5.22 X-ray transfer showing CTE measurement limit.

Figure 5.23 shows a "clock reversal" method that effectively adds more pixel transfers to the measurement. In this test, a 1024 × 1024 frame-transfer CCD is exposed to x-ray events. After the exposure, x-ray events are partially transferred out of the CCD, leaving approximately 650 lines of x rays in the array. Next the transfer direction is reversed to transfer charge back into the upper region of the array by 374 lines. Then transfer is reversed again, moving charge back toward the lower horizontal register by 374 lines. This sequence is repeated, moving the charge back and forth within the array several times. The total number of line transfers is recorded after the 650 lines of x rays are read out into a computer. Figure 5.23(a) shows a baseline response where charge is read out without a vertical clock reversal. Note that the single-pixel-event line is tiltless, showing CTE > 0.99998. In Fig. 5.23(b) charge is moved back and forth ten times, equivalent to 5000 pixel transfers. The single-pixel-event line exhibits a slight reduction in CTE at the edges of the response compared with the baseline response. In Fig. 5.23(c) the CCD is clocked back and forth 40 times, equivalent to 20,000 pixel transfers. Additional CTE loss is seen at the edges. Charge transfer is worse at the edges because the traps responsible for CTE degradation are empty at the boundaries (we will discuss this effect in Section 5.4.3). CTE is measured at the edges where loss is greatest.

Example 5.13

Estimate the vertical CTE for the CCD tested in Figs. 5.23(b) and 5.23(c).

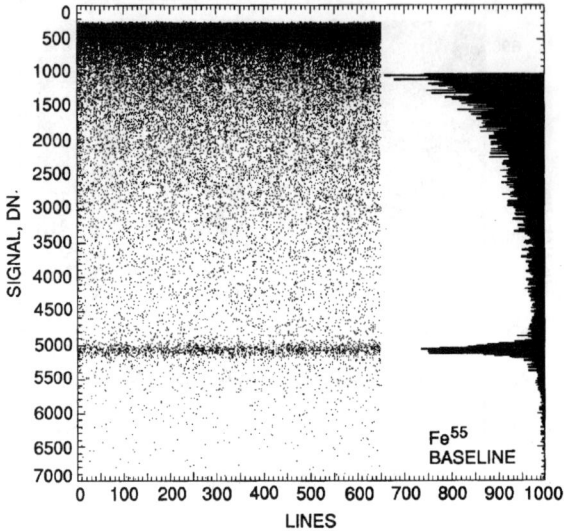

Figure 5.23(a) Baseline x-ray transfer plot for clock reversal experiment.

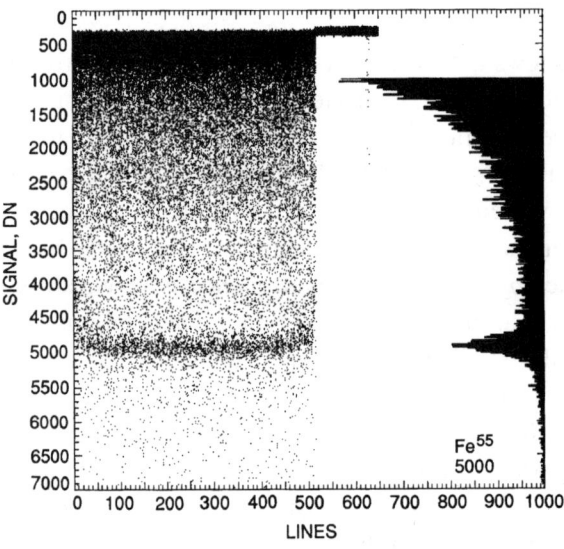

Figure 5.23(b) Response after 5000 pixel transfers.

Solution:
The electrical offset for both figures is approximately 300 DN. The baseline x-ray level is 5100 DN corresponding to a camera gain constant of

$$K = 1620/(5100 - 300),$$

$$K = 0.3375 \text{ e}^-/\text{DN}.$$

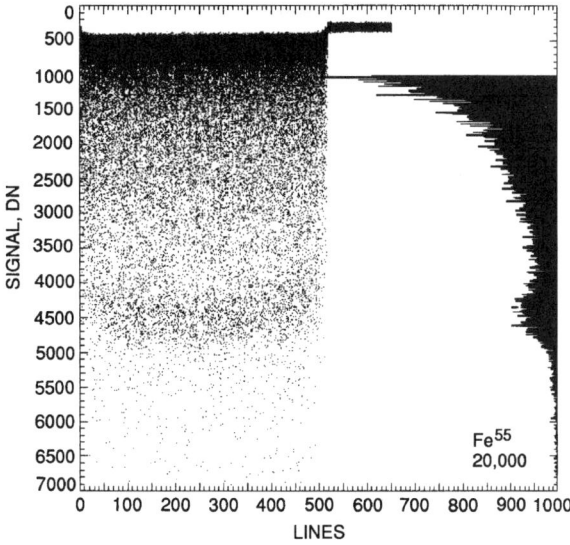

Figure 5.23(c) Response after 20,000 pixel transfers.

CTE loss in Figs. 5.23(b) and 5.23(c) is approximately 400 DN (135 e^-) and 800 DN (270 e^-), respectively, taken from the edge of each response. Corresponding CTEs are

$$CTE_{5000} = 1 - 135/(5000 \times 1620),$$
$$CTE_{5000} = 0.9999833,$$
$$CTE_{20000} = 1 - 270/(20000 \times 1620),$$
$$CTE_{20000} = 0.9999917.$$

5.3.2 EXTENDED PIXEL EDGE RESPONSE (EPER)

The extended pixel edge response (EPER—pronounced "e-per") is a popular CTE measurement technique.[6] The EPER response begins with a flat-field exposure. CTE is estimated by measuring the amount of deferred charge found in the extended pixel or line region. For example, Fig. 5.24 presents an EPER plot showing extended pixels at the end of a horizontal register in response to a 380-e^- flat field. Several lines are averaged together to reduce noise and improve the S/N in the extended pixel region. For this plot, 1500 lines are averaged, which reduces the read noise from 6 e^- to 0.15 e^- rms. Note in the expanded plot that a deferred charge tail following the last column emerges (the expanded plot is shifted to the right by 10 pixels for the illustration). A total of 38 e^- of deferred charge is detected.

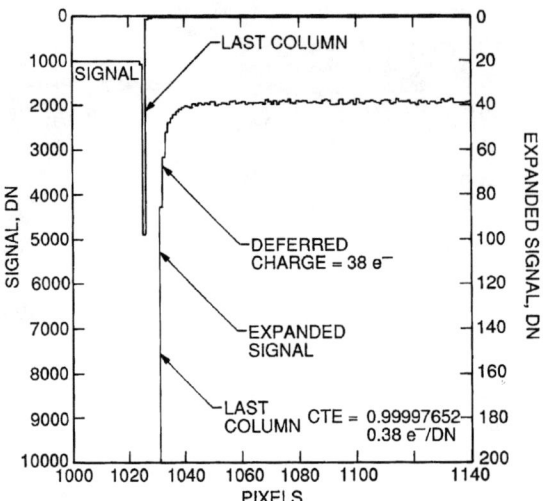

Figure 5.24 Horizontal EPER showing 38 e⁻ of deferred charge.

CTE from an EPER plot is defined as

$$\text{CTE}_{EPER} = 1 - \frac{S_D}{S_{LC}(\text{e}^-)N_P}, \tag{5.21}$$

where S_D is the total deferred charge measured in the extended pixel region (e⁻), $S_{LC}(\text{e}^-)$ is the charge level of the last column (e⁻) and N_P is the number of pixel transfers for the CCD register. Note that the last column is specified because it often contains more charge than other columns. This is because charge from a flat-field exposure diffuses from neutral material from outside the array. This characteristic is not seen for CCDs that have "guard columns" on each side of the device (i.e., n^+ drains).

Example 5.14

Find the horizontal CTE for the CCD tested in Fig. 5.24. Assume $S_D = 38$ e⁻, $S_{LC}(\text{e}^-) = 1900$ e⁻ and $N_P = 1024$ pixel transfers.
Solution:
From Eq. (5.21),

$$\text{CTE}_{EPER} = 1 - (38/1900/1024),$$

$$\text{CTE}_{EPER} = 0.99997692.$$

EPER can also be used to estimate vertical CTE performance. Here extended lines are averaged to determine the amount of vertical deferred charge.

CHARGE TRANSFER 425

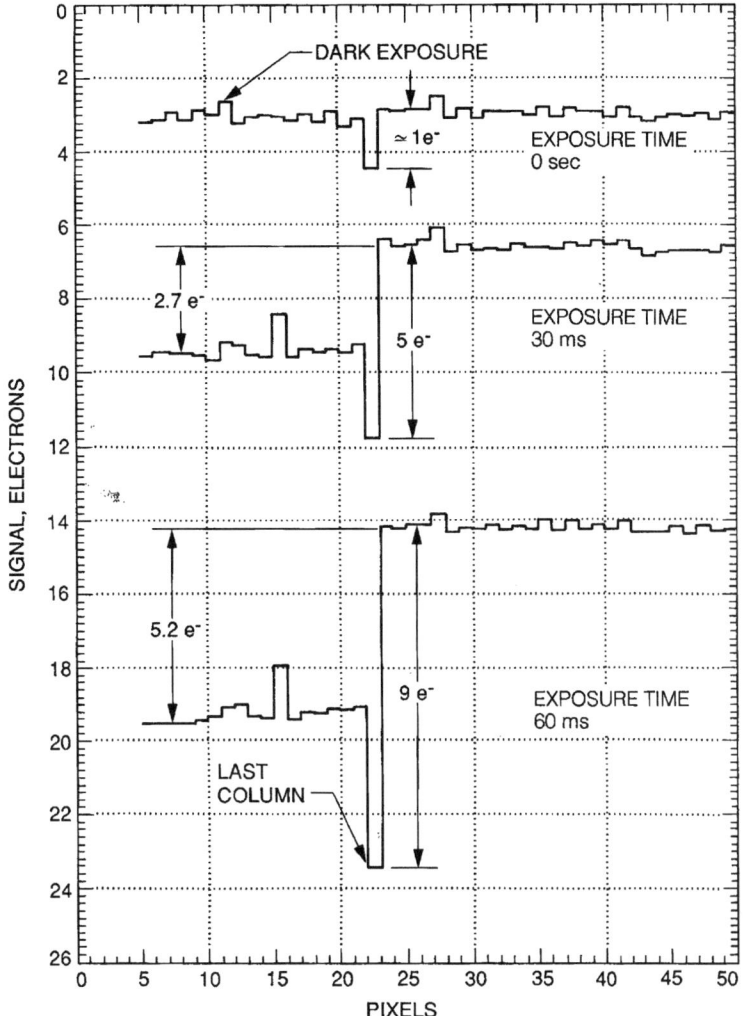

Figure 5.25(a) Ultralow-level EPERs.

Ultralow-level linearity characteristics can also be measured using EPER. For example, Fig. 5.25(a) shows horizontal EPERs generated by a 516×516 CCD. The light intensity was adjusted to the levels indicated (2.7 e^- and 5.2 e^-). Shot noise and read noise are reduced to the levels seen by averaging many lines together. A dark current EPER trace is included at the top of the figure showing a 1 e^- dark level in the last column. Figure 5.25(b) is the linearity response for the last column at several different light levels.

The EPER technique must be used with caution because it is a relative CTE test tool. It and many other relative CTE measurements do not give an absolute measurement for CTE as does the x-ray test. Frequently a CCD with a CTE problem will not show deferred charge as measured by EPER. Such results will overesti-

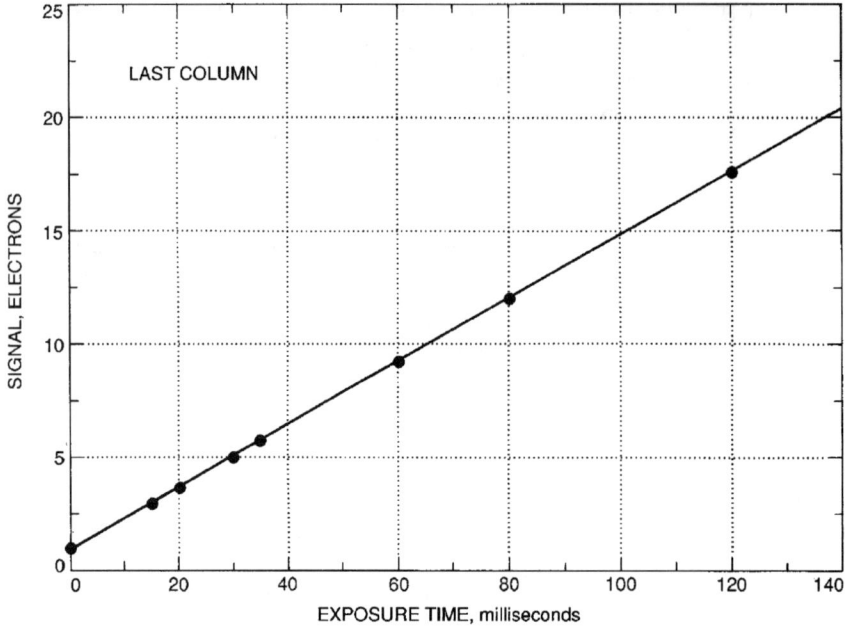

Figure 5.25(b) Low-level EPER linearity curve.

mate CTE performance, sometimes significantly. Figure 5.26(a) demonstrates the problem, showing a CCD that was intentionally damaged with neutrons that generate deep level traps and CTE degradation. The region shown is a 50 × 50-pixel area taken at the top of a 1024 × 1024 array opposite to the lower on-chip amplifier used for readout. Charge is transferred down and to the left. The region contains several dark spikes also induced by neutrons and used as test signals. Note that deferred charge tails follow each dark spike, indicating a CTE problem. Figure 5.26(b) is taken from the same region when the frame readout time is reduced by a factor of ten (i.e., from 23 to 2.0 frames/sec). Note that the deferred charge tails nearly disappear, suggesting that CTE has been enhanced by clocking the CCD faster.

The apparent CTE improvement seen is further investigated in Fig. 5.26(c) by plotting a single column trace that contains a dark spike (charge is transferred from right to left). The dark spike and deferred tail are again measured at different frame rates (23 sec, 19 sec, etc.) and displayed in the figure. As expected, a deferred tail is seen following the dark spike for the 23 second frame. As the vertical clock rate is increased and frame time reduced, the deferred tail becomes smaller. When the frame time is 1.99 sec the tail is nearly eliminated. Note that the amplitude of the dark spike remains the same size independent of clock rate. This behavior indicates that CTE is not improving, or the dark spike amplitude would increase. Instead the deferred charge tail is spreading over more pixels and at the 1.99 sec rate is nearly lost in the read noise floor. This is because signal charge escapes the traps at a fixed rate, whereas the time for transfer decreases with frame readout.

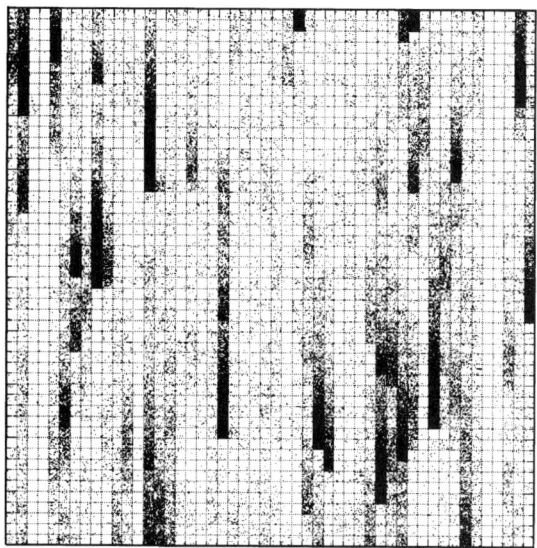

Figure 5.26(a) Dark spike image showing deferred charge tails at 23 sec frame readout rate.

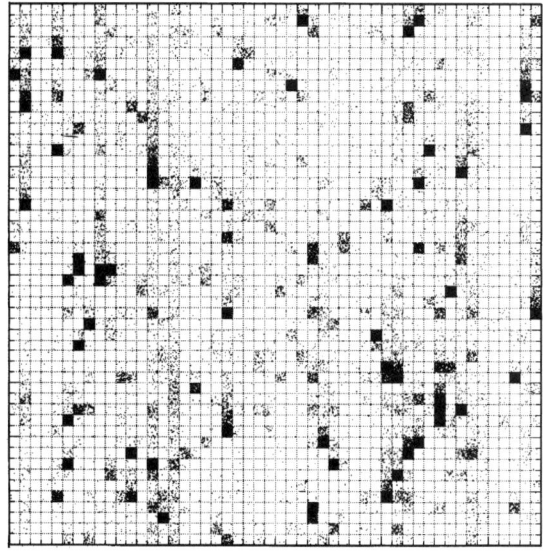

Figure 5.26(b) Dark spike image at 2 sec frame readout rate.

Figure 5.26(d) shows the true amplitude of the dark spike without CTE loss. This is done by transferring charge toward the CCD's upper on-chip amplifier, where very few pixel transfers take place. Comparing the dark spike amplitude in Fig. 5.26(c) to Fig. 5.26(d) shows that approximately 2/3 of the charge generated by the dark spike is in the form of deferred charge independent of vertical clock rate.

As demonstrated in Fig. 5.26, relative CTE tests can be very misleading. Many relative CTE techniques have been invented (e.g., EPER, optical point, edge stim-

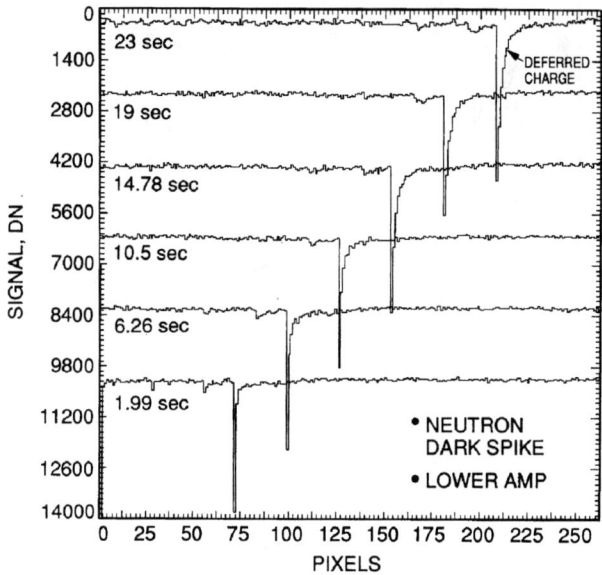

Figure 5.26(c) Column trace taken from lower amplifier showing an individual dark spike and deferred charge tail at different readout rates.

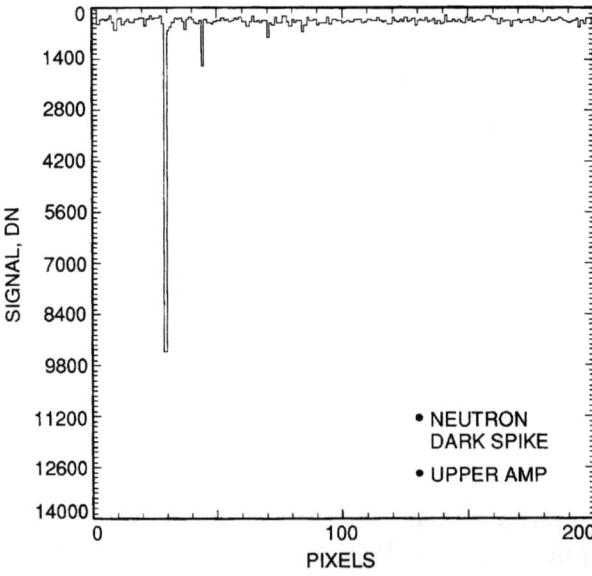

Figure 5.26(d) Column trace of the same dark spike as Fig. 5.26(c) but taken from the upper amplifier.

ulus, bar targets and electrical charge injection via input diode/gate, etc.) that compare the target pixel amplitude to the deferred charge level. Also, operating temperature and pixel rate influence the amount of deferred charge that is measured. On the other hand, x-ray stimulus is an absolute measurement and does not require the user to find the deferred level. EPER will be used to characterize other CTE effects in this chapter that do not require absolute calibration.

5.3.3 First Pixel Edge Response

A technique closely related to EPER is called first pixel edge response (FPER), invented by Michael Jones. The absolute CTE measurement technique requires that vertical and horizontal registers of the CCD be split and independently clocked (e.g., frame transfer layout). For the case of vertical CTE measurement, the CCD is first exposed to a flat-field source of light. Next the storage region is erased several times while charge in the image region remains. Then the image region is read out through the storage region in a normal fashion. The first lines transferred through the storage array will lose charge to any active traps present. The total amount of charge lost is measured and used to determine CTE through the equation

$$\text{CTE}_{FPR} = 1 - \frac{S_D}{S(\text{e}^-)N_P}, \tag{5.22}$$

where S_D is charge lost from the first few lines read out (e$^-$), $S(\text{e}^-)$ is the average charge level (e$^-$) and N_P is the number of pixel transfers.

FPER is a very useful test tool for measuring absolute CTE as a function of signal level. For example, Fig. 5.27 shows CTI measurements as a function of

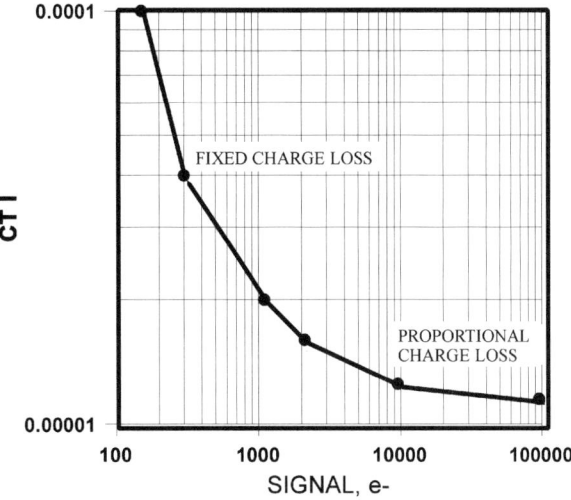

Figure 5.27 First pixel edge response.

signal level for an 800 × 800 CCD. Note that CTI improves as signal increases and becomes nearly constant at high signals. This CTE effect is referred to as "proportional CTE loss" characteristic of all high-performance CCDs. We will look at this effect closer below.

5.3.4 POCKET PUMPING

The technique of "pocket pumping" is used to characterize various charge trap types that degrade CTE performance.[7] The method maps the trap's position, size and even its location within a pixel (i.e., phase location). Traps as small as one electron can be characterized, including "statistical traps" which may or may not trap an electron. Pocket pumping is performed by first exposing the CCD to a flat field of approximately 100–1000 e^- depending on the size of the trap to be detected. After the exposure, the CCD is clocked backwards a specified number of lines and then run forward the same number of lines to the original image position. The number of lines transferred forward and backward is called a "line sector." The clocking scheme is repeated for many cycles before the CCD is read out in the normal fashion. During clocking, signal charge builds or "pumps" up at a trapping site. The rate of charge buildup is directly proportional to the trap size. It should be noted that pocket pumping can only be done on bidirectional CCDs.

PIXEL NO.

SHIFT	1	2	3	4	5	6	7	8	9	10	MODE
1	10	10	10	10	10	10	10	10	10	10	EXPOSURE
2	10	10	10	3	17	10	10	10	10	10	REVERSE
3	10	10	3	10	17	10	10	10	10	10	REVERSE
4	10	3	10	10	17	10	10	10	10	10	REVERSE
5	3	10	10	10	17	10	10	10	10	10	REVERSE
6	10	3	10	10	10	17	10	10	10	10	FORWARD
7	10	10	3	10	10	10	17	10	10	10	FORWARD
8	10	10	10	3	10	10	10	17	10	10	FORWARD
9	10	10	10	10	3	10	10	10	17	10	FORWARD
10	10	10	10	0	13	10	10	17	10	10	REVERSE
11	10	10	0	6	17	10	17	10	10	10	REVERSE
12	10	0	6	10	17	17	10	10	10	10	REVERSE
13	0	6	10	10	24	10	10	10	10	10	REVERSE
14	10	0	6	10	10	24	10	10	10	10	FORWARD
21	0	0	9	10	31	10	10	10	10	10	REVERSE
22	10	0	0	9	10	31	10	10	10	10	FORWARD
29	0	0	2	10	38	10	10	10	10	10	REVERSE
30	10	0	0	2	10	38	10	10	10	10	FORWARD
37	0	0	0	5	45	10	10	10	10	10	REVERSE
38	10	0	0	0	5	45	10	10	10	10	FORWARD
45	0	0	0	0	50	10	10	10	10	10	REVERSE
46	10	0	0	0	0	50	10	10	10	10	FORWARD

Figure 5.28 Pocket-pumping computer simulation assuming a 7-e^- trap.

A computer simulation shows how pocket pumping works. Figure 5.28 presents a 10-pixel segment of a column and shows a typical charge pumping sequence. The pumping sequence involves clocking the device backwards four clock cycles and then forward four cycles (i.e., a four-line sector). The sequence is then repeated many times. In the first line of the figure, each pixel is filled with 10 e$^-$. This represents the state of the CCD following a low-level flat-field exposure. A hypothetical trap of 7 e$^-$ is assumed to be in pixel number 5. In line two, the device is clocked backward one pixel, filling pixel 5 with 17 e$^-$. The 3 e$^-$ excess that is not trapped is transferred to subsequent pixels (steps 3–5). During forward clocking, charge is clocked away from the trap as indicated in the sequence steps 6–9. In the next reverse cycle, the shifted buildup of charge is moved back to the trap, which has trapped 7 additional electrons, resulting in the formation of a spike of 24 e$^-$ as shown in step 13. Since charge in this spike has come from other pixels, a dark tail begins to form behind the trap. The pumping process is repeated several more times until cycle 45 is reached when all charge initially contained in pixels 1–4 has been pumped into the pixel containing the trap. The signature of this trap is a spike of charge, much above the background, followed by many pixels that contain no charge.

Experimentally, we find two trap types using pocket pumping. These are called "forward" and "reverse traps." Figure 5.29(a) shows two column traces that contain forward and reverse traps for a three-phase CCD. The forward trap accumulates charge within the pixel containing the trap site. Its dark tail always precedes the spike. For the reverse trap, charge accumulates at the opposite end of the pumping sequence. Its dark tail always lags behind the spike.

Figure 5.29(b) plots the signal pumped by a 440 e$^-$ trap as a function of pumping cycles at four different exposure times. The device is clocked backwards and forwards 25 lines in one cycle. As demonstrated, the signal pumped increases linearly until all charge within the 25-line sector is pumped out, at which point the response becomes flat. The amount of charge pumped by a trap is given by

$$S_{PP} = N_{PP} S_T, \quad (5.23)$$

where S_{PP} is the charge pumped by the trap (e$^-$), N_{PP} is the number of pumping cycles and S_T is the trap size (e$^-$).

The saturation level where all the charge is removed from the line sector is

$$S_{PPS} = S(\text{e}^-) L_S, \quad (5.24)$$

where S_{PPS} is the saturation level (e$^-$), $S(\text{e}^-)$ is the flat-field level (e$^-$) and L_S is the line sector size (pixels).

The number of pumping cycles required to reach saturation is

$$N_{PPS} = \frac{S_{PPS}}{S_T}. \quad (5.25)$$

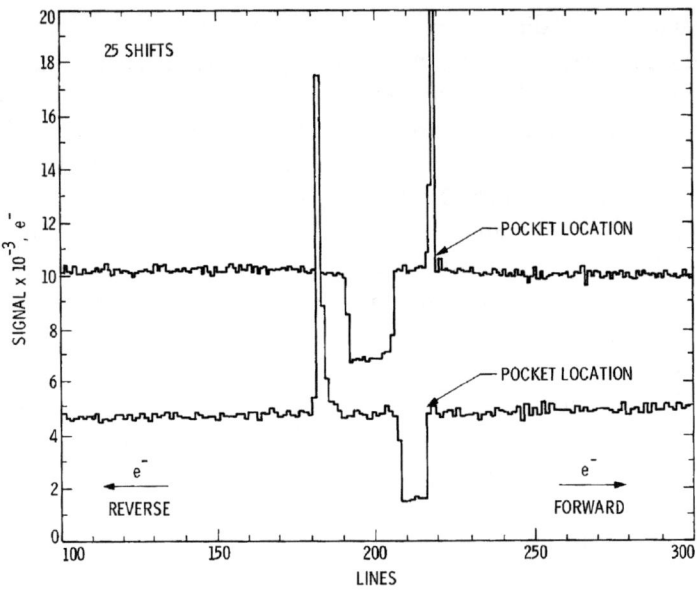

Figure 5.29(a) Pocket-pumping response showing forward and reverse traps.

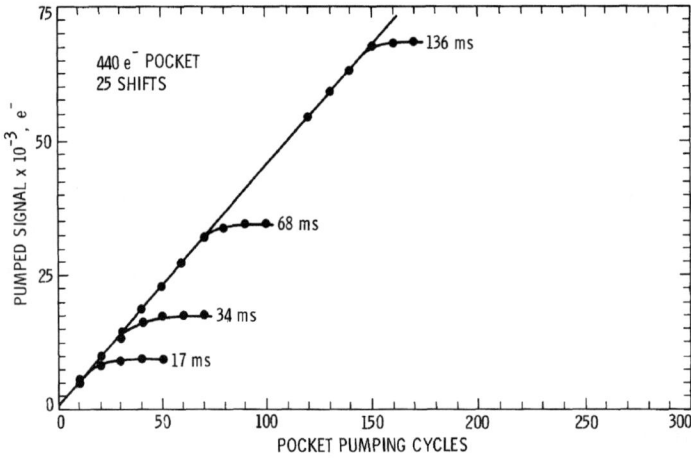

Figure 5.29(b) Charge taken from a trap as a function of pumping cycles.

Example 5.15

Find the total charge pumped by a 3-e^- trap assuming a flat-field level of 500 e^- over a 10 line sector. Also, find the number of pumping cycles to reach the saturation point.

Solution:
From Eq. (5.24),

$$S_{PPS} = 5 \times 10^2 \times 10,$$

$$S_{PPS} = 5 \times 10^3 \text{ e}^-.$$

From Eq. (5.25),

$$N_{PPS} = 5 \times 10^3 / 3,$$

$$N_{PPS} = 1666 \text{ cycles}.$$

We will come back to pocket pumping below and present a CTE formula. The technique represents the most sensitive measurement technique in determining absolute CTE performance.

5.3.5 CHARGE INJECTION

Absolute CTE can also be measured by electrically injecting charge into the CCD as described in Chapter 1.[8] The technique is only applicable to multiphase transfer devices that are bidirectional. This method consists of injecting charge into a single pixel site and moving the charge packet a few pixels into the horizontal register by clocking it backwards. The horizontal clocks are then switched forward to read and calibrate the absolute amount of charge that was injected (in electrons). Then, charge is injected again at the same level and sent to the other end of the horizontal register. At this point, charge is read by an amplifier, assuming one is located there. If not, the charge packet is sent back to the original amplifier that injected the charge. In either case, the charge loss from the charge packet is measured and Eq. (5.22) is applied for the number of total pixel transfers made. Vertical CTE is measured similarly by injecting charge a few pixels into the horizontal register and then taking the charge packet into a vertical register up the array and into the upper horizontal register. Assuming the CCD has dual horizontal registers, the charge packet is read by an output amplifier located there. If not, the charge packet is sent back to the original amplifier that injected the charge for charge measurement.

5.4 TRAPS

As reviewed above, diffusion and self-induced and fringing field drift theoretically set CTE performance. CTE is also limited by traps located in the signal channel. As shown in Fig. 5.30, traps are classified into four main groups: (1) design traps, (2) process traps, (3) bulk traps and (4) radiation traps.

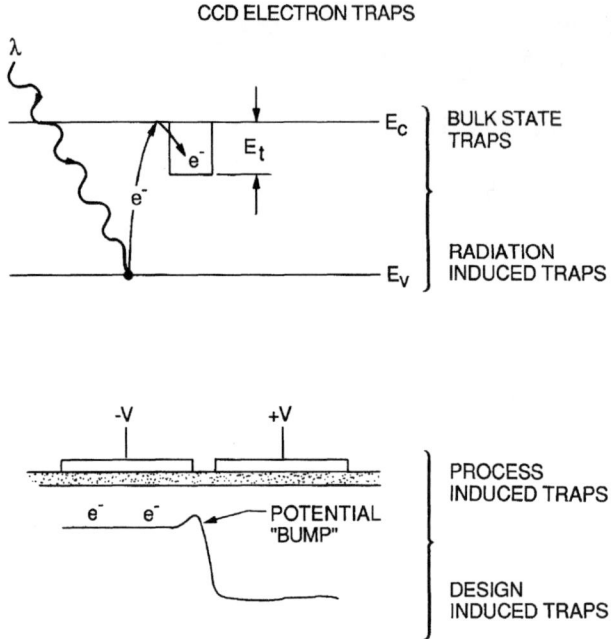

Figure 5.30 Common traps found in CCDs.

5.4.1 DESIGN TRAPS

"Design traps" are created when the CCD is designed. There are numerous causes of these traps; however, the traps of greatest concern are those that affect CTE on a local level. These defects usually operate by extracting a fixed quantity of charge as a signal packet is clocked past the problem site, later to be released into trailing pixels (i.e., deferred charge). Large design-induced traps are capable of trapping thousands of electrons. Consequently, even though the global CTE may be adequate, a single trap, if it is located in the horizontal register, can be disastrous to the CCD.

Design traps are usually associated with an improper design feature that produces a small potential barrier that holds signal charge back. When a design trap exists, it is typically found in the region where charge is forced to transfer from a wider region of the channel into or through a constriction. An excellent example of a design trap was identified on the Hubble WF/PC I CCD project that exhibited the following characteristics. The amount of deferred charge varied from device to device, from a few hundred to over a thousand electrons. A point-source response showed a deferred charge tail in the vertical direction. The tails were roughly the same length from column to column. Their length grew with cooling and vertical clock rate. The tail length was independent of where the array point sources were imaged. For example, Fig. 5.31(a) is an x-ray image showing deferred charge lagging behind each event (charge moves to the top and to the left). Figure 5.31(b)

CHARGE TRANSFER

Figure 5.31(a) Fe^{55} x-ray image showing deferred charge tails.

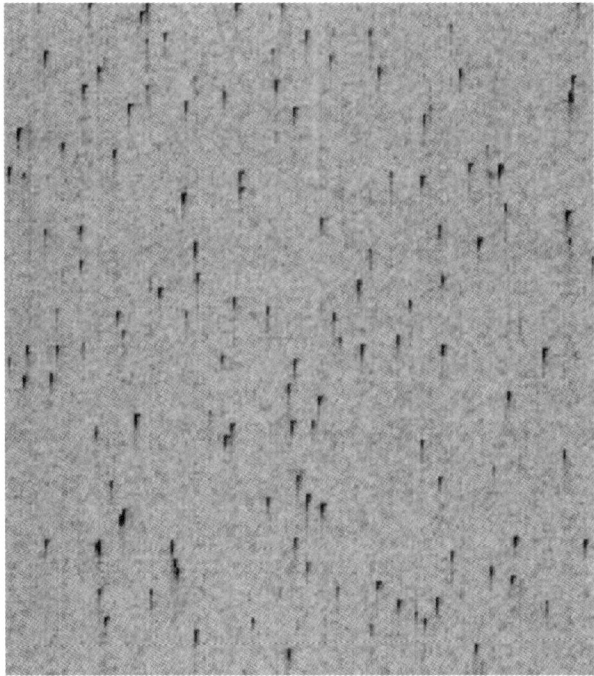

Figure 5.31(b) Magnified view of Fig. 5.31(a) showing deferred tails.

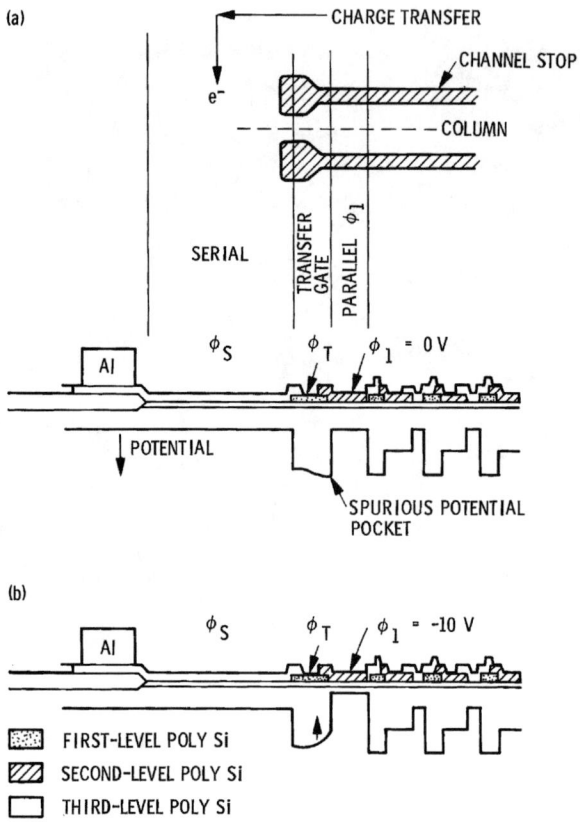

Figure 5.31(c) Transfer gate design-induced trap problem for the Hubble WF/PC-I CCD.

magnifies a region to show the deferred tails. These characteristics together implied that a design trap problem was located between the vertical and horizontal registers (i.e., the array transfer gate).

Further investigation into the problem showed that the deferred tails were caused by a design problem, which is illustrated in Fig. 5.31(c). Note in the upper illustration that the channel stops for the WF/PC I CCD were designed wider in the middle of the array transfer gate to properly direct charge into the horizontal register. However, constricting the signal channel reduced the potential because of the fringing fields produced by the channel stop. Although the potential difference through the transfer gate was small (about 0.1–0.3 V depending on the sensor), the trap that was produced held back a thousand electrons or more.

The discovery next led to a solution to the problem. Driving the neighboring phase negative (i.e., phase-1) caused the fringing fields from that phase to extend under the transfer gate region and collapse the trapping site as shown in the lower illustration. Figure 5.31(d) shows a 400 e^- "star" image with and without the fix applied. The star is absent in the upper image because the design trap swallows it

Charge transfer

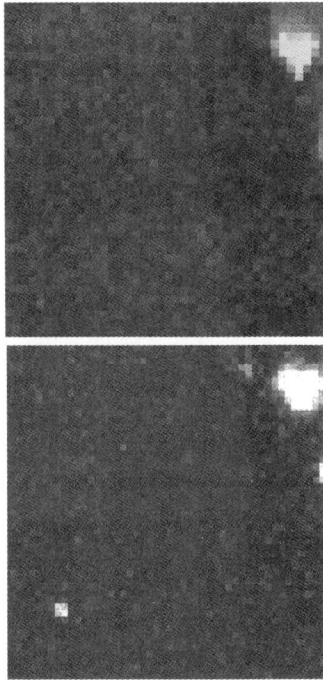

Figure 5.31(d) Star image with and without array transfer gate fix.

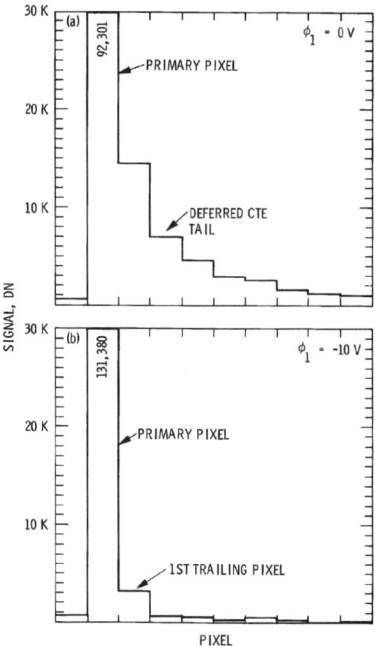

Figure 5.31(e) Fe^{55} x-ray stacking plots that show improved CTE from the fringing field of vertical phase-1.

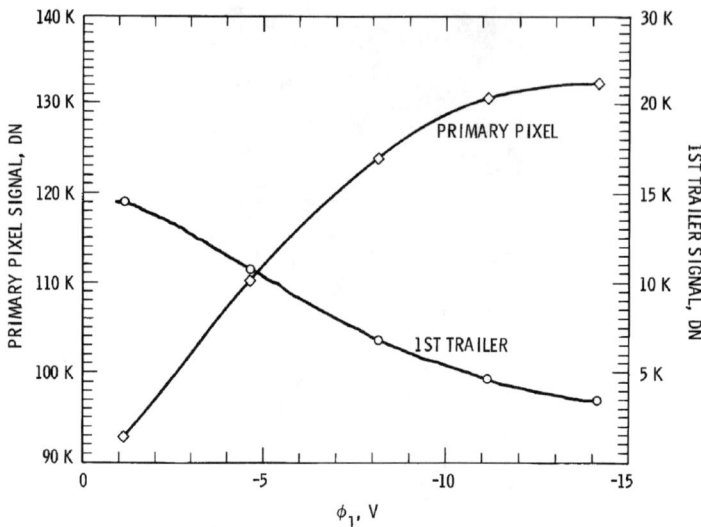

Figure 5.31(f) Charge in the target and first trailing pixels as a function of negative drive on phase-1.

completely. The lower image shows the same region after a -10-V clock drive is applied. The star now is allowed to transfer through the transfer gate. The bright region in the upper left hand corner is a cosmetic defect. A CTE tail is seen following the defect in the top image where the lower image exhibits no significant CTE loss. Charge is transferred toward the bottom in both images.

Figure 5.31(e) shows 2000 Mn single-pixel x rays that are taken randomly from the array and stacked together. Stacking events magnify the deferred tail and reduce the read noise floor by the square root of the number of x rays stacked. For example, the read noise in the plot was reduced from 13 e^- to 0.3 e^- rms. The top response shows the full extent of the deferred charge tail when phase-1 is driven to ground potential. The lower response shows the CTE improvement when phase-1 is driven to -10 V. The deferred tail is nearly eliminated (approximately 50 e^- out of 1620 e^- is deferred). Figure 5.31(f) plots x-ray charge contained in the target pixel and first trailing pixel as a function of the negative level of phase-1. Note that the majority of signal charge is contained in the target pixel with -10 V drive. The remaining 50 e^- problem was fixed on Hubble by injecting 100 e^- of optical fat-zero (refer to Section 5.4.6).

The trapping problem for Hubble could have been prevented if the CCD was designed slightly differently. Today, CCDs are designed so that the signal carrying channel is constricted at the end or beginning of the transfer gate as opposed to midway in the phase. Designed this way, potential bumps or pockets will be dominated by the fringing fields at the interface. The same design philosophy is employed at other interfaces within the signal channel. For example, careful attention is given to the region where charge is transferred from horizontal register into the sense node region where the signal channel is typically narrowed.

It should be pointed out that the Hubble deferred charge problem remained unsolved for several years until x rays came onto the scene. During this period EPER tests pointed WF/PC I in the wrong direction in locating and solving the problem. "Good" sensors using EPER were solely selected on tail length characteristics. For example, those sensors that showed a small or no tail were claimed to be good sensors (especially for cold operating temperatures where kT was not effective in releasing the charge from the trap). This criterion was selecting inferior detectors as devices were screened. A small tail implied a deeper trap which would hold more charge and release it over a greater time period. If the trap was large enough, no tail would be observed because trapped charge would be completely lost in the read noise floor. EPER claimed perfect CTE in this case. X rays, on the other hand, measure the absolute loss from the target pixel independent of tail characteristics. This is a very important distinction between CTE measurement techniques.

5.4.2 PROCESS TRAPS

The second type of CTE defect is the "process-induced trap." These traps can themselves be classified into two categories. The first trap type is uniformly distributed along the signal channel and causes a global CTE effect. These traps are called "spurious potential pockets." The second trap is randomly distributed and isolated to individual pixels. These traps are simply called "localized traps."

5.4.2.1 Spurious Potential Pockets

Spurious potential pocket is a term used to describe charge loss during charge transfer due to improper potential well shape and/or depth within the signal channel. These effects can sometimes arise through polysilicon edge lifting, reticle misalignment and lateral implant encroachment, typically occurring at the edge of a phase.[9] For example, when poly oxide is grown it grows faster at its edge, decreasing the channel potential in this region. This forms a small barrier between the phases. All CCDs exhibit channel potential nonuniformities to some degree. Fortunately, small potential differences like this can usually be overridden by fringing fields between gates.

The spurious trap is more pronounced for large CCD imagers where many transfers are executed. For example, an Fe^{55} x-ray event (1620 e^-) will lose 1000 e^- to spurious pockets over 1000 pixel transfers if only 1 e^- is lost to a trap during each transfer cycle. Figure 5.32(a) shows the effect of spurious pockets distributed through the horizontal register for an early AXAF 1024 × 1024 virtual-phase CCD. The plot presents five point-source horizontal line traces operating at six different operating temperatures. Note that CTE performance degrades with decreasing temperature, as shown by the deferred charge tails to the left of each point of light. The CTE loss is attributed to a lack of thermal energy (i.e., kT) required

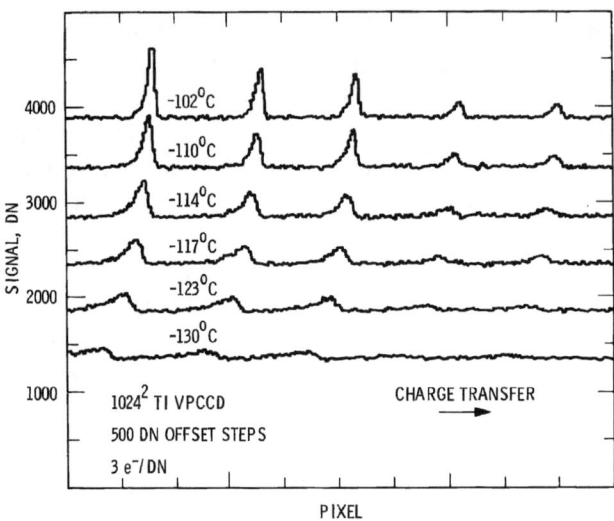

Figure 5.32(a) Point source responses for a 1024 × 1024 virtual-phase CCD that exhibits a global CTE problem as the operating temperature is reduced.

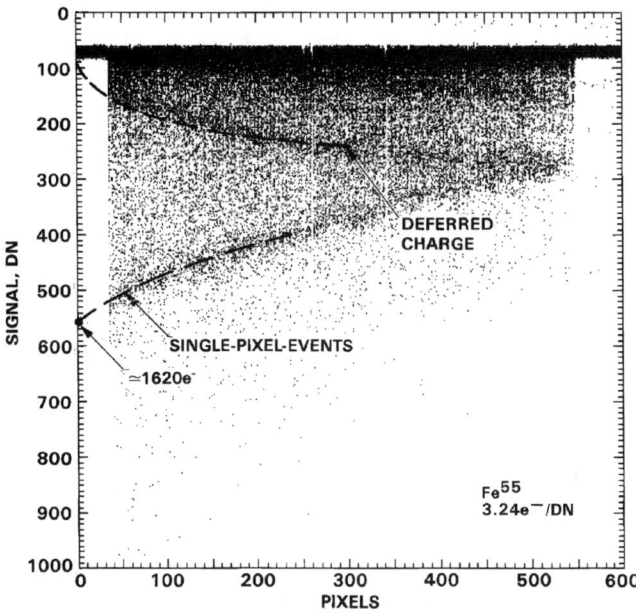

Figure 5.32(b) X-ray transfer showing a CCD with global CTE problem.

to assist emission of signal electrons from the spurious potential pockets during charge transfer. The problem here is likely associated with the alignment and encroachment of the various implants that are employed in virtual-phase technology.

Figure 5.32(b) shows a global CTE problem exhibited by a 512 × 512 three-phase CCD. Note that the target pixel level decreases whereas the first trailing pixel increases. The response shows that most deferred charge ends up in the first trailing pixel. This indicates that the emission from the pockets is fairly fast. We will see other global CTE problems where trailing charge is not seen, because the time constant is longer in comparison to the pixel or line transfer time. This characteristic indicates a deep level trapping problem.

A CCD that exhibits spurious potential pockets can often be restored by four means. The first method involves elevating clock drive to the sensor to increase the fringing fields between phases, which may collapse spurious potential pockets. The second method employs the highest operating temperature possible. This technique decreases the emission time from the pockets. Third, clocking the CCD at a slower rate allows more time for charge to escape from the traps. Fourth, the addition of fat-zero charge fills the pocket and allows signal charge to transfer through the problem region unimpeded.

5.4.2.2 Local Traps

The most serious type of process trap is the "localized trap." Defects like this affect CTE on a local, random level. Localized traps have been seen in devices from every manufacturer reflecting a variety of differing fabrication technologies. These defects are present in devices fabricated in epitaxial as well as bulk material and are present independent of the starting resistivity of the p-substrate. To complicate the issue, the frequency of traps observed in an array varies significantly (from a few to hundreds for any given CCD) from lot to lot even though the same fabrication recipe is employed. The traps are capable of capturing a wide range in charge, from a few electrons to several hundred thousand electrons. Traps are usually confined to a single pixel, and detailed tests show them to be localized to a single level of poly-silicon within a pixel. The presence of traps is independent of the gate structure (1-, 2-, 3- and 4-phase CCDs all have exhibited the trap problem). The filling and emptying of these traps can be strongly affected by the clock voltage.

In-depth physical examination of the pixel in question has generally failed to yield a strong correlation between structured defects and pixels with traps. In testing various CCDs from different manufacturers it appears that the simple gate-oxide dielectric exhibits fewer traps than CCDs fabricated with an oxide/nitride insulator system. Several theories have been given to explain the apparent connection between localized traps and the oxide/nitride layer. One explanation attributes the problem to oxide pinholes, small holes in the oxide that would allow the nitride layer to reside very close to the n-channel. Similar in structure to an EPROM device, it would then be possible for hot electrons to be injected from the silicon through the thin oxide layer and stored at the oxide/nitride interface. The stored charge would cause a voltage shift within that pixel possibly resulting in the formation of a potential bump in the channel. As with EPROMs, the charge retention time at the oxide/nitride interface would be very long (i.e., the traps would be long-lived). Figure 5.33 illustrates another mechanism that produces traps. Here a pinhole in the gate insulator allows phosphorus into the signal channel when the polysilicon layer is doped. The addition of phosphorus increases the channel potential, forming a localized trap.

The localized trap was, much more severe problem in the past than it is today. The dilemma restricted the physical size of the CCD to arrays no larger than

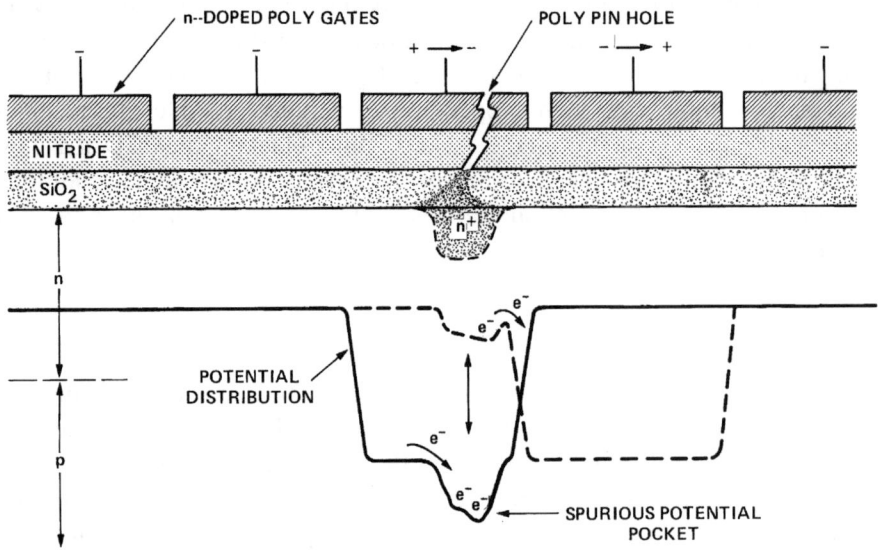

Figure 5.33 Formation of a spurious potential pocket.

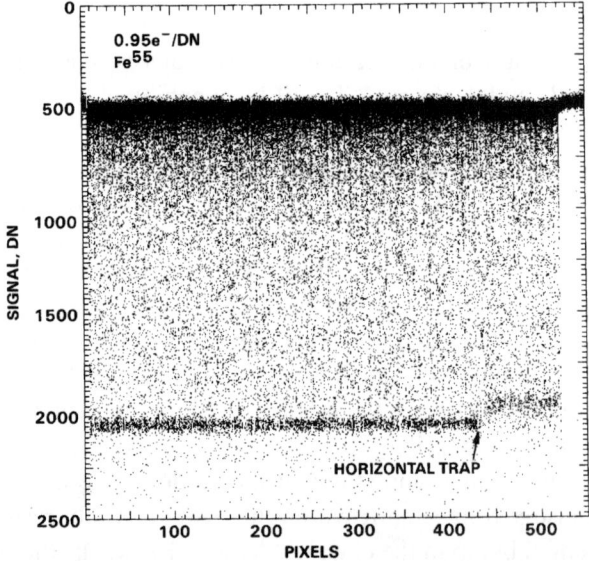

Figure 5.34(a) Horizontal x-ray transfer plot showing a small horizontal 100-e^- process trap.

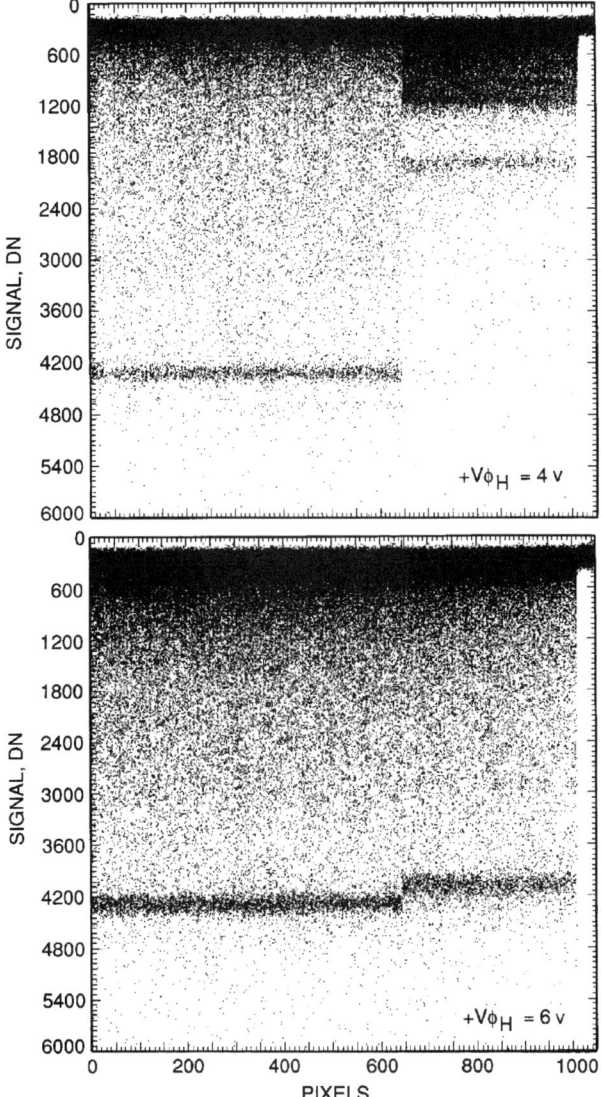

Figure 5.34(b) A horizontal process trap that is sensitive to applied clock drive.

1024 × 1024 formats for many years. Even today, fabricating a large CCD that is completely trap-free is difficult to achieve.

Although the presence of a localized trap in a vertical column is undesirable, its occurrence in the horizontal register can make the CCD totally worthless. Figure 5.34(a) shows a small local trap found at the end of a horizontal register. Note that the trap consumes approximately 100 e^- from the Fe^{55} single-pixel-event line. This trap is small and may be tolerable to the user depending on the application. Figure 5.34(b) shows a much larger trap found in the horizontal register of

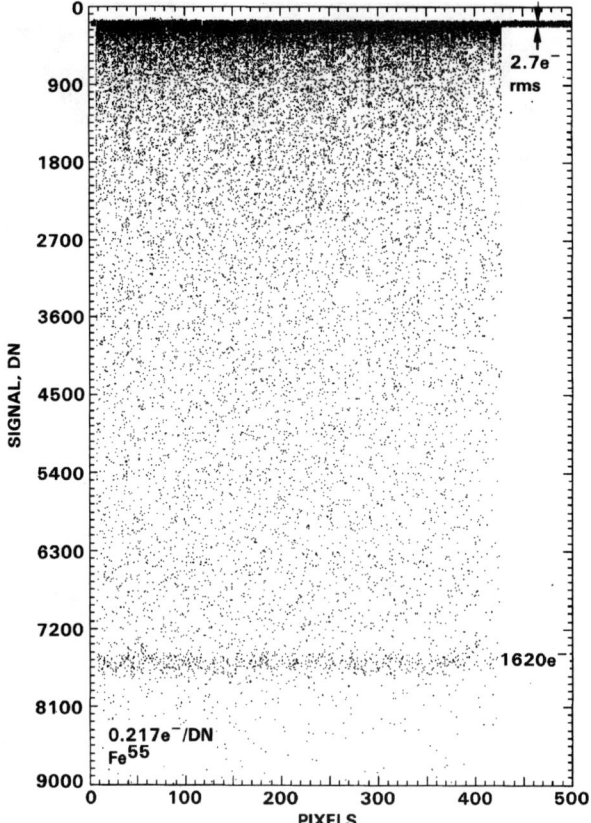

Figure 5.35(a) A horizontal x-ray transfer plot for a CCD that exhibits numerous vertical traps.

a 1024 × 1024 CCD. This trap is highly dependent on clock drive voltage. The top trace was generated by driving the horizontal register with clocks that switch between +4 and −8 V. In the bottom trace the positive rail is increased to +6 V, improving CTE performance dramatically.

Isolated traps are found both in horizontal and vertical registers of the CCD. X-ray stimulation is the best way to find horizontal traps. Vertical traps can also be found with x rays, although other techniques will be discussed below. Figure 5.35(a) presents a horizontal Fe^{55} x-ray response taken from a 420 × 420 CCD. When the response is magnified as in Fig. 5.35(b), we see numerous vertical traps. The white blemishes embedded in the 2.7-e^- noise floor show those columns that contain traps. Traps offset the read noise by the amount of deferred charge that is generated as the sensor is continuously stimulated with x-ray photons under constant readout conditions.

Vertical traps can be found rapidly by simply exposing the CCD to a flat-field light source. This technique is performed by first erasing the CCD under dark conditions to completely empty traps of charge. Next, the CCD is exposed to a low-

CHARGE TRANSFER 445

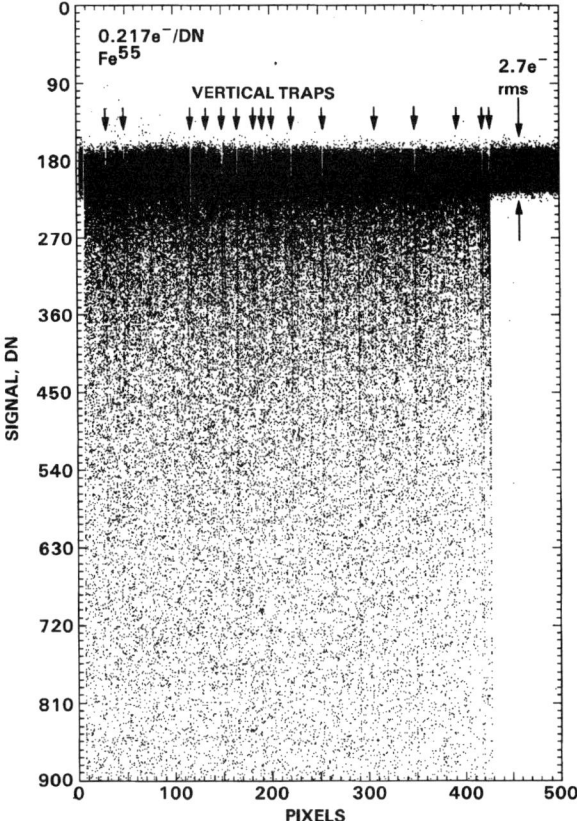

Figure 5.35(b) Magnified view of Fig. 5.35(a).

level flat field. Then the CCD is read out, filling traps. Small traps may require only a few line transfers before they are satisfied whereas larger traps may require hundreds of line transfers before reaching saturation. The resultant image after readout exhibits dark vertical lines running behind each trap location (i.e., charge that is removed and stored by the trap). Figure 5.36 shows a flat-field trap map taken by a 520×520-pixel CCD (charge is transferred down and to the left). Note that the array contains several large isolated traps and many smaller ones.

A very sensitive test for locating vertical traps uses the EPER technique. For example, Fig. 5.37(a) shows the first four extended lines (lines # 1025, 1026, 1027 and 1028) that are stored into the computer after the last line of a 1024×1024 array is read out. The array exhibits numerous traps. Ideally, a trapless CCD would show a flat response without deferred charge contained in the extended lines. The response only locates columns that contain traps. A column trace is required to locate the trap vertically. For example, Fig. 5.37(b) characterizes a single vertical trap that was found in column #188 in response to a 90-e^- flat-field signal. The trap location is recognized by a reduction in signal and requires the passage of seven lines to fill the trap. Note that the signal level does not entirely go to zero as the

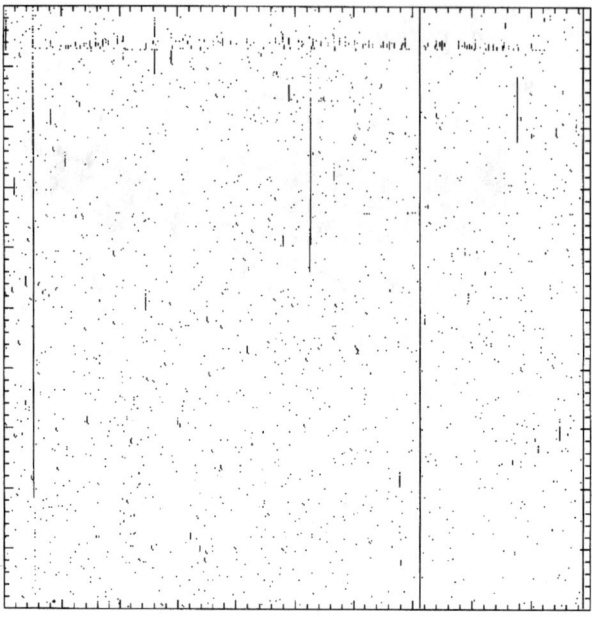

Figure 5.36 A flat-field trap map showing several vertical process-induced traps.

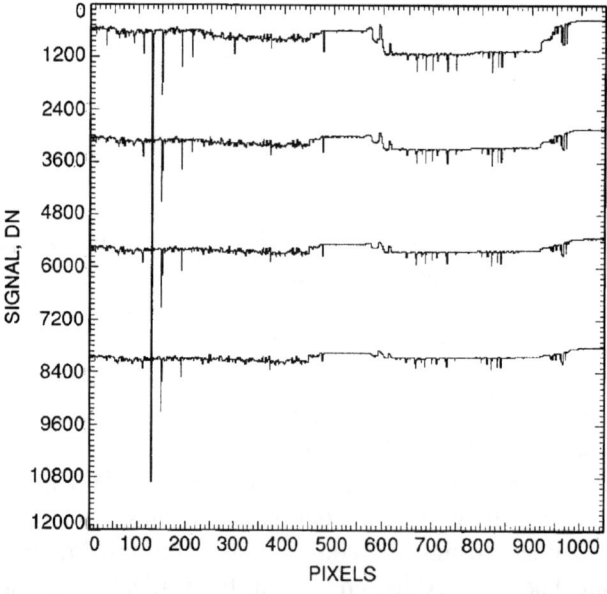

Figure 5.37(a) Vertical EPER showing numerous process traps.

trap consumes charge. This behavior is interpreted as being a result of the interplay between the filling and emptying of the trap, i.e., the capture and emission rates that

CHARGE TRANSFER 447

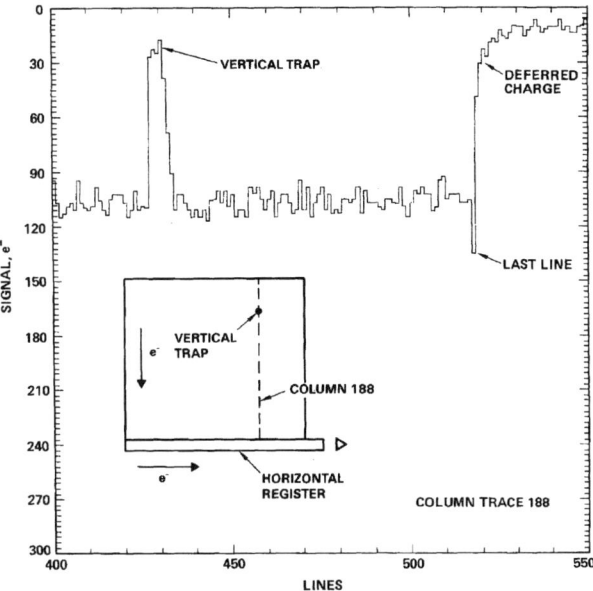

Figure 5.37(b) A vertical process trap located near the end of a column.

characterize the trap. Following the last line of the array an exponential deferred charge tail is observed as charge is slowly released from the trap as the device is overscanned in the vertical direction (i.e., within the extended line region).

Figure 5.38 shows the response of a CCD when stimulated with a small slit of light that is four lines wide (charge transfer is down and to the left). The slit of light is positioned above a vertical trap so charge must transfer through it. Note that the trap consumes the entire signal and then slowly redistributes it into deferred charge in subsequent lines above.

Figure 5.39 presents an image generated by a CCD in response to a uniform light field of 650 e^-. The partial image (13[H] × 7[V] pixels) displays the upper corner of the sensor, a region furthest from the on-chip amplifier (charge is transferred down and to the left in this image). The last vertical column and the last horizontal line of the array are brighter (i.e., displayed here as completely black) due to charge that diffuses in from the sides of the sensor during the exposure period (approximately 1400 e^-). The CCD is overscanned both horizontally and vertically to inspect for deferred charge in the extended pixel and line regions. Deferred charge is seen in the ninth column from the end of the array. The deferred charge is due to the presence of a large localized vertical trap located midway down in the array. Although not completely displayed, nearly 300 additional line transfers are required before the deferred charge level equals that of the noise floor, indicating that the vertical trap is very large. Also note that a small amount of deferred charge (a few electrons) is seen in the overscanned region of the horizontal register (i.e., pixels following the last column), showing a finite CTE problem for the horizontal register.

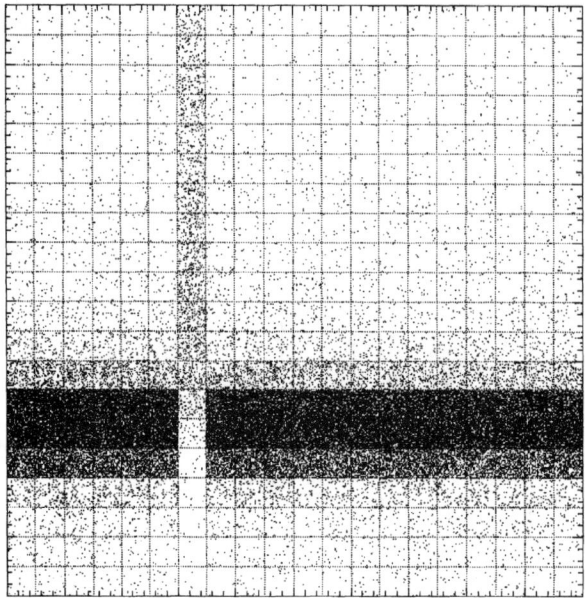

Figure 5.38 A vertical process trap in response to a horizontal slit of light.

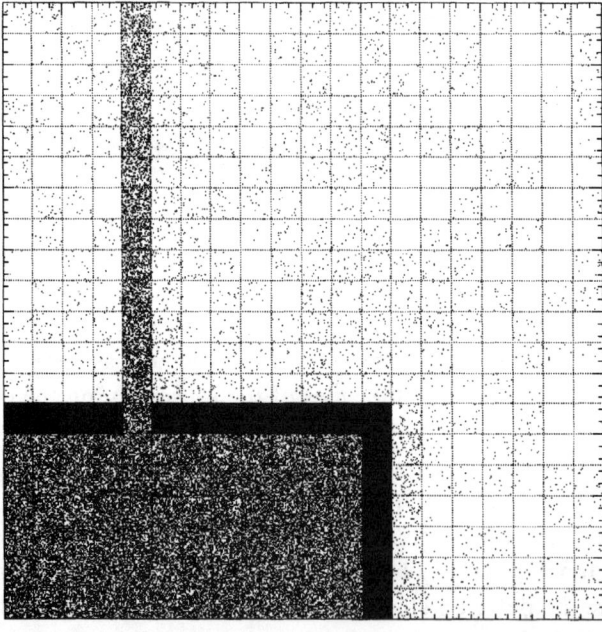

Figure 5.39 A region furthest removed from the on-chip amplifier showing horizontal and vertical extended pixels. A vertical process trap is located in the ninth column from the end.

Figures 5.40(a) and 5.40(b) are pocket pumping trap maps that show numerous trapping sites for a 1024 × 1024 CCD. The tails seen in each image show both forward and reverse traps and are about equal in number. In Fig. 5.40(b) the trap is located by a heavy dot followed by a tail indicating the forward or reverse nature of the trap. Localized traps as small as 10 e^- are measured in this plot.

Figure 5.41 shows a horizontal x-ray response taken from a 512 × 512 four-phase CCD that exhibits a serious CTE flaw. The problem is associated with a single pixel that is partially shorted between phases-2 and -4 (approximately 1 MΩ with 10 V applied). The intralevel short causes a reduction in clock drive voltage and local CTE degradation. Since the CCD was driven from both sides of the array, the line drop-out signature observed is symmetric about the shorted pixel. The problem can be avoided by using the sensor's upper on-chip amplifier because the short is two lines up from the lower horizontal register. When charge is clocked to the upper amplifier only the last two lines are affected, as opposed to 510 lines when using the lower amplifier. Cosmetic problems similar to this can be reduced by selecting the amplifier for best charge transfer results.

Figure 5.42 presents a flat-field response also generated by a four-phase CCD that exhibits a "line dropout" problem. Here charge transfer direction is down and to the left. The "tear" in the image is caused by a single intralevel short located 15 lines up from the lower horizontal register. Poor CTE is exhibited for all lines above the short. As the image shows, approximately six lines are required to fill the troubled line before charge from other lines above can transfer through. Although the apparent CTE improves for these lines, a shot noise reduction is evident in the image, showing that CTE is not well behaved. Line dropouts cause the CTE to vary with signal level (i.e., CTE hysterisis).

The line dropout problem shown in Figs. 5.41 and 5.42 is common to four-phase CCD technology (i.e., double-poly gate technology). In this technology an intralevel short between phases-1 and -3 or between phases-2 and -4 can occur because of poly stringers. These are regions where incomplete etching and patterning of the poly gates occur, resulting in a bridge of poly between the two phases (typically near the channel stop regions). Three-phase CCD technology eliminates the intralevel short problem. This is because each phase is deposited and patterned separately. With three-level gate technology, failure to completely etch a poly level only results in poor CTE at a single pixel site, rather than resulting in a fatal defect that can affect many pixels and lines. Although additional process steps are required compared with four-phase, it is well worth the yield improvement. Therefore, three-phase technology is preferred among very large CCD area array manufacturers.

Figure 5.43(a) presents a bar target image generated by a three-phase CCD (charge moves up and to the left). Note that a group of shorted pixels on the left causes charge to transfer out ahead by three pixels. Charge through this region transfers without clocking, producing a negative delay effect in the region. Also, a small 100-e^- trap is seen on the right that exhibits a positive delay effect. Figure 5.43(b) shows a column trace through this trap. The modulation is disturbed

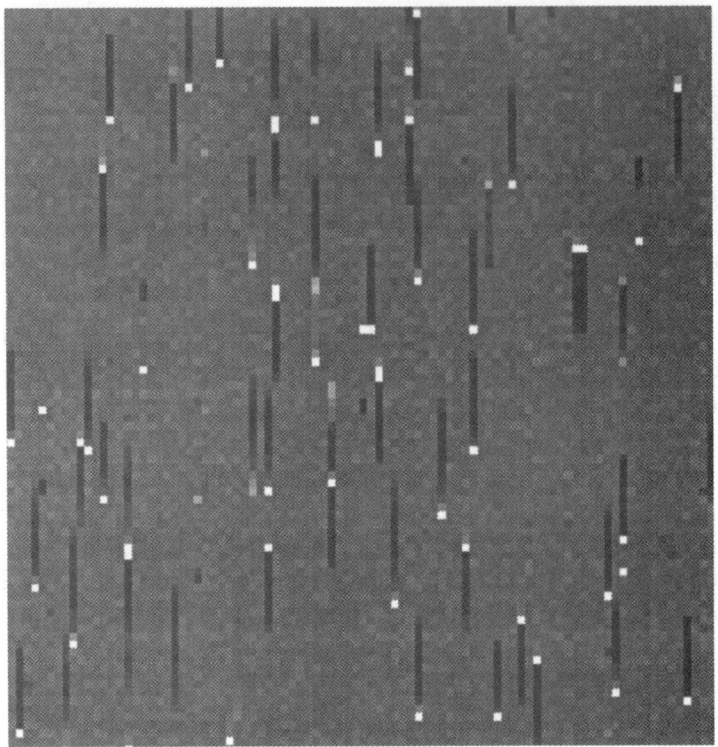

Figure 5.40(a) Pocket pumping response showing forward and reverse traps.

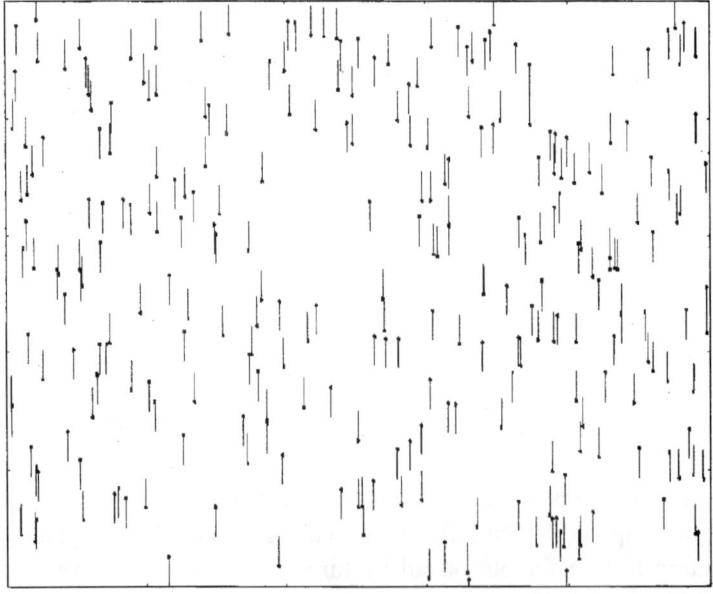

Figure 5.40(b) Pocket pumping response showing process traps as small as 10 e$^-$.

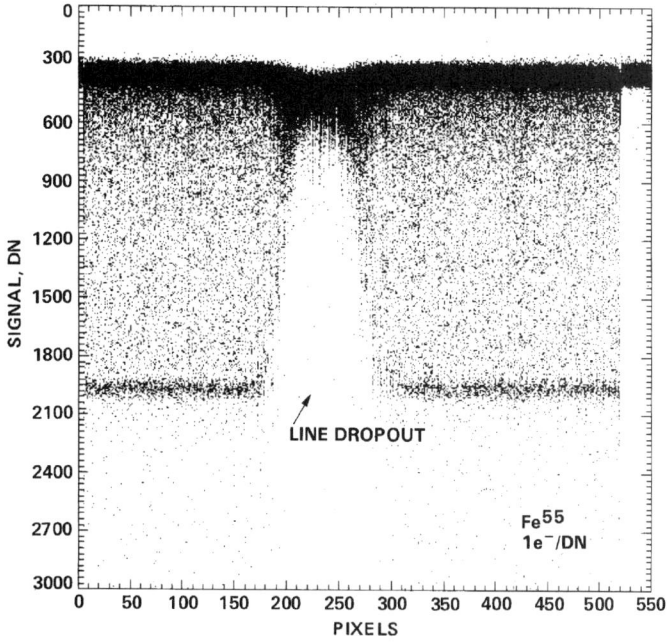

Figure 5.41 Line dropout trap caused by an intralevel short that is located 2 lines up from the lower horizontal register.

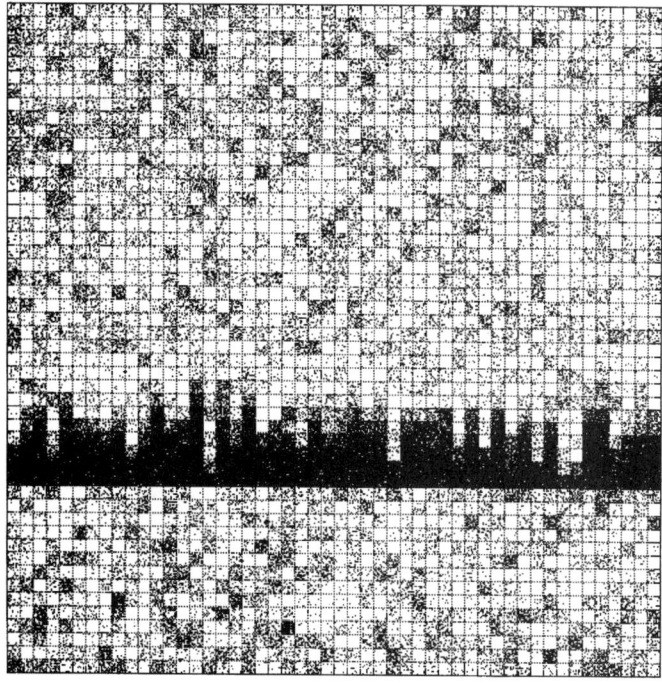

Figure 5.42 A line dropout trap in response to a flat-field stimulus.

Figure 5.43(a) Bar target showing two process-induced traps.

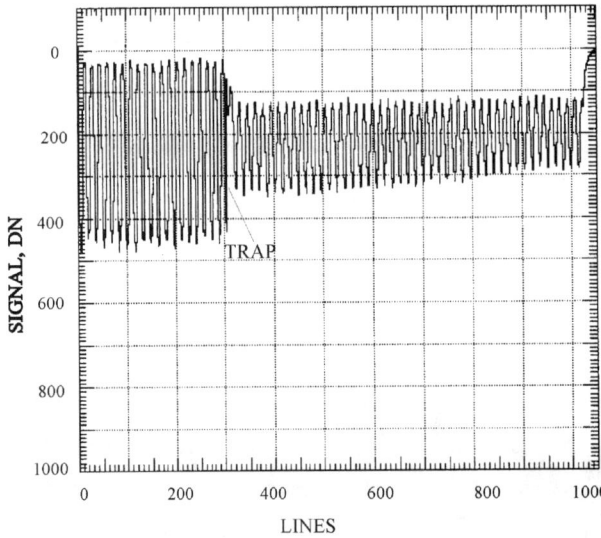

Figure 5.43(b) Bar target column trace taken through the small trap in Fig. 5.43(a).

as charge is captured and released by the trap. Also note that deferred charge is released within the extended line region by the trap. It should be mentioned that

Fig. 5.43(a) is a very important image product when screening CCDs for cosmetic defects.

5.4.3 BULK TRAPS

Bulk traps are associated with impurities (e.g., gold) or lattice defects found in silicon wafers. Bulk traps that happen to lie within the charge transfer channel can trap charge, typically a single electron. They collectively affect the global CTE performance. Bulk traps are not generally a problem for most CCD applications because silicon material used in fabricating CCDs is very high quality. However, as we will discuss below, bulk traps ultimately limit CTE performance. Under optimum conditions CTEs > 0.999999 have been measured on selected "bulk-state-limited" CCDs.

The capture and emission of electrons by bulk states and their influence on CTE is dependent on several factors. Pixel clock rate, operating temperature and trap depth are important variables to consider here. These factors have been analytically described by Choong-Ki Kim[10] and included in his equation,

$$\text{CTI} = qN_T V_V \exp\left(-\frac{t_{PT}}{3\tau_e}\right)\left[1 - \exp\left(-\frac{t_{PT}}{3\tau_e} - \frac{N_Z t_{PT}}{\tau_e}\right)\right], \quad (5.26)$$

where N_T is the bulk trap density (traps/cm^{-3}) located in the silicon band gap, V_V is the volume the charge packet occupies in the potential well (cm^3), t_{PT} is the pixel transfer period (sec), N_Z is the average number of empty pixels between x-ray events (pixels/x-ray event) and CTI is the charge loss for an x-ray event in a single pixel transfer. The equation assumes three-phase operation with symmetric clock overlaps between phases of 1/3 the clock period (i.e., clock overlaps $= t_{PT}/3$). Equation (5.26) is also used to estimate vertical CTE performance. For example, in Chapter 8 t_{LT} will represent the line shift time for deep trap studies.

The variable τ_e in Eq. (5.26) is the emission time constant, which describes the rate at which electrons are emitted from the traps. This constant is

$$\tau_e = \frac{\exp[E_T/(kT)]}{\sigma_n v_{th} N_C}, \quad (5.27)$$

where N_C is the effective density of states in the conduction band (2.8 × 10^{19} states/cm^3) and E_T is the trap energy level below the conduction band edge (eV). Here v_{th} is the thermal velocity of the electrons (cm/sec) where $v_{th} = (3kT/m_e^*)^{1/2} \approx 10^7$ cm/sec where m_e^* is the effective mass of an electron. The quantity σ_n is the electron cross-section which is physically related to the ability of the trap to trap an electron (cm^2). It is a measure of how close the electron has to come to the center to be captured. This distance is on the order of atomic dimensions, approximately 10^{-15} cm. Therefore, $v_{th}\sigma_n N_C$ is approximately equal to 2.8 × 10^{11}/sec.

Example 5.16

Using Eq. (5.26) calculate and plot CTI as a function of τ_e and N_Z assuming a horizontal pixel period of $t_{PT} = 20$ μsec. Normalize the plot to $qN_T V_V$.

Solution:
Take an example data point at $\tau_e = 10^{-3}$ sec and $N_Z = 1$.

$$\exp-(t_{PT}/3\tau_e) = \exp-[20 \times 10^{-6}/(3 \times 10^{-3})] = 9.93 \times 10^{-1},$$

$$\exp-(t_{PT}/3\tau_e + t_{PT}/\tau_e) = \exp-[20 \times 10^{-6}/(3 \times 10^{-3}) \\ + (20 \times 10^{-6})/10^{-3}] = 9.737 \times 10^{-1}.$$

From Eq. (5.26),

$$\text{CTI} = 9.93 \times 10^{-1}(1 - 9.737 \times 10^{-1}),$$

$$\text{CTI} = 2.6 \times 10^{-2}.$$

Figure 5.44 plots CTI for different N_Z.

A wealth of information is contained in Fig. 5.44 which is used to describe many bulk state (and radiation induced) CTE problems. For a given pixel period, bulk traps on the left-hand side of the curve are called "shallow traps" whereas those on the right side of the curve are called "deep traps." For the case of shallow traps, CTE improves as the emission time becomes shorter because more time is allowed to escape bulk states. For deep traps, CTE improves as the emission time constant increases because charge is trapped long enough to allow other x-ray events to transfer through traps without capture. Note that as the separation of x-ray events increases (i.e., N_Z increases) CTE degrades because traps are not supplied with charge as often.

It is important to point out from Fig. 5.44 that the clock overlap time is critical to the CTE measured. For shallow traps, CTE is worse when the emission time constant is approximately equal to the shortest clock overlap period employed. Clock overlap periods on either side of the emission time constant will improve CTE (assuming a finite N_Z). For deep traps, CTE is worse for long clock overlap periods because charge is allowed to escape the traps during that time. For multiphase CCDs, horizontal and vertical clock overlaps do not normally have the same time period. For example, the vertical clocks have one phase that remains high during the horizontal transfer time. It is during this time that charge from deep traps can escape. Adjusting operating temperature and/or clock overlap times are techniques to solve trap problems. CTE degradation caused by radiation-induced

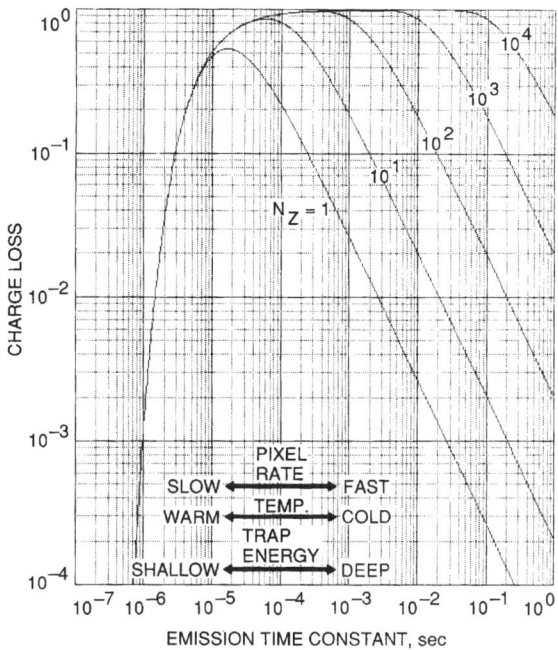

Figure 5.44 Charge loss as a function of bulk trap emission time constant for different N_Z.

traps can also be minimized using these methods. This topic is discussed in detail in Chapter 8.

As indicated by Eq. (5.26), operating temperature and the clock overlap period are two parameters adjusted to circumvent bulk state problems. Figure 5.45(a) presents a horizontal Fe^{55} x-ray transfer response that was generated by an 800 × 800 three-phase CCD at an operating temperature of −130°C. The horizontal register is clocked with a clock overlap period of 1 µsec. Note that the single-pixel event line (i.e., target pixel) is completely annihilated by the bulk traps present. Deferred charge is seen in several trailing pixels. This indicates the presence of a shallow trap because charge escapes the traps quickly enough to see the trailing pixels. Figure 5.45(b) is an approximate "curve fit" to the response using Eq. (5.1) assuming a $CTE_H = 0.996$.

It is interesting to note from Fig. 5.45(a) that vertical CTE is well behaved for this CCD ($CTE_V > 0.99999$). The single-pixel-event line would be much broader if there was a vertical CTE problem, because lines are stacked vertically. The difference in CTE between the two registers is because the vertical registers are clocked at a much slower rate. This allows trapped charge to escape during the transfer period (the vertical clock overlap period is 60 µsec compared with a horizontal clock overlap of 1 µsec). In other words, the trap emission time constant is much shorter than the vertical clock overlap period.

The horizontal CTE problem observed in Fig. 5.45(a) can be improved either by increasing the horizontal clock overlap period or operating temperature. Figure 5.45(c) shows an improved response when the operating temperature is increased to −110°C. In Fig. 5.45(d) the horizontal clock overlap is increased to 4 μsec at −110°C, significantly improving CTE performance even more. Figure 5.45(e) plots the middle region of the sensor at different operating temperatures, leaving the clock overlap at 1 μsec. For operating temperatures greater than −80°C the CTE is well behaved. Below −80°C the CTE rapidly degrades as the trap emission time constant becomes comparable to the horizontal clock overlap period. As long as the clock-overlap period is longer than 1 μsec and the operating temperature is > −80°C, the silicon wafers here are satisfactory for use.

Figure 5.46 shows horizontal CTE characteristics for a three-phase CCD measured using the EPER technique as a function of operating temperature. Note that the greatest CTE loss occurs at −70°C, indicating a specific trapping site is at work. Above −70°C CTE improves as the emission time constant becomes shorter than the clock overlap period. Below −70°C the EPER measurement technique is not accurate because charge spreads over many extended pixels hiding deferred

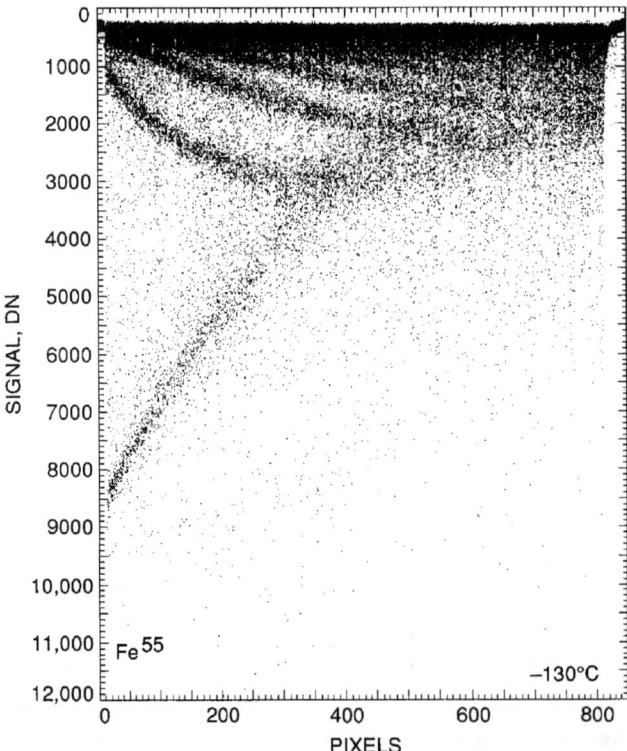

Figure 5.45(a) An Fe^{55} x-ray transfer plot taken from a CCD fabricated on silicon with bulk traps that become active for a clock overlap of 1 μsec and at < −80°C.

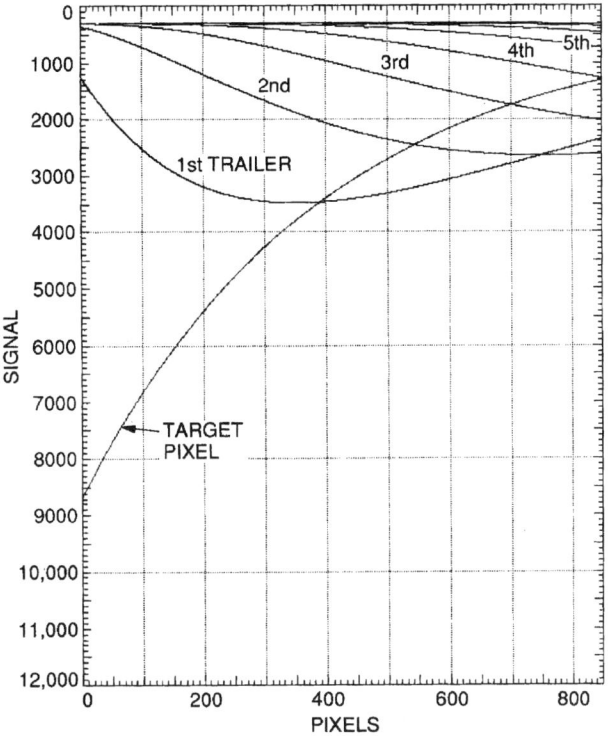

Figure 5.45(b) Theoretical CTE curves that fit data shown in Fig. 5.45(a).

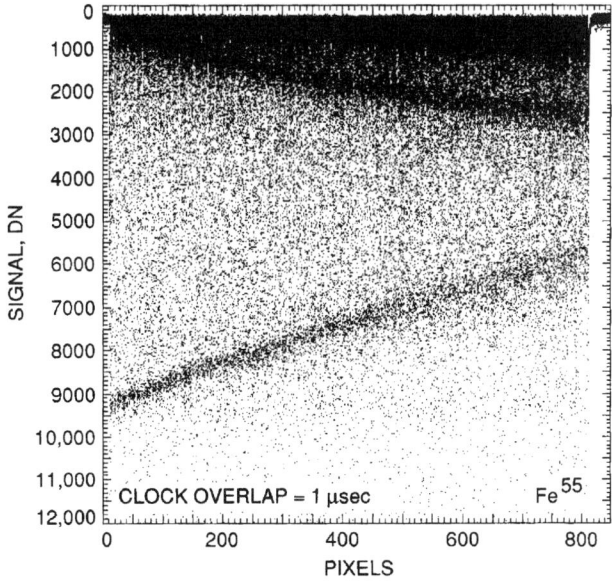

Figure 5.45(c) Improved x-ray transfer response taken at an elevated operating temperature.

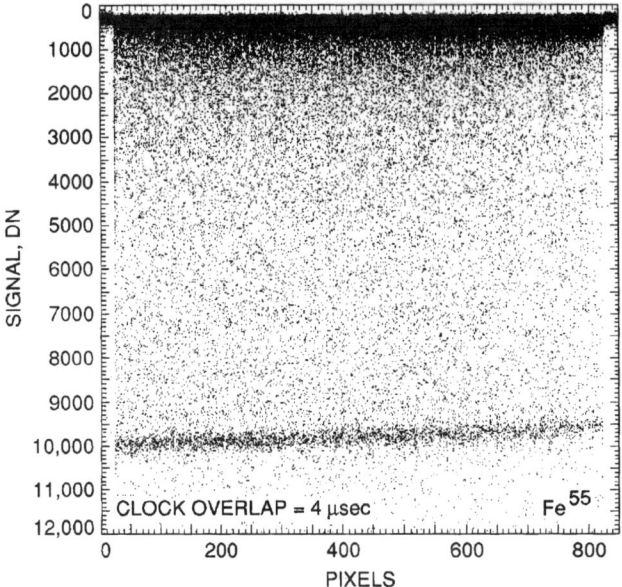

Figure 5.45(d) Improved x-ray transfer response taken at a clock overlap of 4 μsec.

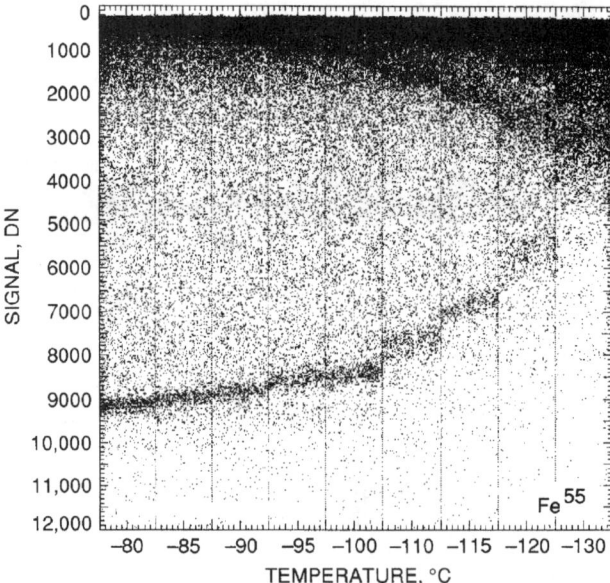

Figure 5.45(e) Several x-ray responses plotted together showing rapid CTE degradation with decreasing operating temperature.

charge in the read noise floor. Nevertheless, the breakpoint is a usefull feature of the plot telling us that absolute CTE improves on either side of −70°C.

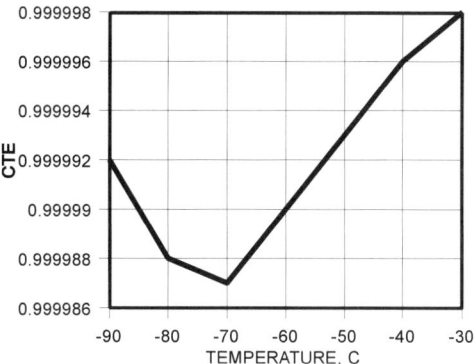

Figure 5.46 Horizontal EPER as a function of operating temperature.

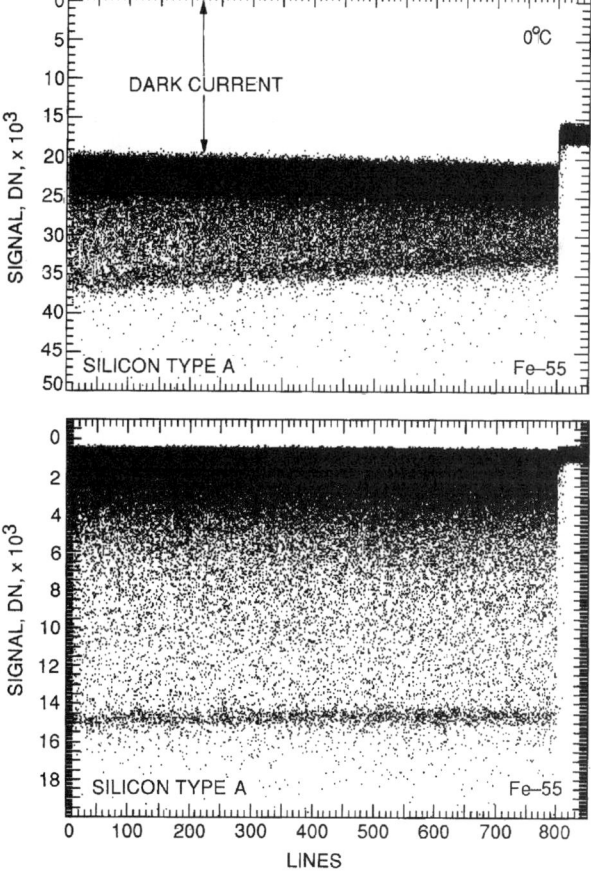

Figure 5.47(a) X-ray transfer responses for two different types of silicon at an operating temperature of 0°C.

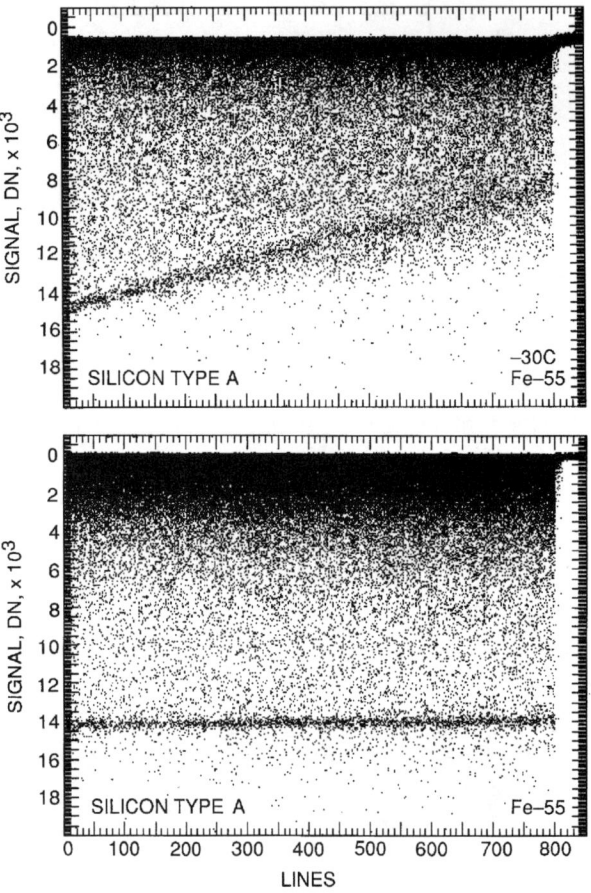

Figure 5.47(b) X-ray transfer responses at an operating temperature of −30°C.

Deep traps exhibit the opposite behavior compared to shallow traps in reaction to operating temperature and clock rate. That is, CTE improves when the operating temperature is lowered or the clocking rate is increased. For example, Fig. 5.47(a) shows vertical x-ray transfer plots for the MPP Hubble WF/PC II CCDs built on two different types of silicon material (called "Type-A" and "Type-B"). Note that silicon Type-A exhibits more traps than Type-B based on CTE_V performance. Horizontal CTE is perfect for both sensors, indicative of deep traps at work ($CTE_H > 0.99999$). Again, this characteristic is because the trap time constant is much greater than the horizontal clock period. Note also that Type-A silicon exhibits much higher dark current. In Chapter 7 we will discuss bulk dark current characteristics. It is typical to find that CTE and dark current track each other. That is, if a bulk CTE problem exists, then bulk dark current will also be higher. Figure 5.47(b) shows the response when the operating temperature is reduced to −30°C. CTE_V improves because charge is held longer in the traps, allowing other

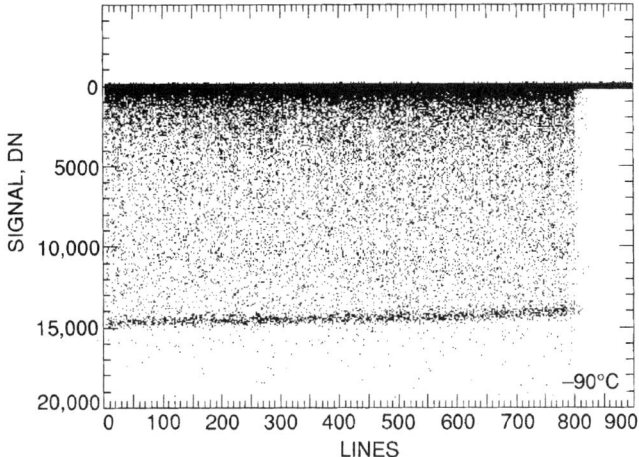

Figure 5.47(c) An improved x-ray transfer response for silicon Type B when the operating temperature is reduced to −90°C.

x rays to transfer through the array. Reducing the operating temperature to −90°C for Type-A material significantly improves CTE_V, as shown in Fig. 5.47(c).

It should be noted that none of the CTE conclusions made above could have been made if the EPER CTE method was used. When deep traps are present, all relative CTE measurement techniques will be in error because deferred charge tails will not be present. This conclusion was very important to Hubble. Unfortunately, silicon Type-A material was flown because Type-B was not available due to schedule problems (a few months too late). Initially after launch the operating temperature for WF/PC II was set to −70°C to allow the instrument to outgas and avoid ice buildup on the CCDs. Calibrating the CCDs using standard stars at this temperature showed an absolute photometry loss of approximately 20% at the top of the array furthest from the output amplifier. The stars looked whole without deferred charge tails present. EPER calibration would have concluded perfect CTE at this operating temperature. The observation would have caused panic in the WF/PC II science team if it weren't for the precalibration x-ray data taken before flight (i.e., Fig. 5.47). Later, after sufficient out-gassing took place and the sensors were cooled to the nominal operating temperature of −90°C, CTE improved to the level shown in Fig. 5.47(c).

Pocket pumping allows individual bulk traps to be located and size measured. For example, Fig. 5.48(a) shows a pocket-pumping response generated by a 1024 × 1024 CCD in response to 1750-e^- flat-field stimulus. The sensor was pumped 1400 times over a five-line sector at a vertical line rate of 2500 Hz. After the pumping sequence was done, the CCD was then read out and the first 500 lines were stacked together into the single composite trace shown. The strong signal level of approximately 7000 DN shows the presence of 1-e^- bulk traps.

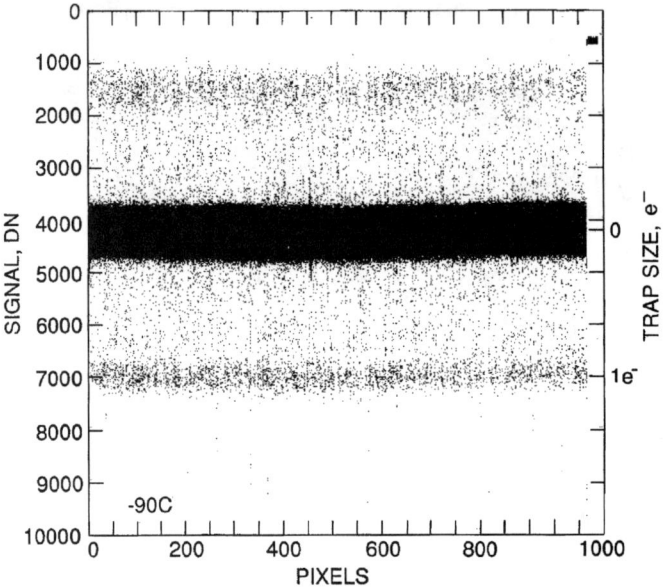

Figure 5.48(a) Pocket pumping response showing 1-e⁻ bulk state traps characterized at an operating temperature of −90°C.

Example 5.17

Determine the trap size required to produce the 7000 DN level measured in Fig. 5.48(a). Assume that the signal difference between the flat-field and trap level is 2800 DN (i.e., 7000 − 4200 DN) and a camera gain of $K = 0.54$ e⁻/DN.

Solution:
From Eq. (5.23),

$$S_T = 2800 \times 0.54/1400,$$

$$S_T = 1.08 \text{ e}^-.$$

Figure 5.48(a) also show traps below the 1-e⁻ level. These are called "statistical traps." Traps of this type may at times capture an electron and release it during the pumping cycle. A uniform distribution of statistical traps is seen in the figure.

The average number of traps per pixel, as measured by pocket pumping, is used to estimate CTE performance through the equation

$$\text{CTE}_{PP} = 1 - \frac{N_{PIXT} T_S}{S(\text{e}^-)}, \tag{5.28}$$

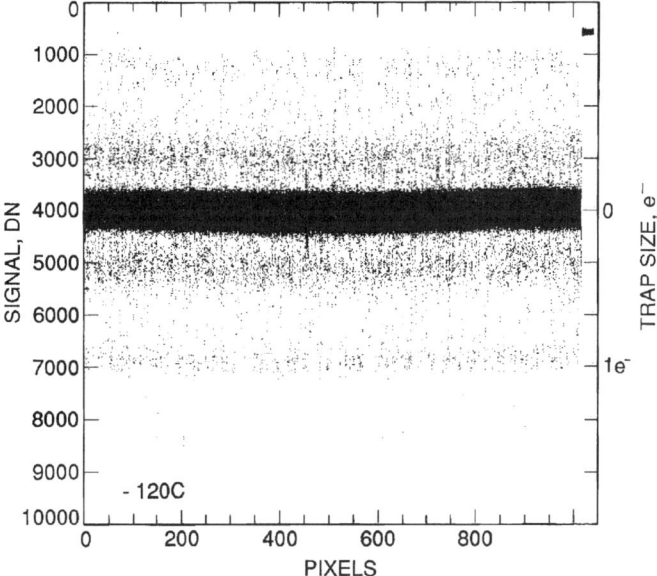

Figure 5.48(b) Pocket-pumping response at −120°C showing two different types of bulk traps.

where N_{PIXT} is the average number of traps/pixel and T_S is the average trap size (e⁻).

Example 5.18

The number of traps counted for the CCD in Fig. 5.48(a) is $N_{PIXT} = 0.002$ traps/pixel (only traps above the 3-sigma shot noise flat-field level are counted). The average trap size is $T_S = 0.81$ e⁻. Calculate the CTE. Assume a flat-field level of $S(e^-) = 1750$ e⁻.

Solution:
From Eq. (5.28),

$$\text{CTE}_{PP} = 1 - (2 \times 10^{-3} \times 0.81/1750),$$

$$\text{CTE}_{PP} = 0.9999987.$$

This example shows that the pocket-pumping technique is very sensitive in measuring high CTE values.

The pocket-pumping technique locates bulk traps that are only active for a given operating temperature and clock rate. Figure 5.48(b) shows a pocket-pumping response for the same CCD as Fig. 5.48(a) at a reduced operating temperature of

−120°C. Note that the 1-e⁻ trap level splits into two lines at approximately the 1 and 1/3 electron respectively. The response indicates that at least two types of bulk traps are present, each exhibiting a different activation energy. Reducing the temperature further below −120°C causes the 1/3 electron trap line to completely disappear into the shot noise floor. At this operating temperature the emission time constant of these traps is comparable to the pocket-pumping cycle period. In other words, the traps remain filled over many pumping cycles and are therefore not seen. The remaining 1-e⁻ trap line at −120°C is unaffected; however, below −140°C it too begins to disappear.

Figure 5.49 presents a bulk trap map generated by the pocket-pumping response taken from Fig. 5.48(a). The map shows a 240(V) × 200(H) region for traps that are above the 3× shot noise level. The 1-e⁻ bulk traps are randomly distributed throughout the array, some columns having more than others. Vertical CTE is therefore column dependent.

Bulk-state-limited CTE performance allows single electrons to be transferred efficiently. For example, Fig. 5.50(a) shows a raw single line trace generated by a 512 × 512 CCD that exhibits a noise floor of 5 e⁻ rms (or about 40 e⁻ peak-to-peak). Although not apparent, an ultralow-level square-wave image is imbedded within the read noise. Figure 5.50(b) shows a 5-e⁻ signal after 1500 images are averaged together, yielding a read noise = 0.13 e⁻ rms. A 6-e⁻ column blemish also emerges, not seen in Fig. 5.50(a). The single electrons that make up the square wave signal are well separated, approximately 1 e⁻ per 300 pixels, indicating that single electrons are being transferred.

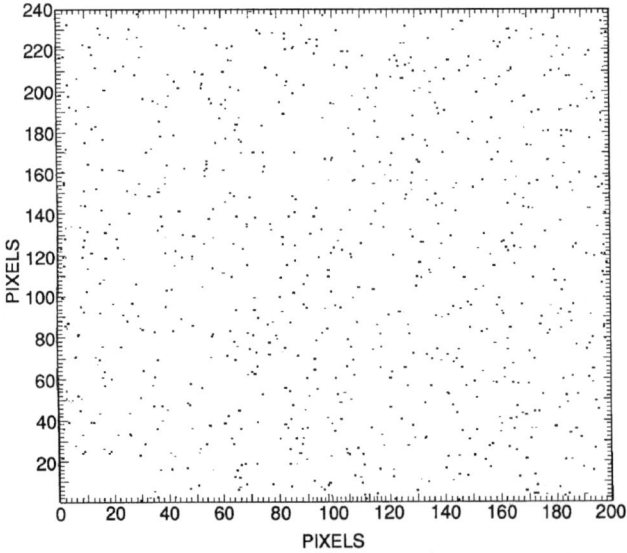

Figure 5.49 Trap map showing single electron traps for a 240(V) × 200(H)-pixel region.

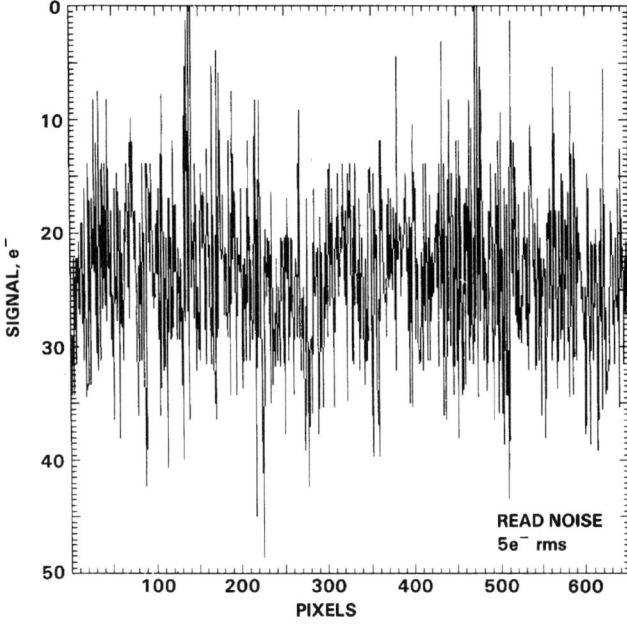

Figure 5.50(a) Line trace showing random read noise at 5 e⁻ rms that hides a bar target image.

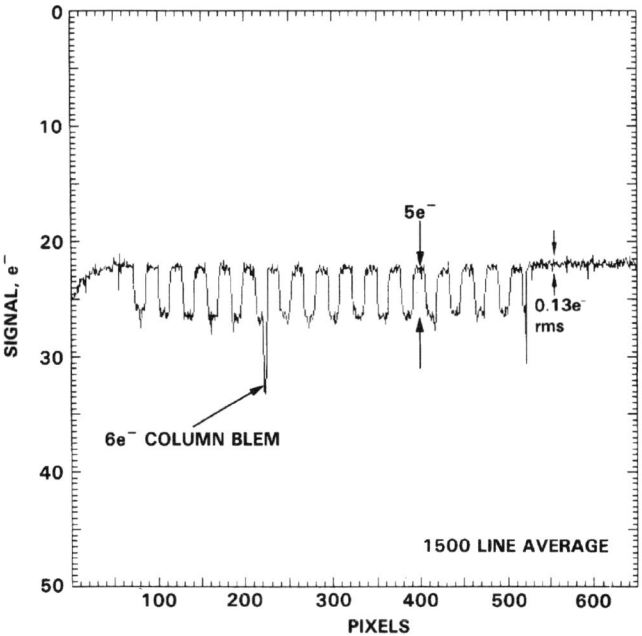

Figure 5.50(b) Average response of 1500 lines showing the 5-e⁻ square-wave signal.

5.4.4 Radiation-induced Traps

Radiation-induced traps are discussed in considerable detail in Chapter 8.

5.4.5 Proportional and Fixed Loss

CTE is dependent on the amount of charge being transferred. The physical reason behind this behavior is best understood by recognizing the difference between "proportional" and "fixed" CTE loss. Fixed CTE loss occurs when a fixed amount of charge is lost to trapping sites. Fixed loss therefore translates to a CTE that increases with charge packet size. Design and process traps behave in this manner by taking a fixed amount of charge from the charge packet. For example, the design trap in Fig. 5.31 consumed approximately 1000 e^- independent of charge packet size. Proportional loss (also called fractional loss) is proportional to the size of the charge packet transferred. Unlike fixed loss, CTE is constant when proportional loss dominates because the traps seen by the charge packet increase proportionally to its size. Proportional loss is characteristic of bulk- and radiation-induced traps, single electron traps that uniformly occupy the signal channel.

For very small charge packets, fixed CTE loss can also apply to bulk and radiation traps. This behavior is caused by the "flat bottom well effect," where the number of bulk traps seen by the charge packet is nearly constant and independent of signal level. Only when the potential well collapses significantly will new traps be open for trapping. Figure 5.51 shows this effect by plotting deferred charge loss as a function of signal level for a bulk-state-limited CCD using the EPER method. Note that for signals > 20,000 e^- deferred charge increases proportionally with signal and follows a CTE slope of 0.999998. Below 20,000 e^- CTE rapidly decreases due to the flat bottom well effect. Figure 5.27 shows similar characteristics.

The overall CTE when proportional and fixed losses are combined is

$$\text{CTE}_{NET} = 1 - \frac{S_{PL} + S_{FL}}{S(e^-)N_P}, \tag{5.29}$$

where S_{PL} is the deferred charge due to proportional loss (e^-) and S_{FL} is fixed loss (e^-). This equation can be written as

$$\text{CTE}_{NET} = \text{CTE}_{PL} - \frac{S_{FL}}{S(e^-)N_P}, \tag{5.30}$$

where CTE_{PL} is proportional CTE loss.

Example 5.19

Estimate CTE performance for the CCD data presented in Fig. 5.51 for a signal level of 1625 e^-. Assume $S_{FL} = 13$ e^-.

CHARGE TRANSFER

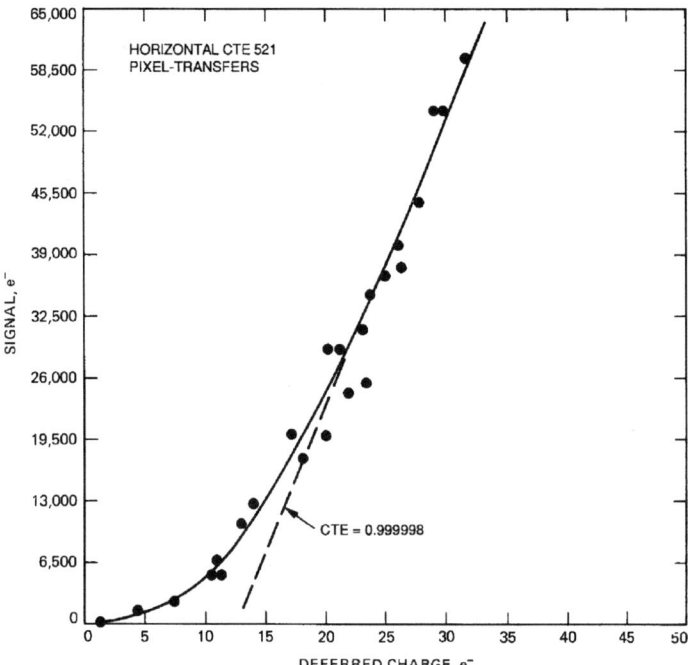

Figure 5.51 Deferred charge as a function of signal showing fixed and proportional loss regimes.

Solution:
From Eq. (5.30),

$$\text{CTE}_{NET} = 0.999998 - 13/(1625 \times 521),$$

$$\text{CTE}_{NET} = 0.9999826.$$

Figure 5.52(a) shows a vertical x-ray transfer plot for a proton-damaged CCD stimulated with a variety of x-ray line sources. X rays were generated by fluorescing a target made of Al (1.48 keV, 407 e^-), S (2.3 keV, 632 e^-), Cl (2.62 keV, 718 e^-), Ca (3.69 keV, 1011 e^-), Ti (4.5 keV, 1235 e^-) and Mn (5.9 keV, 1620 e^-) with Mn x rays. CTI is estimated by measuring the tilt of each single-pixel-event line. These results are plotted in Fig. 5.52(b) as a function of x-ray charge packet size (e^-). Note that CTI decreases with signal level, indicating fixed charge loss is involved.

Figure 5.53 shows a pocket-pumping response generated by a 1024 × 1024 CCD when stimulated with a ramp light stimulus that covers a signal range of 1000 e^- to 18,000 e^-. The band of dots about the response shows the presence

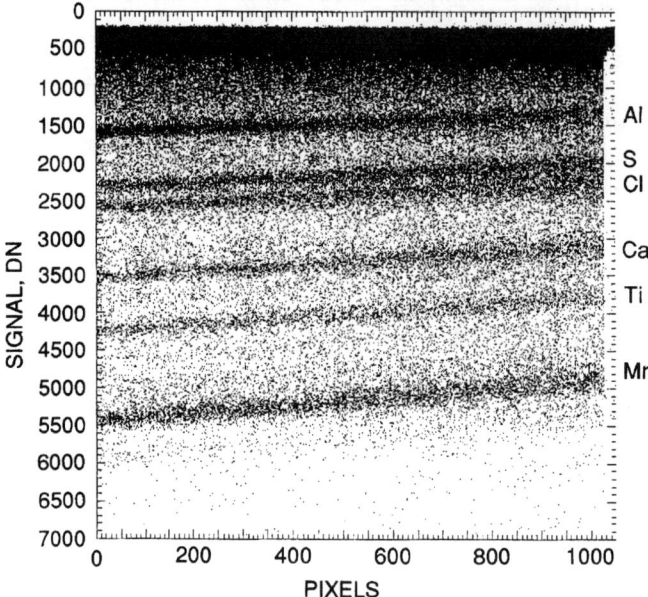

Figure 5.52(a) Vertical x-ray transfer response used to measure CTE as a function of signal.

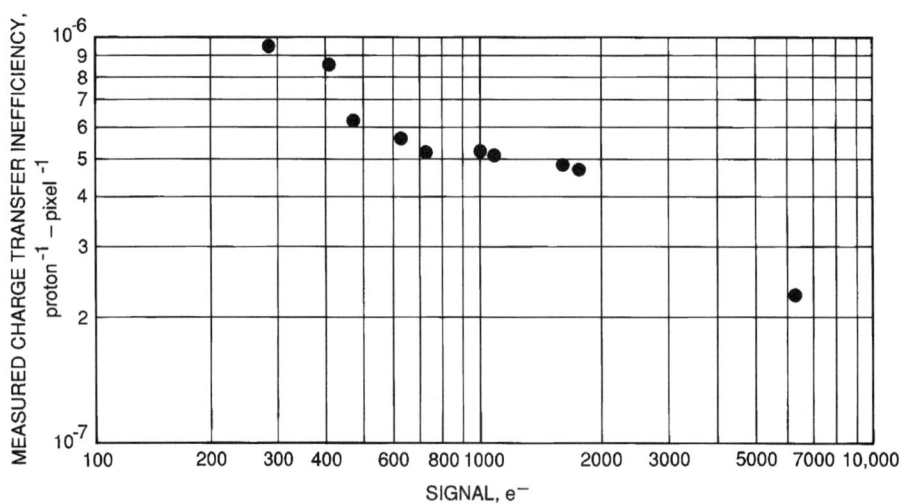

Figure 5.52(b) Vertical CTI as a function of signal using data shown in Fig. 5.52(a).

of single-electron bulk traps. Counting the traps shows that the density of traps is fairly constant up to the point where trap count increases proportionally with signal level.

CHARGE TRANSFER

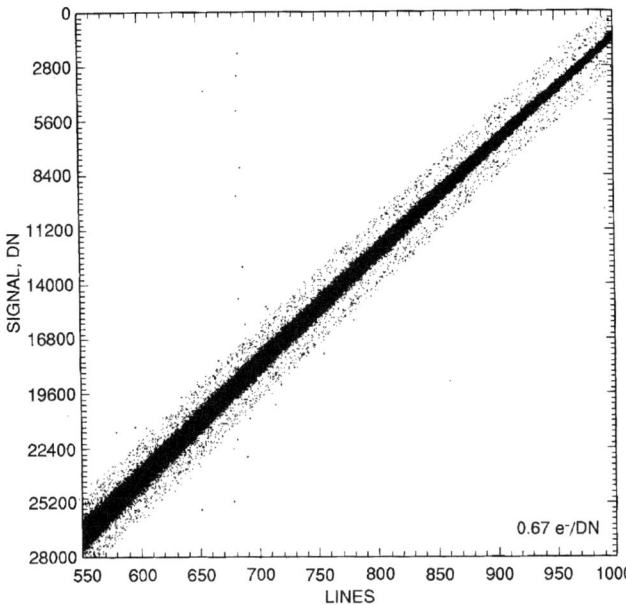

Figure 5.53 Pocket-pumping response for a ramp stimulus showing that bulk trap density increases with signal.

5.4.6 FAT-ZERO

"Fat-zero" is employed as a remedy to rectify some CTE difficulties when design or process traps are present. For example, the Hubble design trap described in Fig. 5.31 was cured by introducing 100 e^- of optical fat-zero. Fat-zero keeps the trap satisfied, which in turn allows signal charge to transfer through the region unimpeded without significant deferred loss. However, the penalty paid for injecting fat-zero is the addition of shot noise to the read noise floor (10 e^- in the case of Hubble). Fat-zero is usually applied optically by a flat-field source. However, fat-zero charge can also be injected electrically using input diode and gate structures (refer to Chapter 1 on charge injection). Although a very popular CTE fix in the past, fat-zero is rarely required today.

Fat-zero does not significantly improve CTE performance for bulk or radiation trap problems (i.e., single-electron traps that are uniformly distributed throughout the signal channel). This is because fat-zero will increase the volume that the charge packet occupies in the potential well, opening up new traps (i.e., CTE is proportional). Figure 5.54 shows the ineffectiveness of adding fat-zero to a radiation-damaged 1024 × 1024 CCD. Figure 5.54(a) presents an Fe^{55} x-ray transfer response without fat-zero, showing a vertical CTE = 0.9999. In Fig. 5.54(b) 300 e^- of fat-zero is injected optically. In comparing the two figures the single-pixel-event line has nearly the same tilt with or without fat-zero. Also note that the extended line region shows a higher noise level because of the shot noise generated by fat-

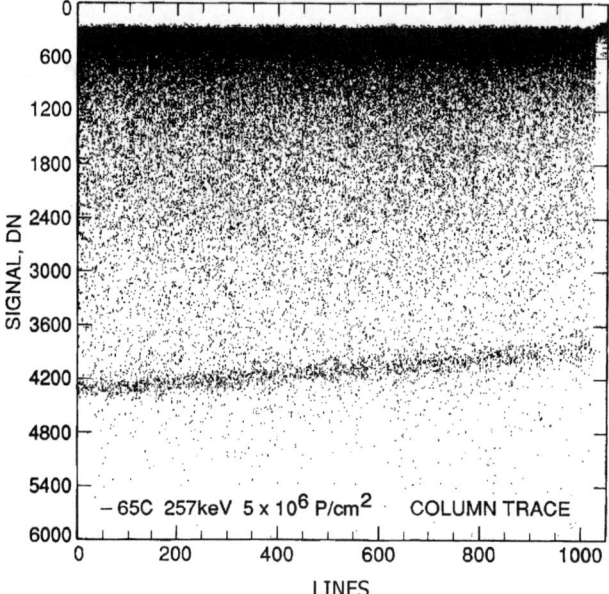

Figure 5.54(a) X-ray transfer plot without fat-zero.

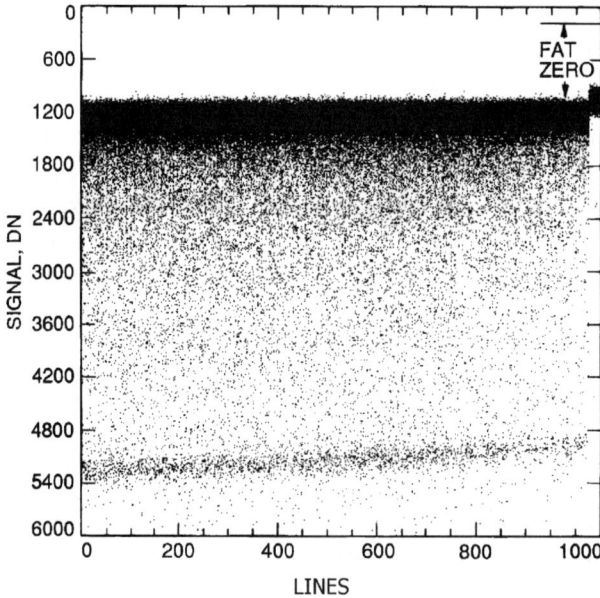

Figure 5.54(b) X-ray transfer plot with fat-zero.

zero (from 5 e$^-$ rms to 17 e$^-$ rms). This example shows that adding fat-zero only degrades sensor performance.

CHARGE TRANSFER 471

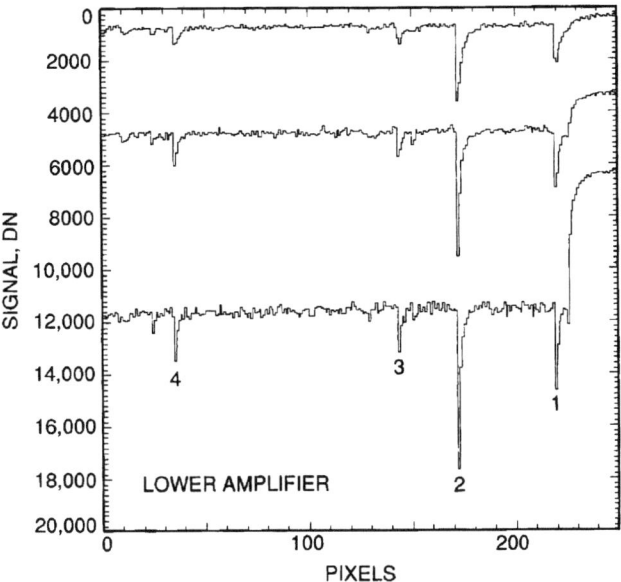

Figure 5.55(a) Deferred charge tails as a function of fat-zero level.

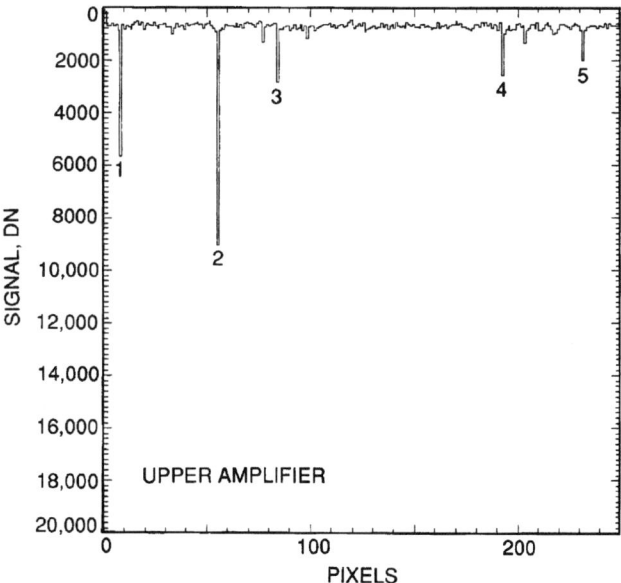

Figure 5.55(b) Dark spikes as read by upper amplifier.

Figure 5.55(a) shows column traces taken from a CCD damaged by 2-MeV neutrons. The traces shown were taken near the top of the array and readout by the lower on-chip amplifier to maximize CTE degradation. Deferred charge tails are

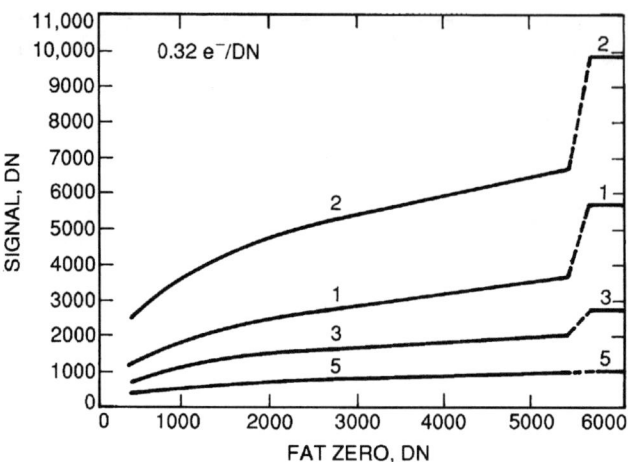

Figure 5.55(c) Dark spike signal as a function of fat-zero.

seen following each dark spike (charge moves from left to right). The top trace was taken without fat-zero whereas the two bottom traces include optical fat-zero levels of 300 and 600 e^-. The true amplitudes of the dark spikes, without significant CTE loss, are shown in Fig. 5.55(b) read by the sensor's upper output amplifier. Although the amplitude of the dark spikes increases slightly with the addition of fat-zero the difference is not significant. Figure 5.55(c) plots dark spike amplitude as a function of fat-zero when transferring spikes to the lower amplifier. Also, shown to the right of the plot is the true amplitude of each spike based on upper amplifier results. Small spikes benefit more from fat-zero than the larger spikes because of the flat bottom well effect discussed above.

5.4.7 NOTCH CHANNEL CCD

Notch channel technology is employed to reduce charge interaction with traps in the buried channel.[11] A notch channel is formed by doping a narrow region within the main signal channel typically with phosphorus. This arrangement is shown in Fig. 5.56. The supernotch, also shown, is the main buried-channel implant. It is sometimes used to isolate the signal channel from the channel stop regions (e.g., refer to Chapter 8 where channel isolation is advantageous in solving radiation damage problems). The small notch that is formed is approximately 2 V deeper than the normal channel potential. With notch technology employed, small charge packets are confined to the narrow notch and encounter fewer traps compared to transferring in the normal wider channel.

Figure 5.57 shows a modeling result for a 3-μm notch created by doping the channel with phosphorus at a concentration of 3×10^{11} cm^{-2} into the main channel doping of 1.6×10^{12} cm^{-2}. For this doping level the potential depth of the notch is approximately 1.5 V. Full well for this notch limits at approximately 12,000 e^-, at

CHARGE TRANSFER

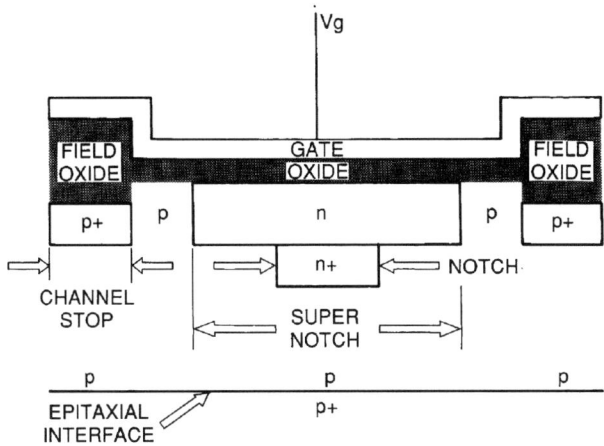

Figure 5.56 Notch channel shown in the middle of the main signal channel.

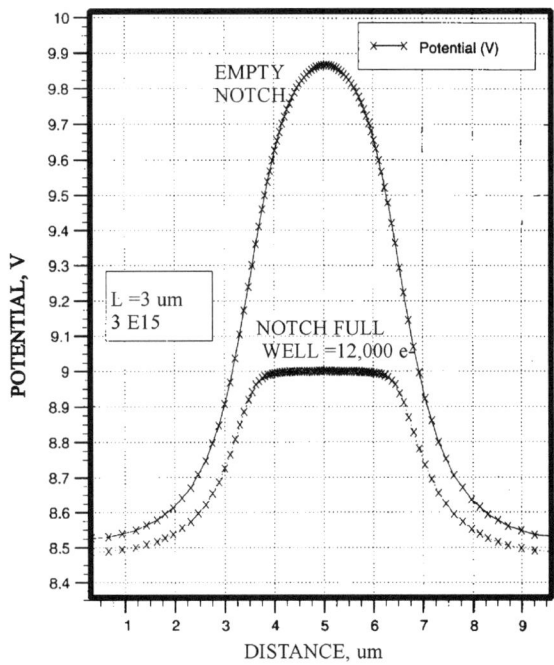

Figure 5.57 Potential plot for a 3-μm notch with and without signal charge.

which point electrons escape into the main channel because of kT.[12,13] Note that fringing fields extend entirely across the notch region limiting the potential depth.

Figure 5.58 shows vertical and horizontal 3-μm notches for a test 12-μm-pixel Cassini CCD. The vertical notch starts at the lower array transfer gate and extends upward to the upper array transfer gate. We end the notch channel at the transfer

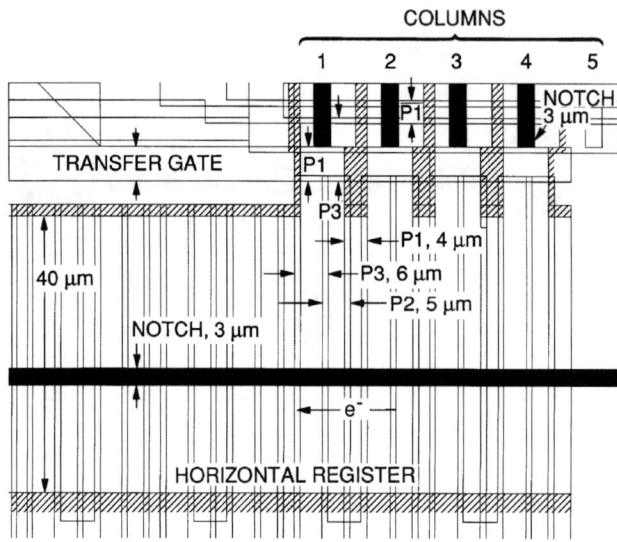

Figure 5.58 Horizontal and vertical registers that employ 3-μm notch channels.

gate to avoid potential pockets that could occur in the interface region. The first experimental notch CCDs fabricated incorporated notches in half the array and the lower horizontal register (the upper register was notchless). This was done to compare CTE performance with and without notch. For a 40-μm-wide horizontal register, a 3-μm notch in theory should improve horizontal CTE performance by a factor of 13 (i.e., 40/3) for small charge packets. The vertical registers would exhibit less benefit because the channel width for the vertical registers is smaller (12 μm vs 40 μm). Channel stops for the CCD occupy approximately 5 μm after processing; therefore, a CTE improvement of only 2.3 (7/3) is expected for these registers.

Figure 5.59 presents horizontal EPERs for the notched CCD described above. The sensor was purposely damaged with alpha particles to show the advantage of a notched CCD. The EPER plot shown in Fig. 5.59(a) is for the notched lower horizontal register, whereas Fig. 5.59(b) is an EPER for the upper unnotched register. The lower register shows less deferred charge compared to the upper register. Figure 5.59(c) plots deferred charge for each horizontal register as a function of signal level. For low signals the notched register is an order of magnitude better than the unnotched register. Note that the notch reaches full well at approximately 10,000 e^-. Beyond this signal level the deferred charge level increases rapidly as electrons interact with traps in the main channel.

Notch is advantageous for all CCD applications that anticipate bulk and radiation trapping problems. It is usually customary to employ notch channels in large CCD arrays (> 1024 × 1204). CTEs as high as 0.9999995 have been measured for bulk-state-limited notch CCDs. For example, Fig. 5.60(a) shows an x-ray transfer plot for a 1024 × 1024 notched CCD that exhibits excellent CTE performance.

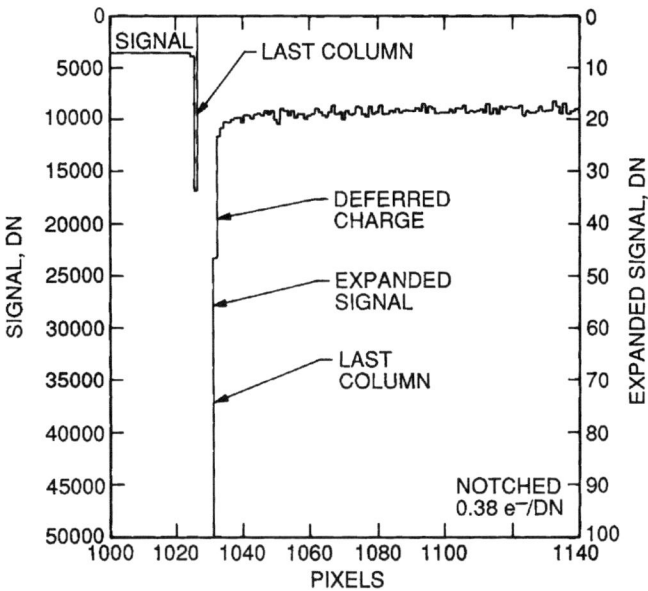

Figure 5.59(a) EPER for a notched horizontal register.

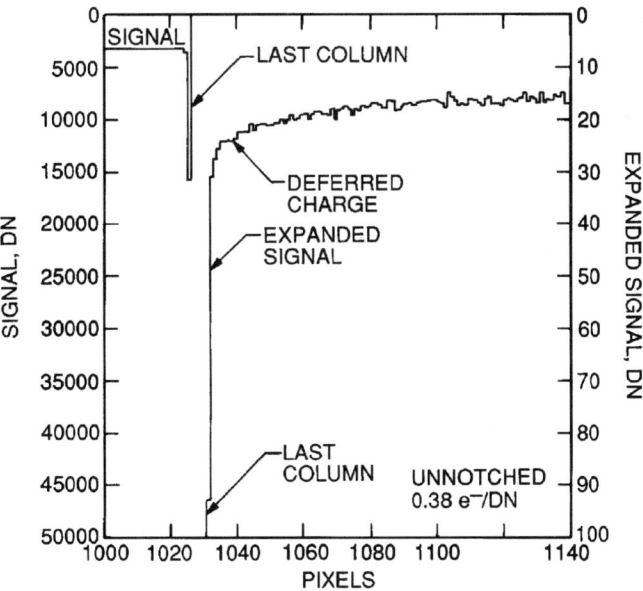

Figure 5.59(b) EPER for an unnotched horizontal register.

Figure 5.60(b) is a corresponding EPER response which shows approximately 115 extended pixels that follow the last column of the array. The DN values shown can be converted to electrons by using a gain constant of 0.32 e$^-$/DN. The signal

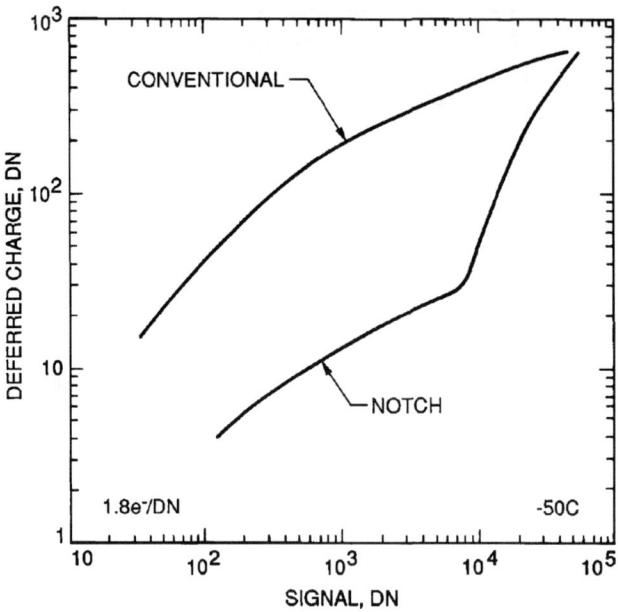

Figure 5.59(c) Deferred charge as a function of signal for a notched and unnotched CCD.

contained in the last column is approximately 1600 e^- with a read noise in the extended pixel region of 0.15 e^- rms (1.2 e^- peak to peak). The magnified trace exhibits no sign of deferred charge, demonstrating very high CTE performance. A CTE of 0.9999993 would be calculated if 1 e^- of deferred charge is assumed. CTE performance at this level is very difficult to measure.

As indicated in Chapter 4, additional phosphorus to the buried channel increases full well performance. Therefore, notch CCDs will exhibit higher charge capacity compared with a notchless device. However, the effective threshold also increases, requiring greater clock drive. Figure 5.61 shows two shutterless photon transfer curves for a very deep (6 V) notch CCD relative to a notchless CCD. The charge capacity is double that compared to the notchless device (this data opened the door for higher full well sensors by increasing the overall doping concentration of the buried channel).

5.5 Transfer Power

As the CCD transfers charge it consumes power.[14] For slow-scan operation the power dissipated is small except for the power generated by the on-chip amplifier (we will discuss this source of power in Chapter 6). For example, the power taken by the WF/PC CCDs amounts to less than 10 mW, most of which is supplied to the output amplifier. Although small, this level of power was still important because

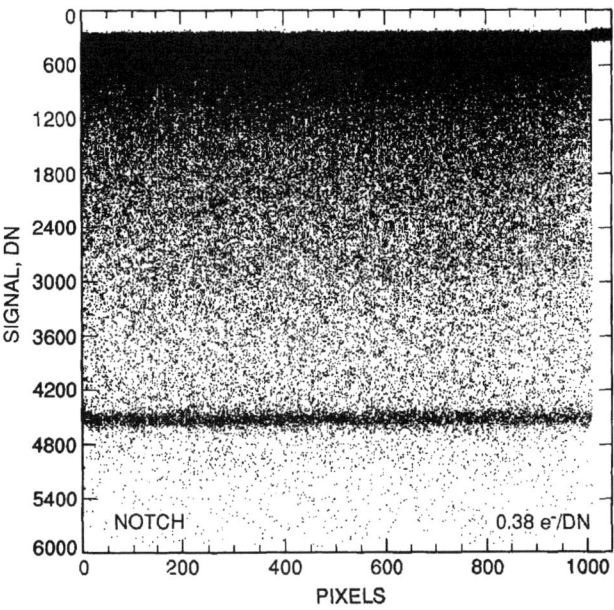

Figure 5.60(a) X-ray transfer response for a notched CCD.

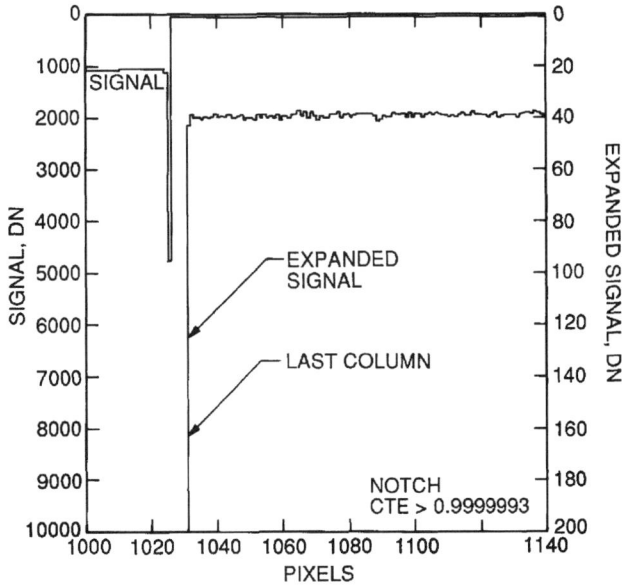

Figure 5.60(b) EPER for a bulk-state-limited notched CCD exhibiting CTE > 0.999999.

the WF/PC CCDs are cooled to $-100°C$ with thermoelectric coolers, which are not efficient cooling devices.

There are many mechanisms internal to the CCD responsible for power dissipation. Discussions below focus on three main sources, which are referred to as 1) charge motion power, 2) potential power and 3) reactive power.

5.5.1 Charge Motion Power

Power is dissipated by friction as electrons move through the signal channel from pixel to pixel. This power is equal to[15]

$$P_F = \frac{nq(Lf_c)^2}{\mu_{SI}}, \qquad (5.31)$$

where P_F is the frictional lattice power (W), L is the length of a pixel (cm), n is the number of electrons transferred, f_c is the clock frequency (Hz) and μ_{SI} is the mobility of electrons within the buried-channel region (cm^2/V-sec).

We will run into the mobility term several times in this book, and, therefore, it deserves some physical explanation. When an electric field is present, an electron will be accelerated by Newton's second law as

$$a = \frac{qE}{m_e^*}, \qquad (5.32)$$

where E is electric field (V/cm) and m_e^* is the effective mass of the electron. The effective mass is not the real mass, but a quantity given to the electron that describes its behavior when moving in the silicon lattice, and thus is variable depending on doping characteristics.

As the electron drifts, its motion is interrupted by collisions yielding an average drift velocity of

$$v_d = \frac{qEt_C}{m_e^*}, \qquad (5.33)$$

where t_C is the time interval between collisions. The drift velocity is dependent on the applied electric field until it becomes comparable to the thermal velocity of the electrons, which is about 10^7 cm/sec. At this point the velocity saturates.

Electrons will collide and scatter with ionized impurity atoms or with the lattice atoms, both of which impede the carriers movement. Mobility quantifies this process and is defined as

$$\mu_{SI} = \frac{v_d}{E} \qquad (5.34)$$

or equivalently using Eq. (5.33)

$$\mu_{SI} = \frac{qt_C}{m_e^*}. \qquad (5.35)$$

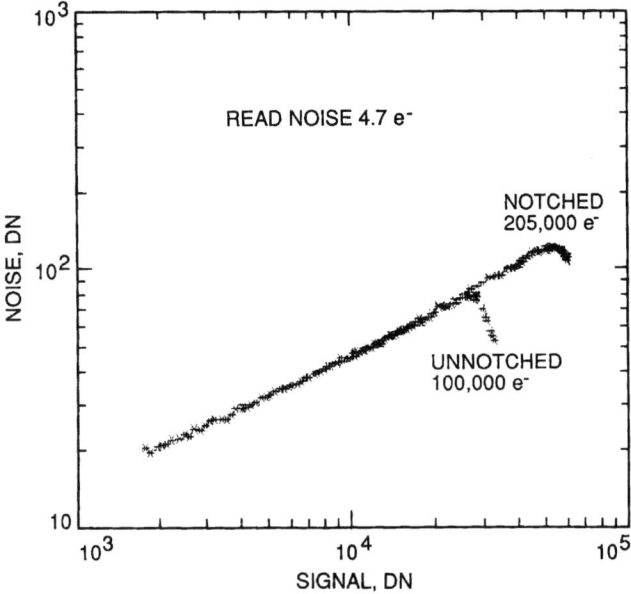

Figure 5.61 Photon transfer curves for a notched and unnotched CCD showing full well improvement.

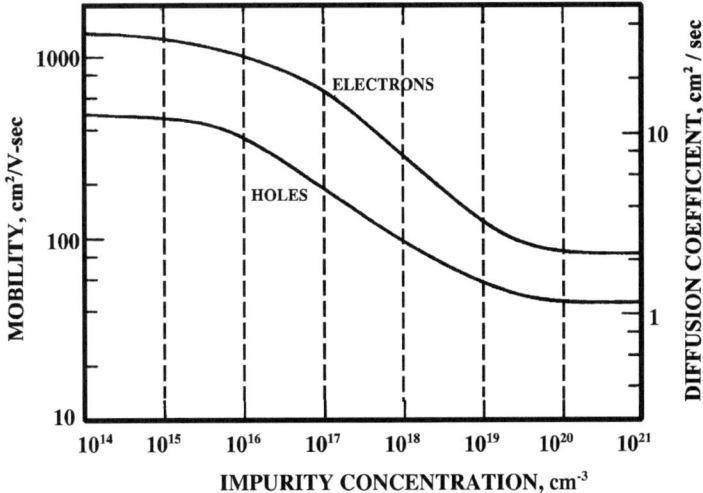

Figure 5.62 Electron and hole mobility as a function of impurity concentration.

Note the unit of mobility is cm^2/V-sec. Figure 5.62 plots electron and hole mobility as a function of doping.[16] For low doping, mobility is dependent primarily on lattice scattering. As the impurity concentration increases, mobility and resistivity decrease.

Example 5.20

Estimate the frictional power required to move 10^6 e^- from one pixel to the next at a 10^7 Hz rate. Assume a 15-μm pixel and a mobility of 1350 cm^2/V-sec.

Solution:
From Eq. (5.31),

$$P_F = (1.6 \times 10^{-19}) 10^6 [(15 \times 10^{-4})(10^7)]^2 / 1350,$$

$$P_F = 2.6 \times 10^{-8} \text{ W (26 nW)}.$$

Compared to other power effects at this speed, electron motion power can usually be neglected.

Power is also dissipated when the p-depletion region expands and collapses as holes move in and out of the device. This power is generated in the epitaxial layer between the depletion region and epitaxial interface. It is estimated by

$$P_{EPI} = \frac{2q \Delta x_p A_{PIX} N_A (x_{ff} f_c)^2}{\mu_{SI-h}}, \quad (5.36)$$

where P_{EPI} is the frictional power generated by a single pixel (W/pixel), Δx_p is the change in depletion depth caused by clocking (cm), A_{PIX} is the area of a pixel (cm^2), N_A is the doping concentration of the epitaxial layer (cm^{-3}), x_{ff} is the field-free distance between the depletion edge and epitaxial interface (cm), and μ_{SI-h} is the mobility of holes within the epitaxial layer (cm^2/V-sec). The factor of 2 is due to the power being generated on both the rising and falling edges of the clock.

Example 5.21

Estimate the power generated in the epitaxial region, assuming

$\Delta x_p = 4 \times 10^{-4}$ cm,
$A_{PIX} = 1.44 \times 10^{-6}$ cm^2 (12-μm pixel),
$N_A = 10^{15}$ cm^{-3},
$x_{ff} = 5 \times 10^{-4}$ cm,
$\mu_{SI-h} = 500$ cm^2/V-sec,
$f_c = 1 \times 10^6$ Hz.

Also calculate the total epitaxial power for a 1000 × 1000 CCD.

CHARGE TRANSFER 481

Solution:
From Eq. (5.36),

$$P_{EPI} = \left\{ 2 \times \left(1.6 \times 10^{-19}\right) \times \left(5 \times 10^{-4}\right) \times \left(1.44 \times 10^{-6}\right) \right.$$
$$\left. \times 10^{15} \times \left[\left(5 \times 10^{-4}\right) \times 10^6\right]^2 \right\} / 500,$$
$$P_{EPI} = 1.15 \times 10^{-10} \text{ W/pixel}.$$

Total power for a 1000×1000 CCD is 1.15×10^{-4} W.

This example shows that epitaxial power is usually negligible compared to other power dissipaters on chip. Note there is also a finite amount of power generated in the substrate layer. The power equation for this region is similar to Eq. (5.36) above except the mobility is considerably lower because it is doped much greater (0.01 Ω-cm is typical). However, the distance the holes must travel is also longer (approximately 500 μm). Therefore, the power dissipation for the two regions is about the same.

Equation (5.36) assumes noninverted operation. For inverted clocking, additional power is dissipated as holes come and go from the channel stops. This power is equal to

$$P_{INV} = \frac{2q A_{PIX} C_{OX} (V_{GL} - V_{INV})(f_c d_h)^2}{\mu_{SI-h}} \quad (5.37)$$

where P_{INV} is the power generated by the inversion layer (W/pixel), C_{OX} is the gate oxide capacitance (F/cm^2), $(V_{GL} - V_{INV})$ is the drive voltage past inversion (V) and d_h is the average distance the holes travel from channel stops (cm).

Example 5.22

Estimate the additional power generated for the pixel described in Example 5.21 when the device is clocked into inversion. Assume $(V_{GL} - V_{INV}) = 4$ V, $C_{OX} = 3.45 \times 10^{-8}$ F/cm^2 and an average distance of 3 μm that holes travel to and from the channel stops.

Solution:
From Eq. (5.37),

$$P_{INV} = \left\{ 2 \times \left(1.6 \times 10^{-19}\right) \times \left(12 \times 10^{-4}\right)^2 \times \left(3.45 \times 10^{-8}\right) \right.$$
$$\left. \times 4 \times \left[10^6 \times \left(3 \times 10^{-4}\right)\right]^2 \right\} / 500,$$
$$P_{INV} = 3.58 \times 10^{-11} \text{ W/pixel}.$$

Total power for a 1000×1000 CCD is 3.58×10^{-5} W.

This example shows that inverted power is also usually negligible. Power is also dissipated when holes move in the channel stops. This power source can be important to backside-illuminated CCDs unless the channel stops are strapped.

5.5.2 POTENTIAL POWER

Clocks must sometimes deliver power to move electrons from one potential height to another. For example, when the collecting phase switches into barrier phase, it is possible that an electron is elevated over a potential difference before it finally transfers. Power consumption to do this operation is[17]

$$P_\Delta = nqNf_c \Delta V_H, \tag{5.38}$$

where P_Δ is the power required to lift n electrons over a barrier height ΔV_H.

For multiphase CCDs this power requirement is much smaller than Eq. (1.53) implies, because charge starts to move before the potential difference between phases becomes large. The situation is different for two-phase CCDs where a potential barrier must be overcome before a small charge packet can move over the barrier region. Nevertheless, power consumption in moving charge over any barrier height is usually negligible.

Example 5.23

Estimate the power required to lift 1 e⁻ through a potential change of 10 V in a three-phase cycle at a 10^7 Hz rate. Assume the electron stays under a phase until the phase has completely switched from a collecting to barrier phase.

Solution:
From Eq. (5.38),

$$P_\Delta = 2 \times (1.6 \times 10^{-19}) \times 10 \times 10^7,$$

$$P_\Delta = 3.2 \times 10^{-11} \text{ W}.$$

5.5.3 REACTIVE POWER

Large displacement currents flow in charging and discharging the capacitance associated with the gate electrodes. This charging mechanism usually generates most of the power in a CCD (excluding the on-chip amplifiers). Reactive power is given by

$$P_R = N_{PH} C_{PIX} \Delta V_{CB}^2 f_c N_{PIX}, \tag{5.39}$$

CHARGE TRANSFER

where P_R is the reactive pixel power (W), N_{PH} is the number of phases, N_{PIX} is the total number of pixels being clocked, V_{CB} is the gate potential difference between the collecting and barrier phases (V) and C_{PIX} is the pixel capacitance (F/pixel/phase). Frequency, f_c, will be specified below as the average lines/sec for the vertical register or average pixels/sec for the horizontal register.

Figure 5.63 shows the various capacitances related to a pixel: (1) the poly gate series oxide (C_{OX}) and depletion capacitance (C_{DEP}), (2) the poly-to-field oxide capacitance (C_F) and (3) the poly-to-poly overlap (C_{OL}) and sidewall gate capacitance (C_{SW}) (also refer to Fig 1.14). The net sum of these capacitors represents the drive capacitance, C_{PIX}, as

$$C_{PIX} = C_T + C_F + C_{OV}, \tag{5.40}$$

where the series oxide and depletion capacitance, C_T, is

$$C_T = L(W - x_{CS})\left(\frac{d}{\varepsilon_{OX}} + \frac{x_d}{\varepsilon_{SI}}\right)^{-1}, \tag{5.41}$$

where W is the width of the phase (equal to pixel pitch [cm]), L is the length of a phase (cm), x_d is the depletion depth and x_{CS} is the width of the channel stop (cm).

The field oxide is

$$C_F = \frac{x_{CS}\varepsilon_{OX}L}{d_F}, \tag{5.42}$$

where d_F is the thickness of the field oxide (cm).

The overlap and associated sidewall capacitance between phases usually represents the largest capacitance and is minimized as much as possible for high-speed CCDs. The overlap dimension varies considerably depending on the manufacturer's design rules (from 0.25 to 2.5 µm). Most CCDs tested in this book are based on a 1-µm phase overlap. The poly-to-poly overlap capacitance, including its sidewall, is

$$C_{OV} = \frac{2\varepsilon_{OX}W(x_{OV} + d_{POLY})}{d_{OX}}, \tag{5.43}$$

where x_{OV} is the phase overlap length (cm), d_{OX} is thickness of the oxide between phases (cm) and d_{POLY} is the poly thickness (cm). The first term represents the overlap capacitance whereas the second term is the sidewall capacitance.

Example 5.24

Estimate the individual capacitances shown in Fig. 5.63 for a 16-µm three-phase pixel with the following parameters:

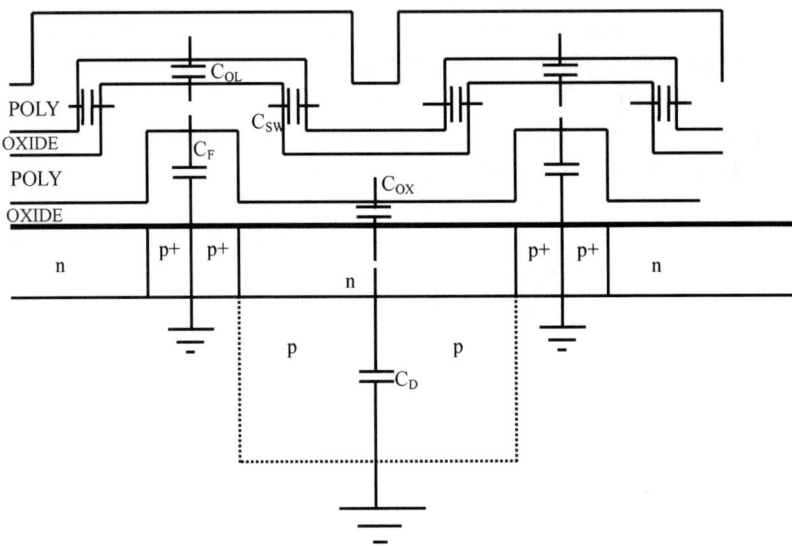

Figure 5.63 Pixel drive capacitances.

$W =$ width of a phase $= 16 \times 10^{-4}$ cm,
$L =$ length of a phase/pixel $= 5.3 \times 10^{-4}$ cm,
$x_{CS} =$ channel stop width $= 4.0 \times 10^{-4}$ cm,
$x_d =$ depletion depth $= 7 \times 10^{-4}$ cm,
$d =$ gate oxide thickness $= 1 \times 10^{-5}$ cm,
x_{OV} is the phase overlap $= 1 \times 10^{-4}$ cm,
d_{POLY} is poly thickness $= 5 \times 10^{-5}$ cm,
$d_{OX} =$ oxide thickness between poly gates $= 1 \times 10^{-5}$ cm,
$d_F =$ field oxide thickness $= 10^{-4}$ cm.

Depletion and oxide capacitance
From Eq. (5.41),

$$C_T = (5.3 \times 10^{-4})[(16 \times 10^{-4}) - (4 \times 10^{-4})][(1 \times 10^{-5})/ (3.45 \times 10^{-13}) + (7 \times 10^{-4})/(1.04 \times 10^{-12})]^{-1},$$

$C_T = 9.06 \times 10^{-16}$ F/pixel/phase.

Field capacitance
From Eq. (5.42),

$$C_F = (4 \times 10^{-4}) \times (3.45 \times 10^{-13}) \times (5.3 \times 10^{-4})/10^{-4},$$

$C_F = 7.31 \times 10^{-16}$ F/phase/pixel.

Overlap and sidewall capacitance
From Eq. (5.43),

$$C_{OV} = 2 \times (3.45 \times 10^{-13}) \times (16 \times 10^{-4}) \times [(1 \times 10^{-4}) + (5 \times 10^{-5})]/(1 \times 10^{-5}),$$

$$C_{OV} = 1.66 \times 10^{-14} \text{ F/pixel/phase}.$$

The total drive capacitance from Eq. (5.40),

$$C_{PIX} = 9.06 \times 10^{-16} + 7.31 \times 10^{-16} + 1.66 \times 10^{-14},$$

$$C_{PIX} = 1.83 \times 10^{-14} \text{ F/pixel/phase (18.3 fF/pixel/phase)}.$$

The calculations above were verified experimentally for a 512(V) × 1024(H) × 16-μm-pixel three-phase CCD. Drive capacitance was measured by placing a series resistor between a clock driver and a CCD phase (say, phase-1). Phases-2 and -3 are grounded in the measurement to eliminate clock feed-through from these lines. The RC time constant of the drive pulse is then measured at the input of the CCD. For example, the CCD tested exhibited a time constant of 20 μsec for a 2-kΩ resistor. The total drive capacitance for phase-1 is therefore 10^{-8} F (i.e., $20 \times 10^{-6}/2 \times 10^3$). The capacitance per pixel is 19 fF/pixel/phase, in good agreement with the calculations performed above. It should be mentioned that this measurement is best done on strapped CCDs without poly resistance. Capacitive measurements taken on large arrays that aren't strapped do not give good comparative results.

Example 5.25

Find the vertical register reactive power for a split-frame 4096 × 4096 × 16-μm CCD running at 60 frames per second. Assume a pixel drive capacitance in Example 5.24 and a clock swing of 17 V. Also find the power for a frame-transfer CCD.

Solution:
Frame Transfer Power
This power is dissipated when charge is transferred 1024 lines from the imaging to the storage regions.
From Eq. (5.39),

$$P_R = 3 \times (1.83 \times 10^{-14}) \times 17^2 \times (1024 \times 60) \times 4096^2,$$

$$P_R = 16.4 \text{ W}.$$

Frame Readout Power
This power is dissipated when the storage region is readout.

$$P_R = 3 \times (1.83 \times 10^{-14}) \times 17^2 \times 1024 \times 60 \times 2048 \times 4096,$$

$$P_R = 8.2 \text{ W}.$$

Note that the frame readout power is $1/2$ that of the frame transfer power because only half the array is being clocked (i.e., the imaging region is static during readout).

The total power dissipated is $16.4 + 8.2 = 24.6$ W.

Note also that split frame transfer is half the power compared to a frame transfer architecture for both transfer operations (i.e., frame transfer and frame readout). The total power for a frame transfer CCD is $32.8 + 16.4 = 49.20$ W.

The above calculation assumes that the CCD driver output impedance is negligible compared to the strapping resistance. In this case all power is dissipated on-chip. If the driver resistance is not negligible then some of this power would be generated off-chip. In theory all power can be generated off-chip if the strap impedance is truly zero ohms. In that case power would be dissipated in the driver circuit. However, CCD bus resistance is always finite.

REFERENCES

1. D. Buss, A. Tasch and J. Barton, "Applications to signal processing," In *Charge-coupled Devices and Systems*, M. Howes and D. Morgan, eds., Chapter 3, pp. 105–107, John Wiley and Sons (1979).
2. E. Banghart, J. Lavine, E. Trabka, E. Nelson and B. Burkey, "A model for charge transfer in buried-channel charge-coupled devices at low temperature," *IEEE Trans. Electron Device* Vol. 38, No. 5 (1991).
3. J. Carnes, W. Kosonocky and E. Ramberg, "Free charge transfer in charge-coupled devices," *IEEE Trans. Electron Devices* Vol. ED-19, No. 6, pp. 798–808 (1972).
4. D. Barbe, "Imaging devices using the charge-coupled concept," *Proc. IEEE* Vol. 63, pp. 38–67 (1975).
5. J. Esser, "The peristaltic charge-coupled device for high speed charge transfer," *1974 IEEE International Solid-State Circuits Conference of Technical Papers*, pp. 38–29 (1974).
6. J. Janesick, T. Elliott, R. Bredthauer, C. Chandler and B. Burke, "Fano-noise-limited CCDs," *Proc. SPIE* Vol. 982, pp. 70–85 (1988).
7. J. Janesick, "Locating electron traps in a CCD," *NASA Tech. Briefs* Vol. 17 No. 9, pp. 51 (1993).

8. J. Orbock, D. Seawalt, W. Delamere and M. Blouke, "Charge transfer efficiency measurements at low signal levels on STIS/SOHO TK1024 CCD's," *Proc. SPIE* Vol. 1242 (1990).
9. G. Taylor and P. Chatterjee, "An evaluation of submicrometer potential barriers using charge-transfer devices," *IEEE Journal of Solid State Circuits* Vol. SC-18, No. 4 (1980).
10. C.-K. Kim, "Physics of charge coupled devices," In *Charge-Coupled Devices and Systems*, M. Howes and D. Morgan, eds., p. 58, John Wiley and Sons (1979).
11. J. Janesick, "Notch charge-coupled devices," *NASA Tech. Briefs* Vol. 16, No. 7, p. 28 (1992).
12. S. Kawai, N. Mutoh and N. Teranishi, "Thermionic-emission-based barrier height analysis for precise estimation of charge handling capacity in CCD registers," *IEEE Trans. Electron Devices* Vol. 44, No. 10 (1997).
13. J. Lavine, E. Banghart and J. Pimbley, "The escape of particles from a confining potential well," *Mat. Res. Soc. Symp. Proc.* Vol. 290 (1993).
14. T. Graeve and E. Dereniak, "Power dissipation in frame-transfer charge-coupled devices," *Optical Engineering* Vol. 32, No. 35 (1993).
15. R. Strain, "Properties of an idealized traveling wave charge coupled device," *IEEE Trans. Electron Devices* Vol. ED-19, pp. 1119–1130 (1972).
16. E. Conwell, "Properties of silicon and germanium," *Proc. IRE* Vol. 46, p. 1281 (1958).
17. C. Sequin and M. Tompsett, *"Charge coupled devices,"* Academic Press, p. 139 (1975).

CHAPTER 6

CHARGE MEASUREMENT

6.1 CHARGE MEASUREMENT

Charge measurement is the last major operation performed by the CCD. This process is accomplished by dumping signal charge onto a small sense capacitor located at the end of the horizontal register. The capacitor is connected to an output amplifier which delivers a buffered output voltage for each pixel. Other than photon shot noise, previous CCD operations of generating, collecting, and transferring charge are noiseless processes. In theory, the charge packet can arrive at the sense node without uncertainty to the single electron! However, there are a host of noise problems that the user may confront. For example, shot noise from thermal dark current will add uncertainty to the charge packet being measured. However, this unwanted source of noise can be completely eliminated by cooling the detector. In Chapter 7, we will discuss dark current and many other on-chip and off-chip noise sources that disturb the measurement process. We will see that each noise source can be controlled and removed.

This chapter discusses the charge measurement process performed by the CCD output amplifier and off-chip signal processor. Unfortunately, the output amplifier adds noise to the charge packet measured. Although the uncertainty is very small, the noise induced cannot be fundamentally reduced to zero. It is interesting to note that the output amplifier read noise has settled to one electron rms (slow-scan). As far as anyone knows, this unique level has happened by coincidence and is not set by any fundamental law of nature. However, it has taken 30 years and millions of dollars of CCD amplifier development to reduce the noise down to this level. In the last 10 years, amplifier noise has been reduced by no more than a factor of two, showing that the optimization process has slowed down and that major advancements to further enhance charge detection are not likely. Also, the need for lower noise for most CCD applications is not warranted. Recall that a 1 e^- signal will generate a shot noise uncertainty of 1 e^-, yielding a S/N of unity independent of amplifier read noise (an exception to this argument is when point or line sources are measured where lower read noise is always better).

Figure 6.1 shows a scanning electron microscope (SEM) image taken of a WF/PC I CCD output amplifier region. Shown are the horizontal register, output summing well, output transfer gate, sense node (also referred to as the floating diffusion or output diode), reset MOSFET, and output amplifier MOSFET. Figure 6.2 shows a cross-sectional diagram and an electronic schematic of the same region. These figures will be referred to often in this chapter as the charge measurement process is described. The output MOSFET amplifier is the most critical element shown. Many different performance characteristics are associated with the device.

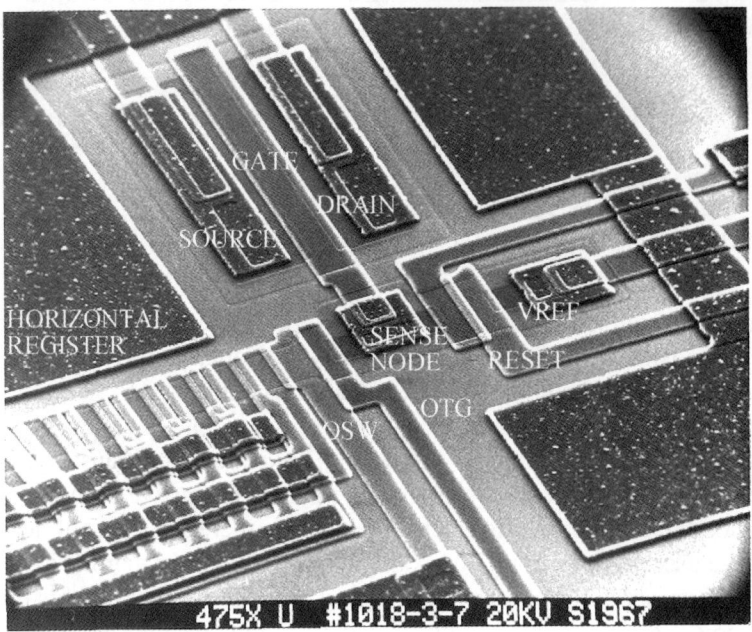

Figure 6.1 SEM image showing the output amplifier region for a Hubble WF/PC I CCD.

Voltage gain, output impedance, frequency response, sensitivity, linearity, power consumption and noise are important areas of discussion below.

6.2 Output Amplifier Characteristics

6.2.1 Operation

This section presents a brief discussion on MOSFET theory and operation. The knowledge gained here will be used to bias the output MOSFET for optimum performance for the charge measurement process.

6.2.1.1 Linear Region

Figure 6.3(a) plots output MOSFET drain current (I_D) as a function of drain-to-source voltage (V_{DS}) for different gate-to-source voltages (V_{GS}). Figure 6.3(b) plots current as a function of gate voltage for different V_{DS} for the same MOSFET. Note in Fig. 6.3(a) that I_D initially increases linearly with V_{DS} as defined by the channel resistance between the drain and source. This operation region is called the "linear region," where the drain current is approximated by

$$I_D = \frac{W}{L} \mu_{SI} C_{OX} (V_{AEFF} + V_{GS}) V_{DS}, \tag{6.1}$$

CHARGE MEASUREMENT

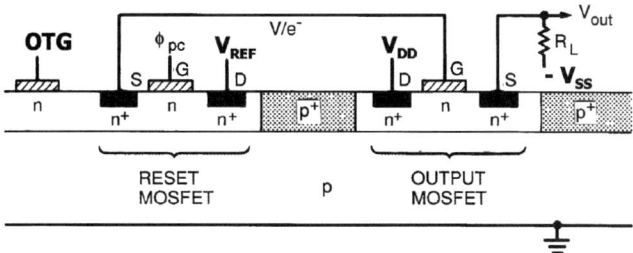

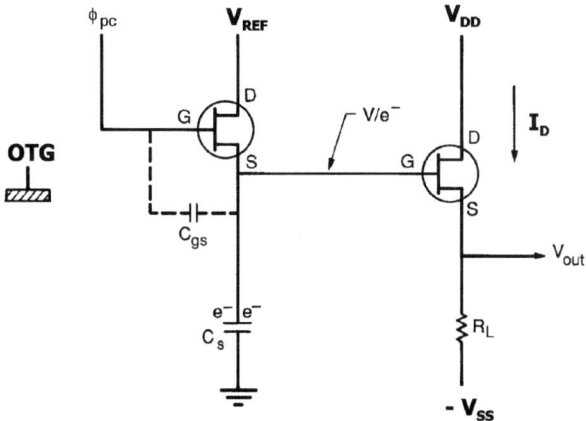

Figure 6.2 Cross-section and electronic schematic of output amplifier region.

where I_D is the drain current (A), W is the width of the gate (cm) and L is the length of the gate (cm), μ_{SI} is the mobility (cm^2/V-sec), C_{OX} is gate capacitance (F/cm^2), V_{GS} is the gate-to-source voltage (V), V_{DS} is the drain-to-source voltage (V), and V_{AEFF} is the effective threshold of the MOSFET that will be defined below. This equation assumes that $V_{DS} \ll V_{AEFF}$.

Example 6.1

Estimate the current within the linear region of a MOSFET with the following characteristics and bias conditions: W = 35 µm, L = 6 µm, $V_{AEFF} + V_{GS} = 8$ V, $C_{OX} = 3.45 \times 10^{-8}$ F/cm^2, $V_{DS} = 1$ V and a mobility of 1000 cm^2/V-sec.

Solution:
From Eq. (6.1),

$$I_D = (35 \times 10^{-4}) \times 1000 \times (3.45 \times 10^{-8}) \times 8 \times 1/(6 \times 10^{-4}),$$

$$I_D = 1.6 \text{ mA}.$$

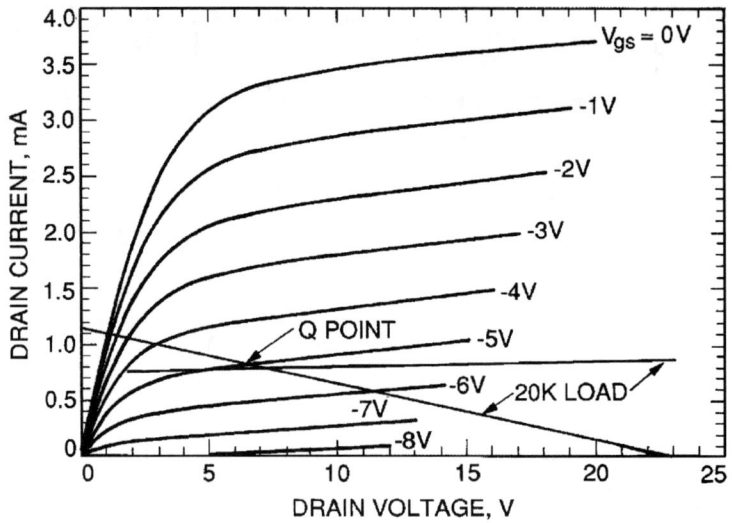

Figure 6.3(a) MOSFET I_D vs. V_{DS} transfer curve.

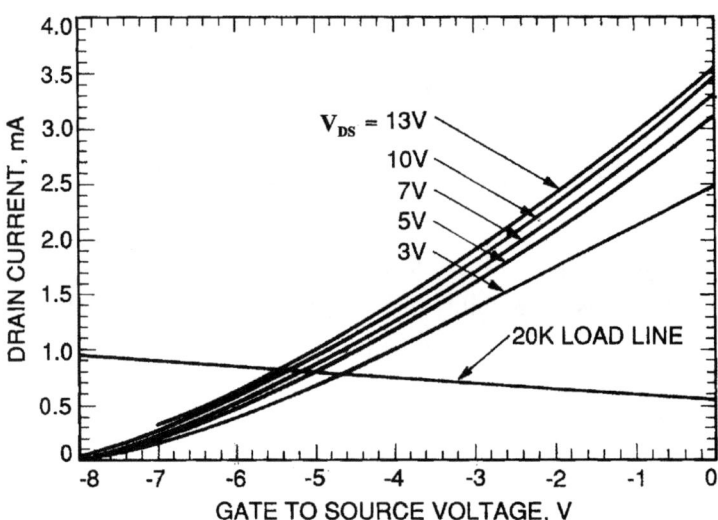

Figure 6.3(b) MOSFET I_D vs. V_{GS} transfer curve.

6.2.1.2 Saturation Region

The depletion region at the drain end of the MOSFET is wider than at the source because of an IR voltage drop that develops across the channel (i.e., the gate voltage relative to the channel is more negative near the drain than at the source, promoting greater depletion in the drain region). At some high V_{DS} the gate and junction depletion regions meet near the drain, a condition called "pinch-off," which is illus-

CHARGE MEASUREMENT 493

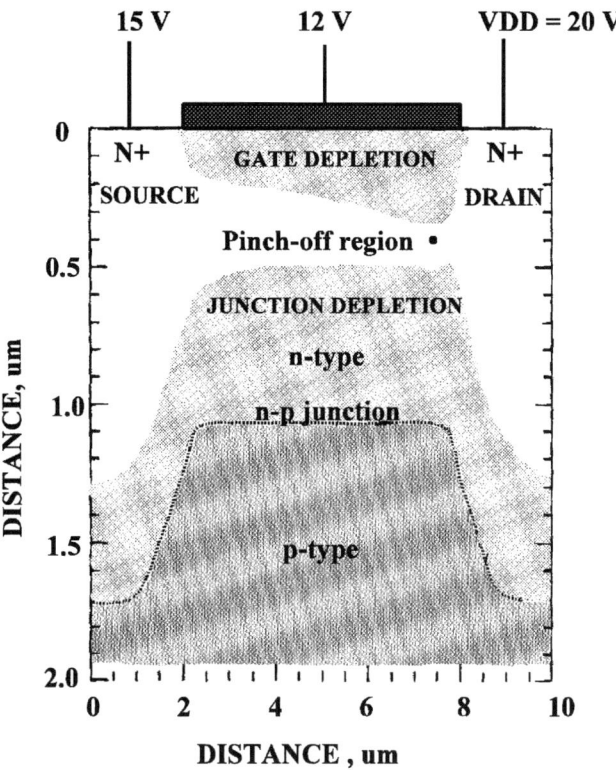

Figure 6.4 Cross-section of output MOSFET showing depletion and pinch-off regions.

trated in Fig. 6.4. When pinch-off occurs, the drain current saturates and becomes constant with V_{DS}. We call this the saturation region, and it is within this operating range the MOSFET is biased for use. The saturation current through the MOSFET is

$$I_{DS} = \frac{W}{2L}\mu_{SI}C_{OX}(V_{AEFF} + V_{GS})^2, \quad (6.2)$$

where I_{DS} is the saturation current (A).

Note from this equation that current varies by the square of gate voltage. It is interesting to note that a current must flow in the MOSFET to sustain the pinch-off condition. Current that flows through depleted material (i.e., the pinch-off region) may be contrary to our understanding about depletion regions where majority carriers are absent. However, carriers still exist within depleted material because of intrinsic carrier concentration. A forward biased diode with a built-in depletion region is a good example of a large current flowing through a depletion region.

Example 6.2

Estimate the saturation current for the MOSFET described in Example 6.1. Assume $V_{AEFF} + V_{GS} = 3$ V.

Solution:
From Eq. (6.2),

$$I_{DS} = (35 \times 10^{-4}) \times 1000 \times (3.45 \times 10^{-8}) \times 3^2 / [(6 \times 10^{-4}) \times 2],$$

$$I_{DS} = 0.9 \text{ mA}.$$

Increasing the drain voltage beyond pinch-off promotes further depletion toward the source. This reduces the resistance between the source and drain, causing the drain current to gradually increase with V_{DS}. This effect causes a slight slope in current as seen in Fig. 6.3(a). The slope becomes more pronounced when the length of the pinch-off region becomes a substantial portion of the conductive channel (i.e., small-gate-length devices as CCDs use).

When V_{GS} is made negative, depletion will come more from the gate side. Therefore, a lower V_{DS} is required to reach the pinch-off state. The saturation current, I_{DS}, is also smaller because there is less conductive volume for current to flow through. A positive V_{GS} will discourage depletion and increase the saturation current level. A higher V_{DS} is also required to reach pinch-off. These characteristics are seen in Fig. 6.3 as the gate voltage is varied.

6.2.1.3 Gate Pinch-off

At some negative V_{GS} bias it is possible to deplete the entire channel even when $V_{DS} = 0$ V. When this occurs, the channel is completely cut off by the gate, and majority current flow is zero for any V_{DS}. This voltage condition is called "gate pinch-off." The "turn-on" voltage for the MOSFET is when V_{GS} is slightly greater than gate pinch-off. This voltage, by definition, is equal to the MOSFET's threshold voltage V_{AEFF}. A small but finite current flows through the channel for gate voltages below cutoff. This current is called the "subthreshold current" and is made up of minority carriers generated in the channel (i.e., dark current).

Gate pinch-off is closely associated with depletion and the buried-channel effective threshold discussions given in Chapter 1. Recall that full depletion under the output transfer gate (OTG) took place when $V_{OTG} = V_{REF} - V_{EFF}$. Similarly, gate pinch-off for a MOSFET occurs when $V_G = V_S - V_{AEFF}$ or, equivalently,

$$V_{GS} = -V_{AEFF}. \tag{6.3}$$

CHARGE MEASUREMENT

This relationship becomes more physically apparent if we think of the OTG/V_{REF} arrangement as half of a MOSFET. In other words, the source of the MOSFET is analogous to V_{REF} and the gate is akin to the OTG.

Example 6.3

Determine the effective threshold for the MOSFET in Fig. 6.3.
Solution:
Gate pinch-off occurs at approximately $V_{GS} = -8$ V. Therefore, the effective threshold for the amplifier is $V_{AEFF} = 8$ V.

6.2.1.4 Drain Pinch-off

Similarly, drain pinch-off occurs when $V_G = V_{DD} - V_{AEFF}$ or, equivalently,

$$V_{GD} = -V_{AEFF}, \tag{6.4}$$

where V_{GD} is the gate-to-drain voltage. In that $V_G = V_{REF}$, the above equation can be written as

$$V_{DD} = V_{AEFF} + V_{REF}. \tag{6.5}$$

This equation is very important in biasing the output MOSFET into pinch-off and is used as a setup equation defined in Chapter 2.

6.2.1.5 Threshold Requirements

Recall from Chapter 1 that the effective threshold (V_{EFF}) for the array is determined by channel doping and thermal drive. Similarly, amplifier threshold (V_{AEFF}) is set by doping and drive made to the amplifier. It would first appear that no channel threshold is required. In fact, many CCD amplifiers in use today are not buried and do not exhibit a significant threshold. We refer to this class of MOSFETs as "surface-channel" or "enhancement" or "normally off" devices. The channel conductance for surface-channel devices is very low when the gate-to-source voltage is zero. A positive gate voltage must be applied to form the n-channel (an inversion layer) for conduction. MOSFETs described in this book are "normally on" and are called depletion mode or buried-channel MOSFETs. These devices require a negative gate voltage to deplete carriers in the channel to turn the device off (refer to Fig. 6.3). There are also depletion and enhancement mode p-channel MOSFETs. They work the same way, except that all voltage polarities to the device are reversed. p-channel MOSFETs are used with p-channel CCDs that generate and transfer holes instead of electrons.

The main preference for surface-channel MOSFETs is low-voltage and low-power operation. This is because the threshold voltage is negligible, requiring a low drain voltage for pinch-off conditions. However, surface-channel MOSFETs exhibit high $1/f$ noise compared with buried-channel devices. We will discuss how much buried-channel implant is required to control $1/f$ noise below.

6.2.1.6 Short Channel Effect

For very small gate lengths, and some high V_{DD}, the drain and source depletion regions can meet under the channel.[1] This allows majority carriers from the drain region to be directly injected into the source region, resulting in a sudden increase in drain current. This undesired condition is referred to as "punch-through." Figure 6.5 presents MOSFET characteristics for a 2-µm gate length that show the punch-through effect. Note that when $V_{DS} > 16$ V, the current increases abruptly, indicating punch-through. CCD processes limit the gate length to approximately 2 µm because of the effect.

6.2.1.7 Transconductance

Transconductance is defined as the ratio of the change in drain current to an incremental change in gate-to-source voltage (expressed in units of A/V or mhos). That is,

$$gm = \frac{\Delta I_{DS}}{\Delta V_{GS}}. \tag{6.6}$$

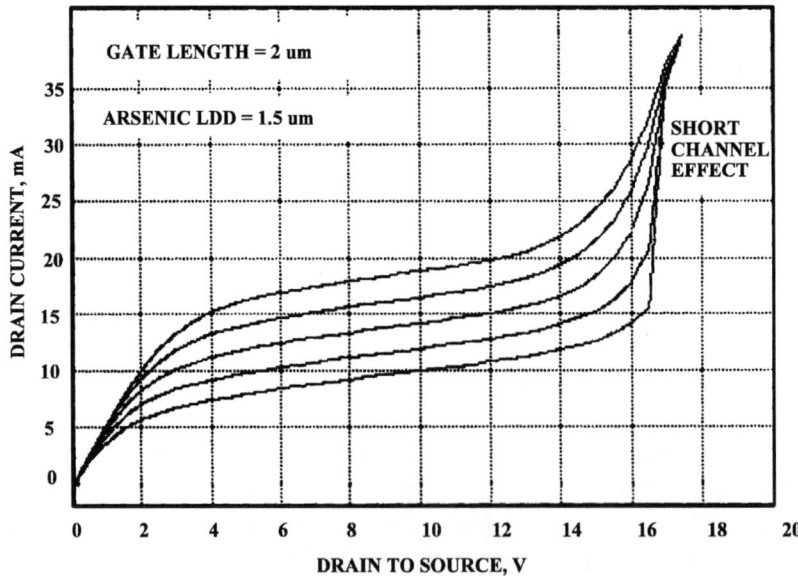

Figure 6.5 MOSFET short channel effect.

Example 6.4

Find the transconductance for the MOSFET in Fig. 6.3(a). Assume a saturation drain current of 2.0 mA.
Solution:
For a gate-to-source voltage change of 1 V, the current changes by 0.5 mA. From Eq. (6.6),

$$gm = (0.5 \times 10^{-3})/1,$$

$$gm = 500 \text{ μmho}.$$

The transconductance in the linear region is approximately

$$gm = \mu_{SI} C_{OX} \frac{W}{L} V_{DS}. \tag{6.7}$$

The transconductance in the saturation region is approximately

$$gm = \mu_{SI} C_{OX} \frac{W}{L} (V_{GS} + V_{AEFF}). \tag{6.8}$$

From Eqs. (6.2) and (6.8), the transconductance in terms of saturation current is

$$gm = \left(2\mu_{SI} C_{OX} \frac{W}{L} I_{DS}\right)^{1/2}. \tag{6.9}$$

Note that the transconductance varies by the square-root of drain current. This MOSFET characteristic is very important for low-noise and high-speed as will be discussed below.

Example 6.5

Find the transconductance for a 35(W) × 6(L)-μm MOSFET that is operating in the saturation region. Assume $C_{OX} = 3.45 \times 10^{-8}$ F/ cm^2, $\mu_{SI} = 1000$ cm^2/V-sec and $(V_{GS} + V_{AEFF}) = 3$ V.
Solution:
From Eq. (6.8),

$$gm = 1000 \times (3.45 \times 10^{-8}) \times (35 \times 10^{-4}) \times 3/(6 \times 10^{-4}),$$

$$gm = 600 \text{ μmho}.$$

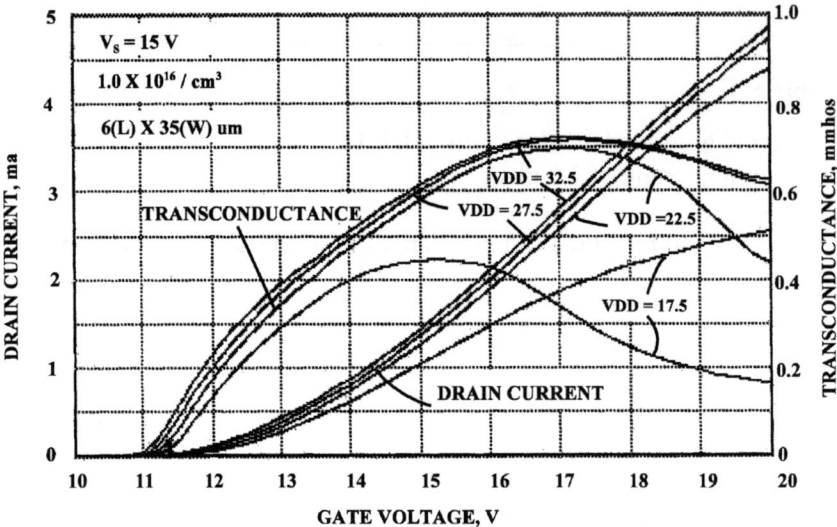

Figure 6.6 MOSFET drain current and transconductance as a function of gate voltage.

As discussed below, transconductance is an important MOSFET parameter that influences many charge measurement parameters, including output voltage gain, frequency response, noise, and linearity.

6.2.1.8 Mobility Behavior

Figure 6.6 plots drain current and transconductance for a 35(W) × 6(L)-μm MOSFET as a function of gate and drain voltages (i.e., V_{REF} and V_{DD}). A wealth of information about MOSFET behavior is seen in the plot. First note that the drain current begins to increase at a gate voltage of 11 V. In that the source was biased at 15 V in the measurement, the threshold for the MOSFET is approximately $V_{AEFF} = 4$ V (i.e., 15 − 11 V). Above threshold the drain current increases by the square of gate voltage [i.e., Eq. (6.2)]. At some high gate voltage the carriers in the channel reach a saturation velocity because of a transverse electric field that is generated.[2] In this bias region the drain current varies linearly with gate voltage. The "mobility effect" reduces saturation current compared to the square-law region. Also, the transconductance becomes constant and even decreases at high gate voltages, as seen in the figure. There is no benefit to biasing the MOSFET in this range.

6.2.2 Voltage Gain

The voltage gain, A_{CCD}, for the source follower output amplifier shown in Fig. 6.2 is given by

$$A_{CCD} = \frac{gmR_L}{1 + gmR_L}, \qquad (6.10)$$

Charge Measurement

where R_L is the output load resistance (Ω). Note that when $gmR_L \gg 1$, $A_{CCD} \simeq 1$ V/V.

Example 6.6

Find the voltage gain, assuming a transconductance of 500 µmho and a load resistance of 20 kΩ.
Solution:
From Eq. (6.10),

$$A_{CCD} = (5 \times 10^{-4}) \times (2 \times 10^4)/[1 + (5 \times 10^{-4}) \times (2 \times 10^4)],$$

$$A_{CCD} = 0.909 \text{ V/V}.$$

The transconductance in Eq. (6.10) can be measured by injecting a low-frequency sinusoidal signal into the drain/gate of the reset transistor and measuring the response at the source of the output amplifier. The ratio of the output voltage to the input voltage represents the gain of the stage. Once A_{CCD} is known, the transconductance can be calculated through Eq. (6.10).

Example 6.7

Find the transconductance for an output amplifier that exhibits a voltage gain of 0.8 V/V with a load resistance of 20 kΩ.
Solution:
First solve for *gm* from Eq. (6.10):

$$gm = A_{CCD}/(R_L - A_{CCD}R_L).$$

Substituting the parameters above yields

$$gm = 0.8/[(2 \times 10^4) - 0.8 \times (2 \times 10^4)],$$

$$gm = 200 \text{ µmho}.$$

6.2.3 Loading

There are several ways to load the output amplifier. The most popular method is to simply connect a load resistor to the CCD and ground the other end. The current through the load resistor under this loading condition is

$$I_{DS} = \frac{V_{GS}}{R_L} + \frac{V_{REF}}{R_L}. \qquad (6.11)$$

This equation can be plotted on the I_{DS}–V_{GS} MOSFET transfer curve as illustrated in Fig. 6.7(a). Note that the equation is a line with a slope of $1/R_L$. The x-intercept (i.e., $I_{DS} = 0$) is equal to V_{REF} and the y-intercept (i.e., $V_{GS} = 0$) is equal to $I_{DS} = V_{REF}/R_L$. We refer to the line as the "load line." The load line intercept with the I_{DS}–V_{GS} curve represents the current through the MOSFET. Figure 6.7(a) shows two load lines at $V_{REF} = 12$ V and $V_{REF} = 16$ V. Note that a large change in the gate voltage (i.e., ΔV_{REF}) represents a very small change in the gate-to-source voltage. This is characteristic of a source follower circuit because of negative feedback between the source and the gate.

Example 6.8

Plot the "load line" for the MOSFET characterized in Fig. 6.3(b). Assume $R_L = 20$ kΩ, $V_{REF} = 10$ V and a $V_{DS} = 13$ V. Then, determine the current through the load resistor and the voltage at the source.

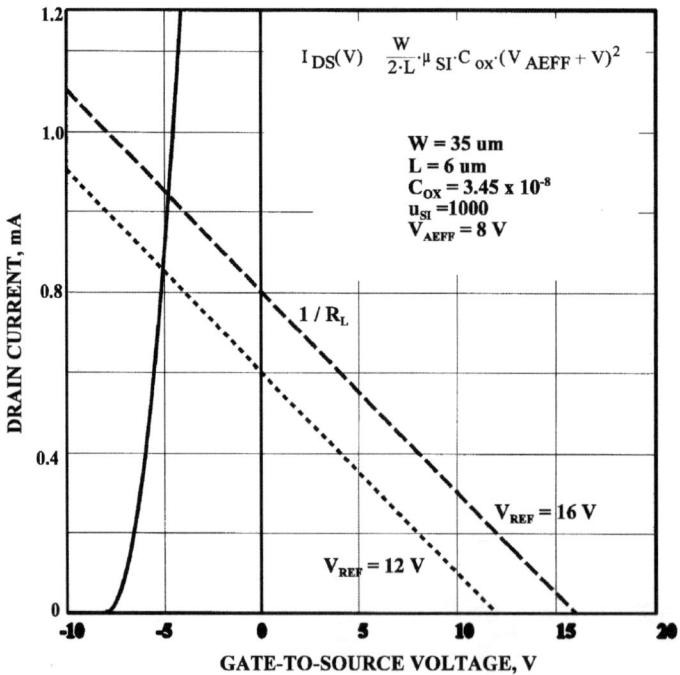

Figure 6.7(a) Source follower load lines.

Solution:
The load line can be plotted by calculating the slope (i.e., $= 1/R_L$) and y-intercept.
The y-intercept is

$$I_{DS} = V_{REF}/R_L,$$
$$I_{DS} = 10/(20 \times 10^3),$$
$$I_{DS} = 0.5 \text{ mA}.$$

The load line is plotted in Fig. 6.3(b). The current through the MOSFET from the load intercept with the I_{DS} curve is 0.8 mA. The voltage at the source is

$$V_S = (0.8 \times 10^{-3}) \times (20 \times 10^3),$$
$$V_S = 16 \text{ V}.$$

A grounded load resistor does not yield the highest output voltage gain or the optimum current flow for low-noise operation. The last requirement will become apparent as we go along in this chapter. One loading technique that helps satisfy both requirements is to bias the load resistor with a negative power supply, as

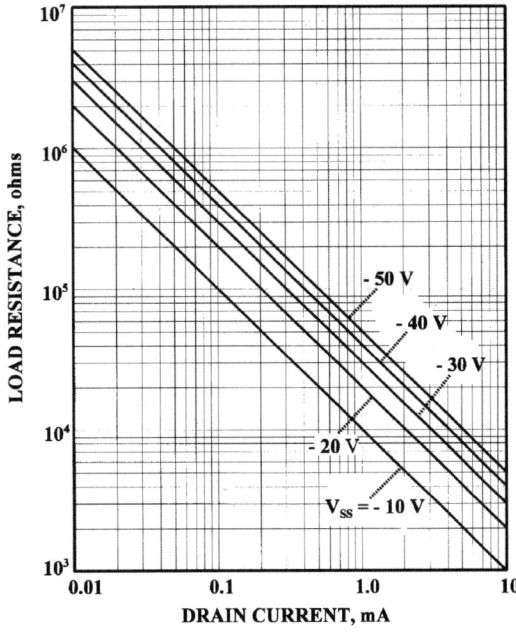

Figure 6.7(b) Load resistance as a function of drain current and V_{SS}.

shown in Fig. 6.2. The current through the load resistor and MOSFET for this arrangement is

$$I_{DS} = \frac{(V_{DD} - V_S) - V_{SS}}{R_L}, \qquad (6.12)$$

where V_S is the source voltage of the MOSFET and V_{SS} is the negative supply. The source voltage is

$$V_S = V_{REF} - V_{GS}. \qquad (6.13)$$

Substituting Eq. (6.5) into this equation yields

$$V_S = V_{DD} - V_{AEFF} - V_{GS}. \qquad (6.14)$$

Taking this equation into Eq. (6.12) yields

$$I_{DS} = \frac{(V_{AEFF} + V_{GS}) - V_{SS}}{R_L}. \qquad (6.15)$$

For normal operation, the output MOSFET will be biased with a negative gate-to-source voltage. Therefore, the first term will be negligible if V_{SS} is made high enough. With this assumption, the current through the MOSFET is simply

$$I_{DS} = \frac{V_{SS}}{R_L}. \qquad (6.16)$$

This equation shows that drain current can be controlled by the V_{SS} independent of V_{GS}.

Example 6.9

Plot the load resistance as a function of drain current for a family of V_{SS} using Eq. (6.16).
Solution:
Figure 6.7(b) is the required plot. Ideally, $R_L > 100$ kΩ for high-voltage gain. The output MOSFET will be typically biased at a few hundred microamps requiring $V_{SS} > 10$ V.

It is also possible to bias the substrate of the CCD with a positive voltage and leave the load resistor at ground potential. This will increase the voltage across the resistor, permitting a larger load resistance. However, elevating the substrate voltage implies that all other voltages to the CCD must be increased by the same amount. This can in itself be advantageous, because it allows single-level clock drive to the CCD. For example, a CCD that is clocked from −8 V to +4 V can also be clocked from 0 V to 12 V if the substrate is elevated to 8 V.

The CCD can also be loaded with a constant current diode. This device is simply a MOSFET that has its gate and source tied together (i.e., $V_{GS} = 0$ V) and its drain biased into pinch-off for constant current operation. The amount of current that flows is determined by the MOSFET gate width-to-length ratio [i.e., Eq. (6.2)]. The diode will give a constant current that is independent of drain voltage as long as the device has sufficient voltage across it to be pinched off. The dynamic input impedance (i.e., $\Delta V_{DS}/\Delta I_{DS}$) is very high, producing high-voltage gain for the output amplifier. However, constant current diodes can be noisy because the device typically operates deep into the pinch-off region (see discussions below on pinch-off luminescence problems). Also, the amount of current delivered is built into the device, making fine bias adjustments impossible. Multistage CCD amplifiers often employ on-chip constant current diodes for loading purposes, as discussed below.

6.2.4 OUTPUT IMPEDANCE

The output impedance, R_{OUT}, for a source follower amplifier is given by

$$R_{OUT} = \frac{R_L}{1 + gmR_L}. \tag{6.17}$$

When $gmR_L \gg 1$, this equation reduces to

$$R_{OUT} = \frac{1}{gm}. \tag{6.18}$$

Example 6.10

Determine the output impedance for the MOSFET characterized in Fig. 6.6. Assume $gm = 500$ μmho (i.e., $V_{GS} = -1$ V or $I_{DS} = 0.8$ mA) and $R_L = 20$ kΩ.
Solution:
From Eq. (6.18),

$$R_{OUT} = (2 \times 10^4)/[1 + (5 \times 10^{-4}) \times (2 \times 10^4)],$$

$$R_{OUT} = 1.8 \times 10^3 \, \Omega.$$

Note from Eqs. (6.9) and (6.18) that the output impedance decreases by the square-root of drain current (I_{DS}). This characteristic is important to high-speed operation where low output impedance is required.

6.2.5 TIME RESPONSE

The video time constant is that time period required for the output signal to reach 63% of its final value (i.e., $1 - 1/e$). The time constant is defined as

$$\tau_{CCD} = R_{OUT} C_L, \qquad (6.19)$$

where C_L is external load capacitance (F).

Example 6.11

Determine the video time constant for the amplifier in Example 6.10. Assume $C_L = 32$ pF.
Solution:
From Eq. (6.19),

$$\tau_{CCD} = (1.8 \times 10^3) \times (3.2 \times 10^{-11}),$$

$$\tau_{CCD} = 5.76 \times 10^{-8} \text{ sec (58 nsec)}.$$

Figure 6.8 plots the video time constant as a function of drain current for the MOSFET characterized in Fig. 6.6. Note that τ_{CCD} decreases by the square-root of I_{DS} until the mobility effect is encountered (approximately $\tau_{CCD} = 80$ nsec at 4 mA). Also plotted is the time constant calculated from Eq. (6.19) using transconductance values measured in Fig. 6.6 (triangle data). Theory is faster than experiment by about a factor of two.

6.2.6 FREQUENCY RESPONSE

The decibel (dB) is a logarithmic unit expressing the ratio of two powers, given by $10 \log(P_{OUT}/P_{IN})$. The unit is used to express power gain or loss for a circuit. Also, the decibel is used to express voltage or current ratios. For example, the definition for voltage gain or loss is given as $20 \log(V_{OUT}/V_{IN})$. For our work, the 3-dB frequency, $f_{3\text{-}dB}$, is where the sinusoidal voltage gain of the output amplifier is reduced by the square-root of two or a factor of two for the half-power frequency. The 3-dB bandwidth frequency of the output amplifier is related to the video time constant through

$$f_{3\text{-}dB} = \frac{1}{2\pi \tau_{CCD}}. \qquad (6.20)$$

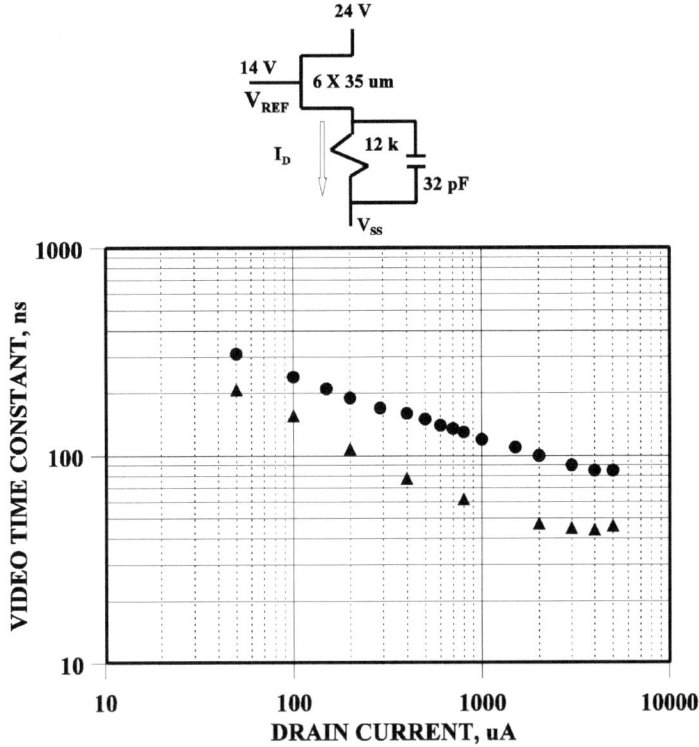

Figure 6.8 Video time constant as a function of drain current.

The frequency response is dependent on the source follower output impedance, which, as mentioned above, decreases by the square-root of the drain current. Figure 6.9 shows white noise plots generated by a 7(L) × 130(W)-μm MOSFET. The data were taken with a spectrum analyzer which measures frequency components that make up a noisy signal. Spectral noise is given in units of rms volts per root-Hz. Note from the plot that the 3-dB frequency is encountered at a lower frequency as the drain is reduced (from 1.8 mA to 4 μA). This is because the output impedance also decreases by the square-root of current as Eqs. (6.9) and (6.18) dictate. The noise spikes seen in this plot are generated by background AM radio frequencies that enter the passband of the measurement.

Example 6.12

Estimate theoretically the bandwidth for a 130(W) × 7(L)-μm MOSFET measured in Fig. 6.9. Assume $C_{OX} = 3.45 \times 10^{-8}$ F/cm^2, $\mu_{SI} = 1000$ cm^2/V-sec, $I_{DS} = 500$ μA and a load capacitance of $C_L = 30$ pF.

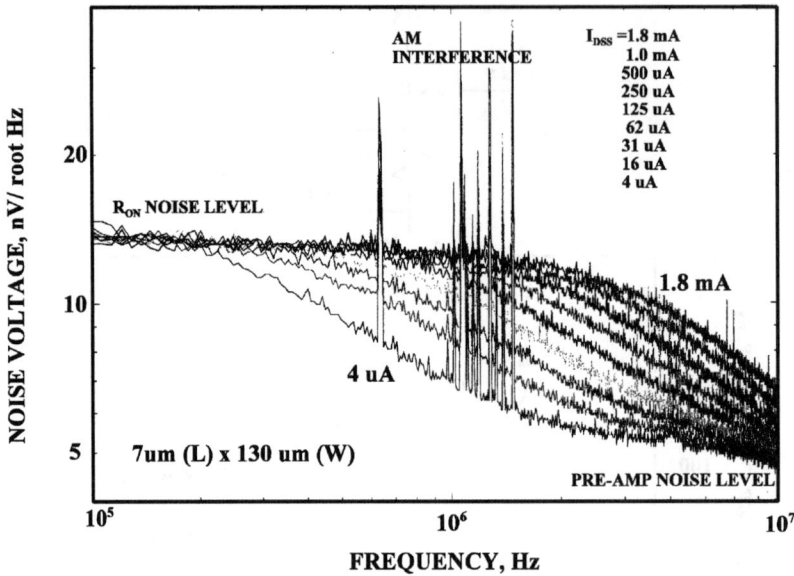

Figure 6.9 Output amplifier spectral noise as a function of drain current.

Solution:
From Eq. (6.9),

$$gm = [2 \times 1000 \times (3.45 \times 10^{-8}) \times (130 \times 10^{-4}) \\ \times (5 \times 10^{-4})/(7 \times 10^{-4})]^{1/2},$$

$gm = 8 \times 10^{-4}$ mho.

The output impedance from Eq. (6.18) is

$$R_{OUT} = 1/(8 \times 10^{-4}),$$

$$R_{OUT} = 1.25 \times 10^3 \ \Omega.$$

The video time constant from Eq. (6.19),

$$\tau_{CCD} = (1.25 \times 10^3) \times (3 \times 10^{-11}),$$

$$\tau_{CCD} = 3.75 \times 10^{-8} \text{ sec.}$$

The 3-dB frequency from Eq. (6.20) is

$$f_{3\text{-}dB} = 1/(2 \times 3.14 \times 3.75 \times 10^{-8}),$$

$$f_{3\text{-}dB} = 4.24 \times 10^6 \text{ Hz}.$$

This frequency is about double that measured in Fig. 6.9.

The 3-dB frequency response is approximately equal to the fastest pixel frequency at which the on-chip amplifier can operate. This rule of thumb includes the time required to shift and convert charge to a digital format (i.e., overhead time periods).

Example 6.13

Determine the maximum pixel rate for the amplifier characterized in Fig. 6.8. Assume $I_{DS} = 1$ mA.

Solution:

The video time constant at $I_{DS} = 1$ mA is approximately 120 nsec. The 3-dB frequency from Eq. (6.20) is

$$f_{3\text{-}dB} = 1/[2 \times 3.14 \times (1.2 \times 10^{-7})],$$

$$f_{3\text{-}dB} = 1.3 \times 10^6 \text{ Hz}.$$

The fastest pixel rate for the CCD is therefore approximately 1.3 Mpixels/sec. Note that the pixel rate can be increased by lowering the output drive capacitance (32 pF used here). The load capacitance for well-designed preamplifiers and layouts can usually be kept below 10 pF.

6.2.7 Bias

Optimum bias for the output MOSFET is found through an important noise transfer curve called I_D transfer. The curve plots read noise (in e⁻ rms!) as a function of drain voltage at different drain currents. For example, Fig. 6.10 shows a set of I_D transfer curves for a CCD output amplifier that is fabricated three ways with varying doping and channel depth. During optimization it is important to maintain high voltage gain for good linearity and low noise reasons (i.e., the MOSFET must be kept in the pinch-off state). Therefore, gain characteristics are also plotted for each curve as the drain voltage is varied. If the drain voltage and current are too low, the voltage gain degrades, causing the noise to increase. On the other hand, if the drain voltage and current are too high, then excess shot and $1/f$ noise are

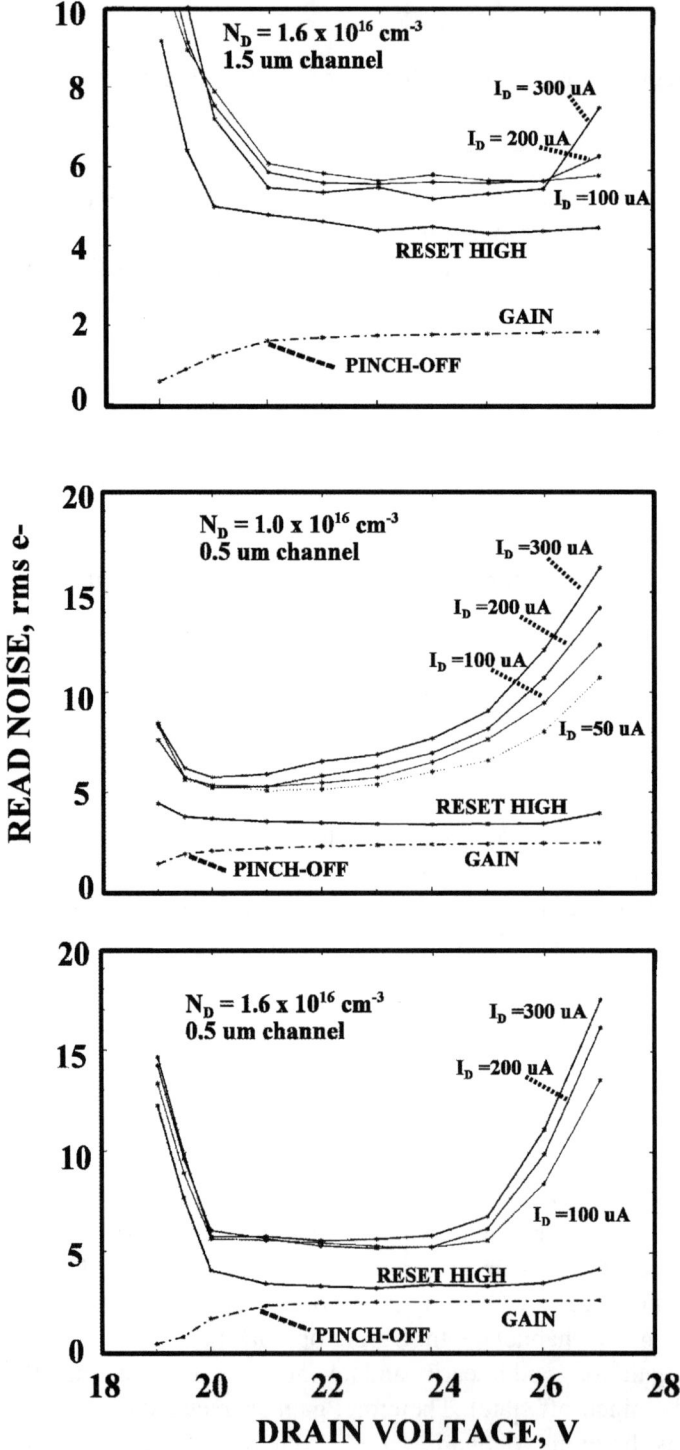

Figure 6.10 Output amplifier read noise I_D transfer curves.

generated (a subject discussed below). Optimum bias is between these two noise breakpoints.

Also shown in Fig. 6.10 are noise levels when the reset MOSFET is biased in the "on" state. Under this bias condition the gate of the MOSFET is not floating, but connected directly to V_{REF}. This arrangement will eliminate those noise sources that originate in the channel and feedback onto the gate of the output transistor. In all cases shown, the read noise is lower when the reset gate is ac shorted to ground. This may not always be the case if the reset "on" white noise resistance dominates (see discussions below on reset noise). Although normal operation must have sense node floating for charge measurement, ideally the read noise floor should be comparable to the reset "on" condition.

6.2.8 Sensitivity

The sensitivity of the output amplifier is defined by

$$S_V = \frac{q}{C_S}, \tag{6.21}$$

where S_V is the sensitivity (V/e$^-$) and C_S is sense capacitance (F) given by

$$C_S = C_{MOS} + C_{FD}, \tag{6.22}$$

where C_{MOS} is the gate capacitance associated with the output MOSFET and C_{FD} is the capacitance related to the floating diffusion and neighboring parasitic capacitances. These capacitors will be defined below.

Example 6.14

Find the node sensitivity for a sense capacitance of 8 fF.
Solution:
From Eq. (6.21),

$$S_V = 1.6 \times 10^{-19} / (8 \times 10^{-15}),$$

$$S_V = 20 \, \mu V/e^-.$$

Sensitivity can be readily measured using photon transfer data and the equation

$$S_V = \frac{1}{K A_{CCD} A_1 A_2}, \tag{6.23}$$

where K is the camera gain (e$^-$/DN), A_{CCD} is the gain of the CCD output amplifier (V/V), A_1 is the signal chain gain from the CCD to the ADC (V/V), and A_2 is the ADC gain (DN/V) that is specified by the manufacturer.

Example 6.15

Find the sensitivity, given $K = 2$e$^-$/DN, $A_{CCD} = 0.8$ V/V, $A_1 = 100$ V/V, and $A_2 = 3.3 \times 10^3$ DN/V (e.g., 16 bits over a 20-V range). Also, calculate the sense capacitance.
Solution:
From Eq. (6.23),

$$S_V = 1/[2 \times 0.8 \times 100 \times (3.3 \times 10^3)],$$

$$S_V = 1.89 \times 10^{-6} \text{ V/e}^-.$$

From Eq. (6.21),

$$C_S = 1.6 \times 10^{-19}/1.89 \times 10^{-6},$$

$$C_S = 8.47 \times 10^{-14} \text{ F (85 fF)}.$$

6.2.8.1 Floating Diffusion Capacitance

Figure 6.11(a) shows the floating diffusion or the sense node region that is bounded by the output transfer gate, the reset gate and surrounding p$^+$ field material. We define the floating diffusion capacitance as

$$C_{FD} = C_{OTG} + C_R + C_{DEP} + C_{SIDE} + C_{TAB}, \quad (6.24)$$

where C_{OTG} is output transfer gate edge capacitance (F), C_R is the reset edge capacitance (F), C_{DEP} is the depletion capacitance below the floating diffusion (F), C_{SIDE} is the diffusion's side-wall capacitance (F), and C_{TAB} is the tab capacitance (F).

The tab capacitance is related to the capacitance of the metal trace that connects the sense node to the output amplifier (see Fig. 6.1). Although this interconnect runs mostly over thick field oxide, it can be significant in some CCD designs. Sidewall capacitance is the lateral capacitance that exists between the diffusion and surrounding material. It can be reduced significantly by isolating n-channel material from p$^+$ field material. This is accomplished by leaving low doped p-epitaxial

Charge Measurement

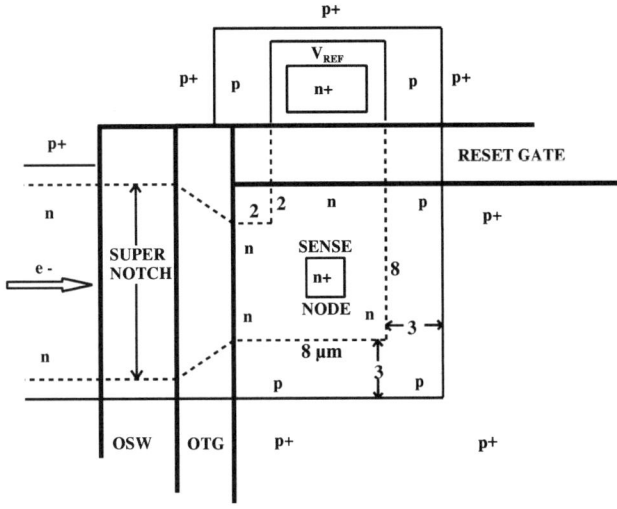

Figure 6.11(a) Sense node region.

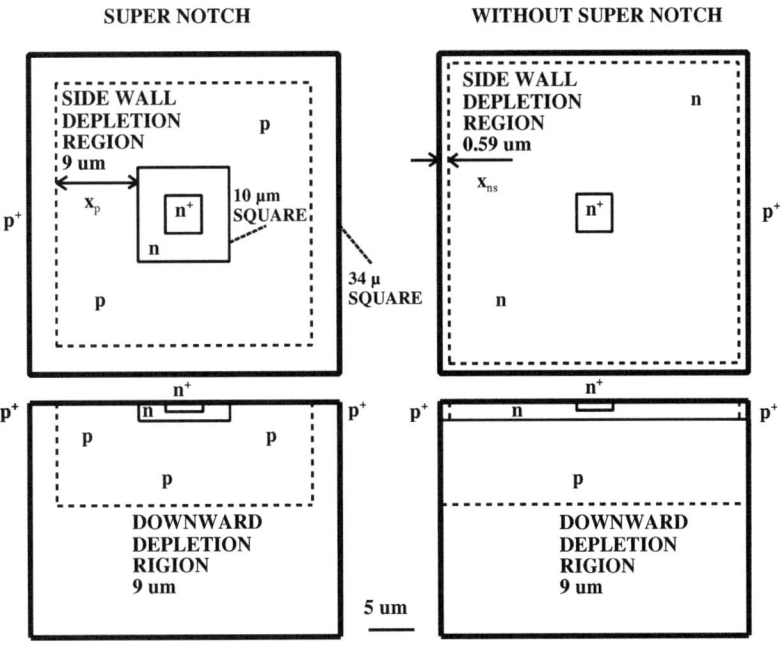

Figure 6.11(b) Sense node with super notch isolation.

material between the two regions. This arrangement is shown in Fig. 6.11(a), a special design feature referred to as a super notch. We will calculate the capacitance difference with and without super notch in the following example.

Example 6.16

Compare the sidewall capacitances for a 34 × 34-μm floating diffusion region with and without super notch isolation [refer to Fig. 6.11(b)]. Assume an n-channel depth of 1 μm doped at $N_D = 5 \times 10^{16}$ cm^{-3}. Also assume a p-epitaxial doping of $N_A = 2 \times 10^{14}$ cm^{-3} with the floating diffusion biased at $V_{REF} = 14$ V.

Solution:
Sidewall without super notch

Without super notch, the n-channel lies directly against p$^+$ field material as shown. Greatest depletion will occur on the n-side, with a depletion width of

$$x_{ns} = \left[\frac{2\varepsilon_{SI}}{qN_D}(V_{REF})\right]^{1/2}, \quad (6.25a)$$

where V_{REF} is the reference voltage (V) that biases the n-channel and N_D is the n-donor concentration of the buried channel (atoms/cm^3).

From Eq. (6.25a),

$$x_{ns} = \left[2 \times \left(1 \times 10^{-12}\right) \times 14 / \left(1.6 \times 10^{-19}\right) \times \left(5 \times 10^{16}\right)\right]^{1/2},$$

$$x_{ns} = 0.59 \text{ μm}.$$

The sidewall capacitance area is $4 \times 33 \times 1 = 132$ μm^2. The sidewall capacitance from Eq. (1.6) is

$$C_{SIDE} = 10^{-12} \times \left(132 \times 10^{-8}\right) / \left(5.9 \times 10^{-5}\right),$$

$$C_{SIDE} = 2.2 \times 10^{-14} \text{ F (22 fF)}.$$

Sidewall with super notch

With super notch the n-channel is isolated from field material with p-epitaxial dopant. A depletion region will extend in the p-material with a depletion width of

$$x_p = \left[\frac{2\varepsilon_{SI}}{qN_A}(V_{REF})\right]^{1/2}. \quad (6.25b)$$

From [Eq. 6.25b],

$$x_p = \left[2 \times \left(1 \times 10^{-12}\right) \times 14 / \left(1.6 \times 10^{-19}\right) \times \left(2 \times 10^{14}\right)\right]^{1/2},$$

CHARGE MEASUREMENT

$x_P = 9$ μm.

The sidewall capacitance area is $4 \times 10 \times 1 = 40$ μm². The sidewall capacitance is

$$C_{SIDE} = 10^{-12} \times (40 \times 10^{-8})/(9 \times 10^{-4}),$$

$$C_{SIDE} = 4.4 \times 10^{-16} \text{ F } (0.44 \text{ fF}).$$

This example shows the importance of minimizing sidewall capacitance through channel isolation with super notch.

Example 6.17

Find the floating diffusion capacitance for Fig. 6.11(a). Assume an n-channel depth of 1 μm doped at $N_D = 5 \times 10^{16}$ cm^{-3} and an epitaxial doping of $N_A = 2 \times 10^{14}$ cm^{-3} with the sense node biased at $V_{REF} = 14$ V. Also, assume an edge gate capacitance of 0.2 fF/μm and a tab capacitance of 3.45×10^{-9} F/cm² with a corresponding area of 21 μm².

Solution:

Downward capacitance

The depletion depth is largest on the p-junction side. It was calculated above as 9 μm. The area associated with the sense region is 60 μm² ($2 \times 6 + 6 \times 6 + 2 \times 6$). The downward capacitance is

$$C_{DEP} = (6.0 \times 10^{-7}) \times 10^{-12}/(9 \times 10^{-4}),$$

$$C_{DEP} = 0.666 \text{ fF}.$$

Sidewall capacitance

The 3 μm of super notch sidewall isolation from field doping will be totally depleted. The side-wall area related to two sides of 8-μm length is $2 \times 8 \times 1 = 16$ μm². The side-wall capacitance is

$$C_{SIDE} = 10^{-12}(16 \times 10^{-8})/(3 \times 10^{-4}),$$

$$C_{SIDE} = 0.533 \text{ fF}.$$

Reset edge capacitance

The edge capacitance related to the reset gate of width 6 μm is

$$C_R = 6 \times 0.2,$$

$$C_R = 1.2 \text{ fF}.$$

Output transfer gate edge capacitance
The edge capacitance related to the output transfer gate of width 6 μm is

$$C_{OTG} = 6 \times 0.2,$$

$$C_{OTG} = 1.2 \text{ fF}.$$

Tab capacitance
The tab capacitance is

$$C_{TAB} = (3.45 \times 10^{-9}) \times (21 \times 10^{-8}),$$

$$C_{TAB} = 0.724 \text{ fF}.$$

Floating diffusion capacitance
From Eq. (6.24), the floating diffusion capacitance is the sum of the above capacitances:

$$C_{FD} = 0.66 + 0.533 + 1.2 + 1.2 + 0.724,$$

$$C_{FD} = 4.3 \text{ fF}.$$

It should be mentioned that the sidewall and downward capacitances calculated above are approximate. The design dimensions shown in Fig. 6.11(a) will vary depending on processing details. For example, the sidewall dimensions will be less than given because dopants will diffuse during processing.

6.2.8.2 MOSFET Capacitance

The MOSFET gate capacitance given in Eq. (6.22) is estimated by

$$C_{MOS} = C_{GS} + C_{GD} + C_G, \tag{6.26}$$

where C_{GS} is the gate-to-source edge capacitance (F) given as

$$C_{GS} = W C_{EDGE}(1 - A_{CCD}), \tag{6.27}$$

C_{GD} is the gate-to-drain edge capacitance (F) given by

$$C_{GD} = C_{EDGE} W, \tag{6.28}$$

and C_G is the gate area capacitance, C_G, given as

$$C_G = \frac{LWC_{OX}}{2}, \quad (6.29)$$

where C_{EDGE} is the MOSFET's gate-edge-to-channel capacitance (F/cm), W is the gate width (cm), and L is the gate length (cm). The term $(1 - A_{CCD})$ in Eq. (6.27) is because the gate and source voltages "follow" each other, reducing edge capacitance. If the source follower gain was perfect (i.e., unity), there would be no voltage change between the source and gate, and thus, no source-to-gate capacitance.

Example 6.18

Find C_{GS}, C_{GD}, and C_G for a 4(L) × 15(W)-μm MOSFET. Assume $C_{EDGE} = 1 \times 10^{-12}$ F/cm, $C_{OX} = 3.45 \times 10^{-8}$ F/cm², and $A_{CCD} = 0.8$ V/V. Also find the sense capacitance and sensitivity when assuming a floating diffusion capacitance of 5.23 fF.

Solution:
Gate-to-source edge capacitance
From Eq. (6.27),

$$C_{GS} = (15 \times 10^{-4}) \times (1 \times 10^{-12}) \times (1 - 0.8),$$

$$C_{GS} = 0.3 \text{ fF}.$$

Gate-to-drain capacitance
From Eq. (6.28),

$$C_{GD} = (15 \times 10^{-4}) \times (1 \times 10^{-12}),$$

$$C_{GD} = 1.5 \text{ fF}.$$

Gate capacitance
From Eq. (6.29),

$$C_G = (3.45 \times 10^{-8})(15 \times 10^{-4})(4 \times 10^{-4})/2,$$

$$C_G = 10.3 \text{ fF}.$$

MOSFET capacitance
From Eq. (6.26),

$$C_{MOS} = 0.3 + 1.5 + 10.3,$$

$$C_{MOS} = 12.1 \text{ fF}.$$

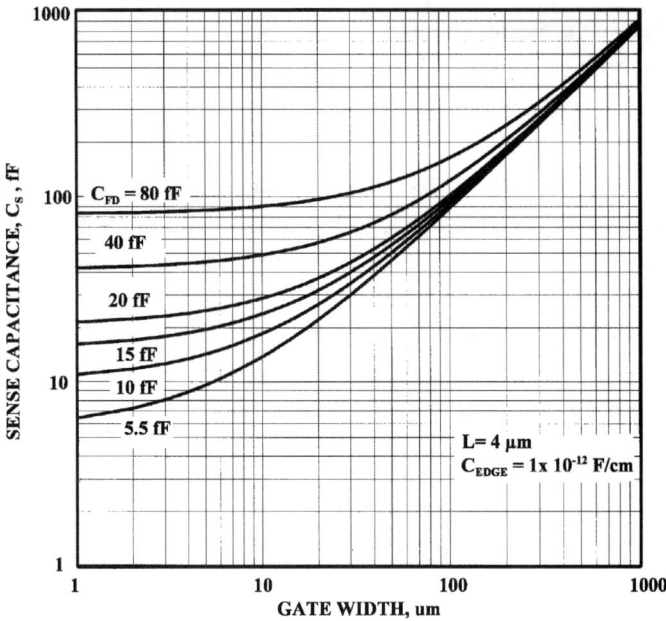

Figure 6.12(a) Sense capacitance as a function of output amplifier gate width and floating diffusion capacitance.

Sense capacitance
From Eq. (6.22),

$$C_S = 12.1 + 5.23,$$

$$C_S = 17.3 \text{ fF}.$$

Sensitivity
From Eq. (6.21),

$$S_V = (1.6 \times 10^{-19})/(1.73 \times 10^{-14}),$$

$$S_V = 9.2 \text{ μV/e}^-.$$

Figure 6.12(a) plots sense capacitance, C_S, as a function of gate width and floating-diffusion capacitance C_{FD} for a fixed gate length (L = 4 μm). Figure 6.12(b) shows the corresponding sensitivities.

6.2.9 LINEARITY

Charge that is generated, collected, and transferred to the sense node should be proportional to the number of photons that strike the CCD. However, the sense

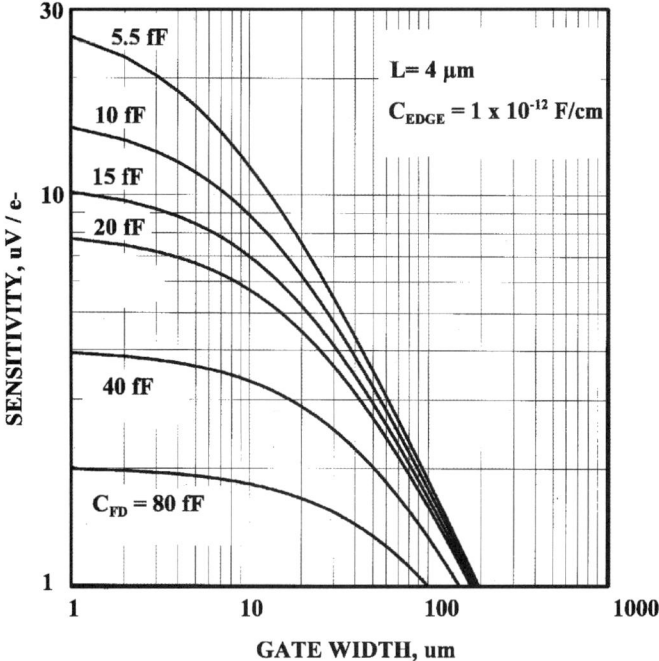

Figure 6.12(b) Sensitivity as a function of output amplifier gate width and floating diffusion capacitance.

node does not convert charge to a working voltage with perfect linearity. Also, the on-chip amplifier does not amplify different signal levels exactly one-to-one. These two sources of nonlinearity are referred to as floating diffusion nonlinearity and MOSFET nonlinearity.

6.2.9.1 Floating Diffusion Nonlinearity

The floating diffusion capacitance is not fixed, but varies depending on sense node bias (i.e., V_{REF}) and electrons transferred to the sense node [i.e., $S(e^-)$]. The impressed voltage generates a depletion capacitance under the diffusion defined by Eq. (6.25b). The depletion width decreases when charge is transferred onto the sense node and, in turn, increases the node capacitance and lowers the sensitivity. The fractional change in sensitivity caused by a change in depletion capacitance induced by a full well signal is

$$\frac{\Delta S_V}{S_V} = \frac{q C_{DEP} S_{FW}}{2 V_{REF} C_S^2}, \qquad (6.30)$$

where S_{FW} is the full well level (e^-).

Example 6.19

Determine the percent change in sensitivity at full well, given $C_{DEP} = 2.6 \times 10^{-15}$ F, $C_S = 12.6 \times 10^{-15}$ F, and $S_{FW} = 100{,}000$ e$^-$.
Solution:
From Eq. (6.30),

$$\Delta S_V/S_V = (1.6 \times 10^{-19}) \times (2.6 \times 10^{-15}) \times 10^5 / \left[2 \times 12 \times (12.6 \times 10^{-15})^2\right],$$

$$\Delta S_V/S_V = 0.011.$$

This represents a nonlinearity in response of 1.1% over the dynamic range of the CCD.

6.2.9.2 MOSFET Nonlinearity

When charge is transferred to the sense node, the gate-to-source voltage will decrease, causing the drain current through the MOSFET to decrease. A reduction in drain current will reduce the transconductance, lowering the voltage gain. This action produces a nonlinearity in the output amplifier. Nonlinearity is estimated by taking Eq. (6.8) and substituting it into Eq. (6.10) and differentiating in respect with V_G, yielding

$$\frac{\Delta A_{CCD}}{A_{CCD}} = \frac{\alpha S_V S_{FW}}{[1 + \alpha(V_{GS} + V_{AEFF})]^2}, \quad (6.31)$$

where ΔA_{CCD} is the voltage gain change of the output amplifier for a full well signal (V/V), and α is

$$\alpha = \frac{W \mu_{SI} C_{OX} R_L}{L}. \quad (6.32)$$

Example 6.20

Find the output amplifier gain change when a full well level of 100,000 e$^-$ is transferred onto the sense node. Assume $V_{AEFF} + V_{GS} = 4$ V, a sensitivity of 3 µV/e$^-$, $R_L = 10^4$ Ω, L = 6 µm, and W = 35 µm.
Solution:
From Eq. (6.32),

$$\alpha = (35 \times 10^{-4}) \times 1000 \times (3.45 \times 10^{-8}) \times 10^4 / (6 \times 10^{-4}),$$

$$\alpha = 2.01.$$

From Eq. (6.31),

$$\Delta A_{CCD}/A_{CCD} = 2.01 \times 10^5 \times (3 \times 10^{-6})/(1 + 2.71 \times 4)^2,$$

$$\Delta A_{CCD}/A_{CCD} = 4.3 \times 10^{-3}.$$

This represents a nonlinearity of 0.43%. Note that diffusion and MOSFET nonlinearity work in the same direction (i.e., the nonlinearities add).

High-dynamic-range (i.e., large full well) CCDs are most prone to the nonlinear effects mentioned above. There are many interesting amplifier designs invented that have dealt with the problem. For example, two amplifiers with "high" and "low" sensitivities that share the same signal channel have been used to control nonlinearity problems. Signal charge is directed to the amplifier of choice with control gates activated by the user. However, the penalty paid for reducing sensitivity by adding capacitance to the sense node is a proportional increase in read noise (i.e., e^- rms).

WF/PC I CCD utilized a dual reset gate as shown in Fig. 6.1, which by accident was found to be a good method to control sensitivity. The original intent of the dual reset gate was to suppress the reset clock feed-through by applying a dc voltage onto the second gate nearest the sense node. Although the reset feed-through was suppressed, it was discovered that the dc gate also decreased sensitivity (approximately by a factor of 2). A lower sensitivity was seen because electrons are allowed under the dc gate if they are to transfer through the reset gate when pulsed high. Electrons will see the full capacitance of the dc gate, thus lowering sensitivity. When both gates are used to reset the sense node, only the gate edge capacitance will be seen, and thus, full sensitivity is realized (the way WF/PC I ended up using the dual gate).

6.2.10 TEMPERATURE CHARACTERISTICS

Output amplifier characteristics improve with decreasing operating temperature because mobility and transconductance both increase. However, gain is not significantly influenced by temperature because of the inherent negative feedback provided by the source follower circuit. For example, measurements show that gain characteristics for the Hubble CCDs are stable to within 8% over a 200 K to 300 K temperature range. This characteristic assumes that the output MOSFET is well into the pinch-off state.

6.2.11 LIGHTLY DOPED DRAIN

Lightly doped drain (LDD) is a special process often employed when fabricating high-performance output amplifier designs. LDD technology significantly reduces edge gate capacitance (i.e., C_{GS} and C_{GD}), thereby increasing node sensitivity compared to non-LDD devices. Figure 6.13 shows a LDD amplifier for a WF/PCII/Cassini design. Note that the n^+ source and drain regions are isolated from the gate by the n-buried-channel dopant (separated by 1 µm in this case). In contrast, the source and drains for a non-LDD amplifier are self-aligned to the gate using a n^+ "blanket" implant (i.e., no mask is required). Subsequent high-temperature processing will cause the n^+ dopant (typically phosphorus) to laterally diffuse under the gate. Therefore, the gate edges reside over highly doped material resulting in high depletion capacitance (recall that depletion width is inversely proportional to the square-root of doping). Figure 6.14 shows depletion regions under the gate for a LDD MOSFET compared to a non-LDD MOSFET. In the case of the LDD device, the depletion region is maintained under the entire gate. For the non-LDD device, phosphorus encroaches under the gate pulling the depletion region along with it. Sensitivity (V/e$^-$) for the WF/PC II/Cassini amplifier designs was 50% greater when using the LDD option.

Test LDD MOSFETs show that the distance between the gate and LDD implant is critical and should be designed as small as possible without degrading sensitivity significantly. This distance is on the order of 1 µm for phosphorus source/drains. Separation more than this results in a parasitic series channel resistance which

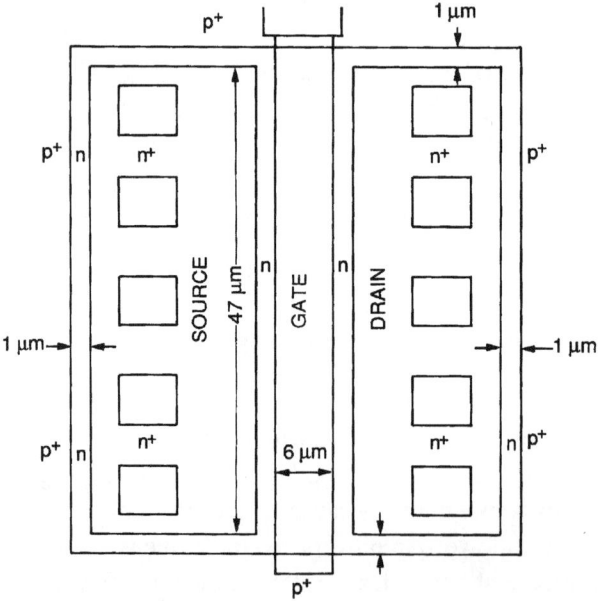

Figure 6.13 Lightly doped drain (LDD) output amplifier.

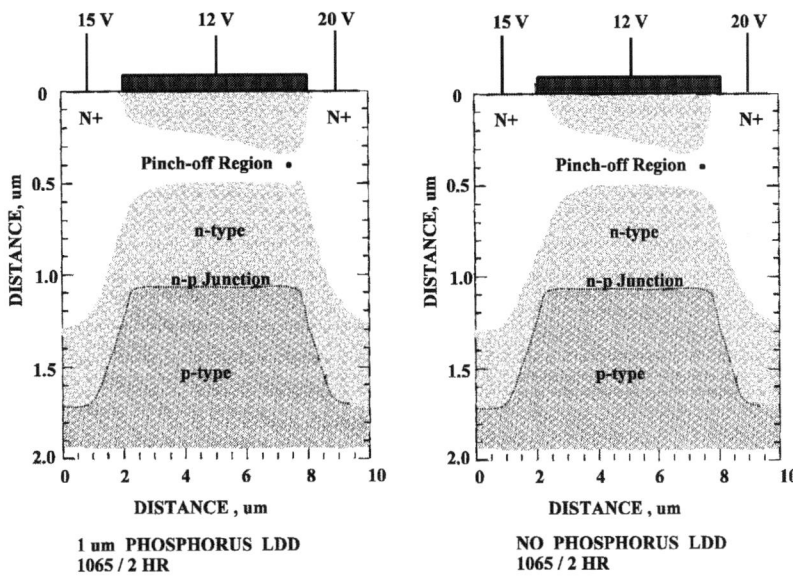

Figure 6.14 Gate depletion regions for an LDD and a non-LDD output amplifier.

reduces transconductance. This series resistance is particulary important to high-speed amplifiers. To circumvent the problem, the gap can be lightly doped to reduce the series effect. Arsenic doping is particularly attractive because lateral diffusion under the gate compared to phosphorus is considerably less. In fact, the LDD structure can be eliminated if arsenic source and drains are employed, which is a common process. In this case, arsenic will only encroach under the gate a couple tenths of a micron, where, by comparison, phosphorus will spread under the gate typically by a couple microns.

LDD processing has a second advantage of improving breakdown characteristics for the output amplifier. Breakdown for an n-p junction varies inversely with doping for the lightly doped side. Figure 6.13 shows that the highly doped source and drains are isolated from highly doped p^+ field material with n-buried-channel dopant (isolation is 1 μm in this case). The $p^+/n/n^+$ sandwich increases breakdown characteristics by a factor of two or more, compared to a non-LDD amplifier.

Figure 6.15 shows V_{DD} transfer characteristics for a non-LDD amplifier which demonstrate the breakdown problem. The plot measures noise under dark conditions as a function of the output drain voltage (V_{DD}) and the reference voltage (V_{REF}). Note that there are two noise breakpoints for each curve. The first noise increase is caused by an increase in gain as the MOSFET pinches off. For good linearity performance, V_{DD} is set in this operating region (e.g., $V_{REF} = 9$ V and $V_{DD} = 21.5$). The second noise breakpoint is caused by avalanche breakdown between the n^+ and p^+ junctions. Figure 6.16 shows a V_{DD} transfer curve for an

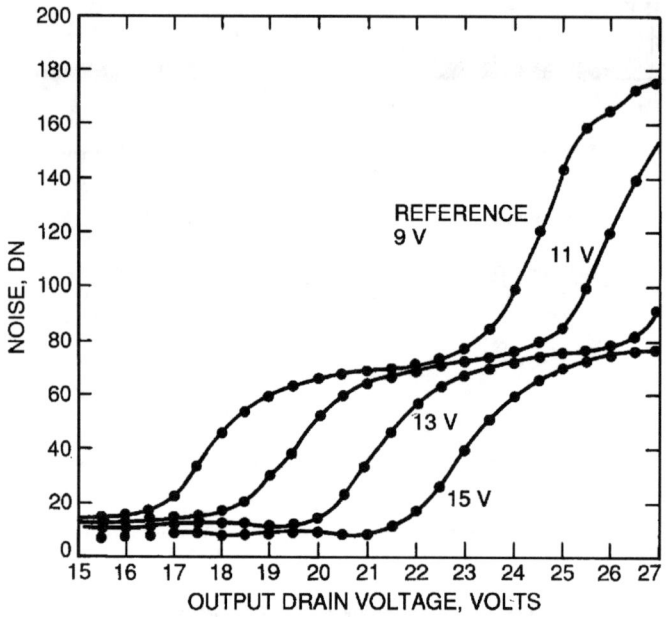

Figure 6.15 V_{DD} transfer curve for a non-LDD output amplifier.

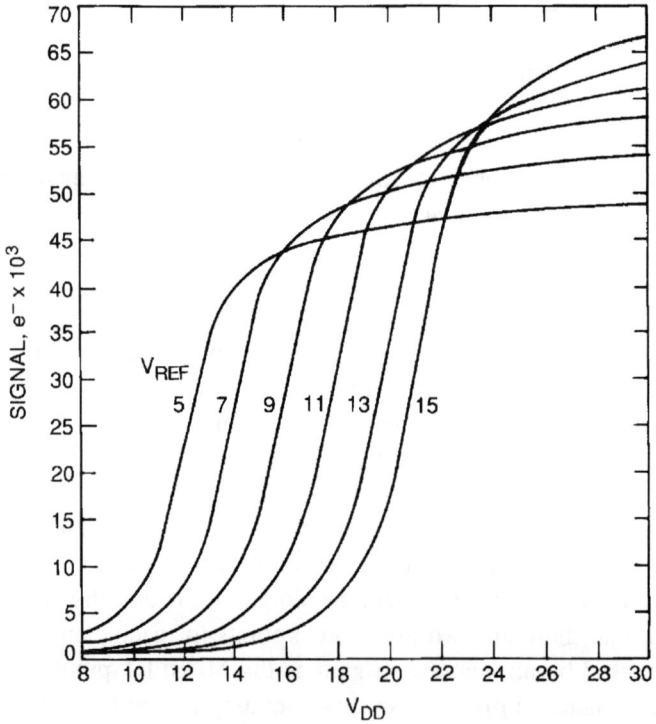

Figure 6.16 V_{DD} transfer curve for an LDD amplifier.

CHARGE MEASUREMENT

LDD output amplifier. Note that the second noise breakpoint is absent compared to the non-LDD MOSFET. The breakdown voltage for this LDD MOSFET is greater than 45 V because of LDD isolation.

6.2.12 AMPLIFIER LUMINESCENCE

The most common CCD luminescence problem is related to the output amplifier (additional luminescence sources are discussed in Chapter 7). For example, Fig. 6.17 is a dark-current image showing a "glowing" corner. When the problem was first observed in 1974, it was believed that the source of the charge was related to high dark-current generation (i.e., the amplifier was heating the silicon). However, thermal studies discounted this possibility because the heat required was too high compared to the actual power dissipated. This fact was later confirmed by maintaining a constant power level to the amplifier while adjusting the drain voltage. The test showed that the luminescence signature was only dependent on applied voltage and not on power consumption.

Today we know that diffusion and pinch-off luminescence are two sources of amplifier luminescence. Pinch-off luminescence is related to the pinch-off region. High fields found here can result in a weak avalanche condition promoting impact ionization and NIR light emission. The electrons generated by the process leave the drain, whereas the holes leave the substrate as a substrate current (substrate current is a direct indicator that luminescence is occurring). Diffusion luminescence is generated in the drain/source regions that are under high bias conditions at the point where avalanche and breakdown occur. As discussed above, LDD technology can

Figure 6.17 Output amplifier luminescence.

eliminate this source of diffusion luminescence by increasing breakdown characteristics. However, many non-LDD amplifiers in use today exhibit the problem. Users of these CCDs often turn off the drain voltage during a long integration period, thereby suppressing the bright corner effect. However, the drain voltage must be turned on to read the CCD, which adds shot noise to the CCD signal.

Luminescence can be observed by taking a CCD image of an operating CCD through a high-powered microscope.[3] Figure 6.18 shows a pinch-off luminescence image taken from a non-LDD MOSFET operating several volts into pinch-off. Luminescence is seen within the pinch-off region near the drain side. Figure 6.19 is a computer-generated image for the same MOSFET under higher bias conditions. Both pinch-off and diffusion luminescence are observed in this image. Note that diffusion luminescence is generated in the corners of the drain where electric fields are highest. The wavelength of emission seen is approximately 10,000 Å (i.e., bandgap emission).

The high-electric fields responsible for pinch-off luminescence are shown in Fig. 6.20. Note that the fields are greatest above the pinch-off region very near the surface at the gate/drain edge. It is in this vicinity where impact ionization will first take place. Figure 6.21 plots electric field strength as function of distance into the channel through this region. The onset of impact ionization approximately takes place at a field strength of 5×10^5 V/cm (i.e., near silicon breakdown).[4]

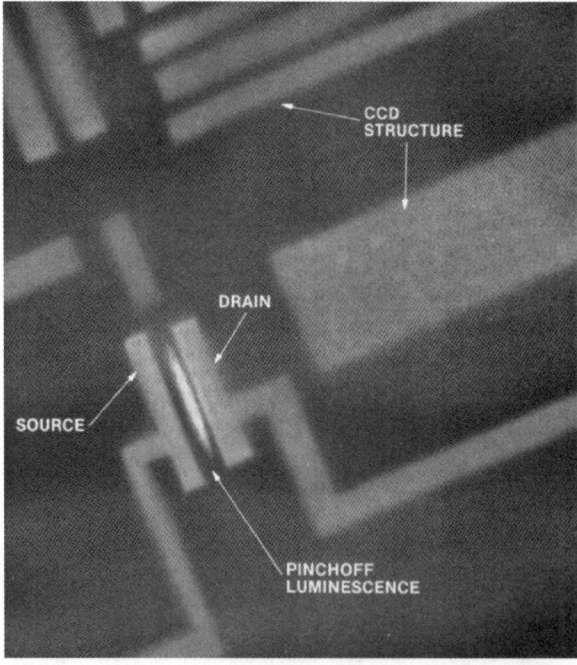

Figure 6.18 Microscope CCD image showing pinch-off luminescence.

CHARGE MEASUREMENT

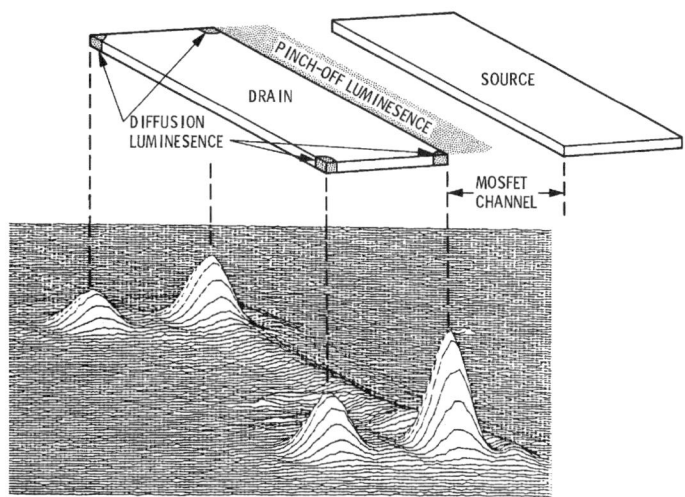

Figure 6.19 Computer-generated image showing pinch-off and diffusion luminescence.

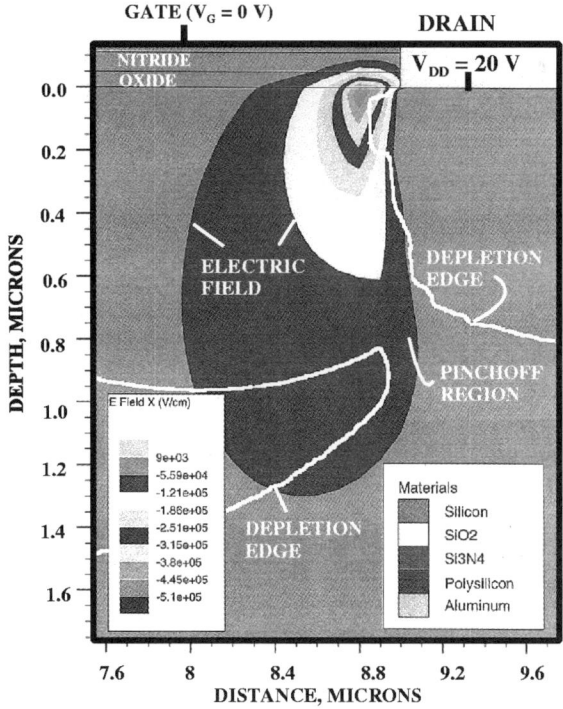

Figure 6.20 Electric field concentration above the pinch-off region.

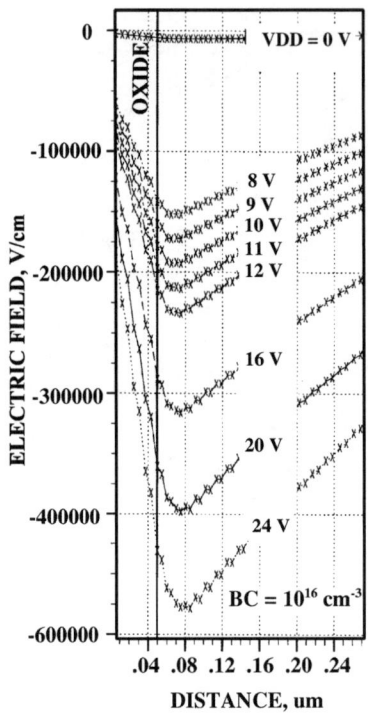

Figure 6.21 Electric field as a function of drain voltage near the drain region.

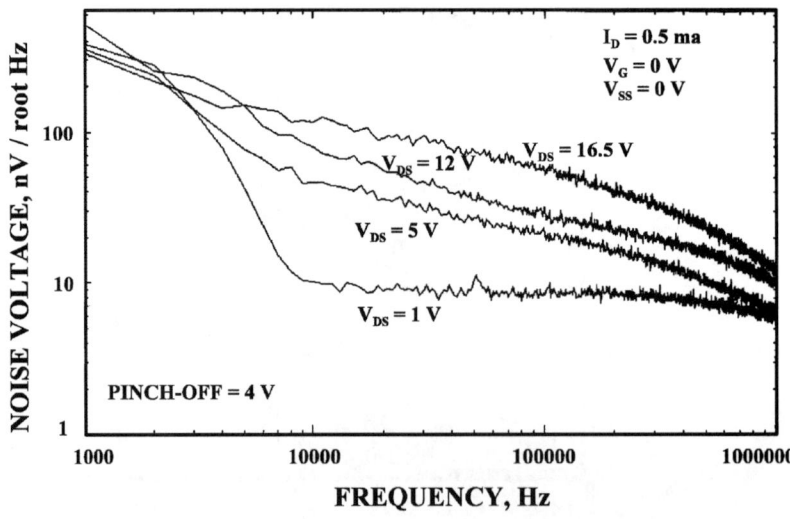

Figure 6.22 Spectral noise characteristics for a MOSFET that is biased into pinch-off.

Figure 6.22 is a spectral noise plot showing how low-frequency noise increases as the device is biased deep into pinch-off. This particular MOSFET exhibits a threshold of $V_{AEFF} = 4$ V ($N_D = 1.0 \times 10^{16}$ cm^3) and a pinch-off point of $V_{DS} = 4$ V for $V_{GS} = 0$ V. Note that excess noise dramatically increases past pinch-off where the onset of luminescence takes place (i.e., $V_{DS} = 4$ V).[5]

Pinch-off noise is also seen for the LDD amplifiers characterized in Fig. 6.10 as a function of drain voltage. The plot also shows that noise generated in the channel by impact ionization modulates the gate voltage when the gate is floating. Note that leaving the reset switch in the "on" state shorts out this feedback source.

6.2.13 MULTISTAGE AMPLIFIERS

6.2.13.1 Design

As discussed above, the MOSFET amplifier is limited to a maximum pixel rate of approximately 2 Mpixels/sec, depending on drain current. For faster pixel rates, multistage amplifiers must be employed. For example, triple-stage amplifiers are used at pixel rates greater than 70 Mpixels/sec (e.g., HDTV CCDs). The purpose of additional amplifier stages is to provide low-output impedance to drive the external load capacitance at the desired pixel rate. Figure 6.23 presents a dual-stage amplifier design showing the reset switch [4.5(L) × 9(W)], first-stage amplifier [2(L) × 17(W)], an active first stage MOSFET load [19(L) × 21(W)], and second-stage amplifier [2(L) × 200(W)].

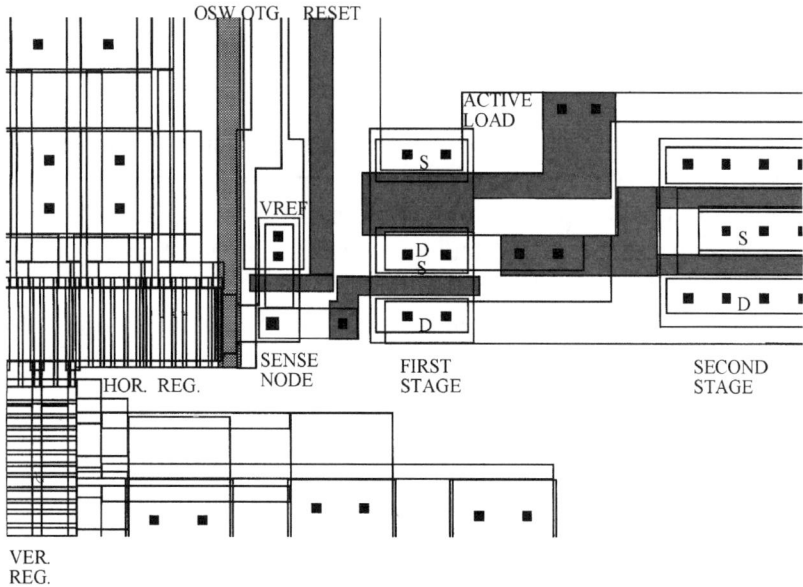

Figure 6.23 Dual-stage amplifier with active MOSFET load.

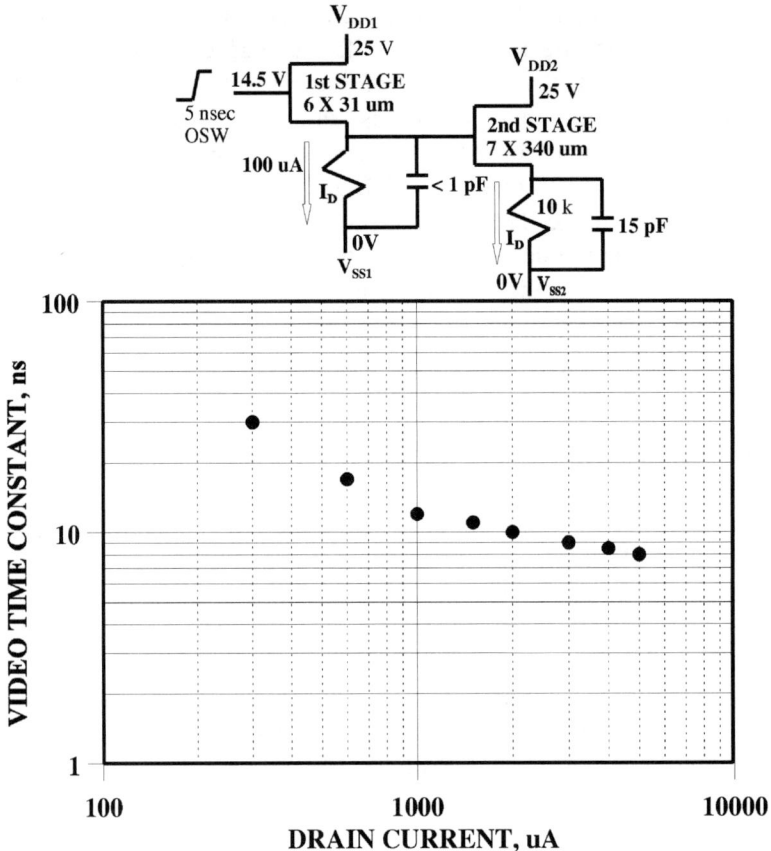

Figure 6.24 Dual-stage amplifier video time constant as a function of drain current.

Figure 6.24 plots the video time constant as a function of drain current for the dual-stage amplifier illustrated. A video time constant of approximately 10 nsec is achieved for a load capacitance of 15 pF at $I_D = 2$ mA. This data compared with Fig. 6.8 for a single-stage amplifier shows a significant speed improvement because the second stage has a lower output impedance.

Eqs. (6.9), (6.18), and (6.19) can be combined to determine the gate width required for the second-stage amplifier for a specified load capacitance. The resultant equation is

$$W = \frac{LC_L^2}{2\tau_{CCD}^2 \mu_{SI} C_{OX} I_{DS}}, \qquad (6.33)$$

where W is the second-stage gate width (cm), L is the second-stage gate length (cm), C_L is the second-stage load capacitance (F), and τ_{CCD} is the second-stage video time constant (sec).

Example 6.21

Estimate the minimum gate width for a second-stage output amplifier, assuming the parameters $L = 3 \times 10^{-4}$ cm, $\tau_{CCD} = 10^{-8}$ sec, $I_{DS} = 1$ mA, $\mu_{SI} = 1000$ cm^2/V-sec, $C_{OX} = 3.45 \times 10^{-8}$ F/cm^2, and $C_L = 20 \times 10^{-12}$ F.

Solution:

The required channel width, from Eq. (6.33), is

$$W = (3 \times 10^{-4})(20 \times 10^{-12})^2 / [2(10^{-8})^2 \times 1000 \\ \times (3.45 \times 10^{-8}) \times (1 \times 10^{-3})],$$

$$W = 1.73 \times 10^{-2} \text{ (173 μm)}.$$

The gate width is usually fabricated larger than calculated for an additional gain-bandwidth margin.

6.2.13.2 Active Load

The first-stage amplifier shown in Fig. 6.23 uses a MOSFET load to control the drain current through this stage. The MOSFET's gate-to-length ratio determines the current flow [i.e., Eq. (6.2)]. The channel doping and drive defines its threshold and pinch-off voltage. To reduce package pin count, some CCD manufacturers elect to connect the gate and source together and reference both to substrate potential. However, this arrangement does not allow one to optimally set the current through the first-stage MOSFET. Ideally, the MOSFETs gate and source should be pinned out and biased independently. This arrangement allows the gate voltage to set the pinch-off voltage, which is important for luminescence reasons (i.e., the load MOSFET should not be biased too far beyond pinch-off). The source voltage sets the current through the MOSFET, which is important for $1/f$, shot-noise and white-noise reasons, as will be discussed below. Figure 6.25 plots current supplied by a load MOSFET as a function of drain-to-source voltage and source voltage. Note that for a gate voltage of $V_G = 0$ V, a pinch-off potential of approximately 16 V is obtained. Current is adjusted by the source voltage.

The high-voltage drop that will develop across the MOSFET load can take the device deep into the pinch-off regime producing luminescence and noise problems. For this reason, a load MOSFET is doped heavier for high-threshold characteristics (e.g., $V_{EFF} = 16$ V at $N_D = 2.3 \times 10^{16}$ for the MOSFET shown in Fig. 6.25). This allows the device to take a larger voltage drop before entering the pinch-off state, thus controlling the luminescence.

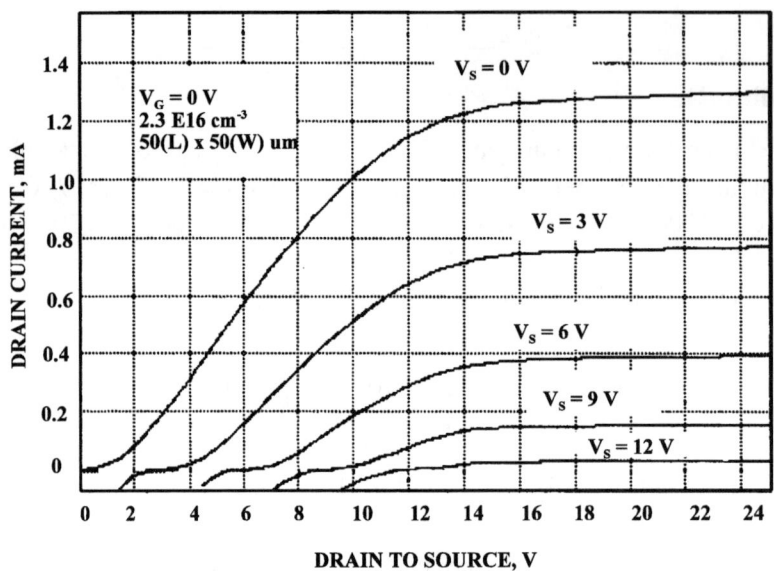

Figure 6.25 Drain current as a function of drain-to-source voltage and source voltage.

Table 6.1 Selected dopants for a 8 × 100-μm channel resistor.

Dopant	Concentration-cm^{-2}	kΩ-sq	8 × 100 μm
Arsenic	2.5×10^{13}	1	12.5 k
Arsenic	5×10^{14}	0.16	2.0 k
Arsenic	5×10^{15}	0.02	250
Phosphorus	1×10^{12}	6.0	75 k
Phosphorus	1.6×10^{12}	3.7	46.25 k
Phosphorus	1.9×10^{12}	3.2	40 k
Phosphorus	2.9×10^{12}	2.2	27.50 k
Phosphorus	3.5×10^{12}	1.9	23.75 k

6.2.13.3 Passive Load

Excess noise seen in active loads has forced some low-noise CCD manufacturers to use simple passive on-chip load resistors. On-chip resistors can be fabricated many different ways. For example, a "poly resistor" can serve as a load resistor for the first stage.[6] An 8 × 100-μm short poly line doped to approximately 2×10^{14}/cm^2 will yield a resistance of approximately 50 kΩ. However, poly resistors are noisy and their resistance varies considerably with operating temperature. A "channel resistor" can also be fabricated by making two n$^+$ contacts within a narrow channel that is selectively doped with arsenic or phosphorus. Table 6.1 shows the resistance produced for an 8 × 100-μm doped region at various implant levels for arsenic and phosphorus. Channel resistors are stable with operating temperature and exhibit low noise characteristics. However, poly resistors have the advantage over channel resistors in that bias can be positive or negative in adjusting drain current to the first stage. Channel resistors are limited to only positive bias because of the n^+ diffusion. A negative voltage would forward bias the n-p junction, resulting in luminescence (all n-p junctions luminesce when forward biased). A small metal film resistor can also be surface mounted close to the CCD in the sensor's package. The size and location of the resistor are important to minimizing the drive capacitance seen by the first-stage amplifier. Lastly, metal film resistors can also be deposited on-chip at the same time the CCD is fabricated.

6.2.13.4 Bias

The first-stage amplifier is responsible for all sensitive performance parameters (i.e., linearity, sensitivity, and noise), whereas the main purpose of the second stage is to drive the outside world. Therefore, bias to the second stage is not as critical. As discussed below, the first stage will require a negative gate-to-source voltage for noise reasons. The gate voltage impressed on the second stage will therefore be greater than V_{REF} and will require a higher drain voltage to pinch-off. To satisfy both output stages, two drain voltages are usually required for independent bias (i.e., V_{DD1} and V_{DD2}). Some CCD manufacturers will ac couple the first stage to the second stage to remove the dc level between stages. However, this arrangement requires two reset transistors and reference supplies (i.e., V_{REF1} and V_{REF2}).

6.2.14 Power Consumption

Power consumption for the dual-stage amplifier configuration shown in Fig. 6.24 with on-chip load is given by

$$P_{AMP} = (V_{DD1} - V_{SS1})I_{D1} + V_{DS2}I_{D2}, \qquad (6.34)$$

where P_{AMP} is the power dissipated on-chip, I_{D1} is the drain current through the first stage, I_{D2} is the current through the second stage, and V_{DS2} is the drain-to-source voltage for the second stage. The first term represents the power dissipated by the first-stage MOSFET and its load, and the second term is the power dissipated by second-stage MOSFET.

Recall from equations above that $V_{AEFF} + V_{REF} = V_{DD1}$ and $V_{AEFF} + V_{GS2} = V_{DS2}$. Substituting these equations into Eq. (6.34) yields

$$P_{AMP} = (V_{AEFF} + V_{REF} - V_{SS1})I_{D1} + (V_{AEFF} + V_{GS2})I_{D2}. \qquad (6.35)$$

Example 6.22

Calculate the power dissipated by a dual-stage amplifier. Assume the following bias conditions:

$$I_{D1} = 0.5 \text{ mA}$$
$$I_{D2} = 2 \text{ mA}$$
$$V_{AEFF} = 6 \text{ V}$$
$$V_{REF} = 14 \text{ V}$$
$$V_{GS2} = -2 \text{ V}$$
$$V_{SS1} = 0 \text{ V}.$$

Solution:
From Eq. (6.35),

$$P_{AMP} = (6 + 14 - 0) \times (5 \times 10^{-4}) + (6 - 2) \times (2 \times 10^{-3}),$$

$$P_{AMP} = 1.8 \times 10^{-2} \text{ W}.$$

Figure 6.24 shows that amplifier speed increases with drain current. Therefore, power consumption also increases with speed. Keep in mind that speed is limited by the mobility effect illustrated in Fig. 6.6. Beyond this point power may be dissipated unnecessarily.

6.3 OUTPUT AMPLIFIER NOISE

There are several noise sources related to the output amplifier. They are reset noise, white noise, flicker noise, shot noise, contact noise, and popcorn noise. Before these noise sources are discussed, we will first review Johnson (white) noise characteristics and important background material.

6.3.1 JOHNSON NOISE

At the atomic level, current flow through any material is erratic. Current is composed of free electrons that each contribute to the flow. When electrons collide and stop, we say the current flow decreases. The degree of stopping and accelerating of a carrier is dependent on the thermal energy that it possesses. Phonon collisions and lattice vibrations impart thermal energy to carriers and their movement. The erratic motion of carriers is measured as thermal noise (i.e., Brownian movement of carriers).

Johnson experimentally recognized characteristics associated with thermal noise. He described thermal noise in terms of random noise power (described in his classic paper "Thermal agitation of electricity in conductors," *Phys. Rev.*, July, 1928). Johnson showed that noise power exists in all conductors with a magnitude proportional to operating temperature. Nyquist, in the same year, theoretically described thermal noise by thermodynamic reasoning and derived the celebrated expression

$$W = \sqrt{4kTBR}, \qquad (6.36)$$

where W is the rms noise voltage (V), k is Boltzmann's constant (1.38×10^{-23} J/K), T is the absolute temperature (K), and R is the resistance of the conducting path (Ω). B is the equivalent frequency bandwidth or noise power bandwidth (Hz), an important noise parameter discussed below.

Example 6.23

Find the rms voltage generated by a 1-MΩ resistor at room temperature (293 K), assuming an effective bandwidth of 200 kHz.
Solution:
From Eq. (6.36),

$$W = \left[4 \times \left(1.38 \times 10^{-23}\right) \times 293 \times \left(200 \times 10^3\right) \times 10^6 \right]^{1/2},$$

$$W = 56.9 \ \mu V \ rms.$$

Figure 6.26 plots thermal noise voltage as a function of bandwidth and resistance. For example, the noise generated by a 10-kΩ resistor is 13 nV/Hz$^{1/2}$.

Due to the fact that carriers can rapidly react to thermal agitation, thermal noise is frequency independent over a broad range of frequencies. Thermal noise is also referred to as "white noise," because of its similarity to white light, which is composed of all the frequencies in the optical spectrum.

It should be mentioned that different resistor types generate more noise than others. Three main types of resistors are used in CCD camera systems: (1) wire wound, (2) composition, and (3) thin film. Of these, the wire wound resistors are best at generating a noise voltage no greater than Johnson noise. Composition resistors exhibit contact noise because they are made of many individual particles molded together. When current flows through this resistor type, low-frequency noise is generated (i.e., $1/f$ noise). The amount of noise generated depends on current flow and resistor manufacturer. Thin film resistors also exhibit contact noise but are considerably lower because the material of the resistor is homogeneous. Wire wound or thin film resistors should be used for the CCD load resistor and

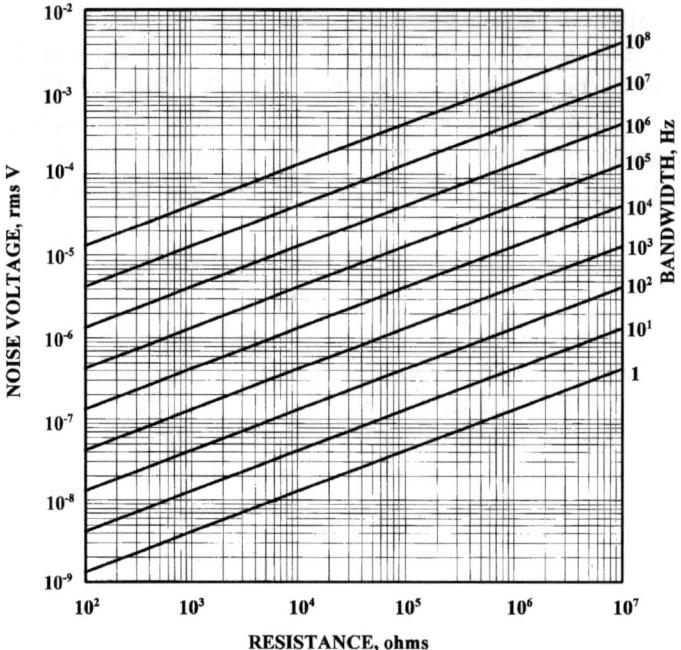

Figure 6.26 White noise voltage and bandwidth as a function of resistance.

sensitive preamplifier circuits. Composition resistors can be used in nonsensitive circuits.

White noise exhibits a Gaussian probability distribution given by

$$\tilde{\omega} = \frac{1}{(2\pi)^{1/2} V_{rms}} e^{-(V^2/2V_{rms}^2)}, \qquad (6.37)$$

where V_{rms} is the rms value that is equal to one standard deviation (or the square-root of the variance).

Example 6.24

Find the probability of occurrence for a thermal noise spike at least 4 times greater than the rms level.

Solution:

The probability of finding an instantaneous noise spike between two values is equal to the integral of the probability density function between the two values. Integrating Eq. (6.37) from four to infinity and doubling it yields the desired probability. Figure 6.27 plots the noise spike probability as a function of standard deviation. Therefore, the probability of occurrence is 0.006%.

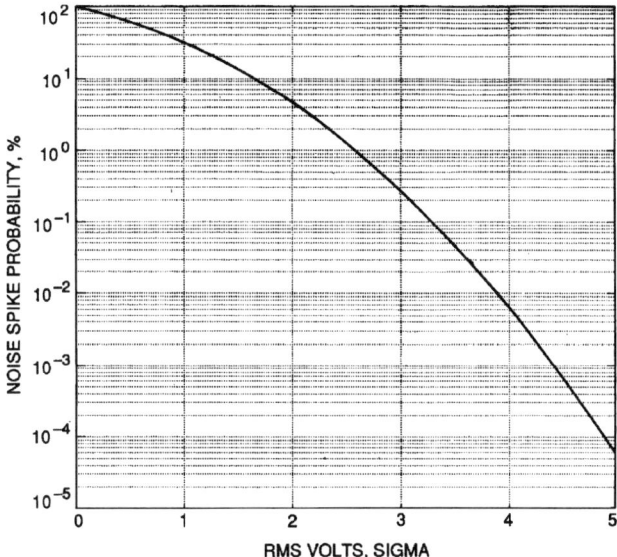

Figure 6.27 Noise spike probability as a function of standard deviation.

Noise is also specified as a peak-to-peak value as measured with an oscilloscope. A peak-to-peak measurement can be converted into rms reading by dividing by a factor of roughly 8. That is,

$$V_{RMS} = \frac{V_{P\text{-}P}}{8}, \qquad (6.38)$$

where V_{RMS} is the rms noise level (V), and $V_{P\text{-}P}$ is the peak-to-peak noise level (V).

Example 6.25

A 5-mV peak-to-peak noise level is measured with an oscilloscope at the output of a preamplifier circuit. Find the rms level.
Solution:
From Eq. (6.38),

$$V_{RMS} = 5 \times 10^{-3}/8,$$

$$V_{RMS} = 6.25 \times 10^{-4} \text{ V}.$$

The noise bandwidth B in Eq. (6.36) is defined for a system that exhibits uniform gain throughout a passband of frequencies, but zero gain outside of this passband. Practical circuits do not exhibit such ideal characteristics, and therefore an "equivalent" bandwidth is used. Theoretically, the equivalent bandwidth is the width of a normalized rectangle with the same area as the normalized power response of the circuit in question. This property can be analyzed for a simple RC lowpass network, a circuit often encountered in CCD noise analysis. The input to output voltage transfer function for this circuit is

$$H(f) = \frac{1}{1 + j(2\pi f RC)}, \tag{6.39}$$

where R is the resistance (Ω), C is the capacitance (F), and f is the frequency (Hz).

The power transfer function is

$$|H(f)|^2 = \frac{1}{1 + (2\pi f RC)^2}. \tag{6.40}$$

The equivalent bandwidth is defined as

$$B = \int_0^\infty |H(f)|^2 df. \tag{6.41}$$

Substituting Eq. (6.40) into (6.41) and integrating yields the effective noise bandwidth,

$$B = \frac{1}{4RC}, \tag{6.42}$$

where the half-power frequency is defined as

$$f_{3\text{-}dB} = \frac{1}{2\pi\tau_D}, \tag{6.43}$$

where

$$\tau_D = RC \tag{6.44}$$

is the dominant time constant of the circuit. Recall that the half-power frequency is equal to voltage gain when down by $2^{1/2}$.

Example 6.26

Find the equivalent noise bandwidth, the half-power frequency, and the time constant for an RC circuit. Assume $R = 10 \text{ k}\Omega$ and $C = 16 \text{ pF}$.

Solution:
From Eq. (6.42),

$$B = 1/[4 \times 10^4 \times (16 \times 10^{-12})],$$

$$B = 1.56 \text{ MHz}.$$

From Eq. (6.43),

$$f_{3\text{-}dB} = 1/[6.28 \times 10^4 \times (16 \times 10^{-12})],$$

$$f_{3\text{-}dB} = 995 \text{ kHz}.$$

From Eq. (6.44),

$$\tau_D = 10^4 \times (16 \times 10^{-12}),$$

$$\tau_D = 0.16 \text{ μsec}.$$

6.3.2 RESET NOISE

Reset noise is generated by the periodic resetting of the sense node by the reset MOSFET. Therefore, the reference level is different from pixel to pixel (this noise is easily seen at the preamplifier output with an oscilloscope). The noise induced on the sense node is caused by thermal noise generated by the channel resistance of the reset transistor. Reset noise at the sense node is

$$N_R = \sqrt{4kTBR_R}, \tag{6.45}$$

where N_R is the noise voltage on the sense node (rms V), R_R is the effective channel resistance of the reset switch (Ω), and B is the noise power bandwidth (Hz).

Substituting the equivalent bandwidth [i.e., Eq. (6.42)] into Eq. (6.45) yields

$$N_R = \sqrt{\frac{kT}{C_S}}. \tag{6.46}$$

Note that reset noise is dependent only on the sense node capacitance and independent of reset channel resistance.

Reset noise, $N_R(\text{e}^-)$, can also be expressed in units of rms noise electrons as

$$N_R(\text{e}^-) = \frac{\sqrt{kTC_S}}{q}, \tag{6.47}$$

where q is the electronic charge (1.6×10^{-19} C). Reset noise expressed this way is called "kTC noise." Reset noise can also be expressed as a function of node sensitivity as

$$N_R(e^-) = \sqrt{\frac{kT}{qS_V}}. \qquad (6.48)$$

At room temperature (300 K), kT equals 4.14×10^{-21} W/Hz, reducing the above equation to

$$N_R(e^-) = \frac{0.158}{\sqrt{S_V}}. \qquad (6.49)$$

This equation implies that node sensitivity can be determined by simply measuring reset noise. However, we will show below that $1/f$ noise also present from the output amplifier prevents us from doing this.

Example 6.27

Find the reset noise at room temperature associated with a sense node capacitance of 0.1 pF.
Solution:
From Eq. (6.47),

$$N_R(e^-) = \left[(4.14 \times 10^{-21}) \times 10^{-13}\right]^{1/2} / (1.6 \times 10^{-19}),$$

$$N_R(e^-) = 127 \text{ e}^-.$$

The effective impedance at the floating diffusion is very large when the reset switch is "off." In theory, the resistance would be infinite if it weren't for dark current that is collected by the sense node during the "off" state. The leakage current defines the "off" resistance by the equation

$$R_{OFF} = \frac{V_{REF}}{I_{DC} A_{SN}}, \qquad (6.50a)$$

where R_{OFF} is the off resistance, I_{DC} is the dark current (A/cm^2) generated in the sense node region, and A_{SN} is the effective area of the sense node (cm^2).

The "off" resistance is experimentally determined by measuring the change in sense node voltage between reset clocks as dark current collects. In order to see a significant change in sense voltage, the reset clock rate must be slow (e.g., resetting

CHARGE MEASUREMENT

the reset switch with a vertical clock is one way of doing this). The output voltage change measured during this time period defines the "off" resistance through

$$R_{OFF} = \frac{V_{REF}\Delta t}{\Delta V S_V q}, \qquad (6.50b)$$

where ΔV is the change in voltage seen at the output of the CCD for a reset time period of Δt seconds.

Example 6.28

Determine the "off" resistance given $\Delta V = 1.0$ V, $\Delta t = 10^{-3}$ sec, $S_V = 5 \times 10^{-6}$ V/e$^-$, and $V_{REF} = 12$ V.
Solution:
From [Eq. (6.50b)],

$$R_{OFF} = 12 \times 10^{-3}/\left[1 \times \left(5 \times 10^{-6}\right) \times \left(1.6 \times 10^{-19}\right)\right],$$

$$R_{OFF} = 1.5 \times 10^{22} \ \Omega.$$

The "on" resistance can be estimated through

$$R_{ON} = \frac{R_S L}{W t}, \qquad (6.51)$$

where R_S is the channel resistivity (Ω-cm), L is the reset gate length from the V_{REF} contact to the sense node contact (cm), W is the reset gate width (cm), and t is the depth of the buried channel (cm).

Example 6.29

Determine the "on" resistance for a 12(L) × 4(W)-μm reset MOSFET. Assume a channel resistivity of 0.2 Ω ($N_D = 1.6 \times 10^{16}$/cm^{-3}) and a channel depth of 0.5 μm.
Solution:
From Eq. (6.51),

$$R_{ON} = 0.2 \times 12 \times 10^{-4}/\left[\left(4 \times 10^{-4}\right) \times \left(0.5 \times 10^{-4}\right)\right],$$

$$R_{ON} = 12{,}000 \ \Omega.$$

This resistance may seem high for resetting purposes. However, the sense node capacitance is very small. For example, the reset time constant associated with a 20 fF node capacitance would only be 0.24 nsec.

540   CHAPTER 6

The reset "on" resistance can be measured from an output amplifier spectral noise plot when the reset switch is constantly left "on" (i.e., the gate of the reset MOSFET is tied to V_{REF}). For example, Fig. 6.9 was taken with the reset MOSFET in the "on" condition. For this CCD, the flat portion of the curve represents the white noise generated by the reset "on" resistance. An "on" resistance of 10 kΩ corresponds to the 14-nV/Hz$^{1/2}$ white noise figure measured (Fig. 6.26 is used to convert white noise to an equivalent resistance). Note that the white noise level is constant with changing drain current, indicating that the output MOSFET white noise floor is less than the reset "on" resistance.

As indicated above, reset noise power is identical whether the reset switch is "on" or "off." However, the time variance of the noise between the two states is radically different. The high resistance in combination with small sense node capacitance generates a noise voltage that changes very slowly with time. On the other hand, when the reset transistor is "on," the noise voltage changes very quickly, typically exhibiting a bandwidth of several hundred megahertz. Figure 6.28 illustrates reset "on" and "off" noise characteristics in the frequency and time domains. Note that when the reset switch is "on," the noise voltage changes very quickly, leaving an undefined level on the sense node after the switch is turned off. However, after the reset switch has turned off, the noise voltage will not significantly change over a pixel period. This very important characteristic allows one to completely remove reset noise by using the technique of "correlated double sampling" (CDS), an important CCD signal processing technique described below. CDS eliminates reset noise by differencing a sample taken from the reference level from a sample taken after signal charge has been transferred to the sense node. We can

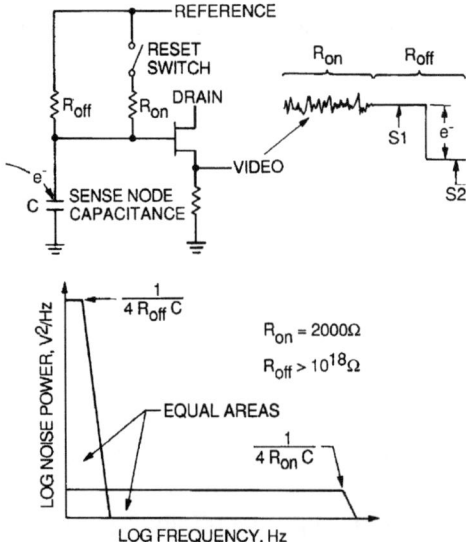

Figure 6.28 Sense node noise frequency spectrums when reset switch is on and off.

eliminate reset noise very effectively because the noise is "correlated" between the two samples (i.e., the noise amplitude does not change over the sample period).

Example 6.30

Find the noise power bandwidth at the sense node when the reset switch is "on" and "off." Assume $R_{ON} = 2$ kΩ, $R_{OFF} = 10^{18}$ Ω, and a node capacitance of 0.1 pF.

Solution:
From Eq. (6.42),
"on" state:

$$B = 1/[4 \times (2 \times 10^3) \times 10^{-13}],$$

$$B = 1.25 \times 10^9 \text{ Hz},$$

"off" state:

$$B = 1/(4 \times 10^{18} \times 10^{-13}),$$

$$B = 2.5 \times 10^{-6} \text{ Hz}.$$

It should be mentioned that there is another component of reset noise, referred to as partition noise.[7] This noise is generated when carriers under the reset gate split between the floating diffusion and V_{REF} as the reset switch goes off. That is, charge does not exactly divide between the two regions during this time, but varies from reset to reset. Partition noise is generated by stochastic thermal diffusion and is proportional to the reset channel capacitance. It too is removed by the CDS processor.

6.3.3 WHITE NOISE

Thermal or white noise generated by the output amplifier MOSFET is given by the Johnson white noise equation,

$$W_{CCD}(V) = \sqrt{4kTBR_{OUT}} \tag{6.52}$$

or equivalently, from Eq. (6.18),

$$W_{CCD}(V) = \sqrt{\frac{4kTB}{gm}}, \tag{6.53}$$

where $W_{CCD}(V)$ is the MOSFET white noise (rms V) and $gmR_L \gg 1$.

White noise in terms of amplifier, geometry from Eq. (6.8) is

$$W_{CCD}(V) = \sqrt{\frac{4kTBL}{\mu_{SI} C_{OX} W(V_{GS} + V_{AEFF})}}. \qquad (6.54)$$

The white noise referred to the sense node in rms e^- as

$$W_{CCD}(e^-) = \frac{W_{CCD}(V)}{S_V A_{CCD}}, \qquad (6.55)$$

where $W_{CCD}(e^-)$ is the MOSFET white noise (rms e^-).

Example 6.31

Determine the thermal noise generated by a CCD source follower amplifier with the following parameters: $gm = 100$ µmho, $T = 293$ K, and $B = 125{,}000$ Hz. Also calculate the noise referred to the sense node, given $S_V = 10^{-6}$ V/e^- and $A_{CCD} = 0.8$ V/V.

Solution:
From Eq. (6.53),

$$W_{CCD}(V) = \left[4 \times (1.38 \times 10^{-23}) \times 293 \times (1.25 \times 10^5)/10^{-4}\right]^{1/2},$$

$$W_{CCD}(V) = 4.5 \text{ µV rms}.$$

From Eq. (6.55),

$$W_{CCD}(e^-) = (4.5 \times 10^{-6})/(0.8 \times 10^{-6}),$$

$$W_{CCD}(e^-) = 5.6 \text{ } e^- \text{ rms}.$$

Example 6.32

Find the noise generated by the amplifier in Example 6.31 when the operating temperature is reduced to $-100°C$ (173 K).

Solution:
From Eq. (6.53),

$$W_{CCD}(V) = \left[4 \times (1.38 \times 10^{-23}) \times 173 \times (1.25 \times 10^5)/10^{-4}\right]^{1/2},$$

$$W_{CCD}(V) = 3.5 \, \mu V \, \text{rms}.$$

From Eq. (6.55), the noise at the sense node is

$$W_{CCD}(e^-) = (3.5 \times 10^{-6})/(0.8 \times 10^{-6}),$$

$$W_{CCD}(e^-) = 4.37 \, e^- \, \text{rms}.$$

Note that white noise is not significantly dependent on operating temperature because of the square-root relation of Eq. (6.52). Noise is reduced only by a factor of 1.28 as compared with room-temperature operation.

6.3.4 FLICKER NOISE

The second noise source associated with the output amplifier is flicker or $1/f$ noise.[8] As will be shown below, $1/f$ noise will ultimately limit the CCD's read noise floor. Figure 6.29 shows the spectral noise characteristics taken from a Hubble WF/PC I CCD amplifier at different operating temperatures. The noise spectral plots were obtained with the CCD amplifier biased under nominal operating conditions. The CCD clocks (i.e., vertical, horizontal, and reset clocks) were inhibited

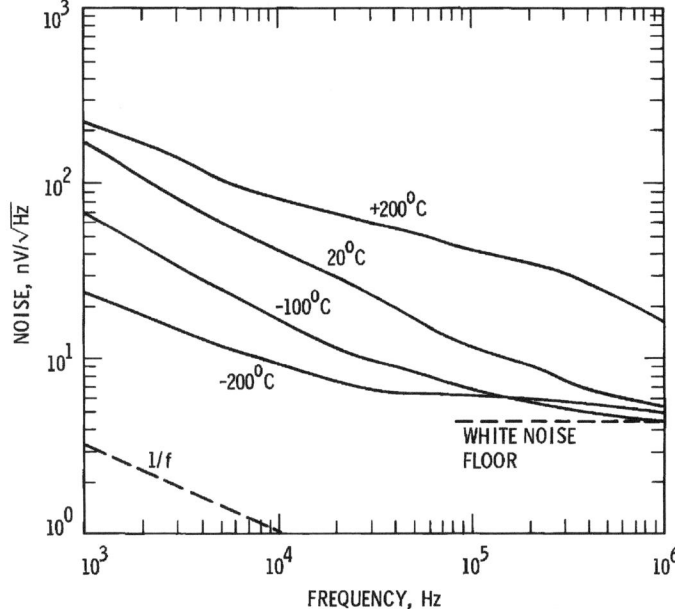

Figure 6.29 Output MOSFET noise frequency spectrum showing $1/f$ and white noise as a function of temperature.

for the measurement to avoid overloading the wave-analyzer by the large clock feed-through pulses present at the output of the CCD when clocked. Note that $1/f$ noise decreases by a factor of 3.16 for each decade increase in frequency. The noise power (i.e., volts squared per Hz) decreases by a factor of 10 for each decade increase in frequency characteristic of a $1/f$ noise spectrum. Flicker noise falls at this rate until the white noise floor is encountered (approximately 4.5 nV per root-Hz for this output MOSFET).

It is generally accepted that flicker noise is generated by surface interface states and to the tunneling of electrons into the oxide.[9] These states arise because of a disruption of the silicon lattice at the surface. This interface problem leads to energy levels within the forbidden gap. In theory, there should be approximately one surface state for every exposed silicon atom. This leads to a density of about 10^{15} states/cm^2. However, silicon does immediately oxidize when exposed to air. The SiO$_2$ and silicon atoms bond together, reducing interface state density (on the order of 10^{11}–10^{12} states /cm^2). High-temperature thermal oxides used in processing CCDs reduce the count to even lower densities (10^9–10^{10} states/cm^2). Silicon wafers with (100) orientation have a lower interface density than (111) material, and are therefore usually preferred for CCDs.

When electrons become trapped in these states, they are later released with a wide range of emission times. Carriers are taken away from the channel current by the ratio τ_e/t_r, where τ_e is the emission time constant of the interface state and t_r is the total transit time that carriers are in the channel. The $1/f$ spectrum arises because fluctuations in the long time constant states (i.e., low frequencies) contribute a greater fluctuation in the output than the shorter time constant states (i.e., high frequencies) because of the τ_e/t_r gain effect.[10]

Flicker noise generated by a single interface state can be approximated by the equation[11]

$$E_n^2 = \frac{K_1 \tau_e}{1 + (2\pi f \tau_e)^2}, \quad (6.56)$$

where E_n^2 is noise power (volts-square per Hz), K_1 is a constant, f is frequency (Hz), and τ_e is the interface state emission time constant [Eq. (5.27)].

The half-power $1/f$ noise frequency is

$$f_{3\text{-}dB} = \frac{1}{2\pi\tau_e}. \quad (6.57)$$

Note from Eq. (6.56) that the noise power increases proportionally with the trap emission time constant, which is characteristic of the model just described. The cutoff frequency is also determined by the trap time constant. Deeper traps generate low-frequency noise, whereas shallow traps exhibit higher cutoff frequencies. The distribution of energy levels of the traps at the Si–SiO$_2$ interface are for the most part uniformly distributed throughout the silicon bandgap.[12] Therefore, the states

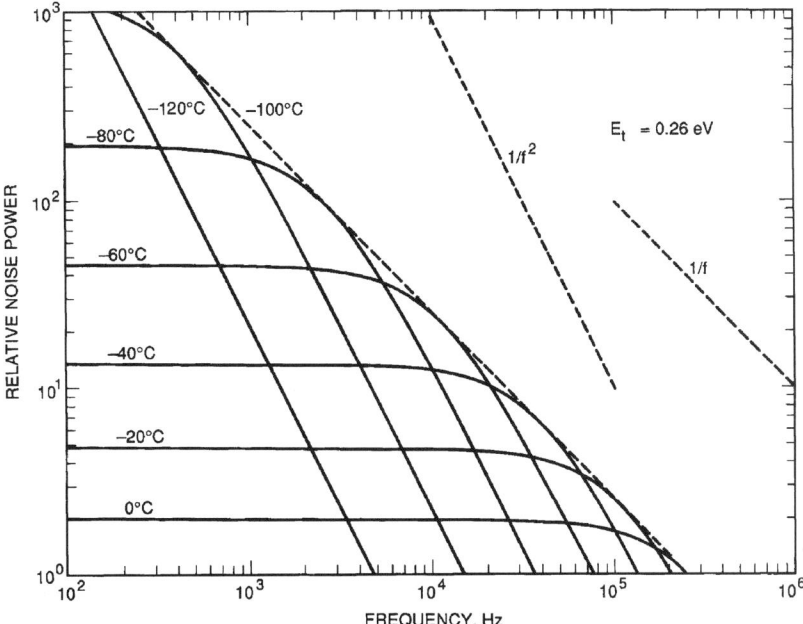

Figure 6.30 Spectral noise characteristics for a 0.26-eV trap at different operating temperatures.

exhibit a uniform distribution of time constants, producing a wide distribution of frequencies and the $1/f$ signature seen in noise spectrums.

Figure 6.30 plots Eq. (6.28) for a single trap level, assuming an activation energy of 0.26 eV at different operating temperatures. As the temperature is reduced, the emission time constant increases, generating a lower-frequency noise. Note that the noise voltage slope exhibits a characteristic $1/f^2$ slope. Many interface states with different activation energies together produce a noise spectrum with a $1/f$ slope. This characteristic is seen in Fig. 6.31, where noise power is plotted for four different traps ($E_T = 0.26, 0.3, 0.35, 0.4$ eV).

Example 6.33

From Fig. 6.30, find the interface noise power at 10^4 Hz for the following operating temperatures: $0, -20, -40, -60, -80, -100°C$.

Solution:

Reading the noise values directly from Fig. 6.30,

$$0°C = 2 \text{ V}^2/\text{Hz},$$
$$-20°C = 5 \text{ V}^2/\text{Hz},$$
$$-40°C = 12 \text{ V}^2/\text{Hz},$$
$$-60°C = 24 \text{ V}^2/\text{Hz},$$

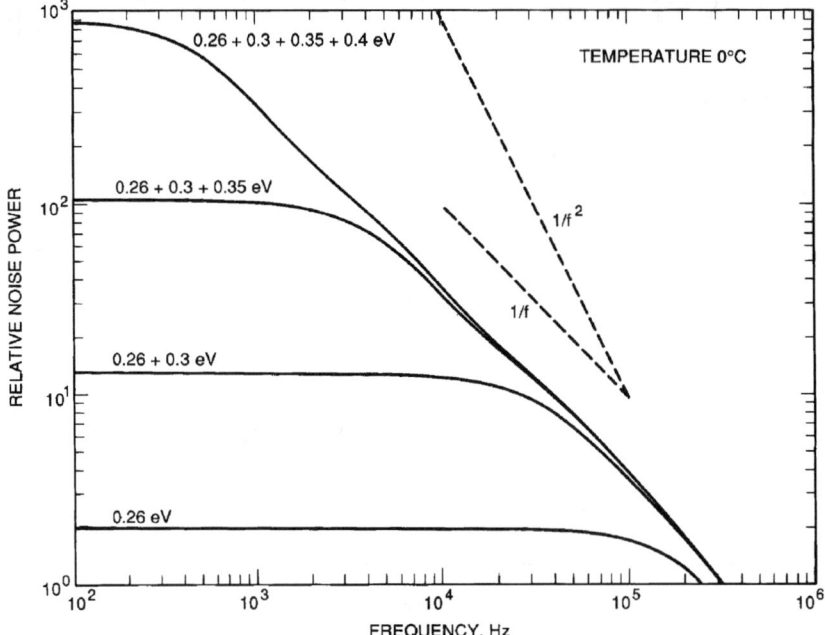

Figure 6.31 Flicker noise generated by several traps at fixed operating temperature.

$$-80°\text{C} = 12 \text{ V}^2/\text{Hz},$$
$$-100°\text{C} = 2.4 \text{ V}^2/\text{Hz}.$$

The noise power peaks at approximately $-60°$C. At this temperature, carriers are emitted approximately at a 10-kHz rate. For operating temperatures above $-60°$C, the detrapping rate of carriers is faster; and for lower temperatures the rate is slower.

Note in Fig. 6.29 that $1/f$ noise decreases as the operating temperature is lowered. Here, the $1/f$ noise is presumably pushed to lower frequencies because the trap time constants increase with decreasing temperature. However, this strong dependence with operating temperature is not normally seen when CCD MOSFET noise spectrums like this are generated. As mentioned above, the trap energies across the bandgap are for the most part uniform.[12] Assuming this is the case, there will be energies present to generate every possible frequency at any operating temperature. The data in Fig. 6.29 suggests that the states are not uniform for this MOSFET or that surface interaction is reduced as the temperature is lowered.

Many models have been invented to describe flicker noise behavior. One equation that fits CCD output amplifier noise data is[13]

$$F_{CCD}(\text{V}) = \frac{K_2 I_D}{(L_{\text{eff}} W f)^{1/2}}, \tag{6.58}$$

where $F_{CCD}(V)$ is the $1/f$ noise voltage (V/Hz$^{1/2}$), K is constant, I_D is the drain current (A), W is the gate width (cm), f is frequency (Hz), and L_{eff} is the effective gate length or that length of the channel where current is influenced by surface states. We will discuss the physical meaning of L_{eff} below. For a fixed current, $1/f$ noise voltage decreases as the square-root of area (W × L_{eff}). This characteristic can be physically described by two properties. First, the noise voltage generated is equal to the square root of the number of traps involved (i.e., $N_t^{1/2}$) which is proportional to the square-root of gate area, $A^{1/2}$. The square-root relation is required because the individual noise sources must be added in quadrature. Second, the noise generated is also dependent on the current density, which is inversely proportional to gate area (if there were no current, no $1/f$ noise would be generated). Therefore, the net noise with area is proportional to $A^{1/2} \times A = 1/(A)^{1/2}$, indicated by Eq. (6.58).

It would appear that $1/f$ noise would not be seen in buried-channel MOSFETs, because the channel current is maintained in the bulk away from surface states; for the most part, this is true. Figure 6.32 shows noise spectral data taken from buried- and surface-channel output amplifiers. Flicker noise for the surface-channel device is considerably greater because of surface interaction. However, there is substantial

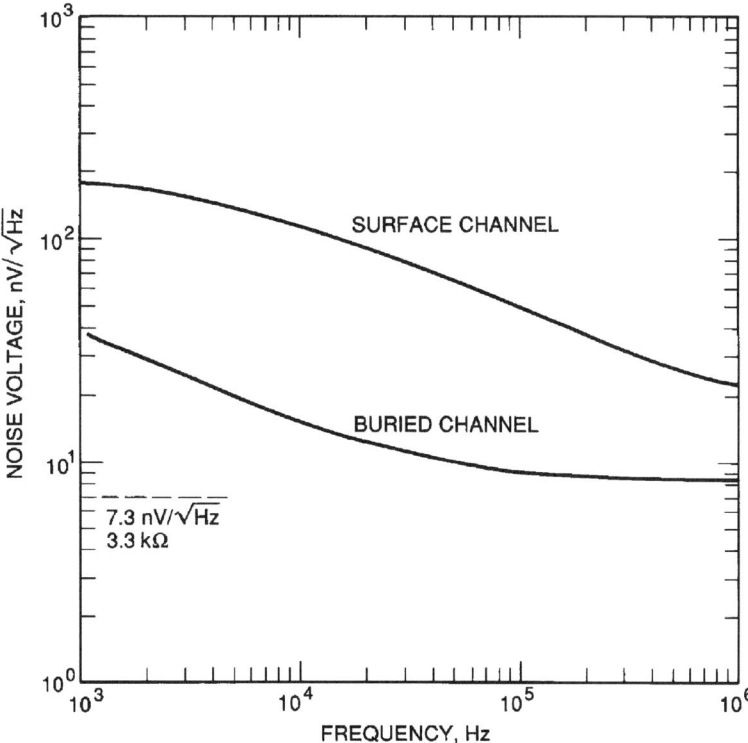

Figure 6.32 Spectral noise plots for surface and buried channel output amplifiers.

$1/f$ noise seen in the buried-channel device as well. Although buried, some surface interaction still takes place.

One theory for why buried-channel devices exhibit $1/f$ noise is shown in Fig. 6.33. Here we plot channel potential for a MOSFET as a function of distance from the gate. The top trace is taken near the source end, whereas the bottom trace is taken at the drain end. The drain, source, and gate were biased at $V_{DD} = 20$ V, $V_{REF} = 14$ V, and $V_G = 15$ V for this model. The doping and drive used in the

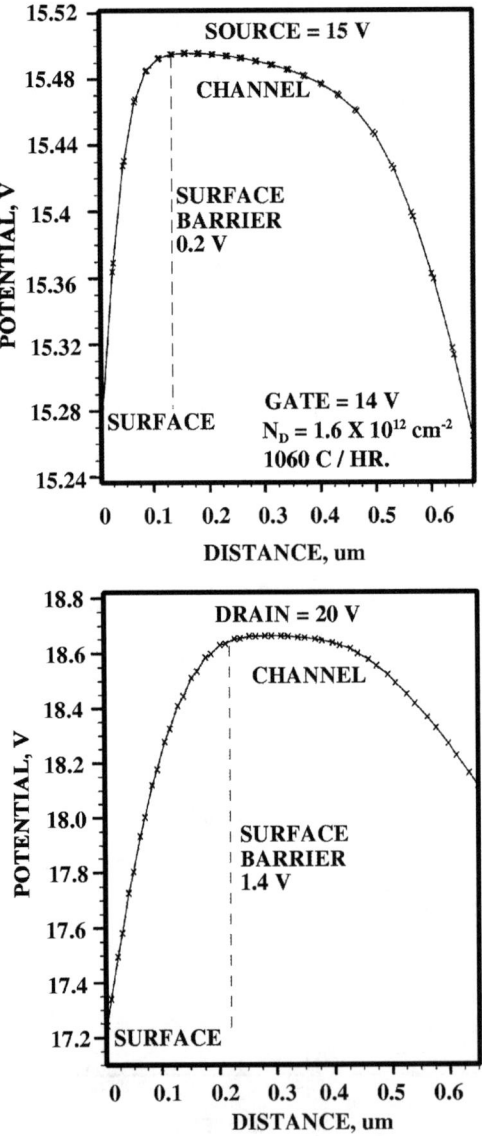

Figure 6.33 Surface barrier potentials near the source and drain regions.

model produces a threshold of $V_{AEFF} = 6$ V. Note that the barrier potential between the channel and surface near the source is small, approximately 0.2 V. The potential on the drain side is greater, about 1.5 V. Recall from Chapter 4 on surface full-well discussions that a thermal barrier height of greater than 0.5 V is required to discourage carriers from interacting with the surface. Similarly, a small barrier at the source of the MOSFET is not sufficient to keep carriers away from the surface. Those electrons that penetrate the barrier have the opportunity to generate $1/f$ noise. The surface barrier height can be increased by increasing the channel doping and thermal drive. However, only a small improvement can be achieved. The MOSFET modeled in Fig. 6.33 was doped at $N_D = 1.6 \times 10^{12}$ cm^{-2} with a drive of 1060°C for one hour. The source barrier height increases to 0.3 V when the drive is extended to 1120°C for one hour.

Figure 6.34 shows modeling data for the same MOSFET as Fig. 6.33 with an elevated bias ($V_{DD} = 29$ V, $V_S = 21$ V, and $V_G = 14$ V). In this case the thermal barrier near the source is 0.75 V, because more depletion from the gate side occurs when a larger negative gate-to-source voltage is applied. Therefore, less surface

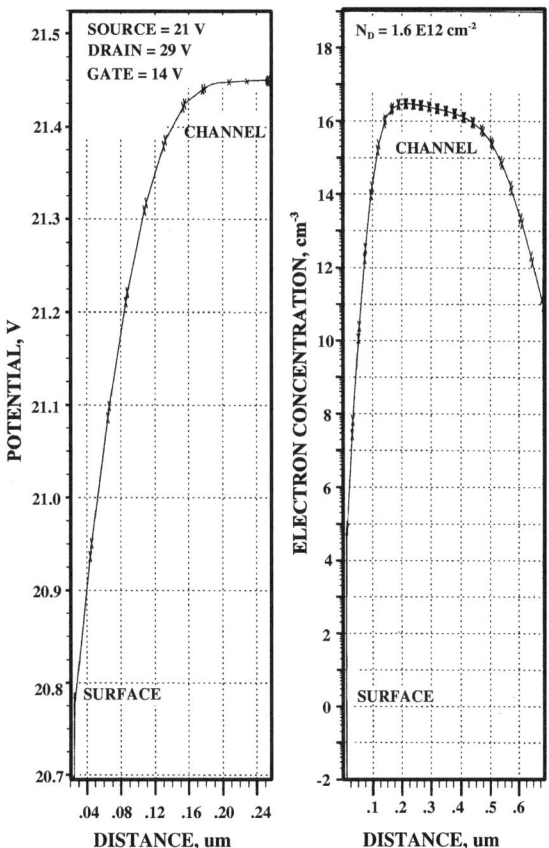

Figure 6.34 Surface barrier potential and electron concentration.

interaction will occur, producing less $1/f$ noise. Shown to the right of this plot is the electron current density near the source. Note that the electron density is high within the channel and decreases exponentially as we move toward the surface into depleted material. The finite electron density at the surface will contribute to $1/f$ noise.

Reducing the gate voltage has the greatest effect on $1/f$ noise because the drain current is reduced [i.e., Eq. (6.58)]. Also, the thermal barrier height and the depletion depth at the source increase, reducing surface interaction. Spectral noise measurements taken with a spectrum analyzer show this behavior. For example, Fig. 6.35 plots spectral noise as a function of frequency and load resistance for a WF/PC I output amplifier. The gate and drain voltages are fixed in this measurement (i.e., $V_{DD} = 29$ V, $V_{REF} = 16.5$ V). Increasing the load resistance reduces the gate-to-source voltage, which, in turn, reduces the drain current and increases the surface barrier height. Both effects will reduce the amount of $1/f$ noise, as seen. However, there are penalties paid when increasing the load resistance too much. As the load resistance is increased, the transconductance is also lowered. Loss of transconductance will increase the white noise to the point where it can dominate the $1/f$ floor. A reduction in transconductance will also lower the output voltage gain, also increasing the read noise (rms e$^-$). The output impedance also increases,

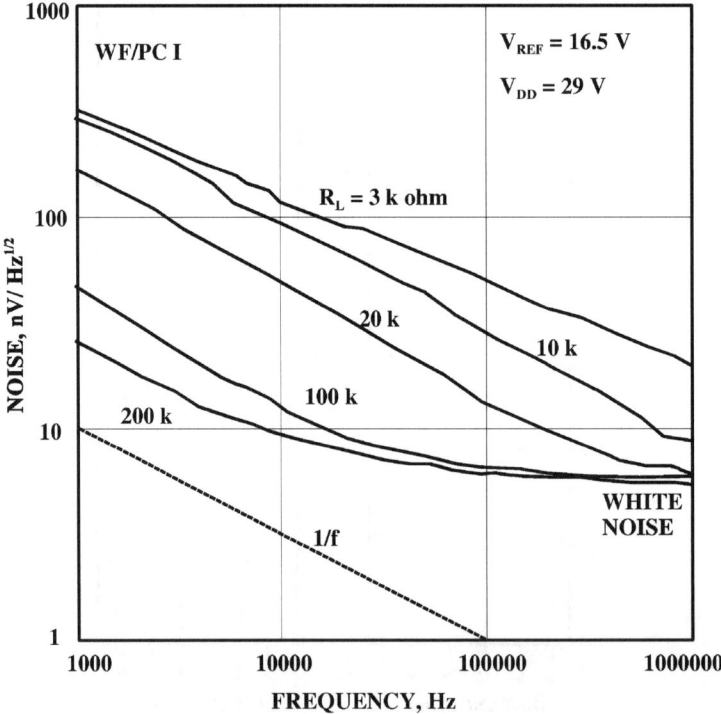

Figure 6.35 Spectral noise as a function of load resistance.

reducing drive capability and high-speed operation. Lastly, linearity performance also becomes an issue [Eq. (6.31)].

Not explained by the foregoing discussion is just how the trapping and untrapping of electrons in surface states translates into $1/f$ noise with the MOSFET drain current, or equivalently, with the MOSFET drain-to-source conductance. The simplest picture is based on the fact that when electrons are trapped, they are removed from the current flow, and when untrapped, they are added back into it.[8] An alternate explanation, that the origin of $1/f$ noise was due to mobility fluctuations, was proposed empirically by Hooge,[14] although a connection between mobility fluctuations and Couloumb scattering by charge trapped in surface states was not made until years later by Surya and Hsiang.[15] The conductivity σ of a semiconductor is given by $q\mu N$, where μ is the mobility and N is the free carrier number density. As can be seen from this equation, either a number density fluctuation or a mobility fluctuation will result in a change in conductivity and related noise generated.

A unique insight into the physical mechanism responsible for the observed $1/f$ conductivity fluctuations came in the mid-1980s, when for the first time it was possible to fabricate and study silicon MOSFETs with submicron gate sizes (e.g., 0.1-μm gate length × 1.0-μm gate width). Recall from the earlier discussion that surface state number densities down to a few times 10^9/cm^2 are possible; the area of a 0.1 × 1.0-μm gate is 10^{-9} cm^2, leading to only a few traps expected per device! By cooling submicron MOSFETs to cryogenic temperatures, investigators were able to observe the noise behavior of a single electron being trapped and untrapped.[16–18] These and later studies all revealed a "random telegraph signal" (RTS) behavior; that is, the drain current (or conductance) switches randomly between a "high" level and a "low" level, and the time spent in either state varies with temperature in a manner consistent with electron occupancy of a single surface state trap. Interestingly, from device to device the RTS amplitude (the high/low level difference) also varies over a large range.

It is this last observation that is most difficult to account for using the number density fluctuation model (recall that the RTS signal is due to the addition or subtraction of one electron, but the free carrier concentration from device to device varies over a much smaller range than the RTS amplitude). Under the trap-induced mobility fluctuation model, the RTS amplitude is expected to depend strongly on exactly where the trap is located with respect to the three-dimensional current flowing through the MOSFET. Imagine inducing RTS fluctuations in a current of water by randomly inserting and removing an obstacle such as a boulder into the stream. The size of the fluctuation will obviously depend upon whether the obstacle is inserted and removed from center of the stream (a large-amplitude RTS) or alternatively inserted and removed from a "trap" location along one of the banks (a small-amplitude RTS current fluctuation).

The flicker noise power of a single RTS signal is Lorentzian, having the form given earlier by Eq. (6.56). For a MOSFET of the size used as a CCD amplifier, a few hundred or more traps are involved, and the total noise power has a $1/f$ spectrum arising as shown in Fig. 6.35. But, the constant K_1 in Eq. (6.56) for a

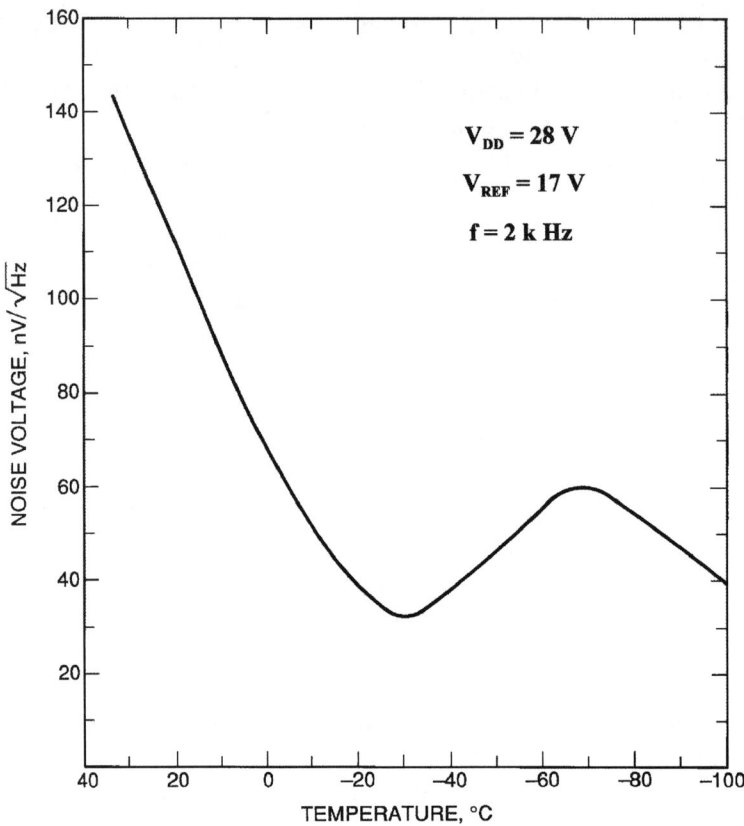

Figure 6.36(a) Bulk trap noise peak at 2 kHz and −70°C.

single trap is in fact proportional to $(\Delta I)^2$, where ΔI is the RTS current amplitude. The strength of the total $1/f$ noise power $S_i(I)$ is then proportional to the sum over all N_t traps (i.e., $S_i(I)/I^2 \propto \sum E_n^2/I^2 \propto N_t(\Delta I/I)^2$ where I is the current through the MOSFET and E_n^2 is given by Eq. (6.56). Following Kirton and Uren, we have chosen to normalize by $1/I^2$ throughout.)[19,20]

For an assumed uniform trap density, the total number of traps N_t increases in direct proportion to the MOSFET gate area A. Conversely, for a uniform current flow, the influence of each trap when expressed as the dimensionless ratio $\Delta I/I$ decreases proportional to $1/A$. Therefore, the normalized $1/f$ noise power $S_i(I)/I^2$ is proportional to $A \times (1/A)^2$, or simply $1/A$. Uren investigates the scaled function $A \times S_i(I)/I^2$ as a function of frequency for MOSFETs ranging in area from $0.4\,\mu m^2$ to $350\,\mu m^2$, and find excellent $1/f$-noise agreement independent of device area when scaled in this way. For CCD applications, one need just remember that $1/f$ noise power spectral density $S_i(I)$ is proportional to I^2/A [i.e. Eq. (6.58)]. There are many interesting papers for further reading on the physics of $1/f$ noise.[21–27]

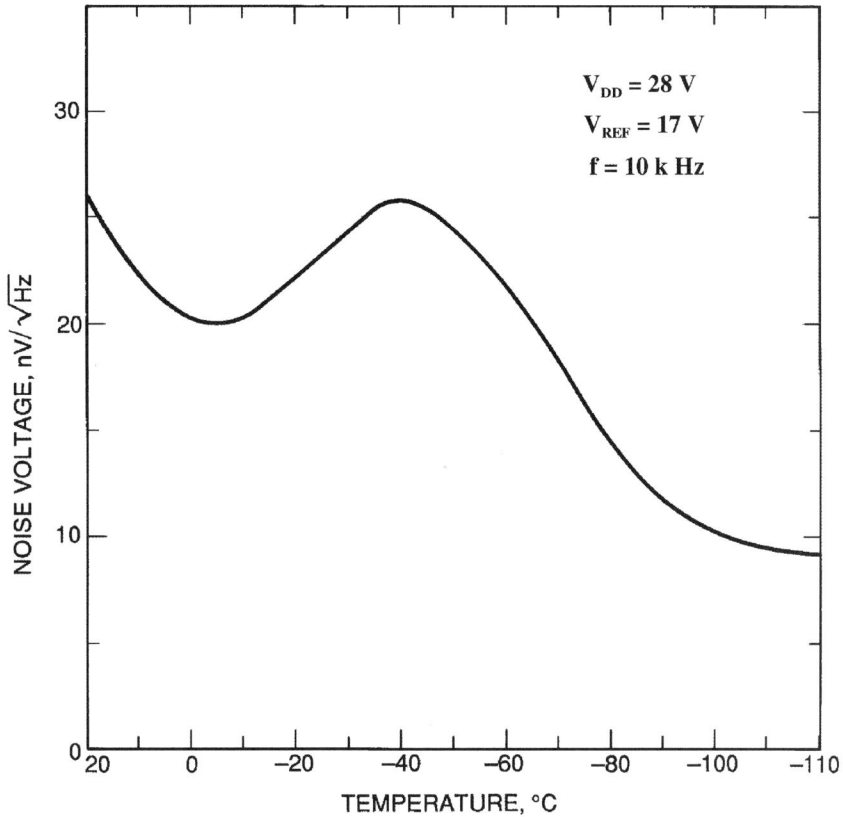

Figure 6.36(b) Bulk trap noise at 10 kHz and −50°C.

It should be mentioned that flicker noise is also generated by bulk states located in the buried channel (e.g., impurities in the silicon).[28,29] A noise peak is observed at specific frequency as the temperature is varied, indicating that a bulk trap is present. A noise peak is seen when $\tau_e = 1/2\pi f_1$, where f_1 is the approximate noise frequency peak. Figure 6.36(a) shows a spot noise measurement taken at 2 kHz for a Hubble WF/PC I MOSFET as a function of operating temperature. The data exhibit a noise peak at −70°C, indicating that a specific trap is at work. Figure 6.36(b) shows the same trap taken at 10 kHz. Note that the peak moves toward a higher operating temperature (−50°C).

6.3.5 SHOT NOISE

Shot noise exhibits a flat frequency spectrum, as does white noise. The primary difference between the two sources is that thermal noise is produced by erratic movement of free carriers, while shot noise is generated when carriers suddenly appear and disappear. Thus, thermal noise is independent of current flow, whereas shot noise is directly dependent on current flow. In semiconductors, shot noise is typically produced when carriers pass through a potential barrier. An example of a device that generates shot noise is a reverse biased junction diode. Here, spontaneous creation of carriers in the depletion region occurs, resulting in a reverse saturation current (i.e., dark current). A bipolar transistor also generates shot noise because minority carriers are randomly injected from the emitter into the base.

Shot noise varies widely for electronic devices. The term "full shot noise" was defined in the days of the vacuum tube. When electrons in a tube are fully emitted by the cathode, a maximum shot noise is seen. However, if a space charge is present around the cathode such that a constant supply of electrons is available, then shot noise is reduced considerably. Shot noise was theoretically analyzed by W. Schottky in 1918. His basic equation for full shot noise is

$$I_{SN} = \sqrt{2qI_D B}, \tag{6.59}$$

where I_{SN} is the full shot noise (rms amps), I_D is the drain current (A), and B is the equivalent noise bandwidth (Hz).

Example 6.34

Find the full shot noise produced for a drain current of 1 mA in a CCD output MOSFET. Assume an effective bandwidth of 200 kHz. Also, find the noise voltage across a 20-kΩ load resistance and the equivalent noise in rms e$^-$.

Solution:
From Eq. (6.59),

$$I_{SN} = \left[2 \times \left(1.6 \times 10^{-19}\right) \times 10^{-3} \times \left(2 \times 10^5\right)\right]^{1/2},$$

$$I_{SN} = 8 \times 10^{-9} \text{ A rms}.$$

The noise voltage across a 20-kΩ resistor is 1.6×10^{-4} V, or 160 e$^-$, assuming a node sensitivity of $S_V = 10^{-6}$ V/e$^-$. Obviously, this noise level is significantly greater than what is typically measured for a CCD (i.e., a few electrons by slow scan). This example serves to show that full shot noise is not generated by the MOSFET output transistor. Fortunately, very few carries are generated or involve generation-recombination processes. The current conduction process in a field-effect device is by majority carriers, where the channel can be considered as a variable conductance.

Shot noise is seen when carriers disappear into surface states, resulting in $1/f$ noise. Shot noise is also generated in the pinch-off region of the MOSFET, induced by impact ionization. Both noise sources are drain current dependent. Therefore, MOSFET current is reduced as much as possible for this reason without significantly affecting gain, output impedance, and linearity characteristics.

6.3.6 Contact and Popcorn Noise

Contact noise is occasionally observed in the output amplifier. Contact noise is generally caused by fluctuating conductivity caused by a subtle defective bond wire connection between the MOSFET and package or aluminum contact on the CCD. The noise generated is usually dependent on operating temperature and bias voltage applied. Contact noise occurs in bursts and results in intermittent offset voltage shifts in the video signal. The period of the bursts may vary between a few micro-seconds to several minutes in an unpredictable manner. In extreme cases, the video signal from the CCD may completely switch on and off, indicating a severe contact or package connection problem. Popcorn noise is similar to contact noise and is usually related to the drain and source diffusions. Aluminum "spiking" in these regions is one possible mechanism that generates popcorn noise. This problem is associated with aluminum metal diffusing into the p-n junction, causing its premature breakdown.

6.3.7 Output Amplifier Noise Equation

Discussions above show that white and $1/f$ noise are the two dominant noise sources generated by the output amplifier. The combination of these noise sources is given by

$$N_{CCD}(f)^2 = W_{CCD}(f)^2 (1 + (f_c/f)^m), \tag{6.60}$$

where $N_{CCD}(f)^2$ is the MOSFET noise power (V²/Hz), $W_{CCD}(f)$ is the white noise voltage (V/root-Hz), and f_c is the $1/f$ noise corner frequency (Hz) defined where the $1/f$ noise power is equal to the white noise power (i.e., when $N_{CCD}(f)^2 = 2W_{CCD}(f)^2$). The constant m characterizes the slope of the $1/f$ spectrum, which can vary within the range $1 < m < 2$, depending on drain bias. For well-biased MOSFETs, the $1/f$ slope is usually close to unity.

Example 6.35

Determine the noise voltage generated by a CCD output amplifier at 10 kHz, 1 kHz, and 100 Hz. Assume a white noise floor of 10 nV/root-Hz, a $1/f$ corner frequency of 50 kHz, and a $1/f$ noise slope of unity.
Solution:
From Eq. (6.60),

$$N_{CCD}(f) = \{(10^{-8})^2 \times [1 + (5 \times 10^4)/10^4]\}^{1/2},$$

$$N_{CCD}(f) = 24.5 \text{ nV/root-Hz } (10 \text{ kHz}),$$

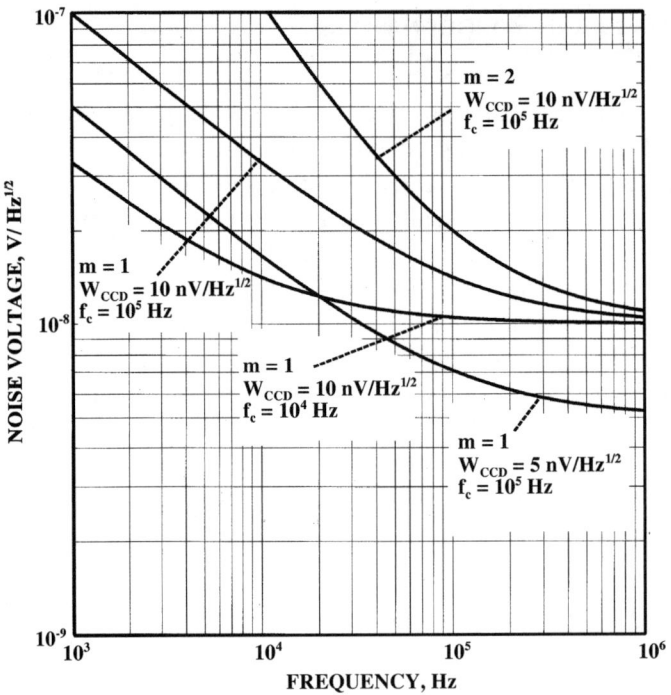

Figure 6.37 Output amplifier spectral noise $|N_{CCD}(f)|^2$.

$$N_{CCD}(f) = 71.4 \text{ nV/root-Hz (1 kHz)},$$

$$N_{CCD}(f) = 245 \text{ nV/root-Hz (100 Hz)}.$$

Note that the noise voltage at 100 Hz is a factor of 10 times greater than at 10,000 Hz, characteristic of $1/f$ noise. Flicker noise dominates white noise within this frequency range.

Figure 6.37 plots Eq. (6.60) as a function of frequency for different parameters (i.e., $W_{CCD}(f)$, f_c, and m). Equation (6.60) is important and will allow us to predict the CCD output read noise below.

6.4 Correlated Double Sampling

Correlated double sampling (CDS) is used to process CCD video signals.[30] The first and primary function of this processor is to achieve optimum signal-to-noise performance (i.e., minimum read noise in units of e^- rms). A properly designed processor will completely eliminate kTC reset noise and optimally filter white and $1/f$ noise generated by the output amplifier. Second, the processor provides sufficient amplification so that each pixel value can be digitized over the sensor's dynamic range. Third, the CDS circuit restores the dc level of each pixel by subtracting an offset generated by the CCD and off-chip electronics. Only the video portion of the signal is digitized (a few DN of offset accompanies the video for noise-encoding reasons). We will review basic CDS circuit elements before its transfer function is derived.

6.4.1 CORRELATED DOUBLE SAMPLING CIRCUIT ELEMENTS

Figure 6.38(a) shows a circuit diagram for a basic CDS circuit. There are five CDS circuit elements: (1) preamplifier, (2) postamplifier, (3) clamp, (4) sample-and-hold, and (5) ADC. The preamplifier provides low-noise voltage gain for the processor. The post-amplifier also furnishes some gain and defines the circuit's bandwidth that is set by a single RC pole with time constant, τ_D. The clamp switch and clamp capacitor restores the video signal to ground level (i.e., a new reference level for each pixel is established). When video is dumped onto the sense node, the clamp switch opens, allowing signal to pass through to the sample-and-hold circuit. The video is then sampled and held after a specified time period. This very important time period is referred to as the "clamp-to-sample time," or the "sample-to-sample time," t_s. After the video level is held, the ADC converts the pixel voltage level to a digital number (DN).

Figure 6.38(b) is a timing diagram for the CDS circuit just described.. There are six critical time periods in processing a pixel: (1) reset, (2) reference, (3) clamp, (4) video dump, (5) sample and hold, and (6) DN conversion. The start of a pixel readout sequence begins when the reset switch is turned "on," setting the sense node to the reference voltage (also refer to discussions on Fig. 1.24). The resetting action also generates a large clock feed-through pulse that accompanies the video signal (refer to point "A") [i.e., Eq. (1.28)]. A high impedance exists on the floating diffusion when the switch goes off, maintaining the reference level until signal charge is dumped. The clamp switch S1 is also "on" at this time, forcing the refer-

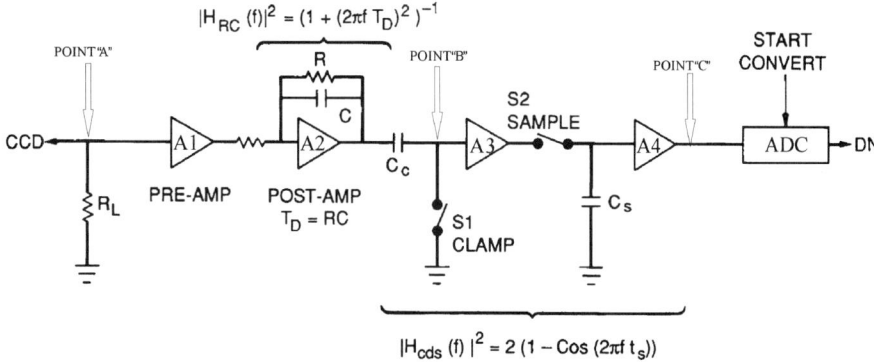

Figure 6.38(a) CDS circuit used to process CCD signals.

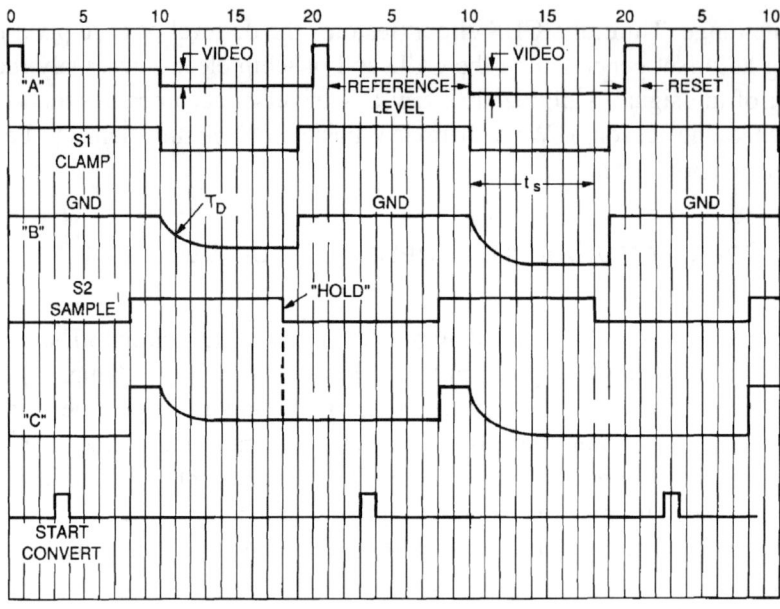

Figure 6.38(b) CDS timing diagram.

ence level to ground potential (refer to point "B"). This first sample reestablishes a precise and known reference level for each pixel before video dump occurs (i.e., reset noise is eliminated at this time). The clamp switch also suppresses the reset feed-through pulse in order to not overload the processor. After the video signal has settled, it is sampled and held by the sample switch S2 (refer to point "C"). The sample switch is turned on before the video is dumped, thereby allowing the circuit to track the signal during the clamp and release process. This assures that video waveform will always start at the same voltage level for each pixel. This timing is important for good linearity performance through the sample-and-hold circuit because it has a finite bandwidth. After the signal level is held, the ADC converts the signal to a digital number (DN).

6.4.2 CORRELATED DOUBLE SAMPLING CIRCUITS

Figure 6.39 shows the Hubble WF/PC preamplifier circuit. The front end uses a 2N 5564 JFET transistor for low-noise amplification. The JFET exhibits a white noise of < 2.5 nV/root-Hz, which is lower than CCD white noise (> 10 nV/root-Hz). The preamplifier gain is 22 V/V ($1 + R_2/R_1 = 1 + 1960/90.9$). The bandwidth (i.e., $f_{3\text{-}dB}$) is approximately 800 kHz with a phase margin of 30 deg. Figure 6.40 shows a high-speed/low-noise preamplifier circuit used for high-speed amplification. The preamplifier works comfortably up to 20 Mpixels/sec at a gain of 3 V/V. The noise floor for the amplifier (OPA-655) is 6 nV/Hz$^{1/2}$.

CHARGE MEASUREMENT

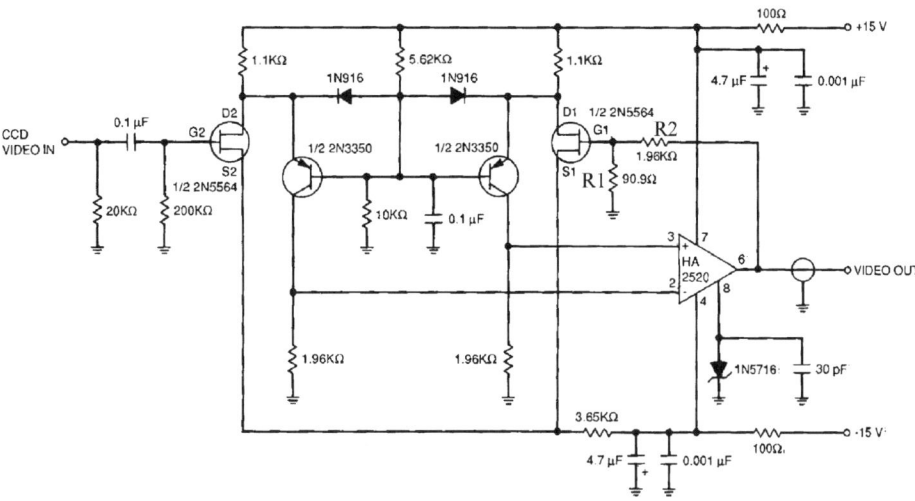

Figure 6.39 WF/PC preamplifier.

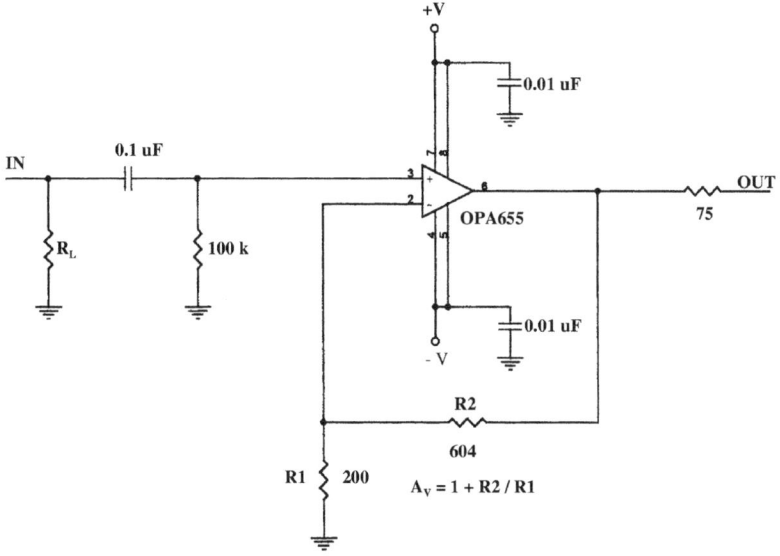

Figure 6.40 High-speed preamplifier.

Figure 6.41(a) is a Hubble-like CDS circuit diagram showing the postamplifier, clamp, sample and hold, and ADC buffer amplifier. The postamplifier voltage gain is set at 11 V/V (35.7 k/3.16 k). The dominant time constant is set at approximately 1 μsec (i.e., $\tau_D = RC = 35.7\,\text{k} \times 27 \times 10^{-12}$). The final gain adjustment of 3 V/V is made at the ADC buffer circuit (1.96 k/1 k + 1). An optional circuit called the "baseline stabilization" is also shown. Its purpose is to sample extended pixels and generate an offset level that is automatically subtracted at the input of the ADC.

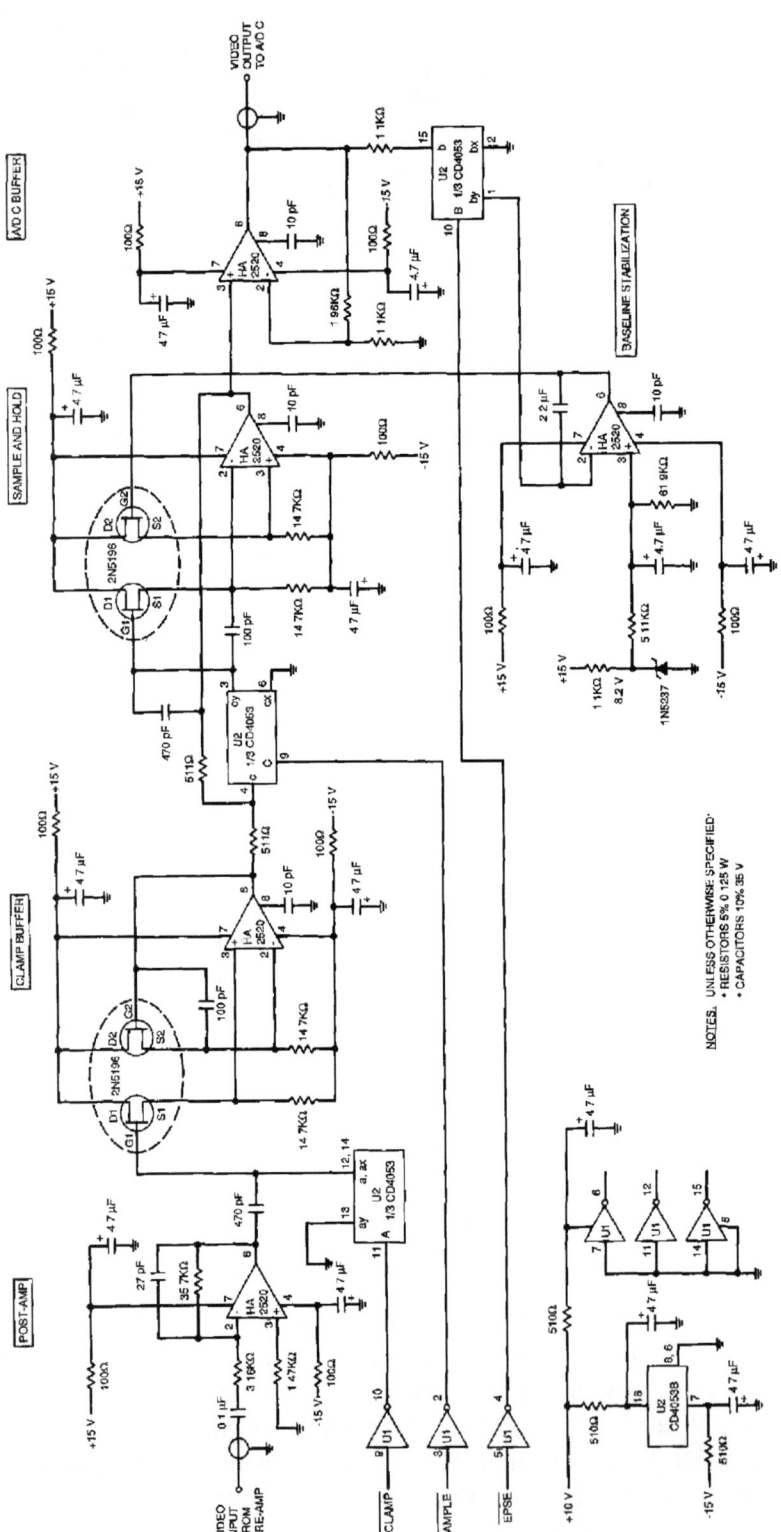

Figure 6.41 WF/PC-like CDS circuit diagram.

CHARGE MEASUREMENT

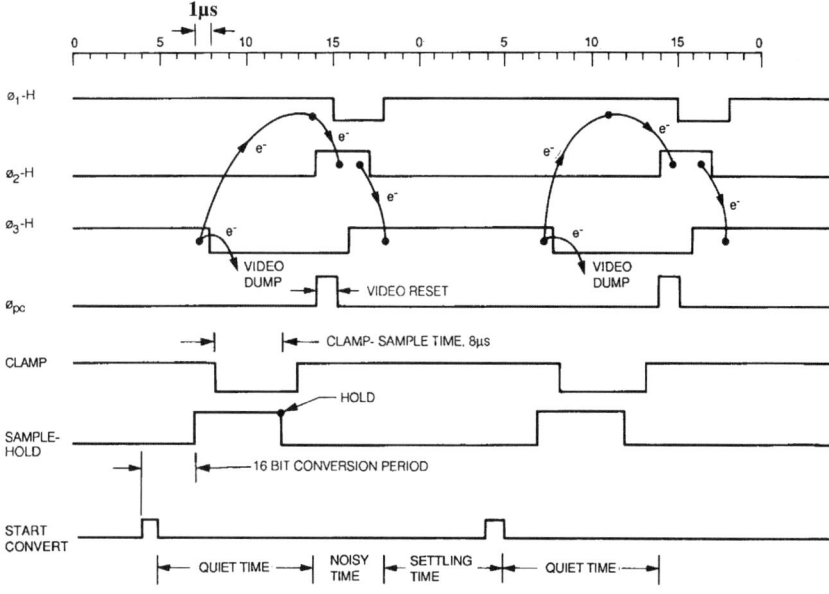

Figure 6.42 WF/PC-like CDS timing diagram.

The circuit is usually required when gain selects are required by the signal chain. Gain selects are employed when the ADC cannot cover the entire dynamic range of the CCD. For example, the Galileo camera utilizes four gain selects with an 8-bit ADC to cover a dynamic range of 5000. The baseline stabilization circuit shown is used by *Galileo* to automatically adjust offset when the gain level is changed (commanded from Earth).

Figure 6.42 shows a Hubble-like timing diagram for a 50-kpixel/sec readout rate. There are 20 time slots to adjust timing pulses, each 1 μsec wide. Note that the horizontal clock edges are tightly grouped into a small time period along with the reset clock. This time period is defined as the "noisy time," or that time when clock activity is allowed. During video dump and ADC conversion, clock activity is stopped. This time period is defined as the "quiet time." There is also a "settling time" to let the reset pulse settle before video is dumped (refer to discussions below on remnant reset noise).

6.4.3 CAMERA GAIN CONSTANT

The required voltage gain through the CCD and the CDS circuit is

$$A_T = S_V A_{CCD} A_1, \tag{6.61}$$

where A_T is the total gain from the sense node to the ADC (V/e$^-$), S_V is the node sensitivity (V/e$^-$), A_{CCD} is the on-chip amplifier gain (V/V), A_1 is the combination of the preamplifier gain (V/V), postamplifier gain (V/V), sample-and-hold gain (V/V), and ADC buffer gain (V/V).

The voltage gain to cover the CCD dynamic range by the ADC is

$$A_T = \frac{V_{ADC}}{S_{FW} + S_{OFF}}, \qquad (6.62)$$

where V_{ADC} is the full input voltage swing at the ADC, S_{FW} is the full well level (e$^-$), and S_{OFF} is the offset required (e$^-$).

The camera gain constant is

$$K(e^-/DN) = \frac{S_{FW} + S_{OFF}}{2^N + 1}, \qquad (6.63)$$

where N is the number of bits used by the ADC.

The camera gain constant can also be found through

$$K(e^-/DN) = (A_T \times A_2)^{-1}, \qquad (6.64)$$

where A_2 is the ADC conversion gain (DN/V) given as

$$A_2 = \frac{2^N + 1}{V_{ADC}}, \qquad (6.65)$$

where N is the number bits for the ADC.

Example 6.36

Determine the total voltage gain required to match a 16-bit ADC that has a 20-V input swing (i.e., $A_2 = 3.27 \times 10^3$ DN/V). Assume a full well level of 100,000 e$^-$ and an offset level of 4000 e$^-$ (i.e., 4% of the dynamic range is given to the offset level). Also, calculate the camera gain constant.

Solution:
From Eq. (6.62),

$$A_T = \frac{20}{1.04 \times 10^5},$$

$$A_T = 1.92 \times 10^{-4} \text{ V/e}^-.$$

From Eq. (6.63),

$$K(e^-/DN) = 1.04 \times 10^5/(2^{16} + 1),$$

$$K(e^-/DN) = 1.58 \text{ e}^-/DN.$$

CHARGE MEASUREMENT

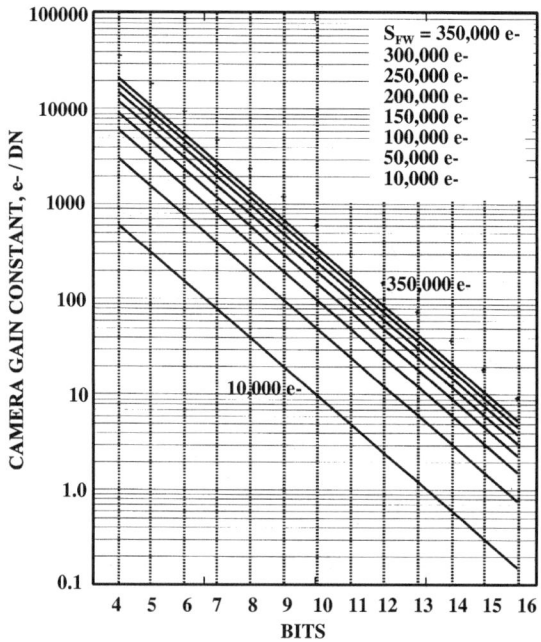

Figure 6.43 Camera gain constant as a function of ADC bits and full well.

From Eq. (6.64),

$$K(\text{e}^-/\text{DN}) = \left[(1.92 \times 10^{-4}) \times (3.27 \times 10^3)\right]^{-1},$$

$$K(\text{e}^-/\text{DN}) = 1.58 \text{ e}^-/\text{DN}.$$

Figure 6.43 plots the camera gain constant required as function of ADC bits and charge capacity.

6.4.4 CORRELATED DOUBLE SAMPLING TRANSFER FUNCTION

The CDS processor is essentially a band-pass filter set by two parameters: τ_D and t_s. Low-frequency noise is rejected when t_s is made small because of noise correlation. The dominant time constant, τ_D, sets the bandwidth to reject high-frequency noise. The center frequency of the CDS filter is tuned by adjusting t_s and τ_D while maintaining a high signal gain. The optimum center frequency is when the lowest noise results (rms e$^-$) for a given CCD input noise-spectrum [i.e., Eq. (6.60)]. These parameters are usually found experimentally using photon transfer. However, the CDS transfer function derived below will give a good estimate in defining t_s and τ_D theoretically.

As for noise, the CDS processor with unity gain generates an output signal, $N_{CDS}(t)$, equal to

$$N_{CDS}(t) = N(t) - N(t - t_s), \tag{6.66}$$

where t_s is the sample-to-sample time (sec), $N(t)$ is the output postamplifier noise as a function of time (V), and $N(t - t_s)$ is the same noise sampled at $t + t_s$ (V).

The Fourier transform of $N_{CDS}(t)$ in the frequency domain is

$$N_{CDS}(f) = N(f)[1 - \exp(-j2\pi f t_s)], \tag{6.67}$$

where f is frequency (Hz) and $N(f)$ is the Fourier transform of $N(t)$.

The magnitude squared response of this quantity is found by using Euler's rule, yielding

$$|N_{CDS}(f)|^2 = |N(f)|^2[2 - 2\cos(2\pi f t_s)]. \tag{6.68}$$

The second term is the CDS transfer function defined as

$$|H_{CDS}(f)|^2 = 2 - 2\cos(2\pi f t_s). \tag{6.69}$$

$|N(f)|^2$ is the filtered input CCD noise given as

$$|N(f)|^2 = |N_{CCD}(f)|^2 |H_{RC}(f)|^2, \tag{6.70}$$

where $|N_{CCD}(f)|^2$ is CCD noise defined by Eq. (6.60) and $|H_{RC}|$ is the single pole filter transfer function of the postamplifier, given by

$$|H_{RC}(f)|^2 = \frac{1}{1 + (2\pi f \tau_D)^2}, \tag{6.71}$$

where τ_D is the dominant time constant for the CDS processor. We assume that the postamplifier and preamplifier gains are unity in our derivation.

Substituting these equations into Eq. (6.68) yields

$$|N_{CDS}(f)|^2 = |N_{CCD}(f)|^2 \frac{1}{1 + (2\pi f \tau_D)^2}[2 - 2\cos(2\pi f t_s)], \tag{6.72}$$

or

$$|N_{CDS}(f)|^2 = |N_{CCD}(f)|^2 |H_{RC}(f)|^2 |H_{CDS}(f)|^2. \tag{6.73}$$

The output noise of the CDS processor is found by integrating Eq. (6.73) with respect to frequency and then taking the square-root, producing

$$N_{CDS}(V) = \left(\int_0^\infty |N_{CDS}(f)|^2 df \right)^{1/2}, \tag{6.74}$$

where $N_{CDS}(V)$ is the output noise (i.e., rms V).

6.4.4.1 White Noise Limited

Equation (6.74) can be explicitly integrated when only white noise is considered [i.e., $N_{CCD}(f) = W_{CCD}(f)$]. The resultant equation after integration is

$$N_{CDSW}(V) = W_{CCD}(f) B^{1/2} [2 - 2\exp(-t_s/\tau_D)]^{1/2}, \tag{6.75}$$

where $N_{CDSW}(V)$ is the output noise (rms V) and B is the equivalent noise power bandwidth (Hz).

Example 6.37

Find the output noise for a CDS processor when $t_s = 4$ μsec and $\tau_D = 1$ μsec ($B = 2.5 \times 10^5$ Hz). Assume $W_{CCD} = 10$ nV/root-Hz. Also calculate the output noise at $t_s = 0.4$ μsec and compare the results.
Solution:
From Eq. (6.75),

$$N_{CDSW}(V) = 10^{-8} (2.5 \times 10^5)^{1/2} \{2 - 2\exp[-(4 \times 10^{-6})/10^{-6}]\}^{1/2},$$
$$N_{CDSW}(V) = 7.02 \text{ μV rms}, \quad (t_s = 4 \text{ μsec}).$$

The noise at $t_s = 0.4$ μsec is

$$N_{CDSW}(V) = 4.06 \text{ μV rms}, \quad (t_s = 0.4 \text{ μsec}).$$

Note that the input noise to the CDS is $(2.5 \times 10^5)^{1/2} \times 10^{-8} = 5$ μV rms, which is smaller than the output noise. This indicates that some white noise frequencies are uncorrelated, causing a higher noise level. In the limit when $t_s = \infty$, the output noise is $2^{1/2}$ times the input noise when all frequencies are uncorrelated. Recall that when uncorrelated noise samples are either added or subtracted, the net noise is equal to the quadrature sum of the samples. The output noise can be made as small as desired by reducing t_s. For example, when $t_s = 0$, the noise is zero because all noise frequencies are correlated.

It is important to note that as t_s is reduced, the signal gain decreases exponentially for a fixed τ_D. For example, when $t_s = \tau_D$, the signal gain through the postamplifier is only 0.63 (i.e., one time constant). The goal of the CDS processor is to also maintain high signal gain along with minimal noise (i.e., maximum S/N or low e^- rms).

The input signal, $S_{CCD}(V)$, to the CDS processor after the postamplifier is

$$S_{CDS}(V) = S_{CCD}(V)[1 - \exp(-t_s/\tau_D)]. \tag{6.76}$$

With Eqs. (6.75) and (6.76), the S/N is

$$\frac{S_{CDS}(V)}{N_{CDSW}(V)} = \frac{S_{CCD}(V)[1 - \exp(-t_s/\tau_D)]^{1/2}}{2^{1/2} W_{CCD}(f) B^{1/2}}. \qquad (6.77)$$

For a fixed bandwidth the equation shows that maximum S/N is achieved when t_s is made as large as possible. This result may seem contrary to above discussions in Example 6.37 where increasing t_s produces more noise. Although the noise does increase, the signal strength increases at a faster rate with t_s, and therefore, a higher S/N is achieved.

Example 6.38

Find the S/N for a CDS processor for the conditions described in Example 6.37. Assume $S_{CCD}(V) = 1 \times 10^{-3}$ V.
Solution:
From Eq. (6.77),

$$S_{CDS}(V)/N_{CDSW}(V) = 10^{-3}[1 - \exp(-4)]^{1/2} / \left[1.414 \times 10^{-8} \times (2.5 \times 10^5)^{1/2}\right],$$

$$S_{CDS}(V)/N_{CDSW}(V) = 140, \quad (4.0 \; \mu\text{sec}),$$

$$S_{CDS}(V)/N_{CDSW}(V) = 81, \quad (0.4 \; \mu\text{sec}).$$

Again, a longer sample-to-sample time is desirable.

Using Eq. (6.77), the CDS output noise referred to the CCD sense node is

$$N_{CDSW}(e^-) = \frac{W_{CCD}(f) B^{1/2} [2 - 2\exp(-t_s/\tau_D)]^{1/2}}{S_V [1 - \exp(-t_s/\tau_D)] A_{CCD}}, \qquad (6.78)$$

where A_{CCD} is the gain of the CCD output amplifier. This equation reduces to

$$N_{CDSW}(e^-) = \frac{2^{1/2} W_{CCD}(f) B^{1/2}}{S_V A_{CCD} [1 - \exp(-t_s/\tau_D)]^{1/2}}. \qquad (6.79)$$

It is very important to note that expressing the noise in rms electrons is equivalent to S/N because the signal at the sense node is also expressed in electrons.

CHARGE MEASUREMENT

Example 6.39

Find the noise at the sense node for the following parameters: $W_{CCD}(f) = 6$ nV/root-Hz, $S_V = 0.5$ μV/e$^-$, $A_{CCD} = 0.8$ V/V, and $\tau_D = 5$ μsec (equivalent noise bandwidth $B = 50$ kHz). Perform the calculation for $t_s = 5$, 10, and 20 μsec.

Solution:
From Eq. (6.79),

$$N_{CDSW}(e^-) = 1.414 \times (6 \times 10^{-9}) \times (5 \times 10^4)^{1/2} /$$
$$((5 \times 10^{-7}) \times 0.8 \times \{1 - \exp[-(5 \times 10^{-6})]\}/$$
$$(5 \times 10^{-6}))^{1/2},$$

$$N_{CDSW}(e^-) = 8.25 \text{ e}^- \text{rms}, \quad (t_s = 5 \text{ μsec}),$$

$$N_{CDSW}(e^-) = 5.09 \text{ e}^- \text{rms}, \quad (t_s = 10 \text{ μsec}),$$

$$N_{CDSW}(e^-) = 4.69 \text{ e}^- \text{rms}, \quad (t_s = 20 \text{ μsec}).$$

Figure 6.44 plots read noise (rms e$^-$) as a function of t_s for $\tau_D = 10^{-4}$, 10^{-5}, and 10^{-6} sec. Read noise decreases by the square-root of t_s until the noise becomes flat, at which point sufficient time has been given to sampling the video signal. It is important to note that the noise does not significantly decrease when t_s is greater than 2 τ_D. The noise level obtains 86% of its final value at that point. Therefore, it is common practice to assume $t_s = 2\tau_D$ for low-noise performance and minimum sample time. With this assumption, Eq. (6.79) can be further simplified, yielding

$$N_{CDSW}(e^-) = \frac{1.075 W_{CCD}(f)}{S_V A_{CCD}(t_s)^{1/2}}. \tag{6.80}$$

This valuable equation is used to calculate read noise for a CCD camera system that is white-noise limited. Note that read noise is proportional to $W_{CCD}(f)$ and inversely proportional to S_V and the square-root of t_s. We will discuss the implications of these parameters below when considering design features for the output amplifier.

Example 6.40

Predict the read noise for a CCD output amplifier that exhibits the following characteristics: $W_{CCD} = 10^{-8}$ V/root-Hz and $S_V = 2 \times 10^{-6}$ V/e$^-$, $A_{CCD} = 0.8$ V/V and $t_s = 10^{-6}$ sec.

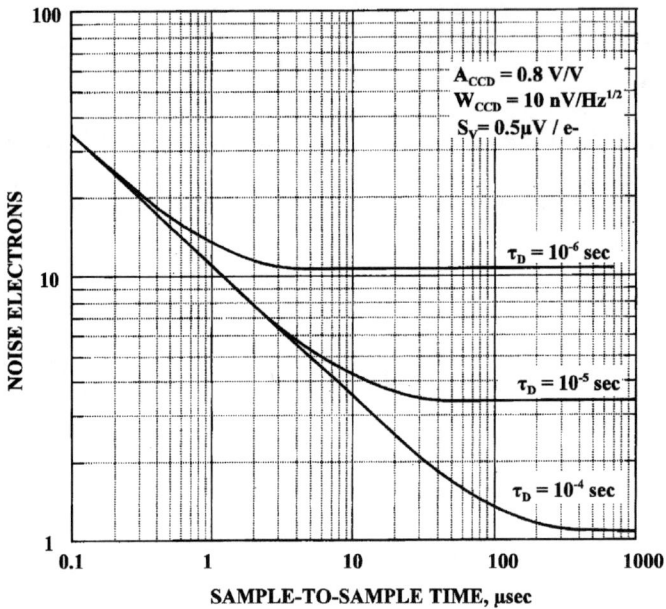

Figure 6.44 Read noise as a function of t_s and τ_D.

Solution:
From Eq. (6.80),

$$N_{CDSW}(e^-) = 1.07 \times 10^{-8} / [0.8 \times (2 \times 10^{-6}) \times (10^{-6})^{1/2}],$$

$$N_{CDSW}(e^-) = 6.7 \; e^- \; \text{rms}.$$

Figure 6.45 plots Eq. (6.80) as a function of sample-to-sample time and node sensitivities. White noise of 10 nV/Hz$^{1/2}$ is assumed in the plot.

Read noise versus pixel rate can also be estimated if an overhead time period is included in Eq. (6.80) (i.e., the time required for ADC conversion, clocking the CCD, and resetting the sense node). Assuming that the overhead time for these functions is equal to the sample-to-sample time, the expression for noise is

$$N_{CDSW}(e^-) = \frac{1.5 W_{CCD}(f) P_R^{1/2}}{S_V A_{CCD}}. \tag{6.81}$$

Example 6.41

Estimate the noise level for a pixel rate of 10 Mpixels/sec. Assume the same parameters as Example 6.40.

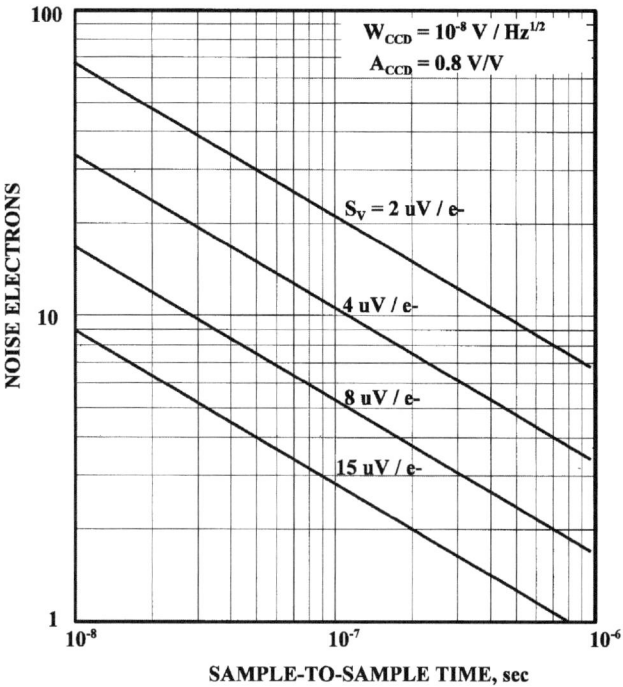

Figure 6.45 Read noise as a function of t_s and output amplifier sensitivity.

Solution:
From Eq. (6.81),

$$N_{CDSW}(e^-) = 1.5 \times 10^{-8} \times (10^7)^{1/2} / [(2 \times 10^{-6}) \times 0.8],$$

$$N_{CDSW}(e^-) = 29 \; e^-.$$

Figure 6.46 plots Eq. (6.81) as function of pixel rate and sensitivity. The data presented is in good agreement with measured noise levels for high-speed CCDs (HSS CCDs) fabricated today for $P_R > 10^6$ pixels/sec and $S_V < 8\,\mu V/e^-$.

Equation (6.80) shows that the node sensitivity, S_V, and white noise $W_{CCD}(f)$ are competing parameters. The trend in the development of the CCD output amplifier has been to make S_V as large as possible while sacrificing white noise performance.[31,32] This is advantageous because sensitivity is inversely proportional to MOSFET gate width, whereas white noise only increases by the square-root of the reduction [i.e., Eq. (6.54)]. This design philosophy has worked very well and has pushed noise levels to a single electron for slow-scan operation. However, there is a limit to how small the output amplifier should be designed. We show

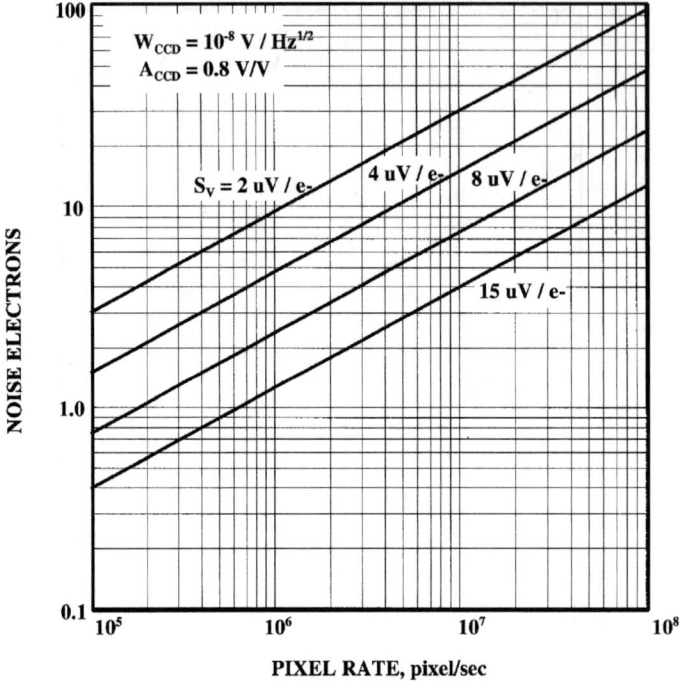

Figure 6.46 Read noise as a function of pixel rate and output amplifier sensitivity.

this limit by substituting the white noise Eq. (6.54) into Eq. (6.80), to produce an equation for read noise as a function of MOSFET geometry. The resultant equation is

$$N_{CDSW}(e^-) = \frac{1.075}{S_V A_{CCD} t_s^{1/2}} \left[\frac{4kT}{\mu_{SI} C_{OX}(W/L)(V_{GS} + V_{AEFF})} \right]^{1/2}, \quad (6.82)$$

where the sensitivity, S_V, as a function of geometry is specified in Fig. 6.12(b).

Example 6.42

Calculate the noise for a MOSFET output amplifier, given

$$t_s = 1 \times 10^{-6} \text{ sec},$$
$$\tau_D = 0.5 \times 10^{-6} \text{ sec},$$
$$B = 1/(4 \times \tau_D) = 5 \times 10^6,$$
$$A_{CCD} = 1 \text{V/V},$$

CHARGE MEASUREMENT

$L = 4 \, \mu m$,
$W = 18 \, \mu m$,
$C_{FD} = 40 \, fF$,
$V_{GS} + V_{AEFF} = 2 \, V$.

Solution:
From Fig. 6.12(b), the sensitivity for a $L = 4 \, \mu m$, $W = 18 \, \mu m$, and $C_{FD} = 40 \, fF$ is $3 \, \mu V/e^-$.

The first term of Eq. (6.82) is

$$1.075 / \left[(3 \times 10^{-6}) \times 0.8 \times (10^{-6})^{1/2} \right] = 4.479 \times 10^8.$$

The second term of Eq. (6.82) is

$$\left\{ 4 \times (1.38 \times 10^{-23}) \times 293 / \left[1000 \times (3.45 \times 10^{-8}) \times 4.5 \times 2 \right] \right\}^{-1/2}$$
$$= 7.2 \times 10^{-9}.$$

The product of these terms yields a read noise of

$$N(e^-) = 3.225 \, e^-.$$

Figure 6.47(a) plots the read noise as a function of gate width. Minimum noise occurs when the MOSFET gate capacitance becomes equal to the floating diffusion capacitance (i.e., $C_{MOS} = C_{FD}$). This point is plotted in Fig. 6.47(b), where the MOSFET gate capacitance is plotted as a function of its width. The intersection of the floating diffusion capacitance with the gate capacitance is where noise is lowest. Note that as the floating diffusion capacitance increases, the optimum gate width must also increase.

6.4.4.2 1/f Noise Limited

Equation (6.80) can be used to calculate read noise only when the $1/f$ corner frequency is substantially below $1/(4t_s)$. When $1/f$ noise becomes important, Eq. (6.74) must be applied. The noise referred to the sense node with $1/f$ noise included is

$$N_{CDS}(e^-) = \frac{1}{S_V A_{CCD}[1 - \exp(-t_s/\tau_D)]} \left[\int_0^\infty |N_{CDS}(f)|^2 df \right]^{1/2}, \quad (6.83)$$

where $|N_{CDS}(f)|^2$ is given by Eq. (6.73).

Equation (6.83) must be numerically integrated because of the $1/f$ term in $N_{CCD}(f)$. The next example will examine each term of this equation.

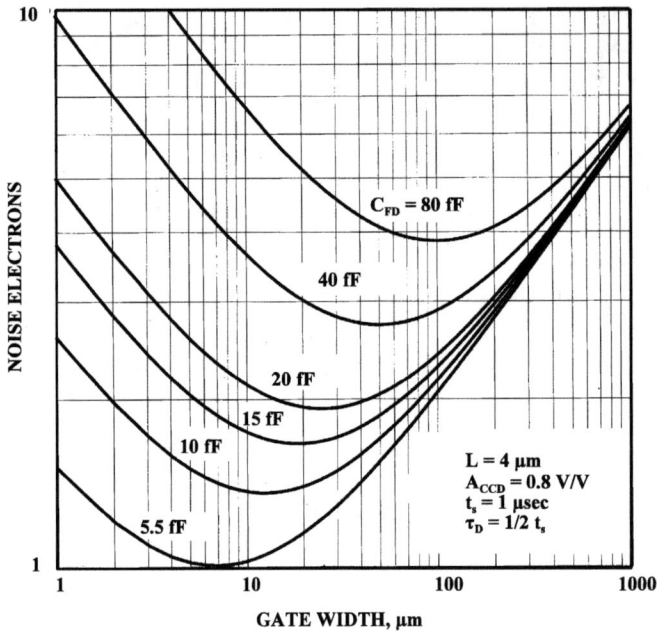

Figure 6.47(a) Read noise as a function of output amplifier gate width and floating diffusion capacitance.

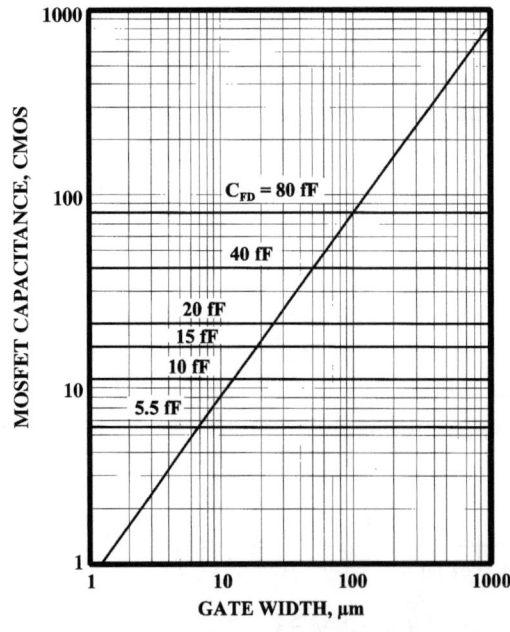

Figure 6.47(b) MOSFET capacitance as a function of gate width and floating diffusion capacitance.

CHARGE MEASUREMENT 573

Example 6.43

Plot the following transfers curves:

The post amp white noise filter $|H_{RC}(f)|^2$,
The correlated double sampling transfer function $|H_{CDS}(f)|^2$,
The product of $|H_{RC}(f)|^2$ and $|H_{CDS}(f)|^2$,
The CCD power noise spectrum $|N_{CCD}(f)|^2$,
The product of $|N_{CCD}(f)|^2$ and $|H_{RC}(f)|^2$ and $|H_{CDS}(f)|^2$.

Assume the output amplifier and CDS parameters are

$$t_s = 1 \text{ μsec}$$
$$t_s = 2\tau_D$$
$$S_V = 4 \times 10^{-6} \text{ V/e}^-$$
$$A_{CCD} = 0.8 \text{ V/V}$$
$$W_{CCD} = 10^{-8} \text{ V/Hz}^{1/2}$$
$$f_c = 100 \text{ kHz}$$
$$m = 1.$$

Solution:
From Eq. (6.71), $|H_{RC}(f)|^2$ is plotted in Fig. 6.48(a).
From Eq. (6.69), $|H_{CDS}(f)|^2$ is plotted in Fig. 6.48(b).
The product of $|H_{RC}(f)|^2$ and $|H_{CDS}(f)|^2$ is shown in Fig. 6.48(c).
From Eq. (6.60), $N_{CCD}(f)|^2$ is plotted in Fig. 6.48(d).
The product of $|N_{CCD}(f)|^2$ and $|H_{RC}(f)|^2$ and $|H_{CDS}(f)|^2$ is shown in Fig. 6.48(e).

Note again from Fig. 6.48(e) that the CDS is a bandpass filter. The low-frequency roll-off characteristics are set by the sample-to-sample time, t_s, whereas high-frequency characteristics are determined by the dominant time constant, τ_D. The area under this curve is proportional to the read noise floor.

Example 6.44

From Eq. (6.83), determine the read noise (e$^-$ rms) as a function of sample-to-sample time for the output amplifier parameters given in Example 6.43. Also, on the same graph, plot the noise for $f_c = 10^3$ Hz and 10^4 Hz.

Solution:
From Eq. (6.83), Fig. 6.48(f) is generated. Note that the noise initially decreases by the square-root of t_s until the $1/f$ noise corner is encountered,

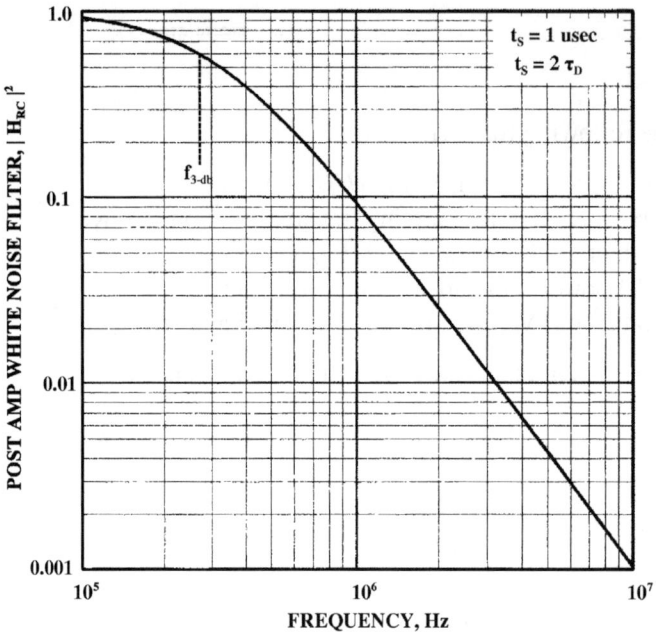

Figure 6.48(a) Postamplifier transfer function, $|H_{RC}|^2$.

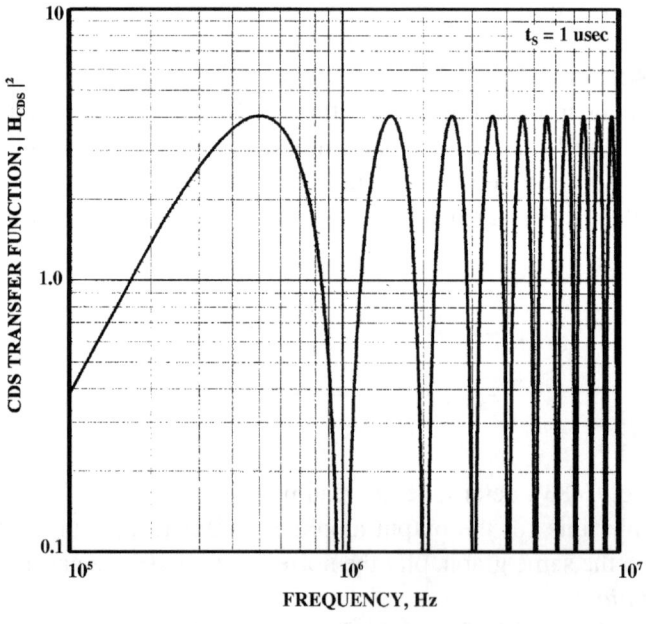

Figure 6.48(b) CDS transfer function, $|H_{CDS}|^2$.

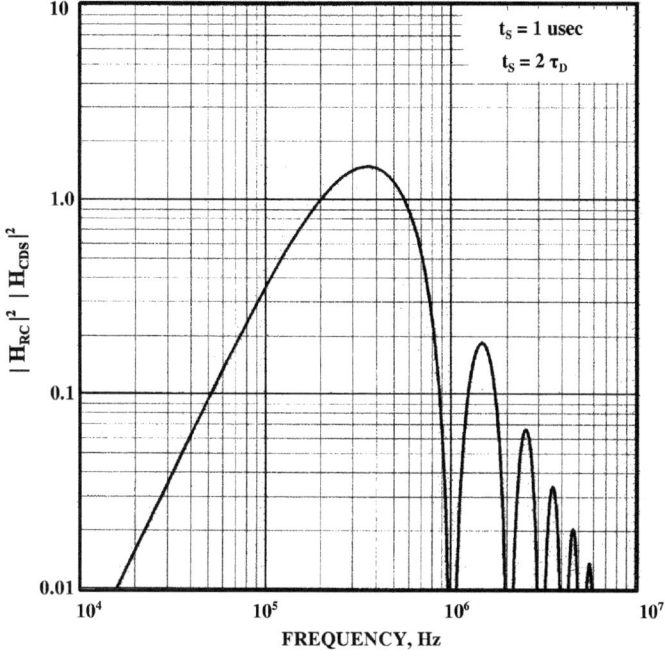

Figure 6.48(c) $|H_{RC}|^2 \times |H_{CDS}|^2$.

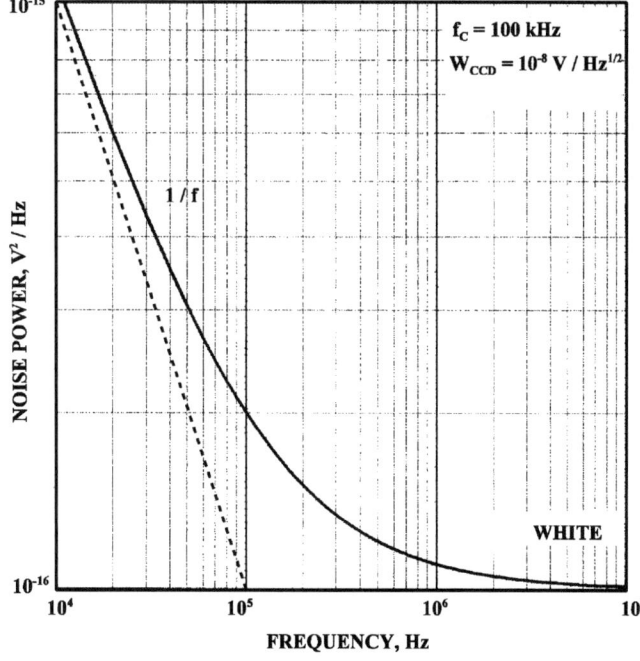

Figure 6.48(d) Output amplifier noise power spectrum, $|N_{CCD}|^2$.

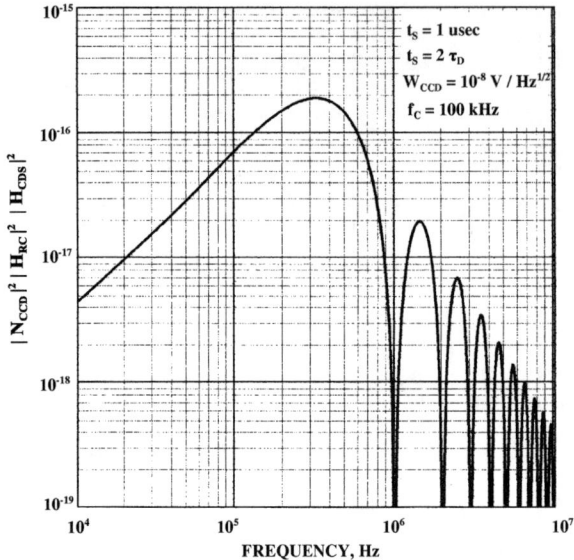

Figure 6.48(e) $|N_{CCD}|^2 \times |H_{RC}|^2 \times |N_{CDS}|^2$.

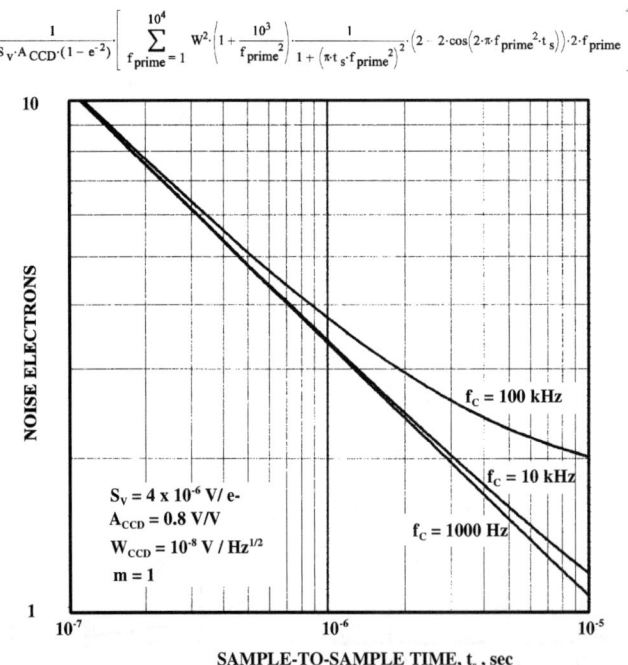

Figure 6.48(f) Read noise as a function of sample-to-sample time and $1/f$ corner frequency.

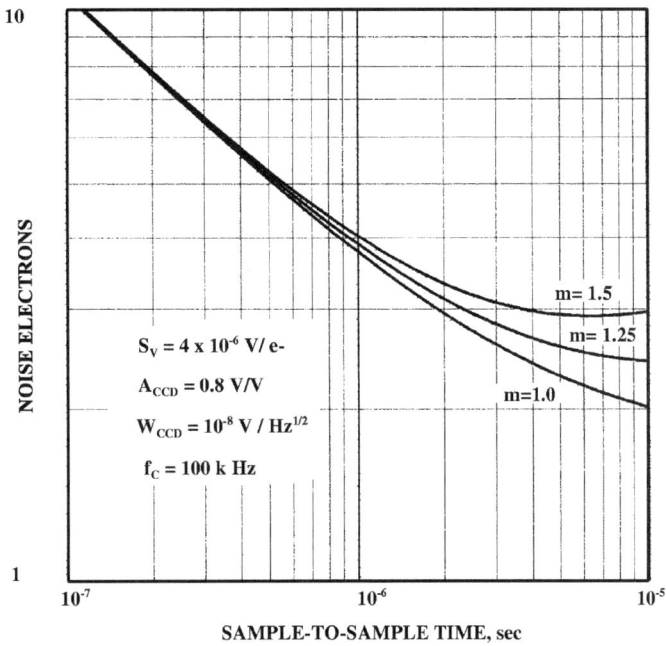

Figure 6.48(g) Read noise as a function of sample-to-sample time and $1/f$ slope.

at which point noise correlation is lost. The gain made in white-noise reduction is exactly canceled by the $1/f$ noise sampled, causing the noise to level off.

An equation is also provided in Fig. 6.48(f) for numerical integration. The term f_{prime} is equal to $(f)^{1/2}$, where f is the frequency (Hz). The limits of integration are set for 1 to 10^8 Hz.

Example 6.45

Using Eq. (6.83), determine the read noise as a function of sample-to-sample time for the output amplifier parameters given in Example 6.43 for $m = 1.0$, 1.25 and 1.5.

Solution:

From Eq. (6.83), Fig. 6.48(g) is generated. CCDs that generate excess $1/f$ noise exhibit a $1/f$ slope greater than unity (i.e., luminescence and shot noise show this characteristic). The read noise (rms e^-) can actually increase at that point where the $1/f$ corner frequency is encountered.

6.5 Dual Slope Processor

6.5.1 Dual Slope Circuit Elements

The dual slope circuit presented in Fig. 6.49(a) is also a popular CCD video processor. As before, the CCD signal is first amplified by the preamplifier. Its output is then fed into an integrating amplifier by way of a three position switch (S1, S2, and S3). One of the three positions is open (S3) and the other two positions, S1 and S2, are connected to a noninverting and an inverting amplifier, each with a gain of

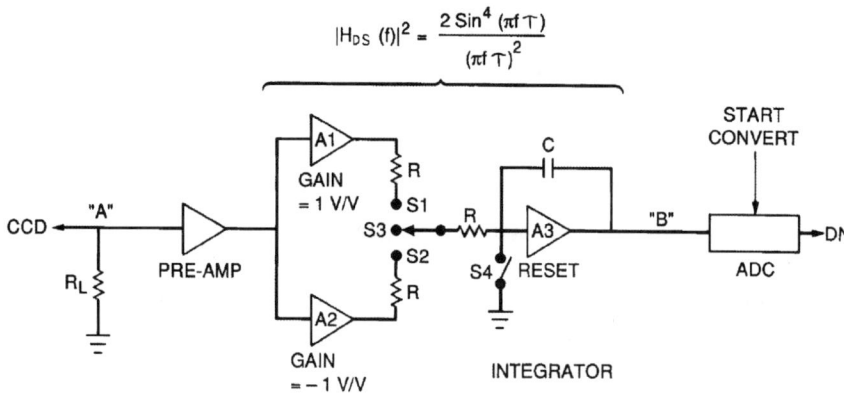

Figure 6.49(a) Dual slope processor.

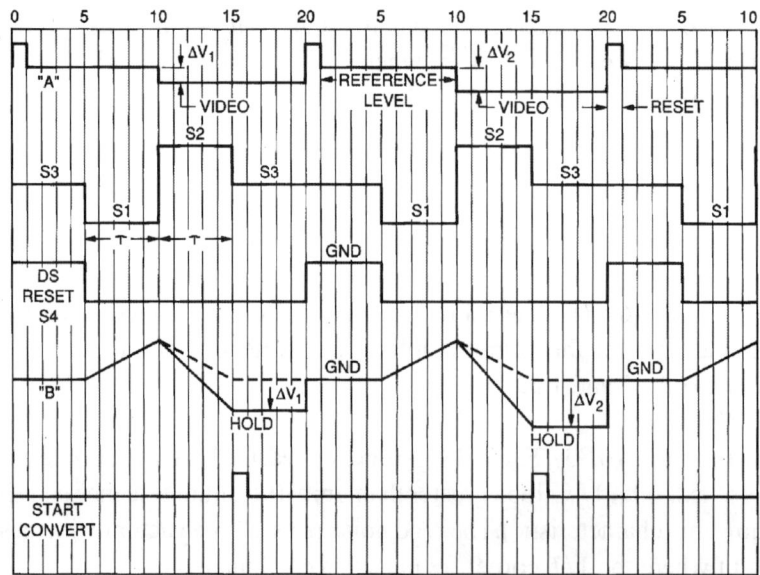

Figure 6.49(b) Dual slope timing diagram.

CHARGE MEASUREMENT

unity. Processing of the CCD video signal begins with resetting the integrator to ground potential by the switch labeled S4 [refer to Fig. 6.49(b)]. Then S4 opens and S1 closes, allowing the reference voltage level to be integrated onto the integrating capacitor C for time τ. Since the integrator begins at ground potential for each pixel, reset noise is eliminated. After integrating for τ seconds, charge is then dumped onto the sense node of the CCD. Simultaneously, S2 closes, connecting the integrator to the inverting amplifier. The reference level plus the video level are integrated for τ seconds when the switch opens to the S3 position. The signal is then held by the integrator while digital conversion takes place.

6.5.2 DUAL SLOPE TRANSFER FUNCTION

The noise referred to the sense node for a dual slope integrator is

$$N_{DS}(e^-) = \frac{2}{S_V A_{CCD}} \left[\int_0^\infty |N_{CCD}(f)|^2 |H_{DS}|^2 df \right]^{1/2}, \qquad (6.84)$$

where $H_{DS}(f)$ is the dual slope transfer function given by

$$H_{DS}(f) = \sin^2(\pi f \tau) \frac{\sin^2(\pi f \tau)}{(\pi f \tau)^2}, \qquad (6.85)$$

where τ is the integration period.

Example 6.46

Plot the following transfers curves for the dual slope processor:

$|H_{DS}(f)|^2$

$|N_{CCD}(f)|^2 |H_{CDS}(f)|^2.$

Assume the output amplifier and CDS parameters

$\tau = 1$ μsec
$S_V = 4 \times 10^{-6}$ V/e$^-$
$A_{CCD} = 0.8$ V/V
$W_{CCD} = 10^{-8}$ V/ Hz$^{1/2}$
$f_c = 100$ kHz
$m = 1.$

Solution:
From Eq. (6.85), $|H_{DS}(f)|^2$ is plotted in Fig. 6.50(a).

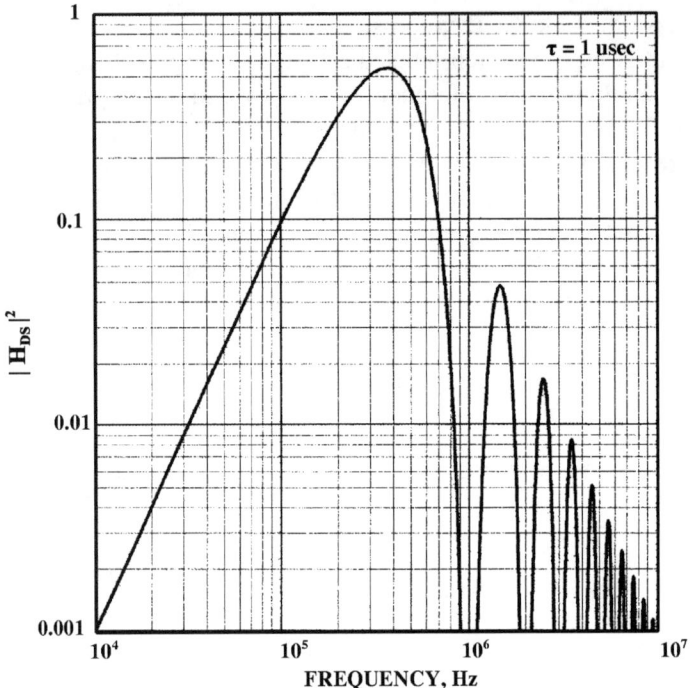

Figure 6.50(a) Dual slope transfer function, $|H_{DS}|^2$.

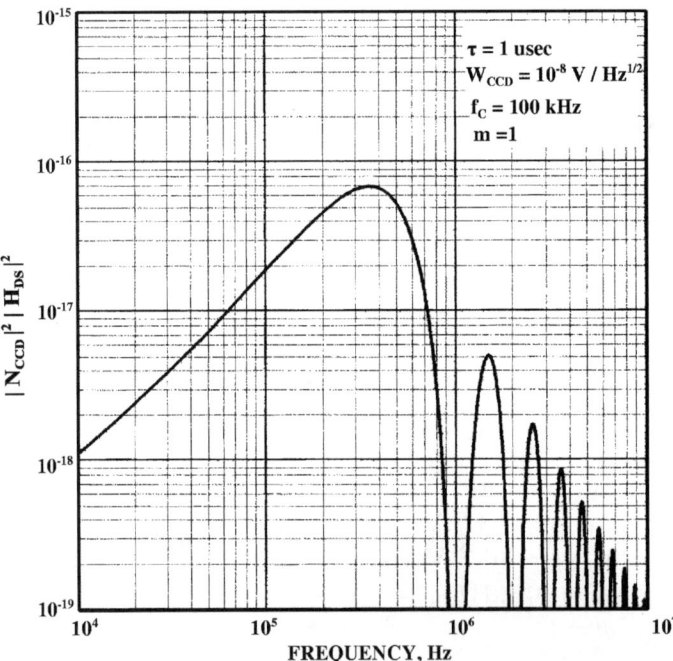

Figure 6.50(b) $|N_{CCD}|^2 \times |H_{DS}|^2$.

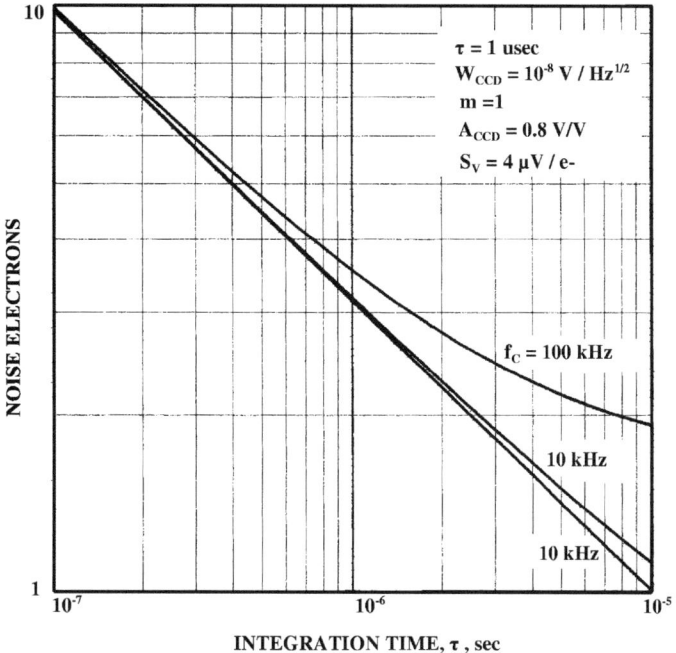

Figure 6.50(c) Read noise as a function of integration time, τ.

Figure 6.48(d) shows the CCD power noise spectrum $|N_{CCD}(f)|^2$. The product of $|N_{CCD}(f)|^2 \times |H_{DS}(f)|^2$ is shown in Fig. 6.50(b).

Example 6.47

Determine the read noise (e⁻ rms) as a function of integrating time for the output amplifier parameters specified in Example 6.46. Also, on the same graph, plot the noise for $f_c = 10^3$ Hz and 10^4 Hz.

Solution:

From Eq. (6.84), Fig. 6.50(c) is generated. As with the dual sampler, noise initially decreases by the square-root of τ until the $1/f$ noise corner is encountered. Although Eqs. (6.84) and (6.83) are remarkably different, the noise filtering action is nearly identical (to within uncertainties generated by numerical integration). Consequently, both processors are used in processing CCD signals. However, when high pixel rates are involved, the dual sampler is the choice circuit because sampling circuits can perform at higher pixel rates than integrating circuits.

6.6 Remnant Signal and Noise

6.6.1 Remnant Signal

The postamplifier filter discharges exponentially and continues to fall even past the sample-to-sample time period. A remnant signal is found in trailing pixels because the filter never completely discharges. In a sense, the filter is a memory element. The problem can also be associated with other filters in the camera's signal chain. For example, the time constant related to the CCD output amplifier can also produce a remnant signal. This problem is sometimes seen when a single-stage output amplifier is used at pixel rates faster than $f_{3\text{-}dB}$ [i.e., Eq. (6.20)]. Normally, the dominant time constant of the camera is defined by one circuit element where the remnant problem can be readily identified (e.g., in the postamplifier). The remnant signal is most noticeable when dark pixels follow a very bright pixel. If the video has not completely settled, a "virtual" deferred charge signal following the bright pixel is seen. For example, EPER is sensitive to this problem and is used as a diagnostic tool in characterizing settling time characteristics for the CDS circuit. Recall from Chapter 5 that EPER can measure signals as small as 1 electron and therefore is very sensitive in measuring remnant signals.

The amount of remnant signal can be determined by superposition signal analysis or simple exponential signal tracking from one pixel to the next. In either case, the ratio of the trailing pixel level to a target signal level is

$$\frac{S_{RS}(e^-)}{S(e^-)} = e^{-i_{th}t_p/\tau_D}\left(e^{t_s/\tau_D} - 1\right), \tag{6.86}$$

where $S_{RS}(e^-)$ is the remnant signal (e^-), $S(e^-)$ is the signal level of the target pixel (e^-), t_p is the pixel period and i_{th} is the ith trailing pixel (e.g., for the first trailing pixel $i_{th} = 1$).

Example 6.48

Find the remnant signal in the first trailing pixel for a pixel period of $t_p = 5$ μsec. Assume a CDS time constant of $\tau_D = 1$ μsec and a sample-to-sample time of $t_s = 2$ μsec.
Solution:
From Eq. (6.86),

$$S_{RS}(e^-)/S(e^-) = \exp\left[-(5.0 \times 10^{-6})/10^{-6}\right]$$
$$\times \left[\exp(2.0 \times 10^{-6}/10^{-6} - 1)\right],$$
$$S_{RS}(e^-)/S(e^-) = 0.043\ (4.3\%).$$

CHARGE MEASUREMENT 583

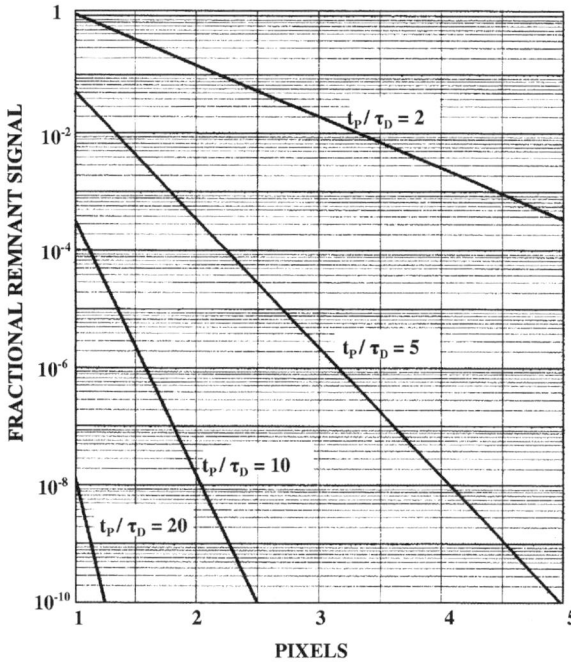

Figure 6.51 Remnant signal as a function of trailing pixels.

Figure 6.51 plots fractional remnant signal as a function of trailing pixels and t_P/τ_D. The plot assumes $\tau_D = t_s/2$.

6.6.2 REMNANT NOISE

Noise variations are also "remembered" on the filter capacitor and produce a noise called remnant noise. For example, a certain amount of time is required after resetting to assure that reset remnant noise is negligible compared to output amplifier noise. This time is the period between the falling edge of the reset pulse and the release of the clamp pulse. The amount of reset remnant noise present is

$$\frac{N_{RRN}(e^-)}{N_R(e^-)} = \left(e^{-t_R/\tau_D} - e^{-(t_R+t_s)/\tau_D}\right), \tag{6.87}$$

where $N_{RRN}(e^-)$ is the reset remnant noise (e^-), $N_R(e^-)$ is the reset noise (e^-) and t_R is the time period between the falling edge of the reset pulse to the end of clamp period (sec). Note that reset remnant noise is independent of reset clock feed-through amplitude (i.e., it can be zero and remnant noise still exists).

Example 6.49

Find the reset remnant noise for a reset noise level of $N_R(\text{e}^-) = 100$ e$^-$. Assume $N = 0.1$, 1, and 2 time-constant periods.

Solution:

Assuming $t_s = 2\tau_D$ and $t_R = N\tau_D$ where N is the number of time constants, Eq. (6.87) reduces to

$$N_{RRN} = N_R(\text{e}^-)\left(e^{-N} - e^{-(N+2)}\right) = N_R(\text{e}^-)e^{-N}\left(1 - e^2\right)$$
$$= N_R(\text{e}^-) \times 0.865 \times e^{-N}.$$

The corresponding reset remnant noise levels using this equation are

$$N_{RRN}(\text{e}^-) = 100 \times 0.865 \times e^{-0.1},$$

$$N_{RRN}(\text{e}^-) = 77 \text{ e}^- \; (N = 0.1),$$

$$N_{RRN}(\text{e}^-) = 32 \text{ e}^- \; (N = 1),$$

$$N_{RRN}(\text{e}^-) = 12 \text{ e}^- \; (N = 2).$$

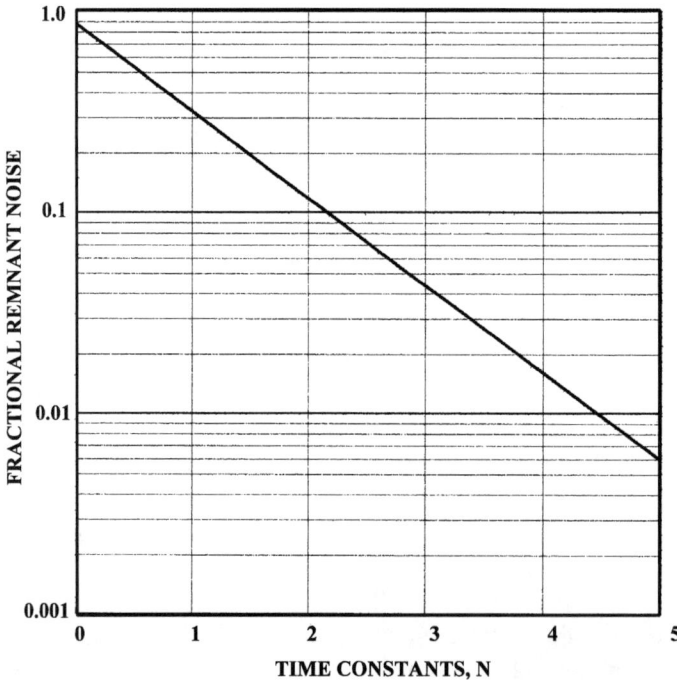

Figure 6.52 Remnant reset noise as a function of dominant time constants, N.

Figure 6.52 plots fractional remnant reset noise as a function of the number of time constants (i.e., $N = t_R/\tau_D$).

Remnant noise and signal are important to high-speed applications where settling time is hard to come by. Both problems can be eliminated by discharging the postamplifier filter with a switch to discharge the capacitor and reduce settling time. Reset remnant noise can also be eliminated by using a floating gate amplifier.[33] Here the gate of the output amplifier is reset once per line, as opposed to once per pixel. We will discuss floating gate amplifiers below. It should also be mentioned that the dual slope processor does not inherently have remnant problems, because the integrating capacitor is reset for each pixel value taken.

6.7 SKIPPER AMPLIFIER

6.7.1 INTRODUCTION

Figure 6.53 shows read noise measurements taken for early and current scientific CCDs. Three critical noise parameters are listed for each curve: the output amplifier white noise, $1/f$ noise corner frequency and sensitivity. Noise improvement has come primarily from improving sensitivity with comparably small increases in $1/f$ and white noise. Current output amplifiers have nearly reached the 1 e⁻

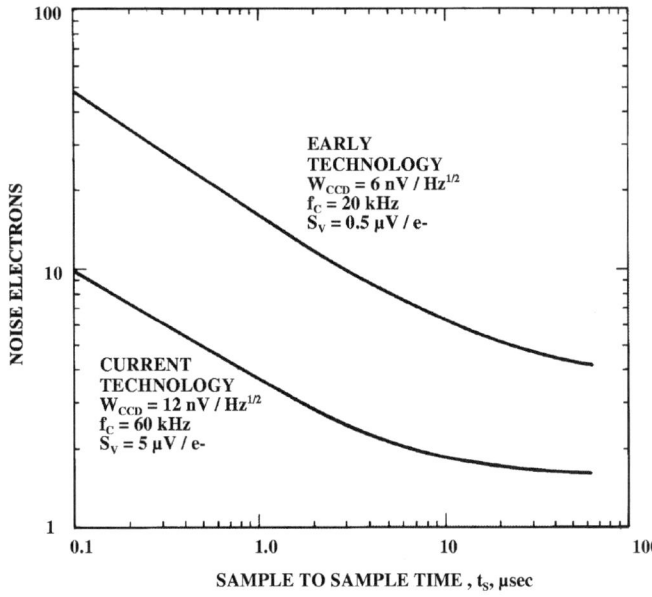

Figure 6.53 Read noise as a function of t_s for early and current output amplifiers.

noise level, with $1/f$ noise being the final limitation. To circumvent the $1/f$ noise limit and break the subelectron noise barrier, a nondestructive multiple-pixel readout amplifier was invented.[34–36] Averaging several samples from the same pixel reduces the read noise by the square-root of the number of samples taken. Multiple sampling is accomplished by shifting the charge packet back and forth between two electrodes, one of which is a floating gate that is connected to the output amplifier. This amplifier arrangement is referred to as a Skipper CCD (the name will be apparent shortly). The sample-to-sample time used by a Skipper amplifier is adjusted to be sufficiently long where $1/f$ noise is encountered (i.e., noise correlation is lost), and further noise performance cannot be achieved unless multiple samples are taken. In this manner, a square-root of noise reduction is realized to any noise level desired. It should be noted that conventional floating diffusion amplifiers only allow one sample per pixel. Moving charge backwards from the output diode back to the summing well is impossible. This is due to the fact that if the potential beneath the output transfer gate ever became equal to the output diode, which is necessary if charge is to move backwards, electrons from the diode would also be injected and would mix with signal electrons.

6.7.2 OPERATION

Figure 6.54 shows the design of the first experimental Skipper amplifier. The layout shows the floating gate electrode and four control gates around it (gates 1–4). The floating gate is connected to a conventional MOSFET source follower amplifier and reset switch. The Skipper sequence begins by clocking the horizontal register one

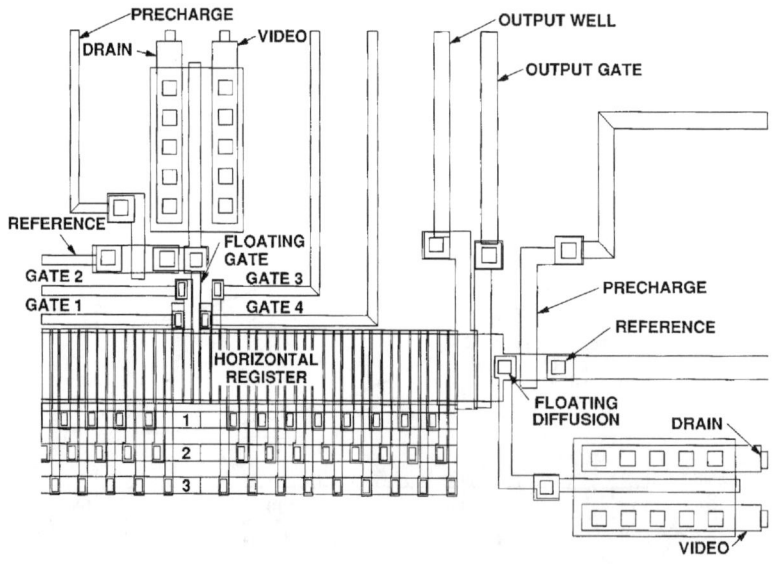

Figure 6.54 Output design of an early Skipper CCD.

pixel and transferring charge under gate 1 (refer to Fig. 6.55). The horizontal clocks are then inhibited. Next, the floating gate is preset to V_{REF}. Gate 1 is then clocked low, forcing charge through gate 2 into the potential well beneath the floating gate. The voltage changes on the floating gate in proportion to the amount of charge transferred. The voltage is then sampled by CDS electronics, producing the first sample for the pixel. Then charge is quickly moved back to gate 1 by clocking gates 2 and 1 high, as shown, completing one sample sequence. The floating gate is reset again, and the above cycle is repeated several times depending on the noise level desired. When all samples for a pixel are collected, gates 3 and 4 are activated, forcing charge back into the horizontal register (i.e., phase-3). As the horizontal register is clocked, the charge packet is then transferred to a conventional floating diffusion MOSFET amplifier (shown). At that point, charge can be sampled again and discarded through the reset MOSFET. The floating diffusion amplifier can also be replaced by a simple drain region to discard charge.

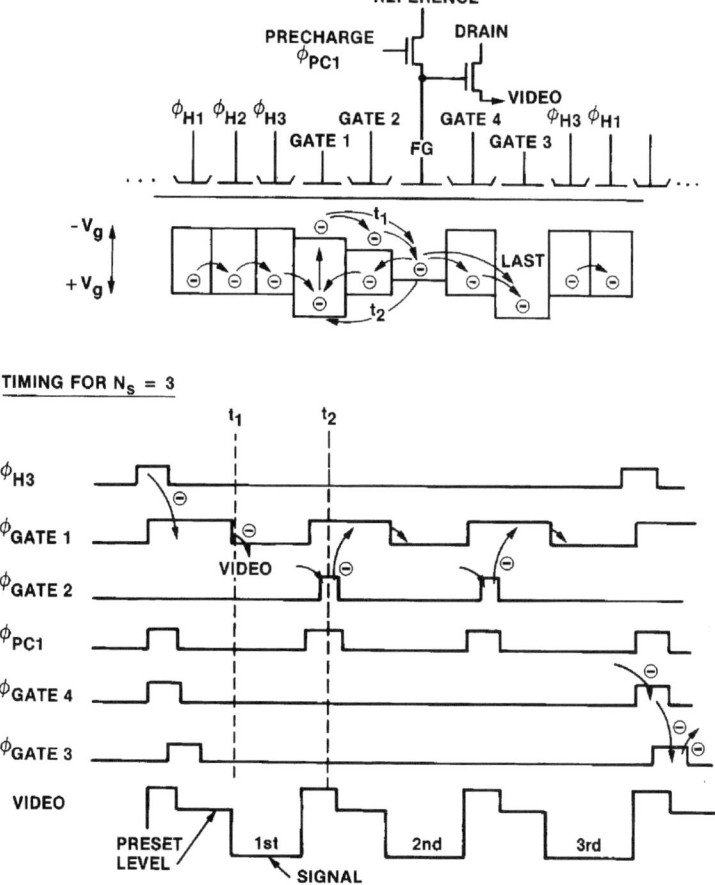

Figure 6.55 Timing diagram for Skipper operation.

6.7.3 PERFORMANCE

Figure 6.56(a) shows a line trace generated by the Skipper CCD in response to a square-wave target. The clocks sample 21 extended pixels followed by 89 video pixels, each of which is sampled 64 times nondestructively. The remaining 167 pixels in a line are rapidly read out ("skipped") by being sampled only once. Figure 6.56(b) magnifies the plot showing the 89th pixel followed by 43 single sampled pixels. Careful examination of this pixel shows random noise variations among the 64 samples generated by noise from the output amplifier. This noise is suppressed by averaging the samples collected into a single value using a computer (noise is reduced by a factor of 8).

Figure 6.57(a) shows a subelectron image produced by the Skipper amplifier shown in Fig. 6.54. The image shows four point sources. Each pixel in the top image is displayed using only the first sample of the 64 sample/pixel set, whereas the lower image uses all 64 samples. Noise levels of 7.6 e$^-$ and 0.97 e$^-$ are measured for the top and bottom images, respectively. The point source of light on the far right side is approximately 3–4 e$^-$. Figures 6.57(b) and 6.57(c) show line traces taken through the point sources for a longer exposure period. It is difficult to detect all four point sources in the 7.6 e$^-$ image; whereas in the 0.97 e$^-$ image each point is clearly seen (the point at the furthest to the right is approximately 14 e$^-$). Figure 6.58(a) shows high-energy electron events generated by a Sr^{90} electron ra-

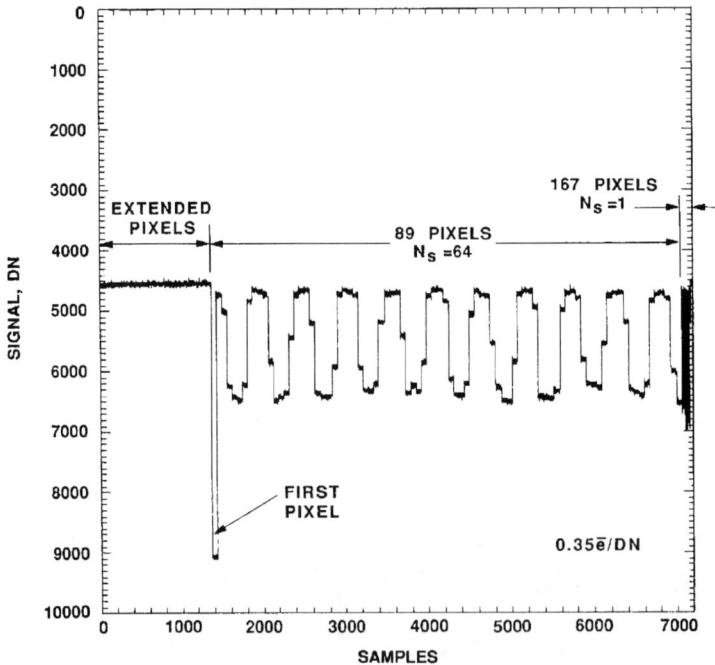

Figure 6.56(a) Skipper video showing 64 samples per pixel.

CHARGE MEASUREMENT 589

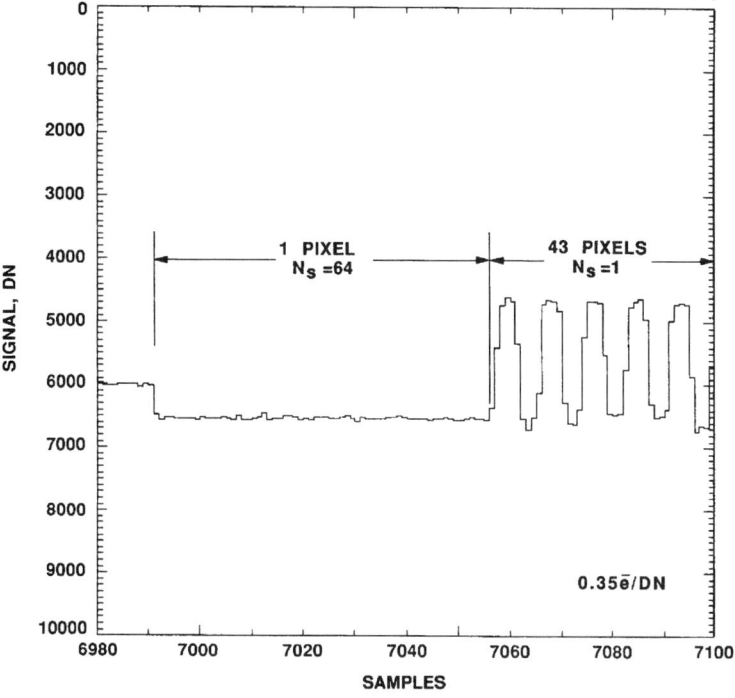

Figure 6.56(b) Magnified view of the 89th pixel showing 64 individual samples.

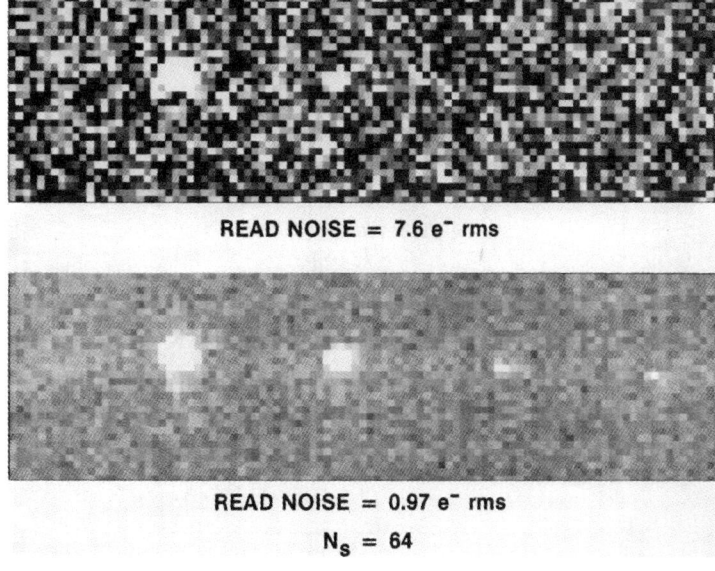

Figure 6.57(a) Image of four point sources showing subelectron noise in the lower image.

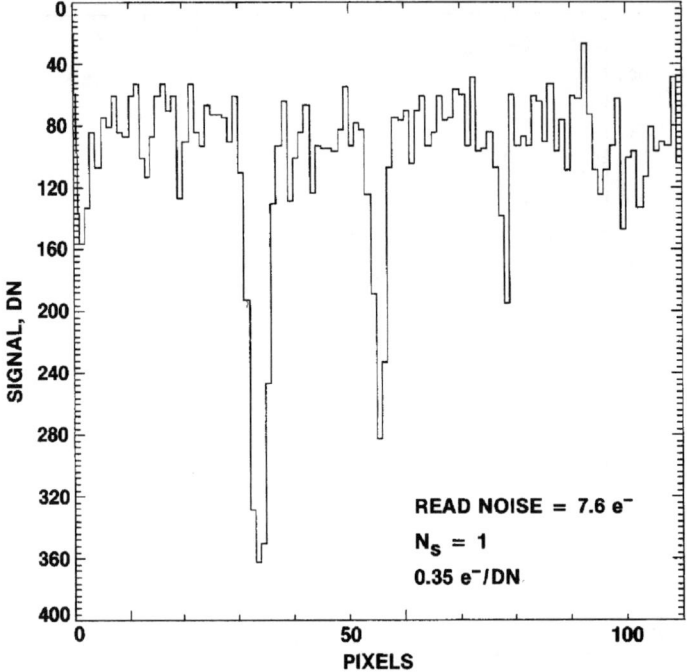

Figure 6.57(b) Line traces of point images using 1 sample/pixel.

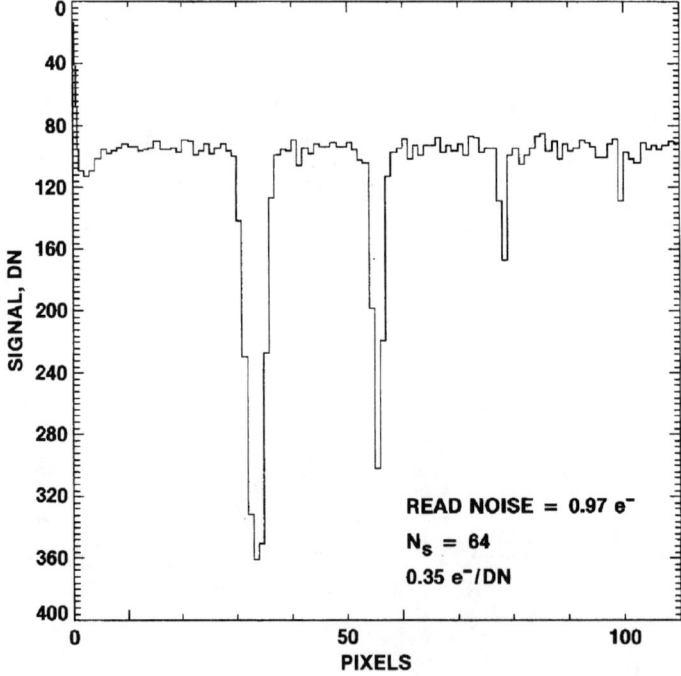

Figure 6.57(c) Line traces of point images using 64 samples/pixel.

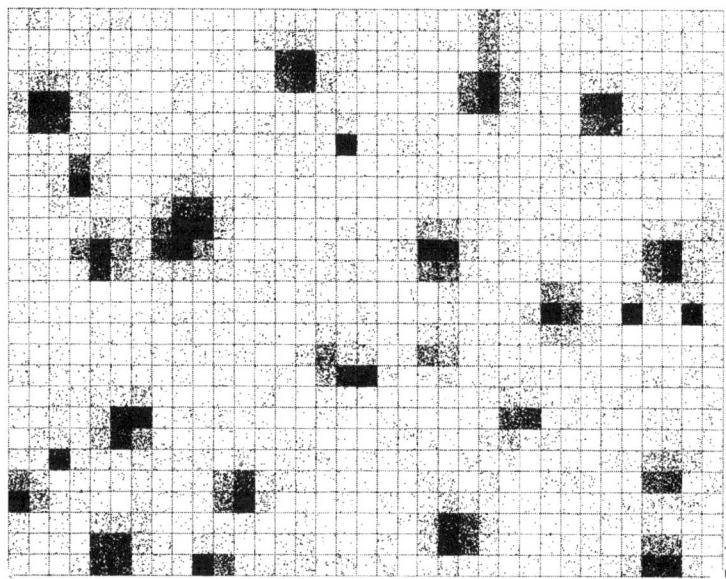

Figure 6.58(a) Electron image using single-pixel sampling.

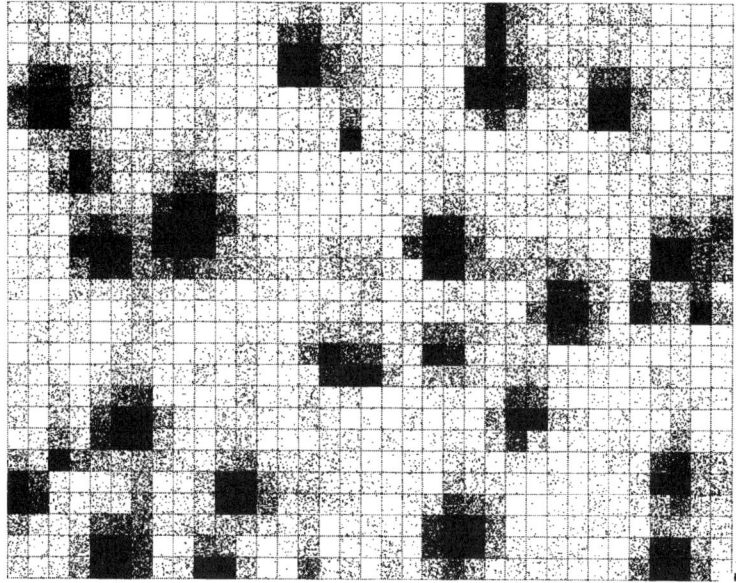

Figure 6.58(b) Electron image using 64 samples/pixel.

diation source for 1 sample/pixel. Figure 6.58(b) shows the S/N improvement when 64 samples are used. Note that the ionization cloud induced by the electrons extends into several pixels unseen in the 7.6 e^- noise image.

Taking multiple samples increases frame time by approximately the number of samples taken per pixel. For many scientific imaging applications, interrogating every pixel on the array multiple times is not necessary. Multiple sampling can be done only in those regions of interest where S/N improvement is desired. Other areas on the array that do not contain signal information (i.e., dark regions) or charge levels that are shot-noise limited can be "skipped" or single-sampled (hence the name Skipper). This process can be done by first taking an image and then zooming into a region of interest on a second take using multiple sampling. The skipping process can also use two Skipper amplifiers incorporated into the horizontal register separated by a fixed number of pixels. The first floating gate is used to detect the incoming signal level to decide if the second gate should perform multiple sampling or not. Read time can also be reduced by fabricating multiple Skipper amps to average a group of pixels simultaneously (similar to multiple output amplifier operation). Outputs from these amps can be averaged, reducing the noise by the square-root of amps used in the horizontal register.

6.7.4 Design

The first Skipper CCD fabricated exhibited a high noise level (7.6 e$^-$), in part due to a non-LDD amplifier configuration. Improvements in amplifier sensitivity have reduced the noise to < 3 e$^-$/sample, allowing subelectron performance in

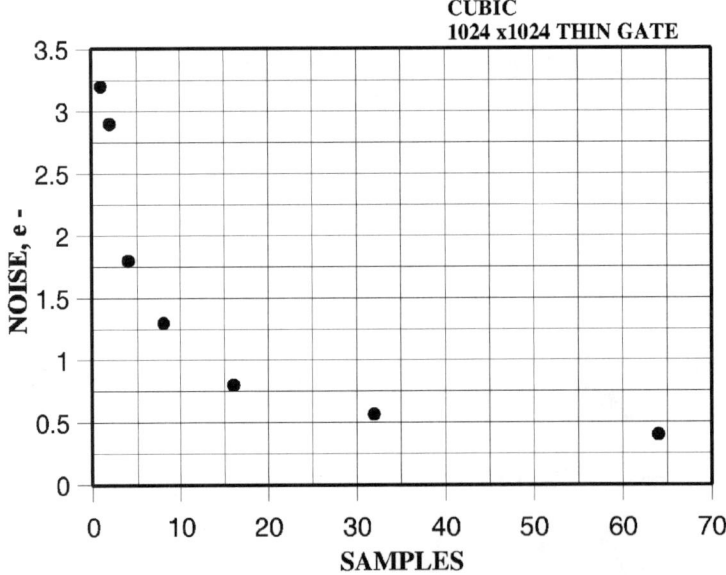

Figure 6.59 Read noise as a function of number of samples for a 1024 × 1024 Skipper CCD.

CHARGE MEASUREMENT

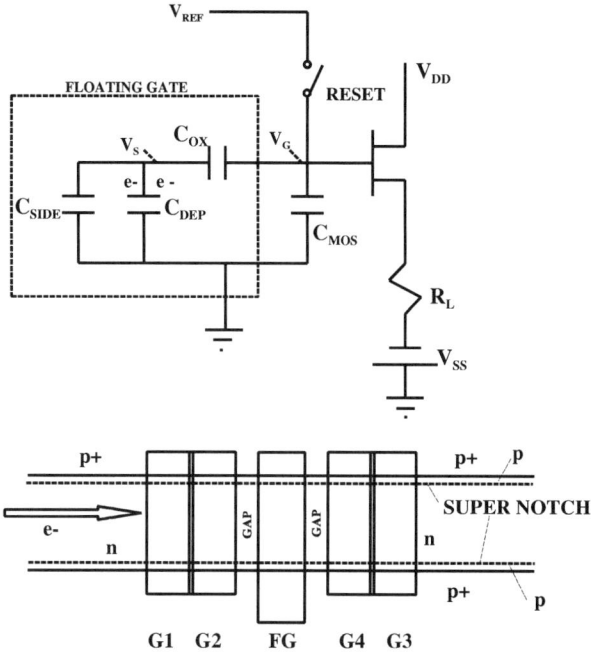

Figure 6.60(a) Skipper amplifier sense capacitances.

16 samples/pixel. For example, Fig. 6.59 shows noise performance for an improved Skipper amplifier that exhibits a sensitivity of 3 μV/e$^-$.

Figure 6.60(a) shows a circuit diagram for a Skipper amplifier and the capacitances that pertain to floating gate charge detection. When charge is dumped under the floating gate, the potential well under it collapses, increasing the depletion capacitance. This in turn generates a voltage change at the Si–SiO$_2$ interface, which is reflected through the gate insulator onto the gate of the output MOSFET. For high sensitivity (V/e$^-$) and low noise (e$^-$ rms), it is important to keep the depletion capacitance under the floating gate as small as possible. This implies that the floating gate area over the channel be small. For small signal (e.g., x-ray CCDs), this area can be designed to be very small.

As already noted in Section 6.2.8.1 on floating diffusion design, sidewall capacitance between the channel and field regions is critical to floating gate sensitivity. This capacitance is minimized by isolating the channel from highly doped field material by a super notch, as shown in Fig. 6.60(a). Also, gate overlap and associated sidewall capacitance contribute to sensitivity loss. This capacitance is reduced by eliminating any overlap between the floating gate and control gates (i.e., gate 2 and gate 4). In fact, some Skipper designs place a small gap between the gates, as shown in the figure. However, there are limits to how much gap and isolation can be employed. The gap represents an undefined potential, which can induce a spurious potential pocket in the signal channel. For example, Fig. 6.60(b) shows a

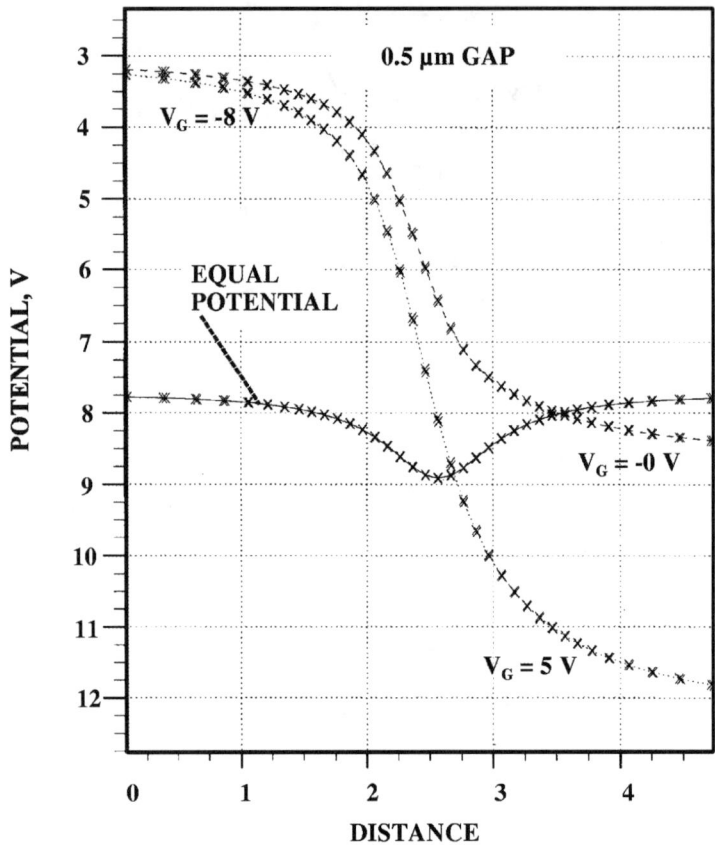

Figure 6.60(b) Potential pocket induced within 0.5 μm gap region.

potential diagram for two gates that are separated by a 0.5-μm gap. When the gates are biased at the same voltage level, a potential pocket of 1 V is created. However, when biased differently, the fringing fields overwhelm and eliminate the pocket as shown. Figure 6.60(c) shows the same gates when a 1-μm gap is employed. Here the fringing fields do not collapse the pocket, and therefore charge transfer problems through the floating gate would result.

Once the floating gate depletion and sidewall capacitances are minimized, the next step to Skipper amplifier design is to make the MOSFET gate capacitance and the source region of reset switch as small as possible. Lowest noise will result when these capacitances are equal to the floating gate capacitance. This is a similar conclusion reached for the floating diffusion amplifier, when the MOSFET gate capacitance is equal to the floating diffusion capacitance. This analysis assumes that the gate oxide capacitance of the floating gate, C_{OX}, is larger than C_{MOS} so that the full voltage change is impressed on the amplifiers gate. If this is not the case, a more detailed analysis is required for optimum design.

CHARGE MEASUREMENT

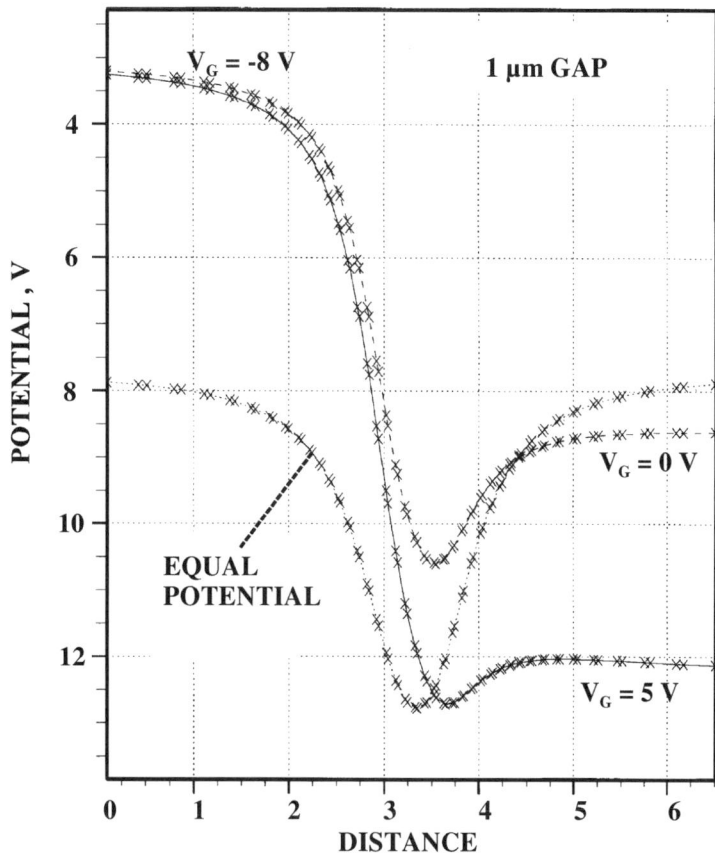

Figure 6.60(c) Potential pocket induced within 1.0 μm gap region.

6.7.5 SIGNAL-TO-NOISE (EXTENDED IMAGES)

For extended images, the benefit of multiple sampling is usually limited over a very small dynamic range. This characteristic is shown by calculating the S/N for a flat-field image that is processed with N_S samples per pixel. That is,

$$\text{S/N} = S(\text{e}^-)\left(\frac{R(\text{e}^-)^2}{N_S} + S(\text{e}^-)\right)^{-1/2}, \qquad (6.88)$$

where $R(\text{e}^-)$ is the read noise (rms e^-), and $S(\text{e}^-)$ is the average signal level of the image (e^-).

Example 6.50

Find the S/N for a flat-field image given, $S(\text{e}^-) = 10\,\text{e}^-$, $R(\text{e}^-) = 10\,\text{e}^-$, and $N_s = 64$ samples. Also perform the same calculation for $R(\text{e}^-) = 1\,\text{e}^-$.

Solution:

$$\text{Read noise} = 10 \text{ e}^-.$$

From Eq. (6.88),

$$\text{S/N} = 10/\left(10^2/64 + 10\right)^{1/2},$$

$$\text{S/N} = 2.94.$$

Without averaging, a S/N of 0.95 is achieved; therefore, multiple sampling is beneficial.

$$\text{Read noise} = 1 \text{ e}^-,$$

$$\text{S/N} = 10/(1/64 + 10)^{1/2},$$

$$\text{S/N} = 3.14.$$

Without averaging, the S/N = 3.05; there is little advantage in multiple sampling because shot noise dominates.

Figure 6.61(a) plots S/N as a function of signal for various N_s with $R(\text{e}^-) = 10 \text{ e}^-$. At high signals, multiple sampling serves no purpose because S/N is limited by shot noise. For low signals, S/N is also limited by shot noise. For a given S/N, the dynamic range of the Skipper circuit is defined for signal levels bounded by $N_S = 1$ to $N_S = \infty$. For example, the figure shows the dynamic range to achieve a S/N = 2 (shown as a horizontal line in the plot). The upper and lower signal levels are 22 e$^-$ ($N_S = 1$) and 4 e$^-$ ($N_s = \infty$) for a dynamic range of 5.5 (22/4). Note that the dynamic range decreases as the S/N requirement increases. For example, the dynamic range is only 2.5 at S/N = 5, as shown in the figure.

The number of samples required to achieve a specified S/N is

$$N_S = \frac{(\text{S/N})^2 R(\text{e}^-)^2}{S(\text{e}^-)[S(\text{e}^-) - (\text{S/N})^2]}. \qquad (6.89)$$

Example 6.51

Find the number of samples required to achieve a S/N = 5, given $S(\text{e}^-) = 40 \text{ e}^-$ and $R(\text{e}^-) = 10 \text{ e}^-$.
Solution:
From Eq. (6.89),

CHARGE MEASUREMENT

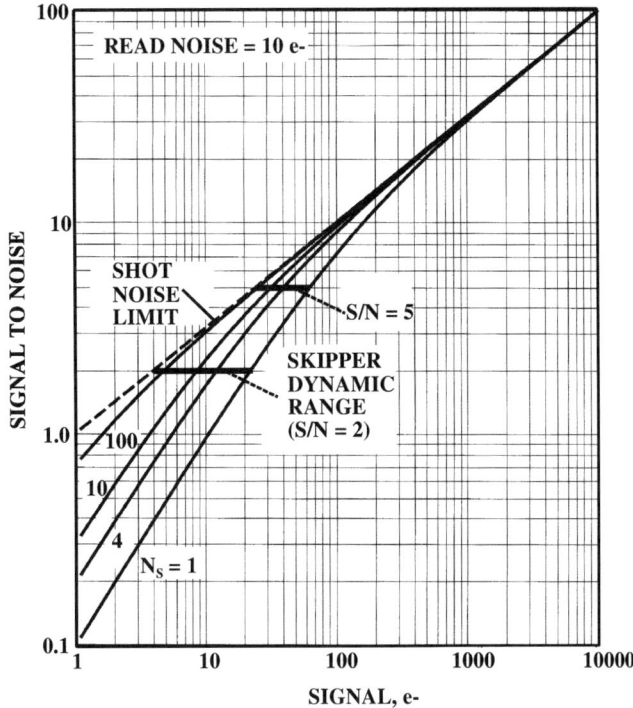

Figure 6.61(a) Signal-to-noise as a function of signal and number of samples for an extended image.

$$N_S = 5^2 \times 10^2 / [50 \times (50 - 5^2)],$$

$$N_S = 2.0.$$

Figure 6.61(b) plots Eq. (6.89) as a function of signal and read noise for a S/N = 5. The horizontal dynamic range line is also shown for a read noise of 10 e$^-$, which corresponds to the line shown in Fig. 6.61(a).

The dynamic range for the Skipper becomes insignificant for read noise levels less than 3 e$^-$, a noise level routinely achieved by high-performance CCDs. There is very little benefit to multiple sampling at noise levels lower than this because of shot noise. It is for this reason that Skipper amplifiers are not popular. If the read noise for the CCD had settled to levels greater than this, Skipper amps would be incorporated much more often (e.g., astronomical applications). Figure 6.62 shows computer-generated bar target images that demonstrate the slight benefit gained when multiple sampling low-read-noise images. The mean signal level in the top image is 1 electron. Multiple sampling has been used to eliminate a 1 electron noise floor also present ($N_S = \infty$). The signal-to-noise for the image is therefore S/N = 1. The mean signal level for the lower image is also 1 electron. Here multiple

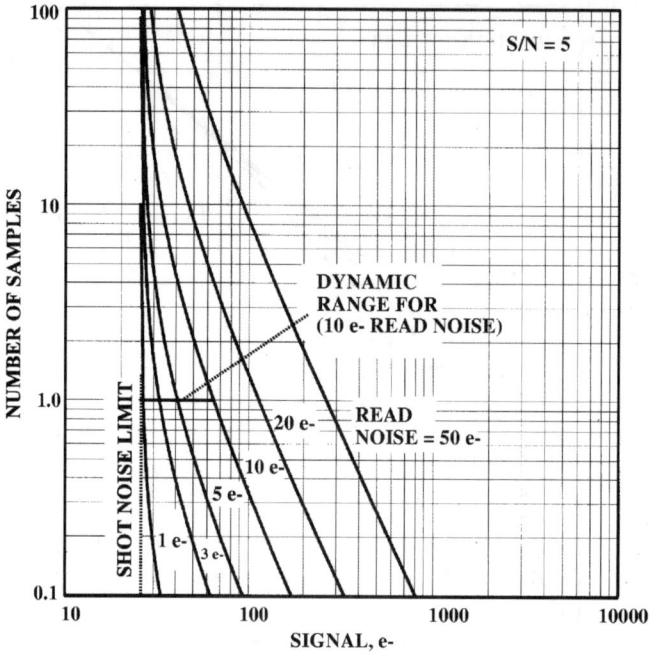

Figure 6.61(b) Number of samples required to reach a S/N = 5 as a function of read noise for an extended image.

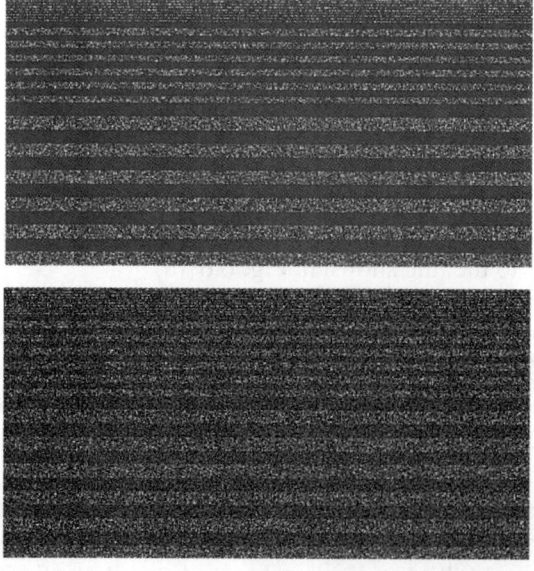

Figure 6.62 Single electron computer-generated images with and without a 1 e⁻ read noise.

sampling is not used, showing the full 1 electron noise floor ($N_S = 1$). The signal-to-noise in this case is S/N = 0.707 [1/(1+1)]. Although the image without noise is better, the improvement gained may not be worth the process time needed for gathering multiple samples.

6.7.6 SIGNAL-TO-NOISE (POINT IMAGES)

Unlike extended images, Skipper readout is very beneficial in measuring point or line sources at any read noise level (e.g., x-ray and spectroscopy applications). S/N for a point source of charge is

$$\text{S/N} = S(\text{e}^-)\frac{N_S^{1/2}}{R(\text{e}^-)}, \tag{6.90}$$

where $S(\text{e}^-)$ is the charge contained in the target pixel. Note that S/N improves by the square-root of the number of samples taken independent of the read noise floor.

Example 6.52

Find the S/N of a point image given, $S(\text{e}^-) = 10$ e$^-$, $R(\text{e}^-) = 10$ e$^-$, and $N_s = 64$ samples. Perform the same calculation for $R(\text{e}^-) = 1$ e$^-$.
Solution:
From Eq. (6.90),

$$\text{S/N} = 10 \times 64^{1/2}/10,$$

$$\text{S/N} = 8, \quad R(\text{e}^-) = 10 \text{ e}^-,$$

$$\text{S/N} = 80, \quad R(\text{e}^-) = 1 \text{ e}^-.$$

When x-ray events are measured by a Skipper amplifier, the Fano-noise must be included. The S/N for x-ray events is

$$\text{S/N} = S(\text{e}^-)\left[\frac{R(\text{e}^-)^2}{N_S} + S(\text{e}^-)F_a\right]^{-1/2}, \tag{6.91}$$

where F_a is the Fano-factor (0.1) and $S(\text{e}^-)$ is the charge generated by the x-ray event.

Example 6.53

Find the S/N for an x-ray event that generates 10 e⁻ of charge. Assume a read noise of 10 e⁻ and 64 samples/pixel. Also, calculate S/N for a read noise of 1 e⁻.

Solution:
From Eq. (6.90),

$$\text{S/N} = 10/\left(10^2/64 + 10 \times 0.1\right)^{1/2},$$

$$\text{S/N} = 3.9.$$

The S/N for a 1 e⁻ noise floor is

$$\text{S/N} = 10/(1/64 + 10 \times 0.1)^{1/2},$$

$$\text{S/N} = 9.92.$$

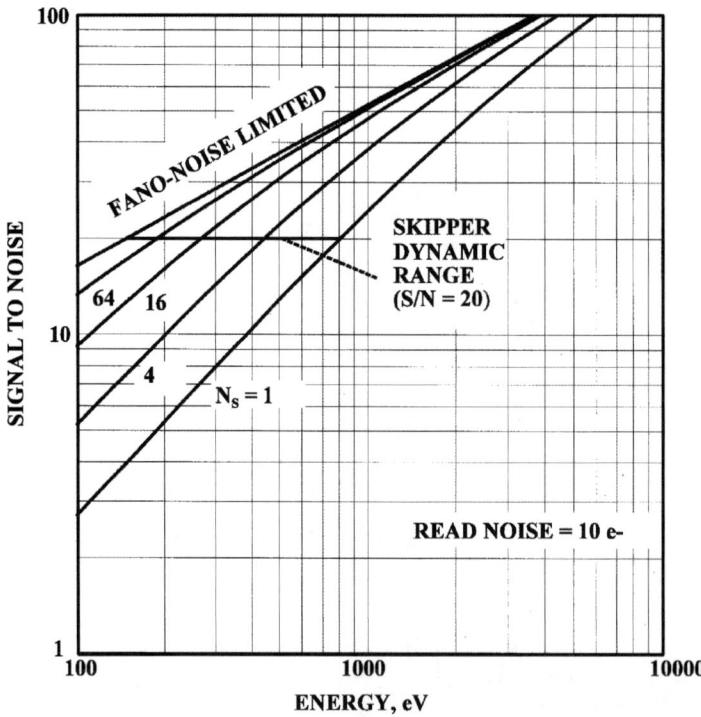

Figure 6.63(a) Signal-to-noise as a function of x-ray energy and number of samples.

CHARGE MEASUREMENT

Figure 6.63(a) plots Eq. (6.91) as a function of x-ray energy and N_S covering the soft x-ray regime. A 10 e⁻ noise floor is assumed in the plot. Recall that the signal generated by an x-ray event is equal to $E(eV)/3.65$, where $E(eV)$ is the energy of the photon. The dynamic range of the Skipper in the figure is shown for a S/N = 20 (i.e., the horizontal line in the plot). Again, the dynamic range for the Skipper is defined by the signal levels that are bounded by $N_S = 1$ to $N_S = \infty$. Note that the lower signal level is bounded by Fano-noise.

The number of samples required to reach a specified signal-to-noise level is

$$N_S = \frac{(S/N)^2 R^2}{S(e^-)^2 - F_a (S/N)^2 S(e^-)}. \tag{6.92}$$

Example 6.54

Find the number of samples required to achieve a S/N = 10, given $E(eV) = 80$ eV and $R(e^-) = 5$ e⁻. Assume a Fano-factor of 0.1.

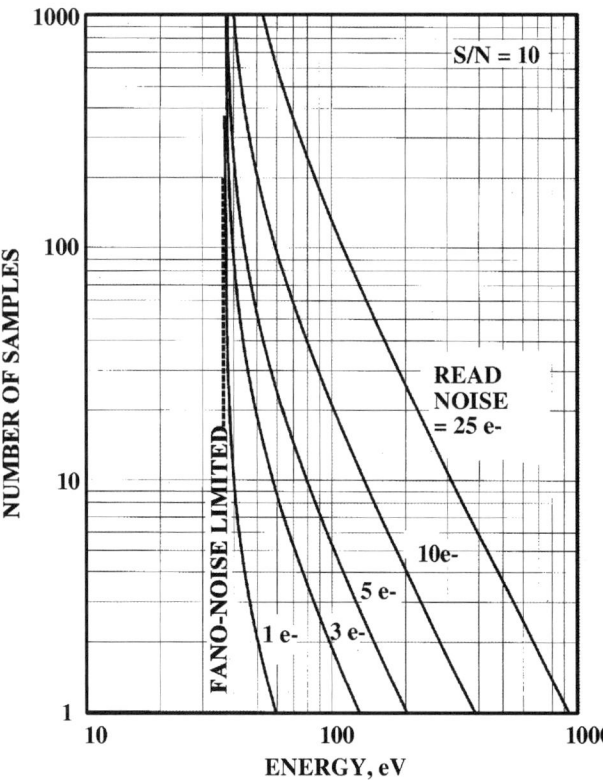

Figure 6.63(b) Number of samples required to reach S/N = 10 as a function of x-ray energy and read noise.

Solution:
The signal generated by an 80-eV photon is

$$S(e^-) = 80/3.65,$$

$$S(e^-) = 21.92 \, e^-.$$

From Eq. (6.92),

$$N_S = 10^2 \times 5^2 / (22^2 - 0.1 \times 10^2 \times 22),$$

$$N_S = 9.57.$$

Figure 6.63(b) plots the number of samples required as a function of x-ray energy and read noise for a S/N = 10.

References

1. S. Sze, *Physics of Semiconductor Devices*, John Wiley and Sons, p. 469 (1981).
2. S. Sze, *Physics of Semiconductor Devices*, John Wiley and Sons, p. 448 (1981).
3. J. Janesick and T. Elliott, "CCD luminescence camera," *NASA Tech. Briefs* Vol. 11, No. 5, p. 34 (1987).
4. A. Grove, *Physics and Technology of Semiconductor Devices*, John Wiley and Sons, pp. 191–192 (1967).
5. J. Hynecek, "Design and performance of a low-noise charge-detection amplifier for VPCCD devices," *IEEE Trans. Electron Devices* Vol. ED-31, No. 12, p. 1718 (1984).
6. T. Jung, H. Guckel, J. Seefeldt, G. Ott and Y. Ahn, "A fully integrated, monolithic, cryogenic charge sensitive preamplifier using channel JFETs and polysilicon resistors," *IEEE Trans. Nucl. Sci.* Vol. 41, No. 4 (1994).
7. N. Teranishi and N. Mutoh, "Partition noise in CCD signal detection," *IEDM—International Electron Devices Meeting*, Washington, DC (1985).
8. A. McWhorter, *Semiconductor Surface Physics*, Philadelphia, University of Philadelphia Press, pp. 207–228 (1957).
9. G. Alers, K. Krisch. D. Monroe, B. Weir and A. Chang, "Tunneling current noise in thin gate oxides," *Appl. Phys. Lett.* Vol. 69, No. 19 (1996).
10. J. Carnes and W. Kosonocky, "Noise sources in charge-coupled devices," *RCA Rev.* Vol. 33, pp. 327–343 (1972).
11. E. Simoen and C. Claeys, "On the flicker noise in submicron silicon MOSFETs," *Solid-State Electronics* Vol. 43, pp. 865–882 (1999).
12. S. Sze, *Physics of Semiconductor Devices*, John Wiley and Sons, p. 385 (1981)

13. K. Hung, P. Ko, C. Hu and Y. Cheng, "A physics-based MOSFET noise model for circuit simulators," *IEEE Trans. Electron Devices* Vol. 37, No. 5 (1990).
14. F. N. Hooge, *Physica* Vol. 60, pp. 130–144 (1972).
15. C. Surya and T. Y. Hsiang, *Phys. Rev. B* Vol. 33, p. 4898 (1986).
16. K. S. Ralls, W. J. Skocpol, L. D. Jackel, R. E. Howard, L. A. Fetter, R. W. Epworth and D. M. Tennant, *Phys. Rev. Lett.* Vol. 52, p. 228 (1984).
17. R. E. Howard, W. J. Skocpol, L. D. Jackel, P. M. Mankiewich, L. A. Fetter, D. M. Tennant, R. W. Epworth and K. S. Ralls, *IEEE Trans. Electron Devices* ED-32, p. 1669 (1985).
18. W. J. Skocpol, *Les Houches Winter School on Physics and Fabrication of Microstructures,* Springer Verlag Physics Lecture Series (1986).
19. M. J. Uren, D. J. Day and M. J. Kirton, *Appl. Phys. Lett.* Vol. 47, p. 1195 (1985).
20. M. Kirton and M. Uren, "Noise in solid-state microstructures: A new perspective on individual defects, interface states and low-frequency $(1/f)$ noise," *Advances in Physics* Vol. 38, No. 4, pp. 367–468 (1989).
21. H. Muller and J. Schulz, *J. Appl. Phys.* Vol. 79, p. 4178 (1996).
22. H. Muller and J. Schulz, *Proc. of the 14th Int. Conf. on Noise in Physical Systems and 1. Fluctuations*, Singapore, World Scientific, pp. 195–200 (1997).
23. R. Hooge, T. Kleinpenning and L. Vandamme, "Experimental studies on $1/f$ noise," *Rep. Prog. Phys.* Vol. 44 (1981).
24. K. Hung, P. Ko, C. Hu and Y. Cheng, "Random telegraph noise of deep-submicrometer MOSFETs," *IEEE Electron Device Letters* Vol. 11, No. 2 (1990).
25. K. Hung, P. Ko, C. Hu and Y. Cheng, "A unified model for the flicker noise in metal-oxide-semiconductor- field-effect transistors," *IEEE Trans. Electron Devices* Vol. 37, No. 3 (1990).
26. H. Mueller and M. Schulz, "Conductance modulation of submicrometer metal-oxide-semiconductor field-effect transistors by single-electron trapping," *J. Appl. Phys.* Vol. 79, No. 8 (1996).
27. C. Surya and T. Hsiang, "Surface mobility fluctuations in metal-oxide-semiconductor field-effect transistors," *Phys. Rev. B* Vol. 35, No. 12 (1987).
28. J. Haslett and E. Kendall, "Temperature dependence of low-frequency excess noise in junction-gate FETs, *IEEE Trans. Electron Devices* (1972).
29. K. Kandiah and F. Whiting, "Nonideal behaviour of buried channel CCDs caused by oxide and bulk silicon traps," *Nuclear Instruments and Methods in Physics Research*, A305, pp. 600–607 (1991).
30. M. White, D. Lampe, F. Blaha and I. Mack, "Characterization of surface channel CCD image arrays at low light levels," *IEEE Journal of Solid-State Circuits* Vol. SC-9, No. 1 (1974).
31. B. Burke, J. Gregory, M. Bautz, G. Prigozhin, S. Kissel, B. Kosicki, A. Loomis and D. Young, "Soft-x-ray CCD imagers for AXAF," *IEEE Trans. Electron Devices* Vol. 44, No. 10 (1997).

32. R. Reich, R. Mountain, M. Robinson, W. McGonagle, A. Loomis, B. Kosicki and E. Savoye, "An epitaxially-grown charge modulation device," *IEEE Trans. Electron Devices* Vol. 44, No. 10 (1997).
33. J. Hynecek, "Low noise and high-speed charge detection in high-resolution CCD image sensors," *IEEE Trans. Electron Devices* Vol. 44, No. 10 (1997).
34. J. Janesick, "Low noise charge-couple device," *NASA Tech. Briefs* Vol. 15, No. 10, p. 20 (1991).
35. C. Chandler, R. Bredthauer, J. Janesick, J. Westphal and J. Gunn, "Sub-electron noise charge coupled devices," *Proc. SPIE* Vol. 1242, pp. 238–251 (1990).
36. J. Janesick, C. Chandler, R. Bredthauer, J. Westphal and J. Gunn, New advancements in charge-coupled device technology – subelectron noise and 4096 × 4096 pixel CCDs," *Proc. SPIE* Vol. 1242 (1990).

CHAPTER 7

NOISE SOURCES

7.1 ON-CHIP NOISE SOURCES

We have already discussed two important array noise sources in Chapter 2 and Chapter 4, shot noise and pixel nonuniformity. Other important on-chip array noise sources that are discussed in this chapter include (1) dark current, (2) spurious charge, (3) fat-zero, (4) transfer noise, (5) residual image, (6) luminescence, (7) cosmic rays and radiation, (8) excess charge, (9) blem spill-over, (10) cosmetic noise and (11) seam noise.

7.1.1 DARK CURRENT

7.1.1.1 Introduction

Dark current is intrinsic to semiconductors and naturally occurs through the thermal generation of minority carriers. We call this source dark current because it is produced when the CCD is in complete darkness. The level of dark current generated determines the amount of time a potential well can exist to collect useful signal charge. This time is not very long, and therefore CCD users must deal with the dark current problem head on. There is only one solution to eliminating dark signal, which is to cool the sensor.

There are three principal regions that contribute to dark current generation: (1) neutral bulk material below the potential well and channel stop regions, (2) depleted material within the potential well and (3) Si–SiO$_2$ interface states (frontside and backside in the case of thinned CCDs). These regions are illustrated in Fig. 7.1. Of these sources, the contribution from surface states is the dominant source of dark current. Bulk dark current without surface generation (i.e., full channel inversion and backside accumulation) can be very low. Levels as low as 3 pA/cm^2 (300 K) have been measured. However, bulk dark current varies considerably depending on the quality of the silicon material and wafer preprocessing employed. Bulk dark current is approximately equal to the epitaxial thickness expressed in units of pA/cm^2 at 300 K. For example, a 10-μm epitaxial CCD, for quality silicon, will exhibit a bulk dark current of approximately 10 pA/cm^2.

Many solid state equations have been derived that determine reverse leakage currents in simple n-p devices. Recall that the buried-channel CCD is based on an n-p junction. We now turn to these equations and apply them to estimate the dark current generated in different regions of the CCD.

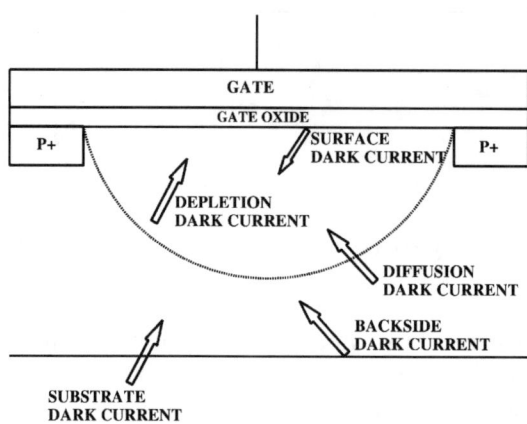

Figure 7.1 Regions of dark current sources in the CCD.

7.1.1.2 Electron and Hole Concentrations

The electron and hole concentrations for a semiconductor under thermal equilibrium are[1]

$$n = N_C e^{-(E_C - E_F)/kT}, \qquad (7.1a)$$

or

$$n = n_i e^{-(E_F - E_i)/kT}, \qquad (7.1b)$$

and

$$p = N_V e^{-(E_F - E_V)/kT}, \qquad (7.2a)$$

or

$$p = n_i e^{-(E_i - E_F)/kT}, \qquad (7.2b)$$

where n is the electron concentration in the n-channel (cm^{-3}) and p is the hole concentration in the p-epitaxial region (cm^{-3}), E_F is the Fermi level (eV), E_C and E_V are the conduction and valence band energies (eV), N_C is the number of effective states in the conduction band (2.8×10^{19} cm^{-3}), N_V is the number of effective states in the valance band (1.04×10^{19} cm^{-3}) and n_i is the intrinsic (i.e., undoped) carrier concentration (cm^{-3}).

The exponential factors in Eqs. (7.1) and (7.2) give the probability that an electron and hole are at the conduction and valence band edges, respectively. The Fermi level is defined by how the semiconductor is doped. The Fermi level moves closer to the conduction band edge when n-doped (e.g., phosphorus), producing more electrons and less holes. On the other hand, doping the semiconductor with p-material (e.g., boron) moves the Fermi level nearer to the valence band edge, producing more holes and less electrons.

The product of electron and hole concentrations from Eqs. (7.1b) and (7.2b) is

$$pn = n_i^2 = N_C N_V e^{-E_g/kT}, \tag{7.3}$$

where N_C and N_V are proportional to $T^{3/2}$, producing

$$n_i^2 = C_1 T^{3/2} e^{-E_g/kT}, \tag{7.4}$$

where C_1 is a constant. This important relationship is the basis for the general CCD dark current equations derived below.

Example 7.1

Determine the intrinsic carrier concentration at room temperature (300 K). Assume $N_C = 2.8 \times 10^{19}$ cm^{-3}, $N_V = 1.04 \times 10^{19}$ cm^{-3}, $k = 8.62 \times 10^{-5}$ eV/K and a bandgap of 1.1 eV. Also, calculate the carrier concentration at 200°C (473 K).

Solution:
From Eq. (7.3),

$$n_i = \left((2.8 \times 10^{19}) \times (1.04 \times 10^{19})\right.$$
$$\left. \times \exp{-\{1.1/[(8.62 \times 10^{-5}) \times 300]\}}\right)^{1/2},$$

$$n_i = 9.9 \times 10^9 \text{ e-h pairs/cm}^3.$$

Similarly, at 200°C,

$$n_i = 2.36 \times 10^{13} \text{ e-h/cm}^3.$$

The number of electrons in the conduction band for intrinsic material is equal to the number of holes in the valence band at any instant of time because they are only created in pairs. The electron-hole pairs are constantly generated and recombined to maintain a constant average. A silicon atom bond that is broken frees an electron, and the missing bond or hole is created. The intrinsic carrier concentration refers to the number of bonds that are broken at a specified operating temperature. The number of bonds that can possibly break is equal to 2×10^{23} bonds/cm^3 (assuming four bonds per atom). Therefore, on the average, less than one bond breaks in 10^{13} atoms at room temperature.

It should be noted that the accepted measure of intrinsic concentration is 1.45×10^{10} e-h/cm^3 at 300 K. The value above is calculated using the values N_c and N_V. We will assume the calculated value in this chapter (i.e., 10^{10} cm^{-3}).

For doped silicon the concentration of majority carriers is

$$n_n = N_D, \qquad p_p = N_A, \tag{7.5}$$

where n_n is the concentration of electrons in n-channel and p_p is the concentration of holes in p-epitaxial layer.

The concentration of minority carriers is

$$p_n = \frac{n_i^2}{N_D}, \qquad n_p = \frac{n_i^2}{N_A}, \tag{7.6}$$

where p_n is the concentration of holes in n-channel and n_p is the concentration of electrons in p-epitaxial layer.

Example 7.2

Determine the number of majority (electrons) and minority (holes) in the n-buried channel where 1 impurity atom for each 10^8 intrinsic atoms has been distributed. Assume 300 K operation and $n_i^2 = 10^{10}$ e-h pairs/cm^3.

Solution:

Silicon contains approximately 5×10^{22} atoms/cm^3. There are 5×10^{14} impurity atoms/cm^3. Assuming each impurity atom gives up an electron, the number of majority carriers is, from Eq. (7.5),

$$n_n = 5 \times 10^{14} \text{ electrons/cm}^3.$$

The number of holes, from Eq. (7.6), is

$$p_n = (10^{10})^2/(5 \times 10^{14}),$$

$$p_n = 2 \times 10^5 \text{ holes/cm}^3.$$

Note that this number is smaller than the number of holes for intrinsic material calculated above. This is because holes have a higher probability to recombine with the vast majority of electrons present, making the count less.

Noise Sources

7.1.1.3 Generation/Recombination

Dark current carriers are generated through intermediate-level centers associated with imperfections or impurities within the semiconductor or at the Si–SiO$_2$ interface. These states introduce energy levels into the forbidden bandgap that promote dark current by acting as "steps" in the transition of electrons and holes between the conduction and valence bands. This process is also referred to as "hopping conduction."

Generation and recombination work by the four bandgap processes shown in Fig. 7.2 (i.e., electron capture, electron emission, hole capture and hole emission). Electron and hole capture are required for recombination. This process rarely occurs in a CCD because signal charge is safe when contained in a potential well (there are no majority carriers with which charge can recombine). Electron and hole emission are required to generate dark charge. The emission of a hole is the transition of an electron from the valence band to the center, leaving a hole in the valence band. The emission of an electron is the transition between the trap and the conduction band. Once in the conduction band, it can be collected by the potential well as dark signal. The hole that is made leaves the CCD through the substrate. Hole and electron emissions must occur nearly simultaneously through the center in order to generate a dark carrier.

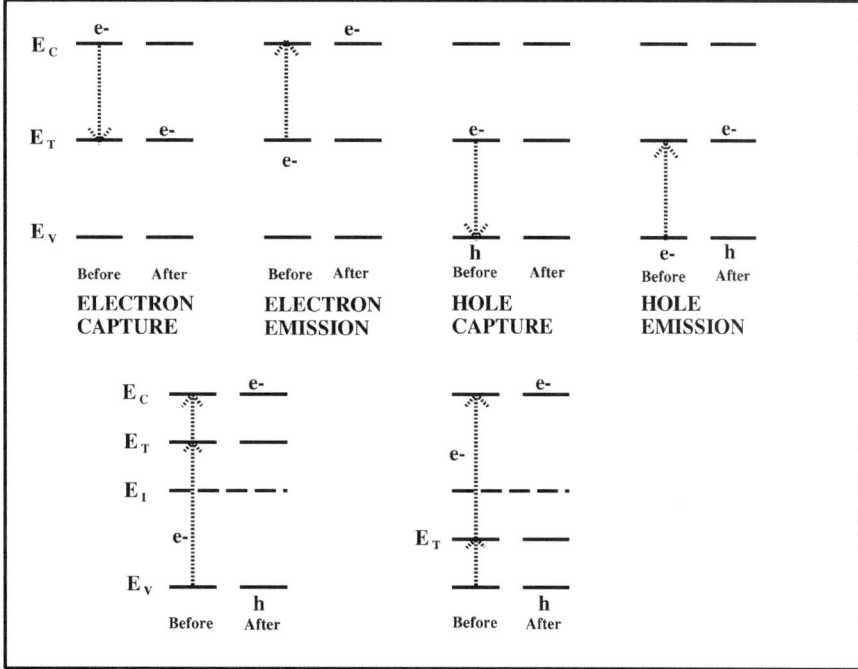

Figure 7.2 Recombination and generation through midband states.

As will be shown below, only those centers whose energy level is near the intrinsic Fermi level contribute significantly to the dark current (i.e., midband states). The generation rate falls off exponentially as the trap level moves away from midgap in either direction. For example, the bottom illustrations in Fig. 7.2 show two trap energies near the conduction and valence bands. A trap near the conduction band readily emits an electron to the conduction band because the energy step is small. However, hole emission from the same trap is not an efficient process. In that both processes must take place simultaneously, the generation rate is much less compared to a midband state transition. The same argument holds true for a trap that is close to the valence edge.

The process of generation/recombination is theoretically described by the famous Shockley/Hall/Read equation,[2-4]

$$U = \frac{\sigma_p \sigma_n v_{th}(pn - n_i^2)N_T}{\sigma_n \left[n + n_i \exp\left(\frac{E_t - E_i}{kT}\right)\right] + \sigma_p \left[p + n_i \exp\left(\frac{E_i - E_t}{kT}\right)\right]}, \quad (7.7)$$

where U is the net carrier generation/recombination (carriers/sec-cm^3), N_T is the concentration of generation/recombination centers (or traps) at energy level E_t (cm^{-3}), $E_t - E_i$ is the energy of the defect level with respect to the intrinsic Fermi level E_i (eV), σ_p and σ_n are the electron and hole cross sections (cm^2) and n and p are the number of free electrons and holes, respectively (cm^{-3}). Note again that the intrinsic energy level is the most probable transition for carrier generation.

For a CCD in thermal equilibrium, $pn = n_i^2$ and $U = 0$, indicating that generation is equal to recombination. A negative U implies a net generation rate of electron-hole pairs, whereas a positive number indicates a net recombination rate of electron-hole pairs. For the dark current equations derived below, we will always come up with a negative result because there will be more generation than recombination. A positive U implies that the CCD is stimulated with light.

7.1.1.4 Depletion Dark Current

The CCD is not in thermal equilibrium when the channel is depleted of majority carriers. Dark current electrons and holes separate by the electric field associated with the depletion region. Therefore, for the depleted state $p = n = 0$ and Eq. (7.7) becomes

$$U_{DEP} = -\frac{\sigma v_{th} n_i N_B}{\left\{\left[\exp\left(\frac{E_t - E_i}{kT}\right)\right] + \left[\exp\left(\frac{E_i - E_t}{kT}\right)\right]\right\}}, \quad (7.8)$$

where U_{DEP} is the carrier-generation rate in the depletion region (carriers/sec-cm^3) and N_B is the bulk state density (cm^{-3}). We have also assumed in the equation that the cross sections for electrons and holes are equal.

NOISE SOURCES

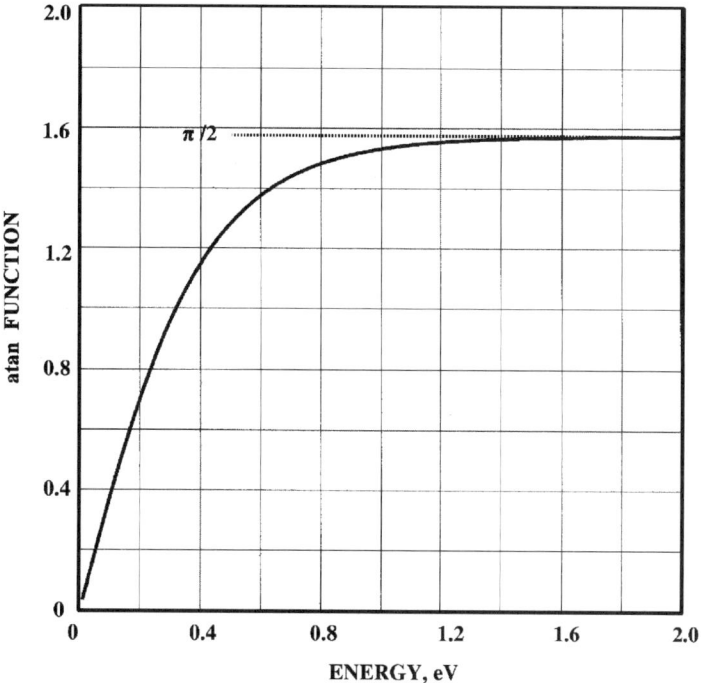

Figure 7.3 Arctan relationship as a function of bandgap energy for Eq. (7.10).

Equation (7.8) can be integrated as a function of energy for a given bulk state concentration. That is,

$$U_{DEP} = -\frac{1}{2}\sigma v_{th} n_i \int_{E_V}^{E_C} \frac{N_B(E_t)}{\cosh\left(\frac{E_t - E_i}{kT}\right)} dE_t, \qquad (7.9)$$

where $N_B(E_t)$ is the density of bulk sites as a function of energy (cm^{-3}-eV^{-1}).

We will now assume that N_B is a uniform concentration of states throughout the bandgap. Integrating Eq. (7.9) with this assumption and using $E_i = (E_C + E_V)/2$ yields

$$U_{DEP} = -\sigma v_{th} n_i N_B kT \left[\arctan\left(e^{E_G/kT}\right) - \arctan\left(e^{-E_G/kT}\right)\right]. \qquad (7.10)$$

Figure 7.3 plots the arctan terms as a function of bandgap energy. It can be seen that for energies greater than 1.2 eV that the function asymptotically approaches $\pi/2$. Therefore, for silicon Eq. (7.10) reduces to

$$U_{DEP} = -\frac{\sigma v_{th} \pi kT n_i N_B}{2}. \qquad (7.11)$$

The effective lifetime of the electrons generated is defined as

$$\tau_{DEP} = \frac{1}{\sigma v_{th} \pi kT N_B}. \tag{7.12}$$

Note that the units for τ_{DEP} are seconds. We can think of the carrier's lifetime as that time it takes to thermally generate an electron-hole pair on the average. The shorter the lifetime, the more dark carriers generated. Substituting this equation into Eq. (7.11) yields

$$U_{DEP} = -\frac{n_i}{2\tau_{DEP}}. \tag{7.13}$$

The depletion dark current is

$$I_{DEP} = \frac{q n_i (x_p + t)}{2 \tau_{DEP}}, \tag{7.14}$$

where I_{DEP} is the depletion dark current (A/cm^2), t is the channel depth (cm) and x_p is the p-depletion depth (cm).

Dark current generated in the depletion region varies considerably because the density of defects and impurities in the starting silicon vary. This difference in silicon quality is why MPP CCDs show large variations in dark current from manufacturer to manufacturer (levels from 3 pA/cm^2 to 100 pA/cm^2 are measured). Therefore, it is not possible to give a typical value for the lifetime. However, the lifetime can be estimated from the amount of dark current measured.

Example 7.3

Determine the lifetime, τ_{DEP}, for a fully inverted CCD that exhibits a dark current rate of 20 pA/cm^2 at 300 K. Assume a total depletion depth of 8×10^{-4} cm and $n_i = 10^{10}$ cm^{-3}. Also, calculate N_B assuming $v_{th} = 10^7$ cm/sec, $\sigma = 10^{-15}$ cm^2 and $kT = 0.0259$ eV.

Solution:
From Eq. (7.14),

$$\tau_{DEP} = (1.6 \times 10^{-19}) \times (8 \times 10^{-4}) \times (10^{10}) / [(20 \times 10^{-12}) \times 2],$$

$$\tau_{DEP} = 0.032 \text{ sec.}$$

From Eq. (7.12),

$$N_B = 1/(10^{-15} \times 10^7 \times 3.14 \times 0.0259 \times 0.032),$$

$$N_B = 3.84 \times 10^{10} \text{ states/cm}^3.$$

Noise Sources

7.1.1.5 Diffusion Dark Current

Diffusion dark current is generated in field-free regions of the CCD below the channel stops and potential wells. For regions deep in the CCD, electrons are uniformly distributed with a concentration n_i^2/N_A. The carriers move around randomly, producing an average diffusion current of zero. This is because the number of electrons moving across a given plane in one direction is equal to the number moving across the same plane in the opposite direction. However, the electron concentration decreases as we move closer to the depletion region. In fact, the concentration of carriers at the depletion edge is zero because any carriers generated at this interface will rapidly move into the potential well. The concentration of carriers increases with distance from depletion region until an equilibrium in carriers results. It can be shown from the general diffusion equation that the distribution of electrons from the depletion edge is[5]

$$n = \frac{n_i^2}{N_A}\left(1 - e^{-x/L_n}\right), \tag{7.15}$$

where the L_n is the diffusion length (cm) and x is the distance from the depletion edge (cm). The diffusion length is the average distance that electrons travel before the total number is reduced by traps (i.e., G/R centers) to $1/e$ of the original number. A region several diffusion lengths beyond the depletion region exhibits a carrier concentration of n_i^2/N_A.

Figure 7.4 plots the carrier concentration as a function of distance and diffusion length. Note that the density of electrons is very low at the depletion edge and n_i^2/N_A in the bulk (i.e., $(10^{10})^2/10^{15} = 10^{15}$ e$^-$/cm^3). The diffusion current as a function of distance from the depletion edge is

$$I_{DIF} = qD_n\frac{dn}{dx}, \tag{7.16}$$

where I_{DIF} is the diffusion current (A/cm^2) and D_n is the diffusion coefficient (cm^2/sec). Evaluating this equation at the depletion edge (i.e., $x=0$) yields

$$I_{DIF} = \frac{qD_n n_i^2}{L_n N_A}. \tag{7.17}$$

Example 7.4

Calculate the diffusion dark current assuming the following parameters: $D_n = 35$ cm^2/sec, $n_i = 10^{10}$ cm^{-3}, $L_n = 0.5$ cm and $N_A = 10^{15}$ cm^{-3}.
Solution:
From Eq. (7.17),

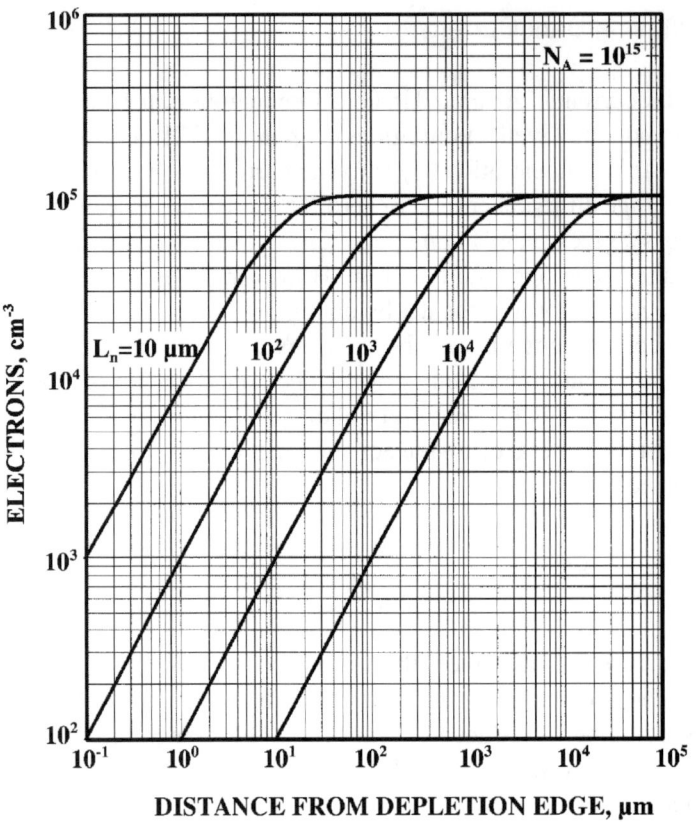

Figure 7.4 Dark current electron density near the depletion edge.

$$I_{DIF} = (1.6 \times 10^{-19}) \times 35 \times (10^{10})^2 / (0.5 \times 10^{15}),$$

$$I_{DIF} = 1.1 \times 10^{-12} \text{ (1.1 pA/cm}^2\text{)}.$$

Recall that the diffusion coefficient is related to mobility through the Einstein relation $D_n = kT\mu_{SI}/q$. Therefore, Eq. (7.17) can also be written as

$$I_{DIF} = \frac{kT\mu_{SI}n_i^2}{L_n N_A}, \tag{7.18}$$

where μ_{SI} is the mobility (cm^2/V-s).

Note that the diffusion current decreases with increasing diffusion length. The diffusion length is a parameter that specifies the quality or trap density of the silicon material. The shorter the diffusion length, the greater the number of traps and the greater the dark current. The diffusion length is also defined by

$$L_n = (D_n \tau_{DIF})^{1/2}, \qquad (7.19)$$

where τ_{DIF} is the lifetime of an electron. Note that a short lifetime implies more trapping centers and a shorter diffusion length. From this definition, Eq. (7.18) can be expressed as

$$I_{DIF} = \frac{q n_i^2 L_n}{N_A \tau_{DIF}}. \qquad (7.20)$$

Example 7.5

Determine the diffusion length, assuming a carrier lifetime of 0.03 sec.
Solution:
From Eq. (7.19),

$$L_n = (35 \times 0.03)^{1/2},$$

$$L_n = 1.025 \text{ cm}.$$

It is important to note that the extent of the field-free region in a CCD is considerably less than the diffusion length. Therefore, only a portion of the dark current generated in a diffusion length can be included. In fact, if there were no field-free material in the CCD there would be no diffusion current. As discussed in Chapter 4, field-free material is minimized as much as possible because of MTF performance. Therefore, Eqs. (7.17) and (7.20) must be modified according to

$$I_{DIF} = \frac{q D_n n_i^2}{L_n N_A}\left(1 - e^{-x_{ff}/L_n}\right), \qquad (7.21)$$

and

$$I_{DIF} = \frac{q n_i^2 L_n}{N_A \tau_{DIF}}\left(1 - e^{-x_{ff}/L_n}\right), \qquad (7.22)$$

where x_{ff} is the field-free width (cm). This is the distance between the depletion edge and epitaxial interface for a frontside-illuminated CCD, or the back surface for a backside-illuminated CCD.

Example 7.6

Determine diffusion dark current generated by a 5-μm field-free region, assuming $L_n = 0.5$ cm, $N_A = 10^{15}$ cm^{-3}, $D_n = 35$ cm^2/sec, $T = 300$ and $n_i = 10^{10}$ e$^-$/cm^3.
Solution:
From Eq. (7.21),

$$I_{DIF} = (1.6 \times 10^{-19}) \times 35 \times (10^{10})^2 / (0.5 \times 10^{15})$$
$$\times \{1 - \exp[-(5 \times 10^{-4})/0.5]\},$$

$$I_{DIF} = 1.1 \times 10^{-15} \text{ A/cm}^2.$$

The depletion and diffusion dark currents calculated above can be compared by dividing Eq. (7.22) by Eq. (7.14), yielding

$$\frac{I_{DIF}}{I_{DEP}} = \frac{\dfrac{qn_i^2 L_n \left(1 - e^{-x_{ff}/L_n}\right)}{N_A \tau_{DIF}}}{\dfrac{qn_i(x_p + t)}{2\tau_{DEP}}}. \tag{7.23}$$

This equation reduces to

$$\frac{I_{DIF}}{I_{DEP}} = \frac{2n_i L_n \left(1 - e^{-x_{ff}/L_n}\right)}{N_A(x_p + t)}, \tag{7.24}$$

where we have assumed that $\tau_{DIF} = \tau_{DEP}$.

Example 7.7

Compare depletion and diffusion currents at room temperature (300 K). Assume $n_i = 10^{10}$ e$^-$/cm^3, $N_A = 1.2 \times 10^{15}$ cm^{-3}, $(x_p + t) = 8 \times 10^{-4}$ cm, $x_{ff} = 2 \times 10^{-4}$ cm and $L_n = 0.5$ cm.
Solution:
From Eq. (7.24),

$$I_{DIF}/I_{DEP} = 2 \times 10^{10} \times 0.5 \times \left(1 - \exp[-2 \times 10^{-4}/0.5]\right)/$$
$$[(1.2 \times 10^{15}) \times (8 \times 10^{-4})],$$

$$I_{DIF}/I_{DEP} = 4.166 \times 10^{-6}.$$

NOISE SOURCES

Equation (7.24) shows that the diffusion dark current is negligible at room temperature because N_A is greater than n_i and field-free material is minimized for MTF reasons. This characteristic is normally the case for CCDs built on epitaxial material.

7.1.1.6 Substrate Dark Current

Substrate dark current is similar to diffusion dark current because the region is also field-free. However, there is one important difference between the substrate and the epitaxial region. In the substrate we find that the lifetime and diffusion length are strongly dependent on a physical process referred to as Auger recombination. The Auger process is dominant if the silicon is very heavily doped, as is the substrate. Auger recombination occurs when an electron recombines directly with a hole. The Auger process takes effect at a doping level of approximately 10^{17} cm^{-3}, depending on the Shockley lifetime defined above. The net lifetime in the substrate region is given by[6]

$$\frac{1}{\tau_{SS}} = \frac{1}{\tau_A} + \frac{1}{\tau_{DIF}}, \qquad (7.25)$$

where τ_{SS} is the lifetime of a carrier in the substrate region and τ_A is the the Auger lifetime given by

$$\tau_A = \frac{1}{G_p N_A^2}, \qquad (7.26)$$

where G_p is the recombination rate (10^{-31} cm^6/sec).

The dark current generated in the substrate region is therefore

$$I_{SS} = \frac{q n_i^2 L_{SS}}{N_A \tau_{SS}}, \qquad (7.27)$$

where L_{SS} is the diffusion length in the substrate (cm) given as

$$L_{SS} = \left(\frac{kT \mu_{SI} \tau_{SS}}{q} \right)^{1/2}. \qquad (7.28)$$

Example 7.8

Find the lifetime and diffusion length in the substrate doped at $N_A = 10^{19}$ cm^{-3}. Assume the lifetime is Auger limited with a mobility of 50 cm^2/V-sec. Also, calculate the dark current generated in the substrate region.

Solution:
From Eq. (7.26),

$$\tau_A = 1/\left[10^{-31} \times \left(10^{19}\right)^2\right],$$

$$\tau_A = 10^{-7} \text{ sec.}$$

From Eq. (7.28),

$$L_{SS} = \left(0.0259 \times 50 \times 10^{-7}\right)^{1/2},$$

$$L_{SS} = 3.6 \text{ μm.}$$

From Eq. (7.27),

$$I_{SS} = \left(1.6 \times 10^{-19}\right) \times \left(10^{10}\right)^2 \times \left(3.6 \times 10^{-4}\right)/\left(10^{19} \times 10^{-7}\right),$$

$$I_{SS} = 5.76 \times 10^{-15} \text{ A/cm}^2.$$

This example shows that the substrate region produces very little dark current.

7.1.1.7 Surface Dark Current

Surface dark current also varies widely among manufacturers, depending on the CCD technology employed and processing details associated with oxide growth and surface passivation. Surface dark currents as low as 60 pA/cm^2 to as high as 10 nA/cm^2 have been measured for noninverted operation. As with bulk centers, an electron can thermally "hop" from interface state into the conduction band, producing a free electron that will be collected by the potential well as dark charge. Noninverted operation maximizes dark current generation because the interface is completely depleted of free carriers.

Surface dark current generation can theoretically be treated in the same fashion as bulk dark current generation, using Eq. (7.7). In that the Si–SiO$_2$ interface is depleted, we can assume that $n \ll n_i$ and $p \ll n_i$. Therefore, the generation rate at the surface per unit area is

$$U_S = -\frac{1}{2}\sigma v_{th} n_i \int_{E_V}^{E_C} \frac{N_{SS}(E_t)}{\cosh\left(\dfrac{E_t - E_i}{kT}\right)} dE_t, \qquad (7.29)$$

NOISE SOURCES

where $N_{SS}(E_t)$ is the surface state density per unit of energy (cm^{-2} eV^{-1}). Integrating yields

$$U_S = -\frac{\sigma v_{th} n_i \pi k T N_{SS}}{2}, \qquad (7.30)$$

where N_{SS} is a uniform distribution of surface states about midband (cm^{-2} eV^{-1}). Surface generation can also be defined as

$$U_S = -\frac{s_o n_i}{2}, \qquad (7.31)$$

where s_o is the surface velocity (cm/sec) given as

$$s_o = \sigma v_{th} \pi k T N_{SS}. \qquad (7.32)$$

We can think of surface velocity as the "speed" it takes to thermally generate an electron-hole pair on the average. The faster the surface velocity, the more dark current is generated.

From these equations the surface dark current for the noninverted operation is

$$I_S = \frac{q n_i s_o}{2}, \qquad (7.33)$$

where I_S is the surface dark current (A/cm^2).

Example 7.9

Determine the surface recombination velocity for a noninverted CCD that generates 1 nA/cm^2. Also, estimate the number of interface states. Assume the parameters $n_i = 10^{10}$ cm^{-3}, $v_{th} = 10^7$ cm/sec, $\sigma_n = 10^{-15}$ cm^2 and $kT = 0.0259$ eV.

Solution:
From Eq. (7.33),

$$s_o = 10^{-9} \times 2/[(1.6 \times 10^{-19}) \times 10^{10}],$$

$$s_o = 1.25 \text{ cm/sec}.$$

The number of interface states is found from Eq. (7.32):

$$N_{SS} = 1.25/(10^{-15} \times 10^7 \times 3.14 \times 0.0259),$$

$$N_{SS} = 1.54 \times 10^9 \text{ states/cm}^2.$$

Depletion dark current can be compared to surface dark current by dividing Eq. (7.14) by Eq. (7.33), yielding

$$\frac{I_{DEP}}{I_S} = \frac{\dfrac{qn_i(x_p+t)}{2\tau_{DEP}}}{\dfrac{qn_is_o}{2}}, \tag{7.34}$$

which reduces to

$$\frac{I_{DEP}}{I_S} = \frac{(x_p+t)}{\tau_{DEP}s_o}. \tag{7.35}$$

Example 7.10

Compare depletion and surface dark current, assuming $s_o = 1.25$ cm/sec, $\tau_{DEP} = 0.03$ sec and $(x_p + t) = 8 \times 10^{-4}$ cm.
Solution:
From Eq. (7.35),

$$I_{DEP}/I_S = (8 \times 10^{-4})/(0.03 \times 1.25),$$

$$I_{DEP}/I_S = 0.021.$$

Depletion dark current for the CCD is usually much lower than surface dark current, as we will experimentally demonstrate below.

Surface dark current can be eliminated by inverting the CCD.[7] The inversion layer can produce a hole population at the surface on the order of 10^{18} carriers/cm^3. The presence of free carriers at the interface can fill the interface states, inhibit hopping conduction and substantially reduce the dark generation rate. The only significant source of dark current remaining is bulk dark current (i.e., depletion dark current). MPP CCDs operate in this mode.

Surface dark current when fully inverted is calculated from Eq. (7.7) assuming $n = 0$, and $\sigma_n = \sigma_p$, yielding

$$U_{SI} = -\sigma v_{th} n_i N_{SS} \int_{E_V}^{E_C} \frac{dE_t}{2\cosh\left(\dfrac{E_t - E_i}{kT}\right) + \dfrac{p}{n_i}}. \tag{7.36}$$

Integrating and assuming $p \gg n_i$,

$$U_{SI} = \frac{4kT\sigma v_{th} n_i^2 N_{SS}}{ip} \arctan\left[\frac{i(e^{E_g/2kT} - 1)}{e^{E_g/2kT} + 1}\right]. \tag{7.37}$$

Noise Sources

Then, using

$$\arctan(z) = \frac{1}{2i} \ln \frac{1+iz}{1-iz}, \quad (7.38)$$

we find

$$U_{SI} = -\frac{\sigma v_{th} n_i^2 N_{SS} E_g}{p}, \quad (7.39)$$

or

$$U_{SI} = -\frac{s_o n_i^2}{p}, \quad (7.40)$$

where we have defined the surface recombination velocity s_o as

$$s_o = \sigma v_{th} E_g N_{SS}. \quad (7.41)$$

Surface dark current under inverted conditions is

$$I_{SI} = \frac{q s_o n_i^2}{p}. \quad (7.42)$$

Example 7.11

Determine the surface dark current under inverted conditions for the parameters, $n_i = 10^{10}$ cm^{-3}, $v_{th} = 10^7$ cm/sec, $\sigma = 10^{-15}$ cm, $N_{SS} = 1.54 \times 10^9$ states/cm^2, $p = 10^{18}$/cm^3 and $E_g = 1.1$ eV.
Solution:
From Eq. (7.41),

$$s_o = 10^{-15} \times 10^7 \times 1.1 \times (1.54 \times 10^9),$$

$$s_o = 1.69 \times 10^1 \text{ cm/sec}.$$

From Eq. (7.42),

$$I_{SI} = (1.6 \times 10^{-19}) \times (1.69 \times 10^1) \times 10^{20}/10^{18},$$

$$I_{SI} = 2.7 \times 10^{-16} \text{ A/cm}^2.$$

This calculations shows that surface dark current is extremely low when the CCD is inverted.

7.1.1.8 Backside Dark Current

Dark current can also be generated at the rear surface of backside-illuminated CCDs. Recall in Chapter 3 that levels as high as 10,000 nA/cm² were observed for recently thinned fully depleted CCDs [refer to Fig. 3.38(a)]. However, when the device is properly accumulated, backside dark current can be completely suppressed, leaving only bulk dark current assuming the frontside is fully inverted. However, for some ion-implantation accumulation processes, backside dark current can dominate. This problem is usually associated with the backside boron implant not being fully activated or annealed (i.e., many bulk states exist).

7.1.1.9 Dark Current Equation

Analysis above showed that surface and depletion dark current, the two main contributors to dark current, exhibit the same temperature dependence as the intrinsic carrier concentration n_i. Although diffusion current has a n_i^2 dependence, this source of dark current is negligible. Therefore, we can use Eq. (7.4) as the general dark current formula for the CCD. That is,

$$D_R(e^-) = CT^{1.5}e^{-E_g/(2kT)}, \qquad (7.43)$$

where $D_R(e^-)$ is the average dark current generated (e⁻/sec/pixel), C is a constant, T is the operating temperature (K), E_g is the silicon bandgap energy (eV), and k is Boltzmann's constant (8.62×10^{-5} eV/K).

The bandgap energy varies with operating temperature following the empirical formula[8]

$$E_g = 1.1557 - \frac{7.021 \times 10^{-4} T^2}{1108 + T}. \qquad (7.44)$$

The constant C can be solved at room temperature (300 K, 27°C), yielding

$$C = \frac{D_{FM} P_S}{qT_{RM}^{1.5} e^{-E_g/(2kT_{RM})}}, \qquad (7.45)$$

where P_S is the pixel area (cm⁻²), T_{RM} is room temperature (300 K) and D_{FM} is called the "dark current figure of merit" at 300 K (nA/cm²). Substituting Eq. (7.45) into Eq. (7.43) produces the final dark current formula,

$$D_R(e^-) = 2.5 \times 10^{15} P_S D_{FM} T^{1.5} e^{-E_g/(2kT)}. \qquad (7.46)$$

This equation is very important in determining operating temperature required to eliminate dark current. The formula agrees precisely with dark current measurements taken. In fact, the operating temperature of the CCD is often determined by simply measuring the average dark current produced after a camera system is calibrated. However, there are a few exceptions where theory

Noise Sources

and experiment do not exactly track. For example, dark current generated by dark spikes may not follow Eq. (7.46). Dark spikes and their temperature behavior is discussed below. Also, the presence of backside dark current may not follow theory. Here we may find that the dark current figure of merit decreases as the temperature is lowered. This is because bulk states responsible for backside dark current will exhibit a longer emission time constant as the temperature is lowered, reducing the dark current rate (i.e., the states are frozen in). Also, the backside well—if present—increases in size as temperature is reduced, thereby lowering the figure of merit.

Example 7.12

Calculate dark current generated at operating temperatures of $-10°C$ (263 K), $-50°C$ (223 K) and $-100°C$ (173 K). Assume a 12-μm pixel and $D_{FM} = 10$ pA/cm^2.

Solution:

The band-gap energy for each operating temperature is found from Eq. (7.44):

$$E_g = 1.1557 - [(7.021 \times 10^{-4})(263)^2/(1108+263)],$$

$$E_g = 1.120278 \text{ eV} \quad \text{at 263 K,}$$

$$E_g = 1.129468 \text{ eV} \quad \text{at 223 K,}$$

$$E_g = 1.139296 \text{ eV} \quad \text{at 173 K.}$$

The area of a 12-μm pixel is equal to 1.44×10^{-6} cm^2. From Eq. (7.46) the dark current generation rates are,

$$D_R(e^-) = (2.55 \times 10^{15}) \times (1.44 \times 10^{-6}) \times (10^{-2}) \times (263)^{1.5}$$
$$\times \exp\{-1.12/[2 \times (8.62 \times 10^{-5}) \times 263]\},$$

$$D_R(e^-) = 2.911 \text{ e}^-/\text{pixel/sec at 263 K,}$$

$$D_R(e^-) = 2.123 \times 10^{-2} \text{ e}^-/\text{pixel/sec at 223 K,}$$

$$D_R(e^-) = 2.149 \times 10^{-6} \text{ e}^-/\text{pixel/sec at 173 K.}$$

Figures 7.5(a) and (b) plot Eq. (7.46) for a 12-μm pixel and different dark current figures of merit ranging from 1 pA/cm^2 to 10 μA/cm^2. Figure 7.5(c) plots dark current at very low operating temperatures (i.e., near absolute zero). It is interesting to point out that the CCD will begin to stop working at liquid nitrogen temperature

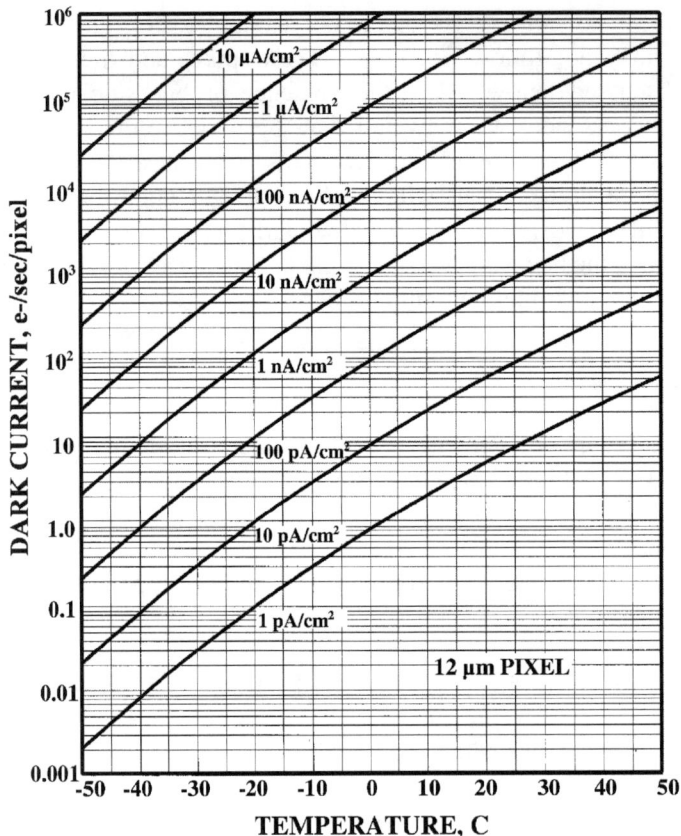

Figure 7.5(a) Dark current as a function of operating temperature and dark current figure of merit.

(77 K or $-196°C$) because of a physical limit known as carrier "freeze-out." At such low temperatures the dopant atoms recombine rather than exist in the lattice in an ionized state. Recall from Chapter 1 that ionization is required to form all depletion regions and potential wells in the CCD. Donor species used in processing CCDs (P, As, Sb) have approximately the same ionization energy requirements (roughly 39–54 meV). This energy becomes large compared to $kT = 6$ meV at 70 K and is why donors cease to ionize at these low temperatures. For the same reason, the resistivity of the poly doped gates increase enormously, resulting in clocking problems at the center of the array. Figure 7.6 shows an image taken at exactly 77 K by a 1024 × 1024 CCD that was completely submerged in LN_2. The picture was acquired by focusing a dollar bill image through the liquid onto the array. Although a picture was successfully obtained, the CCD showed significant CTE loss because of the onset of carrier freeze-out and traps in the signal channel.

Note from Fig. 7.5 that the rate of change of dark current rate increases dramatically as the temperature is reduced (i.e., the slope of the curves). Figure 7.7 plots

NOISE SOURCES 625

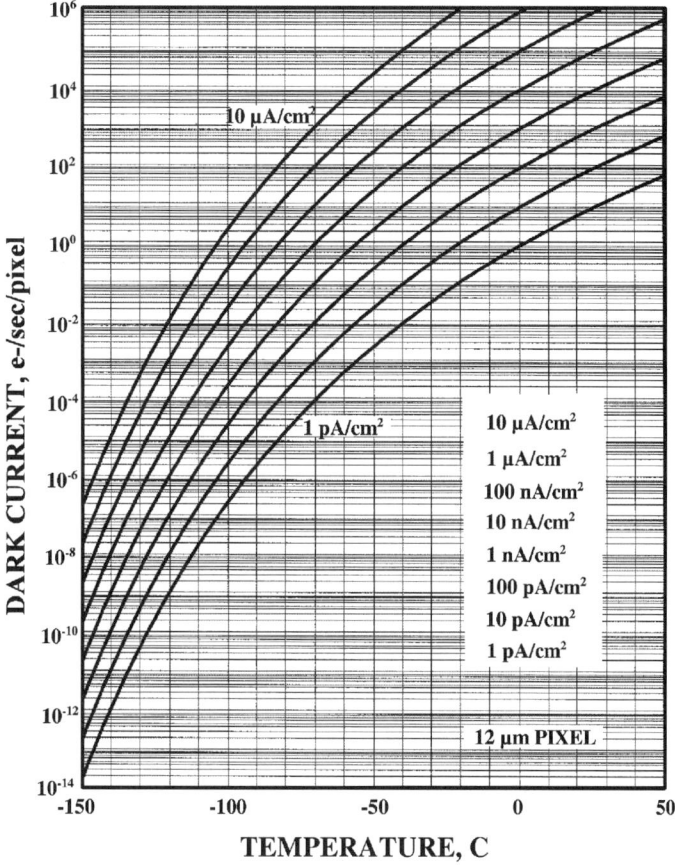

Figure 7.5(b) Dark current as a function of operating temperature and dark current figure of merit.

the change in operating temperature required to reduce the dark current by factors of 1.5, 2, 4, 10 and 20 as a function of operating temperature.

Example 7.13

Using Fig. 7.7, find the temperature change required to reduce the dark current by a factor of two at 20°C, −50°C, −100°C.
Solution:
At 25°C, the operating temperature must be reduced by 8°C. At −50°C a change of 4.8°C is required. At −100°C only 2.9°C is necessary.

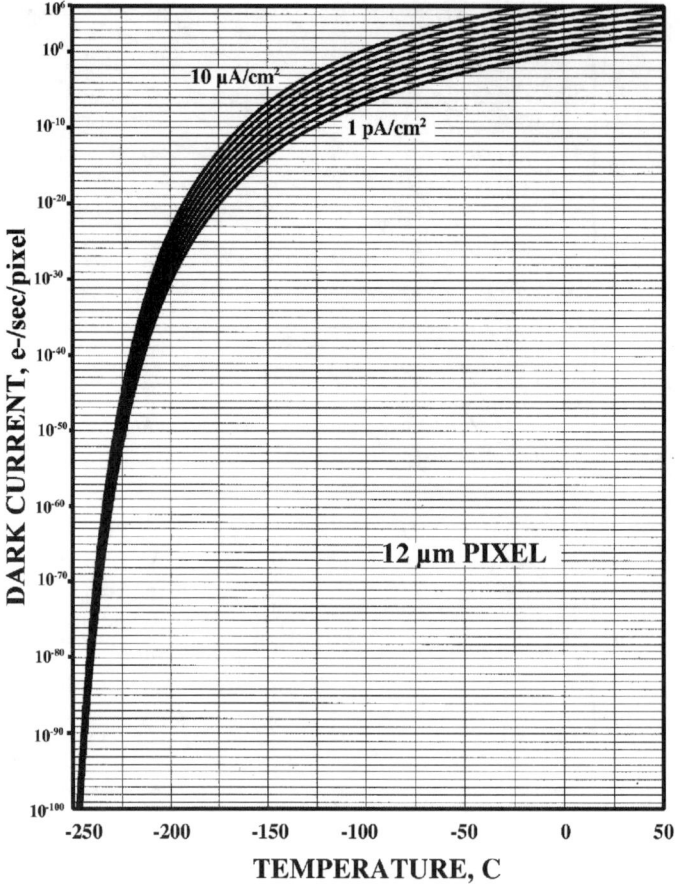

Figure 7.5(c) Dark current over a large operating temperature range.

7.1.1.10 Dark Shot Noise

The shot noise associated with dark current is governed by Poisson's statistics:

$$N_{DSN} = D(e^-)^{1/2}, \qquad (7.47)$$

where N_{DSN} is the dark shot noise (rms e^-) and $D(e^-)$ is the average dark current level (e^-/pixel).

Example 7.14

Determine the dark shot noise generated in a 1-hour exposure. Assume a dark rate of 1 e^-/pixel/sec.

NOISE SOURCES

Figure 7.6 Image taken while CCD was submerged in liquid nitrogen.

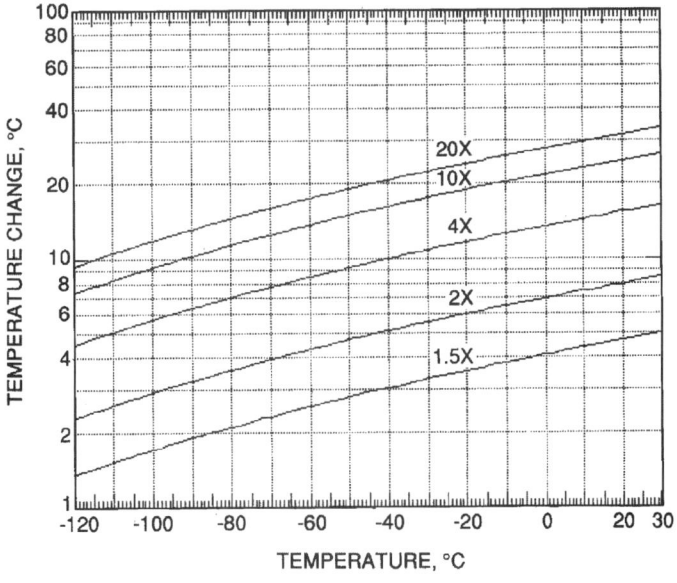

Figure 7.7 Operating temperature change required to change dark current rate.

Solution:
In 1 hour, 3600 e$^-$/pixel of dark current is collected. From Eq. (7.47), the shot noise is

$$N_{DSN} = 3600^{1/2},$$

$$N_{DSN} = 60 \text{ e}^- \text{ rms}.$$

The maximum integration time allowed when the dark shot noise becomes equal to the read noise can be found through

$$R(\text{e}^-) = [I_T D(\text{e}^-)]^{1/2}, \tag{7.48}$$

where I_T is the integration time (sec) and $R(\text{e}^-)$ is the read noise. Solving for I_T yields

$$I_T = \frac{R(\text{e}^-)^2}{D(\text{e}^-)}. \tag{7.49}$$

Example 7.15

Determine maximum integration time before the dark shot noise equals the read noise floor. Assume $R(\text{e}^-) = 10$ e-rms, $P_S = 12$ μm, $T = 263$ K ($-10°$C) and $D_{FM} = 0.4$ nA/cm^2.

Solution:
From Eq. (7.44),

$$E_g = 1.1557 - \left[(7.021 \times 10^{-4}) \times 263^2 / (1108 + 263) \right],$$

$$E_g = 1.120278 \text{ eV}.$$

From Eq. (7.46),

$$D(\text{e}^-) = (2.5 \times 10^{15}) \times (12 \times 10^{-4})^2 \times 0.4 \times 263^{1.5}$$
$$\times \exp\{-1.12/[2 \times (8.62 \times 10^{-5}) \times 263]\},$$

$$D(\text{e}^-) = 115 \text{ e}^-/\text{sec/pixel}.$$

From Eq. (7.49),

$$I_T = 10^2 / 115,$$

$$I_T = 0.87 \text{ sec}.$$

NOISE SOURCES   629

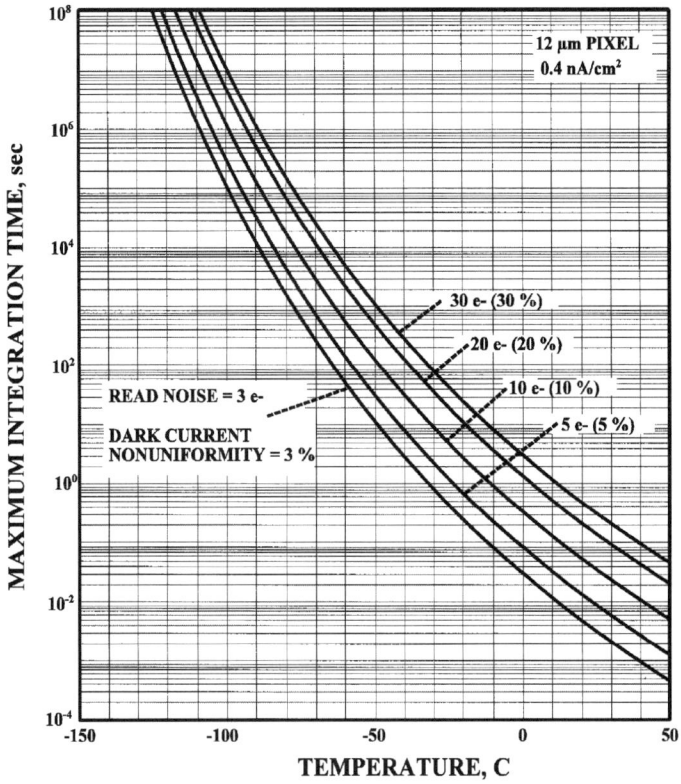

Figure 7.8 Integration time where dark shot noise equals read noise, or dark current nonuniformity noise equals dark shot noise.

Figure 7.8 plots Eq. (7.49) as a function of operating temperature and read noise for $D_{FM} = 0.4$ nA/cm^2 and $P_S = 12$ μm.

7.1.1.11 Dark Current Nonuniformity

Dark nonuniformity is the variation of dark current from pixel to pixel. The noise that is produced is

$$N_{DN}(\text{e}^-) = \sigma_{DN} D(\text{e}^-), \tag{7.50}$$

where N_{DN} is dark current nonuniformity noise (rms e$^-$) and σ_{DN} is the dark current nonuniformity factor. For example, if the average dark current is 10,000 e$^-$, then the rms variation due to nonuniformity is 1000 e$^-$ rms, assuming $\sigma_{DN} = 0.10$.

The total noise generated by dark current is the quadrature sum of dark current shot noise and dark current nonuniformity:

$$N_{DC}(e^-) = \left[(\sigma_{DN} D(e^-))^2 + D(e^-)\right]^{1/2}. \tag{7.51}$$

Note that when the average dark current is low, dark shot noise prevails. As the dark current increases, nonuniformity begins to dominate.

We can find the maximum integration time allowed before the dark current nonuniformity equals the dark shot noise. That is,

$$\sigma_{DN} D(e^-) I_T = [I_T D(e^-)]^{1/2}. \tag{7.52}$$

Solving for I_T yields

$$I_T = \frac{1}{D(e^-)\sigma_{DN}^2}. \tag{7.53}$$

Example 7.16

From Example 7.15, determine maximum integration time allowed before dark current nonuniformity equals the dark shot noise. Assume a dark current nonuniformity of 10% ($\sigma_{DN} = 0.10$) and an operating temperature of 263 K ($-10°$C).
Solution:
From Eq. (7.53),

$$I_T = 1/(115 \times 0.1^2),$$

$$I_T = 0.87 \text{ sec.}$$

Figure 7.8 is also used for Eq. (7.53) as a function of operating temperature and read noise for $D_{FM} = 0.4$ nA/cm^2 and a 12-μm pixel. The values in parentheses specify dark pixel-to-pixel nonuniformity.

Dark current nonuniformity is eliminated by cooling the CCD or removed by image processing techniques (refer to Fig. 7.23 in the discussions on despiking below). Dark current shot noise can only be eliminated by cooling the CCD. For example, the Hubble CCDs require dark rates less than $D_R(e^-) = 0.001$ e$^-$/sec. The devices are cooled by heat pipes and thermoelectric coolers (TEC). The heat pipes to the TECs passively cool the sensors to approximately $-40°$C. The TECs reduce the working operating temperature to approximately $-100°$C. Figure 7.9(a) shows the WF/PC cooling assembly and the five-stage TEC, which is mounted to a copper block for heat transfer. The CCD is supported with a gold rubber band

Noise sources

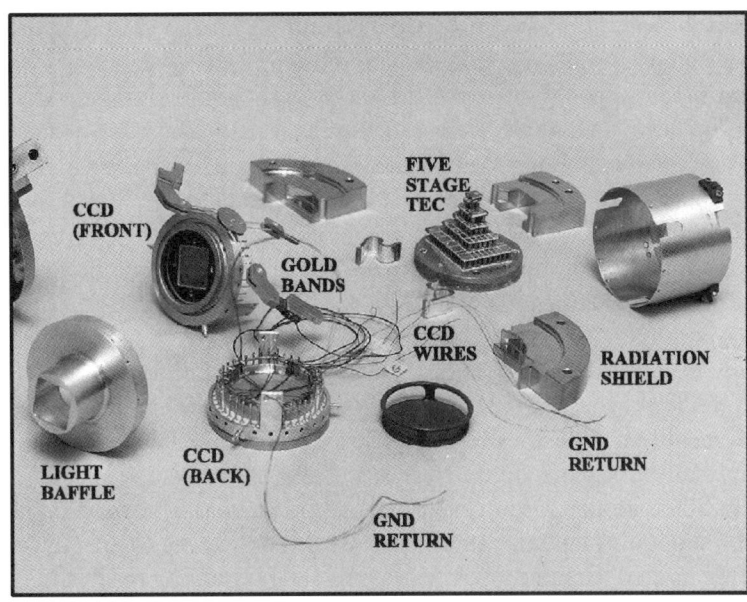

Figure 7.9(a) WF/PC thermoelectric cooling assembly.

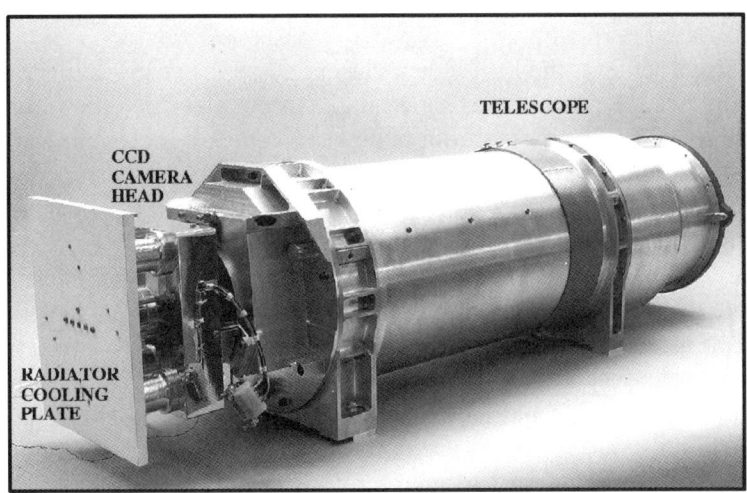

Figure 7.9(b) Galileo camera showing passive radiator plate for cooling the CCD.

assembly to thermally isolate the device from surrounding structures. Constantan wires are used to make electrical connections to the CCD to minimize the heat load. A copper wire is employed for a low-impedance ground return.

The Galileo CCD is cooled passively by a large metal block that stares into outer space to radiate heat from the CCD. The Galileo CCD is cooled to $-120°C$. Figure 7.9(b) shows the Galileo camera system. The round long tube is the telescope that is in front of the CCD electronic assembly. The square aluminum block at the other end of the camera is the radiator cooling plate.

Ground-based applications cool the CCD with liquid nitrogen, dry ice and TECs. Liquid nitrogen cooling is the most popular method used. For example, the camera system shown in Fig. 1.4 cools the CCD by liquid nitrogen gas. The cold nitrogen gas enters the camera head from a LN_2 dewar that circulates through a cold block in which the CCD is mounted. The amount of gas that flows is controlled by a solenoid that is turned on and off by a temperature sensor mounted to the CCD cold block. CCD operating temperatures lower than $-120°C$ can be achieved with stability of approximately $4°C$. The temperature control system usually limits the cooling rate to < 0.1 deg/sec. Cooling a CCD too fast can result in thermal shock and damage the sensor and package. The CCD cooling block usually includes a heating element (e.g., resistor) to warm the CCD from a cooled state. The camera head is normally under vacuum to prevent water condensation on the CCD and cold block. Some groups elect to purge the entire camera head with cold nitrogen gas, thus eliminating vacuum requirements while cooling the CCD. However, temperatures are limited to slightly below freezing by this cooling method. The camera head and window also need to be heated above ambient to prevent condensation and fogging.

7.1.1.12 MPP Dark Current

MPP operation achieves the lowest dark current possible because surface dark current is suppressed. Figure 7.10(a) shows dark current characteristics taken from a four-phase n-MPP CCD biased from noninversion to full inversion. Dark current falls from 1 nA/cm^2 to 27 pA/cm^2 as the device is clocked into inversion, a reduction of 37 times. Note that phases-2 and -4 are the last phases to invert because the phosphorus MPP implant is under these phases.

Figure 7.10(b) plots dark current under each phase for a three-phase p-MPP CCD. Note that the MPP phase (phase-3) inverts first because of the boron MPP implant. The inversion breakpoints for phases-2 and -3 are slightly offset from each other because the gate nitride thicknesses are different (refer to Section 4.2.12). The plot shows that a 2-V barrier exists between phase-3 and phase-2 under fully pinned conditions. Such plots are used determine the MPP barrier height.

Figure 7.11 presents four dark current transfer curves for a 520×520 MPP CCD. Dark current was measured before and after the sensor was subjected to 20 krad of 1.2-MeV gamma rays (i.e., Co^{60} radiation source). Gamma rays induce damage at the Si–SiO$_2$ interface, resulting in an increase in surface dark current generation. As the figure shows, before irradiation, dark current was suppressed from 1 nA/cm^2 to 25 pA/cm^2 by MPP clocking. After irradiation, MPP dark current is three orders of magnitude lower than the noninverted case. The slight increase in dark current for MPP operation after irradiation (a factor of two) is primarily caused by phases coming out of inversion at that time where charge is transferred vertically (120 μsec for each line transfer, in this case). This dark level can be reduced by minimizing the line transfer time as much as possible.

NOISE SOURCES 633

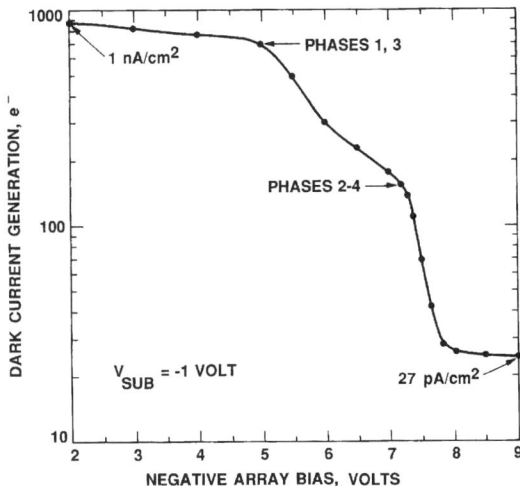

Figure 7.10(a) Dark current generated by a four-phase n-MPP CCD as a function of negative clock voltage, showing inversion breakpoints for each phase.

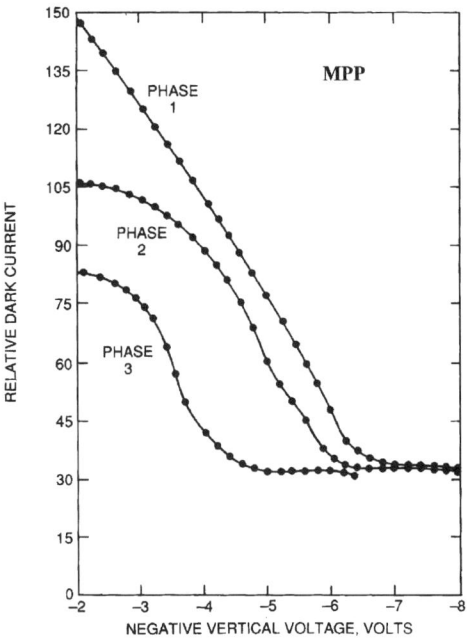

Figure 7.10(b) Dark current generated by a three-phase p-MPP CCD as a function of negative clock voltage showing inversion breakpoints for each phase.

Figure 7.12 shows room-temperature column traces of dark current as a function of the number of lines read out from a 1024 × 1024 MPP CCD driven partially inverted and MPP. The CCD was first rapidly erased, leaving no charge in the ar-

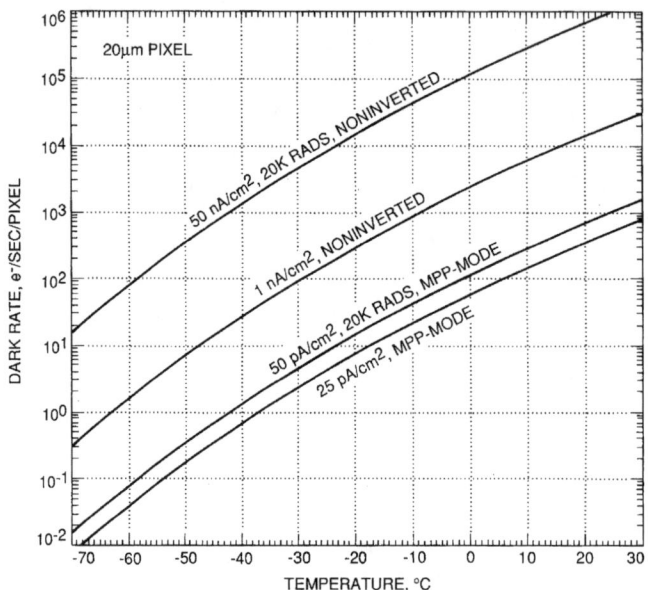

Figure 7.11 Dark current transfer curves for a MPP CCD damaged with Co^{60} radiation.

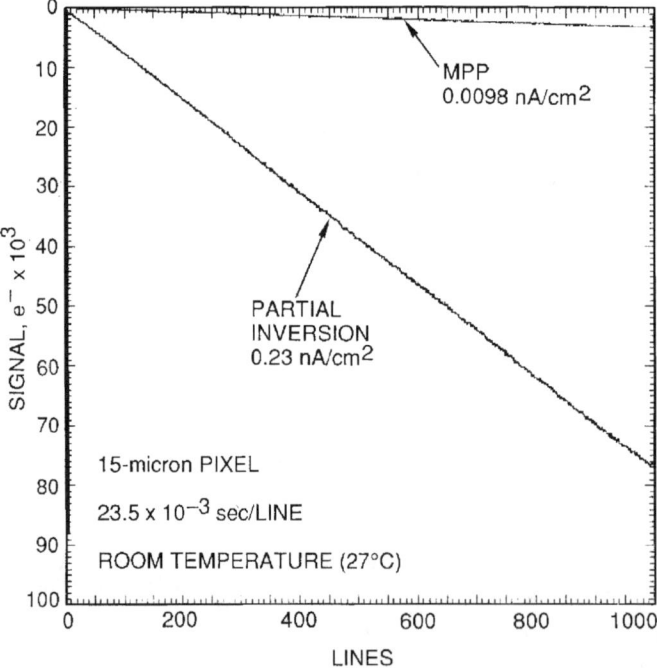

Figure 7.12 Room-temperature MPP and partially inverted dark current buildup during readout after the CCD is erased.

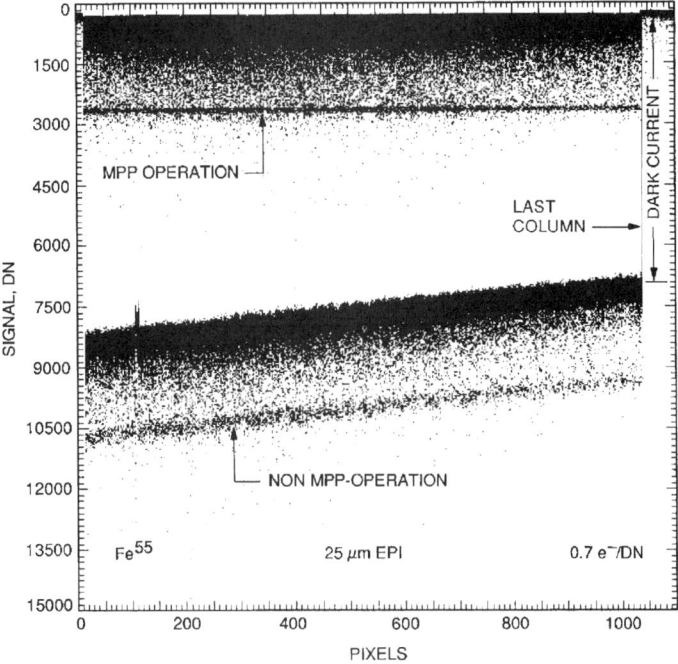

Figure 7.13 X-ray stacking plots for MPP and partial-inversion operation near room temperature.

ray. Then it was read out at a 23.5 msec/line producing the column traces shown. Note that as readout progresses, the dark current increases proportionally with the number of lines read. Dark current generation rates of 0.23 nA/cm^2 (partially inverted) and 0.0098 nA/cm^2 (MPP) are calculated from the slopes of these plots. The charge capacity for the CCD is approximately 100,000 e$^-$. Therefore, the device under partial inverted clocking does not quite reach full well after 1024 lines are read out. For the MPP case, several minutes of integration would be required to saturate the CCD in addition to the 24-sec readout time.

Figure 7.13 shows two Fe^{55} x-ray line traces generated by the same CCD as Fig. 7.12. MPP and noninverted clocking is employed at a $-10°C$ operating temperature. Note that the non-MPP response exhibits dark shot noise, which broadens the single-pixel-event x-ray line. MPP operation suppresses the dark current and improves the response. The tilt observed in the non-MPP case is caused by surface dark current nonuniformity across the array.

7.1.1.13 Dark Current Spikes

A dark current spike is a pixel that generates high dark current, i.e., more than the average. Silicon lattice imperfections or impurities are believed to be responsible for the dark spikes observed. Dark spikes also appear when the CCD is exposed to radiation sources such as protons and neutrons that induce lattice damage (refer to Chapter 8).

The generation rate for a spike can be analyzed by Eq. (7.7) at a specific trap energy level within the bandgap. Assuming $N_B = 1$ and $p = h = 0$, the equation reduces to

$$U_{DS} = -\frac{\sigma v_{th} n_i}{2\cosh\left(\frac{E_t - E_i}{kT}\right)}. \tag{7.54}$$

Example 7.17

Determine the dark current generated by a single midband trap (i.e., $E_t - E_i = 0$). Assume $n_i = 10^{10}$ cm^{-3}, $v_{th} = 10^7$ cm/sec, $\sigma = 10^{-15}$ cm.
Solution:
From Eq. (7.54),

$$U_{DS} = 10^{-15} \times 10^7 \times 10^{10}/2,$$

$$U_{DS} = 50 \text{ e}^-/\text{sec/pixel}.$$

Figure 7.14 plots Eq. (7.54) with Eq. (7.4) as a function of kT. Note that the efficiency in generating dark charge through a trap decreases rapidly as the energy level moves away from midband.

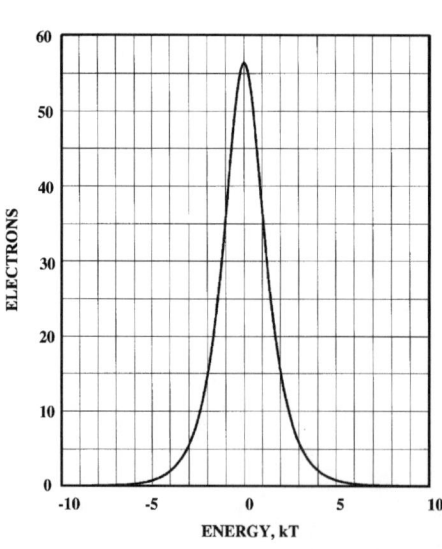

Figure 7.14 Electron-generation rate for traps near midband energy, as a function of number of kT from midband.

NOISE SOURCES 637

The generation rate for the dark spikes seen in CCDs is much greater than calculated above. For example, proton-induced spikes can exhibit rates higher than 10^6 e$^-$/sec/pixel (300 K). This indicates that field-assisted generation is probably playing a role not accounted for by the Shockley generation/recombination equation. We will show experimentally that this effect is responsible for the large dark spikes seen. Also, some dark spikes are associated with more than one bulk state. For example, neutron-induced dark spikes involve cluster damage, which results in very large dark spikes. Chapter 8 discusses these details in more depth.

Figure 7.15(a) shows a 60-sec dark current image generated by a virtual-phase CCD under totally inverted conditions at an operating temperature of $-20°$C. Each dot seen in the image is a dark spike. This is a typical dark current response for this CCD technology. Figure 7.15(b) is a dark current stacking plot where 800 lines have been summed together. The dark spikes seen in the plot exhibit two levels, indicating that there may be two different bulk states at work.[9] Many pixels contain multiple spikes, producing the band structure in the trace. Figure 7.15(c) shows a histogram of the dark spikes from this data.

Figure 7.16 presents a dark current stacking plot for the three-phase MPP CCD characterized in Fig. 7.12. This data shows a uniform distribution of dark spikes. As in Fig. 7.15, the discrete signal levels are produced by the number of dark spikes contained in the pixels. Nearly half of the pixels on the array contain a single dark

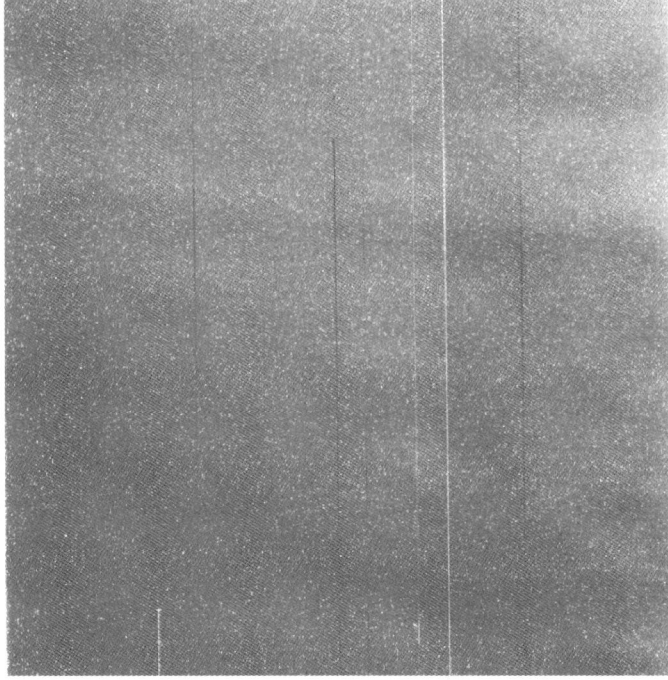

Figure 7.15(a) Dark spike image generated by a virtual-phase Galileo CCD.

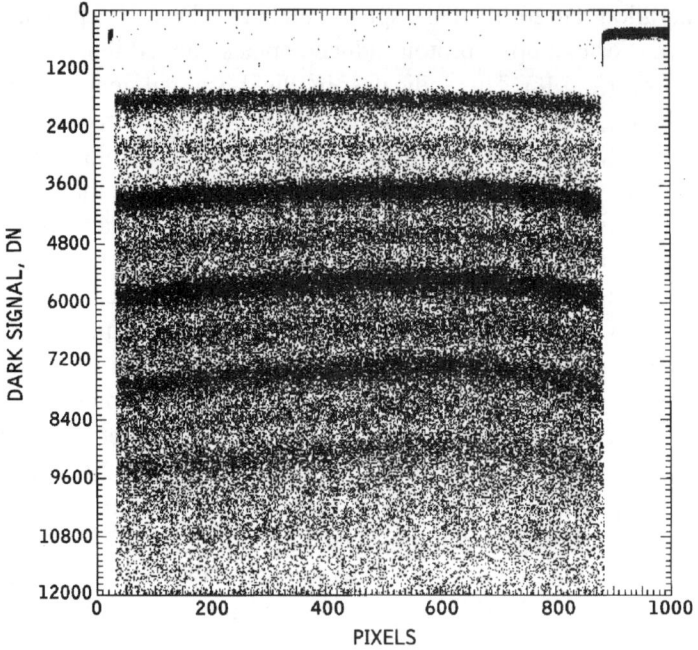

Figure 7.15(b) Virtual-phase dark current spikes showing at least two types of bulk states at work.

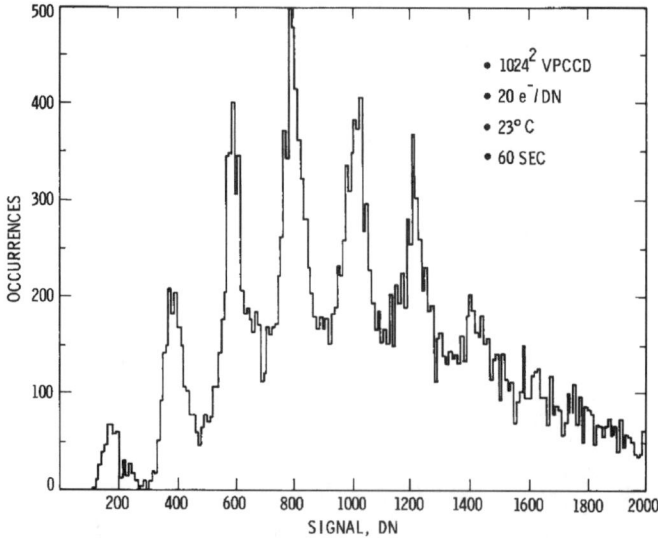

Figure 7.15(c) Virtual-phase dark spike histogram.

spike. Approximately 20% contain two spikes, and very few pixels contain three or more spikes. The average dark floor on which the dark spikes ride is primarily generated at that time when the vertical clocks briefly come out of inversion during

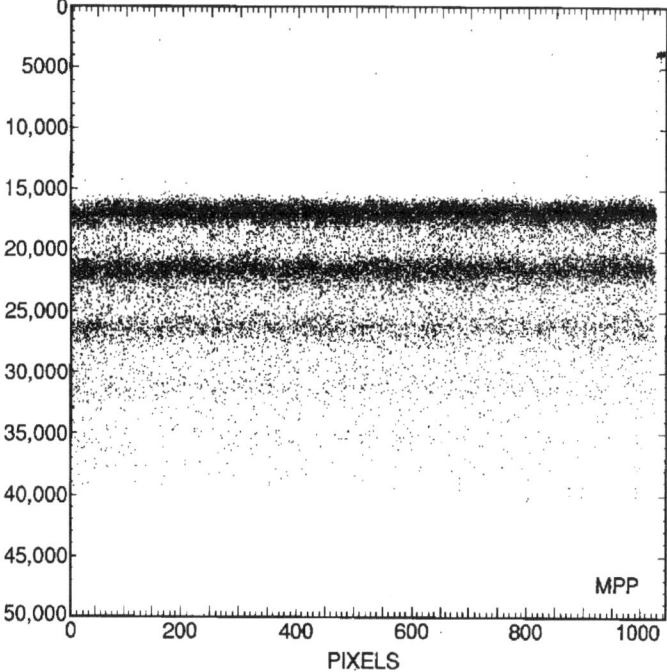

Figure 7.16 Dark current spikes for a three-phase MPP CCD showing multiple dark spikes contained in pixels.

vertical transfer. Most MPP CCDs do not show discrete dark spikes levels as seen in Figure 7.16. Typically, dark spike generation rate is randomly distributed for reasons that will be explained below.

Dark spikes exhibit many interesting characteristics. For example, spikes only occupy a single pixel, implying that they are located within the potential well. Spikes, if generated in field-free material, would show split characteristics among more than one pixel, which is not observed. Also, the dark spike density and amplitude increases with buried-channel implant dose. Again, this property implies that electric fields are influencing the amount of charge that is generated by a spike. The greatest field strength within the potential well is located in the middle of the channel near the Si–SiO$_2$ interface. Figure 7.17 plots the electric field strength as a function of depth from the front surface for three doping concentrations ($N_D = 1 \times 10^{16}$, 2.0×10^{16} and 3.0×10^{16} cm^{-3}). The electric field is greatest at the immediate surface and increases with channel doping. It is in this high field region where dark spikes are likely to be located.

Numerous tests have been performed to verify that dark spikes are in fact generated near the surface of the CCD. Figure 7.18(a) is a horizontal stacking plot showing the dark spikes induced when a 15-μm-thick backside-illuminated CCD is partially irradiated with normally incident 2- and 4-MeV protons, which induce lattice damage and generate dark spikes. Protons of this energy exhibit a range

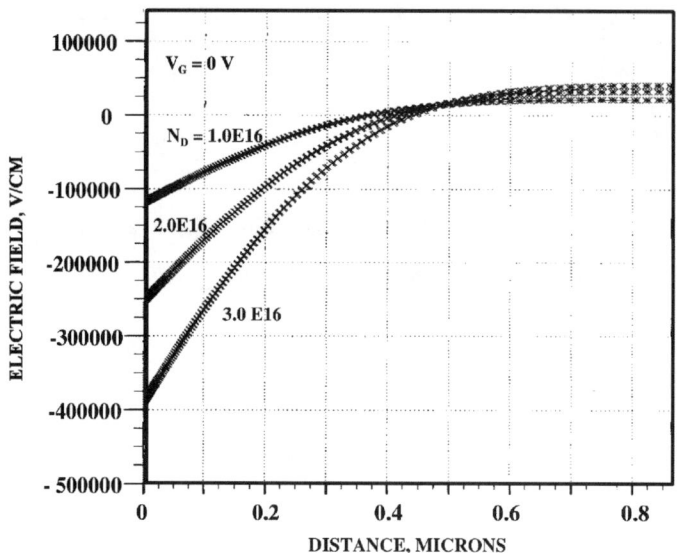

Figure 7.17 Electric field strength as a function of distance from the Si–SiO$_2$ interface and channel doping concentrations.

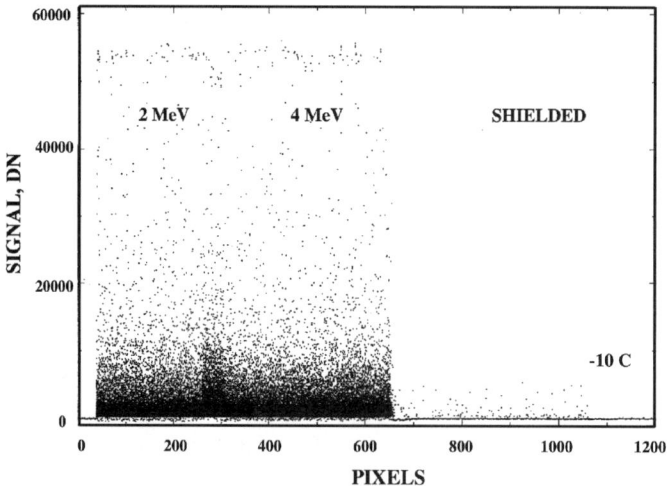

Figure 7.18(a) Proton-induced dark spikes generated by a backside-illuminated CCD (2 and 4 MeV).

that is greater than the membrane thickness of the sensor. For example, 2-MeV protons have a range of 46.8 μm, whereas 4-MeV protons exhibit a range of 146 μm. Therefore, these protons uniformly damage the CCD in depth. Figure 7.18(b) shows a dark response when the sensor is irradiated with 700-keV protons that have a range of only ∼9 μm. Figure 7.18(c) plots the expected damage (i.e.,

NOISE SOURCES 641

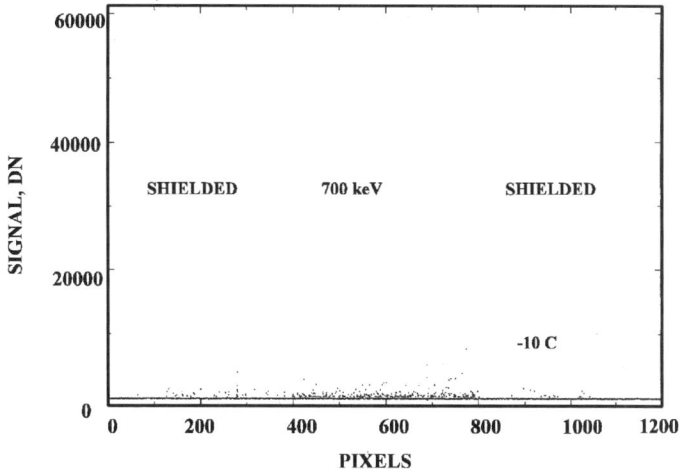

Figure 7.18(b) Dark spikes induced by 700-keV protons.

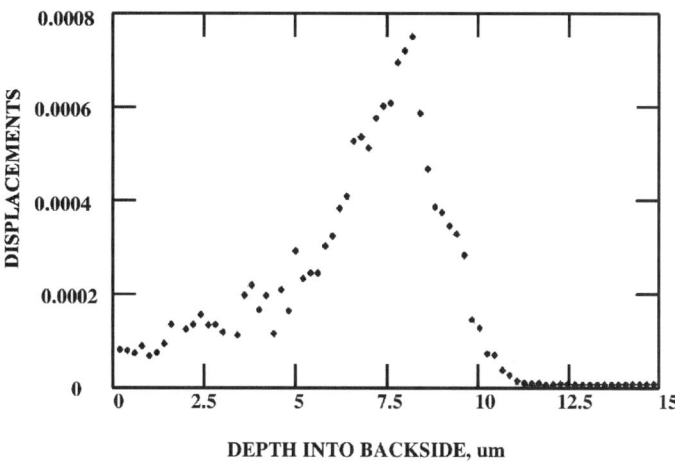

Figure 7.18(c) Expected number of silicon atom displacements produced by 700-keV protons as a function of distance from the backside.

the number of silicon atom displacements per proton) as a function of depth from the backside surface (calculations like this are discussed in Chapter 8). The plot shows no damage at the front surface because the protons cannot penetrate that far into the device. Therefore, dark spikes are absent in Fig. 7.18(b) because the damage is not located in a high field region.

The dark current generated by a dark spike is usually greatest when the potential well is empty because electric fields are stronger. Figure 7.19(a) plots dark signal for several dark spikes as a function of average background dark current level. Note that the generation rate for each spike decreases with average signal level.

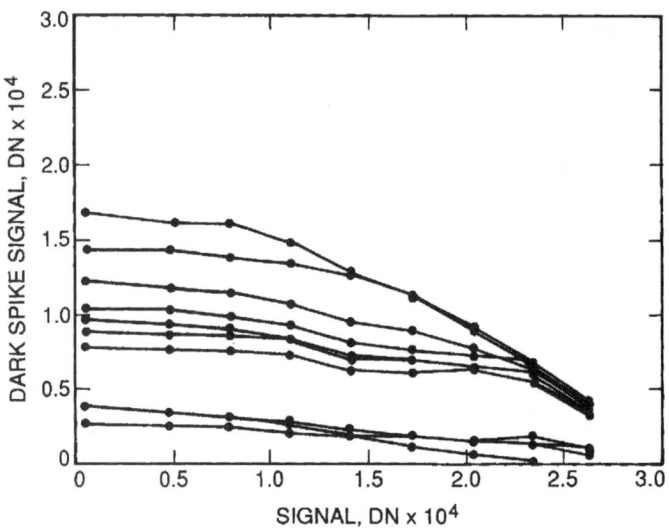

Figure 7.19(a) Dark spike signal as function of average dark current.

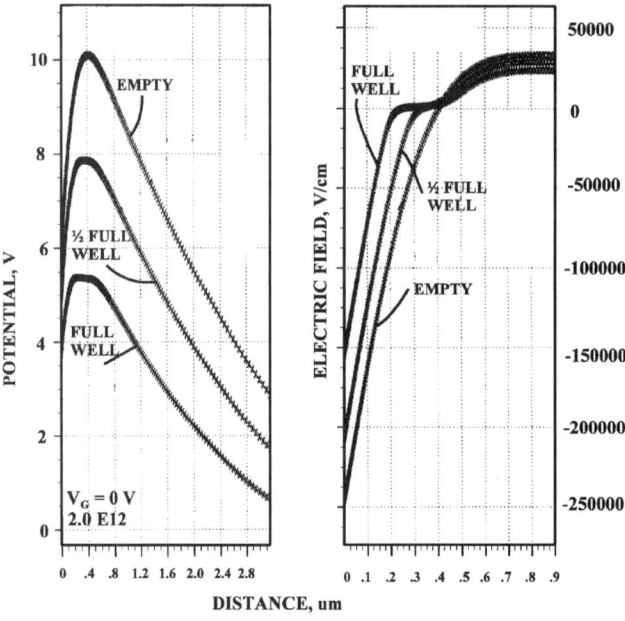

Figure 7.19(b) Potential and electric field distribution as a function of distance and signal level.

Figure 7.19(b) shows modeling results of channel potential and electric fields at three different signal levels (0 e$^-$, half full well and full well). As can be seen, the electric field near the surface decreases with signal level, which in turn reduces field-assisted generation.

NOISE SOURCES

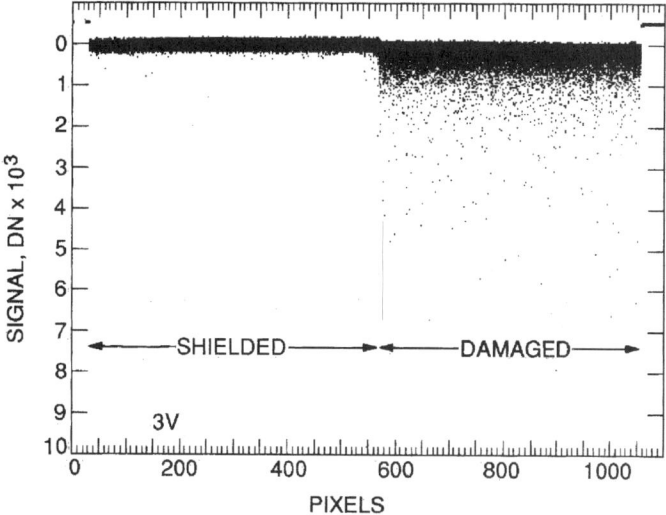

Figure 7.20(a) Dark spike stacking plot taken at a 3-V clock level.

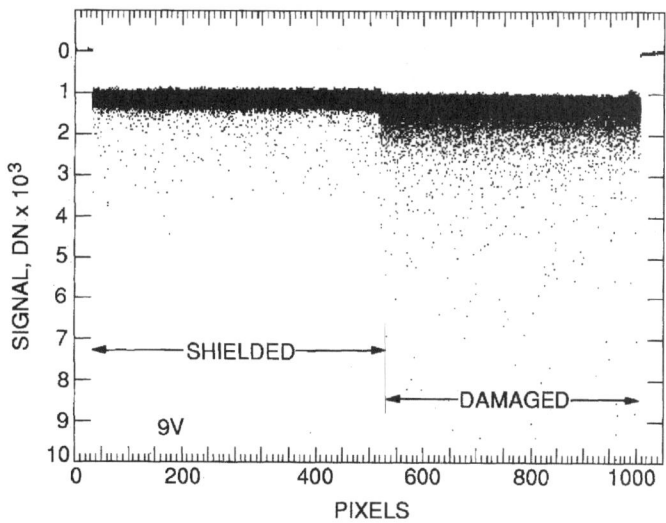

Figure 7.20(b) Dark spike stacking plot taken at a 9-V clock level.

Figures 7.20(a) and 7.20(b) show dark spike stacking plots generated by a partially proton-damaged MPP CCD. Note that the dark spike density increases when the clock voltage is elevated from 3 V to 9 V. The density increase is attributed to higher electric fields. Note that the shielded region also reacts to the increase in gate voltage. Figure 7.20(c) is a differenced plot of Figs. 7.20(a) and 7.20(b) showing that most dark spikes increase with elevated gate voltage.

Dark spikes usually follow the general dark current equation [i.e., Eq. (7.46)]. For example, Fig. 7.21 shows that dark spikes from Fig. 7.16 decrease with operat-

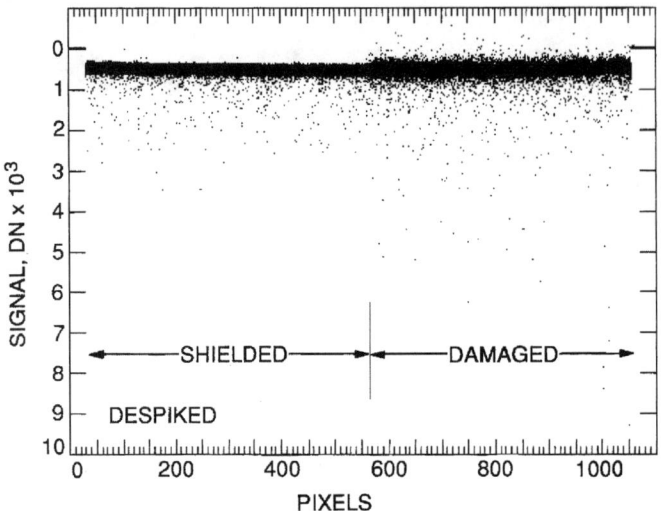

Figure 7.20(c) A differenced stacking plot for Figs. 7.20(a) and (b).

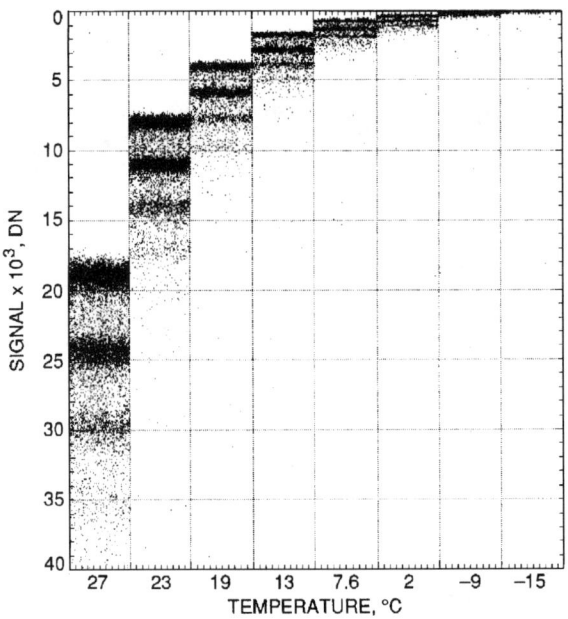

Figure 7.21 Dark spikes and average dark current as a function of operating temperature.

ing temperature at the same rate as the average dark current floor decreases. Highly doped CCDs with surface fields near silicon breakdown (approximately 10^6 V/cm) may not follow Eq. (7.46). For example, dark spikes for the Galileo virtual-phase CCD fall at a slower rate than expected. Virtual-phase technology employs a very shallow and concentrated pinning implant within the virtual region, producing a

NOISE SOURCES

Figure 7.22(a) Dark spike map for the Galileo CCD before launch.

Figure 7.22(b) Dark current after 1.5 years in space.

high field condition. The dark spikes in this region do not follow the dark current equation. These spikes can be described analytically by adjusting the bandgap energy, E_g, specified in Eq. (7.46). For example, high-level neutron-induced dark spikes experienced by the Galileo CCD exhibit an effective bandgap approximately half that of silicon (i.e., 0.6 eV).

Figure 7.22(a) shows a constant-readout dark current image before the Galileo spacecraft was launched in 1991. The image shows two dark spikes in the 800(V) × 400(H) region presented. The dark spikes leave a trail of charge because the imager is first cleared of all charge by a high-speed erase cycle. The spikes generate enough charge from line to line to be seen above the noise floor as readout occurs. Figure 7.22(b) is a similar dark current image taken one and a half years after the CCD was launched into space. Protons from solar flares and neutrons generated by an onboard RTG unit are primarily responsible for the new dark spikes seen. Both images were taken at approximately the same operating temperature and frame readout time (+17°C and 60.67-sec readout time, respectively). Fortunately, dark spikes induced in space were anticipated, and consequently the Galileo

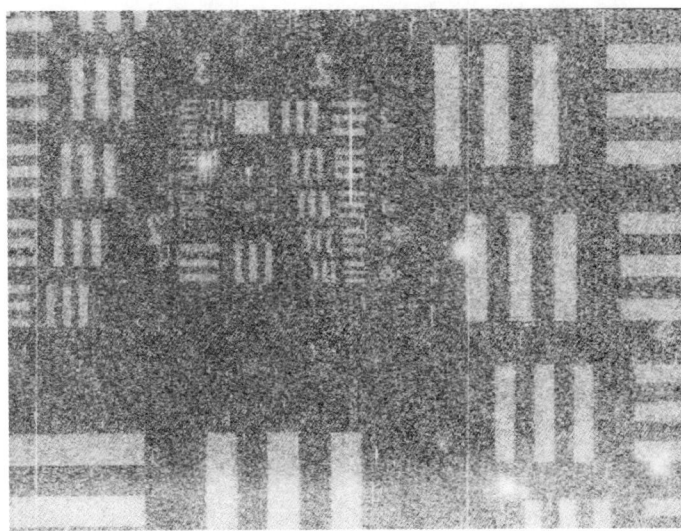

Figure 7.23(a) Two-minute room-temperature bar target image taken with an MPP CCD.

CCD was cooled to $-120°C$ to suppress the problem. Similar images taken after the mission was complete show a CCD that is nearly saturated at $17°C$. Millions of new dark spikes were generated during the mission. Radiation-induced dark spikes are discussed further in Chapter 8.

Dark spikes can for the most part be eliminated by differencing, pixel by pixel, an image with a dark frame image taken at exactly the same operating temperature and integration time. This image-processing scheme is called "despiking." For example, Fig. 7.23(a) shows a 2 1/2-minute bar-target image generated by an MPP CCD operating at room temperature. The majority of the dark current seen in the image is generated by dark current spikes. Figure 7.23(b) shows a despiked image using the process described. Figures 7.23(c) and 7.23(d) show raw and despiked line traces through the bar target. Again, the improvement in image quality with despiking is dramatic. Figure 7.23(e) shows despiking performed on a 6-minute-integration room-temperature image of Saturn. At this exposure time, some spikes bloom several pixels, resulting in small vertical lines in the image after differencing is performed.

Although the quality of the images above is greatly improved, despiking is not a perfect process because the dark spike generation rate is dependent on signal level, as explained above. Therefore, there will be remnant traces of spikes in the image after the process is performed.

Dark current pixel nonuniformity is also removed from an image by the despiking process. Again, the dark image that is subtracted must be taken under exactly the same conditions as the image frame (i.e., operating temperature and integration time). Usually, several dark frames are averaged so as to not add shot noise to the processed frame. The shot noise is reduced by the square-root of the number of frames averaged.

NOISE SOURCES

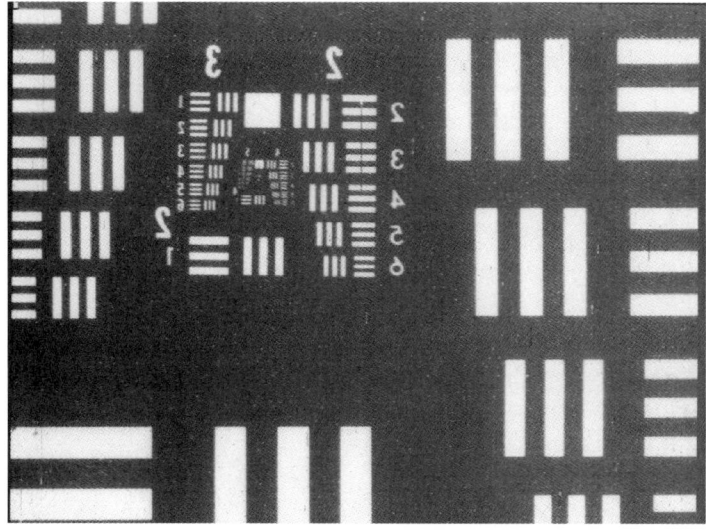

Figure 7.23(b) A despiked image of Fig. 7.23(a).

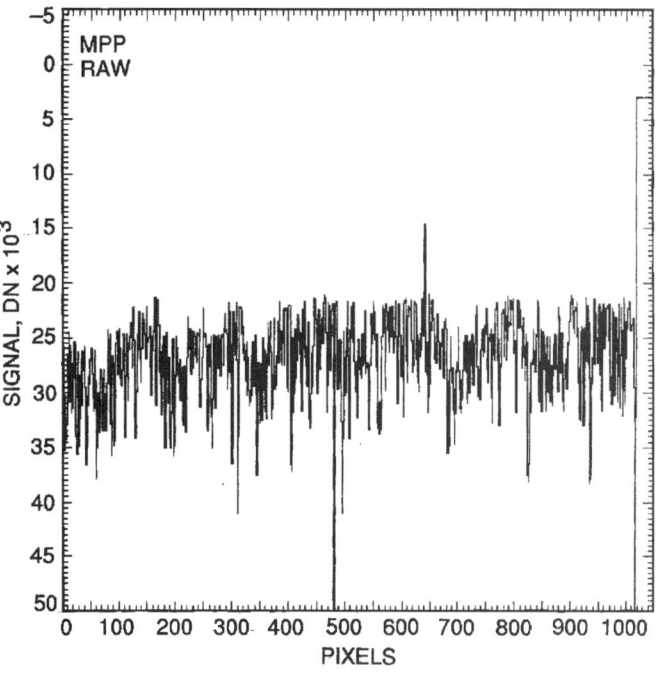

Figure 7.23(c) Line trace through Fig. 7.23(a).

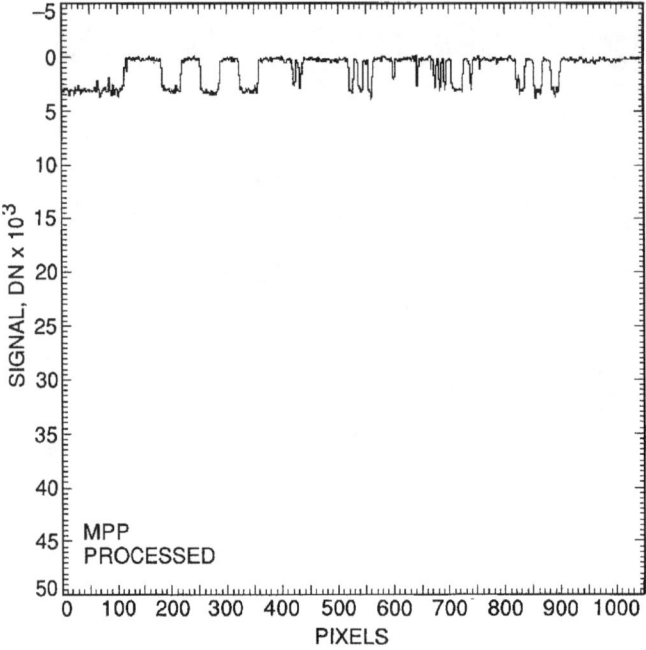

Figure 7.23(d) Line trace through Fig. 7.23(b).

Figure 7.23(e) A partially despiked 6-minute room-temperature image of Saturn.

7.1.2 Spurious Charge

Spurious charge is generated when a CCD is clocked into inversion. When inverting the signal channel, holes from the channel stops migrate and collect beneath the gate (refer to Fig. 7.24). Some holes become trapped at the Si–SiO$_2$ interface. When the clock is switched to the noninverted state, the holes are accelerated from the Si–SiO$_2$ interface with sufficient energy to create electron-hole pairs by colliding with silicon atoms (i.e., impact ionization). These "spurious" electrons are then collected in the nearest potential well.

Spurious charge was first observed when testing the virtual-phase Galileo CCD. The unwanted source of charge limited the read noise floor to approximately 100 e$^-$ rms greater than Galileo's 20 e$^-$ read noise requirement. To meet the specification, an investigation into the physics was undertaken, which revealed many interesting characteristics about spurious charge. First, spurious charge is only generated on the rising edge of the clock when a phase comes out of inversion; the falling edge has no influence on spurious charge generation. Second, spurious charge decreases exponentially with clock rise time and increases exponentially with clock swing. Sending holes back to the channel stops with a fast-moving, high-amplitude clock increases impact ionization and spurious charge generation. Third, spurious charge increases as the clock pulse width increases (i.e., the time the clock spends in the noninverted state immediately after inversion). When the clock width is short, holes will remain in the interface states, reducing impact ionization. This effect is seen for clock widths up to several hundred microseconds, indicating that some traps are long-lived. Fourth, spurious charge increases with decreasing operating temperature. Theory shows that impact ionization is more efficient at cold temperatures because of carrier mobility. Fifth, spurious charge increases linearly with the number of transfers that take place. Sixth—and the most critical to read noise performance—the noise this phenomenon produces can be

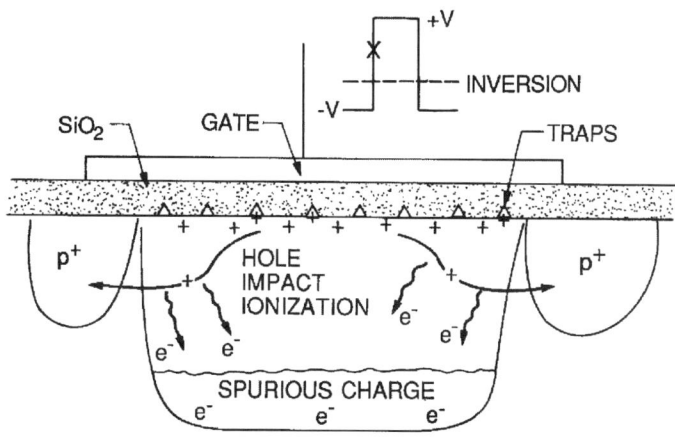

Figure 7.24 Spurious-charge generation.

characterized as shot noise. In other words, the noise increases by the square root of the average spurious charge generated as

$$N_{SC}(e^-) = S_{SC}(e^-)^{1/2}, \qquad (7.55)$$

where $S_{SC}(e^-)$ is the average spurious charge (e^-), and $N_{SC}(e^-)$ is the shot noise produced (rms e^-).

Example 7.18

Determine the spurious charge shot noise generated in a 1024-pixel horizontal register. Assume that 0.1 e^- is generated during each pixel transfer on the average.
Solution:
For 1024-pixel transfers, 102.4 e^-/pixel of spurious charge is collected in each pixel. The corresponding shot noise from Eq. (7.55) is

$$N_{SC}(e^-) = (102.4)^{1/2},$$

$$N_{SC}(e^-) = 10.1 \; e^- \text{ rms}.$$

The question of why spurious charge is only generated on the rising edge and not the falling edge is a curious one. When a phase inverts, holes from the channel stop region populate the surface. This process only takes a few nanoseconds, depending on substrate and channel stop resistivity. The potential difference between the channel stop and surface potential is small when holes begin to move, because each region is near zero volts. Therefore, holes move under a low-field condition. However, the potential difference is greater for trapped holes as a phase comes out of inversion. Figure 7.25 plots lateral surface electric field near the channel stop for various times after the phase is taken out of inversion ($V_G = -10$ to 0 V). Note that equilibrium is reached after approximately 10 ns. The lateral field directly beneath the gate is low because the potential is constant. However, as the holes approach the channel stop region the potential drops suddenly, generating a peak field of 170,000 V/cm. Most holes gradually move back to the channel stop during the clock transition and changing electric field. However, trapped holes will experience the full extent of the field when they are finally released. These holes may gain enough energy to excite other electron-hole pairs through the impact ionization process.

Spurious charge is often mistaken for thermal dark current. The two sources can be easily separated because spurious charge is only generated when the device is clocked and does not depend on integration time as does dark current. Also, spurious charge increases only slightly with decreasing temperature, because hole mobility increases. Dark current, on the other hand, is very dependent on operating

NOISE SOURCES 651

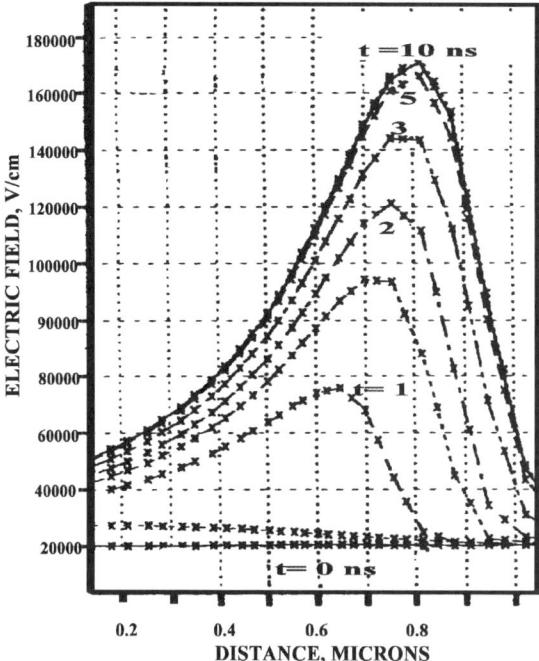

Figure 7.25 Electric fields near the channel stop, the region responsible for spurious-charge generation.

temperature. Therefore, if the dark level increases with clock amplitude or when driven into inversion, then the source is likely to be spurious charge.

It is difficult to detect spurious charge in the horizontal register because it will appear as an offset in each pixel. Horizontal spurious charge is isolated by measuring the read noise as the clocks are adjusted. If the noise level changes as the clock amplitude changes, then spurious charge is a likely source. Ideally, to prevent spurious charge generation, horizontal clocks should not be inverted. Only the vertical registers should require inverted operation. This operation option is possible for multiphase CCDs. However, virtual-phase CCDs must be inverted in order to transfer charge. Therefore, this technology is usually noisier than multiphase CCDs.

Five methods are employed to reduce spurious charge to negligible levels. The main method used is simply to wave-shape the clocks [i.e., see Fig. 5.3(a)]. Wave-shaping allows trapped holes to return to the channel-stop regions under lower field conditions. The second method limits the positive excursion of the clocks (and therefore the electric fields) without degrading full well performance (i.e., for multiphase CCDs it is only necessary to maintain BFW = SFW). Third, trapped holes can also be ameliorated by use of a "trilevel clocking" scheme. This technique uses an intermediate clocking voltage, which lies between the high and low clock rails required for complete charge transfer. The intermediate voltage level is set slightly above the inverted state and remains as long as possible to allow holes

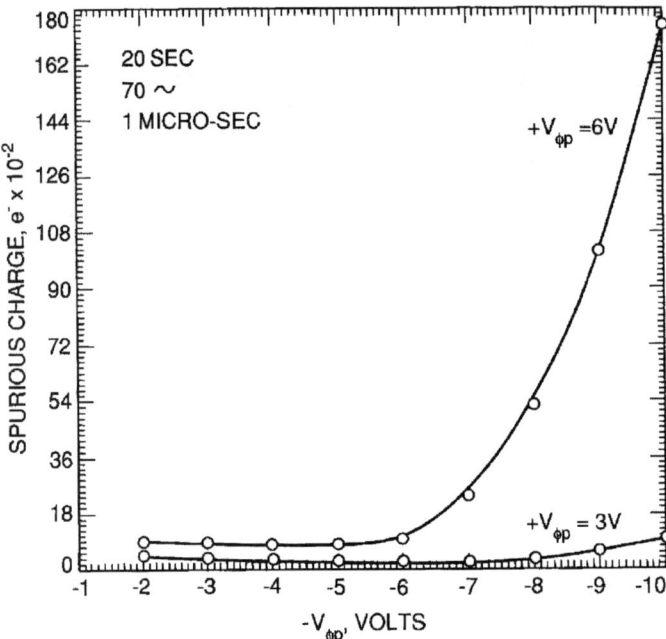

Figure 7.26 Spurious charge as a function of negative and positive clock drive.

to return to channel stops under a low-electric-field condition. Signal charge after this time is transferred by a short positive-going pulse. Trilevel clocking was first implemented in driving the horizontal register for the Galileo CCD in order to meet read noise specification. The fourth method uses very narrow clock pulses. As mentioned above, holes are trapped for a finite amount of time. Therefore, transferring signal charge and quickly returning the phase back to inversion also reduces spurious charge generation (i.e., the holes do not have the opportunity to leave the traps and create charge). This scheme is useful for high-speed applications where clock edges must be fast. The fifth method is not to invert the signal channel. As mentioned above, the horizontal registers should not be inverted. However, there are many advantages to inverting the vertical registers (e.g., low dark current, high full well, residual image suppression, etc.). Fortunately, the vertical registers are inherently slower than the horizontal registers and can be inverted without significant spurious charge generation. What spurious charge remains can usually be eliminated by clock wave-shaping.

Figure 7.26 plots spurious charge generated by a three-phase CCD as a function of negative clock drive. Spurious charge here was collected by switching vertical phases-1 and -2 at a 70-Hz rate for 20 seconds while phase-3 acted as a barrier (pinned at -8 V). The rise time of the vertical clocks was wave-shaped to 1 μsec. Data was collected at $-100°C$ to ensure that thermally generated dark current was absent from the measurement. Note that spurious charge rapidly increases at -6 V at the onset of inversion. Note also that the positive clock level has a dramatic effect on spurious charge. Higher clock levels generate higher fields and more spurious

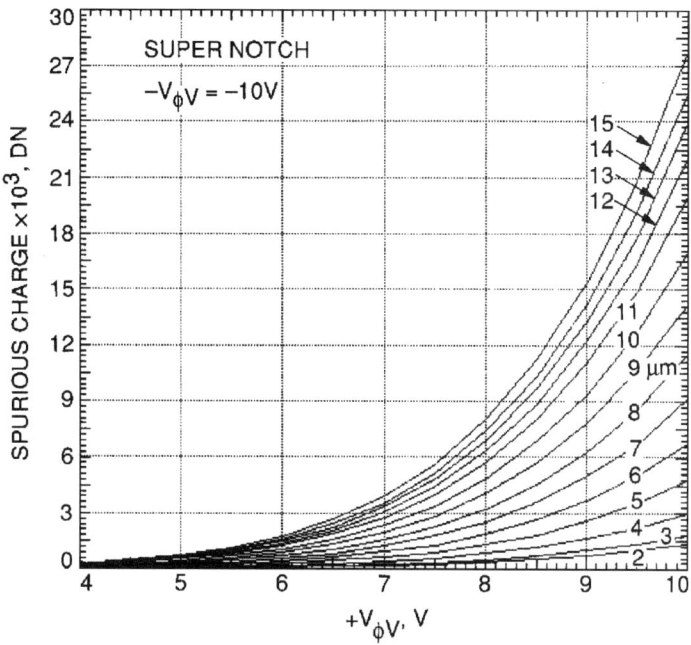

Figure 7.27 Spurious charge as a function of positive clock voltage and channel width.

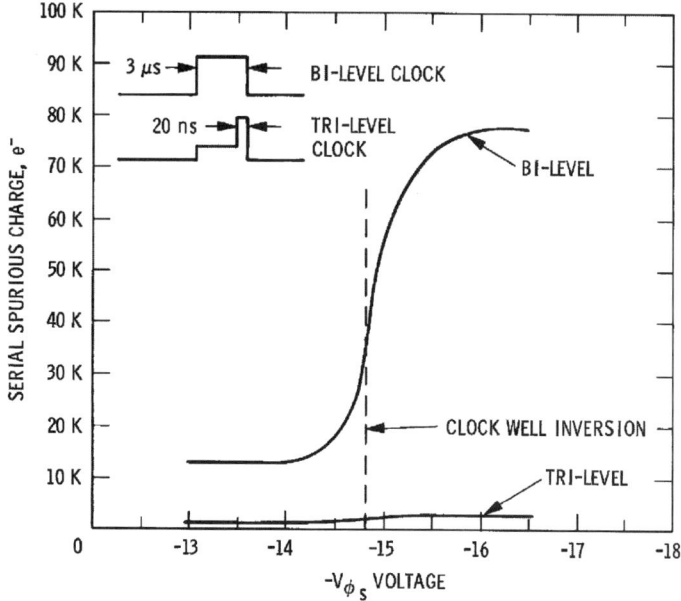

Figure 7.28 Spurious charge as a function of negative clock drive for a virtual-phase CCD driven with bilevel and trilevel clocks.

charge. Figure 7.27 plots spurious charge as a function of positive clock level for a three-phase CCD clocked under partially inverted conditions. The family of curves displayed is for various channel widths, showing that spurious charge increases with surface area under a phase.

Figure 7.28 is a measurement of spurious charge generated by a Galileo virtual-phase horizontal register when clocked into inversion. Note that inversion breakpoint occurs at approximately -14.74 V. Full well for the CCD is 100,000 e$^-$. The amount of spurious charge generated nearly saturates the device when clocked by a 3-μsec pulse with very fast clock edges ($\tau_{WS} = 10$ nsec). The quantity of spurious charge when trilevel clocking is also demonstrated. The objective behind trilevel clocking is to allow trapped holes to escape the interface under a low-field condition. The midvoltage level shown is set slightly out of inversion for this purpose. Although the amount of spurious charge is reduced considerably, 2000 e$^-$ remains, resulting in a shot noise of about 44 e$^-$.

It should also be mentioned that spurious charge generation can be a problem when multiphase CCDs are periodically clocked in and out of inversion during integration to suppress dark current generation.[10] For this technique phases 1 and 3 are held into inversion, while phase 2 is left high to collect dark current (a three-phase CCD is assumed). Charge is then periodically transferred to phase 1 and then back to phase 2. During this time, the surface under phase 2 inverts, suppressing surface dark current. After transfer, the midband states under phase 2 remain filled for a finite time, resulting in low dark current generation. The time constant where states stay quenched is given by $\tau = (n_i v_{th} \sigma)^{-1}$ where $v_{th} = 10^7$ cm/s and $\sigma = 10^{-15}$ cm^2, which yields 10^3 sec at $-70°$C and 10^{-2} sec at $20°$C. Therefore, the dither clock rate between phases 1 and 2 can be slow (e.g., a few hundred hertz at $0°$C). As before, spurious charge generation during clocking can be controlled simply by wave shaping the clocks. However, clocking the CCD in this manner will show trapping sites, because this is the same clocking sequence used by pocket pumping (refer to Section 5.3.4). In addition, this clocking technique is also used to control blooming when the high level of the clock is set into surface full well (refer to Section 4.2.8).

7.1.3 FAT-ZERO

Fat-zero or bias charge is injected when process- and design-induced traps limit CTE performance. Fat-zero fills trapping sites, allowing signal charge to transfer through the troubled regions. Fat-zero is normally introduced optically from a flat-field light source. For example, the Hubble WF/PC I CCD required approximately 100 e$^-$ to correct for the design-induced trap discussed in Chapter 5. However, the penalty for introducing fat-zero is the shot noise that is generated. That is,

$$N_{FZ}(e^-) = F_Z(e^-)^{1/2}, \qquad (7.56)$$

where $N_{FZ}(e^-)$ is the fat-zero noise (e$^-$ rms) and $F_Z(e^-)$ is the average fat-zero injected (e$^-$).

Noise sources

Example 7.19

Find the noise penalty produced when 100 e⁻ of optical fat-zero is introduced. Assume an output amplifier read noise of 13 e⁻ rms.
Solution:
From Eq. (7.56),

$$N_{FZ} = 100^{1/2},$$

$$N_{FZ} = 10 \text{ e}^-.$$

The quadrature sum of the read noise and shot noise yields a net noise of 16 e⁻. Therefore, the addition fat-zero degrades the read noise floor by 6 e⁻ rms.

Electrical fat-zero can also be introduced into the array with the input diode/gate structure described in Section 1.4.3.3. Injecting electrical fat-zero is sometimes advantageous over optical injection because the noise introduced may be less. In this case, the noise generated is given by kTC. That is,

$$N_{CI}(\text{e}^-) = \frac{\sqrt{kTC_I}}{q}, \qquad (7.57)$$

where $N_{CI}(\text{e}^-)$ is the kTC noise (e⁻ rms) and C_I is the input gate capacitance (F). Note that the noise produced is independent of fat-zero level.

Example 7.20

Calculate the noise generated by a 100,000 e⁻ signal that is electrically injected into the horizontal register. Assume an input gate capacitance of $C_I = 0.1$ pF.
Solution:
From Eq. (7.57),

$$N_{CI}(\text{e}^-) = \left[(4 \times 10^{-21}) \times 10^{-13}\right]^{1/2} / (1.6 \times 10^{-19}),$$

$$N_{CI}(\text{e}^-) = 125 \text{ e}^-.$$

A shot noise of 316 e⁻ would be produced if injected optically. Therefore, electrical injection is the optimum method to inject fat-zero at this level. However, when only a small amount of fat-zero charge is required—which is usually the case—optical fat-zero is the best injection method.

Star detection and tracking groups use electrical injection to maintain high CTE performance when working in high-energy radiation environments that produce CCD damage. For example, fat-zero charge is injected into the array and then quickly erased to fill radiation-induced traps before an image is taken. However, the operating temperature must be sufficiently low to not allow fat-zero charge to escape the traps before the frame readout cycle is complete (i.e., the emission trap time constant must be longer than the integration and readout time periods). To reduce temperature requirements, some star detection groups inject a "line of charge" that precedes the guide star as it is read out by a few lines. Therefore, the traps need to hold the fat-zero charge over only a few line periods. However, this mode of operation can be used only if the position of the star is already known, in order to place the fat-zero precisely before the star.

7.1.4 TRANSFER NOISE

Transfer noise is generated when signal charge interacts with surface or bulk states during charge transfer. Transfer noise is related to CTI through

$$N_{CTE}(e^-) = [2 \times \text{CTI} \times N_P \times S(e^-)]^{1/2}, \tag{7.58}$$

where $N_{CTE}(e^-)$ is transfer noise (rms e^-) and N_P is the number of pixel transfers.

Example 7.21

Find the horizontal transfer noise for a 1024 × 1024 CCD when exposed to a 10,000 e^- flat-field signal. Assume a horizontal CTI of 10^{-5}. Also, compare the CTE noise with the signal shot noise.
Solution:
From Eq. (7.58),

$$N_{CTE}(e^-) = \left(2 \times 10^{-5} \times 1024 \times 10^4\right)^{1/2},$$

$$N_{CTE}(e^-) = 14 \, e^- \text{ rms}.$$

The shot noise generated by 10,000 e^- is 100 e^- rms.

This example shows that CTE noise can usually be neglected for high-performance buried-channel CCDs. This may not be the case for surface-channel sensors where CTE performance is much worse. Also, buried-channel CCDs that experience radiation damage exhibit CTE noise problems.

7.1.5 Residual Image

There are two distinct forms of residual image observed in CCDs. The first source is referred to as "residual surface image" (RSI), and is created when signal charge interacts with the Si–SiO$_2$ interface (i.e., SFW). The other form of residual image is referred to as "residual bulk image" (RBI). RBI is associated with trapping centers located at the epitaxial interface.

7.1.5.1 Residual Surface Image

RSI was a very serious problem for early CCDs. Signal charge trapped at the Si–SiO$_2$ interface would be released during readout and appear as deferred charge (e.g., Fig. 4.8). Because the trap lifetimes are long, residual image was seen in subsequent images, sometimes many weeks after the original image was taken. Today, RSI can be completely eliminated by driving the CCD into inversion, allowing trapped charge to recombine with holes.

Surface states at the Si–SiO$_2$ interface occupy the forbidden band between the conduction and valence band edges. Trapped charge will escape these surface states starting with energies nearest to the conduction band edge. The energy level of discharge as a function of time is estimated by

$$E_e = kT \ln(t_d v_{th} \sigma_n N_C), \tag{7.59}$$

where E_e is the energy level discharged to from the conduction band edge (eV), t_d is the time given to discharge the trap (sec), v_{th} is the thermal velocity of an electron (cm/s), σ_n is the electron cross-section (cm^2), and N_C is the effective number of states in the conduction band.

Example 7.22

Find the energy level from the conduction band edge that surface states will discharge over a 60×10^{-6}-sec time period. Calculate at room temperature ($kT = 25$ mV) and 200 K. Assume $v_{th}\sigma_n N_C$ is equal to approximately 10^{11}/sec.

Solution:
From Eq. (7.59),

$$E_e = (2.5 \times 10^{-2}) \ln[10^{11} \times (60 \times 10^{-6})],$$

$$E_e = 0.39 \text{ V } (300 \text{ K}),$$

$$E_e = 0.27 \text{ V } (200 \text{ K}).$$

As mentioned above, RSI has been observed several days at very cold ($-140°$C) operating temperatures showing deep-level trap action.

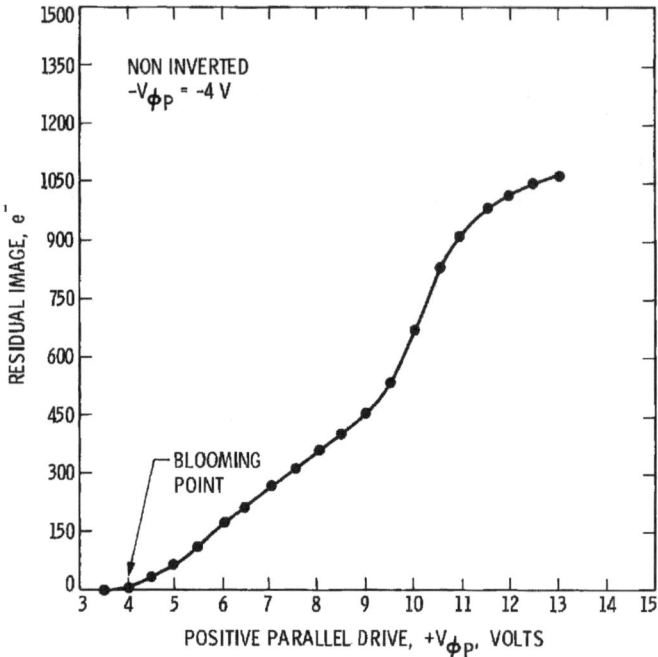

Figure 7.29(a) Residual image as a function of positive clock drive for a three-phase CCD.

Figure 7.29(a) plots RSI after a CCD is overexposed to an intense flat-field source. RSI increases with positive drive because the CCD is driven deeper into surface-channel operation. Note that RSI disappears when BFW = SFW at $V_C = 4$ V. At this point, charge is unable to reach the surface because blooming will occur first. The CCD for this test is clocked noninverted to preserve the RSI signal that otherwise would be lost to hole recombination if inverted.

RSI is also seen when the channel bias (V_{REF}) to the CCD is removed (i.e., the camera is turned off). When this occurs, the channel becomes undepleted, saturating the Si–SiO$_2$ interface with electrons. When V_{REF} is applied again, the interface begins to emit trapped charge as RSI. This form of RSI charge can dominate the dark current floor for many hours, depending on operating temperature. For example, Fig. 7.29(b) plots dark current as a function of operating temperature for a three-phase CCD. Note that dark current rate does not follow theory below −95°C. The additional charge seen is RSI because the camera was powered off and on during the experiment and did not use inverted clocks. Integration times as long as 52 hours were employed in this test. RSI can be eliminated from dark current measurements by inverting the CCD or momentarily elevating the substrate voltage into inversion.

RSI under certain clocking situations may be seen in the horizontal register, since it is normally clocked noninverted (for spurious charge reasons). However, its well capacity is usually much greater than the vertical registers, and therefore, RSI

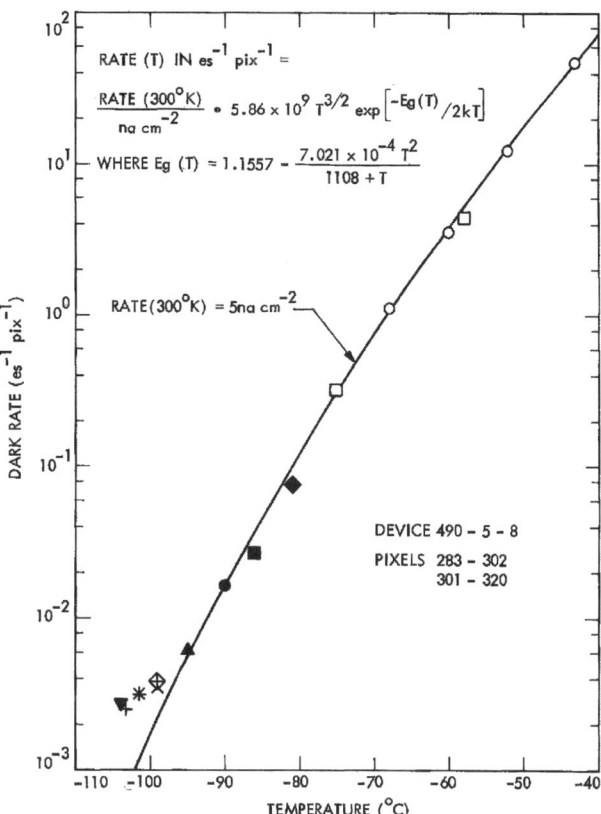

Figure 7.29(b) Dark current transfer showing surface residual charge at very cold operating temperatures.

is not normally seen. Pixel summation could drive the horizontal register beyond SFW, possibly resulting in horizontal RSI problems.

7.1.5.2 Residual Bulk Image

Residual bulk image has an interesting history behind its discovery. Near the end of Galileo CCD camera fabrication, an important radiation damage test was performed. Up to that time, CCDs had only been subjected to Co^{60} gamma-ray radiation, a standard radiation stimulus used to screen and qualify space electronic components. However, it was known that the CCD would be subjected to neutrons because the spacecraft was flying a nuclear energy source (RTG) that emitted this type of particle. Therefore, a test was performed to ensure that neutrons would not be a problem. Unfortunately, after irradiation, the device came back saturated by ultrahigh-level dark spikes. The result was alarming and forced the Galileo imaging team to lower the operating temperature from $-50°C$ to $-120°C$. However, the new temperature opened up the door to a new problem: residual bulk image, or RBI for short.

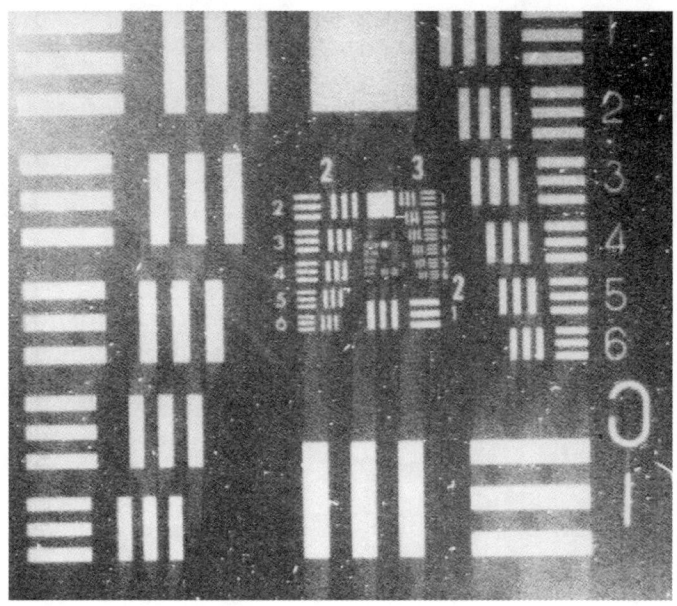

Figure 7.30 Residual bulk image.

RBI is caused by the trapping of signal carriers at the epitaxial/substrate interface. RBI is predominately seen at wavelengths > 7000 Å, where the photons penetrate beyond the depletion region to this interface. Unlike RSI, RBI is detected for exposures below full well, but is most conspicuous when the CCD is overexposed. RBI is independent of how the CCD is clocked. Its signature is the same for SFW, BFW, noninverted and inverted clocking conditions and if the CCD is powered or not. Unfortunately, unlike RSI, there is no special clocking scheme to eliminate RBI. Most CCDs exhibit RBI with only a few exceptions. For example, devices that employ vertical antiblooming do not exhibit RBI because the RBI signal goes in to the substrate. Also, CCDs built on bulk silicon do not show the effect because there is no epitaxial interface. Backside-illuminated CCDs do not exhibit RBI because the epitaxial/substrate interface is completely removed during the thinning process.

Figure 7.30 shows a 9000-Å RBI image taken from a frontside-illuminated CCD. The sensor was exposed to an Air Force target many times above full well; then, the CCD was read out and a dark current image taken, as shown. The exposed regions show dark charge that comes from trapped charge at the epitaxial interface.

Figure 7.31 shows how RSI and RBI sources are separated. The image shown in Fig. 7.31(a) was generated by taking a five-minute dark exposure immediately after the CCD was overexposed ($100 \times$ SFW) with two circular sources of light. The right-hand side of the array was stimulated with 9000-Å light, the left side with 4000-Å light. The CCD array was clocked noninverted to preserve RSI charge. Residual image on the left side is composed only of RSI because 4000-Å photons

NOISE SOURCES

Figure 7.31(a) Bulk and surface residual images taken under noninverted clocking conditions.

do not penetrate any further than the depletion region (recall the absorption length of a 4000-Å photon is 2000 Å). The 100× exposure results in blooming for both images during the integration period. The positive clock drive was adjusted into the SFW regime so that SFW occurs before BFW. Therefore, RSI is observed where blooming occurs. If the clocks had been adjusted into the BFW regime, no RSI would be seen, since charge would not have reached the surface. Residual charge on the right side of the image is composed of both RBI (the white circular region) and RSI (the bloomed part of the image). In Fig. 7.31(b) the same experiment is performed under inverted operation. The RSI signal is erased due to the surface recombination, leaving only the RBI signature.

Figure 7.32 measures the RBI signal for a CCD stimulated with 9500-Å light at five different operating temperatures. The data was generated by flooding the CCD with an intense 100× full well exposure. Next, the CCD was erased for 24 seconds. Then, 2800 lines were read into the computer under dark conditions and a column trace plotted. The test was repeated for different operating temperatures. The data shows that the RBI discharge rate decreases exponentially with decreasing temperature. At −108°C the discharge rate and the RBI signal seen is negligible over a long readout time period (> 68 sec). Figure 7.33 shows the RBI signal produced at different wavelengths at an operating temperature of −80°C. Only those pho-

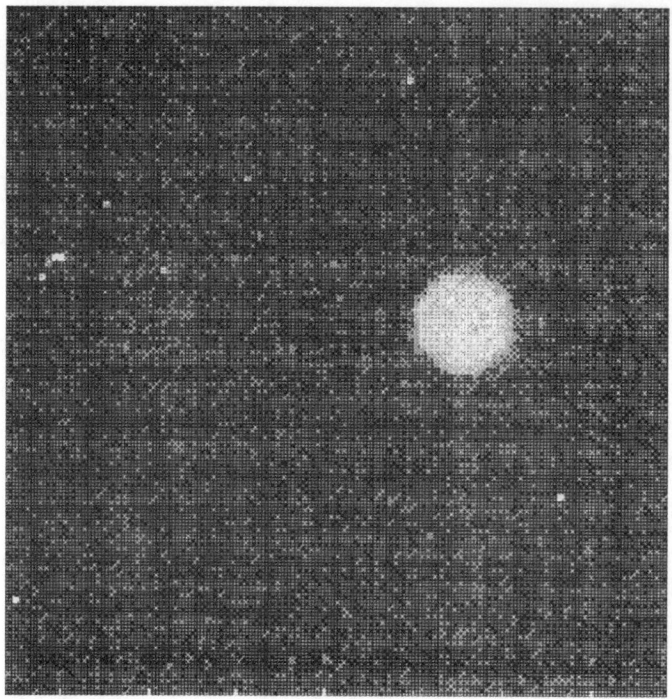

Figure 7.31(b) RBI image with RSI eliminated by inverted operation.

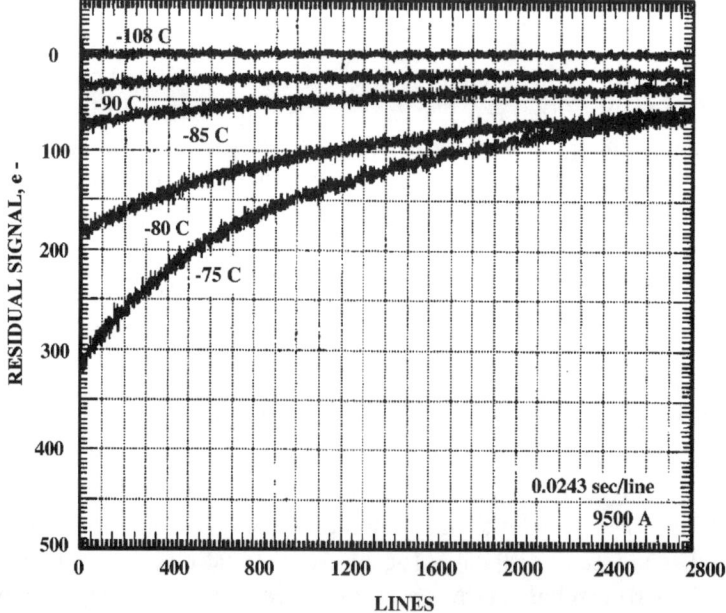

Figure 7.32 RBI as a function of line readout time at different operating temperatures.

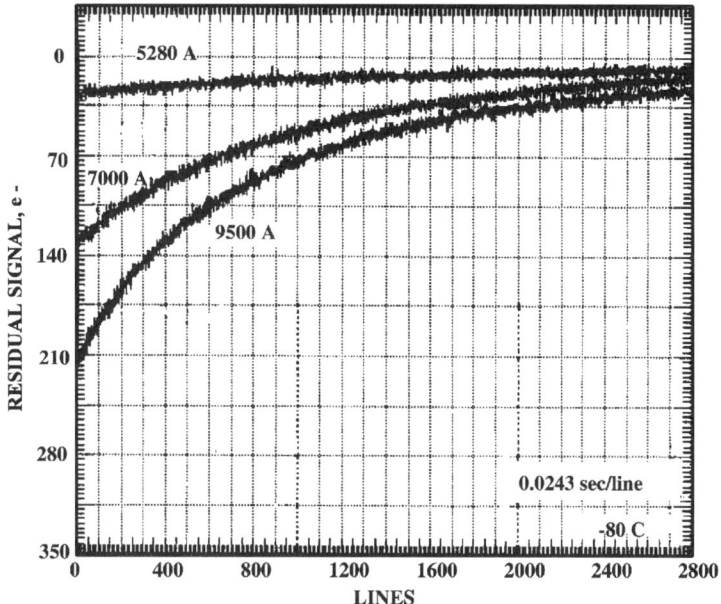

Figure 7.33 RBI as a function of line readout time at different wavelengths.

tons that can reach the epitaxial interface produce the RBI signal. The tests above were performed inverted to guarantee that RSI was completely eliminated from the measurement, leaving only RBI.

As the epitaxial interface bulk states fill, the QE increases, producing an important QE instability problem related to RBI [i.e., quantum efficiency hysteresis (QEH)]. The QEH variation seen depends on the scene being imaged, the wavelength, operating temperature, integration period and readout times involved. As will be shown below, the effect can introduce a photometry error of several percent under some clocking and cooling conditions. Maximum QE occurs when all epitaxial interface bulk states are full, and minimum QE when they are empty. Therefore, bright scenes in the near IR are most vulnerable to the effect. QEH is observed by taking a flat-field image after overexposing the CCD to a bar target using near-IR light (say, 9000 Å). The result is a positive image of the original scene showing locations where QE has increased in the regions of the bars.

Similar to surface states, RBI QEH is exponentially dependent on operating temperature. For slow-scan applications, the QEH problem is not obvious unless the operating temperature is less than approximately $-40°$C. High-speed operation will show the effect at warmer temperatures, because there is less time for the traps to discharge. Maximum QEH is experienced when the discharge time constant is approximately equal to the period between exposures.

One solution to the QEH problem is to allow the CCD to discharge before subsequent images are taken. The time required is dependent on operating temperature. Another technique used to achieve high photometric accuracy (1%) is to use

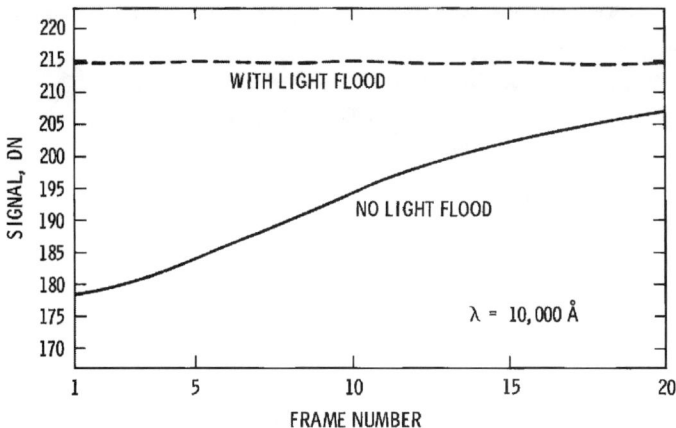

Figure 7.34(a) Near-IR sensitivity as a function of frame number with and without near-IR light flood.

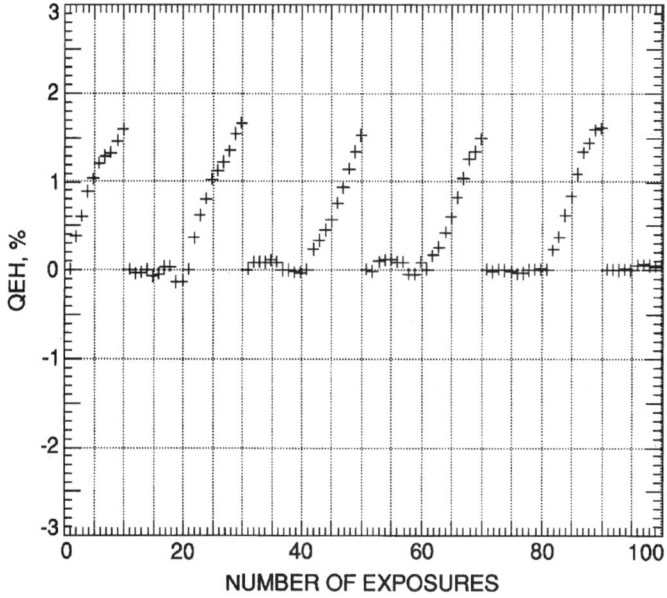

Figure 7.34(b) RBI characteristics without light flood for a Cassini CCD.

a near-IR light flood. For example, the Galileo and Cassini CCDs are preflashed with four IR LEDs to charge the epitaxial interface for maximum sensitivity and stability. The light intensity is adjusted to produce a signal $> 100\times$ full well, thus guaranteeing that all bulk epi traps are filled. After flooding, the CCD is quickly erased and an image is taken. Figure 7.34(a) shows the effectiveness of a light flood in maximizing and stabilizing the near-IR QE for the Galileo CCD. The plot measures signal through a sequence of identically exposed near-IR frames with

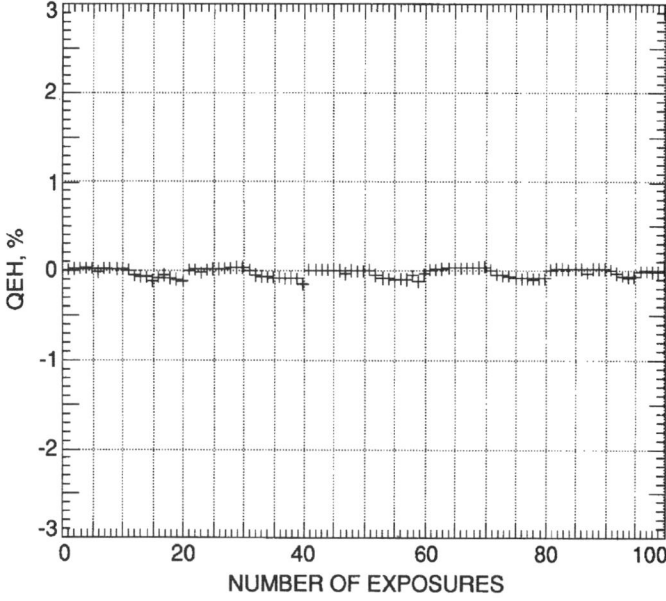

Figure 7.34(c) RBI QEH stabilized after employing light flood.

and without a light flood. Without light flood, the signal progressively increases as greater numbers of trapping sites are filled by previous exposures. When a light flood is applied before each frame, the trapping sites remain filled, and therefore do not influence QE performance. The light flood technique is only effective in stabilizing QE for short integration periods up to 120 sec at an operating temperature of $-120°C$.

Figure 7.34(b,c) shows RBI QEH characteristics for the 1024×1024 Cassini CCD. Figure 7.34(b) shows the QE slowly increasing as 10 flat-field frames of data are taken. For this wavelength and operating temperature (9500 Å and $-80°C$), QE increases by 1.6% as the epi traps fill. For the next 10 exposures the CCD is not exposed, thereby allowing the traps to discharge and bring the QE back to equilibrium. In Fig. 7.34(c) the same experiment is performed using an IR light flood that stabilizes QE performance in addition to eliminating residual images from previous exposures.

7.1.6 LUMINESCENCE

Several luminescence problems have been identified for the CCD. We have already discussed amplifier luminescence and its impact on read noise in Chapter 6. This section will introduce clocking luminescence, pixel luminescence and diode luminescence.

7.1.6.1 Clocking Luminescence

Clocking luminescence can result simply from clocking the gates. The problem appears to be common to those CCDs fabricated with a SiO_2 gate insulator, but is not seen in all cases. Dual insulating systems (e.g., SiO_2/Si-Ni) do not show the problem. Clocking luminescence is observed in the horizontal registers of the Hubble WF/PCI CCD while they are clocked during a long exposure period. As a result, rows in the vertical section nearest the horizontal register acquire additional charge, decreasing exponentially toward the center of the device. The current theory for the phenomenon is that the clocking action generates "hot electrons" in the gate insulator which in turn create IR photons that are reabsorbed in the neighboring pixels in the vertical section. The quantity of charge increases with clock swing, clocking frequency, clock slew rate and decreasing the operating temperature. The effect, as noted above, adds additional background charge and associated shot noise to the image. Early TI and RCA CCDs exhibited horizontal clocking luminescence (both of these devices built with a simple oxide gate dielectric). The problem for Hubble was solved by inhibiting the horizontal clocks for the duration of the integration cycle and running them only when reading an image.

Figure 7.35(a) shows an 1800-sec dark image generated by a TI WF/PC I CCD. Clocking luminescence is observed at the top and bottom of the image near the

Figure 7.35(a) Image showing clocking luminescence from the horizontal registers.

horizontal registers that are clocked during the integration period. The image also shows an Fe^{55} x-ray event field. The tails seen behind each x-ray event are caused by the design-induced trap discussed in Chapter 5. There are also seven column blemishes that have reached SFW. The tails associated with these blemishes are caused by charge being discharged by the surface states (i.e., a form of RSI). These tails disappear when the CCD is clocked into inversion. Also seen in the image is a very bright spot. This defect will be discussed in the next section.

Figure 7.35(b) is a column trace showing horizontal clocking luminescence for an RCA (BIG CID) CCD. The exponential fall-off of light into the vertical registers is used to determine the wavelength of emission (1 μm in this case).

Clocking luminescence is not limited to the horizontal register. For example, the RCA and TI three-phase CCDs both exhibit vertical register luminescence. This phenomenon was first seen when two phases were clocked, leaving the third phase as a barrier. Clocked in this manner, charge that was generated by luminescence did not transfer but collected instead. After clocking the CCD for several minutes, the sensor is then read out. The images show numerous near-IR luminescent pixels because of clocking action. For example, Fig. 7.36 shows a 300-sec WF/PC I CCD dark image in response to the clocking technique just described. Hundreds of luminescence pixels are seen in the image. Also note that the upper horizontal register shows luminescence as it is clocked, whereas the lower register does not because it is not clocked. It is interesting to note that the new WF/PC II CCDs do not show clocking luminescence as do the WF/PC I CCDs. These sen-

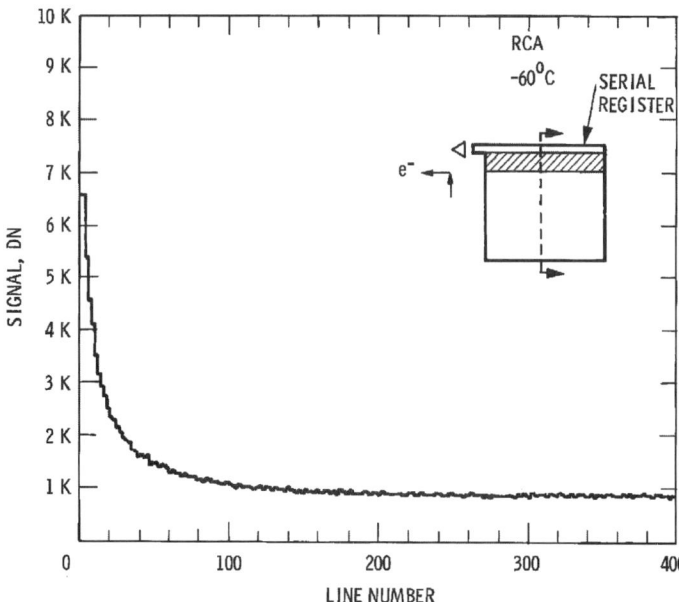

Figure 7.35(b) Column trace showing horizontal register luminescence that extends into the array.

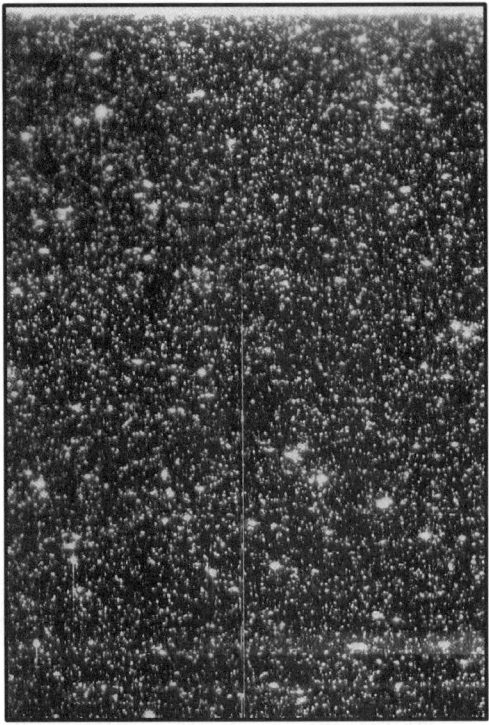

Figure 7.36 Array clocking luminescence.

sors were processed with a oxide/nitride gate dielectric, whereas WF/PC I CCDs were fabricated with a simple oxide dielectric. In flight, the WF/PC II horizontal registers are not inhibited during charge integration. There is no sign of clocking luminescence even for a several-hour integration period.

7.1.6.2 Pixel Luminescence

Pixel or blemish luminescence is the most serious form of all luminescence problems. The problem is usually associated with a single pixel that becomes shorted. Such pixels are referred to as catastrophic pixels when they erupt. The problem can be introduced by ESD damage or by over-stressing the gate insulator momentarily with a high clock voltage. Also, the short may suddenly appear without warning or explanation under normal operating conditions (i.e., latent shorts). Pixel luminescence, when it occurs, can sometimes be controlled by adjusting clock voltages or changing the phase timing (e.g., by switching the normally high phase to a low one when integrating). However, replacing the CCD is the best solution to the problem because catastrophic pixels only worsen with time.

Latent pixel luminescence is a major concern to flight applications. For this reason, all flight CCDs are burned in for several hundred hours to exercise and screen out catastrophic pixels. Burn-in tests typically uses clock levels 2 V beyond

Figure 7.37(a) Image showing two bright luminescent pixels.

inversion to screen weak pixels. Such tests have proved fruitful as some sensors have failed prematurely. Fortunately, to our knowledge, no CCD failures due to latent shorts have occurred in flight.

Figure 7.37(a) shows two glowing catastrophic pixels located in columns #33 and #503 for a TI WF/PC I CCD. The brighter pixel generates enough charge to fill completely several columns even during readout. Note also the image artifact to the left of this bright pixel; this effect is caused by charge transferring back up into the array from the horizontal register through the array transfer gate (an old CCD problem referred to as "transfer gate tunneling"). Charge for this image is read upwards and to the left. Note that horizontal clocking luminescence for this CCD has been suppressed compared with Fig. 7.35. Here, the horizontal clocks have been inhibited during the 100-sec exposure. Figure 7.37(b) also shows a catastrophic blemish related to a single pixel. This image was taken at $-95°C$ using an integration time of one hour.

7.1.6.3 Diode Luminescence

Luminescence is also generated by on-chip diodes used to measure operating temperature. Temperature readings are accomplished by forward biasing the diode and measuring its forward current. By calibrating the diode (a plot of current versus temperature) one can precisely determine the operating temperature of the CCD. If

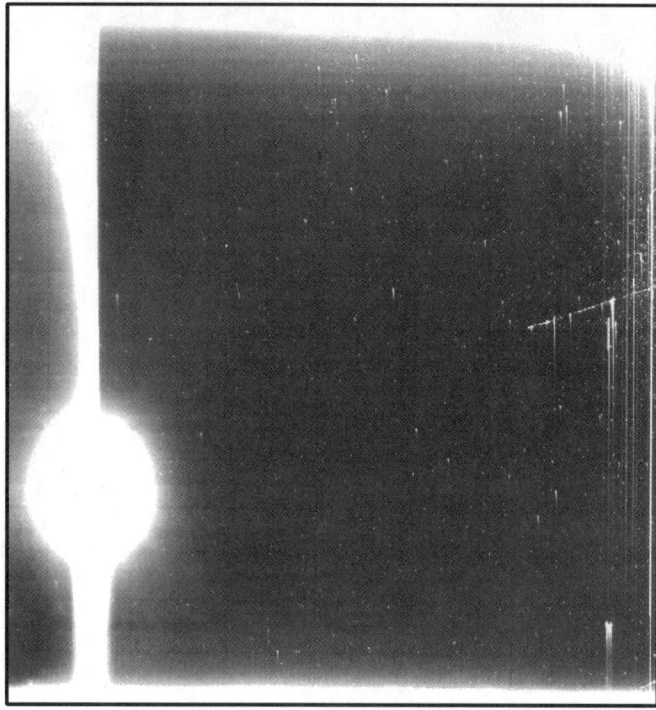

Figure 7.37(b) Severe luminescence generated by a single pixel.

two temperature diodes are available, they can be configured into a bridge circuit to produce an output voltage proportional to operating temperature. Unfortunately, when a diode is forward biased, IR light is generated. The light emission usually saturates the CCD in a few seconds, depending on the applied voltage and location of the diodes. Temperature readings can therefore only be taken when the CCD is not in use.

ESD protection diodes in parallel with the CCD gates exhibit the same problem when forward biased (i.e., when clocks are driven negatively relative to substrate). The diodes prevent inverted clocking, which is detrimental to device performance.

Substrate bounce and clock transients can momentarily forward bias n-p junctions producing light. For example, the source diffusion of an active MOSFET load that is biased to ground can generate luminescence when the vertical clocks go high, causing the substrate to bounce positively.

7.1.7 Cosmic Rays and Radiation Interference

Cosmic-ray events are always seen in CCD images. Cosmic rays originate in space and consist mostly of high-speed protons, the nuclei of hydrogen atoms, with about 15% helium and heavier nuclei and a small percentage of electrons. The earth is

NOISE SOURCES

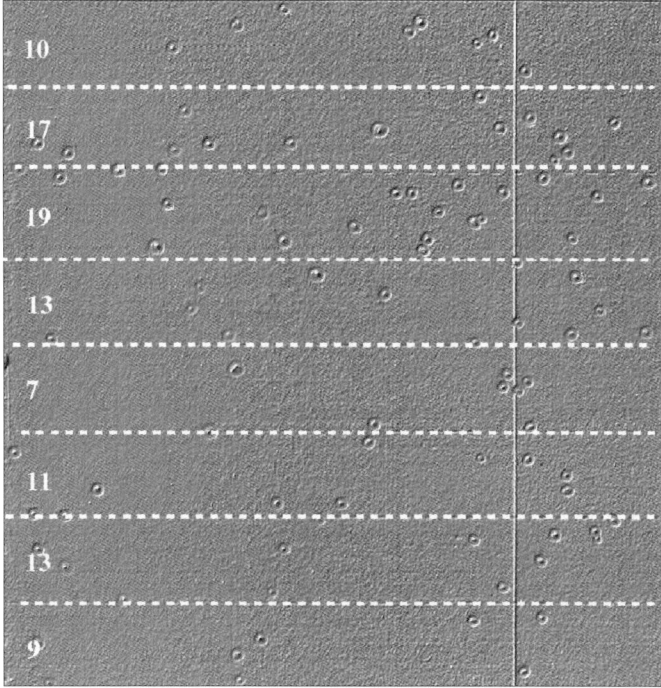

Figure 7.38(a) Cosmic-ray events collected in a 1800-sec exposure.

continually bombarded by these particles, which collide with the stratosphere to produce secondary particles. The secondary rays are comprised of 74% muons, 25% electrons and 1% protons. Muons are short-lived particles with a mass of 207 times that of an electron. They are generated by spallation (i.e., nuclear smashing) when high-energy particles interact with the atoms in the earth's outer atmosphere. Cosmic rays come from the sun, but most rays probably originate elsewhere within our galaxy (e.g., supernovas). A few particles and the most energetic (energies above 10^{17} eV) appear to be coming from outside our galaxy. Both the primary galactic cosmic-ray nuclei and the secondary particles produce ionization trails in the CCD.

The number of cosmic rays seen by the CCD varies with altitude. This effect is familiar to CCD astronomers working at different observatories and altitudes. A nominal fluence at sea-level for cosmic rays is 0.025 $cm^{-2} sec^{-1}$. Therefore, in 1 hour about 90 events are typically seen for a 1-cm^2 detector. For example, Fig. 7.38(a) shows a WF/PC I CCD 1800-sec dark image that collected 99 cosmic events at an altitude of 2000 ft. Figure 7.38(b) shows a cosmic ray that entered a ground-based TI WF/PC I backside-illuminated CCD. The average charge measured in each pixel is approximately 700 e^-.[11] The angle of incidence is determined by length of the ionization track through

$$\tan^{-1}(\theta) = \frac{x_{EPI}}{x_{ion}}, \qquad (7.60)$$

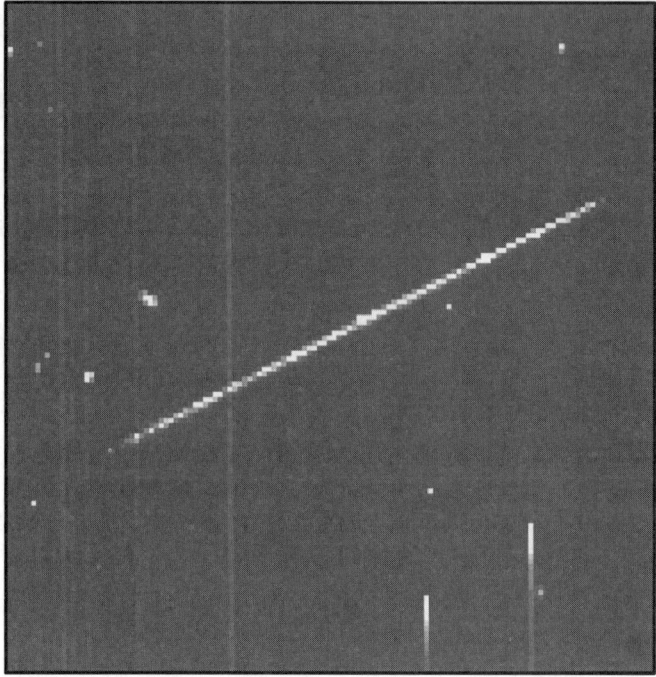

Figure 7.38(b) Cosmic-ray incident on the CCD by 0.249 deg.

where x_{EPI} is the epitaxial thickness (cm) and x_{ion} is the length of the ionization trail (cm).

Example 7.23

Determine the angle of incidence for the cosmic ray shown in Fig. 7.38(b). Assume that 92 pixels are affected. The pixel size is 15 μm and the thickness of the sensor is 6 μm.
Solution:
From Eq. (7.60),

$$\tan^{-1}(\theta) = 6/(92 \times 15),$$

$$\tan^{-1}(\theta) = 0.249 \text{ deg}.$$

Cosmic-ray tracks seen in Fig. 7.38(b) typically involve only a few pixels because their angle of incidence is much greater than 0.249 deg. Some bright pixels seen in Fig. 7.38(b) are dark spikes. The two charge sources can be distinguished, because dark spikes are stationary while cosmic rays are not.

NOISE SOURCES 673

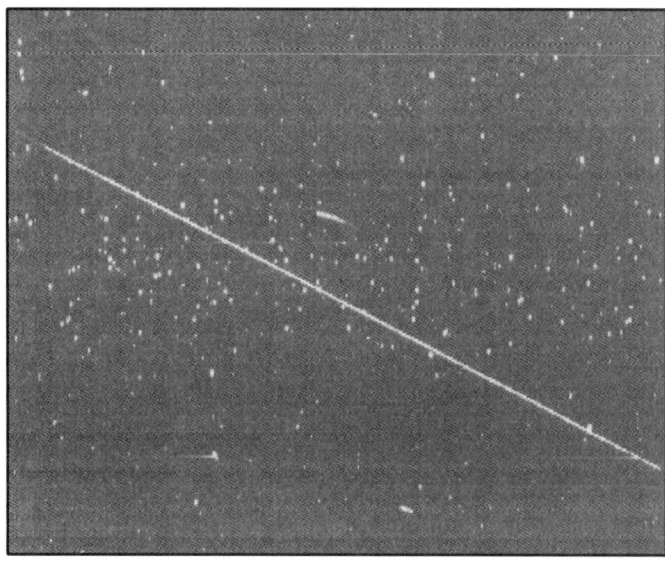

Figure 7.38(c) Energetic proton interacting with a Hubble CCD.

Protons generated by the sun or trapped in the Van Allen belts also represent an interference problem in space. For example, the Hubble WF/PC CCD cameras are not normally used for long exposures when passing through the South Atlantic Anomaly (SAA), a region that is rich in trapped energetic protons [see Fig. 1.6(f)]. Typical event counts for Hubble outside the SAA exhibit a background flux of 0.0385 events/sec/cm^2, producing 800 e$^-$/event on the average. Figure 7.38(c) shows a very shallow angle proton-ionization trial generated by a WF/PC II CCD. Many other proton events are seen in the image which typically interact with a single pixel. The energy of the protons seen in this image are greater than 100 MeV because the Hubble CCDs are shielded with a 1-cm Ta shield that stops proton energies less than this.

Figure 7.39(a) shows a dark image taken in 1992 when the Galileo CCD approached Earth from Venus en route to Jupiter. Numerous proton events are seen in the image. Figure 7.39(b) shows an image from a few hours later when the spacecraft encountered Earth's Van Allen belts. Proton event counts increase dramatically. Pixel readout starts at the bottom of the image and progresses to the top. For these images, the Galileo CCD is quickly erased before integration. Since the read time is greater than the integration time, more events are seen at the top of the image as readout takes place. As with Hubble, the proton events seen in the image are greater than 100 MeV because Galileo also employs a 1-cm Ta shield. The proton activity was so great at this time that the pixel data compression system could not maintain proper throughput, hence the line drop-outs in the upper left-hand corner.

Figure 7.40 shows two Compton electron events generated by a Co^{60} gamma-ray source. These signatures are familiar to the Galileo CCD because the spacecraft

Figure 7.39(a) Galileo image showing interplanetary protons interacting with the CCD.

is powered with a radioactive RTG source that generates gamma rays. The electron path is easily influenced by interactions with silicon valence electrons. The electron event at the top of the image enters the CCD at a steep angle and makes its way beyond the depletion edge where charge diffuses among several pixels. Then the electron heads upward into the depletion region only to return quickly back to field-free material. It then turns upward again and likely leaves the CCD.

It is interesting to note that dark current less than 0.01 e^-/sec is less than the rate of charge generated by cosmic rates at sea level. Therefore, dark currents less than this must subtract this background charge when measurements like this are taken. Ionization tracks and resultant damage to the CCD are discussed in more depth in Chapter 8.

7.1.8 Excess Charge

When inverted, the surface potential at the Si–SiO$_2$ interface is pinned to substrate potential. As the phase is driven more negatively, more holes will populate the interface and maintain the pinned condition. Since the voltage drop across the silicon remains fixed, any further increase in negative gate voltage beyond the pinned condition must be entirely reflected across the gate insulator. As the gate voltage is

NOISE SOURCES

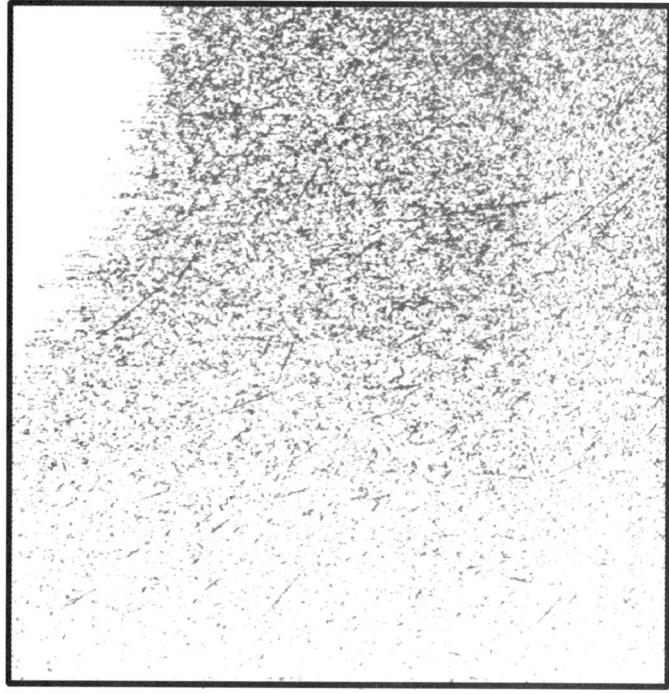

Figure 7.39(b) Galileo image showing proton tracks when encountering Earth's Van Allen belts.

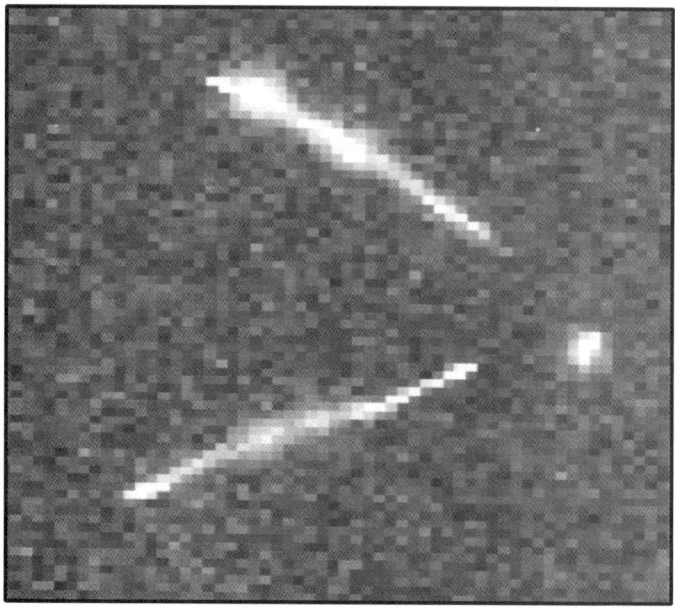

Figure 7.40 Compton electron ionization tracks.

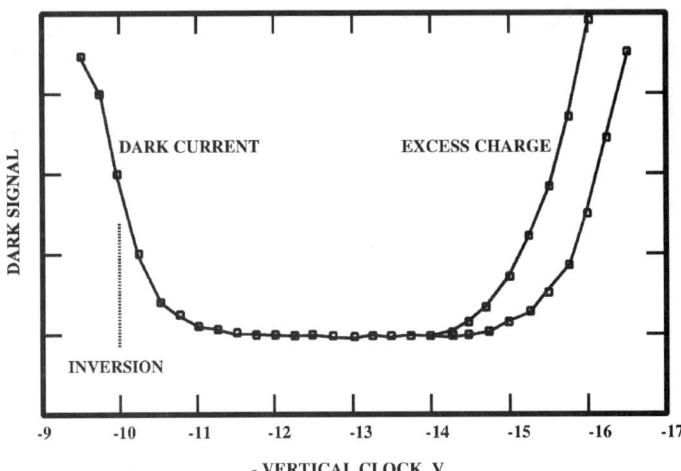

Figure 7.41 Excess charge as a function of negative clock drive, showing hysteresis effect.

driven further negatively, the oxide voltage drop will continue to increase until it shows signs of leakage current, a source of charge called "excess charge."

The onset of excess charge varies considerably among CCD manufacturers because of process differences in growing the gate dielectric. Also, the onset of excess charge usually exhibits a hysteresis effect. For example, if the gate voltage is biased into the excess-charge region, the excess charge measured will slowly decrease with time. A new equilibrium level is established for each new gate voltage applied. Presumably, excess charge carriers fill oxide states, causing leakage current to decline. Exposure to light will also change hysteresis characteristics.

Figure 7.41 plots dark charge as a function of negative clock drive for a Galileo virtual-phase CCD. The decrease in dark charge at -10 V is where the device inverts, suppressing surface dark current. The increase in dark charge at approximately -14 V is caused by excess-charge generation. The curve furthest to the right was generated when the device was initially taken into the excess-charge regime. The second curve was taken several minutes after the first measurement. The excess-charge breakpoint occurs at a lower voltage level because of hysteresis.

Figure 7.42(a) shows an excess-charge image generated by a Galileo CCD. This particular sensor received more implant when the ion beam was mistakenly turned off during one of the implant steps (the dark region in the image). The image was taken when bias to the gate was slightly beyond the onset of excess charge for the lighter region seen. Figure 7.42(b) shows a spurious-charge image for the same CCD. Here the device is clocked with a small voltage swing that does not transfer charge. Instead, spurious charge builds up in the pixels without charge movement. Note that the spurious-charge image is the negative of the excess-charge image.

Noise sources

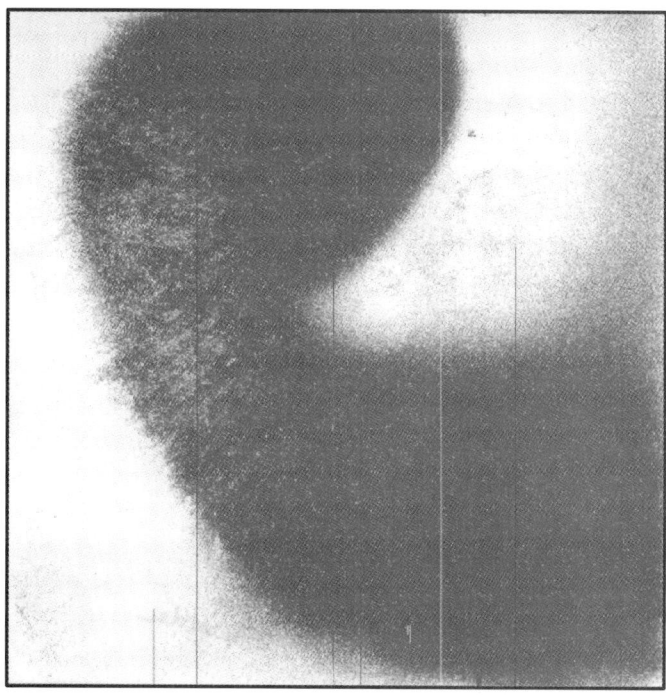

Figure 7.42(a) Excess-charge image generated by a virtual-phase CCD.

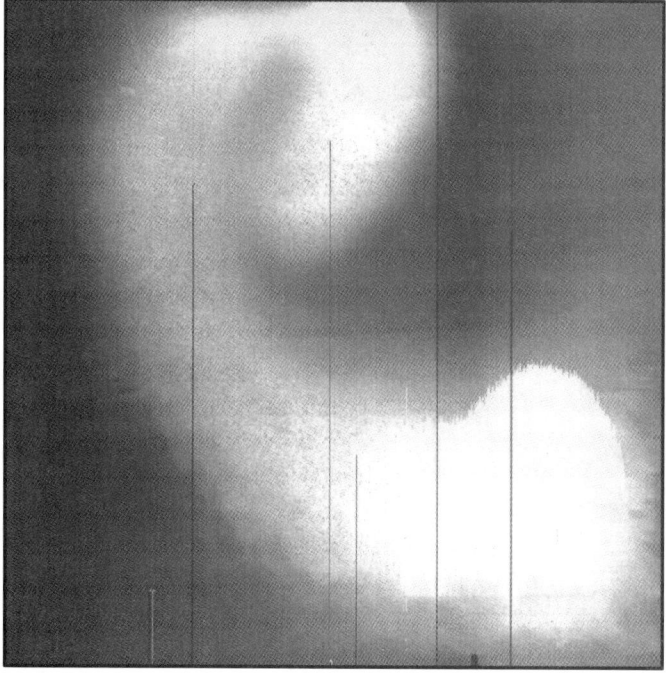

Figure 7.42(b) Spurious-charge image for the same CCD as Fig. 7.42(a).

7.1.9 COSMETIC DEFECTS

Small arrays (< 1000 × 1000) procured today are generally cosmetically clean. This was not the case in the past when the CCD was under development. For example, for many years only a few cosmetically clean CCDs were available to the astronomical community. Most sensors typically exhibited cosmetic defects. For example, the image shown in Fig. 7.43 was taken from a 800 × 800 backside-illuminated CCD that was used extensively for several years in a camera system. Even today, very large area array groups purchase defective sensors with cosmetic problems simply because they are more affordable. Fortunately, most cosmetic defects affect only a column or two and are therefore still useful for scientific purposes. For example, one defective column of a 1000 × 1000 CCD represents only a 0.1% loss in terms of imaging science. Curiously, inexpensive commercial CCD applications are more demanding on cosmetic quality where the goal is to generate "pretty" pictures. Here, cosmetic defects are not acceptable.

White and dark column blemishes are the most common cosmetic defects. White blemishes are produced by a pixel that generates an abnormally high level of dark charge. The problem can be associated with a lattice or process defect or a high impedance substrate short. Figure 7.35(a) shows a 1800-sec 800 × 800 dark current image with several bright column blemishes (for this image, charge moves

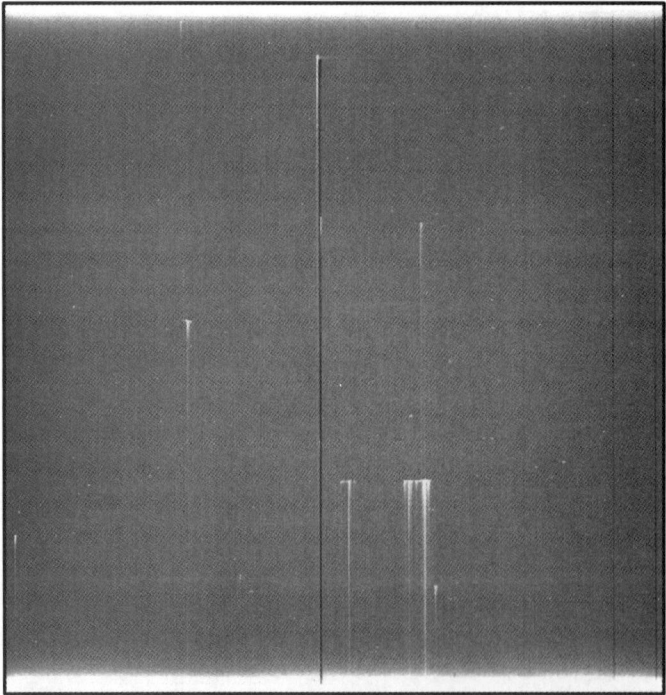

Figure 7.43 Image showing various cosmetic problems.

to the upper left-hand corner). Some bright defects produce enough charge to reach surface full well during integration. When readout commences, a long tail is seen behind the defect as trapped charge is released from the Si–SiO$_2$ interface (the device is not inverted in order to preserve the effect). The column blemish on the right-hand side that extends completely across the image is caused by a pixel defect located on line 600 (i.e., 200 lines from the bottom of the image). The defect generates enough charge to be seen throughout the readout cycle (i.e., the charge level generated during a line shift time is greater than the read noise floor). There is also a weaker column blemish to the right of this defect that extends across the image.

Figures 7.42(b) and 7.43 show dark column blemishes. The defects are referred to as block channels, where charge is either lost or held back by a troubled pixel. Other dark blemishes can extend completely across the CCD. Two of these defects are seen to the far right of Fig. 7.43. Usually this blemish type is related to a trap at the horizontal/vertical register interface region; here, charge is trapped during the vertical-to-horizontal transfer period. Once trapped, charge is slowly emitted during the horizontal shift time, emptying the trap for the next line that comes in. The end result is a dark column that has a lower charge level from line to line.

Figure 7.43 also shows three bright horizontal defects associated with a surface problem that generates dark current. Similar to Figure 7.35(a), the tail that follows behind the defect is charge escaping the Si–SiO$_2$ interface. As with other defects with tails, this form of residual image usually disappears when the device is taken into inversion.

7.1.10 BLEM SPILLOVER

Blem spillover is a term applied when a large column blemish spills charge into the horizontal register during readout. Depending on vertical clocking rate, a blemish can saturate a column in which it is located. Since the blemish charge must go somewhere, charge will "spill over" into the horizontal register. If spillover occurs at a constant rate, the output charge level will appear as an offset in each pixel. However, column blemishes usually generate charge erratically, introducing noise in the horizontal register as pixels are read out.

Figure 7.44 shows the blem spillover problem for a Galileo CCD. At point 1, charge from a large column blemish is transferred into the horizontal register. In region 2, blem charge is held back from the horizontal register. At point 3, charge from the blemish saturates the column and spills over into the horizontal register, increasing the signal and noise levels readout. In region 4, charge continues to spill until the next vertical transfer, completing the cycle. Each line readout shows the same signature.

CCDs with antiblooming protection are protected from blem spillover. CCDs without antiblooming can be somewhat protected by allowing charge to bloom into a unused horizontal register. For example, if the upper horizontal register is not used, then the array transfer gate next to it can be biased at a higher voltage

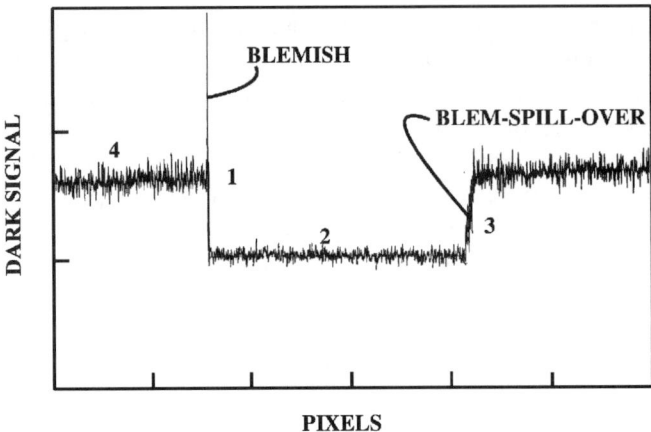

Figure 7.44 Line trace of horizontal blem spillover.

than the lower array transfer gate. This will allow blemish charge to diffuse and dump into the upper register and not disturb the lower register. However, as with catastrophic pixels, it is best to replace the CCD if problems like this develop, because blem spillover will likely worsen with time.

7.1.11 SEAM NOISE

High-speed CCD applications typically use multiple amplifiers for readout. This parallel readout scheme, although very effective, generates a serious problem called seam noise. Each on-chip (and off-chip) amplifier employed exhibits its own offset level, gain constant (i.e., e⁻/DN) and nonlinearity characteristics. Therefore, because the channels are never perfectly balanced, "seams" are seen in the image between each section readout. Unfortunately, the eye is very sensitive to seams, especially when the CCD is stimulated with a flat-field light source. The problem becomes even more pronounced for moving pictures. As we will show below, one can detect a signal difference between channels of only 0.2%. It is therefore important to quantify the variables involved to eliminate the effect below this level. This is accomplished by calculating the S/N for the seam signal generated by two amplifiers [refer to Fig. 7.45(a)].

The seam signal is the difference of signal levels for each channel. That is,

$$S_S(\text{DN}) = [S(e^-)a_1 + O_1] - [S(e^-)a_2 + O_2], \tag{7.61}$$

where $S_S(\text{DN})$ is the seam signal (DN), $S(e^-)$ is the flat-field signal level, a_1 is the gain of channel 1 (DN/e⁻), a_2 is the gain of channel 2 (DN/e⁻), O_1 is the offset for channel 1 (DN) and O_2 is the offset for channel 2 (DN).

Noise sources

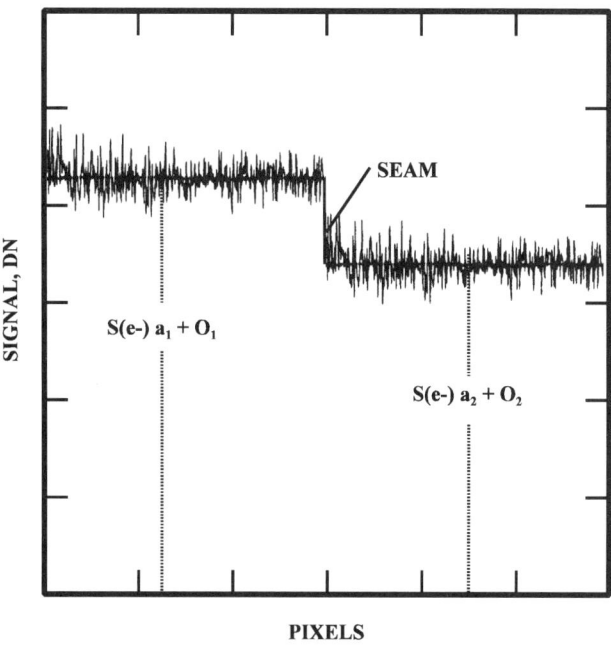

Figure 7.45(a) Illustration showing seam noise for two readout channels.

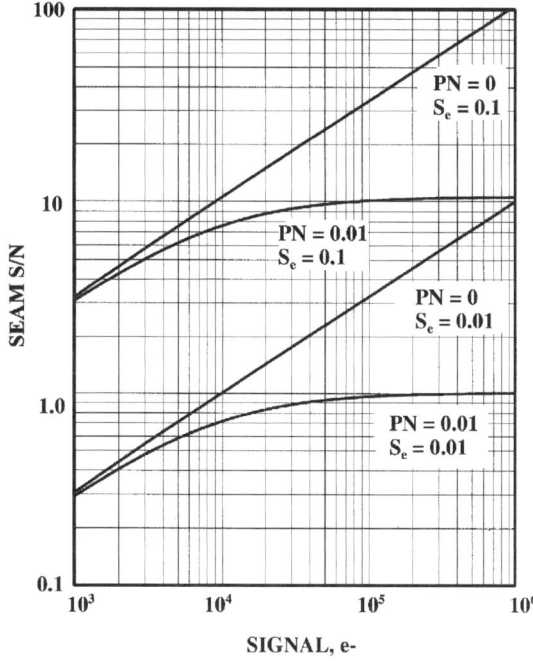

Figure 7.45(b) Seam S/N as a function of signal, pixel nonuniformity and seam factor.

The noise is the average noise level for the regions. That is,

$$N_S(\text{DN}) = \frac{a_1 + a_2}{2}\{S(e^-) + [S(e^-)\sigma_{PN}]^2 + \sigma_R(e^-)^2\}^{1/2}, \qquad (7.62)$$

where the terms under the square root bracket are the shot noise, pixel-nonuniformity noise and read noise respectively.

We now assume that the offset difference between channels is completely removed using extended pixel information. With that assumption we ratio Eqs. (7.61) and (7.62) to calculate S/N, yielding

$$\text{S/N} = S(e^-)\frac{2S_e}{2-S_e}\frac{1}{\{S(e^-) + [S(e^-)\sigma_{PN}]^2 + \sigma_R(e^-)^2\}^{1/2}}, \qquad (7.63)$$

where S/N is the signal-to-noise of the seam signal. S_e is defined as the seam factor and is defined as

$$S_e = 1 - \frac{a_1}{a_2}. \qquad (7.64)$$

When pixel nonuniformity is removed by flat fielding and the read noise is negligible, this equation reduces to

$$\text{S/N} = S(e^-)^{1/2}\frac{2S_e}{2-S_e}. \qquad (7.65)$$

It is important to note from this equation that the seam S/N increases by the square root of average signal. Therefore, high-full-well devices are most vulnerable to the seam noise problem.

Example 7.24

Find the seam S/N given $S(e^-) = 10^6$ e$^-$, $\sigma_{PN} = 0.01$, $\sigma_R(e^-) = 10$ e$^-$ and $S_e = 0.01$ (i.e., 1% gain difference). Also, determine the S/N after pixel nonuniformity is removed.

Solution:

The quadrature sum of shot noise, pixel-nonuniformity noise and read noise is

$$N = \left[10^6 + \left(10^6 \times 0.01\right)^2 + 10^2\right]^{1/2},$$

$$N = 1.004 \times 10^4 \text{ rms e}^-.$$

The S/N of the seam from Eq. (7.63),

NOISE SOURCES 683

$$\text{S/N} = 10^6 \times (2 \times 0.01)/(2 - 0.01) \times 1/1.004 \times 10^4,$$

$$\text{S/N} = 1.00102.$$

The noise when pixel nonuniformity is removed is

$$N = \left(10^6 + 10^2\right)^{1/2},$$

$$N = 10^3.$$

From Eq. (7.63),

$$\text{S/N} = 10^6 \times (2 \times 0.01)/(2 - 0.01) \times 1/10^3,$$

$$\text{S/N} = 10.$$

For this example, pixel nonuniformity essentially hides seam noise (S/N = 1). When nonuniformity noise is removed, the seam is easily seen (S/N = 10).

Figure 7.45(b) plots seam S/N as a function of signal for seam factors of $S_e = 0.1$ and 0.01 with and without a pixel nonuniformity of 1%. Note that seam S/N becomes constant when pixel nonuniformity noise dominates. However, when nonuniformity noise is removed, the S/N increases by the square-root of signal until full well is exceeded.

Seam noise is difficult to see for S/N < 0.2, even when images are contrast enhanced. Figure 7.46 shows a seam image with different S/N; note that the seam begins to disappear for S/N < 0.5.

Seam removal is a challenging image processing problem. As indicated above, offset, gain and nonlinearity differences between channels all contribute to the amount of seam observed. Offset is easiest to remove by using extended pixel information. If desired, offset can be removed real time as images are taken, by averaging and storing extended pixel values into a lookup table that is constantly updated for "on-the-fly" pixel processing.

Several techniques are used to remove gain and nonlinearity differences between channels. One approach is to illuminate the CCD with a flat-field light source and calibrate the gain of each channel. This process is done at different light settings in order to measure gain nonlinearity differences as a function of signal level. Another method is to electrically inject charge into the array. Again, different charge levels are injected to measure nonlinearity characteristics. One can also use dark current to measure gain coefficients of the channels as a function of integration time. The camera system is taken offline to load lookup tables with gain and linearity information for real-time processing.

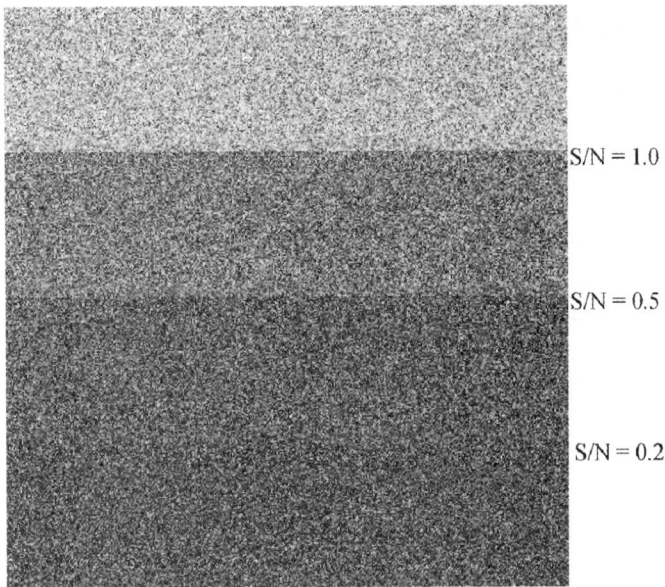

Figure 7.46 Image showing seam noise at sub unity S/N.

7.2 OFF-CHIP NOISE SOURCES

This section briefly reviews off-chip noise sources that can degrade the charge measurement process. We will limit our discussions to five external noise sources: (1) light leak noise, (2) preamplifier noise, (3) ADC quantizing noise, (4) clock jitter noise and (5) electromagnetic interference (EMI).

7.2.1 LIGHT LEAK

Light leak always seems to find its way into CCD images; it is a common problem for all CCD users. For example, Fig. 7.47 shows a light leak source that caused difficulties during a dark current test. The problem was caused by a soldering iron that was left on in the test lab. The source of light was invisible to the naked eye and therefore difficult to locate. After its discovery, the iron was brought up to the CCD camera and photographed. The dark thermal bands seen in the image rapidly move as the soldering iron tries to maintain a constant tip temperature.

7.2.2 PREAMPLIFIER NOISE

It is important that noise generated by pre- and postamplifiers do not add to the CCD read noise floor. When amplifier noise and CCD read noise are added in quadrature, the total noise is

NOISE SOURCES

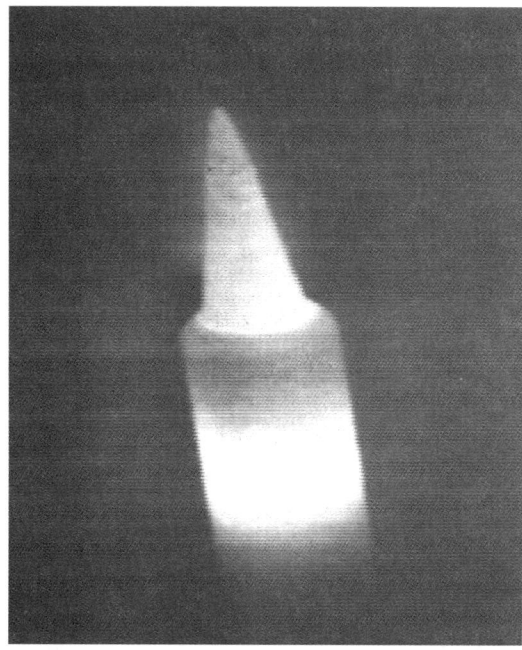

Figure 7.47 Light-leak image of a soldering iron.

$$N_T(e^-) = \frac{B^{1/2}}{S_V}\left[W_{CCD}(f)^2 + W_{A1}(f)^2 + \frac{W_{A2}(f)^2}{A_{PA}}\right]^{1/2}, \qquad (7.66)$$

where A_{PA} is the preamplifier gain (V/V), B is the equivalent noise bandwidth (Hz), S_V is the node sensitivity (V/e$^-$), and $W_{CCD}(f)$, $W_{A1}(f)$, and $W_{A2}(f)$ are the white-noise figures of the CCD output amplifier, preamplifier and postamplifier, respectively (V/root-Hz). We assume the CCD output amplifier gain is unity.

Example 7.25

Find the total noise generated by the CCD, preamplifier, and postamplifier. Assume $W_{CCD}(f) = 10$ nV/root-Hz, $W_{A1}(f) = 2.5$ nV/root-Hz, $W_{A2}(f) = 20$ nV/root-Hz, $B = 2.5 \times 10^5$ Hz, $S_V = 10^{-6}$ V/e$^-$, and $A_{PA} = 22$ V/V. Also, find the noise associated with the CCD, preamplifier, and postamplifier separately.

Solution:
From Eq. (7.66),

$$N_T(e^-) = (2.5 \times 10^5)^{1/2}/10^{-6} \times \left[(10^{-8})^2 + (2.5 \times 10^{-9})^2 + (2 \times 10^{-8})^2/22\right]^{1/2},$$

$$N_T(e^-) = 5.58 \; e^- \; \text{rms}.$$

CCD, preamplifier and postamplifier noise levels referred to the sense node are 5, 1.25, and 0.45 e⁻, respectively. Note that the postamplifier's noise is suppressed by the preamplifier gain factor of 22.

The example above shows that the noise contribution of the off-chip amplifiers can be made negligible. The front-end design of the preamplifier typically utilize a low-noise JFET or bipolar transistor. Off-the-shelf devices with white-noise figures less than 1 nV/root-Hz can be procured, an order of magnitude lower than a typical low-noise CCD amplifier. It is interesting to note that preamplifier noise requirements have been relaxed over the years because CCD white noise has increased in exchange for higher sensitivity, as discussed in Chapter 6.

7.2.3 ADC Quantizing Noise

Figure 7.48 shows the transfer characteristics for an analog-to-digital converter (ADC). Output DN as a function of input voltage is plotted. Each step in the staircase response is equal to

$$A_2 = \frac{N_{ADC}}{V_{ADC}}, \quad (7.67)$$

where A_2 is the conversion gain (DN/V), V_{ADC} is the input voltage range of the ADC (V), and N_{ADC} is the number of DN levels generated by the ADC. This number is

$$N_{ADC} = 2^{N_{BITS}} - 1, \quad (7.68)$$

where N_{BITS} is number of bits provided by the ADC (bits).

Example 7.26

What is the conversion gain for an 8-bit ADC that covers a 0.5-V input range?
Solution:
From Eq. (7.67),

$$A_2 = (2^8 - 1)/1,$$

$$A_2 = 510 \text{ DN/V}.$$

NOISE SOURCES

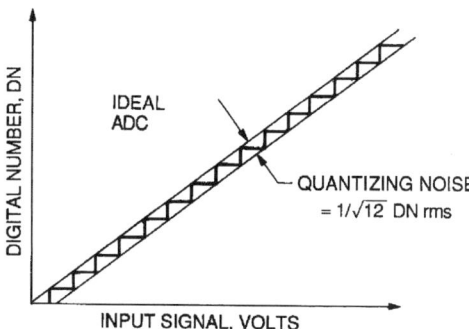

Figure 7.48 ADC transfer function and corresponding quantizing noise.

When a pixel value is digitized, an uncertainty in its value is introduced. This uncertainty is called ADC quantizing noise. Quantizing noise is found by calculating the rms value of the staircase response shown in Fig. 7.48 about a perfect ramp input. In general, the rms value of any cyclic function is so defined that

$$\text{rms} = \left\{ \frac{1}{T} \int_0^T [f(x)]^2 dx \right\}^{1/2}, \quad (7.69)$$

where T is the cycle period.

ADC quantizing noise is found using this equation for a triangular wave with amplitude DN/2:

$$QN = \frac{2}{T} \left[\int_0^{T/4} f_1(x)^2 dx + \int_{T/4}^{T/2} f_2(x)^2 dx \right]^{1/2}, \quad (7.70)$$

where

$$f_1(x) = \left(\frac{2\text{DN}}{T} \right) x, \quad 0 < x < \frac{T}{4} \quad (7.71)$$

and

$$f_2(x) = \left(\frac{-2\text{DN}}{T} \right) x + \text{DN}, \quad \frac{T}{4} < x < \frac{T}{2}. \quad (7.72)$$

Squaring these terms equals

$$f_1(x)^2 = \left(\frac{4\text{DN}^2}{T^2} \right) x^2, \quad (7.73)$$

and

$$f_2(x)^2 = \left(\frac{4\text{DN}^2}{T^2} \right) x^2 - \left(\frac{4\text{DN}^2}{T} \right) x. \quad (7.74)$$

Performing the integration, quantizing noise is found to be

$$QN(\text{DN}) = 12^{-1/2}. \tag{7.75}$$

It is important to note that quantizing noise is constant for a given camera gain constant $K(\text{e}^-/\text{DN})$ when expressed in DN units. It is useful to convert quantizing noise into noise electrons as

$$QN(\text{e}^-) = 0.288675 \, K(\text{e}^-/\text{DN}). \tag{7.76}$$

Example 7.27

Find the quantizing noise for a camera gain of $K = 20 \text{ e}^-/\text{DN}$.
Solution:
From Eq. (7.76),

$$QN(\text{e}^-) = 0.28865 \times 20,$$

$$QN(\text{e}^-) = 5.77 \text{ e}^- \text{ rms}.$$

The total noise when CCD read noise is added in quadrature with quantizing noise is

$$N(\text{e}^-) = \left\{ [QN(\text{e}^-)]^2 + \sigma_R(\text{e}^-) \right\}^{1/2}. \tag{7.77}$$

Example 7.28

Find the total noise generated by a CCD camera that exhibits a gain of $20 \text{ e}^-/\text{DN}$. Assume a CCD read noise of 5 e^-.
Solution:
From Eq. (7.77),

$$N(\text{e}^-) = \left[(0.28867 \times 20)^2 + 5^2 \right]^{1/2},$$

$$N(\text{e}^-) = 7.64 \text{ e}^- \text{ rms}.$$

Figure 7.49 shows a collection of images for a sensor that is optically stimulated linearly (i.e., a ramp stimulus). Each image is generated with a different camera

NOISE SOURCES

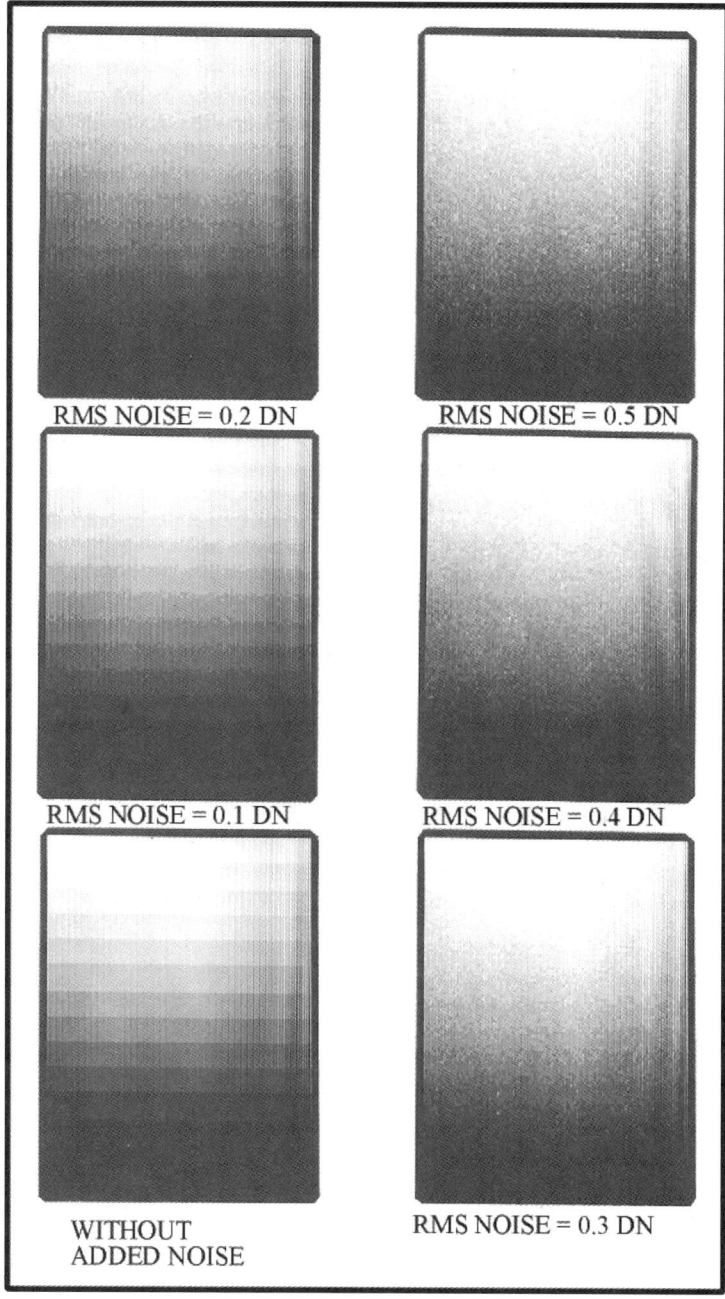

Figure 7.49 Images showing quantizing noise as random read noise increases.

gain setting. The first image in the lower left hand corner uses a low-gain state such that CCD noise is negligible compared to quantizing noise, resulting in a "staircase" of gray levels. The gain for the second image is set such that the read noise is $R(\text{DN}) = 0.1$ DN rms. The net noise in this image is the quadrature sum of the quantizing noise and read noise or 0.30 DN. Quantizing noise is still apparent in the image. Increasing the gain further suppresses quantizing noise until it is completely hidden by CCD noise when $R(\text{DN}) = 0.5$ DN. These results show that $K(\text{e}^-/\text{DN})$ should be about equal to the read noise to make quantizing noise negligible. That is,

$$K(\text{e}^-/\text{DN}) = R(\text{e}^-). \qquad (7.78)$$

Recall that the number of discrete DN levels required to encode a full well signal is

$$N_{ADC} = \frac{S_{FW}(\text{e}^-)}{K(\text{e}^-/\text{DN})}, \qquad (7.79)$$

where $S_{FW}(\text{e}^-)$ is the full well (e$^-$). When the camera gain is set equal to the read noise for quantizing noise reasons, this equation becomes

$$N_{ADC} = \frac{S_{FW}(\text{e}^-)}{R(\text{e}^-)}, \qquad (7.80)$$

which is simply equal to the dynamic range of the CCD.

The number of bits required from an ADC is

$$N_{BITS} = \frac{\log N_{ADC}}{\log 2}. \qquad (7.81)$$

Example 7.29

Find the number of DN levels required to encode a CCD that exhibits a full well capacity of 100,000 e$^-$ and read noise of 3 e$^-$. Also, determine the numbers of bits required.

Solution:
From Eq. (7.80),

$$N_{ADC} = 100{,}000/3,$$

$$N_{ADC} = 33{,}333 \text{ DN}.$$

The number of bits from Eq. (7.81) is

$$N_{BITS} = \log 33{,}333 / \log 2,$$

$$N_{BITS} = 15 \text{ bits}.$$

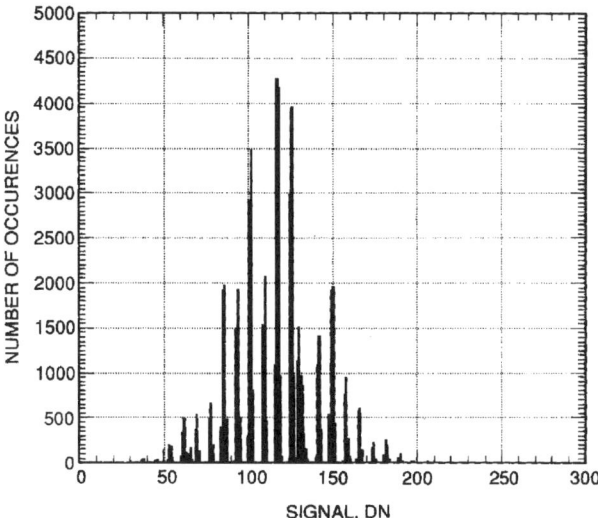

Figure 7.50 ADC noise histogram showing missing bits related quantizing noise.

The discussion above assumes an ideal ADC response (i.e., Fig. 7.48). ADCs can exhibit bit-weighting characteristics where certain DN levels are preferenced more than others. Less frequent but still common are ADCs with missing codes where some DN values are totally absent. Such behavior leads to a distorted staircase transfer function and greater quantizing noise. Bit weighting can be identified quickly by generating a noise histogram as shown in Fig. 7.50. Ideally, a smooth Gaussian distribution should result, in which each bit is equally weighted. A "comb"-like pattern as shown indicates a bit-weighting problem. Bit weighting can be seen even when the ADC is in specification (i.e., $\pm 1/2$ LSB). In this case the problem is usually caused by clock feed-through transients present at the input of the ADC. To avoid this problem, clocks internal to the camera can be inhibited during the conversion period.

7.2.3.1 Square Rooter

Discussions above show that optimum read noise encoding without quantizing noise takes place when $K(\text{e}^-/\text{DN}) = R(\text{e}^-)$. However, because the ADC is a linear circuit, excessive digitization occurs when encoding the signal shot noise. For example, a read noise of 10 e^- rms is equivalent to 1 DN rms, assuming $K(\text{e}^-/\text{DN}) = R(\text{e}^-)$. A shot noise of 1000 e^- rms (i.e., signal of 1 million electrons) is encoded to 100 DN rms, a hundred times greater encoding than the read noise floor. Matching digitization for shot and read noise implies that the camera gain, $K(\text{e}^-/\text{DN})$, should increase by the square-root of the signal. We call this nonlinear encoding processor a "square-rooter"—a powerful compression technique when applied to shot-noise-limited detectors such as a CCD. For example, this encoding scheme is utilized by the Cassini camera to reduce the number of bits transmitted between Saturn and Earth. For Cassini, 12 linear bits for each pixel are compressed to 8 bits using an onboard square-rooter lookup table. We will now look at this processing scheme a little closer.

The output signal from a square rooter is simply

$$S(\text{DN})_{SR} = \left[\frac{S(e^-)}{K(e^-/\text{DN})} \right]^{1/2}, \quad (7.82)$$

where $K(e^-/\text{DN})$ is the camera gain constant (e^-/DN).

The effective gain constant of the square-rooter is therefore

$$K(e^-/\text{DN})_{SR} = \frac{S(e^-)}{[S(\text{DN})]_{SR}} = [K(e^-/\text{DN})S(e^-)]^{1/2}. \quad (7.83)$$

The output noise from the square-rooter is

$$N(\text{DN})_{SR} = \frac{[S(e^-)]^{1/2}}{K_{SR}(e^-/\text{DN})}. \quad (7.84)$$

Using Eq. (7.83), this equation reduces to

$$N(\text{DN})_{SR} = K(e^-/\text{DN})^{-1/2}. \quad (7.85)$$

To avoid quantizing noise problems, the output noise from the square-rooter should be approximately 1 DN rms. Therefore, the camera gain constant is adjusted to yield $K(e^-/\text{DN}) = 1\ e^-/\text{DN}$.

Example 7.30

Determine the output signal from a square-rooter processor for a $10^5\ e^-$ signal. Assume $K(e^-/\text{DN}) = 2\ e^-/\text{DN}$. Also, determine the effective gain constant at this signal level.

Solution:
The square-rooter output signal from Eq. (7.82) is

$$S(\text{DN})_{SR} = \left(10^5/2\right)^{1/2},$$

$$S(\text{DN})_{SR} = 223\ \text{DN}.$$

The square-rooter output noise from Eq. (7.85) is

$$N(\text{DN})_{SR} = (1/2)^{1/2},$$

$$N(\text{DN})_{SR} = 0.707\ \text{DN rms}.$$

NOISE SOURCES

The effective gain constant from Eq. (7.83) at a 10^5 e$^-$ signal level is

$$K(\text{e}^-/\text{DN})_{SR} = \left(10^5 \times 2\right)^{1/2},$$

$$K(\text{e}^-/\text{DN})_{SR} = 447 \text{ e}^-/\text{DN}.$$

Figure 7.51 shows how the square-root conversion process works with the photon transfer curve. The top photon transfer curve represents conventional linear encoding. This simulation assumes a camera gain constant of $K = 1$ e$^-$/DN and a read noise floor of 40 e$^-$ rms. The signal stops at 65,536 DN because the simulation assumes a 16-bit ADC. The middle curve is generated by taking the DN pixel values of the first curve and running them through a square-root lookup table (we will show how this table is generated below). As can be seen, a signal of 65,536 DN is mapped to 256 DN (i.e., $(65{,}536)^{1/2}$). The noise level within the shot noise regime is a constant of 1 DN rms as dictated by Eq. (7.85). The lower curve is generated by taking the square-root curve and mapping it back into the linear domain with the same lookup table. The top and lower curves are identical.

As with linear encoding, it is desirable to maintain a noise level equal to or greater than 1 DN rms to hide quantizing noise. For example, the top curve of Fig. 7.52 is a square-root output for the same data as Fig. 7.51, using a camera gain constant of $K_{LIN} = 3$ e$^-$/DN. This gain forces noise values below the 1 DN rms level into the quantizing noise regime. Note that quantizing noise disrupts the photon transfer process in the shot noise region. The second plot is a magnified view of the top curve, showing the quantizing errors introduced. The square-root plot is then remapped into the linear domain as shown in the third curve, which also exhibits quantizing problems. The bottom curve of Fig. 7.52 includes pixel nonuniformity in the mapping process for a camera gain constant of 4 e$^-$/DN. Again, quantizing noise is seen in the shot noise regime. However, pixel nonuniformity noise hides quantizing noise as the signal increases.

Nonlinear encoding schemes such as a square-rooter are difficult to implement into actual hardware. Analog logarithmic amplifiers do not exhibit a perfect response, leading to camera linearity problems when CCD digital data is mapped back into the linear domain. Also, nonlinear circuits are inherently slow and show instabilities in offset control. The latter characteristic is important because an offset change implies a gain change that must be taken into account when reducing CCD data. A compromise to these problems is to encode the CCD linearly and then employ a digital square-root lookup table to compress bits. This was the scheme used in Figs. 7.51 and 7.52 and by Cassini. For example, the output of a 12-bit converter can in theory be mapped into 6 bits. The following formula is used to generate a square-root lookup table.

$$\text{DN}_{SR} = (\text{DN}_{LIN})^{1/2} 2^{(Y - \frac{N_{BIT}}{2})}, \qquad (7.86)$$

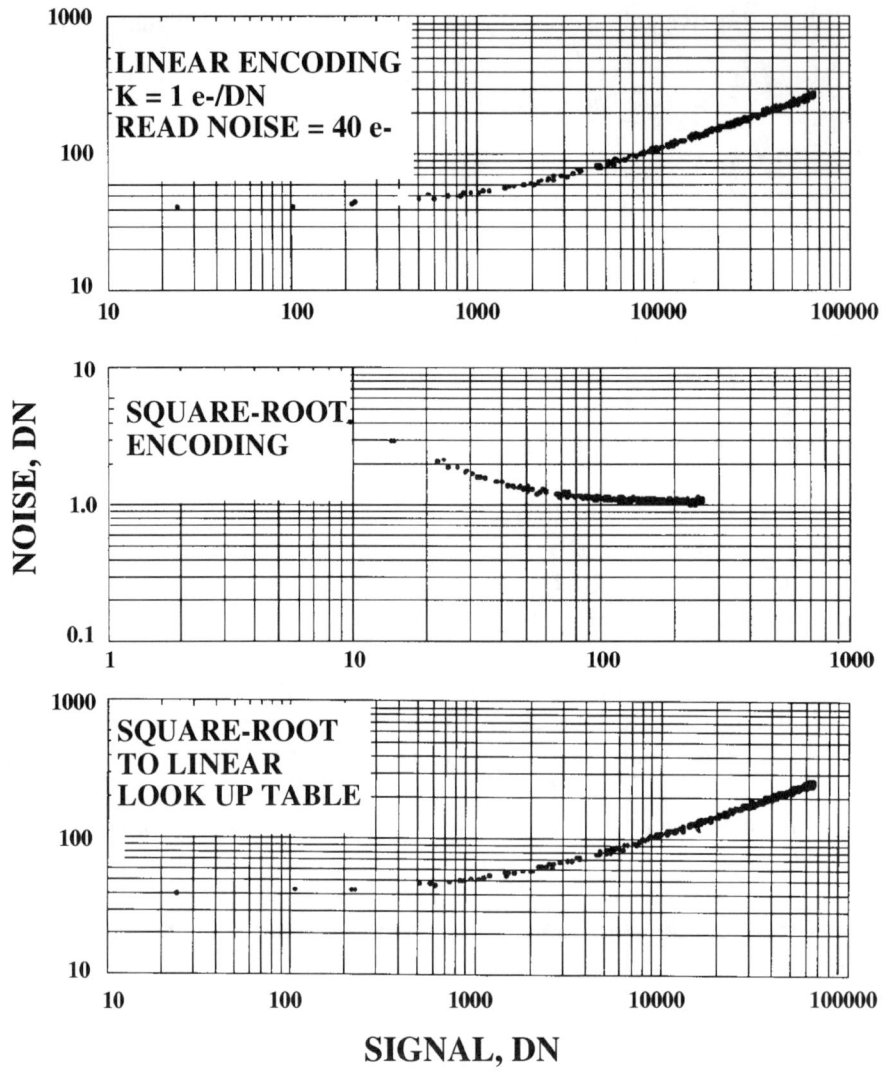

Figure 7.51 Photon transfer curves taken through the square-root process.

where DN_{SR} is the square-root DN number for a Y-bit square-rooter output using a linear N-bit ADC. DN_{LIN} is the linear DN value to be converted. Table 7.1 presents a square-root lookup table in mapping a linear 8-bit ADC to 4, 5, 6 and 7 bits by square-rooting.

Example 7.31

Map the square-root value of 253 DN from a linear ADC into a 7-bit lookup table.

NOISE SOURCES 695

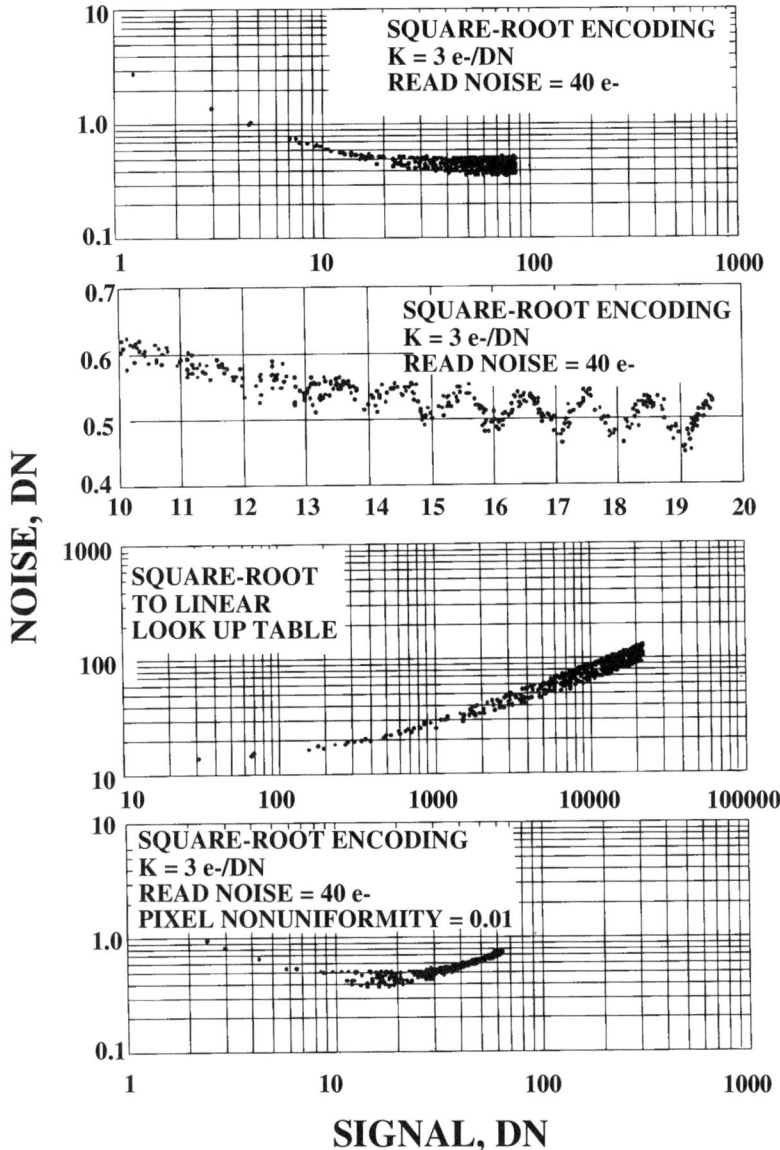

Figure 7.52 Photon transfer curves showing quantizing noise generated by the square-root process.

Solution:
From Eq. (7.86),

$$\mathrm{DN}_{SR} = (253)^{1/2} \times 2^{(7-8/2)},$$

$$\mathrm{DN}_{SR} = 127.$$

Table 7.1 Square-root look up table.

Linear 8 Bit	Square-root 4 Bit	Square-root 5 Bit	Square-root 6 Bit	Square-root 7 Bit
256	16	32	64	128
255	16	32	64	128
254	16	32	64	127
253	16	32	64	127
252	16	32	63	127
251	16	32	63	127
250	16	32	63	126
249	16	32	63	126
...	...	...	...	...
240	15	31	62	124
230	15	30	61	121
220	15	30	59	119
210	14	29	58	116
40	6	13	25	51
30	5	11	22	44
20	4	9	18	36
10	3	6	13	25
5	2	4	9	18
4	2	4	8	16
3	2	3	7	14
2	1	3	6	11
1	1	2	4	8
0	0	0	0	0

The number of DN required by a square-root lookup table is

$$(N_{ADC})_{SR} = \frac{S_{FW}(e^-)}{K(e^-/DN)_{SR}}. \quad (7.87)$$

This equation is equivalent to

$$(N_{ADC})_{SR} = \left[\frac{S_{FW}(e^-)}{K(e^-/DN)}\right]^{1/2}. \quad (7.88)$$

From Eq. (7.81) the number of bits required by the square-rooter is

$$(N_{BITS})_{SR} = \frac{\frac{1}{2}\log\left[\frac{S_{FW}(e^-)}{K(e^-/DN)}\right]}{\log 2}. \quad (7.89)$$

Example 7.32

Determine the number bits required for a linear and a square-root encoder. Assume $S_{FW}(e^-) = 65{,}536\ e^-$ and $K = 1\ e^-/DN$.

Solution:
The number of bits required for linear encoding from Eq. (7.81):

$$N_{BITS} = [\log(65536/1)]/\log 2,$$

$$N_{BITS} = 16 \text{ bits.}$$

From Eq. (7.89), the numbers of bits required by a square-rooter is

$$(N_{BITS})_{SR} = 1/2 \log(65536)/\log 2,$$

$$(N_{BITS})_{SR} = 8 \text{ bits.}$$

Therefore, in theory, the square-rooter reduces the bit count from 16 bits to 8 bits. It is common practice to have more bits than what Eq. (7.89) specifies, for noise margin reasons. For example, Cassini compresses 12 bits into 8 bits.

7.2.4 CLOCK-JITTER NOISE

Jitter on the master clock that generates the clamp, output-summing well and sample pulses can introduce noise. This noise is called clock-jitter noise and is most conspicuous for camera systems whose master clock is not directly run off a crystal oscillator source. For example, phase-lock-loop clocks often exhibit clock-jitter problems. It is critical that the clamp-to-sample time remain fixed from pixel to pixel to a high degree of accuracy. Phase-jitter specifications for crystal-derived clocks are 0.001 (i.e., 0.1% rms clock period variation) and normally do not cause a noise problem.

The noise generated by clock jitter is given by

$$N_{CJ}(e^-) = [S(e^-) + S_{OFF}(e^-)][e^{(-t_s/\tau_D)} - e^{(-t_s(1+J)/\tau_D)}], \quad (7.90)$$

where $N_{CJ}(e^-)$ is the noise generated by clock jitter (rms e^-), $S(e^-)$ is the signal, $S_{OFF}(e^-)$ is the offset signal that accompanies the signal (i.e., clock feed-through), t_s is the sample-to-sample time, τ_D is the CDS dominant time constant and J is the clock jitter of the master clock.

Note that this noise source increases with signal and the amount of clock feed-through present. Clock jitter has become more pronounced in recent years because the clock feed-through level has been increasing due to sensitivity improvements.

Example 7.33

Find the clock-jitter noise assuming $S(e^-) = 10{,}000\ e^-$, $t_s = 2 \times 10^{-6}$ sec, $\tau_D = 10^{-6}$ sec, $S_V = 5\ \mu V/e^-$ and $J = 0.01$. The amount of clock feed-through at the output of the CCD is 100 mV. Also compare the clock-jitter noise to the shot noise.

Solution:
The clock feed-through signal is

$$S_{OFF} = 0.1/(5 \times 10^{-6}),$$

$$S_{OFF} = 2 \times 10^4\ e^-.$$

From Eq. (7.90),

$$N_{CJ} = (10^4 + 2 \times 10^4) \times \{\exp-(2 \times 10^{-6}/10^{-6}) - \exp-[(2 \times 10^{-6}) \times (1 + 0.01)]/10^{-6}\},$$

$$N_{CJ} = 80\ e^-.$$

The shot noise related to the 10,000-e^- signal is 100 e^-, and will therefore hide the clock-jitter noise.

Example 7.34

Find the clock-jitter noise for the example above without signal. Calculate for clock jitter specifications of 0.1, 0.01 and 0.001. Compare to a read noise of 3 e^- rms.

Solution:

$$N_{CJ} = (2 \times 10^4) \times \{\exp-(2 \times 10^{-6}/10^{-6}) - \exp-[(2 \times 10^{-6}) \times (1 + 0.01)]/10^{-6}\},$$

$$N_{CJ} = 53.6\ e^-\ (J = 0.01),$$

$$N_{CJ} = 5.4\ e^-\ (J = 0.001),$$

$$N_{CJ} = 0.54\ e^-\ (J = 0.0001).$$

7.2.5 ELECTROMAGNETIC INTERFERENCE

Figure 7.53 shows the presence of electromagnetic interference (EMI) embedded in the CCD read noise. The peak-to-peak EMI signal contained in the lower image is approximately equal to 6 e$^-$, which is lower than the CCD read noise floor of 10 e$^-$ (i.e., S/N = 0.6). The interference signal is readily apparent. The top image shows a reduction of interference to approximately 2 e$^-$ peak-to-peak. Its presence is still noticeable (S/N = 0.2). Not until S/N < 0.1 will the interference totally disappear. This was a similar conclusion reached for seam noise discussed above (i.e., Fig. 7.46).

EMI and related noise sources are the most difficult and frustrating noise problem to solve by the CCD camera engineer. Solutions are usually panic-oriented and derived by trial-and-error methods without much understanding given to the physical cause and effect. An excellent book that discusses the science of EMI is Henry Otts' *Noise Reduction Techniques in Electronic Systems*.[12] The book is recommended for engineers involved in designing a CCD camera for the first time. Such EMI guidance may prevent needing to tear a new CCD camera system completely apart because of improper EMI design practices.

This section briefly reviews simple methods used to control EMI problems. We refer to these noise control techniques as (1) decoupling, (2) shielding of conductors and (3) structure shielding.

7.2.5.1 Decoupling

Decoupling networks are used to reject interference frequencies that lie within the CDS passband. For example, RC decoupling circuits are used to filter power-supply lines to the CCD and sensitive analog circuits. The RC network is placed in close proximity to these circuits for the highest noise rejection possible. As a rule of practice, the RC time constant, τ_{RC}, of the decoupling filter should be at least 20 times greater than the dominant time constant, τ_D, of the CDS processor. That is,

$$\tau_{RC} = 20\tau_D. \tag{7.91}$$

The frequencies rejected by the decoupling filter are

$$f_{reject} > \frac{1}{2\pi\tau_{RC}}. \tag{7.92}$$

Example 7.35

Using the rule of thumb above, find R and C values to decouple the CCD voltage lines V_{REF}, V_{DD}, and V_{OTG}. Assume a CDS dominant time constant of 0.5×10^{-6} sec.

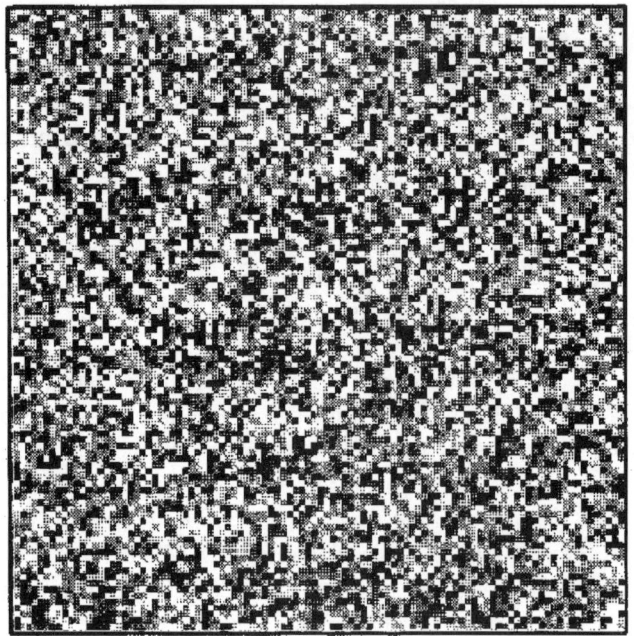

2e⁻ SIGNAL
10e⁻ RMS NOISE

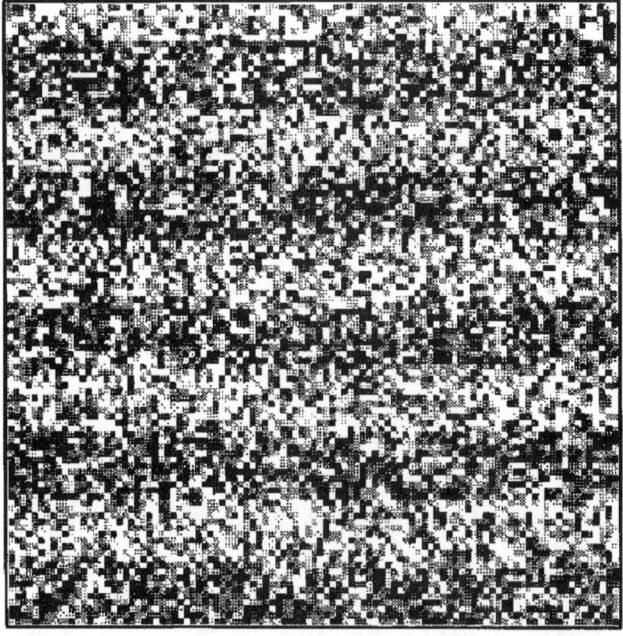

6e⁻ SIGNAL
10e⁻ RMS NOISE

Figure 7.53 Images showing EMI interference at S/N = 0.2 and 0.6.

Solution:
From Eqs. (7.91) and (7.92),

V_{REF}
: V_{REF} draws an insignificant amount of current, and therefore, a large series resistance can be employed in the decoupling network. Assuming 10 kΩ, a capacitance of 1000 pF will provide the desired decoupling time constant. The network will attenuate noise frequencies above 1.59×10^4 Hz.

V_{DD}
: Approximately 1 mA is required from V_{DD} for a single-stage output amplifier. A 500-Ω resistor in conjunction with a 0.02-μF capacitor will provide proper decoupling. The voltage drop across the resistor is 0.5 V for 1 mA. For multiple-output amplifier designs it is best to feed and decouple each amplifier separately to avoid cross talk between amplifiers as discussed below.

V_{OTG}
: V_{OTG} does not draw dc current, and therefore, a high series resistance can be employed. If 10 kΩ is assumed, a capacitance of 1000 pF is sufficient.

The type of decoupling capacitor is an important consideration. The maximum frequency of which a capacitor provides wide-band decoupling is limited by the internal inductance of the capacitor and its leads. For example, aluminum and tantalum electrolytic capacitors are limited to about 25 kHz. On the other hand, mylar capacitors are efficient to 1 MHz. For very high frequency rejection, polystyrene, mica, glass and ceramic capacitors are good to several hundred megahertz. However, high-frequency capacitors are small and cannot alone be used to decouple low-frequency-noise sources. It may be necessary to parallel two types of capacitors for optimum noise rejection.[13] Also, surface-mount capacitors exhibit better frequency characteristics because of their small physical size.

Example 7.36

What parallel filter capacitance arrangement can be used to decouple a 15-V preamplifier power supply voltage? Assume that the voltage drop across the decoupling resistor must be less than 1 V for a 10-mA bias current. Also, assume $\tau_D = 0.5 \times 10^{-6}$ sec.

Solution:
The largest decoupling resistance that can be used is

$$R = 1/0.01,$$

$$R = 100 \ \Omega.$$

The decoupling capacitance from Eq. (7.91) is

$$C_{LOW} = 20 \times (0.5 \times 10^{-6})/100,$$

$$C_{LOW} = 0.1 \; \mu F.$$

An electrolytic capacitor is appropriate here. However, as mentioned above, its inductance is limited to about 25 kHz. Therefore, a high-frequency mylar capacitor can be placed in parallel to reject frequencies above this frequency range. From Eq. (7.92),

$$C_{HIGH} = 0.159/[100 \times (2.5 \times 10^4)],$$

$$C_{HIGH} = 0.0636 \; \mu F.$$

Ferrite beads are also used to filter high-frequency noise from conductors. The small cylindrical-shaped iron beads are simply slipped over the conductor. The resistance presented to the noise is caused by energy loss in the ferrite material. The beads attenuate noise frequencies above ∼1 MHz. However, the reactive impedance of a bead is on the order of 100 Ω. Therefore, decoupling is only effective for low-impedance conductors such as power supply lines. Beads can also be beneficial to damp high-frequency ringing that occurs between the CCD and clock drivers.

7.2.5.2 Conductor Shielding

Wires running to and from the CCD may pick up noise by capacitive or inductive coupling action. Capacitive or electric coupling is due to interaction of electric fields between circuits, whereas inductive or magnetic coupling is due to interaction of magnetic fields. Capacitive coupling is usually the primary noise-coupling problem in CCD work because noise currents are small in comparison to noise voltages.

Although many arrangements can be analyzed to show how capacitive coupling occurs, the simplest circuit to analyze is the coupling that takes place between two parallel wires. It can be shown from Henry Otts' book that the noise voltage generated on a signal conductor from a noisy conductor is

$$V_{cross} = 2\pi f R_{SIG} C_{cross} V_{noise}, \qquad (7.93)$$

where V_{cross} is the cross talk voltage between the two conductors (V), f is the noise frequency (Hz), C_{cross} is the capacitance between the two conductors (F), R_{SIG} is load resistance between the ground and signal conductor (Ω), and V_{noise}

NOISE SOURCES

is the noise voltage on a noisy conductor (V). This equation assumes that the load resistance on the signal conductor is

$$R_{SIG} \ll \frac{1}{2\pi f(C_{cross} + C_{SIG})}, \qquad (7.94)$$

where C_{SIG} is the capacitance between the signal conductor and ground. This is the case for low-impedance sources such as power supply lines. Note that cross talk under this condition only passes high frequencies. We see this effect where only clock edges and high-frequency components are coupled onto a signal line.

For the case of high signal-line resistance, the voltage induced on the signal line is

$$V_{cross} = \left(\frac{C_{cross}}{C_{cross} + C_{SIG}}\right) V_{noise}. \qquad (7.95)$$

High-input lines include the CCD output and inputs to operational amplifier circuits. Note that in this case cross talk is independent of frequency. We see this effect when a full square wave (i.e., all frequencies) is coupled from one conductor to the other.

The capacitance between two conductors is given approximately as

$$C_{cross} = \frac{\pi \varepsilon_o L_{LEN}}{\ln(2L_{DIS}/L_{DIA})}, \qquad (7.96)$$

where C_{cross} is the capacitance between the two conductors (F), L_{LEN} is the length of the two conductors (cm), L_{DIS} is the distance between the two conductors (cm), and L_{DIA} is the diameter of the conductors (cm). The equation assumes $L_{DIS}/L_{DIA} > 3$.

Example 7.37

Find the capacitance between two 22-gauge conductors with $L_{LEN} = 1$ cm, $L_{DIS} = 2.54$ cm and $L_{DIA} = 6.35 \times 10^{-2}$ cm.
Solution:
From Eq. (7.96),

$$C_{cross} = 3.14 \times (8.85 \times 10^{-14})/[\ln(2 \times 2.54)/(6.35 \times 10^{-2})],$$

$$C_{cross} = 6.34 \times 10^{-14} \text{ F}.$$

Example 7.38

Find the noise voltage induced between two conductors. Assume $f = 1$ MHz, $V_{noise} = 1$ V, $C_{cross} = 6.34 \times 10^{-14}$ F and $C_{SIG} = 5 \times 10^{-12}$ F. Calculate for $R_{SIG} = 100\ \Omega$ and 1 MΩ.

Solution:
Low impedance ($R_{SIG} = 100\ \Omega$)
From Eq. (7.94),

$$R_{SIG} = 1/[2 \times 3.14 \times 10^6 (6.34 \times 10^{-14} + 5 \times 10^{-12})],$$

$$R_{SIG} = 3.14 \times 10^4\ \Omega.$$

Therefore, Eq. (7.93) applies, yielding

$$V_{cross} = 6.28 \times 10^6 \times 100 \times (6.34 \times 10^{-14}) \times 1,$$

$$V_{cross} = 3.98 \times 10^{-5}\ V.$$

High impedance ($R_{SIG} = 1$ MΩ)
Eq. (7.95) applies, yielding

$$V_{cross} = (6.34 \times 10^{-14})/(6.34 \times 10^{-14} + 5 \times 10^{-12}),$$

$$V_{cross} = 1.25 \times 10^{-2}\ V.$$

Capacitive noise coupling is suppressed simply by keeping signal lines as short as possible. If further rejection is required, shielding or twisting the signal lines with a ground return is necessary. Fortunately, there are only a few CCD signals that need shields if at all. By far the most sensitive CCD signal is the reset clock because it cannot be heavily RC decoupled, although some wave-shaping can be applied. Noise on the reset line will directly couple onto the sense node and CCD video signal. The sensitivity of the precharge line can be demonstrated by physically touching the signal while looking at the CCD video signal with an oscilloscope. The noise level will dramatically increase when this is done. Therefore, shielding the reset line may prove fruitful for some layout situations. The next potential candidate for shielding is the video signal between the CCD and preamplifier. Again, touching this line makes the video signal jump around erratically because of noise injection.

7.2.5.3 Structure Shielding

Electromagnetic radiation (EMR) sources can interfere with sensitive CCD signal lines. Power and telephone lines, lightning, radio and TV broadcast, car ignition, electric motors, and computers are just some EMR noise sources to contend with. A shielded structure that contains the CCD and sensitive electronics is the best arrangement to reduce EMR noise problems. The shield reflects incoming EMR waves and absorbs the nonreflected part. Absorption loss acts on electric and magnetic fields in a similar fashion. Reflective loss is large for electric fields but relatively small for magnetic fields. Therefore, we primarily rely on absorption loss to shield the CCD and support electronics.

When a electromagnetic wave passes through a shield medium, its amplitude decreases exponentially as

$$W = W_{INC} e^{-x/\delta}, \quad (7.97)$$

where W is the field strength in the shield, W_{INC} is the incident wavefront strength, x is distance into the shield (cm) and δ is the skin depth, that depth where the wave is attenuated to 37% of it original value. Skin depth is defined as

$$\delta = (\pi f \mu_r \sigma_r)^{-1/2}, \quad (7.98)$$

where f is frequency (Hz), μ_r the relative permeability (H/cm) and σ_r relative conductivity (mhos/cm) of the shield structure.

Example 7.39

Find the skin depth for an aluminum-shield structure at 1 MHz. Assume $\mu_r = 4\pi \times 10^{-9}$ H/cm and $\sigma_r = 3.55 \times 10^5$ mhos/cm.
Solution:
From Eq. (7.98),

$$\delta = [3.14 \times 10^6 \times (4\pi \times 10^{-9}) \times (3.55 \times 10^5)]^{-1/2},$$

$$\delta = 8.45 \times 10^{-3} \text{ cm } (84.5 \text{ μm}).$$

Equation (7.97) can be solved for x, yielding

$$x = \delta \ln\left(\frac{W_{INC}}{W}\right). \quad (7.99)$$

This equation gives a good estimate for the required shield thickness to attenuate EMR to a desired level.

Example 7.40

What is the aluminum shield thickness required to attenuate a 1 MHz EMR to 1%? Assume the skin depth is 84.5 μm.
Solution:
From Eq. (7.99),

$$x = [\ln(1) - \ln(0.01)] \times (8.45 \times 10^{-3}),$$

$$x = 3.89 \times 10^{-2} \text{ cm}.$$

7.2.6 GROUNDING

Common impedance noise is generated when signal currents from two different circuits flow through the same ground return. There are two principal grounding philosophies to consider. They are referred to as single-point grounding and multipoint grounding and are illustrated in Fig. 7.54.

Series single-point or "daisy-chain" grounding is undesirable when applied to CCD camera design. Although it is the simplest grounding scheme, the circuits share ground currents, which in turn causes cross talk problems. The parallel ground arrangement is a better choice because it isolates ground currents for

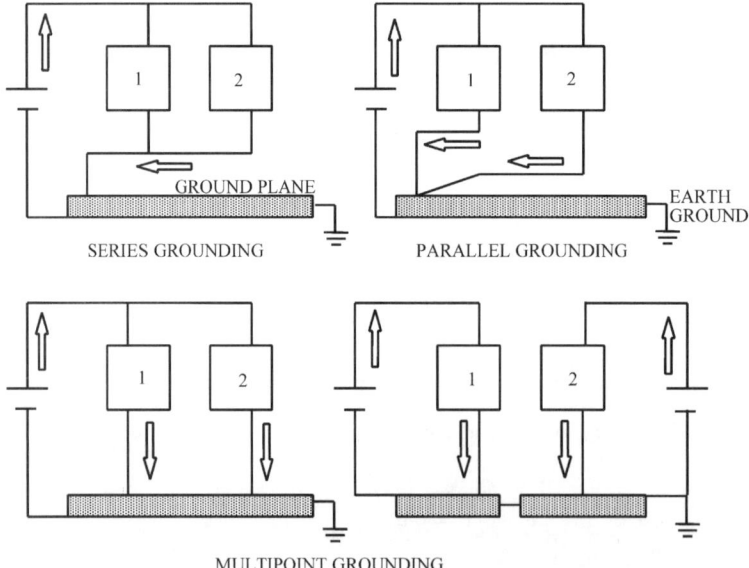

Figure 7.54 Single-point and multipoint grounding schemes.

NOISE SOURCES

each circuit. However, this grounding system is not satisfactory for high-frequency CCD circuits because the long leads to the ground plane exhibit high inductive reactance. Therefore, the ground returns "bounce" when the circuits become active. Also, inductive magnetic coupling and capacitance coupling produce cross talk if the ground returns are in close proximity.

The multipoint grounding system is the best grounding philosophy for CCD work. The arrangement offers good circuit isolation and low inductance returns. Connections are made to a ground plane of negligible resistance and inductance. The CCD shield structure normally acts as the ground plane for this purpose. As shown, the multipoint ground plane can also be divided into different sections to support different electronic functions. For example, the ground plane is usually separated to isolate logic and analog functions. The earth ground for each arrangement indicated is provided for electrostatic reasons and is not intended for circuit current flow.

Inductance associated with ground returns is an important problem for multipoint grounding layouts and must be minimized for high-speed applications. For example, the substrate return for the CCD is very sensitive to inductive reactance as noted in substrate bounce discussions in Chapter 5. The inductance associated with a ground return is given by

$$L_R = 1.96 \times 10^{-9} L_{LEN} \ln\left(\frac{4h}{s}\right), \qquad (7.100)$$

where L_R is the return's inductance (H/cm), h is distance above a ground plane (cm), s is the diameter of the conductor (cm), and L_{LEN} is the length of the conductor.

Example 7.41

Calculate the inductance of a 1-cm length of 22-gauge wire with a diameter of $s = 6.35 \times 10^{-2}$ cm that is located $h = 6.35 \times 10^{-1}$ cm above a ground plane.

Solution:
From Eq. (7.100),

$$L_R = (1.96 \times 10^{-9}) \times \ln[4 \times (6.35 \times 10^{-1})/(6.35 \times 10^{-2})],$$

$$L_R = 7.23 \times 10^{-9} \text{ H}.$$

The inductance reactance is

$$Z_L = 2\pi f_{MAX} L, \qquad (7.101)$$

where Z_L is the inductance reactance (Ω) and f_{MAX} is the maximum working frequency (Hz).

Example 7.42

Find the inductive reactance for the ground return described in Example 7.41 at a frequency of 10 MHz.
Solution:
From Eq. (7.101),

$$Z_L = 6.28 \times 10^7 \times (7.23 \times 10^{-9}),$$

$$Z_L = 0.454 \ \Omega.$$

For comparison, the dc resistance associated with a 22-gauge copper wire is 5.11×10^{-4} Ω/cm. Therefore, inductive reactance dominates at this frequency.

The maximum working frequency, f_{MAX}, for a ground return is defined as 10 times the 3-dB frequency for the fastest wave-shaping time constant, τ_{WS}, applied to the CCD. That is,

$$f_{MAX} = \frac{10}{2\pi\tau_{WS}}. \qquad (7.102)$$

The shortest wave-shaping time constant is typically associated with the horizontal register and reset clock. Recall from Chapter 5 that wave-shaping for a three-phase CCD horizontal register is related to pixel readout rate through

$$\tau_{WS} = \frac{1}{12 P_R}, \qquad (7.103)$$

where P_R is the pixel rate (pixels/sec). Substituting Eqs. (7.102) and (7.103) into Eq. (7.101) yields

$$Z_{MAX} = 120 P_R L. \qquad (7.104)$$

Combining this equation with Eq. (7.100) and solving for L_{LEN} yields

$$L_{LEN} = \frac{1}{2.35 \times 10^{-7} P_R \ln(4h/s)}. \qquad (7.105)$$

This equation estimates the maximum ground return length to maintain low inductive reactance for a specified pixel rate.

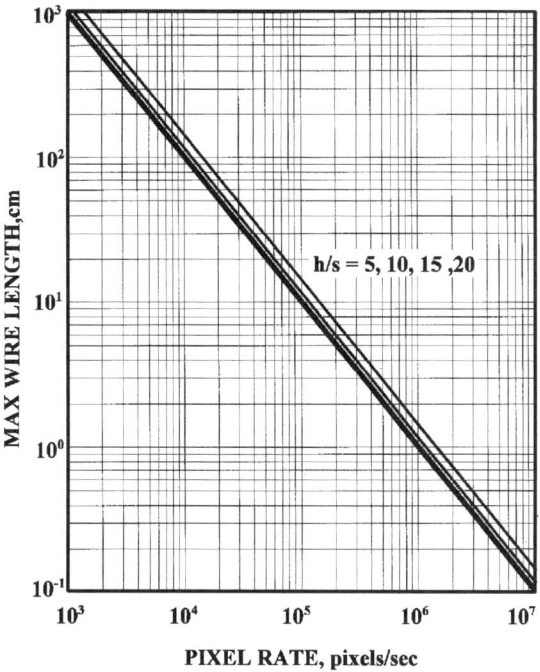

Figure 7.55 Maximum ground return length as a function of pixel rate.

Example 7.43

Determine the length of a ground return at a pixel rate of 10 Mpixels/sec. Assume $h/s = 10$.
Solution:
From Eq. (7.105),

$$L_{LEN} = 1/\left[(2.35 \times 10^{-7}) \times 10^7 \times \ln(4 \times 10)\right],$$

$$L_{LEN} = 0.115 \text{ cm}.$$

Figure 7.55 plots L_{LEN} as a function of pixel rate for various h/s ratios using Eq. (7.105).

The ground bounce voltage generated by inductance reactance is

$$V_{GB} = L\frac{di}{dt}, \qquad (7.106)$$

where V_{GB} is the ground bounce voltage (V), di/dt is the rate of change in current in the ground return when a CCD driver switches state (A/sec) and L is the inductance (H).

The clock driver current requirement in driving the CCD is

$$I = C_D \frac{dV}{dt}, \qquad (7.107)$$

where C_D is the CCD capacitance (F) that is associated with either the horizontal or vertical registers and dV/dt is the voltage change (or slew rate) required.

Example 7.44

Find the ground bounce for the wire described in Example 7.41 that is 20 cm long. Assume the clock driver is driving a 10,000-pF load at a slew rate of 10^8 V/sec (i.e., 10 V in 100 nsec).
Solution:
From Eq. (7.107), the current delivered by the clock driver is

$$I = 10^{-8} \times 10/10^{-7},$$

$$I = 1 \text{ amp.}$$

The inductance for the wire described in Example 7.41 is

$$L = 20 \times \left(7.23 \times 10^{-9}\right),$$

$$L = 1.44 \times 10^{-7} \text{ H.}$$

From Eq. (7.106), the ground bounce induced is

$$V_{GB} = \left(1.44 \times 10^{-7} \times 1\right) \times 1/10^{-7},$$

$$V_{GB} = 1.44 \text{ V.}$$

As discussed in Chapter 5, this amount of ground bounce can significantly influence full well performance for multiphase CCDs.

It is important in the design of a new CCD camera system to include a grounding diagram to show how ground currents originate and flow. Figure 7.56 shows a basic CCD and CDS grounding diagram for a daisy-chain grounding arrangement and a single power-supply line. Numerous grounding and power-supply violations are seen in this diagram that can be corrected by eliminating common impedances through multipoint grounding practices.

NOISE SOURCES

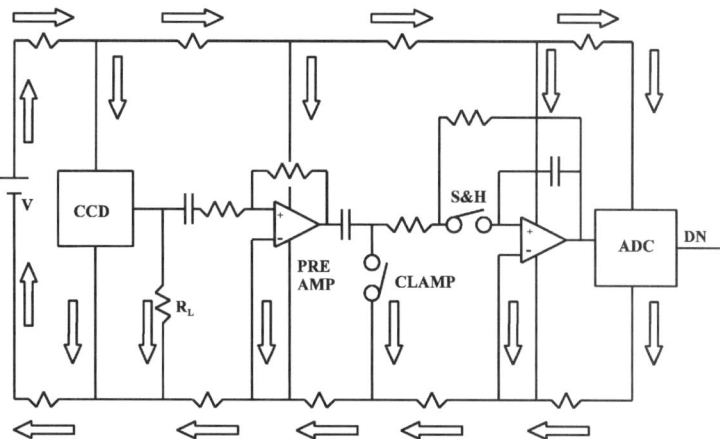

Figure 7.56 Grounding diagram for CCD and CDS processor showing several common impedance paths.

7.2.7 IMAGE CROSS TALK

In addition to seam noise, parallel output amplifier readout also exhibits a problem called image cross talk. This problem develops when two or more channels are reading the CCD simultaneously, producing ghost images between channels. For example, Fig. 7.57 shows a CCD that uses four channels for readout. As channel-1 reads a circular image, cross talk produces a ghost image in channel-2.

There are numerous possibilities for how cross talk develops between channels. Figure 7.58 shows three common configurations that generate ghost images. The first example shows two on-chip CCD amplifiers sharing a V_{DD} voltage supply. As channel-1 reads pixels, its signal couples into channel-2 in the following manner. The drain current for MOSFET-1 decreases when charge dumps onto its sense node. This in turn reduces the voltage drop across the filtering resistor R_{DC}, causing the drain voltage to increase for MOSFET-2. This will result in an increase in source voltage at the time video dumps onto the sense node for MOSFET-2. The cross talk signal will subtract from its own signal, producing a negative ghost image for channel-2. This process can be quantified theoretically. The current change through the load resistor for channel-1 is

$$\Delta I_{OUT1} = -\frac{S_V A_{CCD} S_1(\text{e}^-)}{R_{L1}}, \qquad (7.108)$$

where S_V is the output amplifier sensitivity (V/e$^-$), A_{CCD} is the output amplifier voltage gain (V/V), $S_1(\text{e}^-)$ is the signal for channel-1 (e$^-$) and R_{L1} is the load resistance for channel-1 (Ω).

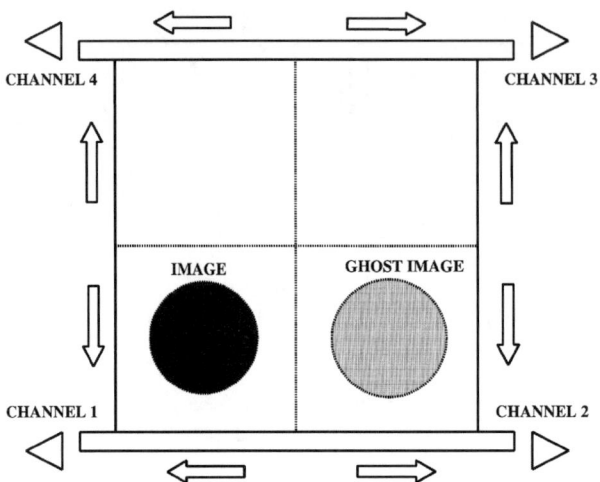

Figure 7.57 Illustration showing a ghost image induced in channel 2 by channel 1.

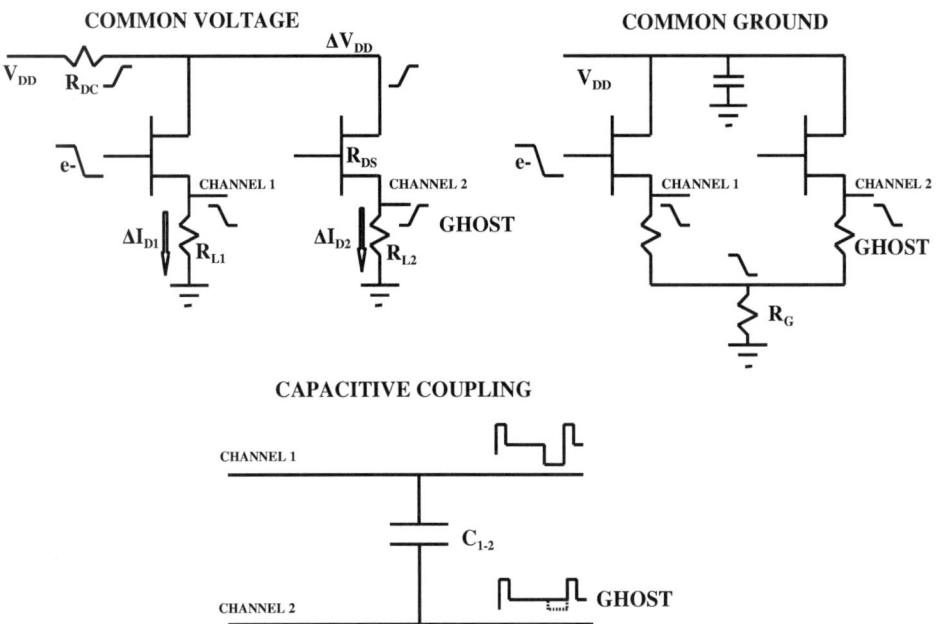

Figure 7.58 Common cross talk problems.

The voltage change across the V_{DD} decoupling resistor R_{DC} is

$$\Delta V_{DD} = R_{DC} \Delta I_{OUT1}. \tag{7.109}$$

The change in current at the output of channel-2 is

$$\Delta I_{OUT2} = \frac{\Delta V_{DD}}{R_{DS}}, \tag{7.110}$$

NOISE SOURCES   713

where R_{DS} is the incremental channel resistance given by

$$R_{DS} = \frac{\Delta I_D}{\Delta V_{DS}}, \qquad (7.111)$$

where ΔI_D is the incremental change in current for an incremental change in drain voltage of ΔV_{DS}. This MOSFET characteristic is seen in Fig. 6.3(a) where R_{DS} is approximately 33 kΩ.

The change in voltage at the output of channel-2 is

$$V_{OUT2} = I_{OUT2} R_{L2}, \qquad (7.112)$$

where R_{L2} is the load resistance for channel-2.

Assuming that $R_{L1} = R_{L2}$, $S_{V1} = S_{V2}$ and $A_{CCD1} = A_{CCD2}$ for channels-1 and -2, the ghost signal induced in channel-2 in electrons is

$$G_2(e^-) = \frac{R_{DC} S_1(e^-)}{R_{DS}}. \qquad (7.113)$$

The negative ghost image that appears in channel-2 can be specified in percent by

$$G_{1\text{-}2} = \frac{100 R_{DC}}{R_{DS}}. \qquad (7.114)$$

Example 7.45

Determine the ghost image generated given $R_{DC} = 100\ \Omega$, $S_1(e^-) = 2.5 \times 10^5\ e^-$ and $R_{DS} = 33$ kΩ.
Solution:
From Eq. (7.113),

$$G_2(e^-) = [100 \times (2.5 \times 10^5)]/(3.3 \times 10^4),$$

$$G_2(e^-) = 757\ e^-.$$

From Eq. (7.114),

$$G_{1\text{-}2} = 100 \times 100/(3.3 \times 10^4),$$

$$G_{1\text{-}2} = 0.3\%.$$

The V_{DD} cross talk problem discussed above can be solved by decoupling each V_{DD} line separately. The same philosophy is used to bias multiple-output transfer gates and V_{REF} voltage lines. For some CCDs, the V_{DD} lines are bussed together on-chip, providing only one external connection to the device. Care must be given to the design of the CCD bus lines to assure that the impedance is sufficiently low to avoid cross talk issues.

The second form of cross talk shown in Fig. 7.58 is associated to a common ground return for the output amplifiers. Here, cross talk occurs as follows. As before, when charge dumps onto the sense node of MOSFET-1, its drain current decreases. In turn, the voltage across the ground impedance R_G decreases, causing the voltage at the source of MOSFET-2 to also decrease. A positive ghost image is produced in channel-2. Cross talk in this case is minimized by using the multipoint grounding scheme shown in Fig. 7.54. In some situations one may need to provide multiple ground planes to isolate ground returns. This can be accomplished by using multilayer printed circuit boards.

The last example shown in Fig. 7.58 shows cross talk that occurs between video lines, which also produces a positive ghost image. Capacitive coupling is minimized by keeping video lines well separated and short. The video lines can also be twisted (ground and video) or shielded to reduce pickup. For extralong runs, the video signal can be sent by differential drivers and receivers.

7.2.8 Noise-Reduction Techniques

Eliminating unwanted noise sources in a CCD camera system can be a time-consuming process. For example, it required nearly one year to solve a 100-e^- WF/PC I noise problem. For this camera system, interference noise patterns seen in images were not stationary but varied from frame to frame. This observation implied that the noise source was not synchronous to the CCD and CDS processing clocks. As it turned out, clocking noise generated by a dc power switcher and the microprocessor circuit used to generate camera clocks was identified as the noise culprit. These circuits generated EMI that coupled into lines that led up to the camera head electronics. Synchronizing clocks was the main solution to the problem. Inhibiting clock activity during sample and ADC conversion "quiet" times was also beneficial. Shielding the reset and video lines to and from the CCD was also required. These efforts paid off and reduced the read noise to the specified level of 13 e^- rms (the noise level of the CCD output amplifier). The same design philosophy was used on WF/PC II, where a noise of level of 3 to 4 e^- rms was achieved.

The location of a cross talk noise problem can be found with the following measuring technique. First the CCD is removed from the camera system. Then, the output summing well (OSW) clock is applied to the preamplifier input of one channel. The OSW signal simulates the video that would normally be generated by the CCD. The OSW pulses are modulated with an arbitrary pattern to better see the cross talk effect. For example, we can gate the OSW "on" for 10 pixels and then "off" for 10 pixels: any pattern will suffice. The rise and fall times of the OSW signal should be comparable to the shortest wave-shaping time constant of the CCD

NOISE SOURCES

system. The amplitude can be adjusted with a simple resistive voltage divider. The other channels remain inactive while stimulating the first channel (i.e., their inputs are grounded). Then, an image is taken and closely examined for ghost images contained in other readout channels. If a ghost is seen, additional decoupling is incorporated or the layout is changed until the ghost disappears.

A powerful test tool was developed for the WF/PC I project to locate random (and coherent) noise sources by using the CDS processor as a test probe. The method begins by calibrating the CDS noise floor without the CCD. This is accomplished by disconnecting the input of the preamplifier from the CCD and reconnecting it to ground potential. Next, the peak-to-peak noise is measured at the input of the ADC using an oscilloscope (a computer can be used to make the measurement if desired). Then the peak-to-peak value measured is converted to an rms reading by dividing the result by 8. Theoretically, the rms noise level at the ADC with the preamplifier shorted to ground is

$$N_{ADC}(V) = 2^{1/2} W_{AI}(f) B^{1/2} A_{CDS}, \quad (7.115)$$

where N_{ADC} is the noise voltage at the input of the ADC (rms V), $W_{AI}(f)$ is the white noise of the preamplifier (V per root Hz) and A_{CDS} is the voltage gain of the CDS processing circuit (V/V). Recall from Chapter 6 that the constant B is the noise power bandwidth and is equal to $1/(4\tau_D)$, where τ_D is the CDS dominant time constant. The square root of two in the equation is because the CDS circuit differences two noise samples, which increases the random noise by this amount.

Example 7.46

Find the ADC input noise for a grounded preamplifier assuming $W_{AI}(f) = 2.5$ nV per root Hz, $A_{CDS} = 100$ V/V, and $\tau_D = 10^{-6}$ sec. Express noise in rms and peak-to-peak units.
Solution:
The noise power bandwidth is

$$B = 1/(4 \times 10^{-6}),$$

$$B = 2.5 \times 10^5 \text{ Hz}.$$

From Eq. (7.115),

$$N_{ADC}(V) = 1.414 \times (2.5 \times 10^{-9}) \times (2.5 \times 10^5)^{1/2} \times 100,$$

$$N_{ADC}(V) = 1.7 \times 10^{-4} \text{ V rms},$$

$$N_{ADC}(V) = 1.4 \times 10^{-3} \text{ V p-p}.$$

This noise level should agree with measurement made at the input of the ADC. If not, the parameters in Eq. (7.115) need to be examined and verified.

We next refer the CDS noise to the sense node and find the equivalent number of noise electrons. This is accomplished through the equation

$$N_{ADC}(e^-) = \frac{N_{ADC}(V)}{(2^{1/2})A_{CDS}A_{CCD}S_V}. \tag{7.116}$$

Example 7.47

Find the equivalent noise calculated in Example 7.46 when referred to the CCD's sense node. Assume $S_V = 4 \times 10^{-6}$ V/e$^-$, $A_{CDS} = 100$ V/V and $A_{CCD} = 0.8$ V/V.
Solution:
From Eq. (7.116),

$$N_{ADC}(e^-) = 1.77 \times 10^{-4}/\left[1.414 \times 100 \times 0.8 \times \left(4 \times 10^{-6}\right)\right],$$

$$N_{ADC}(e^-) = 0.39 \text{ e}^- \text{ rms}.$$

This noise level should be much lower than the CCD noise and represents a level of sensitivity to which interference noise sources can be measured and eliminated.

After the above calibration is made, the input of the preamplifier is disconnected from ground. This input is then used as a test probe to locate noisy lines leading to and from the CCD. Also, other sensitive circuits in the camera can be checked if noise problems are suspect. Ideally, all CCD power and clock lines should exhibit a noise level less than the CDS shorted noise level [i.e., Eq. (7.116)]. This includes voltage bias lines such as V_{DD}, V_{REF}, V_{OTG} and V_{SUB}. In addition, all clock lines should be screened for noise. For this measurement, the CCD clocks are placed into a static condition by pulling the master clock from the logic circuits. Other lines running close to the CCD, such as temperature thermocouples, are also measured. If noise sources are found, additional decoupling is applied. After all lines in the camera head are checked, the input of the preamplifier is reconnected to the CCD and a final system noise level is measured by photon transfer.

7.2.9 Noise-Reduction Summary

The following is a list of noise-reduction techniques that can be reviewed before CCD camera design takes place.

Noise Sources

CCD and Preamplifier

- Keep all voltage and clock lines to and from the CCD as short as possible.
- Use a shield or twisted leads (i.e., signal and ground) between CCD and preamp. Ground both ends of shield. Keep the length of the signal line that extends beyond the shield as short as possible.
- Special attention should be given to the CCD substrate return. This ground return needs to be as short as possible to avoid ground bounce caused by inductive reactance.
- Place the CCD video line close to the chassis ground plane to reduce EMI pick up.
- Limit preamplifier bandwidth to $1/\tau_D$.
- Keep noisy lines (e.g., clocks) well separated from quiet lines (e.g., CCD video).
- Maintain a shorted CDS noise level for all lines entering the camera head [i.e., Eq. (7.116)].

CDS Processor

- The dominant time constant, τ_D, defined by the postamp, should set the overall bandwidth of the system.
- Avoid unnecessary bandwidth for all analog circuits.
- Keep high-impedance lines (and traces) as short as possible (e.g., $\pm$ inputs to the operational amplifiers).
- Decouple each amplification stage separately.
- Keep the ADC input line as short as possible. Decouple the input for quantizing noise reasons.
- Decouple ADC digital output lines that go to the computer to avoid noise from entering the ADC.
- A printed circuit board ground plane is essential for the preamplifier and CDS processor.

Clock Drivers

- Keep all CCD drivers as close as possible to the CCD. Special attention should be given to reset clock driver.
- Keep clock lines close to ground plane to avoid EMI pickup.
- Noise on all clock lines leading to the CCD and CDS should be less than the shorted CDS noise level. Check using measurement technique discussed above.

Wave-shaping

- Wave-shape all CCD clocks including the reset clock.
- Keep wave-shaping networks close to the CCD.

- For RC wave-shaping networks, keep the resistance small and the capacitance large for high-speed applications.

Digital Circuits

- Twist noisy digital signals to reduce noise coupling and EMI. Avoid uniform bundling of logic lines.
- Limit the master clock to the lowest possible frequency to reduce EMI coupling and transient effects.
- Limit rise and fall times. Use the lowest-speed logic family possible. For example, for slow-scan CCD cameras, low-power CMOS logic is a good choice.

Power Supplies

- Avoid common voltage impedances.
- Separate and feed each camera circuit separately with decoupled voltages.
- Filter and decouple voltage supply lines before leaving the power supply.
- Filter and decouple voltage supply lines at the CCD and CDS.
- Noise on any voltage line leading to the CCD and CDS should be less than the shorted CDS noise level [i.e., Eq. (7.116)].
- Do not mix analog and digital power supplies. Use separate power supplies for ADC.
- Bypass electrolytic capacitors with high-frequency capacitors.

Grounding

- A grounding diagram is imperative in the design of a CCD camera system.
- Avoid single-point ground returns.
- Avoid common ground impedances.
- Avoid ground loops.
- Keep ground leads as short as possible to ground plane. When a ground is required, immediately take the lead to the ground plane.
- Isolate ground planes for digital and analog circuits.
- Use multilayer printed circuit boards for multiground planes to isolate ground returns.
- Avoid questionable grounds to the outside world. Isolate camera grounds from surrounding environment. Only one ground between the camera and outside world should exist. An earth ground is normally used for this purpose. Earth ground does not carry circuit currents.
- Use optical couplers to break ground loops between the camera and computer. Optical couplers can also serve as line drivers to drive the cable between the camera and computer. Cables in excess of 1000 feet have been employed in conjunction with simple optical/driver circuits.

Camera System

- Enclose and shield the CCD, preamplifier, CCD drivers and voltage decoupling circuits (i.e., camera head).
- Use printed circuit boards and surface-mount components to reduce signal lead length and ground returns.
- EMI problems are minimized by routing leads to and from the camera head through two different connectors. These groups are referred to as "noisy" and "quiet" connectors. For instance, the quiet lines include dc voltages, preamplifier output, and reset clock. The noisy bundle includes the summing well, horizontal and vertical clocks.
- Enclose noise sources in a shielded enclosure (e.g., master clock and timing logic).
- Carry shields along with signal through connectors. When quiet and noisy signals must go through a common connector, separate them with ground returns.

REFERENCES

1. A. Grove, *Physics and Technology of Semiconductor Devices*, John Wiley and Sons, pp. 100–105 (1967).
2. C. Sah, R. Noyce and W. Shockley, "Carrier generation and recombination in p-n junction and p-n junction characteristics," *Proc. IRE* Vol. 45, p. 1228 (1957).
3. R. Hall, "Electron-hole recombination in germanium," *Phys. Rev.* Vol. 87, p. 387 (1952).
4. W. Shockley and W. Read, "Statistics of recombination of holes and electrons," *Phys. Rev.* Vol. 87, p. 835 (1952).
5. S. Sze, *Physics of Semiconductor Devices*, John Wiley and Sons, p. 53 (1981).
6. S. Sze, *Physics of Semiconductor Devices*, John Wiley and Sons, pp. 145–146 (1981).
7. N. Saks, "A technique for suppressing dark current generated by interface states in buried channel CCD imagers," *IEEE Electron Device Letters* Vol. EDL-1, No. 7 (1980).
8. J. Pankove, *Optical Processes in Semiconductors,* Dover Publications, p. 27 (1971).
9. R. McGrath, J. Doty, G. Lupino, G. Ricker and J. Vallera, "Counting of deep-level traps using a charge coupled device," *IEEE Trans. Electron Devices* Vol. ED-34, No. 12 (1987).
10. B. Burke and S. Gajar, "Dynamic suppression of interface-state dark current in buried channel CCDs," *IEEE Trans. Electron Devices* Vol. ED-38, pp. 285–290 (1991).

11. D. Lumb, G. Berthiaume, D. Burrows, G. Garmire and J. Nousek, "Charge coupled devices (CCDs) in x-ray astronomy," *Experimental Astronomy 2*, Kluwer Academic Publishers, p. 194 (1991).
12. H. Ott, *Noise Reduction Techniques in Electronic Systems*, John Wiley and Sons (1976).
13. Brokaw, "An IC amplifier user's guide to decoupling, grounding and making things go right for a change," *Analog Devices* AN-202.

CHAPTER 8

DAMAGE

8.1 RADIATION DAMAGE

8.1.1 INTRODUCTION

The preceding chapters have shown that the CCD has achieved unprecedented levels of performance in read noise and charge transfer efficiency. Optimizing the sensor to near perfection in these areas has produced a device that is extremely vulnerable to damage induced by high-energy radiation sources.[1-9] CCDs are also more vulnerable because past performance was not as prominent as today. Now, ultralow dark currents (< 5 pA/cm^2) and ultrahigh charge transfer efficiency (> 0.9999995) make the presence of a single defect introduced by radiation important.

The CCD is being used more often in hostile environments where high levels of radiation are encountered (e.g., outer space imaging applications, particle detectors used in beam colliders, nuclear weapon use, plasma physics).[10] The emphasis of this concern reflects upon the author's own experience, because the radiation problem has surfaced on every flight CCD mentioned in this book. Many surprises were encountered, including some show stoppers. For example, the Galileo CCD would have arrived at Jupiter in a saturated state if neutron testing was not carried out before launch. Here, neutrons created very large dark spikes in the array. To circumvent the dilemma, the operating temperature for the CCD was reduced from $-40°C$ to $-120°C$. The SXT CCD would have also failed prematurely because of a serious x-ray-induced ionizing radiation problem. The solution for SXT was to employ a UV flood that neutralized radiation-induced charge in the gate dielectric.[11] The operating temperature for Cassini and Hubble was optimally selected ($-90°C$) to ameliorate a serious trapping problem induced by protons. The Cassini Star Tracker CCD experienced a reverse annealing problem that forced engineers to a new cooling temperature ($-40°C$) and clocking sequence. Unfortunately, the new Chandra mission was not as lucky as these missions. As we will discuss below, this flight camera experienced a radiation problem that seriously degraded CTE performance shortly after being launched into space. The experience gained from these missions clearly shows that the radiation problem cannot be taken lightly and must be thoroughly analyzed to forestall potential crises.

When charge particles pass through the CCD, nearly all energy loss (99.9%) is converted into e-h pairs. We refer to this energy conversion as "ionizing energy loss" (IEL). The remaining energy is given to nonionizing interactions including

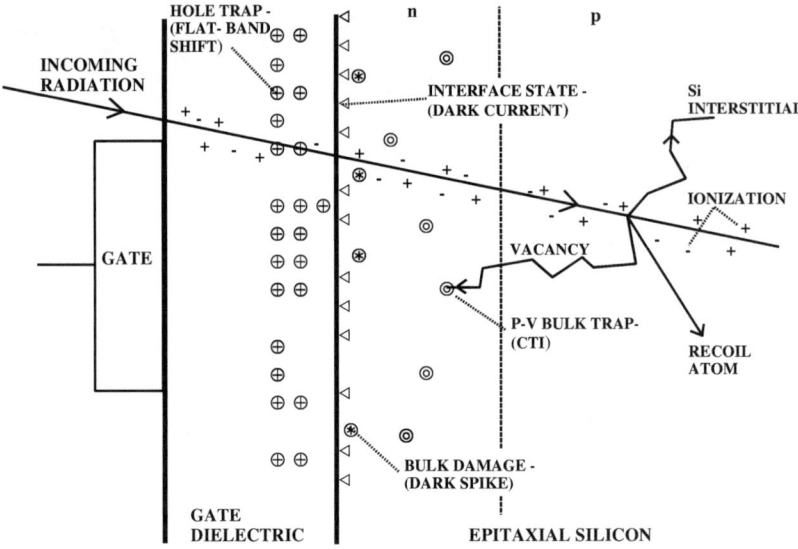

Figure 8.1 Gate dielectric and epitaxial layer where ionization and bulk damage effects take place.

displacing silicon atoms. This energy transformation is called "nonionizing energy loss" (NIEL). These two different forms of energy dissipation translate into two major damage mechanisms for the CCD called ionization and bulk damage, respectively. Figure 8.1 illustrates regions in the sensor that are damaged by high-energy incident particles. Each problem will be discussed in the sections to follow. Here, we briefly introduce the problems.

8.1.1.1 Ionization Damage

The CCD is based on a metal-insulator-semiconductor (MIS) structure, and therefore it is susceptible to ionization damage. The problem is well understood and there are many good references and reviews on the subject.[12–16] Permanent damage to the gate dielectric will occur when e-h pairs are generated in the layer by ionizing radiation. There are two main damaging effects when this happens. First, holes generated in the layer will become trapped, generating a flatband voltage that will effectively shift the clock and output amplifier bias potentials. If the shift becomes greater than the operating windows, the sensor will fall out of specification, usually dramatically. For the scientific CCDs discussed in this book, a typical channel threshold shift caused by ionization damage is approximately -0.1 V per 10^3 rad (Si). The exception is for the Galileo virtual-phase CCD, which is more tolerant to ionizing sources. This sensor exhibits a flatband shift of -1 V per 10^5 rad.[17] For regions near or outside the signal channel (e.g., field oxide), greater shifts can be experienced with corresponding larger effects on performance. For example, shifts greater than -1 V per 10^3 rad will be described below.

The second ionization problem is related to the dielectric's Si–SiO$_2$ interface. Here, ionizing radiation will break weak bonds and create interface states. The newly formed states will generate additional flatband shift and surface dark current through the hopping conduction process discussed in Chapter 7. For some CCDs described, a very small dose (< 5 krad) of ionizing radiation can cause the dark current to increase by two orders of magnitude.

Ionization damage varies widely among CCD manufacturers depending on processing details and the CCD technology employed. Multiphase scientific CCDs discussed in this book are categorized as "soft" CCDs because a noticeable flatband shift begins to appear at 10^3 rad. Radiation "hard" CCDs that use custom processes show shifts of only -1 V per 10^6 rad.[18] Ionizing damage is also very dependent on the type, energy, and rate of incoming radiation. Energetic photons and electrons are the main ionizing radiation sources with which the CCD must deal. Heavier particles, such as protons and heavy ions, can also induce ionization damage; however, bulk damage is the main concern for these particle types. Ionizing damage is very sensitive to operating conditions such as clock bias and operating temperature. These characteristics make it difficult to qualify a new CCD and predict performance for a given working radiation environment. This problem is best handled by irradiating the CCD exactly the way it will be used.

8.1.1.2 Bulk Damage

Bulk damage, or displacement damage, is associated with the epitaxial layer where charge is generated, collected, transferred, and measured. When energetic electrons, protons, heavy ions, and neutrons pass through the CCD, there is the possibility of collisions with silicon target nuclei. These events displace atoms from their lattice positions and create vacancy-interstitial pairs. The permanent defects produced by this process causes degradation in charge transfer efficiency, increased dark current generation, dark current spikes, and effects on the output amplifier.[19] As will be shown below, a small dose of protons (25 rad) will begin to take its toll on these performance parameters.

8.1.2 NEAR-EARTH RADIATION ENVIRONMENT

Three basic mechanisms contribute to the Earth's local space radiation environment: solar events, trapping of particles by the Earth's magnetic field, and the interaction of galactic cosmic rays with the Earth's magnetosphere and atmosphere.[20] Coronal mass ejections and solar flares accelerate protons and heavier ions toward the Earth where they can interact with orbiting spacecraft and CCDs. Protons accelerated toward the Earth by these events have energies in the hundreds of mega-electron-volts; heavier ions have energies in the hundreds of giga-electron-volts. Coronal mass ejections are strongly correlated with increases in proton fluences. These events generate a shockwave that increases the solar wind velocity. Solar flares primarily affect heavier ion fluences. They accelerate the particles directly, increasing the solar wind density rather than its velocity. The frequency of these events, and the ion fluences associated with them, peak during the maximum of the solar cycle.

The magnetosphere acts as a shield, protecting the Earth from the solar wind. About 99% of the particles directed toward the Earth are deflected. The likelihood of penetration is determined by the ratio of a particle's momentum to its charge. The higher this ratio, the more likely it is that the particle can penetrate. Despite the high percentage of particles that are deflected, the increased fluences that accompany solar events can represent a significant portion of the total radiation that an Earth-orbiting spacecraft experiences, especially in high-altitude orbits where the fluences are greatest.

Charged particles can become trapped in the Van Allen belts, where they spiral around the Earth's magnetic field lines, reflecting back and forth between the poles. From within the belts this radiation is omnidirectional and composed of protons (0.4 to 500 MeV), electrons (0.4 to 7 MeV) and heavier ions, which have energies too low to cause a problem for the CCD. The inner belt, which surrounds the Earth at altitudes between 300 km and 25,000 km, is composed primarily of protons. The proton flux for energies above 10 MeV reaches its maximum around 7000 km. For low altitudes, the region of most concern is the South Atlantic anomaly, where the magnetic field lines dip into the Earth's atmosphere because of the misalignment of Earth's magnetic and rotational axes [refer to Fig. 1.6(f)]. The outer belt, between altitudes of about 25,000 km and 50,000 km, is dominated by energetic electrons with a peak flux around 20,000 km for energies above 1 MeV. Electrons are readily absorbed by shielding the CCD with approximately 1 cm of aluminum. On the other hand, high-energy protons are difficult to shield. For example, 3.6 cm of aluminum will shield the CCD only from energies less than 100 MeV.

Near solar maximum, the fluences from trapped electrons peak and fluences from trapped protons are at their minimum. When solar activity is low, trapped proton fluences peak and trapped electron fluences reach a minimum. Average counts of protons and electrons in the belts vary slowly with the solar cycle, but local counts can vary by orders of magnitude during magnetic storms, especially in the outer belt. The danger this radiation represents depends on a spacecraft's orbit and how much time the spacecraft will spend in the belts during its mission life.

Cosmic rays include every element in the periodic table and are found everywhere in interplanetary space. Nuclei of hydrogen, helium, carbon, oxygen, silicon and iron are the most common and particle energies are in the giga-electron-volt range. Most of the nuclei are fully ionized. Like trapped radiation, cosmic rays are an omnidirectional source. Like trapped proton fluences, cosmic ray fluences reach their peak during the solar minimum.

The ionizing dose accumulated by the CCD during a typical satellite mission varies considerably, typically from 1 to 30 krad depending on the lifetime of the mission, type of orbit, time of mission and shielding characteristics around the CCD. Similar dose levels are experienced for interplanetary flight.

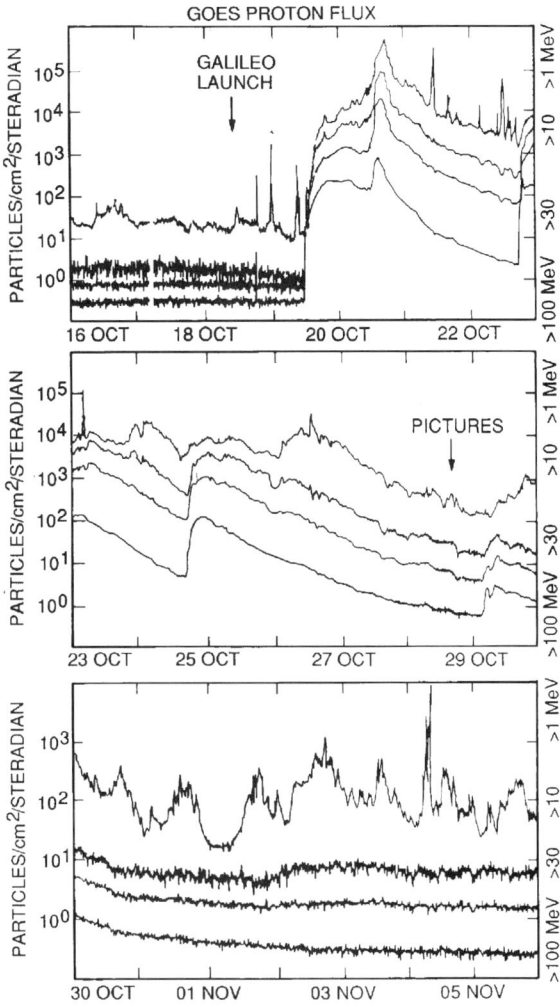

Figure 8.2 GOES proton fluence after the Galileo mission launched into space.

Most institutions involved with radiation analysis have access to computer programs that describe space radiation environments. For example, the Naval Research Laboratory (NRL), the National Aeronautics and Space Administration (NASA) and the National Bureau of Standards (NBS) have extensive radiation programs in wide use today. The analysis performed by these codes usually takes place in four steps. First the program defines the spacecraft position and trajectory. This information is needed by the code if the spacecraft is within the Earth's or planet's magnetosphere. The magnetosphere will act as a shield to certain particles or act as a trap that increases the radiation level (e.g., Van Allen belts). Next, the program will calculate the accumulated radiation exposure at the spacecraft over the mission lifetime. Then, this environment is transferred through various shield structures that normally include the spacecraft and CCD shield. The product of the program lists the working radiation environment immediately at the CCD in terms of total dose and fluence for each particle of concern.

There are also off-the-shelf PC programs available for radiation analysis. These comprehensive software tools model particle radiation effects in space, including solar flares, cosmic radiation and Van Allen belts. The programs simulate CCD exposure to protons, electrons and heavy ions in any specified orbit around Earth or interplanetary flight. The final output from the models typically includes total dose and proton damage in terms of displacements.

Figure 8.2 shows the proton environment that the Galileo spacecraft experienced shortly after launch. Note that the proton flux increases by orders of magnitude on October 19 because of a very large solar flare. Dark current pictures were taken by Galileo several days after the event started, as indicated in the figure. Although the intensity of the radiation decreased significantly at this time, images still showed proton ionization trials much higher than normal.

8.1.3 Radiation Units

Radiation received by the CCD is specified by two different units. For bulk damage effects, incoming radiation is specified as a fluence. For ionization damage, radiation is specified in terms of total dose. Both units are defined below.

8.1.3.1 Total Dose

Total dose received by the CCD is defined as

$$D = \frac{dE}{dM}, \qquad (8.1)$$

where D is the total dose, dE is the average energy imparted by ionizing radiation into the CCD of mass dM. We specify dose as radiation-absorbed dose (rad), which is equivalent to depositing 100 ergs of energy in 1 g of material (or 0.01 J/kg). The silicon unit of absorbed dose is the gray (Gy), which is equal to an absorbed energy of 1 J/kg, or 100 rad.

Example 8.1

Determine the number of electrons generated in silicon and silicon dioxide for a 1-rad exposure. Assume the density of silicon is 2.32 g/cm^3 and silicon dioxide is 2.2 g/cm^3.

Solution:
The number of eV per rad is

$$1\,\text{rad} = [100\,\text{ergs/g}] \times \left[1\,\text{eV}/(1.6 \times 10^{-19}\,\text{J})\right] \\ \times \left[2.32\,\text{g/cm}^3\right] \times [\text{J}/10^7\,\text{ergs}],$$

DAMAGE 727

$$1 \text{ rad} = 1.45 \times 10^{14} \text{ eV/cm}^3 \text{ (silicon)},$$

$$1 \text{ rad} = 1.37 \times 10^{14} \text{ eV/cm}^3 \text{ (silicon dioxide)}.$$

Assuming it takes 3.65 eV of energy to create one e-h pair in the case of silicon, and approximately 18 eV in the case of oxide, the number generated for 1 rad is

$$e - h = 1.45 \times 10^{14}/3.65,$$

$$e - h = 3.97 \times 10^{13} \text{ e-h pairs (silicon)},$$

$$e - h = 7.6 \times 10^{12} \text{ e-h pairs (silicon dioxide)}.$$

When radiation experiments are performed, total dose must be measured to verify that incoming radiation was properly imparted to the CCD. Dosimeters are used for this purpose. There are various types of instruments employed: Si diodes, calorimeter, thermoluminescence, etc. Often it is advantageous to use the CCD to directly measure total dose received. This is accomplished by measuring the signal charge generated by the radiation source and the formula

$$D = \frac{1.6 \times 10^{-14} \, S(e^-) \, E_{e-h}}{\rho_{SI} \, P_S \, x_{EPI}}, \tag{8.2}$$

where D is the dose in the epitaxial layer (rad), $S(e^-)$ is average charge generated by a flat field of radiation events (e^-/pixel), E_{e-h} is the conversion of eV to e-h pairs (3.65 eV/e^-), ρ_{SI} is the density of silicon (2.32 g/cm^3), P_S is the pixel area (cm^2), x_{EPI} is the thickness of the epitaxial layer (cm), and the constant factor 1.6×10^{-14} is the number of ergs per electron volt. The equation assumes that the radiation source uniformity stimulates the epitaxial layer in depth.

Example 8.2

Determine the dose received by a CCD that is exposed to Fe^{55} x rays. Assume $S(e^-) = 4166 \, e^-$/pixel, $x_{EPI} = 15 \times 10^{-4}$ cm and $P_S = 2.25 \times 10^{-6}$ cm^2 (i.e., 15-μm pixel).
Solution:
From Eq. (8.2),

$$D = \left[(1.6 \times 10^{-14}) \times 4166 \times 3.65 \right] /$$
$$\left[2.3 \times (2.25 \times 10^{-6})(15 \times 10^{-4}) \right],$$

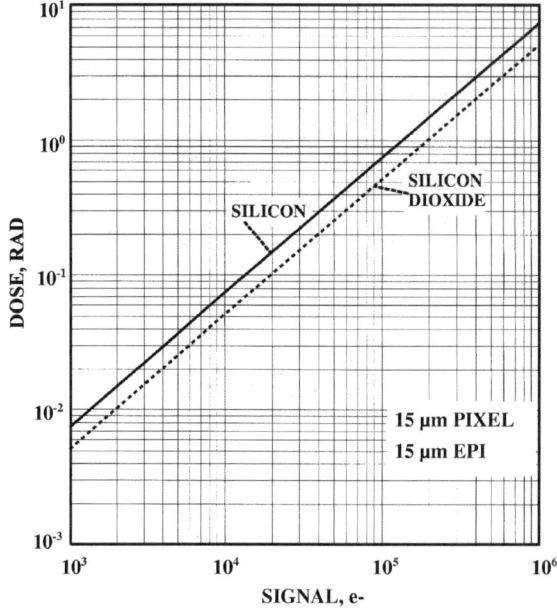

Figure 8.3 Total dose as a function of signal as measured by the CCD.

$$D = 0.0313 \, \text{rad}.$$

Figure 8.3 plots Eq. (8.2) as a function of signal per pixel.

The total dose rad unit must be used with caution when testing CCDs. The bulk damage problem was obscured for many years because early radiation tests were based on total dose experiments using a Co^{60} gamma-ray source. Although Co^{60} was suitable for determining radiation hardness for ionizing damage, the method was unsatisfactory for characterizing the vulnerability of the CCD to bulk damage effects. The same dose of protons and gamma rays will have a completely different outcome for the CCD. Protons will cause severe bulk damage but insignificant levels of ionization damage. On the other hand, Co^{60} gamma rays will induce ionization damage but exhibit insignificant levels of bulk damage. The CCD is very discriminating with respect to the type of particle and energy confronted. Therefore, the ionization (total dose) and bulk (fluence) damage problems must be evaluated separately.

8.1.3.2 Cross-Section and Mean Free Path

Cross-section is a parameter that describes the probability of particle interaction with the target material, and it is usually specified in units of area (cm^2). We can think of cross-section as the effective interacting area that the target presents to the incoming particle. High interaction, or a large cross-section, means that the area around the target particle in which interaction can occur is large compared to the incident particle. On the other hand, low probability of interaction implies the particle effective area is small by comparison. Rutherford calculated cross-section for his famous experiment involving alpha particles being scattered by a foil target. His cross-section gives the probability that an alpha particle is deflected from its path when passing through the target. The cross-section here is defined by the fraction of alpha particles that bounce back from the target, divided by the density of nuclei in the target and target thickness. It is important to note that the cross-section is not just dependent on the geometric area of the particle itself, but also on interaction characteristics around it.

The mean free path of a particle in a medium is the average distance the particle will travel before an interaction occurs. This path is related to the cross-section by the formula

$$\lambda = \frac{A}{N_A \rho \sigma}, \tag{8.3}$$

where λ is the mean free path (cm), A is the atomic weight (g/mole), N_A is Avogadro's number (6.022×10^{23} atoms/mole), ρ is the density (g/cm^3) and σ is the cross-section (cm^2). This formula can be simplified to

$$\lambda = \frac{1}{\rho \sigma}, \tag{8.4}$$

where σ is given in cm^2/g.

Example 8.3

Determine the mean free path length for a 5.9-keV photon (Fe^{55}) in silicon and silicon dioxide. Assume cross-sections of 145 and 106 cm^2/g and densities of 2.32 and 2.2 g/cm^3, respectively.
Solution:
From Eq. (8.4),

$$\lambda = 1/(145 \times 2.32),$$

$$\lambda = 29.7 \, \mu\text{m (silicon)},$$

$$\lambda = 42 \, \mu\text{m (silicon dioxide)}.$$

The mean free path of a soft x ray is equal to its absorption length because the energy of the photon is completely absorbed by a target atom in one stop. For protons and other particles with mass, complete absorption does not usually take place. Energy of absorption occurs in many mean free path steps.

Therefore, the term absorption length does not apply for these particles. Instead, we refer to the interaction depth as the particle's range.

8.1.3.3 Fluence

Total dose and fluence are related by the stopping power through the equation

$$D = K_1 F S, \qquad (8.5)$$

where K_1 is a constant (1.602×10^{-5} ergs/MeV), F is the fluence for a given energy (particles/cm^2) and S is the stopping power of the particle (MeV-cm^2/mg) given as

$$S = \frac{dE}{dx}, \qquad (8.6)$$

where dE is the energy exchange (MeV-cm^3/mg) in a mass depth of dx (cm). Equation (8.5) assumes that the stopping power is constant as the particle moves through the CCD. Note that the greater the stopping power, the greater the energy loss and total dose received by the CCD.

The stopping power in silicon varies widely, depending on the incident particle type and its energy. For example, the stopping power for a 1-MeV electron is 0.0016 MeV-cm^2/mg in silicon. Figure 8.4(a) plots stopping power for electrons as a function of energy.[21] The stopping power for a 2-MeV alpha particle is considerably greater, 1 MeV-cm^2/mg. Figure 8.4(b) plots stopping power for protons as a function of energy (also see Appendix G1).[22] Note that maximum energy exchange between the proton and CCD occurs at an energy of approximately 60 keV, a point referred to as the Bragg peak. Here the stopping power is 0.6 MeV-cm^2/mg, nearly equal to the 2-MeV alpha particle. These stopping power units, MeV-cm^2/mg, can be multiplied by 2.32×10^1 to determine the energy lost by a proton as function of distance (eV/Å) or 2.32×10^9 for eV/cm.

Example 8.4

Determine the 20-MeV proton total dose for a fluence of 8×10^7 protons/cm^2.
Solution:
From Appendix G1, the stopping power for a 20-MeV proton is 0.021 MeV-cm^2/mg.
From Eq. (8.5),

$$D = (1.602 \times 10^{-5}) \times (8 \times 10^7) 0.021,$$

$$D = 26.91 \text{ rad (Si)}.$$

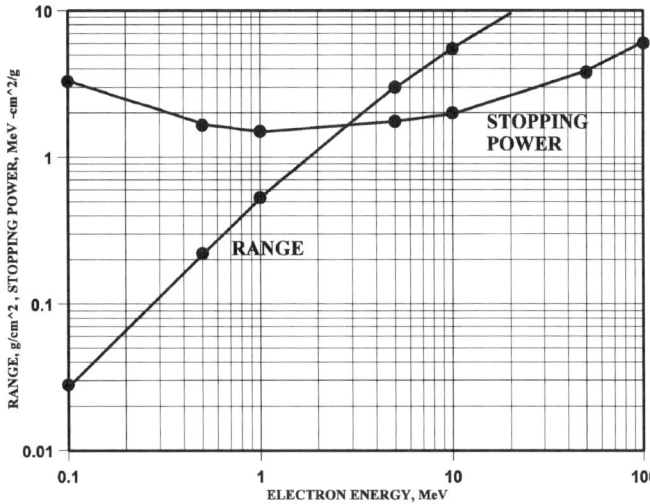

Figure 8.4(a) Electron stopping power and range as a function of energy.[21]

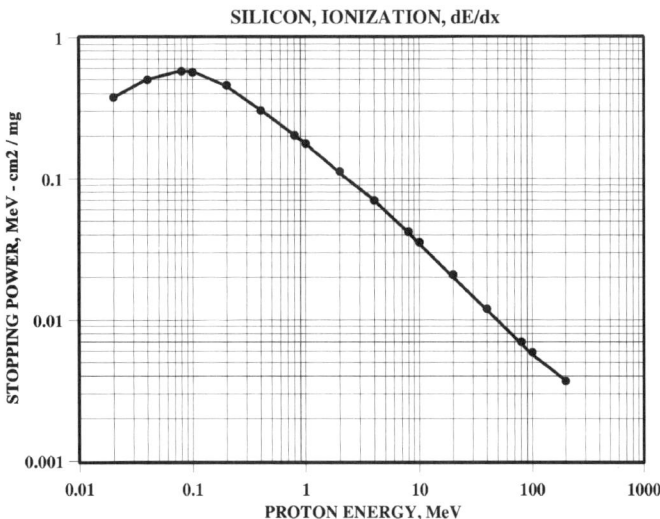

Figure 8.4(b) Proton stopping power as a function of energy.[22] Also see Appendix G1.

Example 8.5

Determine the 1-MeV electron total dose at a fluence of 10^{10} electrons/cm^2. Also calculate the dose for a 1-MeV Co^{60} gamma ray at the same fluence.

Solution:
From Fig. 8.4(a), the stopping power for a 1-MeV electron is 1.6×10^{-3} MeV-cm^2/mg.
From Eq. (8.5),

$$D = (1.602 \times 10^{-5}) \times (1.6 \times 10^{10}) \times 10^{-3},$$

$$D = 256 \text{ rad (Si)}.$$

The stopping power for a 1-MeV gamma ray is 6×10^{-5} MeV-cm^2/mg.
From Eq. (8.5),

$$D = (1.602 \times 10^{-5}) \times (6 \times 10^{10}) \times 10^{-5},$$

$$D = 9.6 \text{ rad (Si)}.$$

Note that the total dose for gamma rays is considerably lower because most penetrate the CCD without interaction.

8.1.3.4 Total Dose Sources

Co^{60} and Fe^{55} are popular ionizing sources used to irradiate the CCD to a given dose level. Each source has its advantages in testing the CCD for radiation hardness. X-ray irradiation is a convenient source because photons can be readily shielded where gamma rays cannot. Therefore, x rays allow several regions on a single device to be irradiated by using appropriate shields. Figure 8.5(a) shows an experimental Fe^{55} x-ray setup that uses a shield with several holes to irradiate the CCD in different places. Between the x-ray source and this shield is a movable shield that can be moved to a different hole location for each radiation experiment performed. Figure 8.5(b) shows a dark current image and seven regions on a 1024 × 1024 array where Fe^{55} x-ray radiation was applied under different operating conditions. The data products generated by this image are used in Fig. 8.18.

Fe^{55} x rays are also monoenergetic, a property that simplifies data analysis when modeling radiation results. Compared to Fe^{55}, gamma rays exhibit a large energy distribution and generate secondary particles that can enhance the dose to the CCD. For example, when gamma rays interact with materials in the vicinity of the CCD (e.g., camera housing, lead bricks, the walls of the lab), soft and hard x rays are also generated. In addition, Co^{60} dosimeters are usually limited to a cutoff energy of approximately 100 keV. Below this energy, soft x rays are not accounted for, underestimating the dose received by the CCD. Consequently, the CCD usually measures a higher dose [Eq. (8.2)] compared to dosimeters. An Fe^{55} source is also convenient to use because it is small and easy to configure and perform radiation tests. Co^{60} sources are usually quite sizeable. For example, Fig. 8.5(c) shows the source used at JPL, which stands over 6 ft tall. These sources are also deadly if

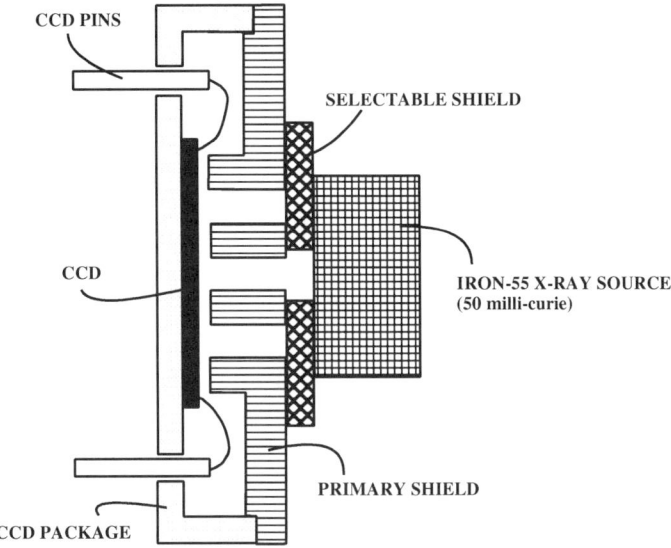

Figure 8.5(a) A Fe^{55} x-ray ionization damage shielding setup to perform multiple radiation experiments on a single CCD.

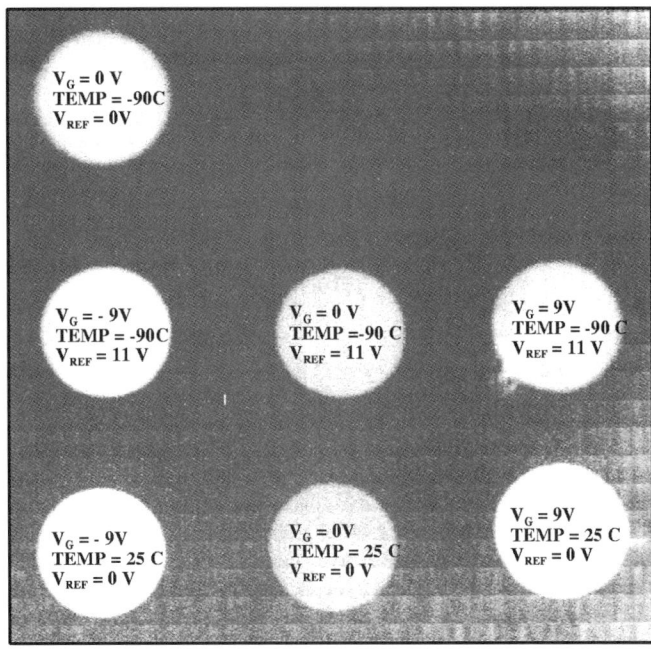

Figure 8.5(b) Dark current image showing multiple damaged regions under different operating conditions.

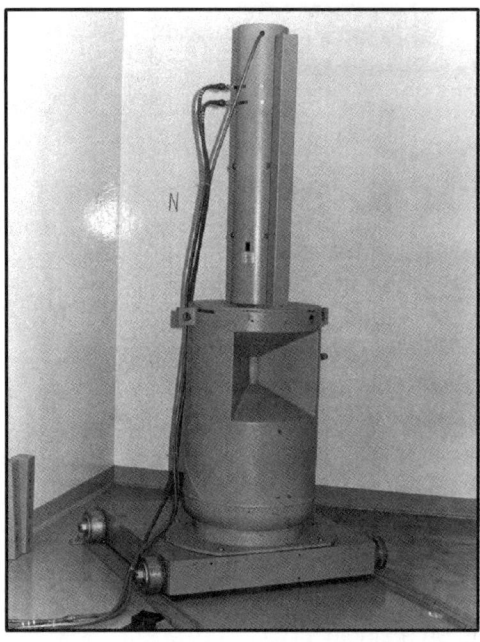

Figure 8.5(c) Co^{60} gamma-ray source.

used mistakenly. The main Co^{60} source at JPL has an intensity of 30,000 Ci. By comparison, a 50-mCi Fe^{55} is sufficient for radiation studies on the CCD.

Equation (8.2) calculates the dose within the epitaxial layer. If the cross-section of the incident particle is independent of target material, then the total dose will be the same for the gate dielectric. This is the case for Co^{60} gamma-ray irradiation for the following reason. Co^{60} nuclei decay by β decay with a half-life of 5.26 years to a stable Ni^{60} atom, producing 1.17- and 1.33-MeV gamma rays in the process. The high-energy photons interact with the CCD by Compton scattering. Here, a large fraction of the energy is transferred to orbiting silicon electrons, with the remainder of the energy going to secondary photons. The electrons produced have an energy that lies in the range of 0.5 to 1.0 MeV. They can be considered free electrons because the energy absorbed is much greater than their binding energy with the target atom. Therefore, the scattering cross-section and total dose for Co^{60} gamma rays are independent of target material. The cross-section for a 1-MeV photon is about 0.06 cm^2/g.

The cross-section for soft x rays is not independent of target material because the energy delivered to the target electrons is comparable to the binding energy. The cross-section depends on this binding energy and therefore on the target material. The cross-section for a 10-keV photon is 32 cm^2/g for Si and 18 cm^2/g for SiO_2. For Fe^{55} 5.9-keV x-ray photons, the cross-sections are 145 cm^2/g and 106 cm^2/g, respectively. The dose relationship between the gate dielectric and epitaxial layer

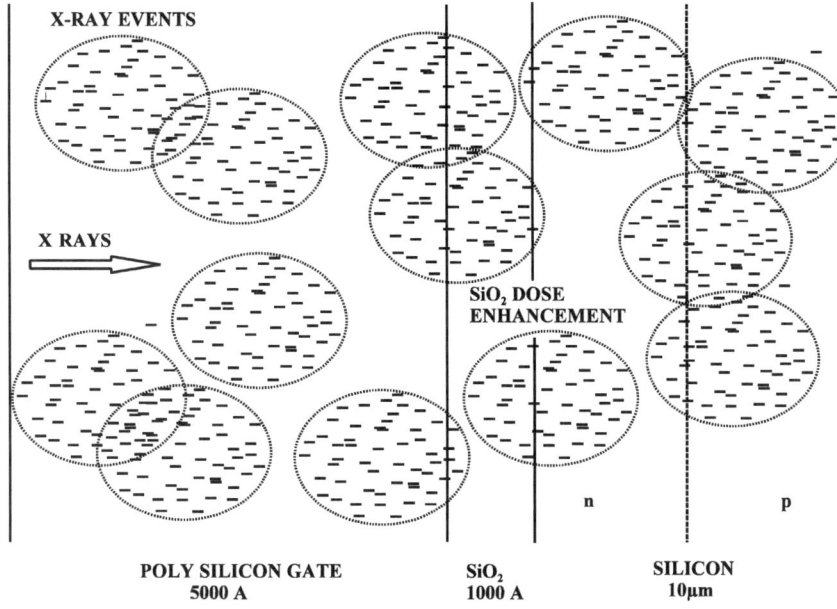

Figure 8.6 Dose enhancement effect in the gate dielectric.

for soft x rays can be approximated using

$$D_{OX} = \frac{\sigma_{OX}\rho_{OX}}{\sigma_{SI}\rho_{SI}} D_{SI}, \tag{8.7}$$

where σ_{SI} is the cross-section for silicon (cm²/g), σ_{OX} is the cross-section for oxide (cm²/g), ρ_{OX} is the density of oxide (2.2 g/cm³) and ρ_{SI} is the density of silicon (2.32 g/cm³).

Example 8.6

What is the ionizing dose in the oxide layer for the CCD irradiated in Example 8.2? Assume $\sigma_{SI} = 145$ cm²/g, $\sigma_{OX} = 106$ cm²/g, $\rho_{OX} = 2.2$ g/cm³, and $\rho_{SI} = 2.32$ g/cm³.

Solution:
From Eq. (8.7),

$$D_{OX} = 106 \times 2.2 \times 0.0313/(145 \times 2.32),$$

$$D_{OX} = 0.0217 \text{ rad}.$$

Figure 8.3 plots the dose for the gate oxide as a function of signal using Eqs. (8.2) and (8.7).

The K-shell electron generated by an Fe^{55} x ray exhibits a range in the silicon lattice of approximately 0.4 μm [see Eq. (4.39)]. Therefore, photoelectrons generated in the poly silicon gates and epitaxial layer will overlap into the gate dielectric (recall the gate oxide layer is approximately 0.1-μm thick). In that more absorption occurs in the silicon layers, the dose received by the gate dielectric by overlapping charge clouds will be greater than expected. The effect is illustrated in Fig. 8.6 and is called "dose enhancement." A less energetic x-ray photon or thicker oxide will reduce the dose enhancement effect. For Fe^{55} events, dose enhancement for the gate dielectric is approximately equal to σ_{SI}/σ_{OX}.

8.1.4 Transient Events

Inelastic collisions with orbital silicon electrons is a predominant mechanism where energetic charge particles lose kinetic energy (i.e., Rutherford scattering or Coulomb scattering). The electrons either experience a transition to an excited state or to an unbound state into the conduction band (i.e., ionization). Conduction band electrons are collected in the nearest potential well, generating a transient event in an image.[23–28] Charged particles leave a e-h track producing approximately 1 e-h pair for every 3.65 eV of energy absorbed. The ionizing trial of charge left behind is not a permanent feature and can be erased simply by reading the CCD.

8.1.4.1 Background Transients

Interference events have been an issue for the CCD since their beginning, in the form of background transients. For example, ceramic headers used in the CCD package were occasionally seen to be radioactive, producing charged particles that interacted with the sensor. Glass on which thinned CCDs were mounted and windows above the sensor sometimes exhibited small amounts of radioactivity. MgF$_2$ windows also exhibit phosphorescent Cherenov events that are induced when energetic particles interact with the window, with signatures that can take many hours to decay. In addition, CCD sockets (e.g., ZIF) and printed circuit boards can show detectable amounts of uranium and thorium, both radioactive elements.

Figure 8.7 shows electron ionization trails generated by a radioactive source used to test Hubble CCD packages for hermeticity. In this test, the sensor was first embalmed with a radioactive gas and then later tested with a radiation detector to find leaks in the package. The gas used was Ra^{225}, a β emitter that generates 0.32-MeV electrons. The test was abandoned because even extremely small microscopic leaks would allow the Ra^{225} gas into the package, levels undetected by the leak radiation monitor but easily seen by the CCD. This method of testing leaks was later replaced by using a helium gas leak detector.

In general, CCD manufacturers (and users) become concerned when background transients are greater than the normal cosmic background level. This quantity is approximately 1.5 events/cm^2/min at sea level.[29]

DAMAGE

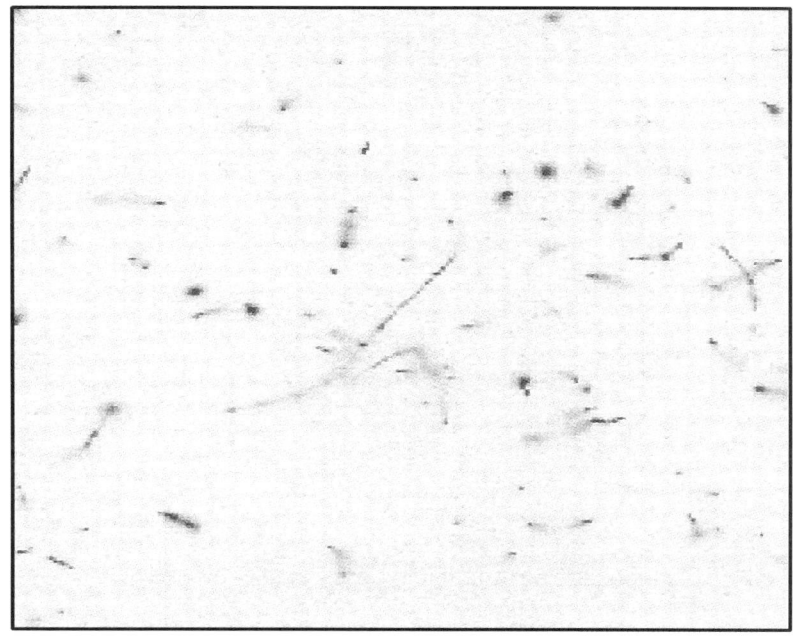

Figure 8.7 Ra^{225} electron transients.

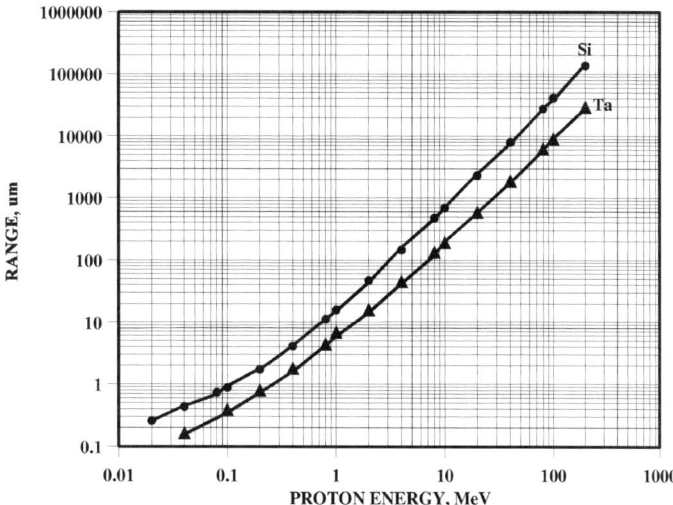

Figure 8.8 Proton range in silicon and tantalum as a function of energy.[22] Refer also to Appendix G1 and G4.

8.1.4.2 Particle Transients

As mentioned previously, protons and heavy ions undergo ionizing loss producing transient events. Figure 8.8 plots the range of a proton as a function of its energy.

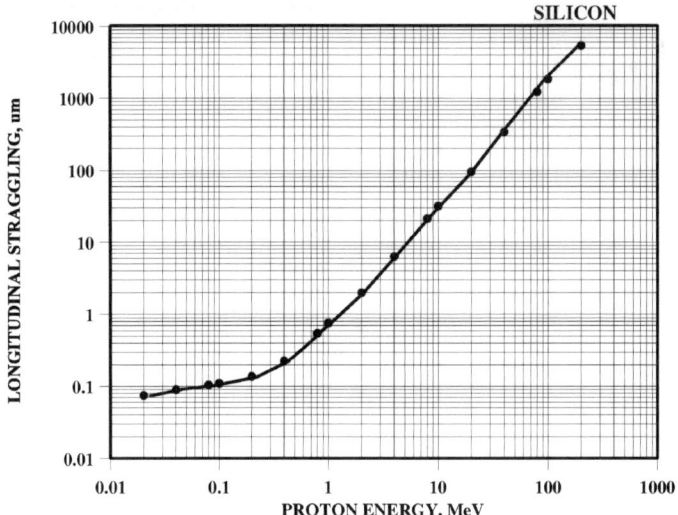

Figure 8.9 Proton longitudinal straggling as a function of energy.[22] Also see Appendix G1.

Figure 8.9 plots the longitudinal straggling, the standard deviation about the average range. The data shown in Figs. 8.4(b), 8.8 and 8.9 were provided by SRIM, which we will discuss more fully in Section 8.1.9.2.[22]

Example 8.7

How far will a 1-MeV proton travel when normally incident on a backside-illuminated CCD? Also, determine the straggling about the range.
Solution:
From Fig. 8.8 or Appendix G1 the range of a 1-MeV proton is 16 μm. The proton will pass through the CCD if thinner than this. From Fig. 8.9, straggling is < 1 μm about its range.

Figure 8.10(a) shows 9-MeV proton ionization trails generated in a 1024 × 1024 × 7.5-μm-pixel × 20-μm epitaxial three-phase CCD. The measured range is 580 μm or 77 pixels, which is in agreement with Fig. 8.8. The images were taken by aligning the surface of the CCD at a slight angle to a beam of protons. Protons enter at the top of the CCD and travel into the 7-μm depletion region, resulting in a trail of charge no wider than a pixel. Because there is a slight angle, most protons pass through the depletion region into field-free material, which for this CCD is 13 μm in extent (i.e., 20 − 7 = 13 μm). This results in charge diffusion, where the charge trail becomes broader thus giving the event a headlike appearance. As can be seen, protons can be useful sources in measuring the point spread and MTF responses.

DAMAGE

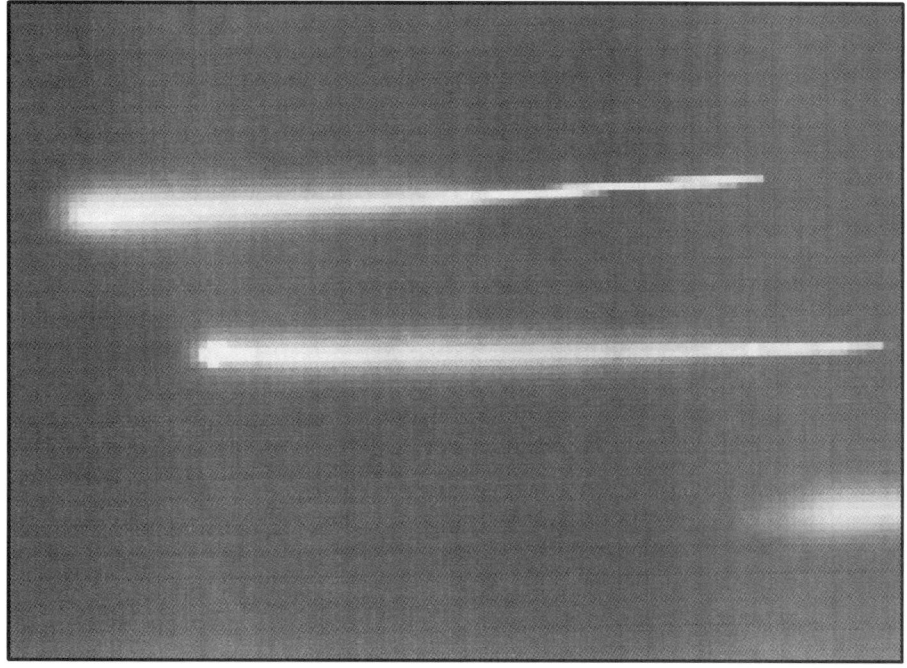

Figure 8.10(a) Grazing incident 9-MeV proton transient events.

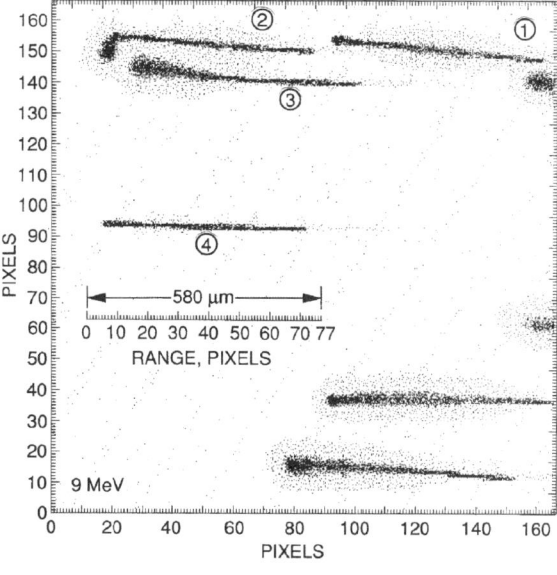

Figure 8.10(b) Several 9-MeV proton transient events.

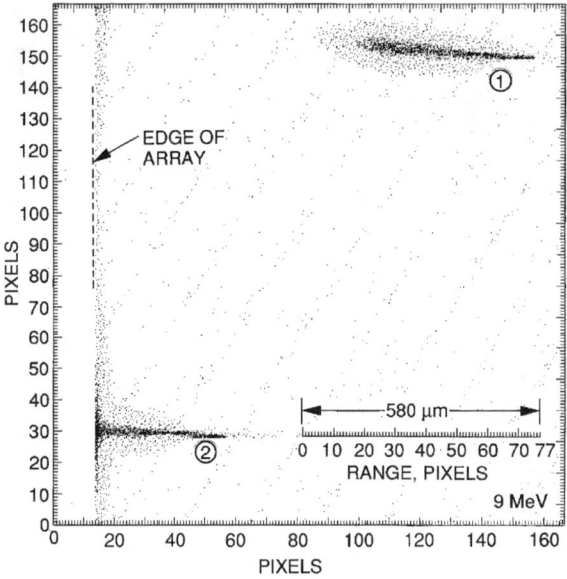

Figure 8.10(c) A 9-MeV proton interacting at the edge of the array.

Each proton track seen is different depending on how it is deflected by silicon lattice atoms. Note that one proton event in Fig. 8.10(b) takes an immediate turn downward, indicating a possible nuclear interaction (event # 2). Another proton in the same image stays near the surface and likely leaves the CCD without leaving a Bragg peak signature (event # 4). Charge generated at the Bragg peak in a single pixel is greater than the charge capacity of the pixel, which is approximately $35,000\,e^-$ for this CCD. This would result in blooming; however, charge diffusion among pixels at the Bragg peak prevents this from happening for most events seen. The proton in Fig. 8.10(c) makes its way to the edge of the CCD into field-free material (event #2). Note that charge diffuses back into the array along the edge.

Figure 8.11 shows ionization trails generated by 6-MeV protons that have a shorter range of 290 µm or 38 pixels. The event labeled #3 in Fig. 8.11(a) penetrates the gate and leaves the epitaxial layer and then comes back, producing a gap in the ionization trail. Figure 8.11(b) is a magnified view of event #2, showing the amount of charge diffusion between pixels as the proton goes deeper into the epilayer.

Figure 8.12(a) plots charge generated as a function of distance for different energy protons. Note that the stopping power and charge generated is nearly constant as the proton travels until the Bragg peak is approached. Figure 8.12(b) is a line trace taken through the top 9-MeV proton event in Fig. 8.10(a), showing the Bragg peak quite clearly. The top trace of Fig. 8.13 plots energy absorbed (eV) per angstrom as a function of depth (microns) into silicon for very low energy protons. The lower figure plots the corresponding energy absorbed as a function of depth.

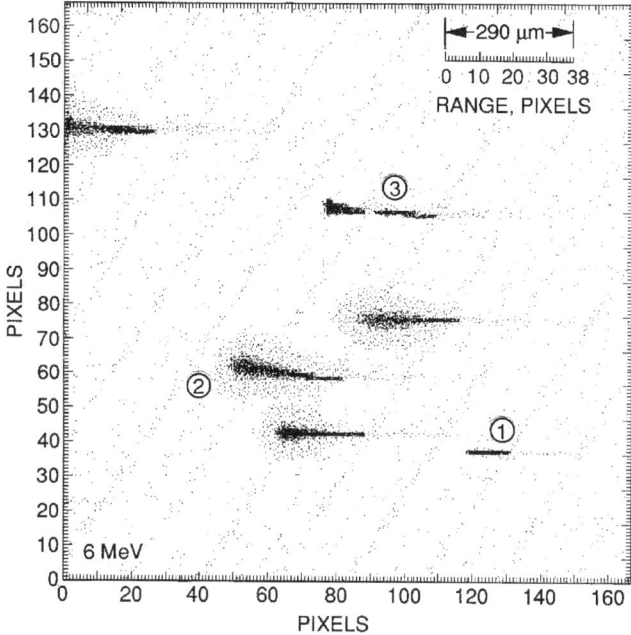

Figure 8.11(a) Several 6-MeV proton transient events.

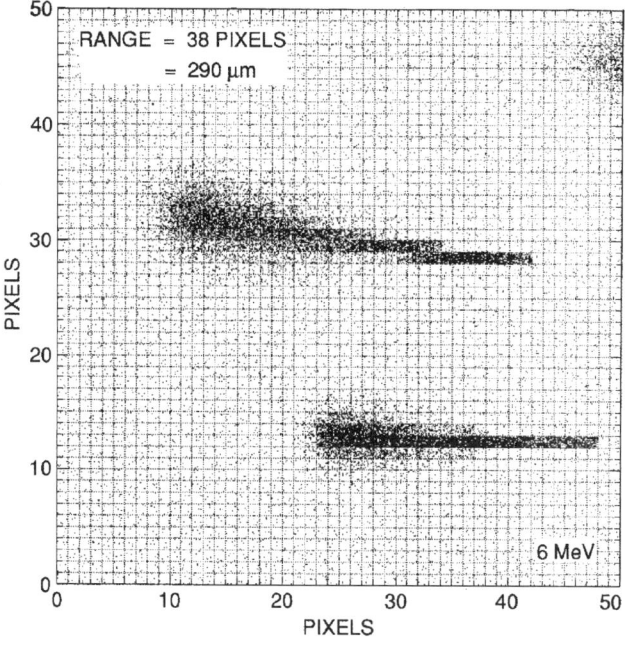

Figure 8.11(b) Magnified view of two 6-MeV proton events.

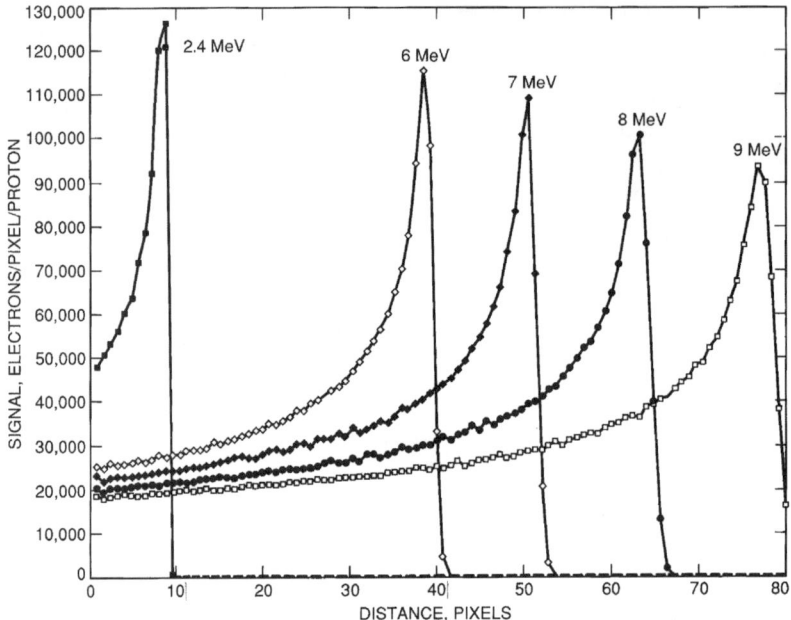

Figure 8.12(a) Charge calculated by grazing incident protons showing the Bragg peak.

For normally incident ions, the number of e-h pairs generated per ion is

$$\text{e-h} = \frac{10^9 S x_{path} \rho}{E_{e-h}}, \tag{8.8}$$

where S is the stopping power on the ion (MeV-cm^2/mg), x_{path} is the path length (cm), ρ is the density of the target material (g/cm^3) and E_{e-h} is conversion of electron-volts to e-h pairs in either the silicon (3.65 eV/e$^-$) or oxide (18 eV/e$^-$).

Example 8.8

Determine the number of e-h pairs generated when a 2-MeV alpha particle (helium nucleus) passes through 1000 Å of gate oxide. Assume a stopping power of 1 MeV-cm^2/mg.

Solution:
From Eq. (8.8),

$$\text{e-h} = \left(10^9 \times 1 \times 10^{-5} \times 2.2\right)/18,$$

$$\text{e-h} = 1.22 \times 10^3 \text{ e-h pairs}.$$

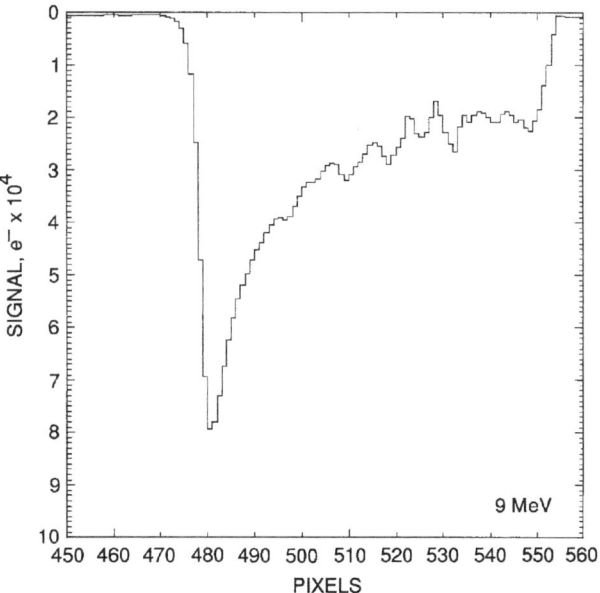

Figure 8.12(b) Charge generated by the top proton event shown in Fig. 8.10(a).

Example 8.9

Determine the number of e-h pairs generated by a 100-MeV proton that passes through a 10-μm epitaxial layer. Assume a stopping power of 5.92×10^{-3} MeV-cm²/mg.

Solution:
From Eq. (8.8),

$$\text{e-h} = [10^9 \times (5.92 \times 10^{-3}) \times 10^{-3} \times 2.32]/3.65,$$

$$\text{e-p} = 3.76 \times 10^3 \text{ e-h pairs}.$$

As an ion passes through the CCD its stopping power will increase, resulting in more charge generation as it slows down. Equation (8.8) assumes that the stopping power is constant and therefore does not give an exact value when the range of the ion is comparable to the thickness of the CCD. The true charge generated is found by

$$\text{e-h} = \frac{E_{ENTER} - E_{EXIT}}{E_{e-h}}, \tag{8.9}$$

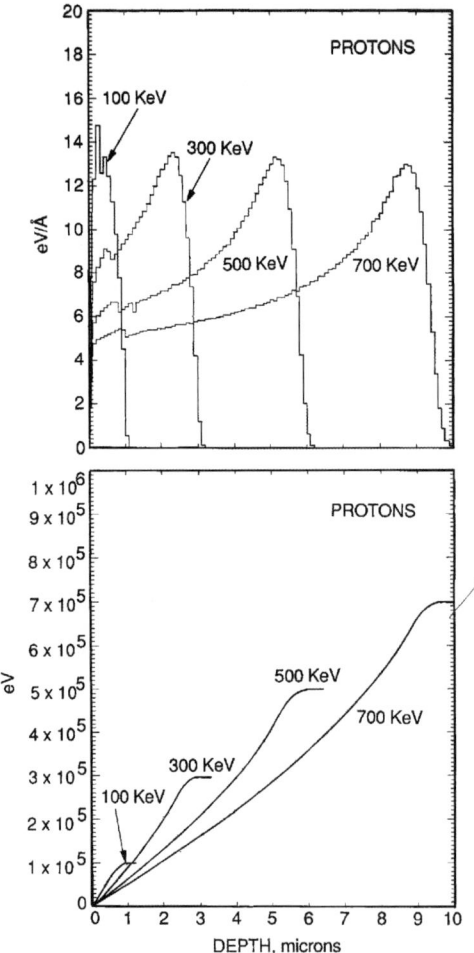

Figure 8.13 Proton ionization energy loss (IEL) as a function of depth for very low proton energies.

where E_{ENTER} is the energy of the incident ion (eV) and E_{EXIT} is the energy of the ion when exiting the epitaxial layer (cm). The next example will show how exit energy is found.

Example 8.10

Determine the exit energy for a 1-MeV proton that passes through a 10-μm-thick CCD.
Solution:
Step 1-Determine the range of the 1-MeV proton in the CCD. From Appendix G1, $R = 15.67$ μm.

Step 2-Subtract the CCD thickness from this range. This is $15.67 - 10 = 5.67$ μm.

Step 3-From Appendix G1 find the energy for range of 5.67 μm. This energy is approximately 0.505 MeV and represents the exit energy.

Example 8.11

Determine the number of electrons created by a 1-MeV proton that passes through a 10-μm CCD. Use Eqs. (8.8) and (8.9) and compare answers. Assume an initial stopping power of 0.177 MeV-cm^2/mg.

Solution:
From Eq. (8.8),

$$\text{e-h} = 10^9 \times 0.177 \times 10^{-3} \times 2.32/3.65,$$

$$\text{e-h} = 1.125 \times 10^5.$$

From Eq. (8.9) and the example above,

$$\text{e-h} = [10^6 - (5.05 \times 10^5)]/3.65,$$

$$\text{e-h} = 1.36 \times 10^5.$$

Although the answers are comparable, Eq. (8.8) produces less signal because it assumes a constant stopping power. For CCD work, Eq. (8.8) can only be used when proton energies are greater than 1 MeV.

Figure 8.14(a) shows an image of normally incident 320-keV proton events taken by a 25-μm epitaxial virtual-phase CCD. The range of a 320-keV proton is approximately 3 μm. Therefore, events are only associated with no more than four pixels because interaction takes place within the 7-μm depletion region. Figure 8.14(b) shows 2.75-MeV proton events that have a range of 80 μm. Charge now occupies several pixels because of charge diffusion within field-free material below the potential well.

Figure 8.14(c) plots experimental and theoretical charge generation as a function of proton energy for two virtual-phase CCDs built on 10- and 25-μm-thick epitaxial material. Maximum charge generation for the 10-μm device occurs at 0.8 MeV (approximately 200,000 e$^-$/proton). The range of a 0.8-MeV proton is approximately 10 μm, which is equivalent to the epitaxial thickness. Therefore, the Bragg peak occurs in the photosensitive volume of the CCD at this energy. Theoretically, 220,000 e$^-$ should be measured (800,000 eV/3.65 eV/e$^-$). The signal level is slightly lower than theory because some ionization takes place in the

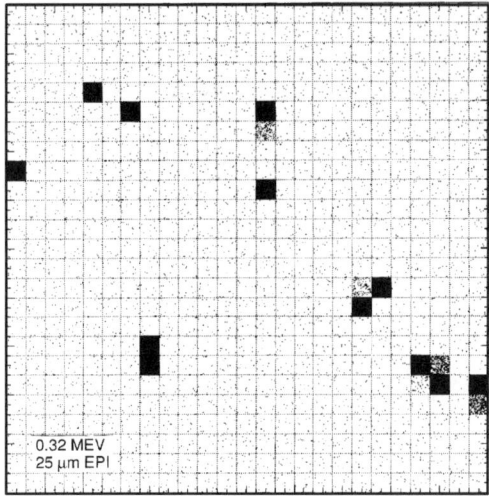

Figure 8.14(a) Image of normal incident 0.32-MeV proton events showing no significant charge diffusion.

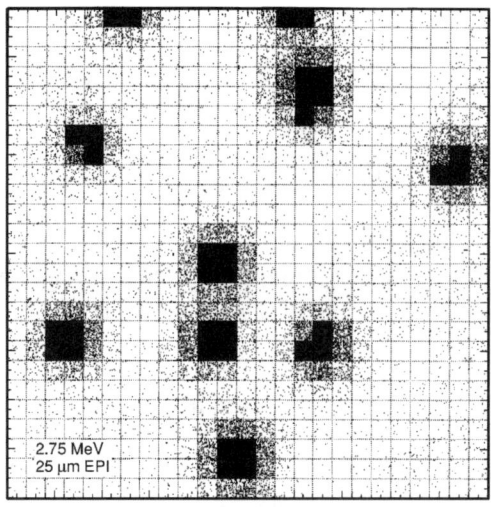

Figure 8.14(b) Image of normal incident 2.75-MeV proton events showing charge diffusion effects.

gate structures of the CCD. Maximum charge generated for the 25-μm CCD occurs at 1.3 MeV. This proton has a range of 24 μm, again about equal to the epitaxial thickness.

Figure 8.14(d) presents a 400-keV proton event histogram taken from the 10-μm epitaxial CCD. Energy resolution is limited by straggling characteristics of the proton source (see Example 8.12 below). Two peaks are seen in the response because the virtual region of the pixel is more transparent than the clocked region,

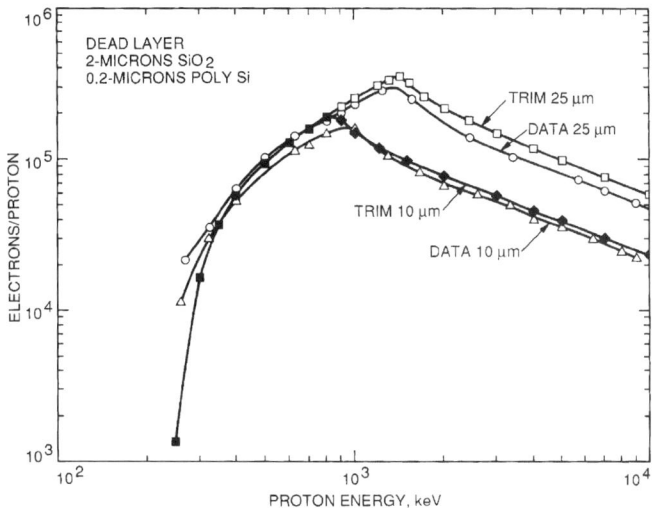

Figure 8.14(c) Experimental and theoretical charge generated by normal incident protons as a function of energy for two epitaxial thicknesses.

resulting in less gate absorption (refer to virtual phase technology discussions in Chapter 4). Figure 8.14(e) shows proton stacking responses as a function of proton energy for the same CCD. Each dot displayed represents the charge generated by a proton event. As in Fig. 8.14(c), maximum charge generation occurs at 800 keV. Again, note that two levels of charge are seen at lower energies. At 100 keV, only the virtual region is sensitive because protons at this energy cannot penetrate the 2000-Å poly gate electrode and 2 μm of glass that covers the clocked region. The range of a 100-keV proton is only 8797 Å in silicon.

Example 8.12

Determine the energy resolution for the 0.4-MeV histogram in Fig. 8.14(d). Express answer in rms e^-, rms eV, e^- FWHM and eV FWHM. Assume an energy resolution of 634 DN rms from Fig. 8.14(e) and camera gain constant of 5.5 e^-/DN. Also, find the ideal energy resolution.
Solution:
Each resolution is

$$\text{rms } e^- = 3487 \, e^- \text{ rms (multiplying 634 DN by 5.5} \, e^-/\text{DN)},$$

$$\text{rms eV} = 12730 \, \text{eV rms (multiplying 3487} \, e^- \text{ by 3.65 eV/} e^-),$$

$$e^- \text{ FWHM} = 8212 \, e^- \text{ FWHM (multiplying 3487} \, e^- \text{ by 2.355)},$$

$$\text{eV FWHM} = 30000 \, \text{eV FWHM (multiplying 12730 eV by 2.355)}.$$

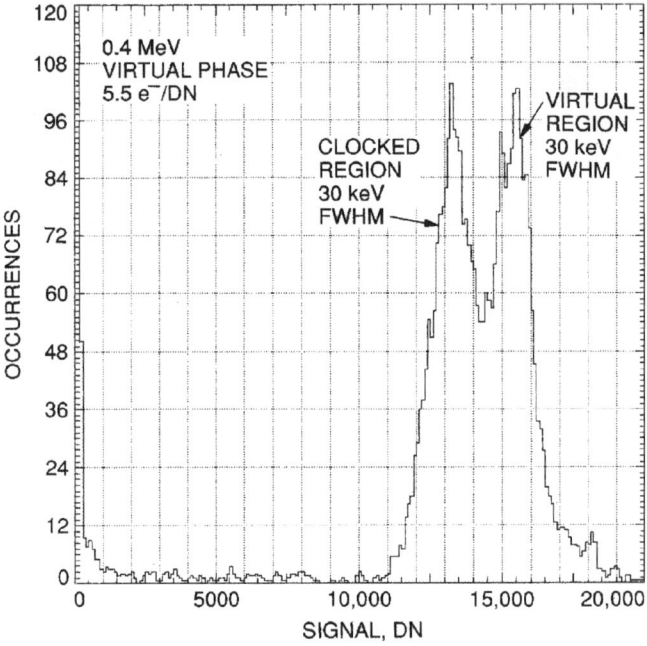

Figure 8.14(d) A 0.4-MeV proton signal histogram generated by a virtual-phase CCD showing virtual and clocked region responses.

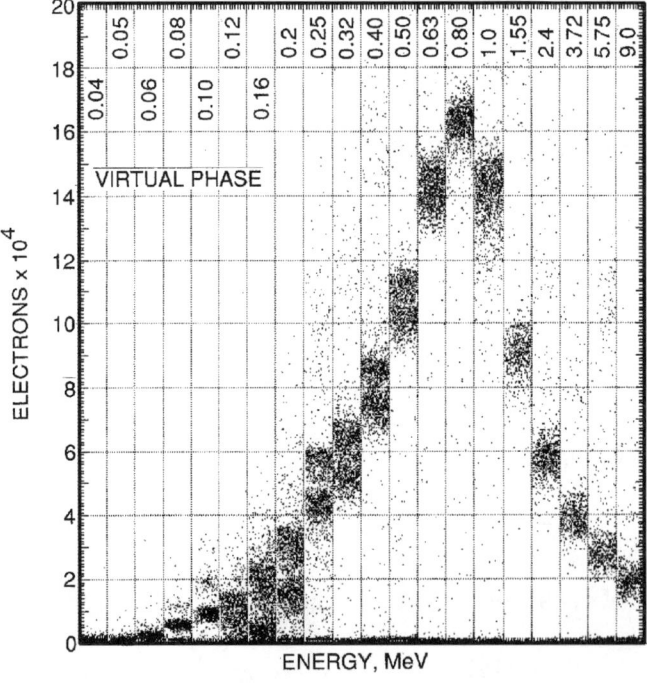

Figure 8.14(e) Proton event stacking plots as a function of energy.

The ideal energy resolution is found by Eq. (2.39),

$$\text{eV FWHM} = 2.355 \times \left[0.1 \times \left(4 \times 10^5\right)\right]^{1/2},$$

$$\text{eV FWHM} = 471\,\text{eV}.$$

8.1.4.3 Photon Transients

Photons also produce transient signals in the CCD. Three different interactions in the CCD are possible: the photoelectric effect, the Compton effect and pair production. In all three cases, the interaction produces energetic free electrons that give rise to a transient signal. Low-energy photons (approximately < 0.1 MeV) interact through the photoelectric effect. Here the incident photon excites and emits a K-shell electron ionizing the target atom. The free electron excites neighboring atoms, and in turn generates e-h pairs. Either a low-energy photon is generated when the K-shell is filled by an outer L-shell orbital electron, or a low-energy Auger electron is emitted from the L shell. This electron can interact photoelectrically, releasing additional e-h pairs. This process is how Fe^{55} x rays generate e-h pairs. The probability of a photoelectric interaction decreases with increasing photon energy and decreasing atomic number.

The minimum energy required to create a single e-h pair in silicon dioxide is approximately 9 eV, its bandgap ($\lambda = 1380$ Å). At this energy the oxide layer becomes opaque to incoming photons (analogous to 1.1-eV photons for silicon). As we will discuss below, direct damage to the gate oxide layer will begin to take place at this energy. The average energy to create multiple e-h pairs in silicon dioxide is 18 eV (analogous to 3.65 eV/e^- for silicon). Electron-hole (e-h) generation occurs in the UV, EUV, soft x-ray, hard x-ray and gamma-ray energy spectrum with decreasing cross-section.

Compton scattering is important for photons with energy greater than 100 keV but less than 10 MeV. For example, Compton scattering occurs when the CCD is irradiated with Co^{60}, 1.25-MeV photons. In comparison to the photoelectric effect, the Compton effect does not totally consume the energy of the interacting photon. Here, the incident photon energy is greater than the binding energy of the K-shell electron. Therefore, the interacting photon transfers its energy to a target electron, and a new photon is created that can itself interact photoelectrically.

Pair production occurs for energies > 10 MeV, although the threshold for interaction is approximately 1 MeV. At this energy the photon collides with the target atom, creating an electron-positron pair. The incident photon is annihilated.

8.1.5 IONIZATION DAMAGE EQUIVALENCE

When working with a complicated radiation environment, it is often convenient to irradiate the CCD with a single energy that induces the same total dose. We refer to this method of testing as ionization equivalence. Table 8.1 shows how this is accomplished for a Hubble like proton radiation spectrum. Here we tabulate the energy bin number, the average energy for that bin (MeV), the fluence (protons/cm^2), the stopping power for SiO$_2$ (MeV-cm^2/mg and eV/Å), the total dose [rad(SiO$_2$)], the energy lost in 1000 Å of oxide (eV) and the number of e-h pairs created assuming 18 eV/pair. The total dose for the entire spectrum is calculated for the dose column along with the total number of e-h pairs generated. We find that an 8-MeV proton is a good energy to simulate space environments such as Hubble. As shown at the end of the table, a fluence of 3.5×10^8 8-MeV protons produces 252 rad or 1.92×10^{10} e-h pairs approximately equivalent to the working spectrum.

8.1.6 IONIZATION DAMAGE

This section reviews some theoretical and detailed aspects of how ionization damage actually occurs. We will address several different physical effects observed immediately after the CCD is subjected to ionizing radiation. The ionization damage process can be divided into eight areas of study: e-h generation, e-h recombination, fractional yield, hole transport, hole trapping, annealing and interface state creation. These mechanisms work together in a complex manner in damaging the CCD's gate dielectric. For purposes of discussion, we assume that the gate insulator is SiO$_2$. However, some comments are given about dual insulating systems such as silicon nitride (Si$_3$N$_4$) on oxide (SiO$_2$).

8.1.6.1 e-h Generation

To our knowledge, visible and near-IR photons do not produce ionization damage. However, as we move into the UV, the first noticeable permanent ionization effect occurs at a wavelength of approximately 2900 Å (4.3 eV). When a CCD is subjected to UV radiation, some photons pass through the poly silicon gates and gate oxide. They then interact with the buried channel immediately next to the Si–SiO$_2$ interface. Some electrons are photoemitted into the gate dielectric where they can get trapped, charging the oxide negatively (i.e., positive flatband shift). For multiphase CCDs, nonuniform charging occurs because the poly electrodes overlap. This can result in spurious potential barriers and traps within the signal channel that degrade CTE performance. An intense UV light source is required to induce the effect because the poly silicon gates absorb and shield most incident radiation. CCDs coated with phosphors do not show this problem because the coating absorbs all incoming UV photons. Backside-illuminated CCDs are not sensitive to near-UV damage because the epitaxial layer acts as a shield, preventing photons from reaching the frontside surface. Curiously, UV photoemission is not always a problem, and can sometimes be used to anneal radiation damage that induces a negative flatband shift. We will discuss an example of this application below.

Shortward of 1800 Å, photons interact directly with the gate oxide layer and generate e-h pairs. True ionization damage is initiated when this happens. If there

Table 8.1 Proton radiation environment that is simulated with single proton energy (8 MeV) for equivalent total dose.

Bin (#)	Energy (MeV)	Fluence (protons/cm^2)	Stopping power (MeV-cm^2/mg)	Dose (rad)	Stopping power (eV/Å)	Energy loss in 1000 Å (eV)	e-h pairs (18 eV/pair)
1	0.18	1.19E + 04	4.85E − 01	9.23E − 02	1.07E + 01	1.27E + 08	7.05E + 06
2	0.223	1.48E + 04	4.51E − 01	1.07E − 01	9.93E + 00	1.47E + 08	8.17E + 06
3	0.282	2.02E + 04	4.08E − 01	1.32E − 01	8.97E + 00	1.81E + 08	1.01E + 07
4	0.359	2.87E + 04	3.61E − 01	1.66E − 01	7.94E + 00	2.28E + 08	1.27E + 07
5	0.442	3.68E + 04	3.21E − 01	1.89E − 01	7.05E + 00	2.60E + 08	1.44E + 07
6	0.579	5.46E + 04	2.80E − 01	2.45E − 01	6.16E + 00	3.36E + 08	1.87E + 07
7	0.706	1.04E + 05	2.49E − 01	4.14E − 01	5.48E + 00	5.70E + 08	3.16E + 07
8	0.897	1.11E + 05	2.13E − 01	3.79E − 01	4.70E + 00	5.21E + 08	2.90E + 07
9	1.1	1.69E + 05	1.88E − 01	5.08E − 01	4.13E + 00	6.99E + 08	3.88E + 07
10	1.4	2.36E + 05	1.60E − 01	6.05E − 01	3.53E + 00	8.32E + 08	4.62E + 07
11	1.8	3.52E + 05	1.35E − 01	7.61E − 01	2.97E + 00	1.05E + 09	5.82E + 07
12	2.31	5.66E + 05	1.14E − 01	1.03E + 00	2.51E + 00	1.42E + 09	7.89E + 07
13	2.79	7.45E + 05	9.89E − 02	1.18E + 00	2.18E + 00	1.62E + 09	9.00E + 07
14	3.68	1.13E + 06	8.23E − 02	1.49E + 00	1.81E + 00	2.05E + 09	1.14E + 08
15	4.58	1.71E + 06	6.97E − 02	1.91E + 00	1.53E + 00	2.62E + 09	1.46E + 08

Table 8.1 (Continued).

Bin (#)	Energy (MeV)	Fluence (protons/cm²)	Stopping power (MeV-cm²/mg)	Dose (rad)	Stopping power (eV/Å)	Energy loss in 1000 Å (eV)	e-h pairs (18 eV/pair)
16	5.45	3.74E + 06	6.00E − 02	3.59E + 00	1.32E + 00	4.93E + 09	2.74E + 08
17	7	3.91E + 06	4.99E − 02	3.12E + 00	1.10E + 00	4.29E + 09	2.38E + 08
18	9.13	5.58E + 06	4.11E − 02	3.67E + 00	9.04E − 01	5.04E + 09	2.80E + 08
19	11.4	8.45E + 06	3.54E − 02	4.78E + 00	7.78E − 01	6.57E + 09	3.65E + 08
20	14	1.27E + 07	2.92E − 02	5.94E + 00	6.43E − 01	8.17E + 09	4.54E + 08
21	18.3	2.16E + 07	2.39E − 02	8.26E + 00	5.26E − 01	1.14E + 10	6.31E + 08
22	22.5	3.37E + 07	2.03E − 02	1.10E + 01	4.47E − 01	1.51E + 10	8.37E + 08
23	28.5	4.67E + 07	1.60E − 02	1.20E + 01	3.52E − 01	1.64E + 10	9.13E + 08
24	35.2	6.61E + 07	1.30E − 02	1.37E + 01	2.86E − 01	1.89E + 10	1.05E + 09
25	45.3	1.02E + 08	1.14E − 02	1.86E + 01	2.51E − 01	2.56E + 10	1.42E + 09
26	57.3	1.35E + 08	9.72E − 03	2.10E + 01	2.14E − 01	2.89E + 10	1.60E + 09
27	69.8	1.66E + 08	8.06E − 03	2.14E + 01	1.77E − 01	2.94E + 10	1.64E + 09
28	90.8	1.81E + 08	6.66E − 03	1.93E + 01	1.47E − 01	2.65E + 10	1.47E + 09
29	113	2.64E + 08	5.74E − 03	2.42E + 01	1.26E − 01	3.33E + 10	1.85E + 09
30	136	2.71E + 08	4.90E − 03	2.12E + 01	1.08E − 01	2.92E + 10	1.62E + 09
31	182	2.61E + 08	4.09E − 03	1.71E + 01	9.00E − 02	2.35E + 10	1.30E + 09
32	221	2.49E + 08	3.60E − 03	1.43E + 01	7.92E − 02	1.97E + 10	1.10E + 09
33	278	1.97E + 08	3.12E − 03	9.83E + 00	6.86E − 02	1.35E + 10	7.51E + 08
34	335	1.07E + 08	3.00E − 03	5.14E + 00	6.60E − 02	7.06E + 09	3.92E + 08
				TOTAL 2.47E + 02		TOTAL 1.89E + 10	
Equivalent Fluence at 8 MeV							
8		3.50E + 08	4.50E − 02	2.52E + 02	9.90E − 01	3.46E + 11	1.92E + 10

were no e-h pairs generated, there would be no ionizing damage to speak of. As discussed below, it will be the hole that is mainly responsible for the flatband shift and the dark current increase experienced.

8.1.6.2 e-h Recombination and Fractional Yield

Once e-h hole pairs are generated, they have an opportunity to recombine. To do so, they must recombine in a very short time period before separating by diffusion and/or field-induced drift. If the CCD is powered, the carriers will separate and move in opposite directions because of the electric fields generated in the insulator. When the CCD is not powered, carriers mainly diffuse. Electrons and holes move through the oxide at much different rates. The mobility for electrons in oxide is 20 to 40 cm^2 V^{-1} s^{-1}, whereas for holes the mobility is considerably less, 10^{-4} to 10^{-7} cm^2 V^{-1} s^{-1}. A slight bias will cause electrons to move near their saturation velocity, about 10^7 cm/sec. Therefore, the recombination process must occur in a very short time period (on the order of a few picoseconds).

Only holes remain in the gate insulator after initial recombination and electron migration occur. The fraction of holes that remain is defined as

$$h_{FY} = \frac{h}{e\text{-}h}, \tag{8.10}$$

where h_{FY} is the fractional yield, h is the number of holes remaining and e-h is the initial number of e-h pairs as calculated by Eq. (8.8).

The fractional yield depends highly on the energy and type of incident particle. When an ion passes through the oxide layer, it generates a column of e-h pairs with a charge density proportional to stopping power of the particle. The faster a particle releases its energy, the greater the charge density and recombination. For example, h_{FY} for a 100-MeV proton is approximately an order of magnitude greater compared with a 0.1-MeV proton. The h_{FY} for incident alpha particles (He nuclei) is < 6%, which is about an order of magnitude less than Co^{60} gamma rays. The charge separation effect is even more dramatic for heavy ions. For Cu ions with energy of 1 MeV, the yield is < 1%.[30]

Fractional yield increases rapidly as bias is applied to the CCD because internal fields generated will separate carriers. For example, 80% of the charge escapes recombination for 12-MeV electrons when a field strength of 10^6 V/cm is applied (approximately equal to the electric fields generated in an operating CCD). If the field is reduced to 0.1 × 10^6 V/cm, only 25% escape recombination. The h_{FY} for Fe^{55} x rays is approximately 50% for a 1-MeV/cm field and about 10% without bias.[30]

It should be noted that e-h recombination is greater for Fe^{55} x-ray events than Co^{60} events. Therefore, the effective dose for soft x rays is less. Dose enhancement and recombination work in opposite directions and tend to cancel each other out. Therefore, when Fe^{55} damage is compared to Co^{60} damage, the net effect is usually less than a factor of two in either direction.

8.1.6.3 Hole Transport

When holes separate from electrons, they can either move to the Si–SiO$_2$ interface or to the gate electrode, The direction of the hole depends on the polarity of the gate voltage. Recall that the voltage drop across the oxide depends on the gate voltage and the voltage applied to the signal channel (i.e., V_{REF}). For buried-channel operation, the surface potential at the Si–SiO$_2$ interface is always greater than the gate voltage. Therefore, as the CCD is clocked, the electric field always directs electrons in the direction of Si–SiO$_2$ interface and holes in the direction of the gate.

Example 8.13

Calculate the electric field strength across a 1000-Å gate dielectric for two bias conditions: (1) $V_G = 5$ V, $V_S = 10$ V, and (2) $V_G = -8$ V and $V_S = 0$ V (i.e., inverted).
Solution:
Case 1
The voltage drop across the gate oxide is 5 V. The corresponding electric field is

$$E = V/d,$$

$$E = 5/(10 \times 10^{-6}),$$

$$E = 0.5 \, \text{MV/cm}.$$

Case 2
The voltage drop across the gate dielectric is 8 V. The electric field in the gate oxide is

$$E = 8/(10 \times 10^{-6}),$$

$$E = 0.8 \, \text{MV/cm}.$$

Note that the electric field is always in the same direction independent of gate potential polarity. However, when the substrate and V_{REF} are intentionally grounded, the voltage drop across the oxide is simply equal to the gate voltage applied. This bias condition is used for experimental purposes in testing both electric field directions.

Although considerably slower than electron migration, hole movement under most operating conditions is still swift and difficult to detect in CCD irradiation experiments. The transit time depends on field strength, operating temperature, and thickness of the oxide layer. For example, at room temperature, hole migration

Damage

is on the order of 10^{-3} sec for a field strength of 1 MeV/cm.[31] However, if the CCD is very cold and unbiased, times on the order of thousands of seconds may be required for complete hole transport.

8.1.6.4 Hole Trapping

Immediately after irradiation, a net negative flatband voltage exists because of new holes in the dielectric. The flatband voltage induced by these carriers at this time is given by[32]

$$V_{FB} = -\frac{1}{\varepsilon_{OX}} \int_0^{x_{OX}} x \rho_{OX}(x) dx, \qquad (8.11)$$

where x_{OX} is the thickness of the dielectric and $\rho_{OX}(x)$ is the distribution of holes in the layer as a function of distance from the gate electrode. Note that positive charge (holes) give rise to a negative flatband shift.

Note also from Eq. (8.11) that the flatband voltage is dependent on where holes are situated. The largest flatband shift will occur when holes become trapped at the Si–SiO$_2$, which is predominantly the case in real life. Equation (8.11) can be integrated by letting $\rho_{OX}(x) = Q_h \delta(x_o)$, where $\delta(x_o)$ is a delta-function located at the Si–SiO$_2$ interface. This produces the equation

$$V_{FB} = -\frac{qQ_h x_{ox}}{\varepsilon_{OX}}, \qquad (8.12)$$

where Q_h is the hole charge density per unit area at the Si–SiO$_2$ interface (hole traps/cm^2).

Example 8.14

Determine the hole trap density for a flatband shift of -1 V. Assume an oxide thickness of 1000 Å.
Solution:
From Eq. (8.12),

$$Q_h = 1 \times (3.45 \times 10^{-13}) / [(1.6 \times 10^{-19}) \times 10^{-5}],$$

$$Q_h = 2.16 \times 10^{11} \text{ holes/cm}^2.$$

It is interesting to note that on the silicon surface there are approximately 7×10^{14} atoms/cm^2. Therefore, for this example there is only 1 trap per 3246 atoms, a small number considering that a -1-V flatband shift takes place.

8.1.6.5 Annealing

Trapped holes are not permanent features but slowly detrap (i.e., anneal). There are two main mechanisms that anneal a hole trap: tunnel annealing and thermal annealing. The former mechanism is when electrons from the silicon tunnel into the oxide and recombine with the hole. Thermal annealing is when a hole escapes a trap thermally. The speed at which these processes work depends on the distance of the trap from the Si–SiO$_2$ interface and the depth of the trap or corresponding emission time constant. Therefore, after irradiation, the flatband voltage may take days to stabilize and require even longer periods at cold operating temperatures.

8.1.6.6 Interface State Creation

Recall from previous chapters that interface traps are allowed energy levels in the forbidden band. Good interfaces exhibit an interface state density that ranges between 10^9 and 10^{11} traps/cm^2-eV at midband (i.e., those states important to dark current generation). The actual density is very dependent on fabrication details and varies significantly from manufacturer to manufacturer. One possible form of an interface state is a silicon atom at the interface, back-bonded to three Si atoms on the silicon side and having a single dangling bond extending into the SiO$_2$ interface (called a P$_b$ center). This structure is illustrated in Fig. 8.15 where we show the Si–SiO$_2$ interface and two unsatisfied dangling bonds.

The incorporation of hydrogen into the CCD process will passivate a significant portion of dangling bonds as shown (i.e., Si •+ H → Si–H). Hydrogen annealing is normally employed throughout the CCD process, especially when high-energy plasma etches are utilized. In this step a mixture of 5% hydrogen and 95% nitrogen is typically used (i.e., a forming gas mixture). Some CCD manufacturers have used

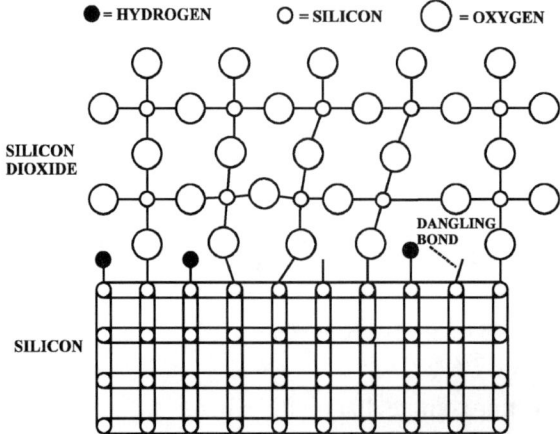

Figure 8.15 Si–SiO$_2$ interface region showing dangling and hydrogen-passivated bonds.

a single anneal cycle at the very end of the process, using 100% hydrogen at 350°C. For oxide/silicon nitride gate dielectrics, the hydrogen gas can only enter through the channel stop/field oxide regions where nitride is absent (hydrogen has great difficulty diffusing through nitride). For implanted channel stops without field oxide, small nitride windows must be provided in the design to allow hydrogen into the silicon.

Unfortunately, passivating dangling bonds with hydrogen will produce a CCD that is vulnerable to ionizing radiation damage. As discussed below, radiation will break these weak bonds and form dangling bonds that will produce a flatband shift and surface dark current. In other words, ionizing radiation undoes what the CCD manufacturer has worked hard to build in. The radiation data presented below is generated by CCDs with soft oxides that rely on hydrogen passivation.

For some CCDs, dark current and flatband may continue to increase for several years after the sensor is irradiated. This long-term reaction is called "reverse annealing" because the flatband shift and dark current increase rather than decrease after the CCD is irradiated.[33,34] Recall from above, the effect cannot be associated with simple hole-transport because hole movement is too rapid, on the order of a few seconds at room temperature. Another physical explanation for reverse annealing is therefore required.

Reverse annealing has been observed in other MOS devices and various models have been presented to explain the phenomenon.[30,35–37] We briefly present two theories here that appear to fit CCD reverse annealing characteristics. Researchers in this field have proposed a multistep process for the creation of new interface traps. The first step in the process is the transport of radiation-induced holes that migrate through the oxide as discussed above. Next, the holes interact and break weak H-bonds with trivalent Si atoms that form during oxide growth. A H^+ ion is released from this reaction, which drifts toward the Si–SiO$_2$ interface. When reaching the interface, the H^+ ion picks up an electron supplied by tunneling from the bulk silicon nearby, forming a neutral hydrogen atom. This atom then reacts with a hydrogen-passivated bond at the Si–SiO$_2$ interface, producing H$_2$ and a dangling Si bond. In equation form, these last two reactions are

$$3H-Si-H + H^+ + e^- = 3H-Si-H + H^0 = 3H-Si-\bullet + H_2,$$

where 3H—Si—H is a hydrogen-passivated trap and 3H—Si—• is the dangling bond.

The diffusion process of the H^+ ion in the oxide is exponentially dependent on operating temperature. It may take several hours for the above equation to stabilize at room temperature. However, an elevated temperature (say, $> 50°C$) may reduce the reaction time to a few minutes.

A second model applies to oxide/nitride gate insulators where the equilibrium time constant can be considerably longer than a few hours. The model assumes that there are positively charged defects within the oxide layer after irradiation. Hydrogen gas that is present in the nitride layer will then slowly diffuse into the oxide region. In turn, the hydrogen molecules will be "cracked" by the charged

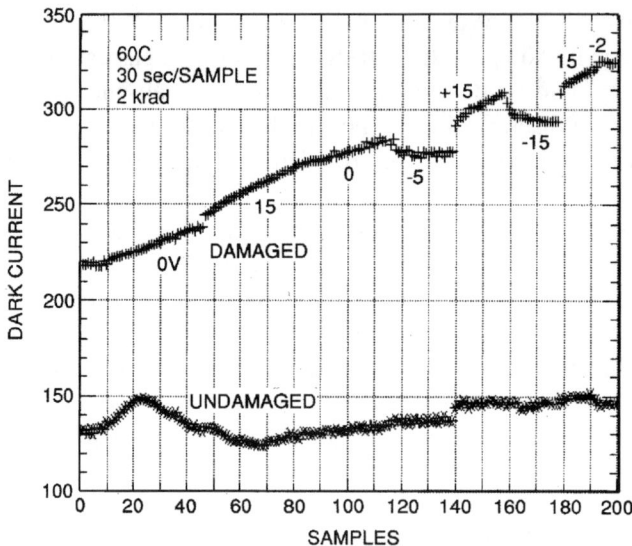

Figure 8.16 Dark current as a function of time showing the reverse annealing effect for different applied gate potentials.

oxide defects, producing H^+ that then starts the first reaction already described above. The time rate limiting mechanism for this reaction is the hydrogen gas that diffuses from the nitride layer into the oxide layer. This process may take several years to complete compared to the diffusion of H_2 and H^+ in the oxide.

It is not surprising that reverse annealing is dependent on bias conditions, because holes and ions are charged. For example, when a negative gate bias relative to the substrate is applied, an electric field is generated that attracts the H^+ ions to the gate instead of to the interface. In theory, this will stop reverse annealing. However, biasing the CCD positively repels the ions to the interface, accelerating the reaction. Figure 8.16 presents dark current data showing how reverse annealing is controlled by bias. In this experiment a test Cassini CCD is first irradiated with 2 krad of 5.9-keV x rays. The CCD is heated to 60°C to speed the reverse annealing reaction. Dark current is measured briefly under normal clocking conditions for each data point shown. A control region shielded from the x rays is also measured to correct for slight temperature differences that occur in the experiment (bottom plot). Each data point represents a bias condition applied to the CCD for a 30-sec time period. For example, note initially that when the gates and V_{REF} are grounded (0 V), dark current increases as positive ions make their way to the Si–SiO$_2$ interface. Biasing the gates to > 15 V promotes the reaction, further increasing the creation of interface states and dark current generation, as shown. However, applying –5 V stops the process completely, as the model above would predict. Returning the bias to > 15 V causes the dark current to increase again. Reverse annealing can be controlled completely by applied bias.

Flatband shift also exhibits reverse annealing characteristics. This is because the new interface states that form also contribute to flatband shift. Here we assume

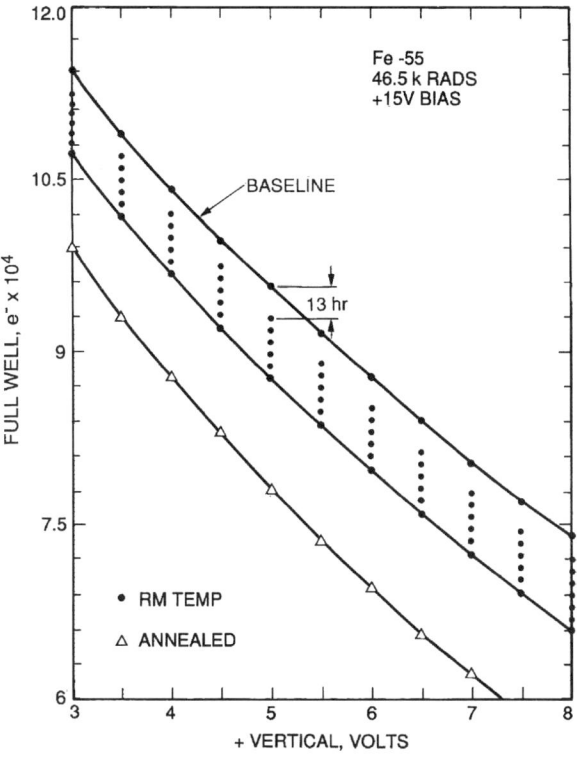

Figure 8.17 Full well transfer curves in response to ionization damage induced by Fe^{55} x rays.

that the states are donorlike in nature and are positively charged when empty, and neutral when filled with an electron. Figure 8.17 plots full well characteristics for a Cassini CCD irradiated with 5.9-keV x rays. The baseline response before irradiation is shown. Every 13 hours, at room temperature, while the device is being constantly irradiated with x rays (at > 15-V bias to maximize radiation damage), a new full well plot is taken. Note that the full well curve slowly shifts toward the left, indicating that new positive states are being created (i.e., it takes less positive clock voltage to reach the same full well level). After exposing the CCD to 46.5 krad, the sensor is annealed at < 50°C for two hours and a new full well curve is generated (triangles). A net flatband shift of approximately 2 V is measured, half of which comes from the high-temperature anneal cycle.

8.1.7 Clock and Bias for Minimum Ionization Damage and Control

Discussions above show that clock bias applied to the CCD strongly affects how ionization damage is induced. For example, minimum damage will result during irradiation if the CCD is not powered. This bias condition is best for e-h recombination in the gate dielectric. On the other hand, after holes are generated it is best to power the CCD and attract holes and H$^+$ ions to the gate electrode, thereby reducing new interface state buildup. Also, operating temperature can be an important parameter. For example, it is best to have the sensor cold for recombination reasons (i.e., e-h pairs will diffuse slower, promoting more recombination). Lowering the temperature also reduces exponentially the reverse annealing process. The Cassini Star Tracker CCD relies and maintains an operating temperature of −40°C to control the reverse annealing problem.

Figure 8.18 plots dark current generated for a CCD subjected to ionizing radiation while gate bias was applied. The substrate and buried channel (i.e., V_{REF}) of the CCD were grounded in this test to study the difference in polarity across the gate dielectric. The plot shows that damage and the corresponding dark current increase with bias. Also note that the damage is polarity independent. Fractional yield is also dependent on operating temperature. Two data points are shown at −20°C and −100°C, with zero gate bias, showing less fractional yield and damage at colder temperatures.

Negative flatband voltage shifts can be negated by lowering the CCD clock voltages by the same amount. For multiphase sensors, the most sensitive clock voltage is the positive vertical clock. Recall from Chapter 4 that optimum full well occurs at a specific operating potential (i.e., where BFW = SFW). A small flatband shift on either side of this point can result in a significant full well loss. For this reason, the multi-phase user may elect to command clock voltage settings when the CCD is used in a radiation environment. This is the case for the Cassini CCD where the positive vertical clock voltage can be set over a large range (approximately 6 V). Other clock and amplifier bias voltages for Cassini are preset and not

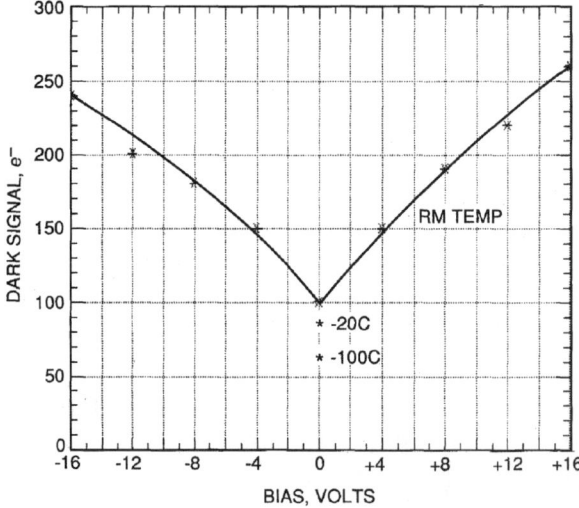

Figure 8.18 Dark current as a function of gate to substrate voltage during irradiation.

commandable. This is because the operating windows for these voltages are well within the expected flatband shift caused by radiation over the mission lifetime (i.e., < 1-V flatband shift in conjunction with an operating window of ±2 V).

8.1.8 IONIZATION DAMAGE MEASUREMENTS

This section presents three example ionizing radiation tests that show subtle effects behind the ionizing damage problem. The tests are called (1) thick and thin oxide charging, (2) UV charging and (3) process-induced charging.

8.1.8.1 Thick and Thin Oxide Charging

Differences in soft and radiation-tolerant hard oxides depend mainly on the thickness of the gate dielectric, the manner in which the oxide is grown (e.g., wet versus dry) and the Si–SiO$_2$ stress that is induced by overlaying gate layers. Of these three, oxide thickness is the most important parameter because the number of e-h pairs generated is proportional to the oxide's volume. In theory, if the oxide thickness were made thin enough, there would be no particle interaction and ionizing damage to speak of. In fact, groups have fabricated MOS devices with ultrathin gate oxides less than 10-nm thick for radiation hardening purposes.[38,39] For thin dielectrics, there is very little flatband shift because the layer is neutralized by tunnel annealing.

The type of gate dielectric is also important. For example, MNOS CCDs with 100 Å of SiO$_2$ and 1200 to 1700 Å of Si$_3$N$_4$ have shown good radiation tolerance.[18,40,41] As Saks quotes, trapping in the silicon nitride layer appears to be negligible because most of the e-h pairs generated in the nitride either recombine or are trapped very close to the point where they are created.

The radiation test discussed in this section compares ionizing damage of a thick-field oxide (10,000 Å) to a thin-gate oxide (500 Å). Field oxide is found over the channel stop regions whereas the gate oxide is over the signal channel. Figure 8.19 shows these regions and where they converge. This important interface is called the "bird's beak." In that the thick field oxide is approximately 20 times thicker than the gate oxide, we would expect more radiation damage to occur there.

Figure 8.20 presents Co^{60} ionizing damage characteristics for a frame transfer MPP CCD processed as Fig. 8.19 shows. The data plots dark current as a function of negative clock drive at different irradiation levels (2 krad, 4 krad and 6 krad). Figure 8.20(a) plots dark current for the imaging registers whose gates were grounded during the exposure. Figure 8.20(b) is a similar plot for the storage registers that were biased at –9 V. A baseline dark current response is also shown in both figures for determining the preirradiated inversion breakpoint (approximately $V_{INV} = -6.0$ V for both regions). Note that inversion breakpoint shifts significantly after irradiation, indicating a negative flatband shift has taken place. The storage region exhibits a greater flatband shift because e-h pairs were allowed to separate by the applied bias. Note that the 6-krad storage region experiences a

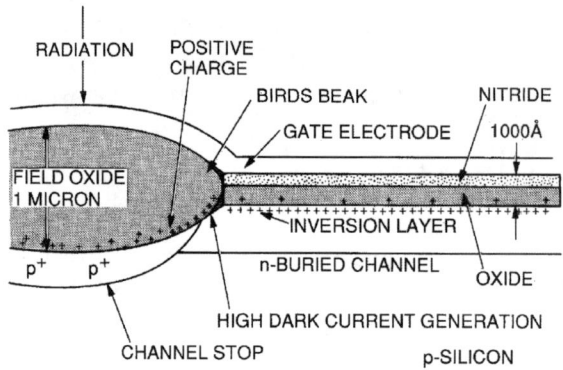

Figure 8.19 Bird's beak region where large flatband voltage shifts take place.

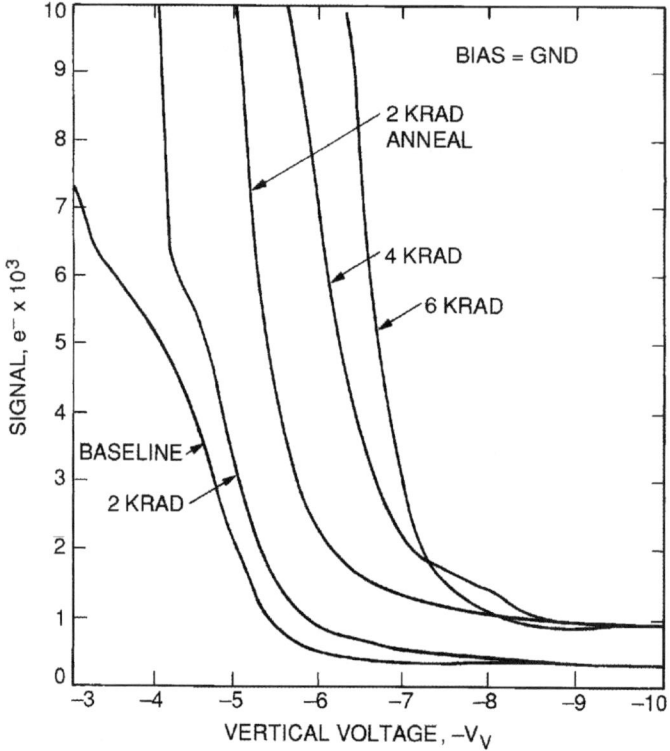

Figure 8.20(a) Dark current as a function of negative vertical clock drive showing flatband shift (unbiased during irradiation).

very large shift, approximately –10 V. After the 2-krad exposure, the CCD was allowed to anneal for two hours at 50°C. The additional flatband shift caused by reverse annealing is readily apparent in both figures. No intentional annealing was attempted for the 4-krad and 6-krad dose levels. The results of Fig. 8.20 show that

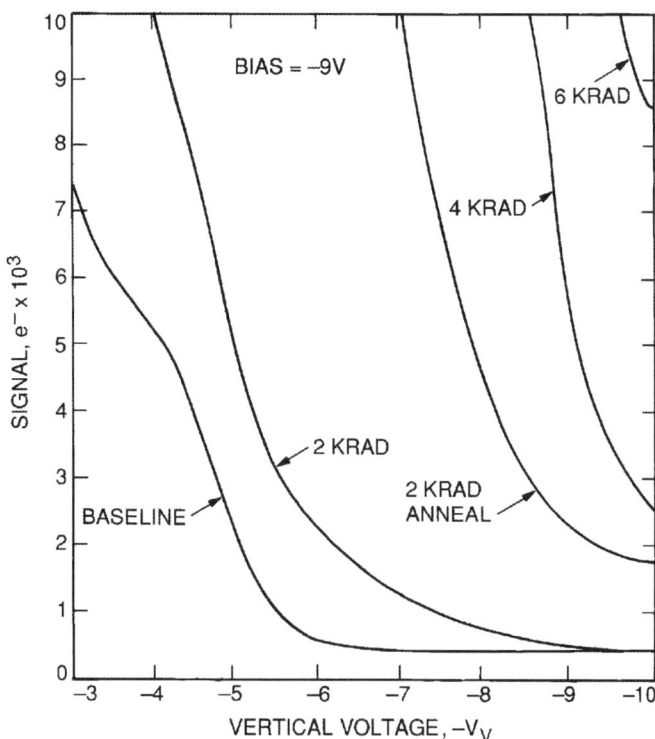

Figure 8.20(b) Dark current as a function of negative vertical clock drive showing large flat-band shift compared to unbiased case.

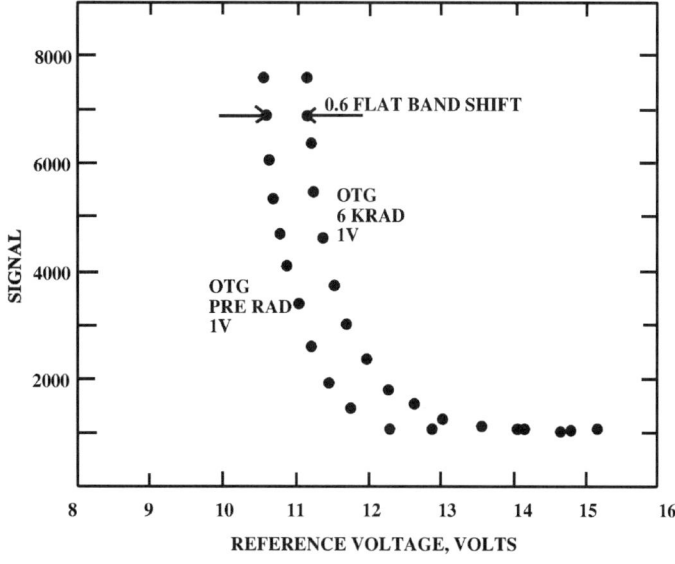

Figure 8.20(c) OTG transfer curve that measures flatband shift in the signal channel.

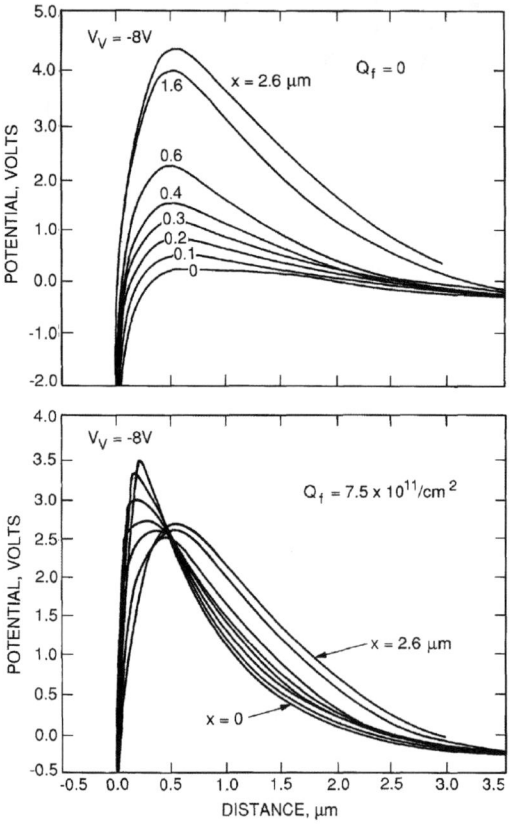

Figure 8.21 Top figure shows potential well plots for non-MPP phases before irradiation at various points in the signal channel and bird's beak region ($x = 0$). Bottom figure shows potential wells for an MPP phase after irradiation, showing a large flatband shift at the bird's beak.

the gate dielectric is charging significantly. The question is, where does such a large flatband shift come from? The channel flatband shift between channel stops is best measured using the OTG transfer curve [refer to Fig. 1.23(b)]. For example, Fig. 8.20(c) shows an OTG transfer curve before and after irradiation with the OTG gate biased at −9 V during irradiation for maximum flatband shift. Note that the charge injection point has only shifted slightly by 0.6 V after the 6-krad exposure. Figure 8.19 helps to explain why the flatband shift for the inversion breakpoint is much greater than the OTG data. As the bird's beak is approached, the oxide thickness increases, which results in more radiation damage and greater flatband shift. Therefore, it takes more negative clock drive to maintain inversion near the bird's beak region compared to the signal channel.

Flatband shift in the bird's beak region also causes full well degradation. Figure 8.21 presents theoretical potential diagrams for a three-phase MPP CCD to show why this happens. The top traces of Fig. 8.21 plot channel potentials before

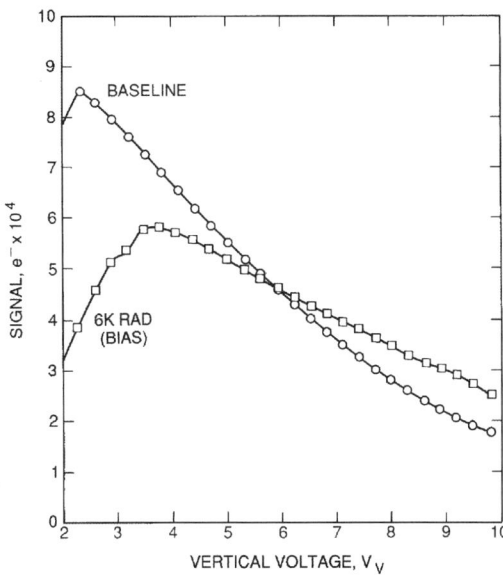

Figure 8.22 Full well transfer curves before and after 6 krad of Co^{60} irradiation for a MPP CCD in the biased state.

irradiation for a non-MPP phase at the center of the channel (labeled $x = 2.6\,\mu\text{m}$) and at various points closer and closer to the bird's beak region (i.e., $x = 0\,\mu\text{m}$). The analysis assumes that all gates are inverted at $V_G = -8$ V. The channel potential is maximum in the center of the channel (i.e., 4.5 V) and falls to zero at the channel stop (i.e., 0 V). We assume there is no trapped positive charge in this case ($Q_f = 0$). The bottom traces of Fig. 8.21 show channel potentials for the MPP phase after irradiation. Note that the potential is lower in the center of the channel because of the boron MPP implant in this phase (approximately 2.7 V, compared to 4.5 V). It is assumed that the trapped charge level is proportional to oxide thickness, with 7.5×10^{11} charges/cm^2 located in the center of the channel. As the bird's beak is approached, the potential increases caused by the positive charge buildup in the field oxide. Recall that full well performance for an MPP CCD is determined by the potential difference between the MPP phase and non-MPP phases. The simulation shows that only a 0.7-V barrier remains after irradiation, compared with 2 V before irradiation; therefore, charge capacity decreases for the CCD. Figure 8.22 shows a full well transfer plot for the same CCD tested in Fig. 8.20(b) for the 6-krad biased case. A significant reduction in full well is experienced after irradiation.

The bird's beak radiation problem described can be alleviated by employing the "super notch" structure shown in Fig. 5.56. The super notch isolates the bird's beak by leaving a small neutral p-region between the signal channel and the 4-μm channel stop to help maintain the inverted state. Figure 8.23 plots dark current generation as a function of negative clock drive and various super notch widths. Data was taken after damaging a 18-μm-pixel test MPP CCD with Co^{60} radiation. As

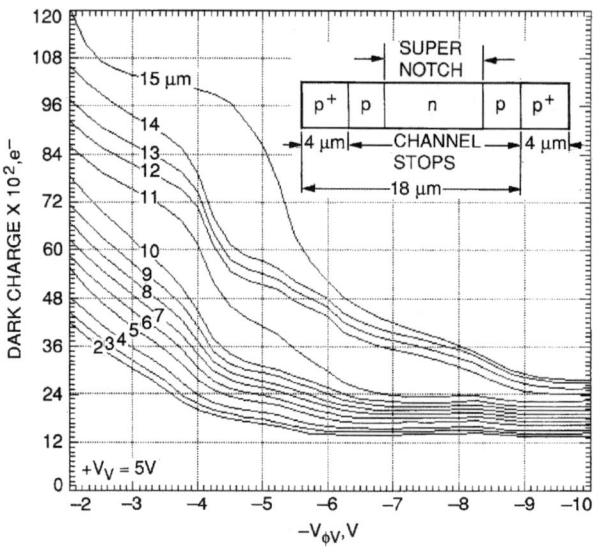

Figure 8.23 Dark charge showing super notch control on the inversion breakpoint.

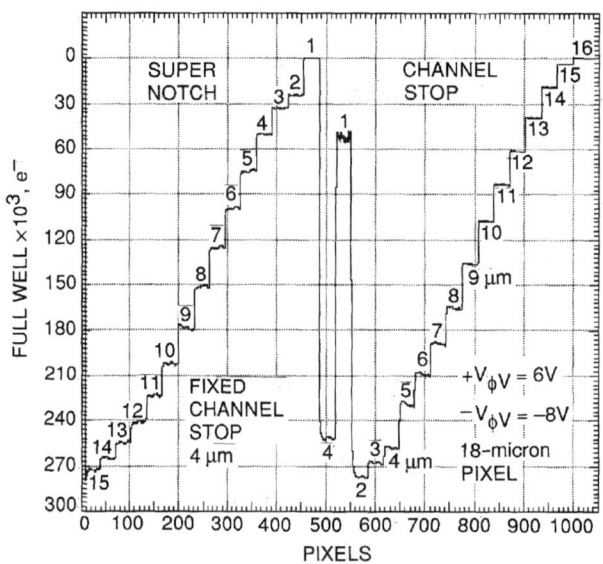

Figure 8.24 Full well as a function of super notch width.

in Fig. 8.20, inverting the CCD is difficult when the signal channel is not isolated from the channel stop region (i.e., 11- to 15-μm super notches). However, when the channel becomes sufficiently isolated, the inversion break point works its way back to the preirradiation level of –6.5 V. Good isolation occurs for a super notch width of 10 μm. The penalty paid for employing super notch is a full well reduc-

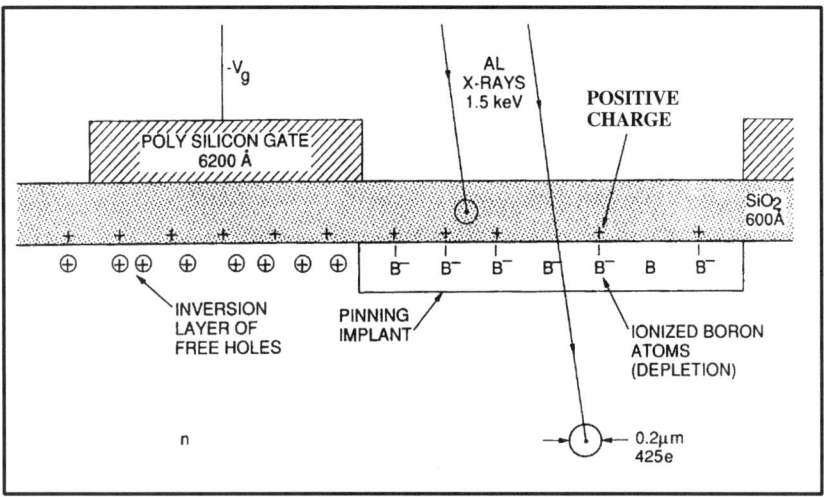

Figure 8.25 Virtual-phase pixel showing the virtual region pinning implant and resultant depletion caused by x-ray damage.

tion, because the signal channel is reduced in width. Figure 8.24 plots well capacity as a function of super notch width. Charge capacity decreases from 270,000 e⁻ to 195,000 e⁻ for the 10-μm super notch. The Cassini Huygen's probe CCD uses a super notch for radiation protection.

8.1.8.2 UV Charging

Negative flatband shifts induced by ionizing radiation may be reversed for some CCD technologies by using a UV flood. This solution was employed by the Solar X-ray Telescope (SXT) to prevent excessive positive charge buildup during the mission lifetime.[11] The SXT camera system uses a 1024 × 1024 × 18.3-μm-pixel virtual-phase CCD. The camera has worked quite successfully, producing more than a million x-ray solar images in the 0.3- to 4-keV x-ray band. However, during qualification tests, the sensor prematurely failed when it was subjected to a dose of x rays that simulated flight conditions. As we will show below, the ionizing damage produced an unacceptably large increase in dark current and severe CTE degradation.

Both the clocked and virtual phase regions of a virtual-phase pixel are sensitive to ionizing radiation (refer to Fig. 3.16). The clocked region is less sensitive to soft x rays because of the shielding effect of the poly gate. On the other hand, the virtual region is completely open to x rays. As noted in Chapter 3, the potential within the virtual region is pinned with a highly concentrated shallow boron p-doped layer. The pinning implant is connected to the channel stop, thereby maintaining ground potential at the surface. It is imperative in virtual-phase technology that the channel potential of the virtual region remain fixed as the device is clocked. Any potential

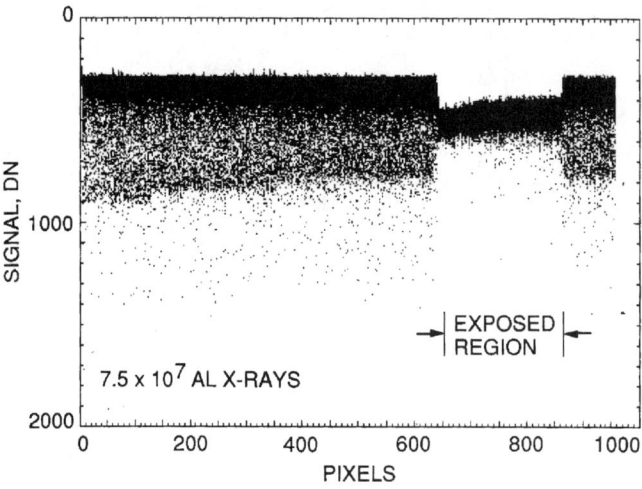

Figure 8.26(a) Fe^{55} CTE response after irradiation to 7.5×10^7 Al x rays.

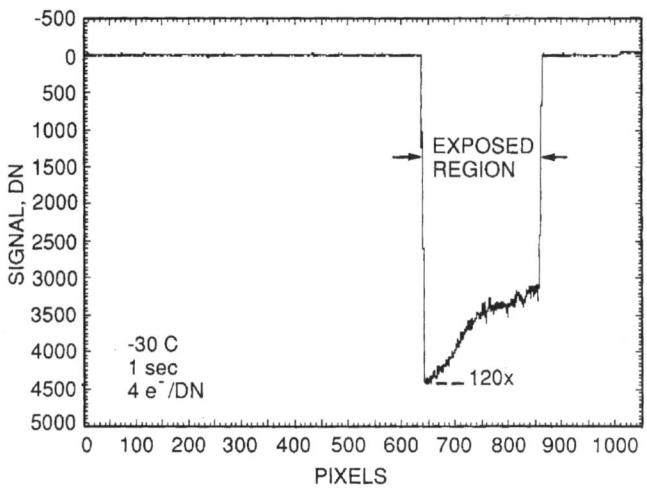

Figure 8.26(b) Dark current response after irradiation.

change under this region will result in CTE problems. In addition, an unpinned surface will generate high surface dark current. Trapped positive charge induced by x rays depletes the boron pinning layer as shown in Fig. 8.25. This action then diminishes the pinning effect, allowing the potential in the region to bounce when clocked.

Figure 8.26(a) shows a horizontal Fe^{55} x-ray transfer response taken with a SXT CCD after being partially exposed to 7.5×10^7 aluminum x rays/pixel (approximately the mission dose). It can be seen that CTE performance within the irradiated region is very poor. Figure 8.26(b) shows a dramatic increase in dark

DAMAGE   769

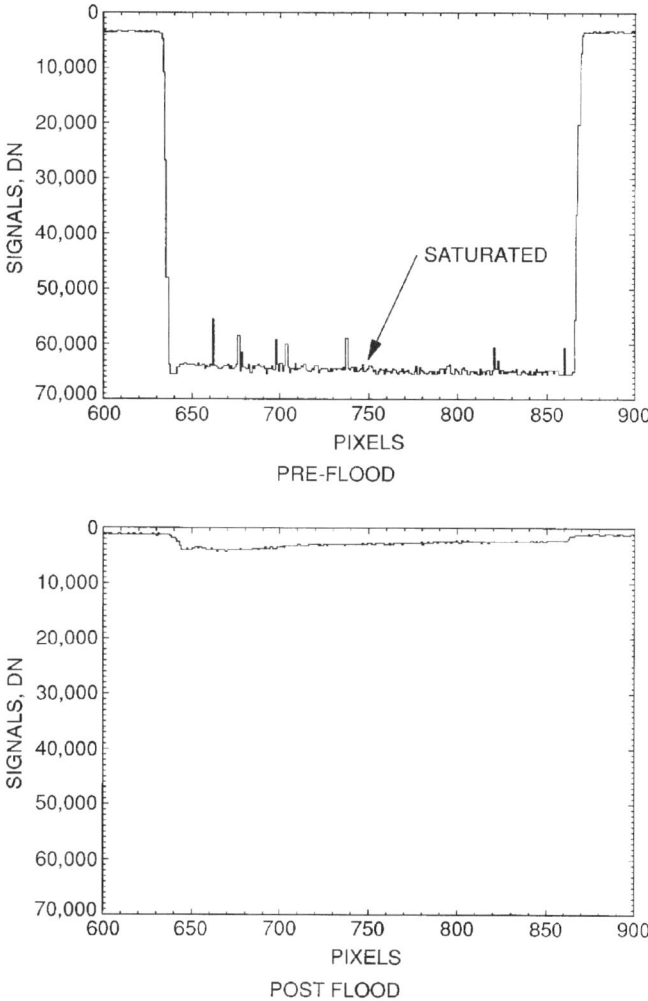

Figure 8.27 Dark current before and after UV flood.

current for the same region. Here an increase of 120 times is measured compared to the nonirradiated region.

In Chapter 3 it was shown that positive oxide charge for backside-illuminated CCDs can be neutralized by employing UV flood (i.e., photoemission of electrons into the native oxide layer). Fortunately, UV light flood was also beneficial for the SXT CCD in neutralizing positive charge buildup in the virtual region. Figure 8.27 shows dark responses before and after UV flood for an x-ray-damaged CCD. Note that dark current is nearly eliminated after UV light flood. Figure 8.28 shows the reduction of dark current as a function of UV flood time. Dark current after several minutes of flooding returns to the preirradiated level. The CTE problem was also cured completely with UV flood.

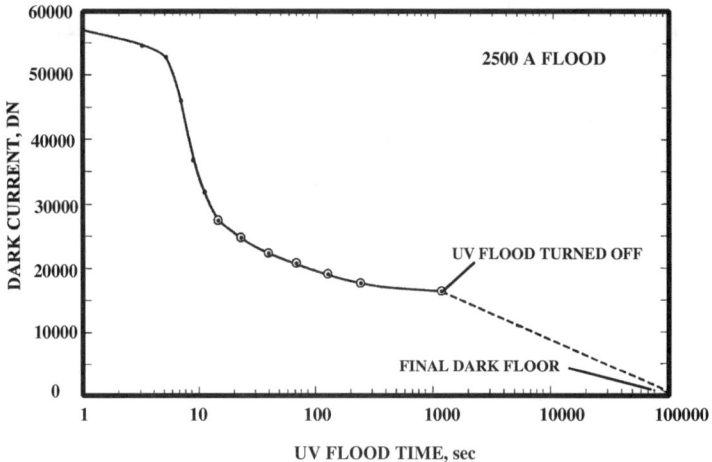

Figure 8.28 Dark current as a function of UV flood time.

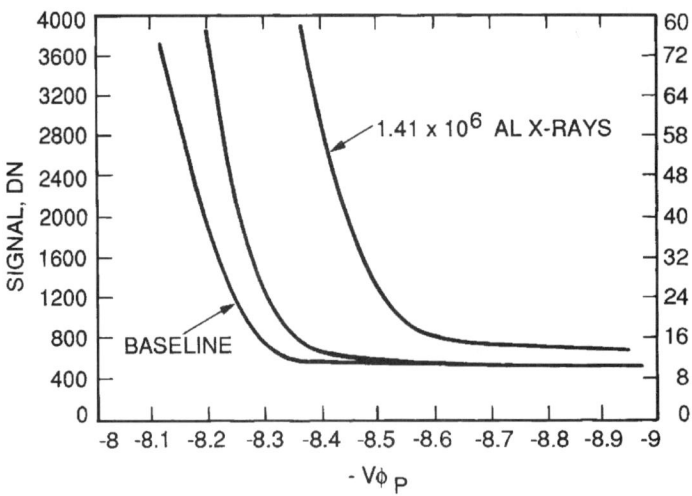

Figure 8.29(a) Inversion breakpoint for the clocked region before and after irradiation with 1.41×10^6 Al x rays.

The flatband shift within the clocked region could not be alleviated with UV flood. This is because UV light could not penetrate the poly silicon gate structure. Fortunately, the x-ray-induced flatband shift for this region was not significant. The small shift that did occur was compensated by increasing the negative clock drive to the CCD, thereby maintaining inversion. Figure 8.29(a) shows the inversion breakpoint for the clocked region after a test virtual-phase CCD was exposed to 1.41×10^6 aluminum x rays per pixel. The flatband shift is less than 1 V. Figure 8.29(b) shows the excess charge breakpoint, which decreases after irradiation. The onset of the new excess charge breakpoint occurs at approximately –18.6 V.

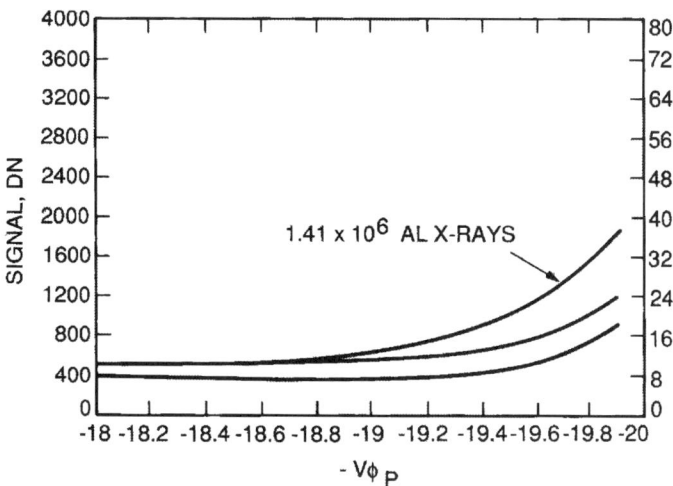

Figure 8.29(b) Excess charge breakpoint for the clocked region before and after irradiation.

The inversion and excess charge breakpoints define the operating window for the negative clock rail. SXT elected to drive the CCD at −13 V, which gave sufficient margin in circumventing the radiation problem along with the UV flood for the virtual region.

It should be mentioned that the UV flood technique used by SXT was very beneficial. However, as mentioned above, multiphase CCDs usually react negatively to UV flood because of nonuniform charging induced by overlapping gate electrodes.

8.1.8.3 Process-induced Charging

Radiation damage can also be induced when fabricating the CCD. High-energy process steps involving plasma etching, ion implantation and sputtering often generate secondary x rays that are damaging. Because of this problem, frame transfer and split frame transfer architectures may show different dark current levels in the image and storage regions. For example, Fig. 8.30 shows dark current nonuniformity observed in four sections of a 1024 × 2048 split-frame transfer CCD. The damage produced here was induced in the last stages of fabrication, at the time when the contacts and metal were plasma etched. In these steps, the poly silicon gates that are patterned float to arbitrary voltage level. Recall from discussions above that ionization damage is very sensitive to gate bias. Therefore, the gates with the highest voltage will suffer the most damage and exhibit highest dark current. Normally, process-induced radiation damage is not severe because H_2-forming gas anneal cycles eliminate most of the ionizing damage that occurs. Manufacturers may elect to combine wet and dry etching processes to reduce ionizing radiation as much as possible. Nevertheless, the user may still find remnants of process-induced damage when dark current images are taken.

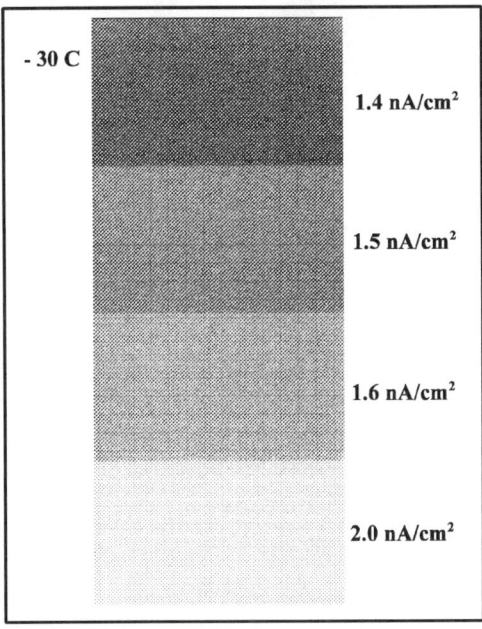

Figure 8.30 Dark current image taken with a split-frame transfer CCD showing process-induced radiation damage.

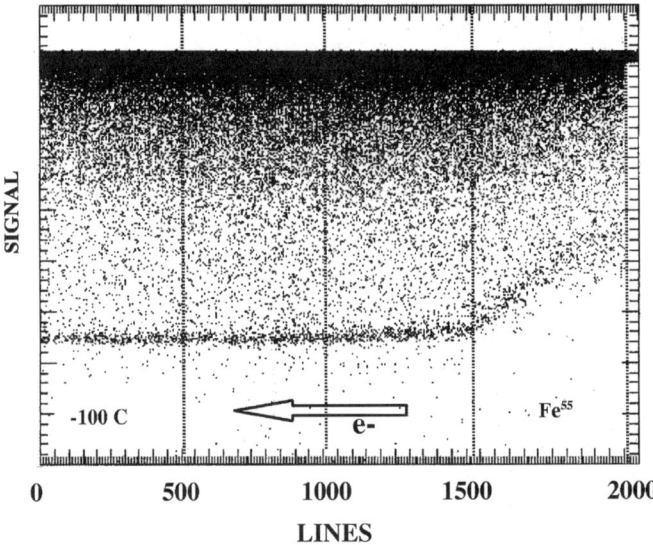

Figure 8.31 X-ray response showing poor CTE in the last quadrant.

Process-induced damage can also degrade CTE characteristics because of nonuniform flatband shifts that develop within a pixel. For example, the flatband shift may be less in the middle of a phase than near the edges of the bird's beak where greater damage occurs. Also, poly gate overlaps result in nonuniform damage and flatband shift over the signal channel. These nonuniformites can induce potential barriers and traps. Figure 8.31 shows a vertical x-ray transfer plot for the same sensor as Fig. 8.30. Note that CTE is worse within the 2-nA/cm^2 region where traps are present. The radiation problems shown in Figs. 8.30 and 8.31 were solved by modifying several process steps involving the plasma-etching processes.

The aluminum light shield used on frame and split-frame transfer CCDs is another process-induced problem. For frontside-illuminated CCDs, dark current is normally lower in the shielded region. There are two explanations for this observation. During aluminum deposition, which typically involves a sputtering process, secondary x rays are generated, causing radiation damage. When the light shield is patterned with a dry plasma etch, radiation is seen by the CCD. Dark current is greater in the image region because aluminum remains over the storage region and acts as a shield. The second and more popular theory is that aluminum is a good catalyzer that creates active hydrogen radicals to help tie up dangling bonds present. The effect produces lower dark current in the aluminum-covered areas.

8.1.9 BULK DAMAGE

8.1.9.1 Physics

Energetic charged particles undergo Rutherford-scattering-type coulombic interactions with the silicon lattice structure.[7] The energy expended by the interacting ion can pull a silicon atom out of its lattice position, forming an interstitial silicon atom and a vacancy (called a Frenkel pair). In order for this interaction to occur, the ion must pass well within the K-shell electron orbit, close enough to the target nucleus. The displaced atom is also referred to as the primary recoil [or primary knock-on atom (PKA)]. The PKA may have sufficient energy to undergo collisions with lattice, producing even more vacancies. The recoil energy and the resultant PKA damage spectrum is dependent on the energy, mass and charge of the incident particle. For example, electrons and photons usually produce single defects, whereas protons and neutrons produce a mixture of isolated and cluster defects.

The interactions described above are elastic collisions involving electromagnetic fields between the incoming particle and silicon nucleus. Inelastic collisions involve those ions that directly interact with the nucleus, leaving it in an activated state. The nucleus can emit nucleons and the recoiling nucleus is displaced from its lattice site, causing further damage. For example, a proton can be absorbed in a target nucleus which can then emit an alpha particle that is very damaging because of its short range. Nonelastic nuclear interactions are important when proton energies are greater than 10 MeV.

Vacancies are mobile above 100 K and are considered unstable defects when first introduced.[42,43] Depending on the energy delivered to the interstitial atom, most vacancy pairs recombine before they form stable defects and cause permanent damage. On the average, 98% of the vacancies recombine.[44] The process of migration and rearrangement is not instantaneous but exhibits a recovery time after vacancies are made. At room temperature the recovery time is on the order of a few seconds to a minute. Some long-term annealing characteristics do occur over a time period of a year or more, especially if the sensor is cold (dark spikes in particular).

The remaining silicon vacancies that do not recombine migrate in the lattice to form stable defects.[45] For example, two vacancies can form a defect known as the divacancy, which is stable up to 300°C[7,46] and higher-order complexes are even possible.[47] Defects also interact with impurities or dopants found in the signal channel of the CCD.[4,48,49] Such defects are termed defect-impurity complexes. These stable defects give rise to states with energy levels within the forbidden bandgap. The most common defect center is the E-center, which for a phosphorus-doped n-channel CCD is called the P-V pair complex (As-V centers form for arsenic-doped channels). The P-V trap exhibits an activation energy of approximately 0.44 eV below that of the silicon conduction band. The defect is electrically active and can become negatively charged by accepting an electron from the conduction band. The vacancy can also tie up with an oxygen impurity, which produces an O-V trap level at 0.17 eV. The P-V pair anneals near 150°C and the O-V trap anneals at 350°C, either by direct disassociation or long-range migration to other defect centers located in the CCD.[50] To our knowledge, bulk trap formation through vacancies is independent of CCD bias and operating temperature above 100 K.

In addition to proton damage, we will also be concerned with neutron damage in our discussions below. As will be shown, the damage produced by neutrons will be different than protons for two reasons. First, the cross-section for a neutron is smaller, on the order of 2.4×10^{-24} cm^2 for a 1-MeV neutron. Therefore, the number of primary displaced silicon atoms is smaller compared to proton interactions. Second, the interaction will not be coulombic, but nuclear. Therefore, more energy is transferred to the recoiling silicon atom (on the average, approximately 70 keV for a 1-MeV neutron). The energetic silicon atom will collide with the lattice, displacing many silicon atoms at one time (approximately 1500 atoms for a 1-MeV neutron). Initial damage produced is unstable, and a considerable amount of lattice reordering takes place (i.e., annealing). Compared to low-energy proton damage, the resultant cluster damage region will produce many traps within a pixel, as will be demonstrated below.

The presence of defect centers in the silicon bandgap can affect the electrical characteristics of the CCD in two major ways: trapping and charge generation. The first process is the temporary trapping of signal carriers at a defect center and emission to the conduction band producing deferred charge. The trapping process can have a profound effect on CTE performance. The second process creates thermally generated dark current typically in the form of a dark spike (i.e., hopping conduction, as explained in Chapter 7). CTE and dark spike problems induced by radiation will be discussed in considerable depth below.

DAMAGE

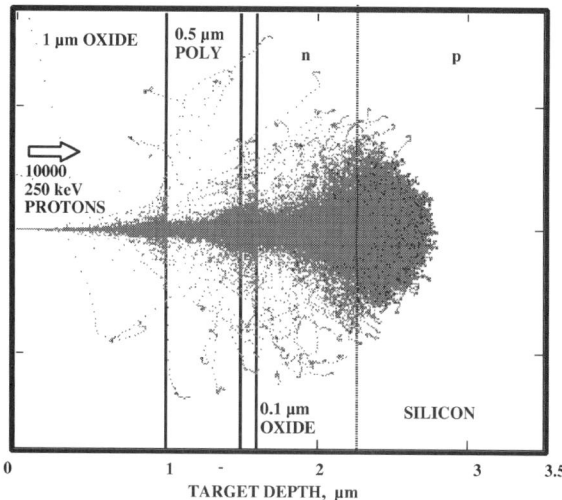

Figure 8.32(a) SRIM output showing 250-keV proton interactions with the CCD.[22]

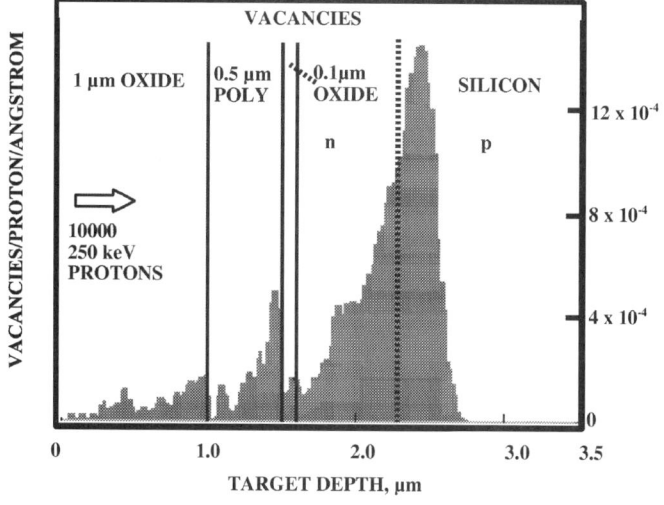

Figure 8.32(b) 250-keV proton vacancies as a function of depth.[22]

8.1.9.2 SRIM

Unlike ionizing radiation damage, the amount of bulk damage introduced into the CCD is predictable in that most sensor families react similarly. Therefore, as we will show below, only a few spot measurements will be necessary to calibrate and predict CTE degradation and dark spike generation for a specified radiation environment. A Monte-Carlo computer simulation called the "Stopping and Range of Ions in Matter (SRIM)" will serve as our theoretical model for

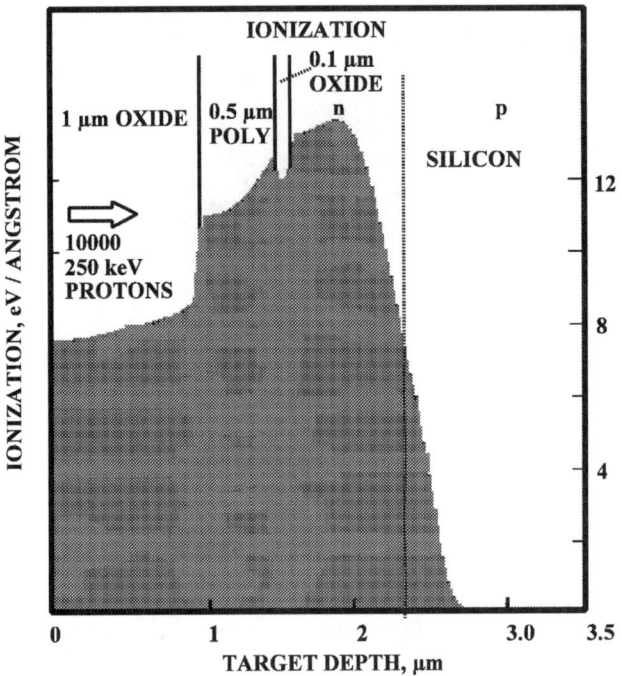

Figure 8.32(c) Ionization energy loss (IEL) as a function of depth.[22]

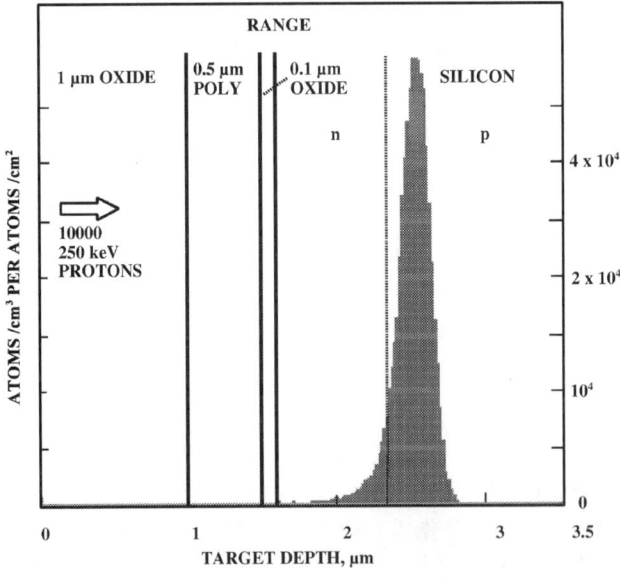

Figure 8.32(d) Range of 250-keV protons as a function of depth.[22]

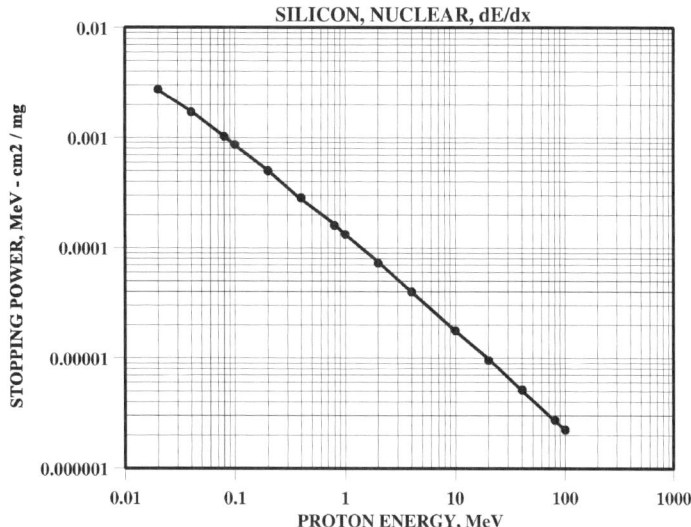

Figure 8.33 Proton stopping power as a function of energy.[22]

bulk damage studies below.[22] SRIM is a program that calculates the stopping and range of ions in the CCD. Outputs from this user friendly but powerful program are numerous (the software is available on the web free of charge at http://www.research.ibm.com/ionbeams/). For example, Fig. 8.32(a) shows a simulation performed when taking a 250-keV hydrogen ion (i.e., proton) into a CCD structure (1-μm overcoat oxide, 5000-Å gate electrode, 1000 Å of gate dielectric and epitaxial silicon layer). The plot shows how 10,000 protons travel and scatter as a function of target depth. Figure 8.32(b) plots the number of silicon vacancies produced as a function of depth. Figure 8.32(c) plots IEL as a function of depth. Figure 8.32(d) plots the final range of the proton (the protons will eventually turn into hydrogen atoms and slowly escape from the CCD). SRIM also tabulates the range, stopping power (ionization and nuclear), longitudinal and lateral straggling as a function of ion energy. For example, Appendix G gives SRIM proton data for silicon, silicon dioxide, aluminum, and tantalum. Figure 8.33 plots NIEL for protons as derived by SRIM. Note that NIEL decreases logarithmically with energy indicating that less lattice damage will occur as the proton energy increases. Units for NIEL are usually given in MeV-cm^2/gm (i.e., nuclear stopping power).

8.1.9.3 Displacements

Displacement reactions are extremely rare as a particle passes through the silicon lattice. The probability of interaction is described by a particle's cross-section, or equivalently, by its mean free path length.

Example 8.15

Determine the mean free path length for a 1-MeV proton. Assume $\sigma = 3.5 \times 10^{-20}$ cm².
Solution:
From Eq. (8.3),

$$\lambda = 28/[(6 \times 10^{23}) \times 2.32 \times (3.5 \times 10^{-20})],$$

$$\lambda = 5.74 \times 10^{-4} \text{ cm } (5.74 \text{ μm}).$$

In that there are 4.977×10^{22} silicon atoms/cm³, the proton passes through 3.68×10^7 atoms before an interaction occurs. Therefore, if there were no ionization loss, a 1-MeV proton would be very penetrating, exhibiting a range in silicon of approximately 27.3 cm. This value assumes that 21 eV is lost in each mean free path length (i.e., 5.74×10^{-4} cm). However, because of ionization loss, the proton only travels 15.7 μm, showing that IEL is much greater than NIEL.

The displacement of an atom from its lattice position involves the formation of a vacancy, an interstitial, and energy losses from other atomic movements. Displacement energy, when induced by a proton, is estimated by[21]

$$E_D = \Lambda E_T, \tag{8.13}$$

where E_T is the threshold energy of the proton (eV) and Λ is defined as

$$\Lambda = \frac{4 A m_p}{(m_p + A)^2}, \tag{8.14}$$

where A is the atomic weight (g/mole, Si = 28) and m_p is the proton mass (1 g/mole).

Example 8.16

Using Eq. (8.13), estimate the displacement energy. Assume a proton threshold interaction energy of 160 eV.
Solution:
From Eq. (8.14),

$$\Lambda = 4 \times 28 \times 1/(1 + 28)^2,$$

$$\Lambda = 1.33 \times 10^{-1}.$$

DAMAGE

From Eq. (8.13),

$$E_D = (1.33 \times 10^{-1}) \times 160,$$

$$E_D = 21.3 \text{ eV}.$$

The total number of displacements is related to the displacement cross-section through

$$D_T = \frac{N_A \sigma \rho N_{PRI} x_{path}}{A}, \tag{8.15}$$

where D_T is the number of displacements/particle, A is the atomic weight of the target material (g/mole), N_A is Avogadro's number (6.022×10^{23} atoms/mole), σ is the particles cross-section (cm^2), x_{path} is the path length of the particle (cm), ρ is the density of silicon (g/cm^3) and N_{PRI} is the average number of displacements produced for every primary knock-on including the knock-on. This number is estimated by[51]

$$N_{PRI} = \frac{1}{2}\left[1 + \ln\left(\frac{\Lambda E_{PRO}}{2 E_D}\right)\right], \tag{8.16}$$

where E_{PRO} is the energy of the proton (MeV).

Example 8.17

Determine the number of displacements for a 1-MeV proton over a path length of 1.5 μm. Assume $\sigma = 3.5 \times 10^{-20}$ cm^2 and $A = 28$ g/mole.
Solution:
From Eq. (8.16),

$$N_{PRI} = \frac{1}{2}\left[1 + \ln(0.133 \times 10^6)/(2 \times 21)\right],$$

$$N_{PRI} = 4.5 \text{ displacements}.$$

From Eq. (8.15), the average number of displacements over a 1.5-μm path length is

$$D_T = (6.023 \times 10^{23}) \times (3.5 \times 10^{-20})$$

$$\times 2.32 \times 4.5 \times (1.5 \times 10^{-4})/28,$$

$$D_T = 1.18 \text{ displacements/proton}.$$

Stopping power from NIEL can also be used to estimate the number of displacements through

$$D_T = \frac{10^9 \rho S x_{path}}{E_D}, \qquad (8.17)$$

where x_{path} is the distance traveled by the ion (cm), E_D is the displacement energy (eV), ρ is the density of silicon (2.32 g/cm^3) and S is the nuclear stopping power (MeV-cm^2/mg).

Example 8.18

Determine the number of displacements created by a 1-MeV proton that passes through 1.5 μm of silicon material. Assume $E_D = 21$ eV and a stopping power of 6×10^{-5} MeV-cm^2/mg.
Solution:
From Eq. (8.17),

$$D_T = 10^9 \times 2.32 \times (6 \times 10^{-5}) \times (1.5 \times 10^{-4})/21,$$

$$D_T = 0.994 \text{ displacements/proton}.$$

This value is close to the number of displacements calculated in Example 8.17 using the cross-sectional assumption (i.e., $\sigma = 3.5 \times 10^{-20}$ cm^2).

Equations (8.15) and (8.17) can be equated to estimate the cross-section from stopping power data yielding

$$\sigma = \frac{10^9 S A}{N_A N_{PRI} E_D}. \qquad (8.18)$$

For silicon this equation reduces to

$$\sigma = \frac{4.65 \times 10^{-14} S}{N_{PRI} E_D}. \qquad (8.19)$$

Example 8.19

Determine the cross-section for a 10-MeV proton. Assume $S = 8.5 \times 10^{-6}$ MeV-cm^2/mg (Fig. 8.49) and $E_D = 21$ eV.

DAMAGE

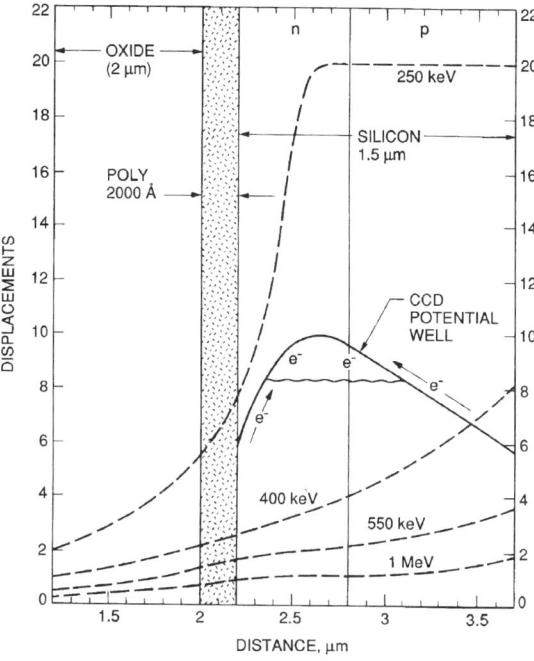

Figure 8.34 Low-energy proton displacements induced into a simple CCD structure as a function of depth.

Solution:
From Eq. (8.16),

$$N_{PRI} = \frac{1}{2}[1 + \ln(0.133 \times 10^7)/(2 \times 21)],$$

$N_{PRI} = 5.68$ displacements.

From Eq. (8.19),

$$\sigma = (4.65 \times 10^{-14}) \times (8.5 \times 10^{-6})/(5.68 \times 21),$$

$$\sigma = 3.31 \times 10^{-21} \text{ cm}^2.$$

Figure 8.34 plots the integral of displacement damage as a function of distance into the CCD for different proton energies based on SRIM. Note that the 250-keV proton induces the greatest damage to the signal channel for the parameters assumed (i.e., 2 μm of oxide and 2000-Å poly gate). At this energy, damage is maximum because the protons stop directly in the signal channel (i.e., coincident

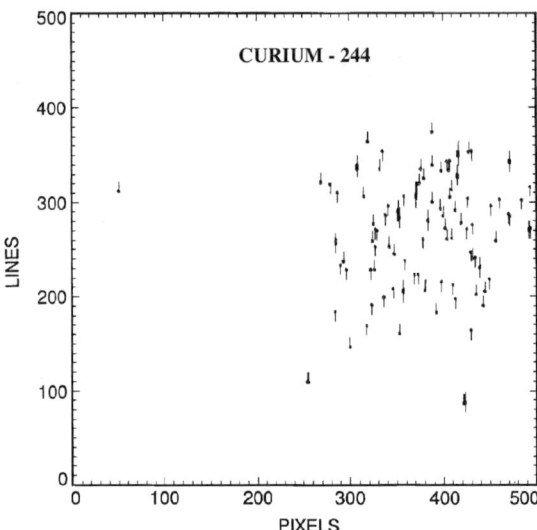

Figure 8.35 Alpha particle trap map generated by pocket pumping.

with the Bragg peak). Approximately 12 displacements/proton result on the average. For energies > 250 keV, an increasing fraction of displacements is created outside of the CCD signal channel. For example, at 1 MeV, less than 1 displacement/proton takes place on the average. The discrepancy between this plot and the above calculations lies in the subtle details of calculation between SRIM and the NIEL curve shown in Fig. 8.49.[7,52,53]

8.1.9.4 Early Bulk Damage Experiments

It is enlightening to review seven early heavy ion, proton and photon experiments that showed the vulnerability of CCD performance to displacement damage. These tests were called (1) alpha particle test, (2) low-energy proton test, (3) dark spike test, (4) Co^{60} test, (5) 487-rad proton test, (6) 25-rad proton test and (7) neutron test.

8.1.9.4.1 Alpha particle test. Exposing a CCD to Cm^{244}, a 5.9-MeV alpha-particle emitter, was one of the first experiments performed to show the extreme sensitivity of the CCD to the displacement problem. Recall from Chapter 2 that low-energy x rays can be fluoresced by using a Cm^{244} source and a desired target material. For this setup, it is critically important that the CCD not see alpha-particles, because a single particle can destroy CTE performance in that pixel of interaction. The range of a 5.9-MeV alpha particle is approximately 30 μm in silicon and does not cause significant damage if it is allowed to pass through the signal channel with its full energy. However, the energy of an alpha particle is attenuated when passing through air. A few inches of air reduces the energy to a level where

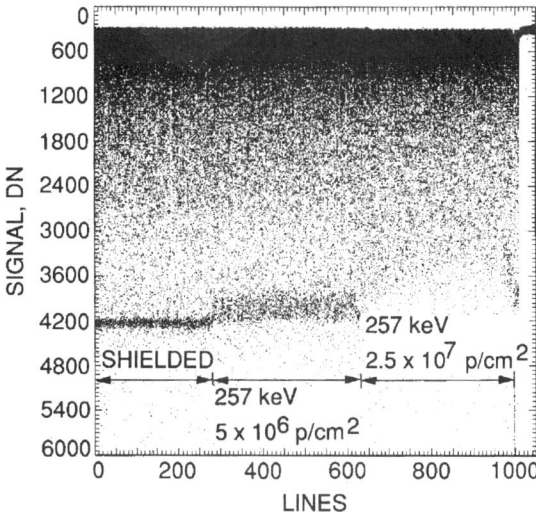

Figure 8.36(a) Horizontal Fe^{55} x-ray response for 257-keV protons at two fluences.

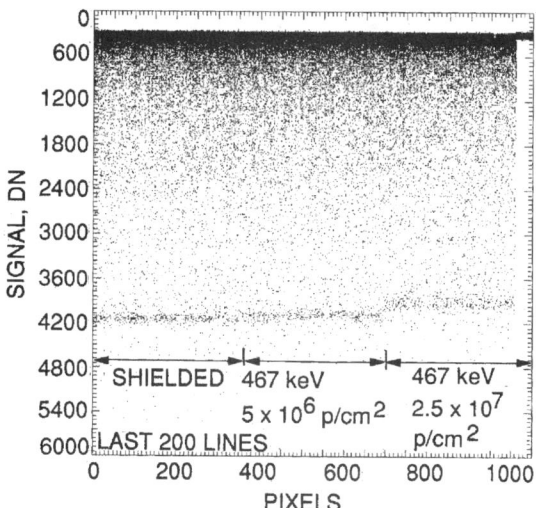

Figure 8.36(b) Horizontal Fe^{55} x-ray response for 467-keV protons at two fluences.

the Bragg peak can coincidently match that of the signal channel depth. In this case, many displacements and traps are produced in a single interaction. For example, a 225-keV alpha particle produces 173 vacancies in a 1.5-μm depth.

Figure 8.35 shows a pocket pumping response taken from a CCD that was exposed to Cm^{244} alpha particles in a very short time period. The source was approximately 4 in. from the CCD, making them very damaging once reaching the CCD.

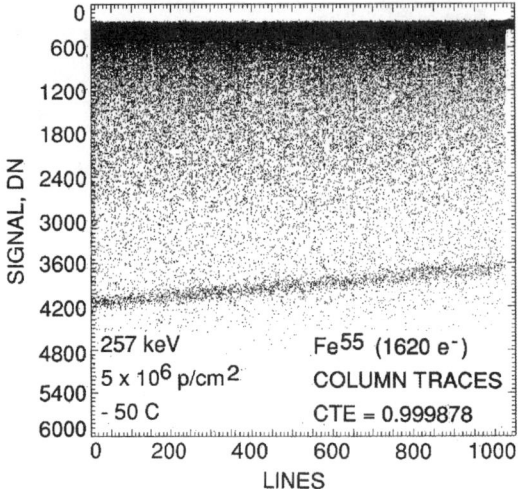

Figure 8.36(c) Vertical x-ray transfer plot for 257-keV protons (fluence = 5×10^6 protons/cm^2).

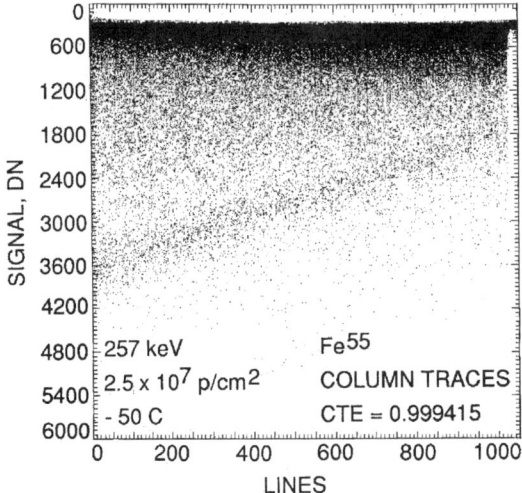

Figure 8.36(d) Vertical x-ray transfer plot for 257-keV protons (fluence = 2.5×10^7 protons/cm^2).

Each trap seen in the plot represents one alpha-particle hit. This small exposure ruined the CCD and it was discarded.

8.1.9.4.2 Low-energy proton test.

This test verified that low-energy protons (< 1 MeV) were very damaging to the CCD. Figure 8.36(a) shows a horizontal x-ray stacking plot taken with a Cassini CCD that was damaged with 257-keV protons at two fluences, 5×10^6 and 2.5×10^7 protons/cm^2. A third region on

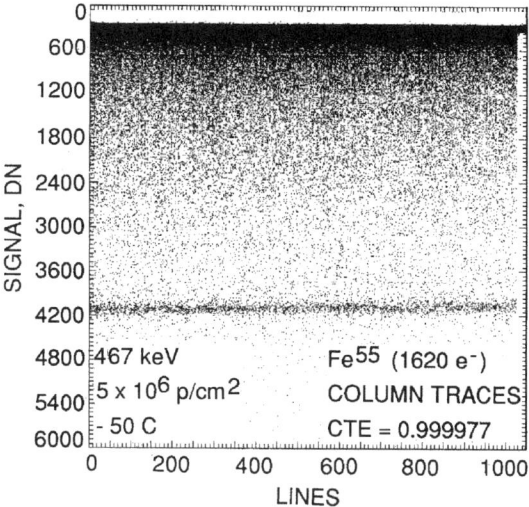

Figure 8.36(e) Vertical x-ray transfer plot for 467-keV protons (fluence = 5×10^6 protons/cm^2).

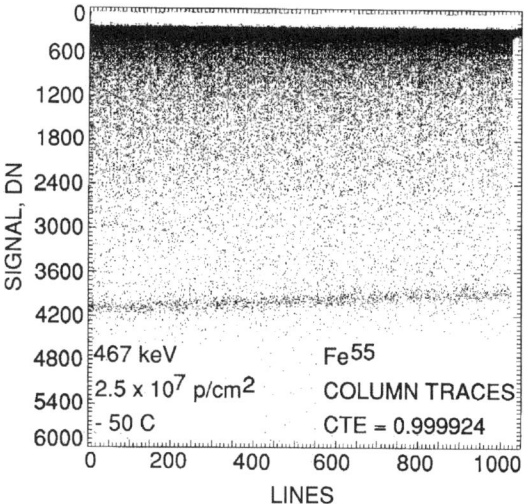

Figure 8.36(f) Vertical x-ray transfer plot for 467-keV protons (fluence = 2.5×10^7 protons/cm^2).

the device was shielded for reference purposes. The plot shows severe vertical CTE degradation in the irradiated regions. The same test was performed by exposing a different CCD with 467-keV protons and same fluences. The results of this test are shown in Fig. 8.36(b). Although the proton energy difference is slight, the 250-keV proton is much more damaging for reasons stated above (i.e., Fig. 8.34). Figures 8.36(c), 8.36(d), 8.36(e) and 8.36(f) show vertical x-ray stacking plots for

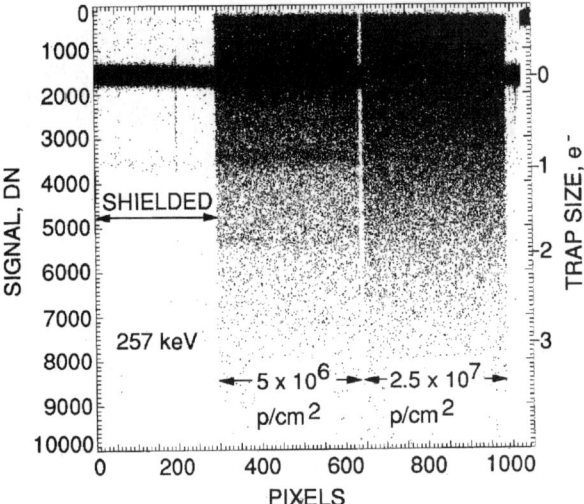

Figure 8.36(g) Pocket pumping response for 257-keV protons at two fluences.

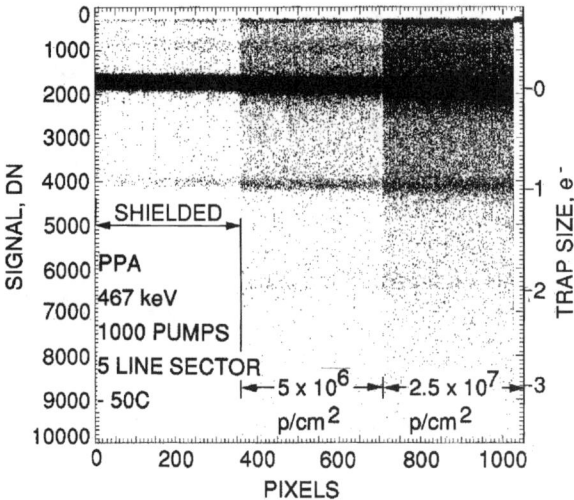

Figure 8.36(h) Pocket pumping response for 467-keV protons at two fluences.

the 257-keV and 467-keV exposures respectively. Again, the difference in CTE performance is striking between the two energies. Figures 8.36(g) and 8.37(h) are the pocket pumping response for each energy. The shielded region shows a few bulk traps, whereas the irradiated regions exhibit thousands of new single electron traps induced by the protons. Again, many more traps are produced by 257-keV protons compared to 467-keV protons. Figure 8.36(i) are SRIM results that plot displacements per angstrom as a function of depth for different proton energies below 1 MeV. The upper plot is the damage produced by recoils, whereas the bottom

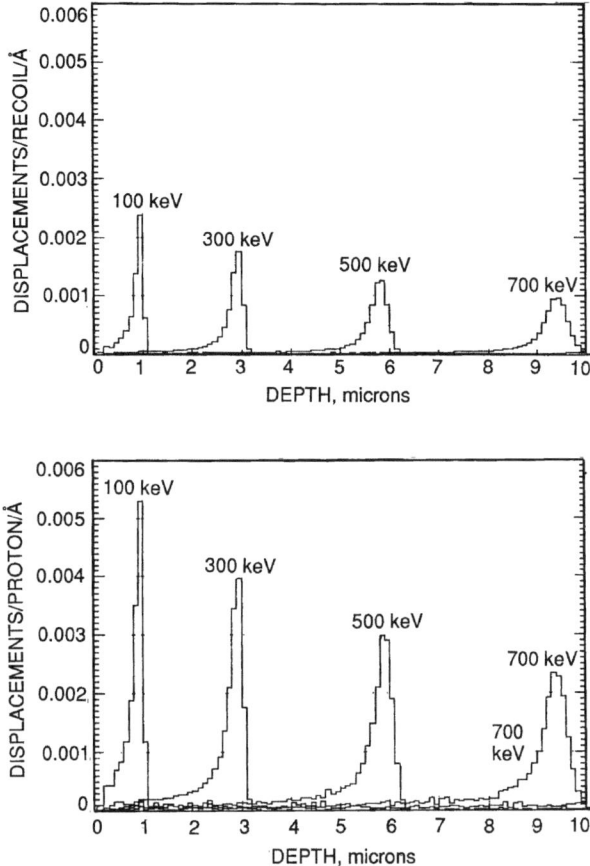

Figure 8.36(i) Recoil and total displacement as a function of depth for different proton energies.

plot is the total displacements per proton. The plots do not include the oxide and poly layers as in Fig. 8.34. Here the protons enter the silicon directly. Figure 8.36(j) integrates the vacancies as a function of depth. Although the total displacements increases with proton energy, the number of displacements created in the signal channel decreases (within the first 0–1.5 μm of the surface).

The proton experiment also showed that vertical CTE degraded dramatically, whereas the horizontal CTE was only slightly affected. The discovery was confusing at first, but later the mystery was resolved because of a relationship between the P-V trap emission time constant and the clocking rate applied to the CCD registers. We will discuss this interesting characteristic below.

Unfortunately, the low-energy proton damage problem was experienced head-on during the first month of the Chandra/ACIS x-ray mission. Immediately after the problem surfaced, it was concluded that protons with energy up to 300 keV could pass through the grazing incident x-ray telescope mirrors used by Chandra.

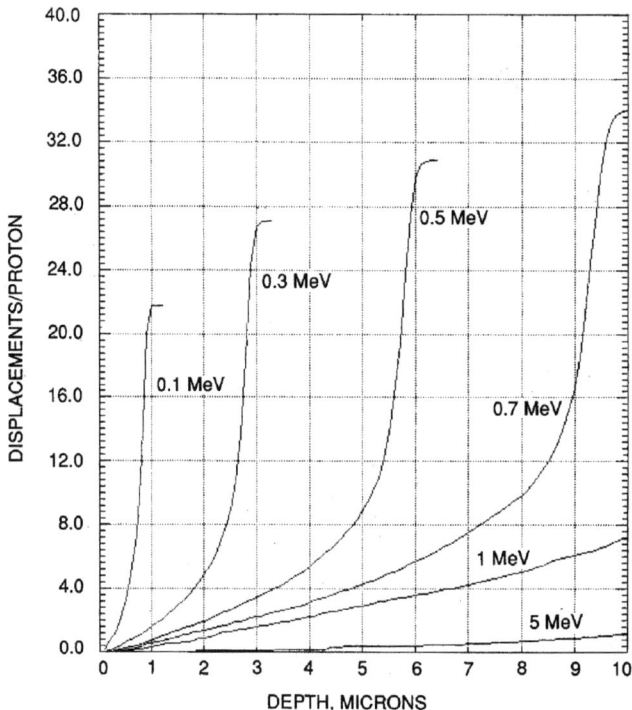

Figure 8.36(j) Displacements per proton as a function of depth for different proton energies.

And below 100 keV, the protons were unable to penetrate the visible light blocking filters above the CCD. Therefore, a very narrow energy range of protons was able to reach the CCD. Unfortunately, this energy range was very damaging to the ACIS sensors. CTI performance degraded in a short period of time, from 7×10^{-7} (CTE = 0.9999993) at launch to 2×10^{-4} (CTE = 0.9998) after a few months, as measured by Fe^{55} at $-90°C$.

Chandra's orbit (10,000 × 140,000-km orbit ~2.7 R_e geocentric) takes the instrument through the Van Allen belts where a high intensity of protons are located. This passage through the belts is where damage was introduced. The single backside-illuminated CCD on board did not show CTE problems as it did for the frontside-illuminated sensors. The backside CCD is 45-μm thick, which protected the signal channel from low-energy protons (the most damaging energy to the backside-illuminated CCD is the 1.8-MeV proton, an energy that could not pass through the telescope). Also, the frame store regions of the frontside-illuminated CCD did not show CTE problems. The 2- to 3-mm Al light shield protected these regions to incoming protons. Stowing the CCD focal plane in a protected position during passages through the belts and geomagnetic storms arrested further rapid CTE degradation for Chandra.

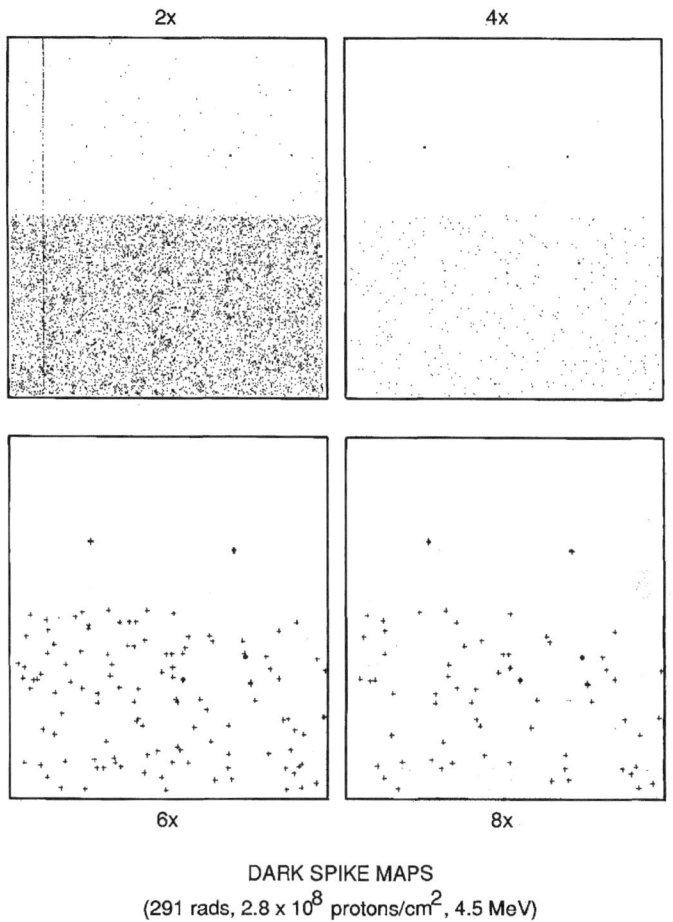

Figure 8.37(a) Proton-induced dark spike maps specified at levels greater than the average dark current floor.

8.1.9.4.3 **Dark spikes.** Figure 8.37(a) shows a dark current image for a $1024 \times 1024 \times 12$-µm pixel Cassini CCD that was partially irradiated with 291 rad (2.8×10^8 protons/cm^2) of 4.5-MeV protons. The upper half of the sensor was shielded for a control region. The figure shows those dark spikes that are above a specified average dark current floor (i.e., 2 times, 4 times, 6 times and 8 times). Only two major dark spikes are seen in the shielded side, possibly induced by neutrons during irradiation. Figure 8.37(b) plots the percentage of pixels that contain dark spikes as a function of dark spike signal level above the mean dark current floor before irradiation. Figure 8.37(c) is a similar plot after irradiation. Figure 8.37(d) shows a partial 3D image of proton-induced dark spikes where charge is transferred toward the reader. Note that a trail of dark charge follows each dark spike as the image is read out.

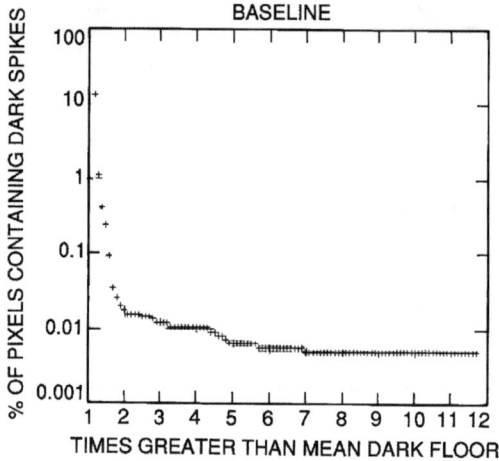

Figure 8.37(b) Pre-irradiation dark spike histogram as a function of signal level greater than the average dark current floor.

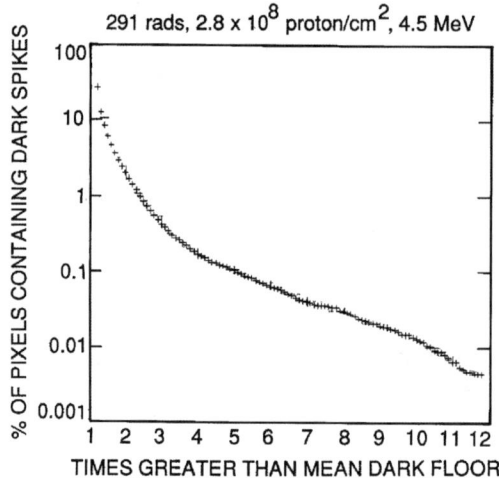

Figure 8.37(c) Post-irradiation dark spike histogram as a function of signal level greater than the average dark current floor.

The density of dark spikes generated for a given proton fluence and stopping power is estimated by

$$D_{SPIKE} = \frac{10^9 \rho S x_{SPIKE} F D_{EFF}}{E_D}, \tag{8.20}$$

where D_{SPIKE} is the density of dark spikes (cm^{-2}), x_{SPIKE} is that depth from the Si–SiO$_2$ interface where dark spikes are generated (recall from Chapter 2 that dark

Figure 8.37(d) Dark image showing trailing charge behind each dark spike during readout.

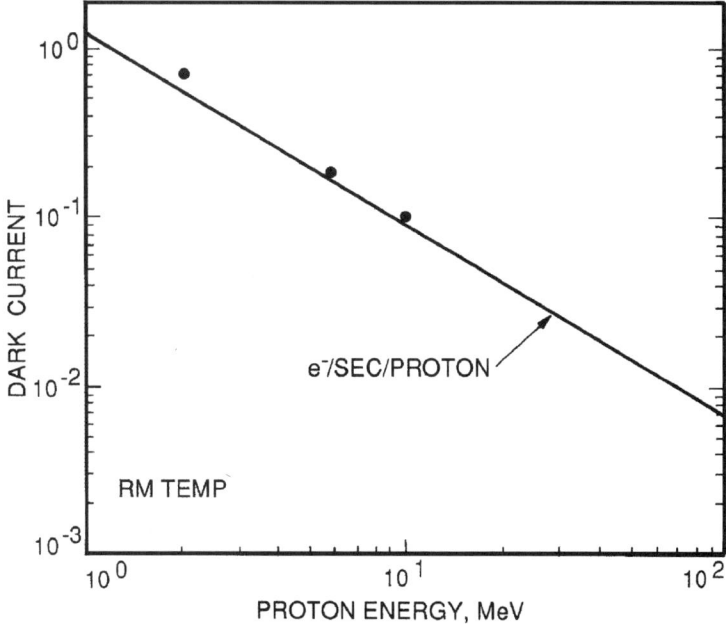

Figure 8.37(e) Average dark spike current as a function of proton energy for a fixed fluence.

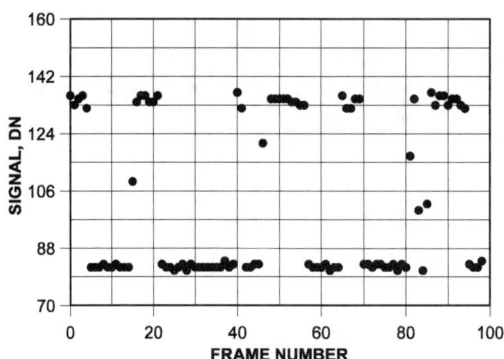

Figure 8.37(f) RTS dark spike (data from Rick Baer of Hewlett–Packard).

spikes are generated very near the surface), F is the proton fluence (protons/cm^2), and D_{EFF} is the dark spike efficiency factor defined as

$$D_{EFF} = \frac{P_{DS}}{P_P}, \qquad (8.21)$$

where P_{DS} is the number of dark spikes per pixel and P_P is the number of protons per pixel.

Example 8.20

Using Eq. (8.20), determine the expected number of dark spikes for the following parameters:
$S = 5 \times 10^{-5}$ MeV-cm^2/mg (4.5-MeV protons),
$x_{SPIKE} = 0.1 \times 10^{-4}$ cm,
$E_D = 21$ eV,
$F = 2.8 \times 10^8$ protons/cm^2
$D_{EFF} = 1$ dark spike/proton.
Solution:
From Eq. (8.20),

$$D_{SPIKE} = (10^9) \times 2.32 \times (5 \times 10^{-5}) \times (0.1 \times 10^{-4})$$

$$\times (2.8 \times 10^8)/21,$$

$$D_{SPIKE} = 1.54 \times 10^7 \text{ dark spikes/cm}^2.$$

Example 8.21

Determine the dark spike efficiency factor D_{EFF} for Example 8.20 for those dark spikes that are two times greater than the average dark current floor. Assume from Fig. 8.37(c) that 1.7% of the pixels on the array are affected.
Solution:
The number of pixels affected by dark spikes is $0.17 \times (1024)^2 = 1.78 \times 10^4$. The number of dark spikes per pixel is therefore

$$P_{DS} = 1.78 \times 10^4 / 1024^2,$$

$$P_{DS} = 0.0172 \text{ dark spikes/pixel}.$$

The number of protons per pixel is $(2.8 \times 10^7) \times (12 \times 10^{-4})^2 = 40.3$ protons/pixel.
From Eq. (8.21),

$$D_{EFF} = 0.0172/40.3,$$

$$D_{EFF} = 4.27 \times 10^{-4} \text{ (2 times greater than average dark current floor)}.$$

The number of dark spikes counted in Fig. 8.37(a) is considerably less than the value calculated in Example 8.20 (assuming $D_{EFF} = 1$). This is because not all displacements translate into dark spikes because vacancies recombine and dark spikes anneal (e.g., refer to Fig. 8.44). Also, dark spikes are very dependent on electric fields within the CCD. Therefore, some CCD technologies are less prone to the dark spike problem than others. In addition, many dark spikes induced will be unnoticed. For example, non-MPP CCDs will exhibit less dark spikes than inverted sensors because they are hidden in the dark current shot and nonuniformity noise. For these reasons, it is impossible to predict the dark spike density for a new CCD and specified radiation environment. However, dark spike population appears to be proportional to the stopping power as a function of energy. Figure 8.37(e) plots average dark current produced by dark spikes as a function of proton energy for a fixed fluence. The predicted NIEL curve by SRIM is shown along with three experimental data points.

It should be mentioned that some radiation-induced dark spikes generate charge erratically. Here, dark current fluctuates between two or more discrete levels with the appearance of a random telegraph signal (RTS).[54–56] Hopkinson suggests that a field is produced when a trap becomes charged, which in turn enhances the emission from neighboring generation centers. The time constant of the RTS signal can

Table 8.2 Davis proton energy spectrum.

Proton energy MeV	Rads	Protons/cm^2
20	26	8×10^7
22	29	1×10^8
26	41	1.5×10^8
30	48	2.0×10^8
35	52	2.5×10^8
39	48	2.5×10^8
48	80	5×10^8
55	86	6×10^8
63	77	6×10^8

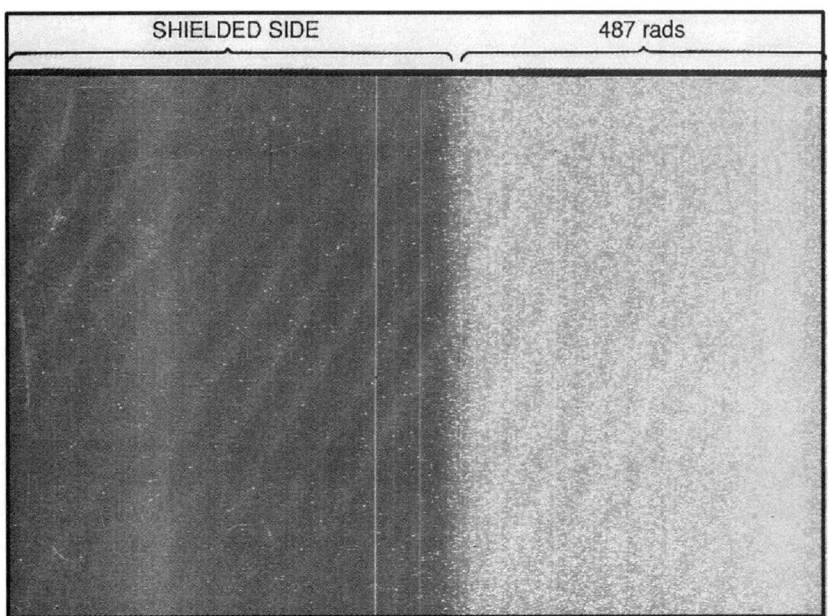

Figure 8.38 Dark current image showing proton-induced dark spikes.

be long (several minutes at room temperature), which implies that an energy barrier is likely to be involved with the capture/emission process. Figure 8.37(f) shows an RTS dark spike that generates dark charge at two discrete levels. Each data point shown represents a 1-sec integration using a camera gain constant of 0.07 DN/e$^-$.

8.1.9.4.4 487-rad proton test. This test demonstrated that a small total dose of protons is very damaging to the CCD. This fact was realized by irradiating a 1024 × 1024-MPP CCD to a dose of only 487 rad using the proton energy spectrum shown in Table 8.2 (protons supplied by Davis University). The CCD was

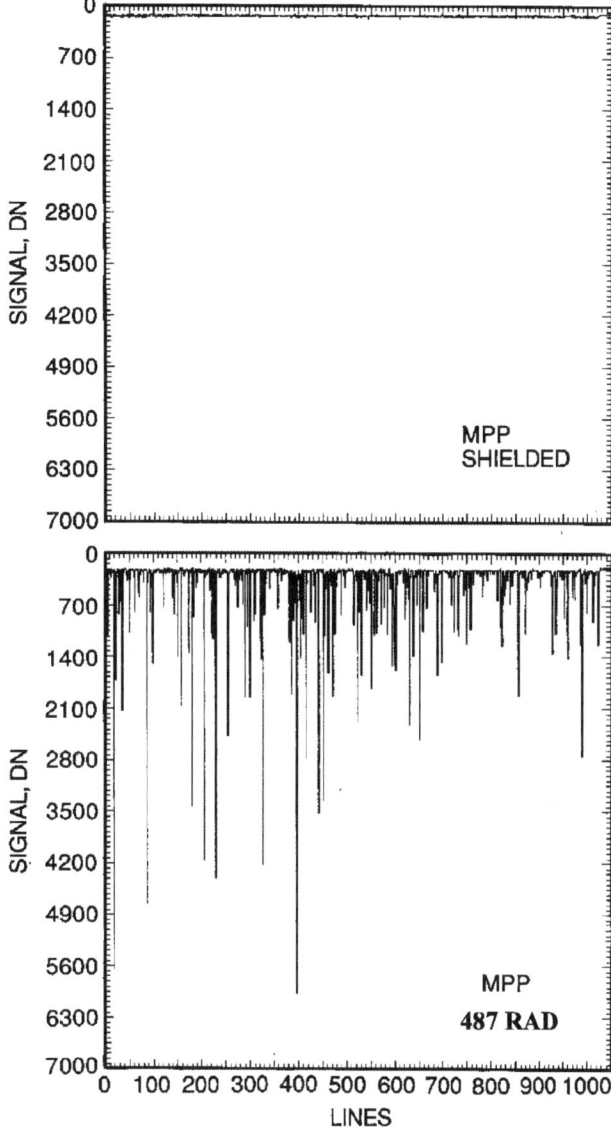

Figure 8.39(a) Line traces showing dark spike density before and after proton irradiation.

partially shielded with a 1-cm tantalum shield to provide a reference region. Figure 8.38 shows a dark current image after irradiation. The irradiated side of the sensor shows thousands of high-level proton-induced dark spikes. Figure 8.39(a) shows dark column traces taken from the baseline and irradiated regions for the test CCD. Dark current in the shielded region is approximately 10 pA/cm^2, standard for a MPP CCD. Some dark spikes in the damaged region exhibit a dark rate of 1000 times greater than this level (i.e., 10 nA/pixel). Figure 8.39(b) presents

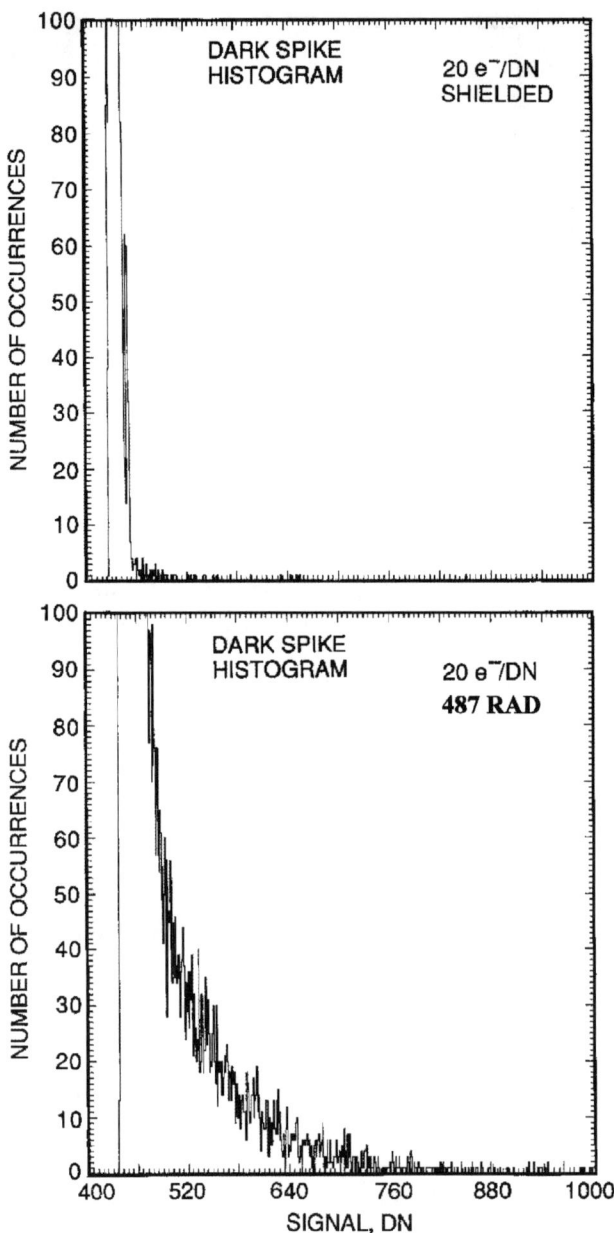

Figure 8.39(b) Dark spike histograms before and after proton irradiation.

dark spike histograms for each region, showing that dark spike density decreases exponentially with signal level.[57] Figure 8.39(c) plots individual dark spikes as a function of operating temperature, showing that the dark spike signal tracks the average dark current floor [Eq. (7.46)]. Figure 8.40(a) presents a horizontal Fe^{55} x-ray stacking plot, showing complete annihilation of the single-pixel event line

DAMAGE

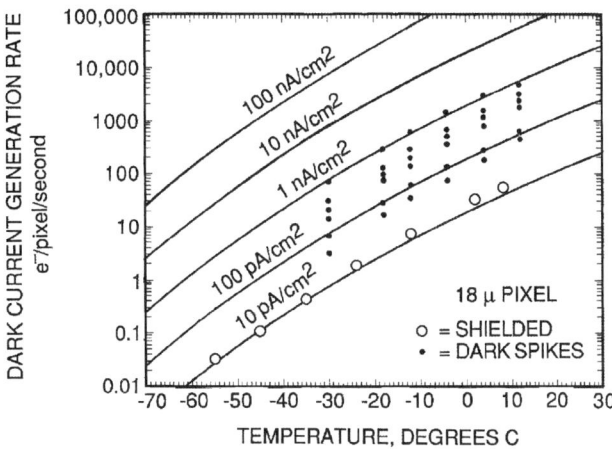

Figure 8.39(c) Selected dark spikes as a function of operating temperature.

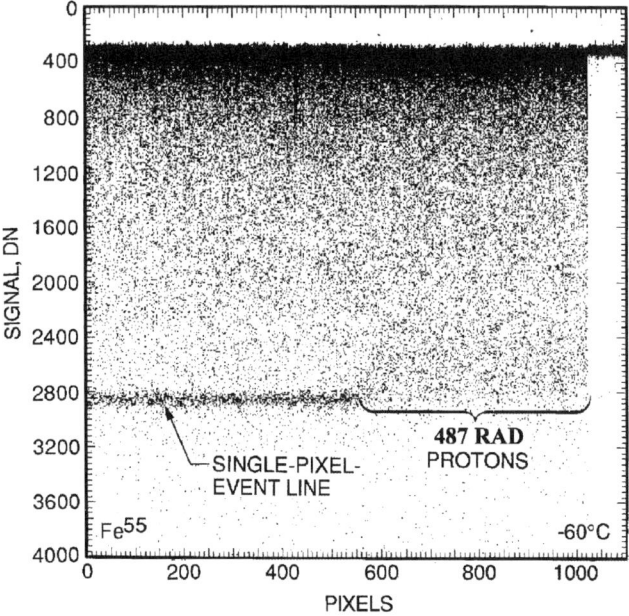

Figure 8.40(a) Horizontal Fe^{55} response showing severe CTE degradation in the nonshielded side.

within the irradiated region. Figure 8.40(b) shows x-ray histograms for the same data set, again showing no evidence of the K-alpha event line.

8.1.9.4.5 Co^{60} test. For many years CCD hardness was evaluated with a Co^{60} source. Although flatband and surface dark current reacted to Co^{60} exposure, no significant bulk damage or CTE degradation was observed. Much higher total dose

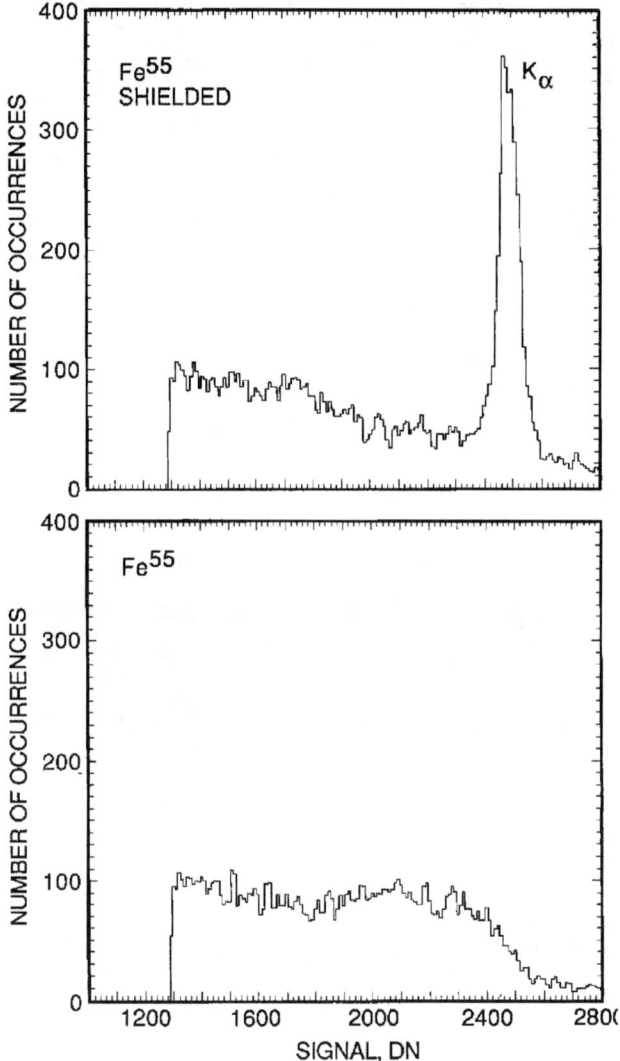

Figure 8.40(b) Fe^{55} histogram for the shielded and nonshielded regions.

levels were required before bulk damage became noticeable. The test performed here showed that CCD damage is very dependent on particle type. For example, Fig. 8.41(a) shows a vertical x-ray response before Co^{60} irradiation. CTE performance is near perfect for the CCD. Figure 8.41(b) shows the vertical response after the CCD was subjected to 5000 rad of Co^{60}. Note that this dose is an order of magnitude greater than the proton dose of 487 rad used above. CTE performance for the gamma-ray exposure degrades slightly to 0.99991, whereas the CTE for the proton exposure is less than 0.995 [Fig. 8.40(a)].

Ionization and bulk damage became separate radiation issues after this test was performed. In this comparative case, NIEL characteristics are dramatically differ-

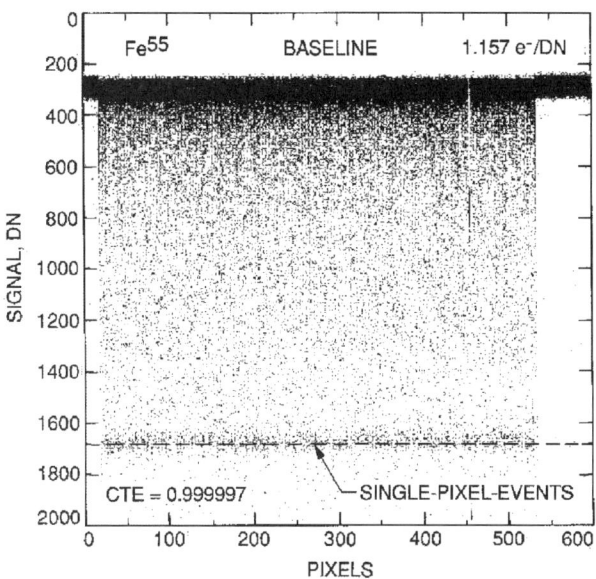

Figure 8.41(a) Baseline vertical Fe^{55} CTE response before Co^{60} irradiation.

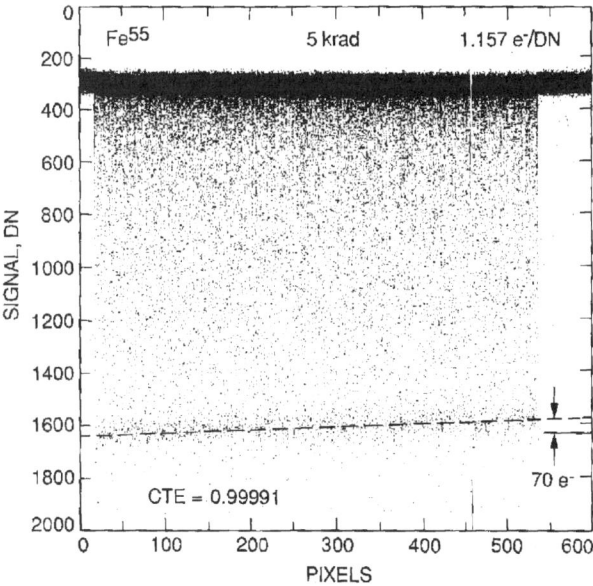

Figure 8.41(b) Vertical Fe^{55} CTE response after 5-krad Co^{60} exposure.

ent for protons and Compton electrons (NIEL curves will be given for many types of particles below). It should be mentioned that Fe^{55} x rays do not exhibit bulk damage characteristics at any dose level. As a rule of thumb, bulk damage only begins to take place for photons energies greater than 300 keV.

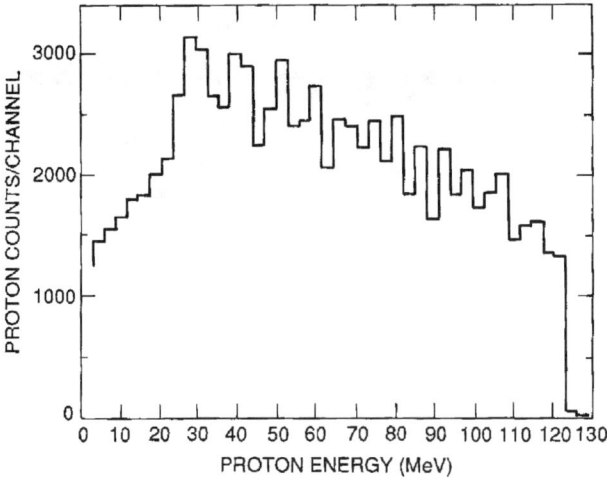

Figure 8.42 Harvard proton energy spectrum.

8.1.9.4.6 25-rad proton test. This test showed that the CCD is sensitive to only 25 rad of protons. For this experiment three regions of a 1024 × 1024 Cassini CCD were exposed to a dose of 25, 100, and 200 rad. Refer to Fig. 8.42 and Table 8.3 for the energy spectrum and total dose breakdown employed in the simulation (protons supplied by Harvard University Cyclotron). The energy spectrum simulated the South Atlantic Anomaly (SAA) at a 400-km altitude. As with all proton tests, one region was shielded for baseline purposes. Figure 8.43 shows partial vertical x-ray responses for each region irradiated. The baseline region exhibits a CTE = 0.999995. The 25-rad region shows some CTE degradation (CTE = 0.999984) whereas 100- and 200-rad regions result in further degradation (CTEs of 0.999969 and 0.999922, respectively). The x-ray stacking plots on the right-hand side of the figure show annealing characteristics at 200°C for 24 hours. Annealing bulk damage (i.e., P-V centers) is usually very difficult to employ in practice because of the high temperatures required.

Figure 8.44 shows proton dark spikes induced for the three regions for the same sensor. Dark spikes also exhibit some annealing characteristics as shown in the lower figure. It is interesting to note that dark spikes exhibit their own activation energy and temperature when annealing takes place. For example, many more dark spikes are seen when the CCD is always kept cold during irradiation and test. Warming the CCD to room temperature after irradiation anneals dark spike density to roughly the same level that would be produced when irradiating at room temperature. There is no magical temperature where all dark spikes will anneal together. This characteristic is used in practice on the Hubble Space Telescope where the operating temperature is regularly elevated to anneal proton-induced dark spikes induced at the nominal operating temperature of −90°C.

Table 8.3 25-rad Harvard proton spectrum.

Proton energy MeV	Rads	Protons/cm^2
4	2.32	2.07×10^6
7	1.59	2.21×10^6
10	1.32	2.36×10^6
13	1.25	2.56×10^6
16	1.08	2.59×10^6
19	0.98	2.86×10^6
22	0.93	3.03×10^6
25	1.08	3.79×10^6
28	1.18	4.50×10^6
31	1.05	4.34×10^6
34	0.85	3.77×10^6
37	0.76	3.64×10^6
40	0.81	4.27×10^6
43	0.75	4.13×10^6
46	0.55	3.20×10^6
49	0.59	3.61×10^6
52	0.65	4.19×10^6
55	0.51	3.41×10^6
58	0.49	3.49×10^6
61	0.52	3.89×10^6
64	0.38	2.94×10^6
67	0.44	3.50×10^6
70	0.41	3.43×10^6
73	0.37	3.17×10^6
76	0.39	3.46×10^6
79	0.34	3.01×10^6
82	0.38	3.54×10^6
85	0.28	2.61×10^6
88	0.33	3.21×10^6
91	0.23	2.33×10^6
94	0.31	3.16×10^6
97	0.26	2.64×10^6
100	0.29	2.93×10^6
103	0.24	2.44×10^6
106	0.26	2.64×10^6
109	0.28	2.87×10^6
112	0.20	2.09×10^6
115	0.22	2.26×10^6
118	0.22	2.29×10^6
121	0.19	1.90×10^6
124	0.18	1.87×10^6

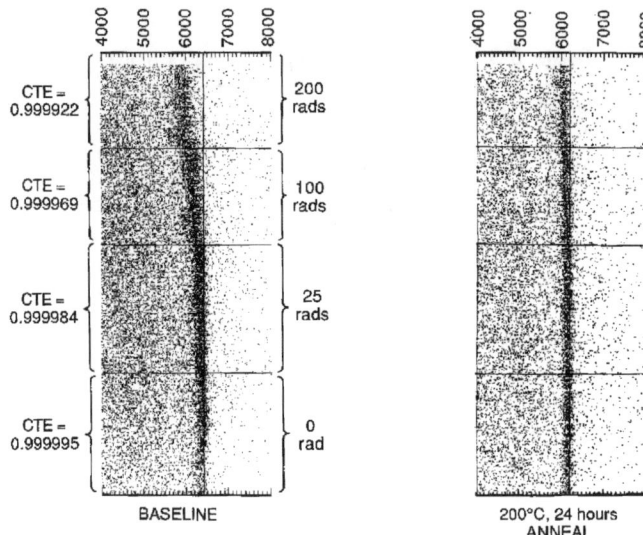

Figure 8.43 Fe^{55} vertical x-ray response after irradiating a CCD to the proton spectrum shown in Table 8.3. Also shown is the response after a high-temperature anneal.

8.1.9.4.7 Neutron test. This test showed the damaging effects of neutrons. Figure 8.45 shows a pocket pumping trap plot taken from a 1024×1024 CCD that was irradiated with 14.5-MeV neutrons. As mentioned above, neutrons interact directly with the silicon nucleus, ejecting high-energy recoil silicon atoms. The atoms are short ranged and induce "cluster" damage at the point of interaction. Cluster and single displacements exhibit different trapping characteristics. For example, the plot shows that neutrons induce a wide distribution of trap sizes compared to 1 e^{-} proton traps shown in Fig. 8.36(h). Figure 8.45(b) shows a pocket pumping response for the same CCD in response to a ramp signal. This data should be compared with the proton response shown in Fig. 5.53. Note again that the distribution in trap size for neutrons is considerably broader. Figure 8.46 shows a pocket pumping response to 2-MeV neutrons showing smaller traps as compared to 14.5-MeV neutrons indicating less displacements per interaction.

Neutrons that become embedded in the CCD will eventually turn into a proton and electron after several minutes (i.e., half-life of a neutron = 10.6 min). When this happens, hydrogen gas can be produced that will slowly diffuse out of the device. However, the neutron may also leave the device entirely before the half-life reaction occurs (i.e., the neutron may roll out of the camera head and onto the ground).

8.1.9.5 Proton Transfer

Space applications usually focus on the damaging effects of the hydrogen ion (i.e., proton) over a very large energy range (100 keV to 1 GeV). Due to the vast

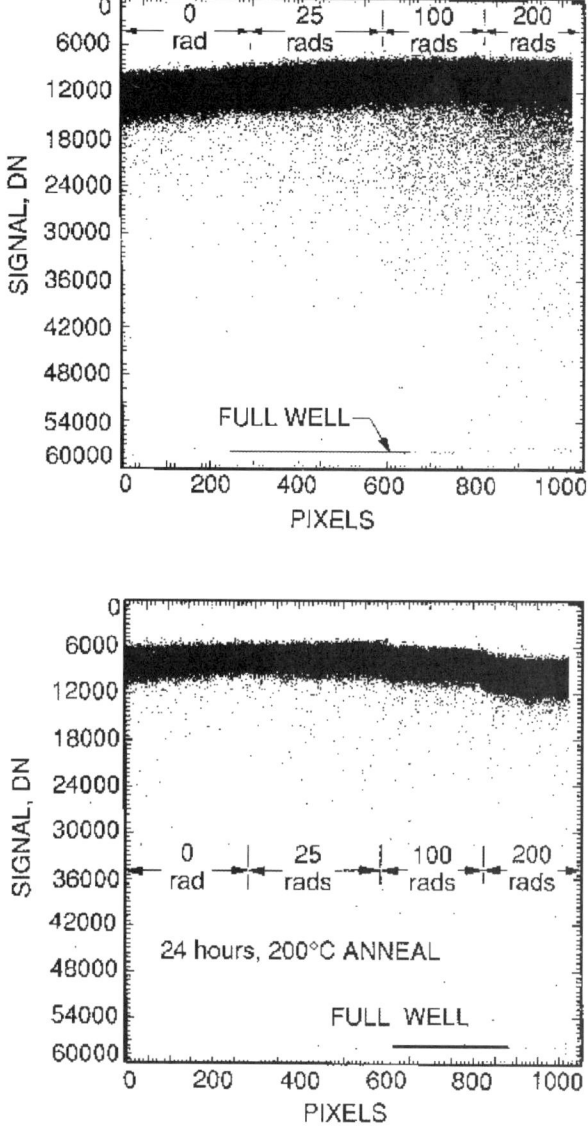

Figure 8.44 Post-irradiation dark current in response to the photon energy spectrum shown in Table 8.3.

range of energies and types of particles involved, it is extremely difficult—if even possible—to accurately simulate a space radiation environment on the ground for test purposes. In addition, mission test requirements (e.g., spacecraft trajectories, shielding arrangements) frequently evolve and change. Characterizing the CCD under all relevant conditions would require prohibitive amounts of time and money. Therefore, it is important that suitable analytical techniques be developed to permit such evaluation and prediction. NIEL and the proton transfer technique discussed

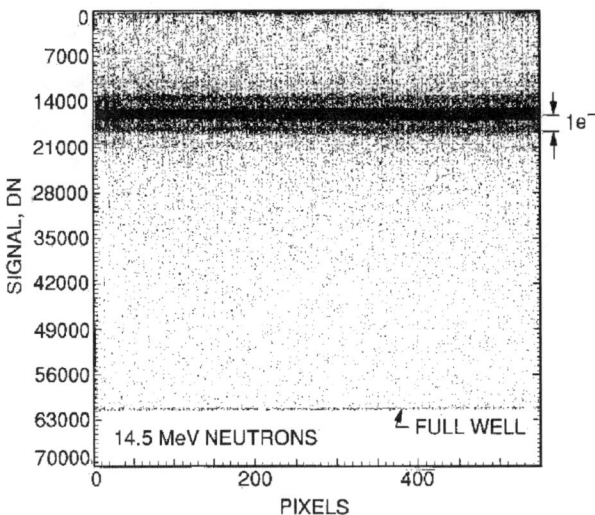

Figure 8.45(a) Flat-field pocket pumping response to 14.5-MeV neutrons.

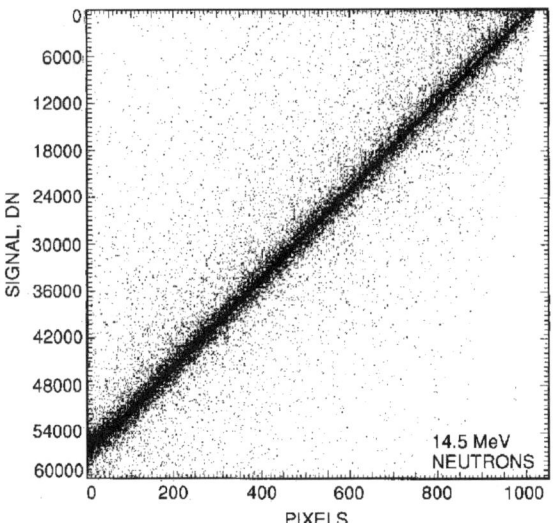

Figure 8.45(b) Ramp-field pocket pumping response to 14.5-MeV neutrons.

below was developed for this purpose.[3,58] This section presents an approximate correlation between experimental results and a prediction of particle damage and explanations for remaining inconsistencies. As a consequence of this agreement, predicting the effect of complicated particle environments upon CCD performance parameters is possible. This prediction requires evaluation of the damage resulting from only a small number of quasi-mono-energetic particle exposures to calibrate the susceptibility of a given device. It is interesting and very important to point out

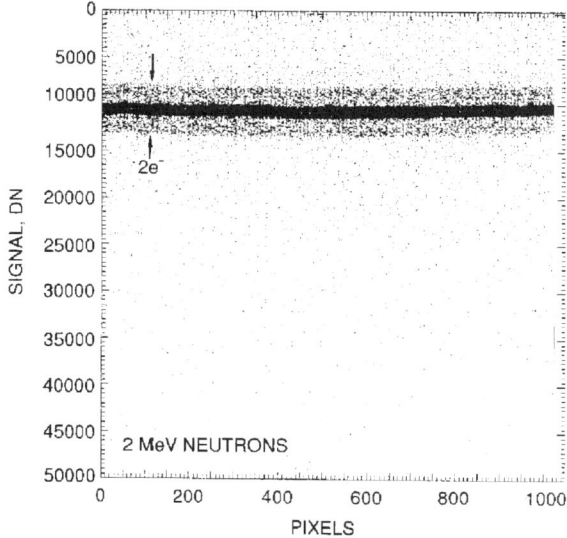

Figure 8.46 Flat-field pocket pumping response to 2-MeV neutrons.

that NIEL and experimental data in the form of CTE and dark current are in good agreement.[3,57,59] This implies that the details between the CCD and interacting particle are not important. For example, it appears that elastic and inelastic reactions are equivalent in making traps and bulk defects for the same energy dissipated. We will emphasize this important point again below.

The relation between experiment and SRIM was verified by irradiating 1024 × 1024 CCDs with approximately mono-energetic beams of protons at energies ranging from 0.25 to 150 MeV. During irradiation, a region of the CCD was covered with an appropriate shield to serve as a control. The results of these tests are shown in Table 8.4. Here we list the energy of the proton (eV), fluence (protons/cm^2), the predicted number of displaced silicon atoms per incident proton (found through SRIM), the measured and predicted CTI after irradiation and the radiation trap inefficiency (RTI) factor, which compares theory to experiment. Data was taken at an operating temperature of $-50°$C and a line shift time of 10^{-2} sec. As will be explained below, this operating temperature and line time was intentionally selected to produce a worse case CTE effect and the full extent of the bulk damage induced.

Figure 8.47 plots the data in Table 8.4 as a function of proton energy: (1) measured CTI/proton/pixel (triangles), (2) displacements/proton based on SRIM (solid line) and (3) the RTI factor. The solid curve is called the proton transfer curve, a key transfer curve used to predict CTE performance for a specified proton energy spectrum. Note the rapid falloff for protons whose energy is < 180 keV, the cutoff energy where the gate and oxide layers absorb all incoming protons. The curve shows that peak damage takes place at approximately 250 keV, the energy where the proton's range is coincident with the signal-carrying channel depth. Beyond this energy, damage drops off steeply as the proton penetrates deeper into

Table 8.4 Proton transfer data generated for Fig. 8.47.

Energy (MeV)	Fluence protons/cm^2	Displacements /proton	Measured CTI	Predicted CTI	RTI
0.250	5.86×10^6	13	9.7×10^{-5}	4.37×10^{-2}	2.22×10^{-2}
0.400	1.48×10^7	5	9.7×10^{-5}	4.38×10^{-2}	2.21×10^{-2}
0.550	3.67×10^7	2	9.7×10^{-5}	4.35×10^{-2}	2.23×10^{-2}
0.700	6.10×10^7	1.2	1.1×10^{-4}	4.33×10^{-2}	2.54×10^{-2}
0.850	9.23×10^7	0.75	1.1×10^{-4}	4.10×10^{-2}	2.68×10^{-2}
1.0	1.37×10^8	0.55	1.1×10^{-4}	4.46×10^{-2}	2.47×10^{-2}
3.07	4.94×10^8	0.15	1.5×10^{-4}	4.40×10^{-2}	3.41×10^{-2}
5.58	9.15×10^8	0.08	1.5×10^{-4}	4.34×10^{-2}	3.46×10^{-2}
8.5	1.45×10^9	0.05	1.5×10^{-4}	4.29×10^{-2}	3.50×10^{-2}
18.5	3.71×10^9	0.02	2.3×10^{-4}	4.39×10^{-2}	5.20×10^{-2}
39.6	8.25×10^9	0.009	2.5×10^{-4}	4.44×10^{-2}	5.63×10^{-2}
67.4	1.4×10^{10}	0.0053	2.6×10^{-4}	4.44×10^{-2}	5.80×10^{-2}
100	1.98×10^{10}	0.0033	4.04×10^{-4}	3.87×10^{-2}	1.04×10^{-2}
150	2.78×10^{10}	0.002	3.94×10^{-4}	3.29×10^{-2}	1.19×10^{-2}

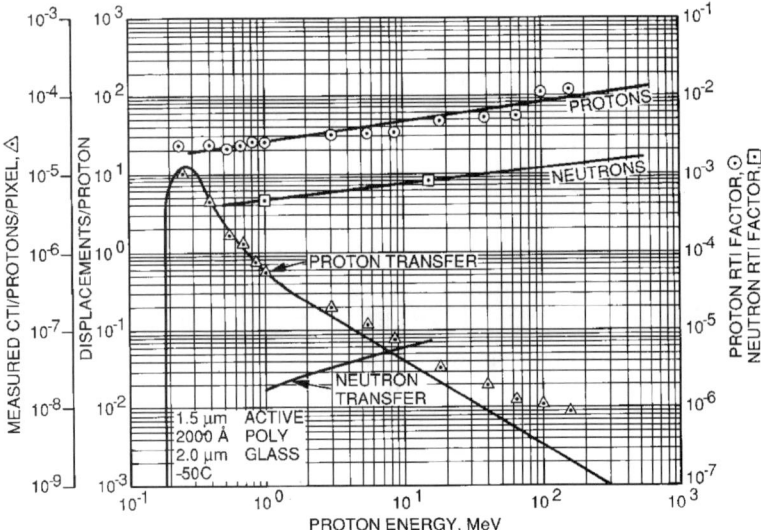

Figure 8.47 Proton and neutron transfer curves with RTI factors.

the CCD. All plots assume the CCD structure in Fig. 8.34 and normal incidence protons.

When bulk radiation experiments are performed, there must be a finite CTI loss induced in the CCD to measure it. If, for example, the fluence is too small, then the tilt in the single-event x-ray line will be difficult to measure. On the other hand, too much damage could destroy the line. The fluence given in Table 8.4 was adjusted to induce a CTI loss of approximately 0.0001 (CTE = 0.9999). For

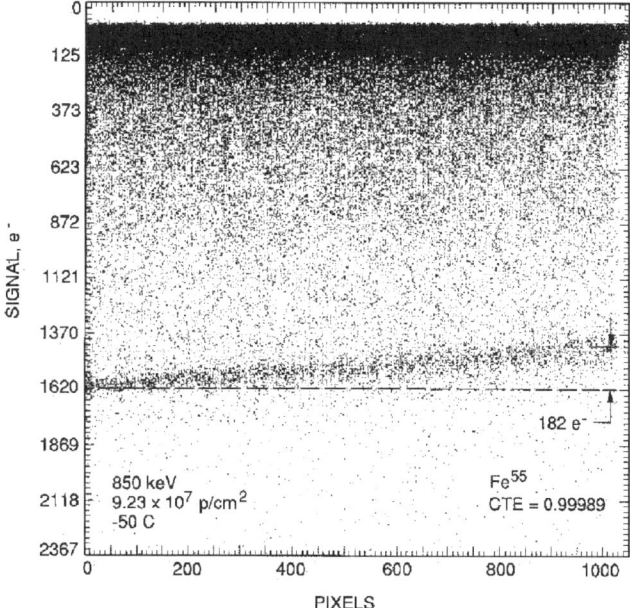

Figure 8.48 An example Fe^{55} vertical CTE response used to generate the proton transfer curve shown in Figure 8.47.

example, Fig. 8.48 shows the tilt that develops when a 1024 × 1024 CCD is exposed to 850-keV protons at a fluence of 9.23×10^7 protons/cm^2 (the fifth data entry in Table 8.4). A CTE of 0.99989 is measured from the tilt in the single-pixel event line (CTI = 0.00011).

Table 8.4 attempts to maintain a constant CTI as a function of energy by adjusting the fluence accordingly. Knowing the CTI for a given fluence and corresponding proton energy allows one to find the fluence at a different energy while producing the same CTI effect. This new fluence level is calculated through

$$F_{NEW} = \frac{F_{BASE} D_{BASE}}{D_{NEW}}, \quad (8.22)$$

where F_{NEW} is the new fluence (protons/cm^2), F_{BASE} is the baseline fluence used to set the original CTI level (protons/cm^2), D_{BASE} is the number of displacements for the baseline fluence (displacements/proton) and D_{NEW} is the number of expected displacements for the new fluence level (displacements/proton).

Example 8.22

Determine the fluence at which 8.5-MeV protons produce the same CTI effect as 3 MeV protons at a fluence of 5×10^8 protons/cm^2.

Solution:
From Table 8.4, $D_{BASE} = 0.15$ and $D_{NEW} = 0.05$ displacements/proton.
From Eq. (8.22),

$$F_{NEW} = (5 \times 10^8) \times 0.15/0.05,$$

$$F_{NEW} = 1.5 \times 10^9 \text{ protons/cm}^2.$$

Predicted CTE given in Table 8.4 is found by assuming that each silicon displacement produces a trap with unity capture probability. For signal charge distributed throughout a specified channel volume, this estimate of CTI is

$$CTI_{SRIM} = \frac{D_{SRIM} A_{PIX} F}{S(e^-)}, \qquad (8.23)$$

where D_{SRIM} is the number of displacements/proton found through SRIM, A_{PIX} is the active area of a CCD pixel (cm^2), F is the fluence (protons/cm^2), and $S(e^-)$ is the signal charge packet being transferred which is normally based on a Fe^{55} test signal (e$^-$).

Example 8.23

From Table 8.4, predict CTI performance for a 12-μm-pixel CCD when exposed to 1.37×10^8 protons/cm^2 1-MeV protons. Assume $A_{PIX} = 9.6 \times 10^{-7}$ cm^2 (12 × 8-μm active pixel area less channel stops), and $S(e^-) = 1620\,e^-$.
Solution:
From Table 8.4, $D_{SRIM} = 0.55$ displacements/proton.
From Eq. (8.23),

$$CTI_{SRIM} = 0.55 \times (9.6 \times 10^{-7}) \times (1.37 \times 10^8)/1620,$$

$$CTI_{SRIM} = 4.46 \times 10^{-2} \ (CTE_{SRIM} = 0.9554).$$

The radiation trap inefficiency (RTI) factor is the ratio between the measured CTI and predicted CTI. That is,

$$RTI = \frac{CTI_M}{CTI_{SRIM}}, \qquad (8.24)$$

where CTI_M is the measured CTI. As we will discuss below, we can look at the RTI factor as a proportionality constant between the damage measured in terms of CTE and vacancies predicted by SRIM.

Example 8.24

Determine the RTI factor for the above example using CTI data provided by Table 8.4 (i.e., $CTI_M = 1.1 \times 10^{-4}$).
Solution:
From Eq. (8.24),

$$RTI = (1.1 \times 10^{-4})/(4.46 \times 10^{-2}),$$

$$RTI = 2.47 \times 10^{-3}.$$

As noted from Table 8.4, theory and experiment for CTI are far from agreement (the RTI factor would need to be unity for full agreement). As it will turn out, the discrepancy will not alter our ability to predict absolute CTE performance for a specified radiation environment. The RTI factor was invented to deal with many uncertainties involved in the prediction, including the shortcomings of SRIM analysis (i.e., it is only a scaling factor). These unknowns are numerous. First, our model assumes that active traps occupy a thickness of 1.5 µm (refer to Fig. 8.34). In fact, signal charge is only allowed to transfer in the n-region, which for the average CCD is approximately 0.5 µm in extent. Therefore, the model overestimates the active transfer region and the number of active traps (although vacancies can diffuse in from other regions of the device). In addition, for small charge packets such as x-ray events, the volume that a charge packet occupies is a small fraction of the n-region. Therefore, the charge packet does not encounter all traps induced in the n-region. For example, it has been estimated that for small charge packets such as soft x rays, only 0.15 µm of the n-region is used (refer to discussions below). Second, many displaced vacancies recombine with interstitials before trapping centers can form. Third, the strength of local electric fields within a pixel affects the trapping that takes place. Traps created between phases see the greatest field strength and therefore may be inactive (i.e., field-assisted transfer). This fact is supported by pocket pumping experiments.

The CTI data presented in Fig. 8.47 show reasonable correlation between experiment and theory, the ratio being expressed by the RTI factors plotted. It is important to note that the RTI factor increases with proton energy. The reason for this behavior is that SRIM only calculates elastic coulomb interactions. This model is inaccurate for energies greater than about 10 MeV where nuclear elastic scattering becomes significant. At still higher energies the inelastic cross-section begins to play a role, again not accounted for by SRIM. At 60 MeV, about half of the displacements are produced by inelastic interactions. Figure 8.49 shows an NIEL curve that includes nuclear inelastic recoil distributions.[53] Note that the number of displacements is nearly constant at energies greater than 100 MeV, considerably different than the SRIM results shown in Fig. 8.47.

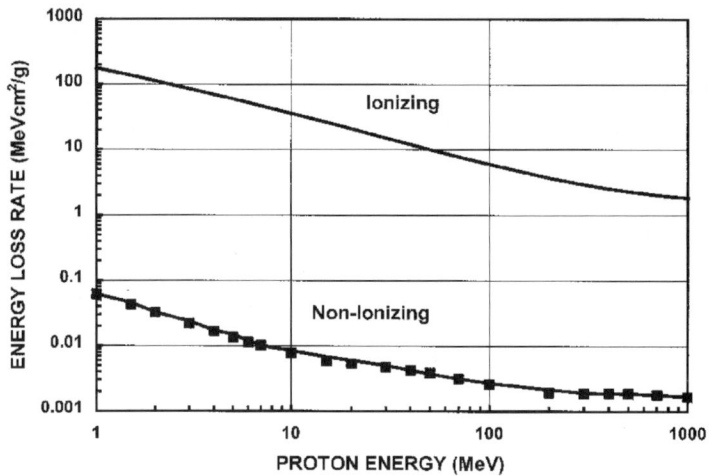

Figure 8.49 NIEL curve that includes nuclear inelastic interactions.[53]

With proton transfer and RTI in hand we can predict CTE loss induced by protons for a given energy spectrum. For example, Fig. 8.50(a) shows the expected proton environment that the Cassini CCD will experience over its mission lifetime. The figure also includes different tantalum shield thicknesses that surround the CCD. For the spectrum illustrated, protons with energies < 50 MeV are primarily generated by the solar flares, whereas protons with energies > 50 MeV are cosmic in origin. The fluence is blocked in 27 energy bins covering 0.1 to 100 MeV. This data is then multiplied by the proton transfer curve producing the plot in Fig. 8.50(b). The energy bins are given in units of displacements/cm^2. The total number of displacements is found by adding up the number of displacements for each bin. For the 1-cm shielding case, this number is 1.7×10^7 displacements/cm^2, or about 24.5 displacements for a 12-μm pixel.

Total displacements can be translated into a CTI figure through

$$CTI_{PRE} = \frac{D_T A_{PIX} RTI}{S(e^-)}, \qquad (8.25)$$

where CTI_{PRE} is the predicted CTI performance and D_T is the total number of displacements/cm^2 supplied by Fig. 8.50(b).

Example 8.25

Predict CTE performance for the Cassini CCD, assuming $A_{PIX} = 9.6 \times 10^{-7}$ cm^2, $RTI = 2.7 \times 10^{-3}$, $D_T = 1.7 \times 10^7$ displacement/cm^2 and $S = 1620\,e^-$.

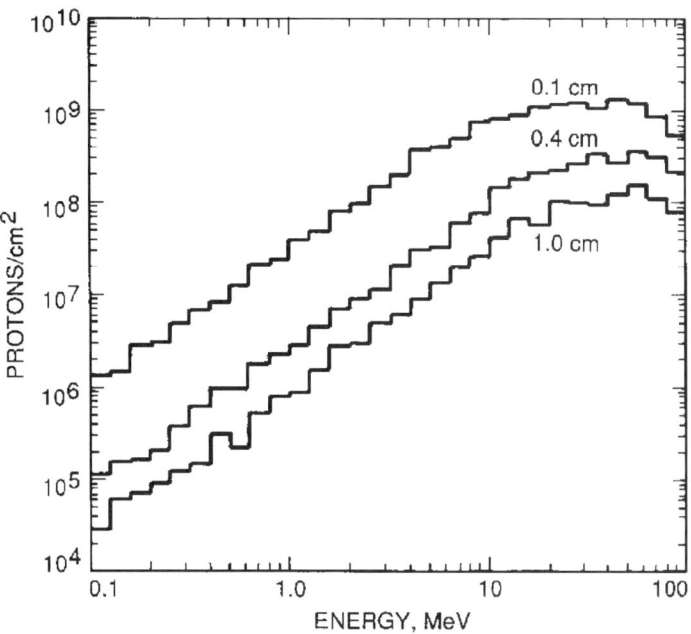

Figure 8.50(a) Cassini proton energy spectrum at the CCD for different Ta shield thicknesses.

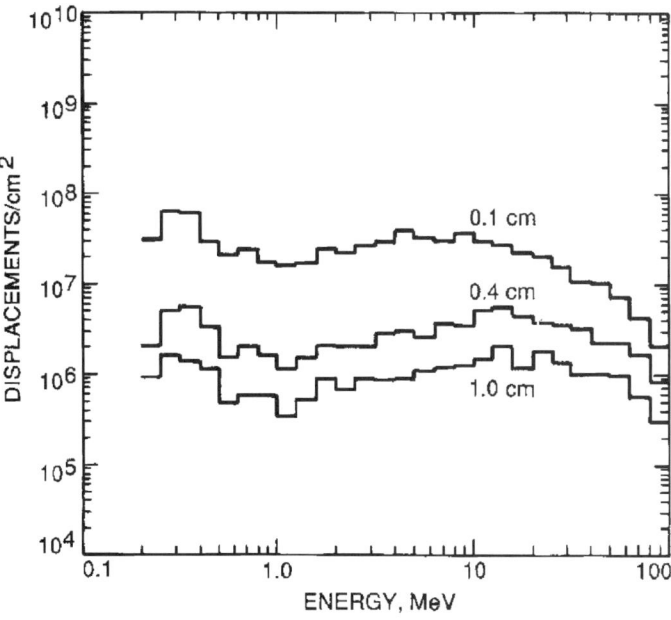

Figure 8.50(b) Cassini displacements as a function of energy.

Solution:
From Eq. (8.25),

$$CTI_{PRE} = (1.7 \times 10^7) \times (9.6 \times 10^{-7}) \times (2.7 \times 10^{-3})/1620,$$

$$CTI_{PRE} = 2.72 \times 10^{-5} \text{ (CTE} = 0.999973).$$

We can summarize the steps taken above in predicting CTI performance for a given proton environment:

1) Specify radiation environment in terms of fluence (protons/cm^2) and energy bins [Fig. 8.50(a)].

2) Multiply each bin of the radiation environment by the proton transfer curve [Fig. 8.50(b)]. This result will only be an estimate, because as mentioned above, inelastic interactions are not taken into account with SRIM. For a more rigorous analysis, experimental data points in Fig. 8.50(b) or the NIEL curve can be used (i.e., Fig. 8.49).

3) From this resultant product, integrate and determine the total number of displacements, D_T.

4) Then, predict CTI using RTI factor [Eq. (8.25)].

8.1.9.6 Bulk Damage Equivalence

It is a truly remarkable outcome that device degradation follows the NIEL/SRIM curves. As discussed above, there are many complicated physical processes that take place when vacancies are converted to active CTE traps and dark current spike defects. Whether this damage comes about through elastic or inelastic collisions or through the form of single or cluster defects doesn't seem to matter. Radiation tests performed on different CCDs (i.e., pixel size, design, technology, manufacturer, etc.) have demonstrated that the NIEL/SRIM curve is reliable in predicting CTI performance using a single equivalent proton energy. We call this concept bulk damage equivalence. For example, Table 8.5 presents a radiation environment that shows the energy bin number, the average energy for the bin (MeV), the fluence for that bin (protons/cm^2), the number of displacements produced from the data points shown in Fig. 8.47 and the number of displacements (displacements/cm^2). The net fluence and number of displacements for the entire environment are shown at the bottom of the table. Using the proton transfer data shown in Fig. 8.47 (or the NIEL curve), the energy spectrum can be simulated by exposing the CCD to 60-MeV protons. Here we will find that a 60-MeV fluence of the same number of particles (i.e., 2.14×10^9 protons/cm^2) will produce the same number of displacements (i.e., 2.56×10^7 displacements/cm^2) for the specified environment. After irradiation to this single energy level, the RTI factor can be measured and CTI calculated.

It may be more convenient to use a different proton energy and fluence than 60 MeV, depending on the proton facility used. For example, Table 8.4 can also

Table 8.5 Proton radiation environment that can be simulated with a single proton energy for equivalent bulk damage.

Bin number	Energy (MeV)	Protons/cm^2	Displacements	Displacements/cm^2
1	0.18	1.19E + 04	5	5.95E + 04
2	0.223	1.48E + 04	8	1.18E + 05
3	0.282	2.02E + 04	15	3.03E + 05
4	0.359	2.87E + 04	8	2.30E + 05
5	0.442	3.68E + 04	3	1.10E + 05
6	0.579	5.46E + 04	1.3	7.10E + 04
7	0.706	1.04E + 05	1.2	1.25E + 05
8	0.897	1.11E + 05	0.73	8.10E + 04
9	1.1	1.69E + 05	0.4	6.76E + 04
10	1.4	2.36E + 05	0.3	7.08E + 04
11	1.8	3.52E + 05	0.23	8.10E + 04
12	2.31	5.66E + 05	0.2	1.13E + 05
13	2.79	7.45E + 05	0.2	1.49E + 05
14	3.68	1.13E + 06	0.15	1.70E + 05
15	4.58	1.71E + 06	0.13	2.22E + 05
16	5.45	3.74E + 06	0.11	4.11E + 05
17	7	3.91E + 06	0.09	3.52E + 05
18	9.13	5.58E + 06	0.07	3.91E + 05
19	11.4	8.45E + 06	0.044	3.72E + 05
20	14	1.27E + 07	0.04	5.08E + 05
21	18.3	2.16E + 07	0.03	6.48E + 05
22	22.5	3.37E + 07	0.026	8.74E + 05
23	28.5	4.67E + 07	0.022	1.03E + 06
24	35.2	6.61E + 07	0.02	1.32E + 06
25	45.3	1.02E + 08	0.015	1.53E + 06
26	57.3	1.35E + 08	0.013	1.76E + 06
27	69.8	1.66E + 08	0.011	1.83E + 06
28	90.8	1.81E + 08	0.01	1.81E + 06
29	113	2.64E + 08	0.009	2.38E + 06
30	136	2.71E + 08	0.0085	2.30E + 06
31	182	2.61E + 08	0.008	2.09E + 06
32	221	2.49E + 08	0.008	1.99E + 06
33	278	1.97E + 08	0.008	1.58E + 06
34	335	1.07E + 08	0.008	8.56E + 05
		TOTAL = 2.14E + 09		TOTAL = 2.60E + 07

be simulated with a 8.5-MeV proton fluence. In that 8.5-MeV protons are more damaging than 60-MeV protons, the fluence needs to be reduced by the ratio of displacements. In this case, a 8.5-MeV proton produces 0.05 displacements/proton, whereas a 60-MeV proton produces 0.012 displacements/proton (see data points in Fig. 8.47). Therefore, a 8.5-MeV fluence of 5.13–10^8 protons/cm^2 will produce equivalent damage as do 60-MeV protons.

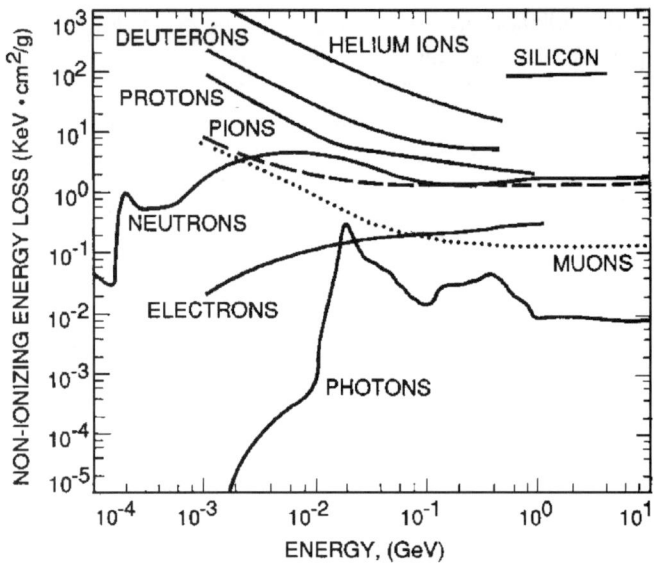

Figure 8.51 NIEL radiation transfer curves for different particles.[7,61,62]

8.1.9.7 Radiation Transfer

Radiation transfer curves can be generated for a variety of particles (neutrons, electrons, heavy ions, photons, pions, etc.).[7,52,60–62] For example, Fig. 8.51 shows a set of NIEL curves as calculated by Burke and Van Ginneken. Note that more massive particles produce more damage than lighter particles because their stopping power is greater. These curves can be used to transfer radiation data from one particle type to another, assuming that the RTI factors are known for each particle. This radiation transfer technique is governed by the linear interpolation equation

$$CTI_{P2} = \frac{F_2 D_{SRIM2} RTI_2}{F_1 D_{SRIM_1} RTI_1} CTI_{M1}, \qquad (8.26)$$

where the subscripts 1 and 2 refer to the particle type 1 and 2. CTI_P is the predicted CTI, CTI_M is the measured CTI, F is the fluence at the CCD (particles/cm^2), D_{SRIM} is the number of displacements predicted by SRIM and RTI is the radiation trap inefficiency factor.

The particle type in Eq. (8.26) can be the same variety. For example, if the CTI is measured for a CCD exposed to a specific proton energy and fluence, the CTI for a different proton energy and fluence can be predicted. The next example will show how this transfer process is performed.

Example 8.26

Determine the CTI for a 550-keV proton at a fluence of 7.34×10^7 protons/cm^2, given

Measured proton energy (5.58 MeV)

$$F_1 = 9.15 \times 10^8 \text{ protons/cm}^2,$$
$$D_{SRIM1} = 0.08 \text{ displacements},$$
$$RTI_1 = 3.46 \times 10^{-3},$$
$$CTI_{M1} = 1.5 \times 10^{-4} \text{ (CTE} = 0.99985).$$

Predicted proton energy (550 keV)

$$F_2 = 3.67 \times 10^7 \text{ protons/cm}^2,$$
$$D_{SRIM2} = 2 \text{ displacements},$$
$$RTI_2 = 2.23 \times 10^{-3}.$$

Solution:
From Eq. (8.26),

$$CTI_{P2} = (3.67 \times 10^7) \times 2 \times (2.23 \times 10^{-3})$$
$$\times (1.5 \times 10^{-4}) / [(9.15 \times 10^8) \times 0.08 \times (3.46 \times 10^{-3})],$$
$$CTI_{P2} = 9.69 \times 10^{-5} \text{ (CTE} = 0.9999031).$$

Example 8.27

Neutrons were also a problem to the Cassini CCD because of an onboard RTG unit used to generate electricity. This radioactive source generated neutrons and gamma rays. Using the neutron transfer curve shown in Fig. 8.47, predict the CTI performance for a neutron fluence of 1.5×10^{10} neutrons/cm^2. Also, from Example 8.25, calculate the net CTE loss for Cassini when proton and neutron damage are combined.

Measured proton energy (8.5 MeV)

$$F_1 = 1.45 \times 10^9 \text{ protons/cm}^2,$$
$$D_{SRIM1} = 0.05 \text{ displacements/proton},$$
$$RTI_1 = 3.5 \times 10^{-3},$$
$$CTI_{M1} = 1.5 \times 10^{-4}.$$

Predicted neutron energy (1 MeV)

$F_2 = 1.5 \times 10^{10}$ neutrons/cm^2 (equivalent 1-MeV fluence),
$D_{SRIM2} = 0.018$ displacements/neutron,
$RTI_2 = 4.54 \times 10^{-4}$.

From Eq. (8.26), the CTI for neutron damage is

$$CTI_{P2} = (1.5 \times 10^{10}) \times 0.018 \times (4.54 \times 10^{-4})$$
$$\times (1.5 \times 10^{-4})/[(1.45 \times 10^9) \times 0.05 \times (3.5 \times 10^{-3})],$$
$$CTI_{P2} = 7.246 \times 10^{-5} \text{ (CTE} = 0.999929).$$

From Example 8.25, the net CTI when proton and neutron damage are combined is

$$CTI = (2.72 \times 10^{-5}) + (7.246 \times 10^{-5}),$$

$$CTI = 9.966 \times 10^{-5} \text{ (CTE} = 0.9999).$$

This value was lower than Cassini's CTE specification of 0.99999. Solutions to meet this specification are discussed in the next section.

8.1.9.8 Bulk Damage Remedies

This section presents six remedies to reduce the impact of the bulk radiation problem on CCD performance: (1) operating temperature, (2) clocking and charge packet density, (3) design and processing, (4) shielding, (5) fat-zero and light flood, and (6) annealing.

8.1.9.8.1 Operating temperature.
Measurements made in Table 8.4 and Fig. 8.47 were taken at an operating temperature of $-50°C$ (223 K). As we will show below, at this temperature the P-V trap emission time constant is approximately equal to the vertical line time (i.e., 10 msec for this test). This timing and operating temperature produces the worst vertical transfer loss. As mentioned above, it is best to keep the P-V trap filled during transfer by lowering the operating temperature. This advantage is seen through Eq. (5.26) when written in the form

$$CTI = \frac{D_T A_{PIX} RTI}{S(e^-)} \exp\left(-\frac{t_{LT}}{3\tau_e}\right)\left[1 - \exp\left(-\frac{t_{LT}}{3\tau_e} - \frac{N_Z t_{LT}}{\tau_e}\right)\right], \quad (8.27)$$

where N_Z is the spacing between charge packets (lines/event), t_{LT} is the line time between vertical shifts (sec) and τ_e is the emission time constant for the P-V trap defined below (sec). A similar expression can be derived for the horizontal register by substituting line time with pixel time.

The first term in Eq. (8.27) represents the worse-case CTI for a selected operating temperature and line shift time (i.e., maximum deferred charge). Therefore, Eq. (8.27) can be simplified to

$$CTI = CTI_{WOR} \exp\left(-\frac{t_{LT}}{3\tau_e}\right)\left[1 - \exp\left(-\frac{t_{LT}}{3\tau_e} - \frac{N_Z t_{LT}}{\tau_e}\right)\right], \quad (8.28)$$

where CTI_{WOR} is the worse-case CTI.

The second and third terms of Eq. (8.28) are used to improve CTE performance through the trap emission time constant, given here as[63]

$$\tau_e = \frac{e^{E_T/kT}}{X_n \sigma_n v_{th} N_C}, \quad (8.29)$$

where E_T is the energy level of the trap below the conduction band (eV); X_n is the entropy factor, which accounts for the entropy change after an electron is emitted from a trap; v_{th} is the thermal velocity (cm/sec); N_C is the number of effective states in the conduction band (cm^{-3}) and σ_n is the electron cross-section of the trap (cm^2). The product $v_{th} N_C$ is determined from the equations

$$N_C = 2\left(\frac{2\pi m_e^* kT}{h^2}\right)^{3/2}, \quad (8.30)$$

and

$$\frac{1}{2}m_e^* v_{th}^2 = \frac{3}{2}kT, \quad (8.31)$$

where m_e^* is the effective electron mass ($0.5 \times 9 \times 10^{-31}$ kg). When combined, these equations yield the product

$$v_{th} N_C = 1.6 \times 10^{21} T^2, \quad (8.32)$$

where $v_{th} N_C$ is given in units of cm^{-2} sec^{-1}.

Figure 8.52(a) measures the emission time constant as a function of $1/kT$ for a Co^{60}-damaged CCD. The measurements are based on the EPER technique described in Chapter 5. Figure 8.52(b) is a typical EPER response, showing the time response at the end of the array as charge detraps in the form of deferred charge. The time constant of the edge response represents the emission time constant for the trapping centers. EPER data was generated at several different operating temperatures to produce Fig. 8.52(a). The slope of the data points represents the activation energy of the P-V trap, which for this measurement works out to be approximately 0.40 eV—close to the accepted value of 0.44 eV.

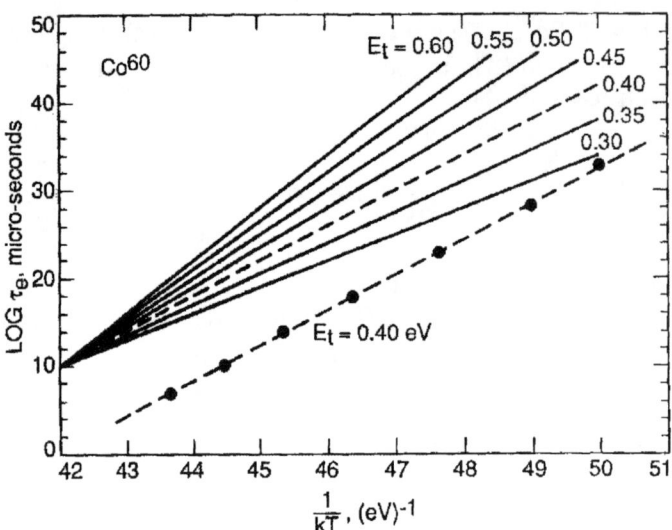

Figure 8.52(a) P-V trap emission time constant as a function of operating temperature.

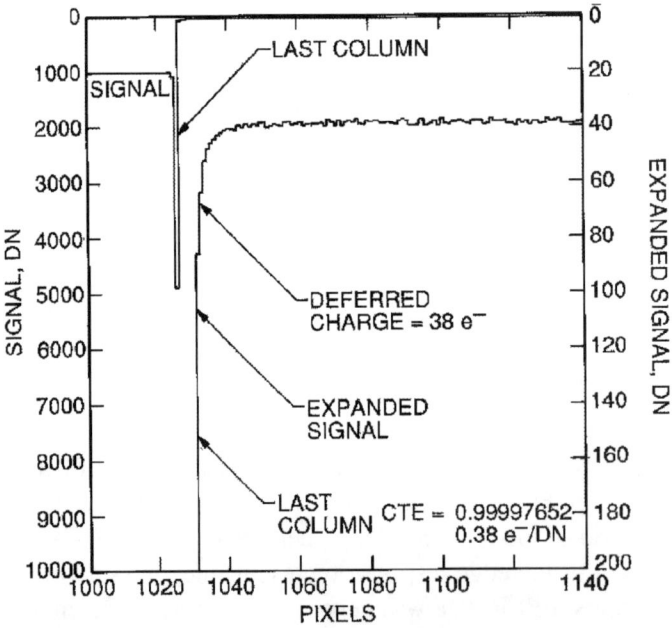

Figure 8.52(b) EPER used to determine emission time constant shown in Fig. 8.52(a).

Example 8.28

Determine the emission time for a 0.44-eV P-V center at room temperature. Assume $X_{PV}\sigma_n = 6 \times 10^{-16}$ cm^2 and $\sigma_n = 3.5 \times 10^{-15}$ cm^2.

Solution:
From Eq. (8.32),

$$v_{th}N_C = 1.6 \times 10^{21} \times 300^2,$$

$$v_{th}N_C = 1.44 \times 10^{26} \text{ cm}^{-2} \text{ sec}^{-1}.$$

From Eq. (8.29),

$$\tau_e = \exp(0.44/0.0254)/[(6.0 \times 10^{-15}) \times (1.44 \times 10^{26})],$$

$$\tau_e = 3.8 \times 10^{-5} \text{ (300 K)}.$$

P-V emission time constants at other operating temperatures are

0°C (273 K) = 1.8×10^{-4} sec,
−50°C (223 K) = 0.018 sec,
−90°C (183 K) = 3.977 sec,
−100°C (173 K) = 22.29 sec,
−110°C (163 K) = 153 sec,
−120°C (153 K) = 1.35×10^3 sec.

Figure 8.53 plots emission time constant as a function of operating temperature for the P-V trap. At −90°C it is possible to keep most traps satisfied during frame readout time by background charge that is present (e.g., cosmic rays, spurious charge, dark current charge). Operating the CCD below −90°C has been the main solution used for addressing the bulk radiation problem in space applications. The CCDs on Hubble, Galileo, Cassini, Chandra, etc., operate at temperatures below −90°C primarily for this reason.

Example 8.29

Using Eq. (8.28), plot vertical CTI performance for the Cassini CCD as a function of operating temperature and line shift times of $t_{LT} = 10^{-2}$, 10^{-3}, 10^{-4}, 10^{-5} and 10^{-6} sec. Assume a worse-case CTI_{WOR} of 9.966×10^{-5} from Example 8.27 and $N_Z = 50$ lines/event.

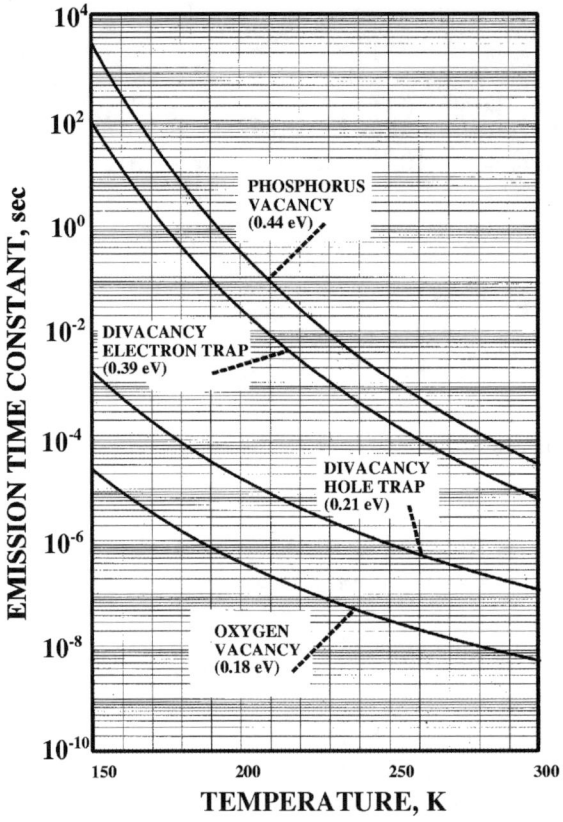

Figure 8.53 Emission time constant for various trap types as a function of operating temperature.

Solution:

Figure 8.54 is the required plot. Note that worse-case CTI occurs at approximately −50°C (223 K) and line time of 10^{-2} sec. To meet specifications (CTI = 10^{-5}), Cassini operates at a temperature of −90°C (183 K).

Also shown in Fig. 8.53 are the emission time constants for the oxygen vacancy (O-V) or A-center ($E_T = 0.170$ eV, $\sigma_n = 1 \times 10^{-14}$ cm^2), the divacancy center ($E_T = 0.390$ eV, $\sigma_n = 4 \times 10^{-15}$ cm^2) and the divacancy hole trap ($E_T = 0.210$ eV, $\sigma_p = 2 \times 10^{-16}$ cm^2). The O-V center is difficult to "freeze out" because its energy is shallow compared to the P-V center. At 170 K its emission time constant is only $\tau_e = 3.7$ μsec. This trap is present in the background of P-V centers because of the presence of oxygen in the silicon. Therefore, the O-V trap prevents us from achieving perfect CTE by simply cooling the CCD. Divacancy traps are produced when P-V and O-V centers are exhausted, and are considered second-order traps.

DAMAGE   821

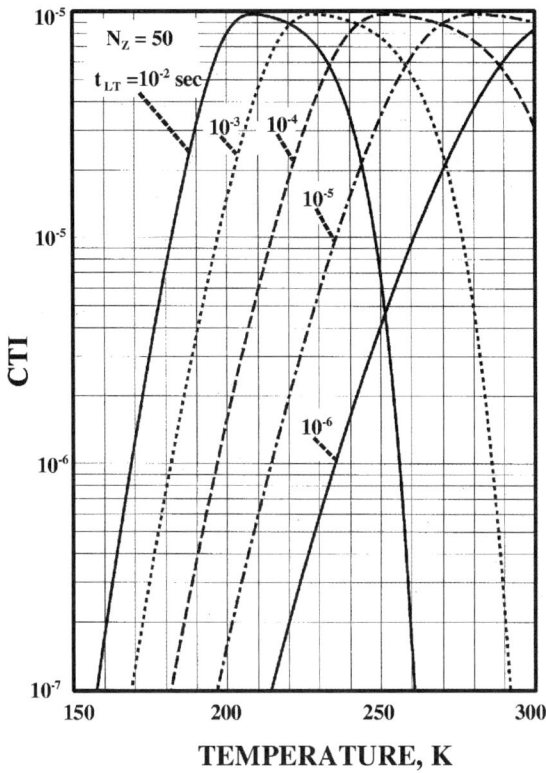

Figure 8.54 Predicted CTI for the Cassini CCD as a function of operating temperature and line shift time.

It should also be mentioned that defects can cluster, which can alter the apparent trap energy and emission time constant further complicating the problem.

Figure 8.55(a) presents a pocket pumping response generated by a 1024 × 1024 CCD that was partially irradiated with protons. The response was generated using a 2250 e^- flat field level and 1000 pump cycles for a 5-line sector. Column traces from the resultant pumped image are accumulated and stacked. Each dot displayed in the figure represents a trapping site. The response uncovers a wealth of information associated with the density and amplitude of radiation traps observed in CCDs. The left-hand side of the sensor was exposed to 280 rad (2.9×10^8 protons/cm^2) of 5-MeV protons, whereas the right side of the CCD was shielded from the proton dose. Within the shielded side of the detector, intrinsic bulk traps of 1 e^- and smaller are seen. This region contains a trap density of 0.002 bulk traps/pixel, limiting the sensor's CTE to approximately 0.999999 (i.e., CTE = 1 − 0.002/2250). For the irradiated side of the CCD, proton-induced traps exhibit discrete trapping lines (1, 2, 3 e^-, etc.) with most of the traps clustered at 1 e^-. The discrete lines imply that some pixels contain more than one active trap. The response relates proton fluence to the number and size of individual trapping centers created. For example, 300 traps greater than 3 e^- are in a 400 × 500 pixel region measured. Assuming a

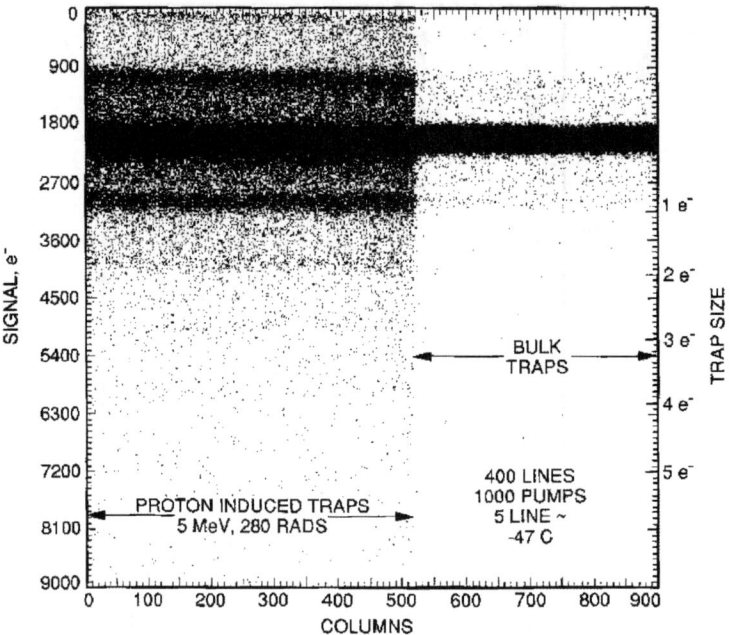

Figure 8.55(a) Flat-field pocket pumping response for a proton-irradiated Cassini CCD at an operating temperature of –47°C.

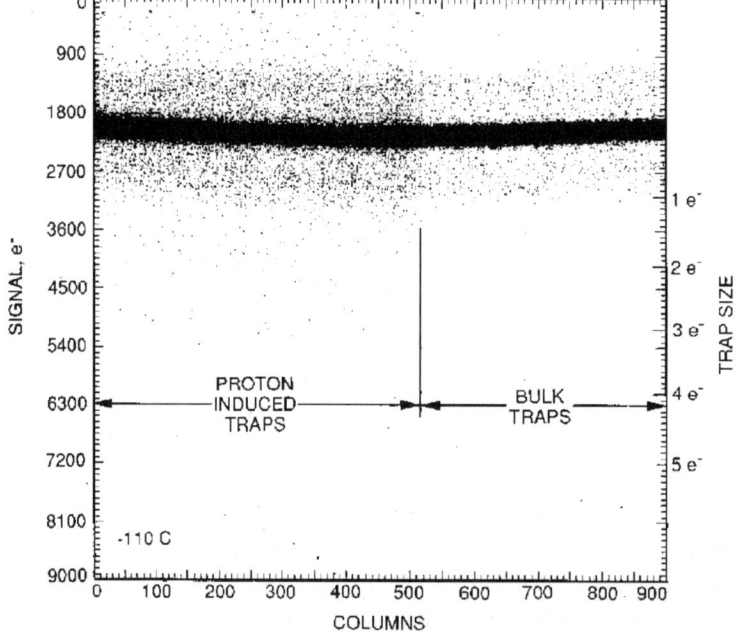

Figure 8.55(b) Flat-field pocket pumping response for a proton-irradiated Cassini CCD at an operating temperature of –110°C.

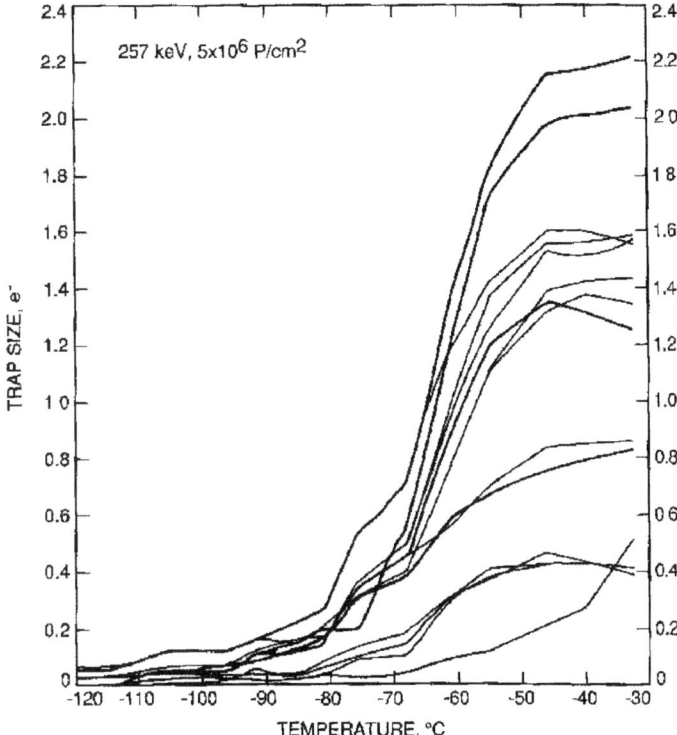

Figure 8.55(c) Individual P-V traps measured by pocket pumping with operating temperature.

fluence of 2.9×10^6 protons/cm^2 (280 rad), the probability of a trap being created above 3 e$^-$ is 0.0359%. CTE for the irradiated region is 0.99996 at the operating temperature of $-47°$C.

In Fig. 8.55(b) the same experiment is performed, except that the operating temperature of the CCD is reduced from $-47°$C to $-110°$C. Note that most proton-induced traps disappear becoming less active at this temperature because they remain filled over a very long time (a few minutes at this temperature). The intrinsic bulk traps do remain active. This result shows that proton-induced traps are deep-level states compared with the intrinsic bulk states. Cold temperature operation improves CTE from 0.99996 (at $-40°$C) to 0.999995 (at $-110°$C) for the irradiated side of the CCD. Reducing the temperature below $-120°$C causes some of the intrinsic bulk states to also freeze out. The remnant traps remaining at $-110°$C for the irradiated side are likely associated with oxygen and divacancy vacancy centers. Much colder temperatures are required to freeze these traps.

Figure 8.55(c) presents the temperature dependence of individual proton-induced traps by plotting their size using the pocket pumping technique. Each radiation-induced trap freezes out at a temperature of $-80°$C, becoming almost totally inactive at $-110°$C. The data in this plot can also be used to determine the activation energy of the traps (approximately 0.44 eV from the conduction band edge).

8.1.9.8.2 Clocking and charge packet density.
Equation (8.28) predicts that CTE improves as the charge packet density increases or when the line shift time decreases. In other words, if the traps can be fed fast enough, or the spacing between events is small enough, CTE will improve to allow other events to transfer through the array without loss.

Example 8.30

> Plot vertical CTI performance for the Cassini CCD as a function of operating temperature using Eq. (8.28). Assume $CTI_{WOR} = 9.966 \times 10^{-5}$ and a line time of 10 msec for $N_Z = 1, 10, 100$ and 1000 lines/event.
> *Solution*:
> Figure 8.56 shows CTI as a function of operating temperature and charge packet spacing. Note again the worse-case CTI is 9.966×10^{-5} and decreases as N_Z becomes smaller (i.e., less separation between events).

The data presented in Fig. 8.57 show the charge packet density effect for a 1024 × 1024 CCD that was partially (one half of the array) damaged by energetic protons working at −50°C (223 K). Figure 8.57(a) shows a vertical x-ray stacking response when there are 100 x-ray events/line (i.e., 100 x-ray events per 1024 pixels or approximately $N_Z = 10$ pixels/event). The first 400 lines from the device were read at a 300-μsec per line shift rate and discarded. The remaining 624 lines were read slow-scan at 20-msec per line and stored into a computer. Stacking plots were generated, producing the tilted x-ray line shown. The shielded region exhibits good CTE performance, as demonstrated by the flat x-ray trace also seen in the plot. Note that the shielded and damaged x-ray lines coincide at that point where slow-scan operation was initiated (again, the first 400 lines were read out quickly and not recorded by the computer). Near-perfect CTE is seen during high-speed transfer because the proton traps are fed at a rate that is faster than charge can escape. During fast-scan, traps are fed at an average rate of 3 msec (10 pixels/event × 300 μsec). From Fig. 8.53, at −50°C, the P-V trap emission time constant is 20 msec. This time is longer compared with 3-msec between events and is why no significant CTE loss is measured. On the other hand, for slow-scan operation the traps are only fed every 200 msec, which is longer than the emission time constant. Also, more trapping is experienced during slow-scan because more time is allowed for signal electrons to find traps that thermally diffuse within the potential well (i.e., we will discuss the capture time constant below).

Figure 8.57(b) shows CTE characteristics when the event density is reduced to 1 event/line ($N_Z = 1000$ pixels/event). In this case, traps during fast-scan are fed every 30 msec and every 2 sec slow-scan. Note that the single-pixel event line for

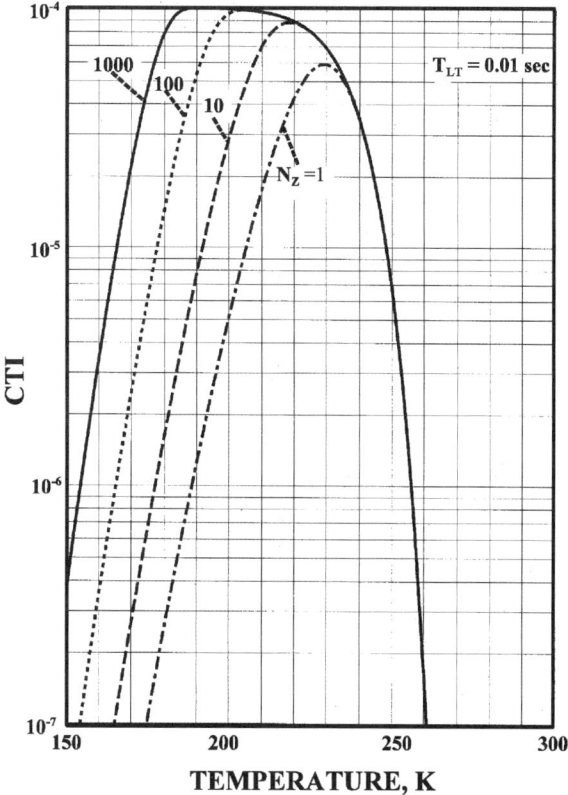

Figure 8.56 Cassini CTI as a function of operating temperature and spacing between charge packets.

the irradiated and reference regions does not coincide as they did in Fig. 8.57(a). This result indicates that some traps are not fed charge fast enough even during fast-scan. The tilt in the single-pixel event line during slow-scan is also steeper compared with Fig. 8.57(a). Also note that the variance in the single-pixel event line is greater for large N_Z. This characteristic is attributed to differences in proton trap density from column to column, which results in a variation in CTE performance across the array. Figure 8.57(c) shows the average CTI as a function of the number of events/line. The operating temperature and transfer time are fixed in the experiment. CTI degrades with the number of events integrated until the count does not affect CTI (< 10 events/line). At this point x-ray events are well separated and uncorrelated with one another (i.e., traps are empty before a new event arrives).

In theory, if the CCD is clocked fast enough, charge would not have enough time to find the traps. The capture time constant defines the time required to trap charge and is given by[4,63]

$$\tau_c = \frac{1}{\sigma_n v_{th} n_s}, \qquad (8.33)$$

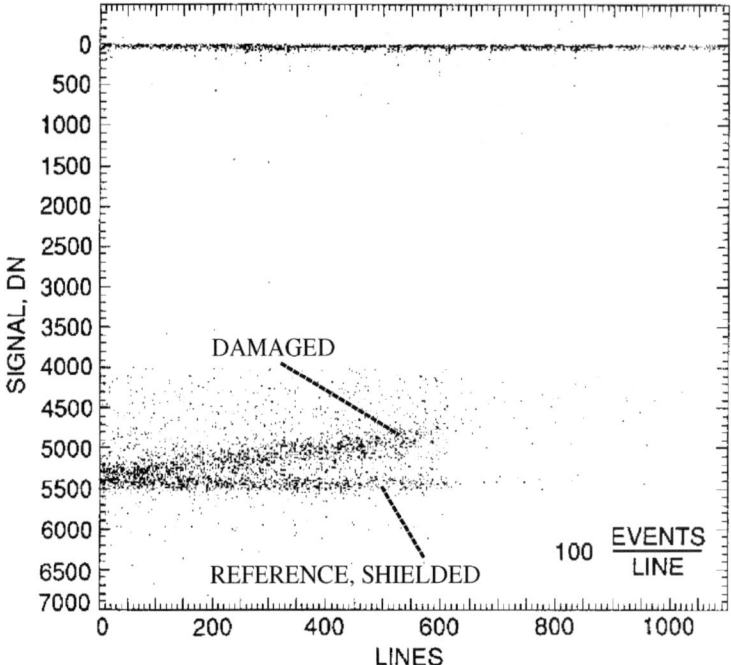

Figure 8.57(a) Vertical Fe^{55} CTE response for 100 events/line.

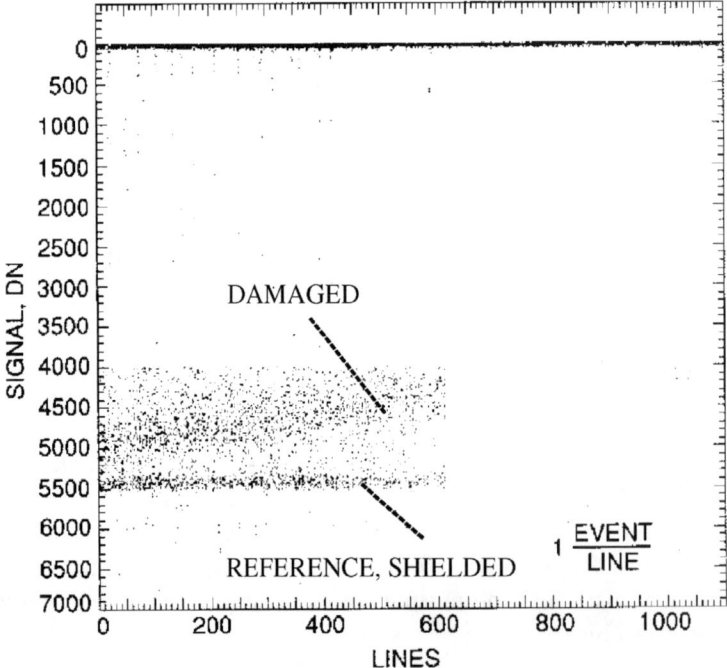

Figure 8.57(b) Vertical Fe^{55} CTE response for 1 event/line.

DAMAGE

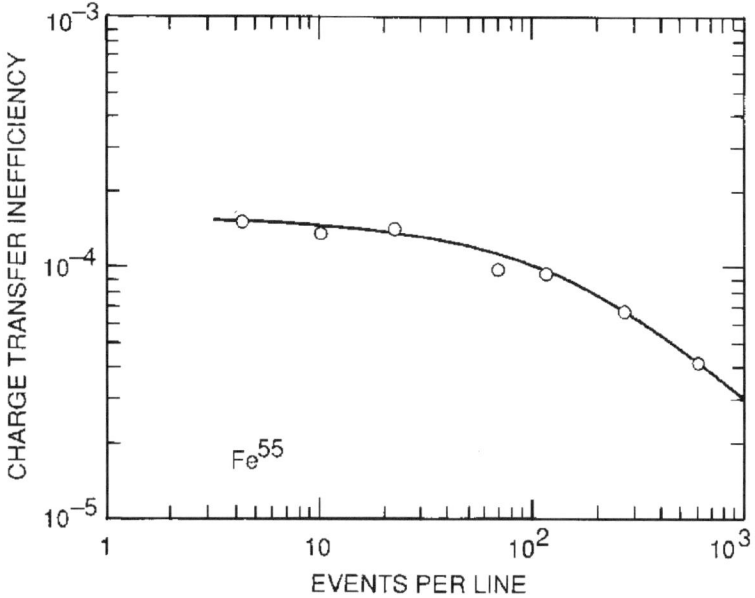

Figure 8.57(c) CTI as a function of events per line.

where n_s signal density (electrons/cm^3). The charge density for a three-phase CCD is approximately

$$n_s = \frac{3S(\text{e}^-)}{x_{con} P_S^{1/2}(P_S^{1/2} - x_{CS})}, \tag{8.34}$$

where $S(\text{e}^-)$ is the charge packet size (e$^-$), P_S is the pixel area (cm^2), x_{CS} is the channel stop width (cm) and x_{con} is the electron confinement range within the potential well (cm). This dimension can be approximated by

$$x_{con} = x_n \left(\frac{S(\text{e}^-)}{S_{FW}(\text{e}^-)} \right)^{1/3}, \tag{8.35}$$

where x_{con} is the confinement distance (cm), x_n is the depth of the buried channel (cm), $S(\text{e}^-)$ is the charge packet size (e$^-$) and $S_{FW}(\text{e}^-)$ is the full well (e$^-$).

Example 8.31

Calculate the trap capture time constant assuming the following:
$S_{FW} = 100,000\,\text{e}^-$,
$S(\text{e}^-) = 1620\,\text{e}^-$,
$x_d = 0.5\,\mu\text{m}$,
$P_S = 1.44 \times 10^{-6}\,\text{cm}^2$ (i.e., 12-μm pixel),

$x_{CS} = 4 \times 10^{-4}$ cm.
Solution:
From Eq. (8.35),

$$x_{con} = (5 \times 10^{-5}) \times (1620/10^5)^{1/3},$$

$$x_{con} = 0.126 \, \mu m.$$

From Eq. (8.34),

$$n_s = 3 \times 1620/[(1.26 \times 10^{-5}) \times (12 \times 10^{-4})$$

$$\times (12 \times 10^{-4}) - (4 \times 10^{-4})],$$

$$n_s = 4.02 \times 10^{14} \text{ electrons/cm}^3.$$

From Eq. (8.33),

$$\tau_c = 1/[(3.5 \times 10^{-15}) \times 10^7 \times (4.01 \times 10^{14})],$$

$$\tau_c = 7.1 \times 10^{-8} \text{ sec}.$$

This example shows that the capture time constant can usually be assumed to be instantaneous for slow-scan operation.

8.1.9.8.3 Design and processing.

There are several radiation damage remedies that are available through proper CCD design and processing.

1) Recall from Chapter 5 that notch technology confines signal charge to a narrow region of the channel, minimizing radiation-induced trap interaction.[64-66] Notch technology is essential for CCDs working in radiation environments.
2) Inverted-mode (MPP) CCDs show negligible ionization-induced surface dark current (refer to Fig. 7.11).
3) CTE performance is improved by reducing the number of pixel transfers. This can be accomplished for the vertical registers by shifting charge in both directions (i.e., split-frame transfer). Horizontal CTE can be improved by using multiple amplifier readout.
4) Multiphase CCDs usually exhibit less radiation dark spikes compared with sensors built with ion-implanted barriers or pinning implants because of high internal electric fields. For example, the Galileo virtual-phase CCD exhibits much larger spikes than multiphase CCDs as used by Hubble and Cassini.

5) For small-pixel devices, the buried channel implant is usually increased to maintain high-charge capacity. Internal electric fields also increase with this implant dose, making the CCD more vulnerable to dark spike generation. Therefore, consideration should be given to trading full well performance for radiation hardness.

6) Custom processes used against radiation-impurity-induced traps is a new area of research. For example, p-channel CCDs are less prone to bulk radiation problems than n-channel devices.[67] The advantage behind the p-channel CCD is that only one stable hole-trap-produced irradiation is known: the divacancy, which has a lower activation energy than the P-V center (0.21 eV compared to 0.44 eV). This trap may be shallow enough to not influence CTE performance significantly depending on operating temperature and clocking rate (refer to Fig. 8.53).[68]

8.1.9.8.4 Shielding.
Shielding is the most direct solution employed to protect the CCD from high-energy radiation sources. For example, Hubble, Galileo and Cassini employ 1-cm tantalum shields to protect the CCD primarily from proton damage and transient events. Figure 8.58 shows the effectiveness of shielding the CCD to a 2-krad proton energy spectrum shown in Fig. 8.42 and Table 8.6. Here, a shield with different thicknesses is placed above the CCD and then irradiated. Figure 8.58(a) shows the resultant CTE performance in each region at an operating temperature of $-80°C$. CTE performance is annihilated without a shield whereas the 3-in. shield shows very little loss in performance. Figure 8.58(b) is a corresponding dark current response showing the dark spike density for each region. As would be expected, the unshielded region shows the greatest density of dark spikes. Figure 8.58(c) shows the resultant CTE when the nonshielded region is cooled to $-130°C$. The cooling has effectively brought the CCD back to life for reasons discussed above.

There are usually strict limits to how much shielding can be employed. For space applications, weight and compactness of the shield are serious considerations. Also, the thickness of some shield materials become ineffective at certain energies.[58,60] For example, tantalum shields thicker than approximately 0.75 cm lose their shielding effectiveness at approximately 90 MeV. For energies greater than this, secondary neutrons are created by the shield. As mentioned above, neutrons can create very large dark spikes. Therefore it may be advantageous to limit shield thickness to keep the neutron count below the proton fluence at the CCD. Low-Z (i.e., low atomic weight) shields do not suffer from this problem as much (e.g., aluminum).

For monoenergetic proton beams, shields may result in more damage to the CCD than an unshielded CCD. For example, Fig. 8.59 plots displacement damage for 50-MeV protons that first encounter a 0.9-cm aluminum shield before exiting into the CCD. Note that the protons lose most of their energy in the shield and enter the CCD at a lower energy level, producing more damage.

It is often necessary for shield analysis to calculate the exit energy of a proton after it passes through a shield. The next example shows how this is accomplished.

Table 8.6 Harvard proton energy spectrum.

Proton energy MeV	Rads	Protons/cm^2
4	170.13	1.52×10^8
7	116.91	1.62×10^8
10	96.80	1.73×10^8
13	91.51	1.88×10^8
16	78.88	1.90×10^8
19	72.08	2.10×10^8
22	67.87	2.22×10^8
25	78.84	2.78×10^8
28	86.59	3.30×10^8
31	77.15	3.18×10^8
34	62.22	2.77×10^8
37	55.48	2.67×10^8
40	59.64	3.13×10^8
43	54.88	3.03×10^8
46	40.40	2.35×10^8
49	43.21	2.65×10^8
52	47.64	3.07×10^8
55	37.06	2.50×10^8
58	35.99	2.56×10^8
61	38.30	2.85×10^8
64	27.97	2.16×10^8
67	32.03	2.57×10^8
70	30.17	2.51×10^8
73	27.24	2.33×10^8
76	28.96	2.54×10^8
79	24.62	2.21×10^8
82	28.18	2.60×10^8
85	20.25	1.92×10^8
88	24.21	2.36×10^8
91	17.13	1.71×10^8
94	22.89	2.32×10^8
97	18.88	1.94×10^8
100	20.93	2.15×10^8
103	17.46	1.79×10^8
106	18.88	1.94×10^8
109	20.52	2.11×10^8
112	14.90	1.53×10^8
115	16.13	1.66×10^8
118	16.33	1.68×10^8
121	13.58	1.39×10^8
124	13.37	1.37×10^8

DAMAGE

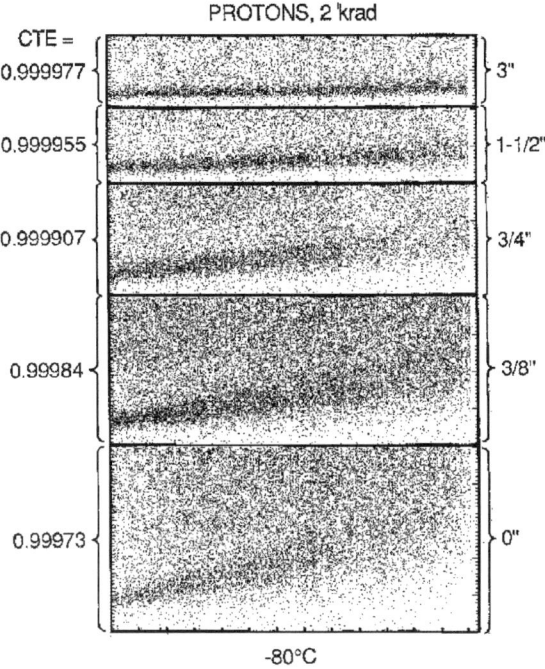

Figure 8.58(a) Vertical Fe^{55} CTE responses for different shield thicknesses.

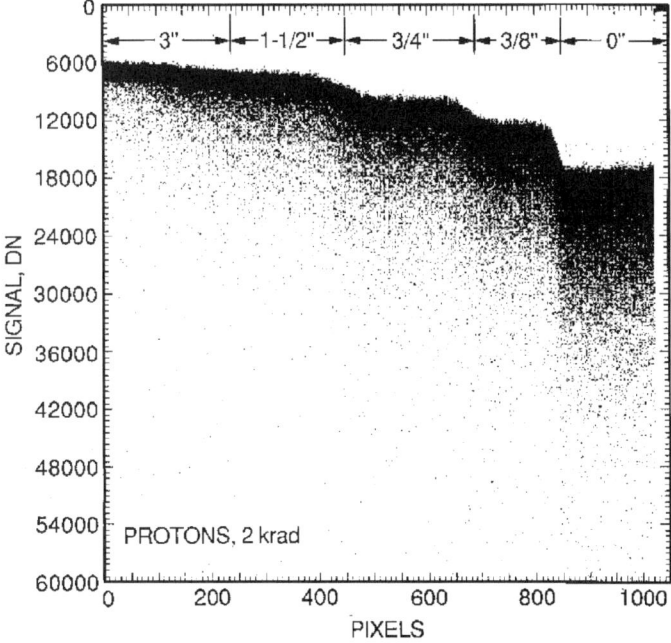

Figure 8.58(b) Dark current response for different shield thicknesses.

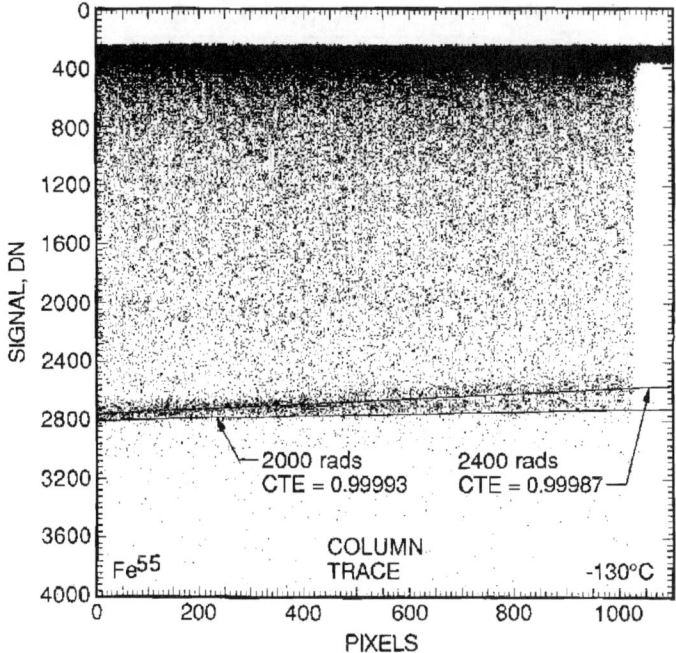

Figure 8.58(c) Vertical Fe^{55} response at an operating temperature of −130°C.

Example 8.32

Determine the exit energy for 7-MeV protons after passing through the aluminum shield arrangement shown in Fig. 8.60.

Solution:

We calculate the exiting energy through the 300-μm portion of the shield here. Other regions are handled in a similar fashion.

Step 1) Determine the range of the proton within the aluminum shield. From Appendix G3, a 7-MeV proton has a range of approximately 337 μm.

Step 2) Subtract the shield thickness from this range. This is $337 - 300 = 37$ μm.

Step 3) From Appendix G3 find the energy that corresponds to a range of 37 μm. This is approximately 1.9 MeV and represents the exit energy.

Figure 8.61 plots exit energy as a function of aluminum shielding thickness and proton energy.

Figure 8.62 shows signal histograms generated by the virtual-phase CCD described in Fig. 8.14, using the shield arrangement shown in Fig. 8.60 in response to 7-MeV protons. Note the proton signal increases as the shield thickness increases because the stopping power increases.

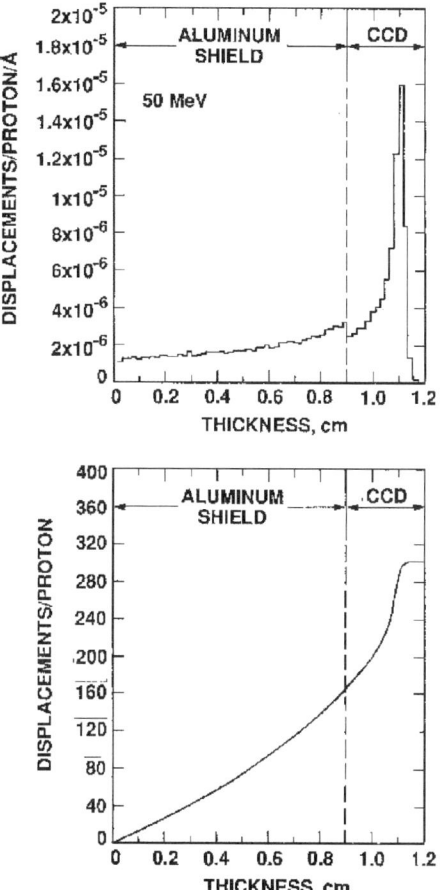

Figure 8.59 Displacements/proton as a function of depth through a shield/CCD arrangement.

A complete proton energy spectrum can also be transferred through the shield to determine the energy spectrum at the CCD. This is accomplished through

$$[F(E_{in})]_{out} = \frac{[F(E_{in})]_{in} S(E_{in})}{S(E_{out})}, \qquad (8.36)$$

where E_{in} is the input energy bin, $[F(E_{in})]_{out}$ is the equivalent output fluence at E_{in} (protons/cm^2), $[F(E_{in})]_{in}$ is the input fluence before the shield at energy E_{in} (protons/cm^2), $S(E_{in})$ is the input stopping power (MeV-cm^2/mg) at energy E_{in} and $S(E_{out})$ is the output stopping power (MeV-cm^2/mg).

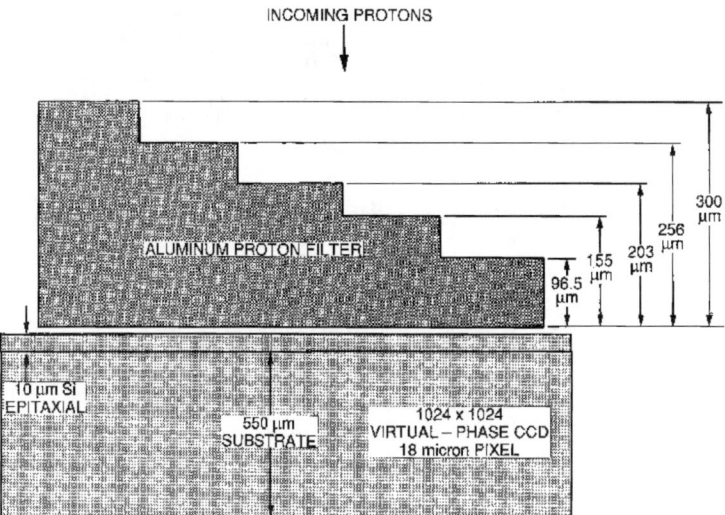

Figure 8.60 Experimental low-energy proton shield.

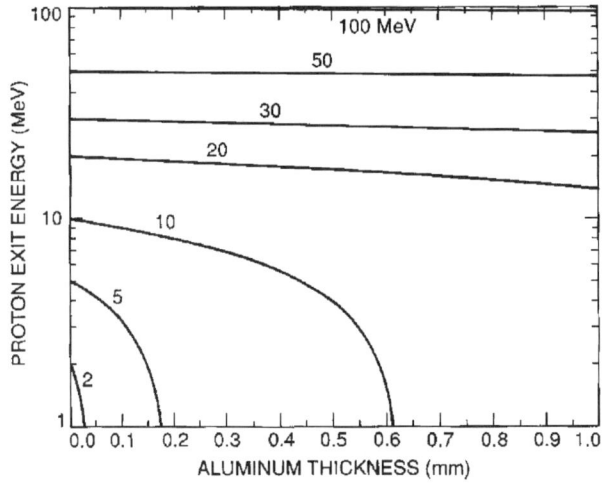

Figure 8.61 Proton exit energy as a function of aluminum shield thickness.

Example 8.33

Determine the output fluence through a 0.5-cm Al shield for an input fluence of $[F(E_{in})]_{in} = 1.21 \times 10^7$ protons/cm^2 at 40 MeV.
Solution:
Step 1) Find the range of a 40-MeV proton in the aluminum shield (0.714 cm).

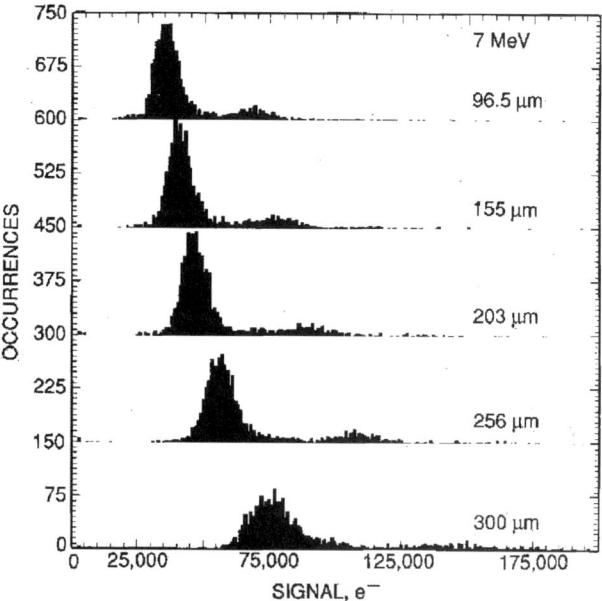

Figure 8.62 CCD signal measured after 7-MeV protons pass through the shield shown in Fig. 8.60.

Step 2) Subtract the shield thickness from this range (0.714 − 0.5 = 0.214 cm).

Step 3) Find the energy of a proton with a 0.214 cm range (20 MeV). This is the exiting energy of the 40-MeV proton.

Step 4) A 40-MeV proton has a stopping power in Al of $S(E_{in}) = 0.01155$ MeV-cm^2/mg.

Step 5) A 20-MeV proton has a stopping power in Al of $S(E_{out}) = 0.02$ MeV-cm^2/mg.

From Eq. (8.36), the equivalent fluence at 40 MeV is

$$[F(E_{in})]_{out} = (1.21 \times 10^7) \times 0.01155/0.02,$$

$$[F(E_{in})]_{out} = 7.0 \times 10^6 \text{ protons/cm}^2.$$

Other energy bins that define the working spectrum are calculated in the same manner.

The resultant spectrum at the CCD through Eq. (8.36) is then taken through the proton transfer or NIEL curves to the find number of displacements:

$$D_T = \sum RTI(E_{in}) D(E_{in})[F(E_{in})]_{out}, \tag{8.37}$$

where D_T is the total number of displacements (cm^{-2}), $RTI(E_{in})$ is the RTI factor at E_{in} and $D(E_{in})$ is the number of displacements per proton at E_{in} (cm^{-2}).

The predicted CTI using Eq. (8.37) is

$$CTI_{PRE} = \frac{D_T A_{PIX}}{S(e^-)}. \tag{8.38}$$

8.1.9.8.5 Fat-zero and light flood.
Chapter 5 discussed the benefits and disadvantages of fat-zero charge in filling potential barrier traps (i.e., designed and processed induced traps). However, fat-zero does not significantly help the bulk trap problem (i.e., single electron traps). Recall that fat-zero only exposes the signal charge packet to new traps contained in the potential wells. However, very small charge packets can show some benefit from fat-zero because of the flat-bottom well effect.

An intense light flood can be introduced to fill radiation-induced traps. The flood level employed should be several times full well to assure that all traps are filled. Immediately after the flood the CCD is quickly erased before an image is taken and read out. Recall from Chapter 7 that the Galileo and Cassini CCD camera systems provide light flood to fill epitaxial interface bulk traps before each image is taken (i.e., to eliminate RBI). The same light flood also helps the radiation-induced trap problem. In the case of the Cassini CCD, the operating temperature ($-90°C$) is not cold enough to keep all traps filled during a frame time. There is some detrapping of light flood charge during readout. However, some of this charge is also retrapped, improving CTE performance compared to no light flood at all. For Galileo, the operating temperature is cold enough ($-120°C$) to keep radiation-induced traps satisfied during the entire frame integration readout cycle.

Star tracking groups also electrically inject a charge packet a few lines before the target star image. Recall charge injection can be provided by the output summing well and output transfer gates (refer to Chapter 1 on charge injection). The injected charge packet fills the traps, allowing the star image to transfer through the array without significant deferred charge loss. The time required to hold fat-zero trapped charge is only a few line transfer times, and therefore, the operating temperature can be higher compared to full-frame light flood requirements above (on the order of 10 msec for slow-scan systems). However, for the method to work, the star's position must be known in order to determine where fat-zero charge should be injected.

8.1.9.8.6 Annealing.
Bulk damage induced by protons and neutrons can be annealed (60%) when heating the CCD for 100 hr at 127°C.[69] Still higher anneal temperatures (200°C) show 85% recovery.[70] For example, the 200-rad proton data shown in Fig. 8.43 show a CTE improvement from 0.999922 to 0.999985 after a 200°C 24-hr anneal. However, this remedy is difficult to implement into practice because of the high temperatures involved. Special heaters and high power usually prohibits this solution in space applications.

8.2 ELECTRICAL, THERMAL AND ESD DAMAGE

8.2.1 ELECTRICAL DAMAGE

The most common CCD failure is when the output amplifier is mistakenly shorted to ground (often with an oscilloscope probe). This unfortunate connection impresses the full drain voltage, V_{DD}, across the output MOSFET, resulting in immediate catastrophic failure (a crater is typically left behind). Sensors with price tags greater than $100,000 have been lost this way. The problem can be avoided by not probing or working near a CCD while power is applied. Also for new camera systems the impedance between the CCD and ground should be verified before powering the camera. One should measure the CCD load resistance. Similar damage can be caused by the decoupling capacitor that is used to connect the CCD to the preamplifier. Excessive current can flow through the MOSFET as this capacitor charges when power is applied. To avoid this problem, the capacitor should be no more than approximately 0.1 μF, depending on the circuit time constant. If a blown output amplifier is suspect, a VOM can be used to measure the impedance of the MOSFET's source and drain diffusions. A diode-like characteristic should be seen to substrate.

Improper grounding to the CCD camera can also cause catastrophic damage. This particular problem was experienced by the author at Kitt Peak National Observatory while installing a 400 × 400 backside-illuminated CCD camera system on a new 4-m telescope. Unfortunately, the camera system did not produce images when power was applied. The problem was traced back to the telescope ground, which was at a substantially lower potential than the camera's ground. The grounding arrangement effectively elevated the operating voltages to the CCD, to a level higher than it would tolerate. The CCD and many other camera components suffered impairable damage when power was applied. The lesson and solution to this electrical problem is to always isolate the CCD camera ground from other foreign ground structures (refer to Chapter 7 on grounding).

Improper wiring to the CCD is another common electrical failure. For example, the CCD can be destroyed if a negative bias voltage (e.g., $-V_{OTG}$) is mistakenly connected to a diffusion (i.e., V_{DD}, V_{REF} and video). Damage can take place if that voltage is not current limited, or is not heavily bypassed with an RC decoupling network. Also, bipolar CCD clocks that are accidently connected to diffusions can result in permanent damage. Fortunately, it is difficult to damage CCD gates with nominal operating voltages of any polarity. However, CCDs with ESD diodes can suffer damage if the gates are connected to negative going clocks.

The CCD is also damaged when the clock and bias voltages become too high. Some pixels will typically show warning signs of leakage currents before this occurs. The onset of excess charge discussed in Chapter 7 is a good indicator that the clock voltage is too high. Luminescence from the output amplifier and on-chip load are indicators when voltages to diffusions are too high.

Luminescence may also indicate that oxide charging is also taking place. As discussed in Chapter 6, a weak avalanche occurs within the pinch-off region, causing pinch-off luminescence. In addition, ionization and hot electron generation are possible. These electrons can gain sufficient energy to overcome the Si–SiO$_2$ energy barrier and be injected and trapped in the oxide layer. Over time, a flatband voltage shift will develop, which can seriously affect the operating windows of the CCD. An oxide charging problem was experienced by two flight Cassini CCDs during burn-in. Here the precharge clock and V_{REF} bias voltage applied to the reset MOSFET were excessive. Over time, a -10-V flatband shift was experienced, which made it very difficult to turn off the reset switch.

8.2.2 Thermal Damage

As far as we know, the CCD cannot be damaged when exposed to visible or NIR radiation unless the intensity level is very high. Destructive tests show that the CCD can take an amazingly high dose of near-IR photons before permanent damage results. For example, a Galileo CCD was subjected to a 10-ns pulse of 1.06-μm laser light at an intensity level of 10^{11} photons focused onto a single pixel. The sensor responded without ill effects taking place. However, the pixel reacted negatively when the exposure level was doubled in intensity. Here, the gate electrodes of the pixel showed signs of meltdown. Figure 8.63 shows the damage produced for a slightly higher photon level than just described.

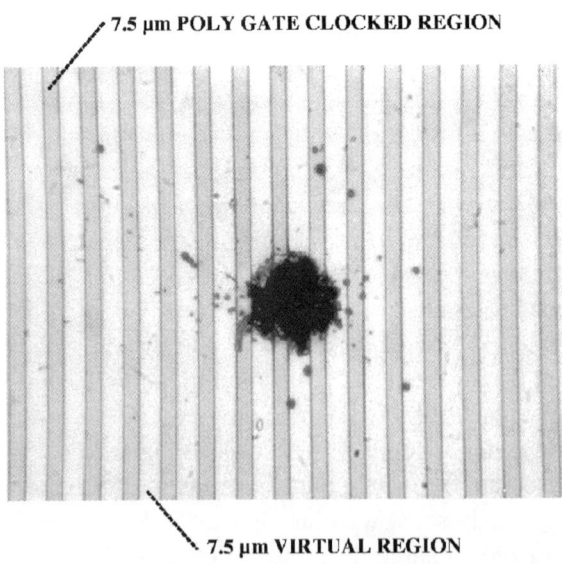

Figure 8.63 A high-intensity near-IR damaged virtual-phase pixel.

8.2.3 ELECTROSTATIC DISCHARGE (ESD) DAMAGE

Electrostatic electricity is an imbalance of positive and negative charges on the surface of objects. Charge can be generated by the contact and quick separation of dissimilar materials. This process is called triboelectric charging (tribology is the study of friction, wear and lubrication). The amount of charge generated depends on the area of contact, the speed of contact and separation, relative humidity and conductivity of the materials involved. For example, if two materials are dry and exhibit high surface resistivity and the charge leakage path to earth is restricted, then high levels of static charge are possible. Essentially all materials can be triboelectrically charged. Charge can also be created on conductors as long as the conductor is well isolated from ground.

A person walking across the floor can generate static charge as the shoes make contact and separate from the floor surface. The human body in this process can charge to hundreds, and in some cases thousands of volts, if the motion is very quick. Once charged, discharge from the fingertip to the CCD is possible, creating an electrostatic discharge event (ESD). The tiniest spark requires about 500 V, which is approximately 10 times what a typical CCD gate dielectric can take before damage results. Human discharge is the most common ESD damage CCD problem. The model used to simulate this event type is called the Human Body Model. Here a 100-pF capacitor acts as the body that discharges through a 1.5-kΩ series resistor to the CCD.

ESD damage sensitivity is defined as the CCD's ability to dissipate discharge energy or withstand the voltage level involved. Unfortunately, the CCD is one of the most ESD-sensitive electronic components manufactured. Most scientific CCDs, without ESD protection diodes, are susceptible to only a 50-V discharge. The ESD event can cause the bus lines to melt, generate ESD craters, diode junction breakdown or insulator failure. Of these problems, dielectric damage is by far the most prevalent.

ESD events come in three varieties: (1) discharge to the CCD from an object that is charged, (2) discharge from a charged CCD to an object and (3) radiation-induced discharge. The first category occurs when a charged material (e.g., human body) discharges to the CCD. The second category is when static charge is transferred from the CCD to a neutral (i.e., ground) material. For example, a CCD might be charged when it is inserted into a camera or conductive foam package, causing a discharge event. Third, the CCD can become charged when working in electrostatic and radiation environments. For example, this problem was a concern for Galileo where radiation sources (electrons, protons, gamma rays, etc.) could potentially ionize and induce static charge on the window of the device. The issue was alleviated by using a special conductive AR coating that discharged any charge buildup on the window.

Damage from ESD may produce catastrophic failure or result in a latent effect. Latent defects are difficult to identify at the manufacturer because it takes time for the damaged region to show its face. The CCD may prematurely fail after the

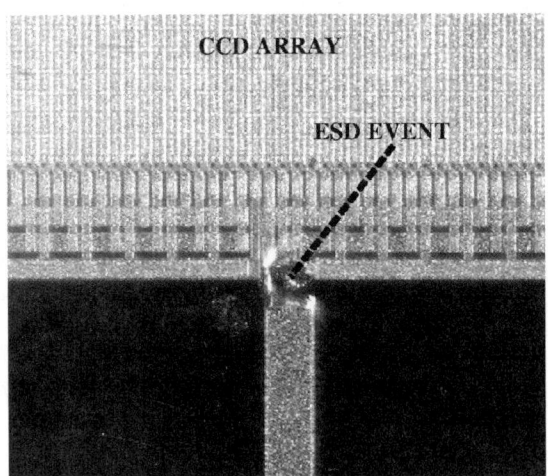

Figure 8.64 ESD event.

user places the part into service. One method used to bring out latent damage is to "burn in" the sensor. Here the device is powered for an extended period of time (e.g., 1000 hr) using elevated clock and bias voltages. Tests may even include temperature cycling to fully exercise the problem. CCD burn-in is important for space applications to ensure that the sensor is latent-free and reliable for flight use.

To our knowledge, no CCD has failed in space because of latent problems. However, the SXT camera system almost fell victim to the problem. Here, only a few weeks before delivery to the launch pad, a new small column blemish emerged in the last column of the array. SXT engineers needed to decide if the CCD should be replaced or gamble and leave it in the system and hope that the problem would not worsen with time. There was a natural tendency from the team to fly the CCD as it was, because of schedule problems. The lead cognizant engineer decided to switch sensors. After removal, the old sensor was characterized in a test set to understand the sudden eruption of the column blemish. In doing so, the blemish became much worse, generating copious amounts of charge. After several hours of testing, under elevated clock voltages, the blemish was categorized as a catastrophic blemish, indicating complete device failure. The sensor was then inspected under a microscope, where a small crater on the side array was found (Fig. 8.64). The damage implied that during camera assembly an ESD event was experienced.

The risks and problems arising from static electricity can be avoided by ensuring that static charge can dissipate through the surfaces of materials to earth ground more quickly than charge can be generated. For normal handling motions, this means the discharge time constant needs to be on the order of one tenth of a second or less. The rate of discharge can be controlled with static dissipative materials that are properly grounded. Wrist straps, ESD floor mats and ESD footwear are advantageous in sending static charge to the ground rather than being discharged into the CCD.

For plastics, clothing and other insulators, grounding may not remove static charge, because there is no conductive pathway to do so. Raising the humidity (i.e., humidifiers) will cause these materials to become slightly conductive. Therefore, as the materials separate, charges can instantly flow back together. Also, grounding becomes more effective. ESD problems usually become nonexistent when the relative humidity is above 60% (one can almost "feel" when it is safe to work with CCDs). Flight CCD work at JPL was usually terminated whenever the humidity of the lab dropped below 30%.

REFERENCES

1. J. Killiany, "Radiation effects on silicon charge coupled devices," *IEEE Transactions on Components, Hybrids and Manufacturing Technology*, Vol. CHMT-1, No. 4 (1978).
2. J. Janesick, T. Elliott and F. Pool, "Radiation damage in scientific charge coupled devices," *IEEE Trans. Nucl. Sci.* NS-36, pp. 572–578 (1989).
3. J. Janesick, G. Soli, T. Elliott and S. Collins, "The effects of proton damage in scientific charge coupled devices," *Proc. SPIE* Vol. 1447, pp. 87–108 (1991).
4. G. Hopkinson, "Radiation effects on solid state imaging devices," *Radiat. Phys. Chem.* Vol. 43, No. 1/2, pp. 79–91 (1994).
5. G. Hopkinson, C. Dale and P. Marshall, "Proton effects in charge-coupled devices," *IEEE Trans. Nuc. Sci.* Vol. 43, pp. 614–627 (1996).
6. G. Hopkinson, "Cobalt 60 and proton radiation effects on large format, 2-D, CCD arrays for an earth imaging application," *IEEE Trans. Nucl. Sci.* NS-38(6) (1992).
7. C. Dale and P. Marshall, "Displacement damage in Si imagers for space applications," *Proc. SPIE* Vol. 1447, pp. 70–86 (1991).
8. R. Murowinski, G. Linzhuang and M. Deen, "Effects of space radiation damage and temperature on CCD noise for the Lyman FUSE Mission," *Proc. SPIE* Vol. 1953, pp. 71–81 (1993).
9. D. Norris, J. Janesick, M. Blouke, P. May and D. McGrath, "Progress in 800 × 800 CCD imager development and applications," *Proc. SPIE* (1981).
10. K. Marsh, C. Joshi, J. Janesick and S. Collins, "Nondispersive x-ray spectroscopy and imaging of plasmas using a charge-coupled device," *Rev. Sci. Instrum.* Vol. 56, No. 5 (1985).
11. L. Acton, M. Morrison, J. Janesick and T. Elliott, "Radiation concerns for the solar-a soft x-ray telescope," *Proc. SPIE* Vol. 1447, pp. 123–139 (1991).
12. F. McLean, H. Boesch and T. Oldham, "Charge generation, transport, and trapping," In *Ionizing Radiation Effects in MOS Devices and Circuits*, T. P. Ma and P. Dressendorfer, Eds., Chapter 3, John Wiley and Sons (1989).
13. P. Winokur, "Radiation induced interface states," In *Ionizing Radiation Effects in MOS Devices and Circuits*, T. P. Ma and P. Dressendorfer, Eds., Chapter 4, John Wiley and Sons (1989).

14. D. Braunig and F. Wulf, "Atomic displacement and total ionizing dose damage in semiconductors," *Radiation Physics and Chemistry* Vol. 43, pp. 105–127 (1994).
15. J. Schwank, "Basic mechanisms of radiation effects in the natural space environment," IEEE Nuclear and Space Radiation Effects Conference, Tucson, AZ (1994), Short Course Notes.
16. J. Srour and J. McGarrity, "Radiation effects on microelectronics," *Proc. IEEE* Vol. 76, pp. 1443–1469 (1988).
17. J. Janesick, J. Hynecek and M. Blouke, "A virtual phase imager for Galileo," *Proc. SPIE* Vol. 290, pp. (1981).
18. N. Saks, J. Killiany, P. Reid and W. Baker, "A radiation hard MNOS CCD for low temperature applications," *IEEE Trans. Nucl. Sci.* Vol. NS-26, No. 6 (1979).
19. R. Murowinski, G. Linzhuang and M. Deen, "Effects of space radiation damage and temperature on the noise in CCDs and LDD MOS transistors," *IEEE Trans. on Nucl. Sci.* Vol. 40, No. 3 (1993).
20. Radiation Environments IEEE NSREC Short Course, Session I, July 21, 1997, Janet Barth, NASA/Goddard Space Flight Ctr.
21. H. Tada, J. Carter, B. Anspaugh and R. Downing, *Solar Cell Radiation Handbook*, JPL Publication 82-69 (1982).
22. J. F. Ziegler, J. P. Biersack and U. Littmark, *The Stopping and Range of Ions in Solids*, Pergamon Press, New York, NY (1985).
23. T. Lomheim, R. Shima, J. Angione, W. Woodward, D. Asman, R. Keller and L. Schumann, "Imaging charge-coupled device (CCD) transient response to 17 and 50 MeV proton and heavy-ion irradiation," *IEEE Trans. Nucl. Sci.* Vol. 37, No. 6 (1990).
24. K. McCarthy, A. Owens, A. Wells, W. Hajdas, F. Mattenberger and A. Zehnder, "Measured and modeled CCD response to photons in the energy range 50 to 300 MeV," *Nuclear Instruments and Methods in Physics Research* A 361, pp. 586–601 (1995).
25. G. Yates, G. Smith, P. Zagarino and M. Thomas, "Measuring neutron fluences and gamma/x-ray fluxes with CCD cameras," *IEEE Trans. Nucl. Sci.* Vol. 39, No. 5 (1992).
26. B. Paczkowski, "Radiation induced noise analysis," Interoffice Memorandum, May 5 (1993).
27. K. Klaasen, M. Clary and J. Janesick, "Charge-coupled device television camera for NASA's Galileo mission to Jupiter," *Optical Engineering* Vol. 23, No. 3, pp. 334–342 (1984).
28. T. Daud, J. Janesick, K. Evans and T. Elliott, "Charge-coupled device response to electron beam energies os less than 1 keV up to 20 keV," *Optical Engineering* Vol. 26, No. 8, pp. 686–691 (1987).
29. M. Roos et al., "Review of particle properties," *Phys. Letters* 111 B, p. 83 (1982).

30. F. McLean, H. Boesch and T. Oldham, "Electron-hole generation, transport and trapping," In *Ionizing Radiation Effects in MOS Devices and Circuits*, T. P. Ma and P. Dressendorfer, Eds., Chapter 3, John Wiley and Sons (1989).
31. R. Pierret, "Field effect devices," In *Modular Series on Solid State Devices, Vol. IV*, Addison-Wesley Publishing Company, pp. 65 (1983).
32. D. Herve et al., "Cumulated dose long term effects in charge coupled devices," CEA, BP 12, 91680, Bruyeres-le Chatel, France.
33. J. Boudenot and B. Augier, "Total dose effects on CCDs—reverse annealing phenomena," Proc. ESA Electr. Comp. Conf., Noordwijk, NL, 12–16 Nov. 1990 ESA SP313, pp. 319–324 (1991).
34. N. S. Saks and D. B. Brown, "Observation of H+ migration during interface trap formation," *IEEE Trans. Nucl. Sci.* NS-37, pp. 1624–1631 (1990).
35. T. P. Ma and P. Dressendorfer, *Ionizing Radiation Effects in MOS Devices and Circuits*, John Wiley and Sons, pp. 234–236 (1989).
36. N. Saks and D. Brown, "Interface traps formations via the two-stage H^+ process," *IEEE Trans. Nucl. Sci.* NS-36, No. 6, pp. 1848–1857 (1989).
37. T. Oldham, F. McLean, H. Boesch and J. McGarrity, "An overview of radiation induced interface traps in MOS structures," In *Semiconductor Science and Technology 4*, pp. 986–999 (1989).
38. T. Chu, J. Szedon and C. Lee, *Solid State Electron.* 10, 897 (1967).
39. C. Perkins, K. Aubuchon and H. Dill, *Appl. Phys. Letts.* 12, 385 (1968).
40. J. Cricchi, M. Fitzpatrick, F. Blaha and B. Ahlport, *IEEE Trans. Nucl. Sci.* NS-24, 2185 (1977).
41. N. Saks, *IEEE Trans. Nucl. Sci.* NS-25, 1226 (1978).
42. J. Walker and C. Sah, "Properties of 1.0 MeV electron-irradiated defect centers in silicon," *Phys. Rev. B* Vol. 7, No. 10, p. 4587 (1973).
43. L. Kimerling, P. Blood and W. Gibson, "Defects states in proton-bombarded silicon at T < 300 K," *International Conference on Defects and Radiation Effects in Semiconductors*, Inst. Phys. Conf. Ser. 46, pp. 273–280 (1979).
44. V. Van Lint, *Nucl. Instr. Meth.* A253, p. 435 (1987).
45. V. Van Lint, *Mechanisms of Radiation Effects in Electronic Materials*, Wiley, (1980).
46. J. Corbett, J. Bourgoin, L. Cheng, J. Corelli and Y. Lee, "The status of defect studies in silicon," *Inst. Phys. Conf. Series*, No. 31, pp. 1–11 (1977).
47. J. Matheson, M. Robbins and S. Watts, "Evidence that type inversion in n-type high resistivity silicon diodes is due to a radiation induced deep acceptor," *IEEE Trans. Nucl. Sci.* Vol. NS-42, No. 6 (1995).
48. G. Watkins and J. Corbett, "Defects in irradiated silicon: electron paramagnetic resonance and electron-nuclear double resonance of the si-e center," *Physical Review* Vol. 134, No. 5A, pp. (1964).
49. H. Stein and F. Vook, "Electrical studies of electron-irradiated n-type Si: impurity and irradiation-temperature dependence," *Physical Review* Vol. 163, No. 3 (1967).

50. H. Mitsuji, H. Masako and H. Saito, "Effect of impurities on the annealing behavior of irradiated silicon," *J. of Appl. Phys.* Vol. 38–6 (1967).
51. G. Kinchin and R. Pease, "Displacement of atoms in solids by radiation," *Report Prog. Phys.* 18, 1 (1955).
52. E. Burke, "Energy dependence of proton induced displacement damage in silicon," *IEEE Trans. Nucl. Sci.* Vol. NS-33, No. 6, pp. 1276–1281 (1986).
53. C. Dale, L. Chen, P. McNulty, P. Marshall and E. Burke, "A comparison of Monte Carlo and analytical treatments of displacement damage in Si microvolumes," *IEEE Trans. Nucl. Sci.* Vol. 41, No. 6, pp. 1974–1983 (1994).
54. I. Hopkins and G. Hopkinson, "Further measurements of random telegraph signals in proton irradiated CCDs," *IEEE Trans. Nuc. Sci.* Vol. 42, pp. 2074–2081 (1995).
55. H. Hopkins and G. Hopkinson, "Random telegraph signals from proton irradiated CCDs," *IEEE Trans. Nucl. Sci.* NS-40(6) (1993).
56. G. Hopkinson, "Radiation-induced dark current increases in CCDs," RADECS93 Conference, St Malo, France (1993).
57. C. Dale, P. Marshall, A. Burke, G. Summers and G. Bender, "The generation lifetime damage factor and its variance in silicon," *IEEE Trans. Nucl. Sci.* NS-36, pp. 1872–1881 (1989).
58. C. Dale, P. Marshall and A. Burke, "Particle induced displacement damage in silicon imagers," *Proc. SPIE* Vol. 1447, pp. 70–86 (1991).
59. C. Dale, P. Marshall, B. Cummings, L. Shamey and D. Holland, "Displacement damage effects in mixed particle environments for shielded spacecraft CCDs," *IEEE Trans. Nucl. Sci.* Vol. 40, No. 6, pp. 1628–1637 (1993).
60. C. Dale, P. Marshall, E. Burke, G. Summers and E. Wolicki, "High energy electron induced displacement damage in silicon," *IEEE Trans. Nucl. Sci.* Vol. NS-35, No. 6, pp. 1208–1214 (1988).
61. G. Summers, E. Burke, C. Dale, E. Wolicki, P. Marshall and M. Gehlhausen, "Correlation of particle-induced displacement damage in silicon," *IEEE Trans. Nucl. Sci.* Vol. NS-34, No. 6, pp. 1134–1139 (1987).
62. A. Van Ginneken, "Non ionizing energy deposition in silicon for radiation damage studies," Fermi National Accelerator Laboratory Report FN-522, (1989).
63. A. Holland, "The effect of bulk traps in proton irradiated EEV CCDs," *Nuclear Instruments and Methods in Physics Research* A326, pp. 335–343 (1993).
64. B. Burke, J. Gregory, R. Mountain, J. Huang, M. Cooper and V. Dolat, "High performance visible/UV CCD imagers for space-based applications," *Proc. SPIE* Vol. 1693, pp. 86–100 (1992).
65. R. Bredthauer, J. Pinter, J. Janesick and L. Robinson, "Notch and large area CCD imagers," *Proc. SPIE* Vol. 1447, pp. 310–315 (1991).
66. J. Janesick, "Notch charge-coupled device," *NASA Tech. Briefs*, July (1992).
67. J. Spratt et al., "The effect of nuclear radiation on p-channel CCD imagers," 1997 Radiation Effects Data Workshop Record, IEEE Catalog No. 97TH8293, (1997) NSREC.

68. L. Kimerlingh, "Defects states in electron-bombarded silicon: capacitance transient analysis," *Radiation Effects in Semiconductors* (1976), *Int. Phys. Ser.* (GB), No. 32, N. Urli and J. Corbett, Eds., London, p. 221 (1977).
69. A. Holland, A. Holmes, B. Johlander and L. Adams, "Techniques for minimizing space proton damage in scientific charge coupled devices," *IEEE NSREC* (1991).
70. A. Holland, A. Abbey and K. McCarthy, "Proton damage effects in EEV charged coupled devices," *Proc. SPIE* Vol. 1344, pp. 378–395 (1990).
71. J. Chubb, "The control of static electricity," J. N. Chubb—John Chubb Instrumentation, Unit 30, Lansdown Industrial Estate, Gloucester Road, Cheltenham, GL51 8PL.

Appendixes

APPENDIX A

Unit prefixes and dimensions

Prefix	Multiple	Symbol
10^{12}	tera	T
10^{9}	giga	G
10^{6}	mega	M
10^{3}	kilo	k
10^{2}	hecto	h
10^{1}	deka	da
10^{-1}	deci	d
10^{-2}	centi	c
10^{-3}	milli	m
10^{-6}	micro	μ
10^{-9}	nano	n
10^{-12}	pico	p
10^{-15}	femto	f
10^{-16}	atto	a

Appendix B

International system of units

Quantity	Unit	Symbol	Dimensions
Capacitance	farad	F	C/V
Current	ampere	A	
Electric Charge	coulomb	C	A-sec
Energy	joule	J	N-m
Force	newton	N	kg-m/sec^2
Frequency	hertz	Hz	1/sec
Length	meter	m	
Mass	kilogram	kg	
Potential	volt	V	J/C
Power	watt	W	J/sec
Resistance	ohm	Ω	V/A
Temperature	kelvin	K	
Time	second	sec	

APPENDIX C

Physical constants

Quantity	Symbol	Value
Avogadro's number	N_A	6.022×10^{23} molecules/g-mole
Boltzmann's constant	k	8.6171×10^{-5} eV/K, 1.3807×10^{-23} J/K
Electron mass	m_e	9.102×10^{-31} kg
Electron charge	q	1.60218×10^{-19} C
Electron volt	eV	1 eV = 1.6022×10^{-19} J
Permeability of free space	μ_0	1.2566×10^{-8} H/cm
Permittivity of free space	ε_0	8.8542×10^{-14} F/cm
Planck's constant	h	6.6262×10^{-34} J-sec
Proton rest mass	m_p	1.6726×10^{-27} kg
Speed of light	c	2.99792×10^{10} cm/sec
Standard atmosphere		1.01×10^5 N/m^2
Thermal voltage at 300 K	kT/q	0.0259 V
Wavelength of 1 eV photon	λ	1.23985 µm

Appendix D

Properties of silicon (300 K)

Property	Symbol	Value
Activation energy, self diffusion		4.8 eV, 7.7×10^{-19} J
Atomic radius		1.18 Å
Atomic weight		28.09
Atoms/cm^3		5.0×10^{22}
Breakdown field		3×10^5 V/cm
Coefficient of expansion		2.5×10^{-6} /K
Conduction band states	N_C	2.8×10^{19} cm^{-3}
Crystal structure		diamond, 8 atoms
Density		2.328 g/cm^3
Dielectric constant	K_{SI}	11.8
Electron affinity		4.05 V
Electron diffusion constant	D_N	35 cm^2/sec
Energy of ionization	E_{e-h}	3.65 eV, 5.76×10^{-19} J
Energy gap (0 K)		1.21 eV, 1.91×10^{-19} J
Energy gap (300 K)	E_g	1.12 eV, 1.78×10^{-19} J
Energy of sublimation		4.9 eV, 7.8×10^{-19} J
Fano factor	F_a	0.1
Hole diffusion constant	D_P	12 cm^2/sec
Index of refraction		3.5–3.7
Intrinsic carrier concentration	n_i	1.45×10^{10} carriers/cm^3
Intrinsic resistivity		2.3×10^5 Ω-cm
Lattice mobility, hole		450 cm^2/V-sec
Lattice mobility, electron	μ_{SI}	1350 cm^2/V-sec
Lattice constant		5.43 A
Melting point		1415°C
Minority carrier lifetime		2.5×10^{-3} sec
Permittivity	ε_{SI}	$\varepsilon_0 K_{SI} = 1.04 \times 10^{-12}$ F/cm
Specific heat		0.7 J/g-K
Thermal conductivity		1.5 W/cm-K
Valence band states	N_V	1.04×10^{19} cm^{-3}
Vapor pressure		10^{-5} Torr at 1250°C

APPENDIX E

Properties of amorphous silicon (SiO_2) dioxide (300 K)

Property	Symbol	Value
Coefficient of expansion		0.55×10^{-6}/K
DC resistivity		10^{14}–10^{16} Ω-cm
Density		2.27 g/cm^3
Dielectric constant	K_{OX}	3.9
Dielectric strength		10^7 V/cm
Energy gap		8 eV, 12.8×10^{-19} J
Etch rate 35 %HF		1000 Å/min
Index of refraction		1.46–1.51
Melting point		1600°C
Molecular weight		60.08
Molecules/cm^3		2.3×10^{22} cm^{-3}
Permittivity	ε_{OX}	$\varepsilon_0 K_{OX} = 3.45 \times 10^{-13}$ F/cm
Specific heat		1 J/g-K
Thermal conductivity		0.014 W/cm-K
Vapor pressure		10^{-3} Torr at 1450°C

APPENDIX F

Properties of amorphous silicon nitride (300 K)

Property	Value
Chemical structure	Si_3N_4
dc resistivity	10^{14} Ω-cm
Density	3.1 g/cm^3
Dielectric constant	7.5
Dielectric strength	10^7 V/cm
Energy gap	5 eV
Etch rate — 35% HF	7.5 Å/min
Index of refraction	2.05

APPENDIX G1

Particle = proton. Target = Silicon. E = energy. (dE/dx)e = electron stopping power (MeV-cm^2/mg). (dE/dx)n = nuclear stopping power (MeV-cm^2/mg). R = Range. $L1$ = Longitudinal Straggling. $L2$ = Lateral Straggling

E	(dE/dx)e	(dE/dx)n	R	$L1$	$L2$
20.00 keV	3.426E−01	2.754E−03	2946 Å	943 Å	946 Å
22.00 keV	3.632E−01	2.587E−03	3158 Å	968 Å	982 Å
24.00 keV	3.827E−01	2.441E−03	3361 Å	989 Å	1014 Å
26.00 keV	4.010E−01	2.312E−03	3556 Å	1009 Å	1044 Å
28.00 keV	4.182E−01	2.199E−03	3745 Å	1026 Å	1071 Å
30.00 keV	4.343E−01	2.097E−03	3927 Å	1042 Å	1096 Å
33.00 keV	4.565E−01	1.962E−03	4192 Å	1063 Å	1130 Å
36.00 keV	4.764E−01	1.846E−03	4446 Å	1082 Å	1162 Å
40.00 keV	4.996E−01	1.713E−03	4774 Å	1104 Å	1200 Å
45.00 keV	5.236E−01	1.574E−03	5167 Å	1128 Å	1242 Å
50.00 keV	5.424E−01	1.459E−03	5548 Å	1150 Å	1280 Å
55.00 keV	5.568E−01	1.360E−03	5919 Å	1169 Å	1315 Å
60.00 keV	5.672E−01	1.276E−03	6285 Å	1186 Å	1347 Å
65.00 keV	5.741E−01	1.202E−03	6646 Å	1201 Å	1378 Å
70.00 keV	5.782E−01	1.137E−03	7004 Å	1216 Å	1407 Å
80.00 keV	5.793E−01	1.028E−03	7722 Å	1245 Å	1461 Å
90.00 keV	5.738E−01	9.401E−04	8445 Å	1272 Å	1512 Å
100.00 keV	5.640E−01	8.672E−04	9181 Å	1297 Å	1562 Å
110.00 keV	5.517E−01	8.056E−04	9934 Å	1322 Å	1610 Å
120.00 keV	5.380E−01	7.530E−04	1.07 μm	1346 Å	1658 Å
130.00 keV	5.238E−01	7.074E−04	1.15 μm	1370 Å	1706 Å
140.00 keV	5.097E−01	6.675E−04	1.23 μm	1393 Å	1754 Å
150.00 keV	4.958E−01	6.322E−04	1.32 μm	1417 Å	1803 Å
160.00 keV	4.825E−01	6.007E−04	1.40 μm	1441 Å	1852 Å
170.00 keV	4.697E−01	5.725E−04	1.49 μm	1466 Å	1902 Å
180.00 keV	4.576E−01	5.471E−04	1.58 μm	1491 Å	1953 Å
200.00 keV	4.354E−01	5.029E−04	1.77 μm	1553 Å	2058 Å
220.00 keV	4.156E−01	4.659E−04	1.97 μm	1617 Å	2167 Å
240.00 keV	3.979E−01	4.343E−04	2.18 μm	1684 Å	2280 Å
260.00 keV	3.821E−01	4.071E−04	2.40 μm	1753 Å	2398 Å
280.00 keV	3.678E−01	3.833E−04	2.63 μm	1825 Å	2520 Å
300.00 keV	3.549E−01	3.624E−04	2.86 μm	1898 Å	2647 Å
330.00 keV	3.375E−01	3.352E−04	3.23 μm	2035 Å	2846 Å
360.00 keV	3.222E−01	3.121E−04	3.62 μm	2175 Å	3054 Å
400.00 keV	3.044E−01	2.862E−04	4.16 μm	2399 Å	3346 Å
450.00 keV	2.854E−01	2.596E−04	4.89 μm	2724 Å	3734 Å
500.00 keV	2.691E−01	2.379E−04	5.66 μm	3052 Å	4146 Å
550.00 keV	2.549E−01	2.197E−04	6.47 μm	3385 Å	4581 Å
600.00 keV	2.425E−01	2.043E−04	7.33 μm	3722 Å	5036 Å
650.00 keV	2.314E−01	1.910E−04	8.23 μm	4066 Å	5513 Å

Particle = proton. Target = Silicon. (Continued)

E	(dE/dx)e	(dE/dx)n	R	L1	L2
700.00 keV	2.215E−01	1.795E−04	9.18 μm	4414 Å	6009 Å
800.00 keV	2.044E−01	1.604E−04	11.18 μm	5502 Å	7060 Å
900.00 keV	1.902E−01	1.452E−04	13.35 μm	6557 Å	8184 Å
1.00 MeV	1.781E−01	1.328E−04	15.68 μm	7602 Å	9377 Å
1.10 MeV	1.677E−01	1.224E−04	18.15 μm	8648 Å	1.06 μm
1.20 MeV	1.586E−01	1.137E−04	20.78 μm	9702 Å	1.20 μm
1.30 MeV	1.506E−01	1.062E−04	23.54 μm	1.08 μm	1.34 μm
1.40 MeV	1.435E−01	9.964E−05	26.46 μm	1.18 μm	1.48 μm
1.50 MeV	1.371E−01	9.391E−05	29.51 μm	1.29 μm	1.63 μm
1.60 MeV	1.314E−01	8.885E−05	32.70 μm	1.40 μm	1.79 μm
1.70 MeV	1.262E−01	8.433E−05	36.02 μm	1.52 μm	1.95 μm
1.80 MeV	1.214E−01	8.028E−05	39.48 μm	1.63 μm	2.12 μm
2.00 MeV	1.131E−01	7.330E−05	46.79 μm	2.00 μm	2.47 μm
2.20 MeV	1.060E−01	6.750E−05	54.61 μm	2.36 μm	2.85 μm
2.40 MeV	9.980E−02	6.259E−05	62.95 μm	2.71 μm	3.24 μm
2.60 MeV	9.442E−02	5.839E−05	71.77 μm	3.06 μm	3.66 μm
2.80 MeV	8.966E−02	5.475E−05	81.09 μm	3.42 μm	4.09 μm
3.00 MeV	8.543E−02	5.156E−05	90.88 μm	3.77 μm	4.55 μm
3.30 MeV	7.987E−02	4.744E−05	106.44 μm	4.50 μm	5.27 μm
3.60 MeV	7.509E−02	4.397E−05	123.05 μm	5.20 μm	6.03 μm
4.00 MeV	6.965E−02	4.010E−05	146.76 μm	6.36 μm	7.11 μm
4.50 MeV	6.399E−02	3.616E−05	178.87 μm	8.03 μm	8.56 μm
5.00 MeV	5.930E−02	3.296E−05	213.67 μm	9.65 μm	10.11 μm
5.50 MeV	5.533E−02	3.031E−05	251.10 μm	11.24 μm	11.77 μm
6.00 MeV	5.192E−02	2.807E−05	291.11 μm	12.84 μm	13.53 μm
6.50 MeV	4.896E−02	2.616E−05	333.64 μm	14.46 μm	15.39 μm
7.00 MeV	4.636E−02	2.450E−05	378.66 μm	16.09 μm	17.35 μm
8.00 MeV	4.199E−02	2.177E−05	475.89 μm	21.69 μm	21.55 μm
9.00 MeV	3.847E−02	1.961E−05	582.62 μm	26.98 μm	26.11 μm
10.00 MeV	3.556E−02	1.786E−05	698.61 μm	32.17 μm	31.03 μm
11.00 MeV	3.329E−02	1.640E−05	823.32 μm	37.33 μm	36.28 μm
12.00 MeV	3.115E−02	1.518E−05	956.56 μm	42.50 μm	41.85 μm
13.00 MeV	2.929E−02	1.413E−05	1.10 mm	47.75 μm	47.75 μm
14.00 MeV	2.765E−02	1.323E−05	1.25 mm	53.09 μm	53.98 μm
15.00 MeV	2.620E−02	1.244E−05	1.41 mm	58.53 μm	60.53 μm
16.00 MeV	2.490E−02	1.174E−05	1.58 mm	64.08 μm	67.41 μm
17.00 MeV	2.374E−02	1.112E−05	1.75 mm	69.73 μm	74.60 μm
18.00 MeV	2.269E−02	1.057E−05	1.94 mm	75.49 μm	82.11 μm
20.00 MeV	2.087E−02	9.613E−06	2.33 mm	95.76 μm	98.06 μm
22.00 MeV	1.935E−02	8.825E−06	2.76 mm	115.22 μm	115.24 μm
24.00 MeV	1.805E−02	8.161E−06	3.22 mm	134.40 μm	133.62 μm
26.00 MeV	1.693E−02	7.594E−06	3.71 mm	153.55 μm	153.18 μm
28.00 MeV	1.595E−02	7.104E−06	4.23 mm	172.82 μm	173.90 μm
30.00 MeV	1.510E−02	6.676E−06	4.79 mm	192.28 μm	195.76 μm
33.00 MeV	1.399E−02	6.126E−06	5.67 mm	233.34 μm	230.65 μm
36.00 MeV	1.305E−02	5.663E−06	6.63 mm	273.54 μm	268.00 μm

Particle = proton. Target = Silicon. (Continued)

E	$(dE/dx)e$	$(dE/dx)n$	R	$L1$	$L2$
40.00 MeV	1.200E−02	5.149E−06	8.00 mm	340.80 μm	321.53 μm
45.00 MeV	1.093E−02	4.629E−06	9.87 mm	438.55 μm	394.25 μm
50.00 MeV	1.006E−02	4.207E−06	11.92 mm	532.81 μm	473.19 μm
55.00 MeV	9.334E−03	3.859E−06	14.14 mm	626.01 μm	558.14 μm
60.00 MeV	8.723E−03	3.566E−06	16.52 mm	719.25 μm	648.88 μm
65.00 MeV	8.199E−03	3.316E−06	19.05 mm	813.11 μm	745.23 μm
70.00 MeV	7.745E−03	3.100E−06	21.75 mm	907.89 μm	846.99 μm
80.00 MeV	6.998E−03	2.745E−06	27.58 mm	1.24 mm	1.07 mm
90.00 MeV	6.407E−03	2.466E−06	34.00 mm	1.56 mm	1.30 mm
100.00 MeV	5.929E−03	2.240E−06	40.97 mm	1.87 mm	1.56 mm
110.00 MeV	5.532E−03	2.054E−06	48.47 mm	2.17 mm	1.84 mm
120.00 MeV	5.199E−03	1.897E−06	56.48 mm	2.47 mm	2.13 mm
130.00 MeV	4.914E−03	1.763E−06	64.98 mm	2.77 mm	2.43 mm
140.00 MeV	4.668E−03	1.648E−06	73.95 mm	3.08 mm	2.76 mm
150.00 MeV	4.453E−03	1.547E−06	83.37 mm	3.38 mm	3.09 mm
160.00 MeV	4.264E−03	1.458E−06	93.22 mm	3.69 mm	3.44 mm
170.00 MeV	4.097E−03	1.379E−06	103.50 mm	4.00 mm	3.80 mm
180.00 MeV	3.947E−03	1.309E−06	114.19 mm	4.31 mm	4.17 mm
200.00 MeV	3.690E−03	1.188E−06	136.70 mm	5.42 mm	4.95 mm
220.00 MeV	3.479E−03	1.089E−06	160.69 mm	6.46 mm	5.77 mm
240.00 MeV	3.302E−03	1.005E−06	186.05 mm	7.46 mm	6.63 mm
260.00 MeV	3.151E−03	9.341E−07	212.69 mm	8.43 mm	7.52 mm
280.00 MeV	3.021E−03	8.726E−07	240.55 mm	9.38 mm	8.44 mm
300.00 MeV	2.909E−03	8.189E−07	269.54 mm	10.31 mm	9.38 mm
330.00 MeV	2.765E−03	7.502E−07	315.00 mm	12.27 mm	10.85 mm
360.00 MeV	2.645E−03	6.924E−07	362.69 mm	14.11 mm	12.36 mm
400.00 MeV	2.513E−03	6.284E−07	429.39 mm	17.12 mm	14.44 mm
450.00 MeV	2.381E−03	5.637E−07	517.27 mm	21.33 mm	17.12 mm
500.00 MeV	2.276E−03	5.115E−07	609.62 mm	25.19 mm	19.88 mm
550.00 MeV	2.191E−03	4.684E−07	705.89 mm	28.81 mm	22.69 mm
600.00 MeV	2.121E−03	4.322E−07	805.63 mm	32.26 mm	25.53 mm
650.00 MeV	2.062E−03	4.014E−07	908.44 mm	35.57 mm	28.41 mm
700.00 MeV	2.013E−03	3.748E−07	1.01 m	38.77 mm	31.30 mm
800.00 MeV	1.935E−03	3.312E−07	1.23 m	49.85 mm	37.11 mm
900.00 MeV	1.877E−03	2.970E−07	1.46 m	59.51 mm	42.92 mm
1.00 GeV	1.833E−03	2.694E−07	1.69 m	68.26 mm	48.68 mm

APPENDIX G2

Particle = proton. Target = Silicon Dioxide. E = energy. $(dE/dx)e$ = electron stopping power (MeV-cm^2/mg). $(dE/dx)n$ = nuclear stopping power (MeV-cm^2/mg). R = Range. $L1$ = Longitudinal Straggling. $L2$ = Lateral Straggling

E	$(dE/dx)e$	$(dE/dx)n$	R	$L1$	$L2$
20.00 keV	3.165E−01	3.085E−03	3239 Å	940 Å	973 Å
22.00 keV	3.330E−01	2.889E−03	3482 Å	966 Å	1012 Å
24.00 keV	3.485E−01	2.719E−03	3717 Å	989 Å	1047 Å
26.00 keV	3.631E−01	2.570E−03	3944 Å	1010 Å	1080 Å
28.00 keV	3.770E−01	2.439E−03	4163 Å	1028 Å	1110 Å
30.00 keV	3.900E−01	2.321E−03	4377 Å	1045 Å	1138 Å
33.00 keV	4.081E−01	2.167E−03	4687 Å	1069 Å	1177 Å
36.00 keV	4.246E−01	2.035E−03	4986 Å	1089 Å	1213 Å
40.00 keV	4.443E−01	1.883E−03	5372 Å	1114 Å	1256 Å
45.00 keV	4.654E−01	1.726E−03	5836 Å	1141 Å	1304 Å
50.00 keV	4.831E−01	1.596E−03	6283 Å	1165 Å	1347 Å
55.00 keV	4.976E−01	1.485E−03	6719 Å	1186 Å	1387 Å
60.00 keV	5.094E−01	1.390E−03	7145 Å	1205 Å	1423 Å
65.00 keV	5.188E−01	1.308E−03	7564 Å	1222 Å	1458 Å
70.00 keV	5.261E−01	1.236E−03	7977 Å	1238 Å	1490 Å
80.00 keV	5.353E−01	1.115E−03	8794 Å	1270 Å	1550 Å
90.00 keV	5.391E−01	1.017E−03	9605 Å	1299 Å	1605 Å
100.00 keV	5.390E−01	9.365E−04	1.04 µm	1326 Å	1657 Å
110.00 keV	5.360E−01	8.688E−04	1.12 µm	1351 Å	1707 Å
120.00 keV	5.310E−01	8.109E−04	1.21 µm	1375 Å	1756 Å
130.00 keV	5.246E−01	7.609E−04	1.29 µm	1398 Å	1803 Å
140.00 keV	5.173E−01	7.172E−04	1.37 µm	1421 Å	1849 Å
150.00 keV	5.094E−01	6.786E−04	1.46 µm	1444 Å	1895 Å
160.00 keV	5.011E−01	6.443E−04	1.55 µm	1466 Å	1940 Å
170.00 keV	4.927E−01	6.136E−04	1.64 µm	1488 Å	1985 Å
180.00 keV	4.842E−01	5.859E−04	1.73 µm	1510 Å	2031 Å
200.00 keV	4.673E−01	5.379E−04	1.91 µm	1565 Å	2121 Å
220.00 keV	4.510E−01	4.977E−04	2.10 µm	1621 Å	2214 Å
240.00 keV	4.354E−01	4.635E−04	2.30 µm	1677 Å	2307 Å
260.00 keV	4.208E−01	4.340E−04	2.51 µm	1735 Å	2403 Å
280.00 keV	4.070E−01	4.083E−04	2.73 µm	1794 Å	2501 Å
300.00 keV	3.941E−01	3.857E−04	2.95 µm	1854 Å	2601 Å
330.00 keV	3.763E−01	3.565E−04	3.29 µm	1968 Å	2757 Å
360.00 keV	3.602E−01	3.316E−04	3.66 µm	2084 Å	2918 Å
400.00 keV	3.410E−01	3.037E−04	4.16 µm	2272 Å	3143 Å
450.00 keV	3.201E−01	2.752E−04	4.84 µm	2547 Å	3440 Å
500.00 keV	3.020E−01	2.519E−04	5.55 µm	2827 Å	3754 Å
550.00 keV	2.862E−01	2.325E−04	6.31 µm	3111 Å	4086 Å
600.00 keV	2.722E−01	2.160E−04	7.11 µm	3400 Å	4434 Å
650.00 keV	2.598E−01	2.019E−04	7.95 µm	3694 Å	4799 Å

Particle = proton. Target = Silicon Dioxide. (Continued)

E	$(dE/dx)e$	$(dE/dx)n$	R	L1	L2
700.00 keV	2.486E−01	1.896E−04	8.82 μm	3992 Å	5179 Å
800.00 keV	2.293E−01	1.692E−04	10.69 μm	4965 Å	5986 Å
900.00 keV	2.132E−01	1.530E−04	12.71 μm	5907 Å	6853 Å
1.00 MeV	1.995E−01	1.398E−04	14.87 μm	6839 Å	7776 Å
1.10 MeV	1.876E−01	1.289E−04	17.18 μm	7770 Å	8754 Å
1.20 MeV	1.773E−01	1.196E−04	19.62 μm	8708 Å	9785 Å
1.30 MeV	1.682E−01	1.116E−04	22.21 μm	9654 Å	1.09 μm
1.40 MeV	1.601E−01	1.047E−04	24.93 μm	1.06 μm	1.20 μm
1.50 MeV	1.529E−01	9.863E−05	27.78 μm	1.16 μm	1.32 μm
1.60 MeV	1.463E−01	9.328E−05	30.76 μm	1.26 μm	1.44 μm
1.70 MeV	1.404E−01	8.850E−05	33.88 μm	1.36 μm	1.57 μm
1.80 MeV	1.350E−01	8.422E−05	37.12 μm	1.46 μm	1.70 μm
2.00 MeV	1.255E−01	7.685E−05	43.98 μm	1.80 μm	1.98 μm
2.20 MeV	1.173E−01	7.073E−05	51.34 μm	2.13 μm	2.28 μm
2.40 MeV	1.103E−01	6.555E−05	59.19 μm	2.45 μm	2.59 μm
2.60 MeV	1.042E−01	6.113E−05	67.52 μm	2.78 μm	2.92 μm
2.80 MeV	9.878E−02	5.729E−05	76.33 μm	3.10 μm	3.27 μm
3.00 MeV	9.396E−02	5.393E−05	85.60 μm	3.43 μm	3.64 μm
3.30 MeV	8.765E−02	4.960E−05	100.37 μm	4.10 μm	4.21 μm
3.60 MeV	8.223E−02	4.595E−05	116.16 μm	4.76 μm	4.83 μm
4.00 MeV	7.606E−02	4.188E−05	138.76 μm	5.85 μm	5.70 μm
4.50 MeV	6.967E−02	3.774E−05	169.45 μm	7.44 μm	6.87 μm
5.00 MeV	6.437E−02	3.439E−05	202.82 μm	8.97 μm	8.14 μm
5.50 MeV	5.989E−02	3.161E−05	238.81 μm	10.48 μm	9.50 μm
6.00 MeV	5.607E−02	2.926E−05	277.37 μm	12.00 μm	10.95 μm
6.50 MeV	5.274E−02	2.726E−05	318.47 μm	13.52 μm	12.48 μm
7.00 MeV	4.984E−02	2.552E−05	362.06 μm	15.07 μm	14.10 μm
8.00 MeV	4.497E−02	2.266E−05	456.49 μm	20.51 μm	17.58 μm
9.00 MeV	4.105E−02	2.040E−05	560.54 μm	25.64 μm	21.39 μm
10.00 MeV	3.782E−02	1.857E−05	674.00 μm	30.68 μm	25.52 μm
11.00 MeV	3.533E−02	1.706E−05	796.30 μm	35.68 μm	29.94 μm
12.00 MeV	3.300E−02	1.578E−05	927.23 μm	40.70 μm	34.65 μm
13.00 MeV	3.098E−02	1.469E−05	1.07 mm	45.80 μm	39.65 μm
14.00 MeV	2.921E−02	1.374E−05	1.22 mm	50.98 μm	44.94 μm
15.00 MeV	2.764E−02	1.292E−05	1.37 mm	56.25 μm	50.51 μm
16.00 MeV	2.625E−02	1.219E−05	1.54 mm	61.62 μm	56.37 μm
17.00 MeV	2.500E−02	1.154E−05	1.71 mm	67.10 μm	62.51 μm
18.00 MeV	2.388E−02	1.097E−05	1.90 mm	72.68 μm	68.92 μm
20.00 MeV	2.193E−02	9.974E−06	2.29 mm	92.72 μm	82.58 μm
22.00 MeV	2.031E−02	9.153E−06	2.71 mm	111.90 μm	97.31 μm
24.00 MeV	1.892E−02	8.462E−06	3.17 mm	130.76 μm	113.11 μm
26.00 MeV	1.774E−02	7.872E−06	3.66 mm	149.57 μm	129.94 μm
28.00 MeV	1.670E−02	7.362E−06	4.18 mm	168.47 μm	147.78 μm
30.00 MeV	1.580E−02	6.917E−06	4.73 mm	187.55 μm	166.63 μm
33.00 MeV	1.462E−02	6.345E−06	5.61 mm	228.20 μm	196.75 μm
36.00 MeV	1.363E−02	5.864E−06	6.56 mm	267.90 μm	229.03 μm

Particle = proton. Target = Silicon Dioxide. (Continued)

E	$(dE/dx)e$	$(dE/dx)n$	R	L1	L2
40.00 MeV	1.253E−02	5.330E−06	7.93 mm	334.64 μm	275.33 μm
45.00 MeV	1.140E−02	4.790E−06	9.80 mm	431.82 μm	338.31 μm
50.00 MeV	1.048E−02	4.352E−06	11.85 mm	525.32 μm	406.76 μm
55.00 MeV	9.724E−03	3.991E−06	14.06 mm	617.64 μm	480.48 μm
60.00 MeV	9.082E−03	3.687E−06	16.44 mm	709.91 μm	559.29 μm
65.00 MeV	8.533E−03	3.428E−06	18.98 mm	802.71 μm	643.01 μm
70.00 MeV	8.058E−03	3.204E−06	21.68 mm	896.33 μm	731.50 μm
80.00 MeV	7.275E−03	2.837E−06	27.52 mm	1.23 mm	922.15 μm
90.00 MeV	6.658E−03	2.547E−06	33.95 mm	1.55 mm	1.13 mm
100.00 MeV	6.157E−03	2.313E−06	40.94 mm	1.85 mm	1.35 mm
110.00 MeV	5.743E−03	2.120E−06	48.47 mm	2.15 mm	1.59 mm
120.00 MeV	5.395E−03	1.958E−06	56.50 mm	2.45 mm	1.85 mm
130.00 MeV	5.098E−03	1.819E−06	65.04 mm	2.75 mm	2.12 mm
140.00 MeV	4.841E−03	1.700E−06	74.04 mm	3.06 mm	2.40 mm
150.00 MeV	4.617E−03	1.596E−06	83.51 mm	3.36 mm	2.69 mm
160.00 MeV	4.420E−03	1.504E−06	93.41 mm	3.66 mm	2.99 mm
170.00 MeV	4.245E−03	1.422E−06	103.74 mm	3.97 mm	3.31 mm
180.00 MeV	4.089E−03	1.350E−06	114.48 mm	4.28 mm	3.64 mm
200.00 MeV	3.821E−03	1.225E−06	137.13 mm	5.39 mm	4.32 mm
220.00 MeV	3.601E−03	1.122E−06	161.26 mm	6.43 mm	5.04 mm
240.00 MeV	3.416E−03	1.036E−06	186.78 mm	7.43 mm	5.79 mm
260.00 MeV	3.259E−03	9.624E−07	213.60 mm	8.40 mm	6.56 mm
280.00 MeV	3.124E−03	8.989E−07	241.65 mm	9.34 mm	7.37 mm
300.00 MeV	3.007E−03	8.435E−07	270.86 mm	10.28 mm	8.20 mm
330.00 MeV	2.857E−03	7.725E−07	316.67 mm	12.24 mm	9.49 mm
360.00 MeV	2.732E−03	7.129E−07	364.74 mm	14.09 mm	10.81 mm
400.00 MeV	2.595E−03	6.468E−07	432.00 mm	17.11 mm	12.64 mm
450.00 MeV	2.457E−03	5.801E−07	520.66 mm	21.35 mm	15.00 mm
500.00 MeV	2.348E−03	5.263E−07	613.85 mm	25.22 mm	17.42 mm
550.00 MeV	2.259E−03	4.819E−07	711.05 mm	28.87 mm	19.89 mm
600.00 MeV	2.186E−03	4.446E−07	811.78 mm	32.33 mm	22.40 mm
650.00 MeV	2.125E−03	4.128E−07	915.64 mm	35.66 mm	24.93 mm
700.00 MeV	2.074E−03	3.854E−07	1.02 m	38.86 mm	27.48 mm
800.00 MeV	1.993E−03	3.405E−07	1.24 m	50.06 mm	32.60 mm
900.00 MeV	1.932E−03	3.053E−07	1.47 m	59.81 mm	37.72 mm
1.00 GeV	1.887E−03	2.768E−07	1.71 m	68.63 mm	42.81 mm

Appendix G3

Particle = proton. Target = Aluminum. E = energy. $(dE/dx)\text{e}$ = electron stopping power (MeV-cm^2/mg). $(dE/dx)\text{n}$ = nuclear stopping power (MeV-cm^2/mg). R = Range. $L1$ = Longitudinal Straggling. $L2$ = Lateral Straggling

E	$(dE/dx)\text{e}$	$(dE/dx)\text{n}$	R	$L1$	$L2$
20.00 keV	3.757E−01	2.648E−03	2213 Å	612 Å	633 Å
22.00 keV	3.890E−01	2.485E−03	2386 Å	630 Å	660 Å
24.00 keV	4.010E−01	2.343E−03	2555 Å	645 Å	684 Å
26.00 keV	4.117E−01	2.219E−03	2720 Å	660 Å	707 Å
28.00 keV	4.213E−01	2.108E−03	2883 Å	673 Å	729 Å
30.00 keV	4.299E−01	2.010E−03	3043 Å	686 Å	750 Å
33.00 keV	4.411E−01	1.880E−03	3279 Å	703 Å	779 Å
36.00 keV	4.504E−01	1.767E−03	3511 Å	719 Å	806 Å
40.00 keV	4.602E−01	1.639E−03	3816 Å	739 Å	840 Å
45.00 keV	4.690E−01	1.505E−03	4191 Å	762 Å	880 Å
50.00 keV	4.747E−01	1.394E−03	4564 Å	783 Å	918 Å
55.00 keV	4.779E−01	1.299E−03	4934 Å	802 Å	953 Å
60.00 keV	4.790E−01	1.218E−03	5304 Å	821 Å	987 Å
65.00 keV	4.786E−01	1.147E−03	5675 Å	838 Å	1020 Å
70.00 keV	4.770E−01	1.085E−03	6048 Å	855 Å	1052 Å
80.00 keV	4.712E−01	9.802E−04	6803 Å	889 Å	1113 Å
90.00 keV	4.633E−01	8.957E−04	7571 Å	923 Å	1173 Å
100.00 keV	4.545E−01	8.258E−04	8355 Å	955 Å	1232 Å
110.00 keV	4.454E−01	7.669E−04	9157 Å	986 Å	1290 Å
120.00 keV	4.362E−01	7.165E−04	9976 Å	1017 Å	1348 Å
130.00 keV	4.273E−01	6.729E−04	1.08 μm	1048 Å	1406 Å
140.00 keV	4.186E−01	6.347E−04	1.17 μm	1078 Å	1464 Å
150.00 keV	4.104E−01	6.010E−04	1.25 μm	1109 Å	1522 Å
160.00 keV	4.025E−01	5.710E−04	1.34 μm	1139 Å	1581 Å
170.00 keV	3.949E−01	5.441E−04	1.44 μm	1170 Å	1640 Å
180.00 keV	3.877E−01	5.198E−04	1.53 μm	1200 Å	1699 Å
200.00 keV	3.743E−01	4.777E−04	1.72 μm	1276 Å	1819 Å
220.00 keV	3.620E−01	4.424E−04	1.92 μm	1352 Å	1941 Å
240.00 keV	3.507E−01	4.123E−04	2.12 μm	1428 Å	2065 Å
260.00 keV	3.402E−01	3.863E−04	2.33 μm	1505 Å	2192 Å
280.00 keV	3.305E−01	3.637E−04	2.55 μm	1582 Å	2321 Å
300.00 keV	3.214E−01	3.437E−04	2.78 μm	1660 Å	2452 Å
330.00 keV	3.088E−01	3.179E−04	3.12 μm	1801 Å	2653 Å
360.00 keV	2.974E−01	2.959E−04	3.49 μm	1941 Å	2859 Å
400.00 keV	2.835E−01	2.713E−04	3.99 μm	2159 Å	3143 Å
450.00 keV	2.682E−01	2.460E−04	4.65 μm	2466 Å	3510 Å
500.00 keV	2.546E−01	2.254E−04	5.36 μm	2771 Å	3893 Å
550.00 keV	2.425E−01	2.081E−04	6.09 μm	3074 Å	4290 Å
600.00 keV	2.317E−01	1.935E−04	6.87 μm	3379 Å	4701 Å
650.00 keV	2.219E−01	1.809E−04	7.68 μm	3686 Å	5127 Å

Particle = proton. Target = Aluminum. (Continued)

E	$(dE/dx)e$	$(dE/dx)n$	R	L1	L2
700.00 keV	2.130E−01	1.699E−04	8.52 μm	3996 Å	5566 Å
800.00 keV	1.974E−01	1.518E−04	10.31 μm	4955 Å	6488 Å
900.00 keV	1.842E−01	1.374E−04	12.24 μm	5880 Å	7465 Å
1.00 MeV	1.729E−01	1.256E−04	14.30 μm	6794 Å	8495 Å
1.10 MeV	1.630E−01	1.158E−04	16.49 μm	7706 Å	9579 Å
1.20 MeV	1.543E−01	1.075E−04	18.81 μm	8623 Å	1.07 μm
1.30 MeV	1.467E−01	1.004E−04	21.25 μm	9549 Å	1.19 μm
1.40 MeV	1.398E−01	9.420E−05	23.82 μm	1.05 μm	1.31 μm
1.50 MeV	1.336E−01	8.878E−05	26.51 μm	1.14 μm	1.44 μm
1.60 MeV	1.281E−01	8.398E−05	29.32 μm	1.24 μm	1.58 μm
1.70 MeV	1.230E−01	7.970E−05	32.26 μm	1.34 μm	1.71 μm
1.80 MeV	1.184E−01	7.587E−05	35.31 μm	1.44 μm	1.86 μm
2.00 MeV	1.102E−01	6.926E−05	41.75 μm	1.76 μm	2.16 μm
2.20 MeV	1.032E−01	6.377E−05	48.66 μm	2.07 μm	2.48 μm
2.40 MeV	9.714E−02	5.913E−05	56.01 μm	2.38 μm	2.81 μm
2.60 MeV	9.183E−02	5.515E−05	63.81 μm	2.69 μm	3.17 μm
2.80 MeV	8.715E−02	5.170E−05	72.04 μm	3.00 μm	3.54 μm
3.00 MeV	8.297E−02	4.868E−05	80.70 μm	3.31 μm	3.93 μm
3.30 MeV	7.749E−02	4.479E−05	94.48 μm	3.95 μm	4.54 μm
3.60 MeV	7.277E−02	4.151E−05	109.19 μm	4.57 μm	5.19 μm
4.00 MeV	6.740E−02	3.785E−05	130.24 μm	5.60 μm	6.12 μm
4.50 MeV	6.181E−02	3.413E−05	158.78 μm	7.08 μm	7.36 μm
5.00 MeV	5.718E−02	3.111E−05	189.76 μm	8.52 μm	8.70 μm
5.50 MeV	5.327E−02	2.860E−05	223.15 μm	9.94 μm	10.13 μm
6.00 MeV	4.991E−02	2.648E−05	258.88 μm	11.36 μm	11.66 μm
6.50 MeV	4.699E−02	2.467E−05	296.93 μm	12.80 μm	13.27 μm
7.00 MeV	4.444E−02	2.311E−05	337.26 μm	14.26 μm	14.97 μm
8.00 MeV	4.015E−02	2.053E−05	424.52 μm	19.30 μm	18.62 μm
9.00 MeV	3.670E−02	1.849E−05	520.56 μm	24.07 μm	22.61 μm
10.00 MeV	3.385E−02	1.684E−05	625.15 μm	28.75 μm	26.93 μm
11.00 MeV	3.173E−02	1.547E−05	737.64 μm	33.40 μm	31.54 μm
12.00 MeV	2.973E−02	1.431E−05	857.68 μm	38.05 μm	36.42 μm
13.00 MeV	2.798E−02	1.332E−05	985.53 μm	42.76 μm	41.59 μm
14.00 MeV	2.644E−02	1.247E−05	1.12 mm	47.53 μm	47.05 μm
15.00 MeV	2.507E−02	1.172E−05	1.26 mm	52.39 μm	52.78 μm
16.00 MeV	2.384E−02	1.107E−05	1.42 mm	57.33 μm	58.78 μm
17.00 MeV	2.275E−02	1.048E−05	1.57 mm	62.36 μm	65.06 μm
18.00 MeV	2.175E−02	9.957E−06	1.74 mm	67.48 μm	71.61 μm
20.00 MeV	2.002E−02	9.058E−06	2.09 mm	85.58 μm	85.52 μm
22.00 MeV	1.857E−02	8.315E−06	2.48 mm	102.93 μm	100.47 μm
24.00 MeV	1.734E−02	7.689E−06	2.89 mm	120.01 μm	116.46 μm
26.00 MeV	1.627E−02	7.154E−06	3.33 mm	137.04 μm	133.45 μm
28.00 MeV	1.534E−02	6.692E−06	3.79 mm	154.16 μm	151.44 μm
30.00 MeV	1.452E−02	6.288E−06	4.29 mm	171.44 μm	170.41 μm
33.00 MeV	1.346E−02	5.770E−06	5.08 mm	207.98 μm	200.66 μm
36.00 MeV	1.256E−02	5.334E−06	5.93 mm	243.71 μm	233.03 μm

Particle = proton. Target = Aluminum. (Continued)

E	$(dE/dx)e$	$(dE/dx)n$	R	L1	L2
40.00 MeV	1.155E−02	4.849E−06	7.15 mm	303.55 μm	279.38 μm
45.00 MeV	1.053E−02	4.359E−06	8.83 mm	390.52 μm	342.30 μm
50.00 MeV	9.691E−03	3.962E−06	10.65 mm	474.32 μm	410.56 μm
55.00 MeV	8.995E−03	3.633E−06	12.63 mm	557.11 μm	483.98 μm
60.00 MeV	8.407E−03	3.357E−06	14.75 mm	639.90 μm	562.38 μm
65.00 MeV	7.904E−03	3.122E−06	17.01 mm	723.19 μm	645.60 μm
70.00 MeV	7.467E−03	2.918E−06	19.42 mm	807.26 μm	733.46 μm
80.00 MeV	6.748E−03	2.584E−06	24.62 mm	1.11 mm	922.58 μm
90.00 MeV	6.179E−03	2.321E−06	30.33 mm	1.39 mm	1.13 mm
100.00 MeV	5.718E−03	2.108E−06	36.54 mm	1.66 mm	1.35 mm
110.00 MeV	5.336E−03	1.933E−06	43.22 mm	1.93 mm	1.59 mm
120.00 MeV	5.015E−03	1.785E−06	50.36 mm	2.20 mm	1.84 mm
130.00 MeV	4.740E−03	1.659E−06	57.93 mm	2.46 mm	2.10 mm
140.00 MeV	4.503E−03	1.550E−06	65.92 mm	2.73 mm	2.38 mm
150.00 MeV	4.296E−03	1.455E−06	74.31 mm	3.01 mm	2.67 mm
160.00 MeV	4.114E−03	1.372E−06	83.09 mm	3.28 mm	2.97 mm
170.00 MeV	3.952E−03	1.298E−06	92.24 mm	3.55 mm	3.28 mm
180.00 MeV	3.807E−03	1.231E−06	101.76 mm	3.83 mm	3.60 mm
200.00 MeV	3.560E−03	1.118E−06	121.82 mm	4.82 mm	4.27 mm
220.00 MeV	3.356E−03	1.024E−06	143.18 mm	5.74 mm	4.98 mm
240.00 MeV	3.185E−03	9.457E−07	165.77 mm	6.63 mm	5.72 mm
260.00 MeV	3.039E−03	8.786E−07	189.50 mm	7.49 mm	6.49 mm
280.00 MeV	2.914E−03	8.207E−07	214.32 mm	8.33 mm	7.28 mm
300.00 MeV	2.805E−03	7.702E−07	240.15 mm	9.16 mm	8.09 mm
330.00 MeV	2.666E−03	7.055E−07	280.65 mm	10.90 mm	9.36 mm
360.00 MeV	2.551E−03	6.512E−07	323.14 mm	12.54 mm	10.66 mm
400.00 MeV	2.423E−03	5.909E−07	382.57 mm	15.22 mm	12.46 mm
450.00 MeV	2.296E−03	5.301E−07	460.88 mm	18.97 mm	14.77 mm
500.00 MeV	2.194E−03	4.809E−07	543.16 mm	22.40 mm	17.14 mm
550.00 MeV	2.112E−03	4.404E−07	628.95 mm	25.62 mm	19.57 mm
600.00 MeV	2.045E−03	4.064E−07	717.84 mm	28.69 mm	22.02 mm
650.00 MeV	1.988E−03	3.774E−07	809.46 mm	31.64 mm	24.50 mm
700.00 MeV	1.940E−03	3.524E−07	903.51 mm	34.48 mm	27.00 mm
800.00 MeV	1.865E−03	3.114E−07	1.10 m	44.35 mm	32.01 mm
900.00 MeV	1.810E−03	2.792E−07	1.30 m	52.95 mm	37.02 mm
1.00 GeV	1.768E−03	2.532E−07	1.51 m	60.74 mm	41.99 mm

Appendix G4

Particle = proton. Target = Tantalum. E = energy. $(dE/dx)e$ = electron stopping power (MeV-cm^2/mg). $(dE/dx)n$ = nuclear stopping power (MeV-cm^2/mg). R = Range. $L1$ = Longitudinal Straggling. $L2$ = Lateral Straggling

E	$(dE/dx)e$	$(dE/dx)n$	R	$L1$	$L2$
20.00 keV	6.770E−02	7.494E−04	893 Å	727 Å	587 Å
22.00 keV	7.057E−02	7.218E−04	981 Å	769 Å	625 Å
24.00 keV	7.333E−02	6.965E−04	1069 Å	809 Å	663 Å
26.00 keV	7.598E−02	6.732E−04	1157 Å	847 Å	699 Å
28.00 keV	7.853E−02	6.517E−04	1244 Å	884 Å	734 Å
30.00 keV	8.097E−02	6.317E−04	1330 Å	918 Å	767 Å
33.00 keV	8.447E−02	6.043E−04	1459 Å	967 Å	815 Å
36.00 keV	8.776E−02	5.796E−04	1587 Å	1013 Å	862 Å
40.00 keV	9.185E−02	5.501E−04	1755 Å	1071 Å	920 Å
45.00 keV	9.650E−02	5.179E−04	1963 Å	1137 Å	990 Å
50.00 keV	1.007E−01	4.898E−04	2168 Å	1197 Å	1055 Å
55.00 keV	1.044E−01	4.650E−04	2371 Å	1254 Å	1117 Å
60.00 keV	1.077E−01	4.430E−04	2572 Å	1306 Å	1176 Å
65.00 keV	1.106E−01	4.233E−04	2771 Å	1356 Å	1233 Å
70.00 keV	1.131E−01	4.056E−04	2969 Å	1403 Å	1287 Å
80.00 keV	1.172E−01	3.748E−04	3363 Å	1490 Å	1391 Å
90.00 keV	1.201E−01	3.489E−04	3755 Å	1569 Å	1490 Å
100.00 keV	1.221E−01	3.269E−04	4148 Å	1644 Å	1584 Å
110.00 keV	1.232E−01	3.078E−04	4542 Å	1715 Å	1675 Å
120.00 keV	1.238E−01	2.912E−04	4940 Å	1782 Å	1764 Å
130.00 keV	1.238E−01	2.764E−04	5341 Å	1847 Å	1852 Å
140.00 keV	1.235E−01	2.633E−04	5747 Å	1909 Å	1938 Å
150.00 keV	1.229E−01	2.515E−04	6158 Å	1970 Å	2024 Å
160.00 keV	1.220E−01	2.409E−04	6576 Å	2030 Å	2109 Å
170.00 keV	1.210E−01	2.312E−04	6999 Å	2089 Å	2194 Å
180.00 keV	1.199E−01	2.224E−04	7429 Å	2147 Å	2278 Å
200.00 keV	1.175E−01	2.069E−04	8308 Å	2262 Å	2448 Å
220.00 keV	1.151E−01	1.936E−04	9214 Å	2376 Å	2619 Å
240.00 keV	1.126E−01	1.821E−04	1.01 μm	2489 Å	2792 Å
260.00 keV	1.102E−01	1.721E−04	1.11 μm	2601 Å	2966 Å
280.00 keV	1.079E−01	1.632E−04	1.21 μm	2714 Å	3143 Å
300.00 keV	1.057E−01	1.553E−04	1.31 μm	2826 Å	3322 Å
330.00 keV	1.027E−01	1.449E−04	1.47 μm	2997 Å	3594 Å
360.00 keV	9.989E−02	1.360E−04	1.63 μm	3168 Å	3872 Å
400.00 keV	9.650E−02	1.258E−04	1.85 μm	3402 Å	4249 Å
450.00 keV	9.273E−02	1.152E−04	2.14 μm	3699 Å	4734 Å
500.00 keV	8.937E−02	1.064E−04	2.45 μm	4000 Å	5230 Å
550.00 keV	8.634E−02	9.903E−05	2.76 μm	4304 Å	5739 Å
600.00 keV	8.359E−02	9.267E−05	3.09 μm	4611 Å	6259 Å
650.00 keV	8.107E−02	8.715E−05	3.43 μm	4922 Å	6790 Å

Particle = proton. Target = Tantalum. (Continued)

E	$(dE/dx)e$	$(dE/dx)n$	R	$L1$	$L2$
700.00 keV	7.875E−02	8.231E−05	3.78 μm	5236 Å	7332 Å
800.00 keV	7.460E−02	7.421E−05	4.52 μm	5913 Å	8449 Å
900.00 keV	7.098E−02	6.768E−05	5.30 μm	6600 Å	9606 Å
1.00 MeV	6.779E−02	6.229E−05	6.11 μm	7299 Å	1.08 μm
1.10 MeV	6.494E−02	5.777E−05	6.97 μm	8009 Å	1.20 μm
1.20 MeV	6.238E−02	5.390E−05	7.87 μm	8732 Å	1.33 μm
1.30 MeV	6.006E−02	5.056E−05	8.80 μm	9466 Å	1.46 μm
1.40 MeV	5.796E−02	4.765E−05	9.77 μm	1.02 μm	1.60 μm
1.50 MeV	5.603E−02	4.507E−05	10.77 μm	1.10 μm	1.73 μm
1.60 MeV	5.425E−02	4.279E−05	11.81 μm	1.17 μm	1.87 μm
1.70 MeV	5.262E−02	4.074E−05	12.89 μm	1.25 μm	2.02 μm
1.80 MeV	5.110E−02	3.889E−05	14.00 μm	1.33 μm	2.17 μm
2.00 MeV	4.837E−02	3.569E−05	16.31 μm	1.51 μm	2.47 μm
2.20 MeV	4.599E−02	3.302E−05	18.76 μm	1.69 μm	2.79 μm
2.40 MeV	4.388E−02	3.074E−05	21.33 μm	1.88 μm	3.12 μm
2.60 MeV	4.201E−02	2.878E−05	24.02 μm	2.07 μm	3.46 μm
2.80 MeV	4.032E−02	2.707E−05	26.83 μm	2.26 μm	3.81 μm
3.00 MeV	3.880E−02	2.557E−05	29.76 μm	2.45 μm	4.18 μm
3.30 MeV	3.676E−02	2.362E−05	34.37 μm	2.78 μm	4.74 μm
3.60 MeV	3.498E−02	2.197E−05	39.22 μm	3.10 μm	5.33 μm
4.00 MeV	3.291E−02	2.012E−05	46.07 μm	3.58 μm	6.16 μm
4.50 MeV	3.072E−02	1.823E−05	55.23 μm	4.23 μm	7.24 μm
5.00 MeV	2.886E−02	1.668E−05	65.01 μm	4.88 μm	8.38 μm
5.50 MeV	2.726E−02	1.539E−05	75.41 μm	5.55 μm	9.58 μm
6.00 MeV	2.586E−02	1.430E−05	86.40 μm	6.23 μm	10.83 μm
6.50 MeV	2.464E−02	1.336E−05	97.96 μm	6.93 μm	12.14 μm
7.00 MeV	2.355E−02	1.254E−05	110.10 μm	7.64 μm	13.49 μm
8.00 MeV	2.169E−02	1.119E−05	135.99 μm	9.45 μm	16.36 μm
9.00 MeV	2.016E−02	1.012E−05	163.99 μm	11.26 μm	19.41 μm
10.00 MeV	1.887E−02	9.245E−06	194.05 μm	13.10 μm	22.64 μm
11.00 MeV	1.794E−02	8.518E−06	225.92 μm	14.97 μm	26.02 μm
12.00 MeV	1.703E−02	7.903E−06	259.49 μm	16.86 μm	29.55 μm
13.00 MeV	1.620E−02	7.375E−06	294.84 μm	18.79 μm	33.22 μm
14.00 MeV	1.544E−02	6.917E−06	331.99 μm	20.76 μm	37.04 μm
15.00 MeV	1.476E−02	6.516E−06	370.91 μm	22.78 μm	41.01 μm
16.00 MeV	1.414E−02	6.161E−06	411.62 μm	24.84 μm	45.13 μm
17.00 MeV	1.356E−02	5.845E−06	454.10 μm	26.96 μm	49.39 μm
18.00 MeV	1.304E−02	5.562E−06	498.34 μm	29.13 μm	53.81 μm
20.00 MeV	1.211E−02	5.074E−06	592.02 μm	34.84 μm	63.08 μm
22.00 MeV	1.132E−02	4.669E−06	692.63 μm	40.59 μm	72.95 μm
24.00 MeV	1.063E−02	4.328E−06	800.07 μm	46.43 μm	83.39 μm
26.00 MeV	1.003E−02	4.035E−06	914.24 μm	52.41 μm	94.41 μm
28.00 MeV	9.502E−03	3.781E−06	1.04 mm	58.53 μm	106.00 μm
30.00 MeV	9.031E−03	3.559E−06	1.16 mm	64.82 μm	118.14 μm
33.00 MeV	8.417E−03	3.273E−06	1.37 mm	76.28 μm	137.38 μm
36.00 MeV	7.891E−03	3.031E−06	1.58 mm	87.91 μm	157.84 μm

Particle = proton. Target = Tantalum. (Continued)

E	(dE/dx)e	(dE/dx)n	R	$L1$	$L2$
40.00 MeV	7.296E−03	2.763E−06	1.89 mm	105.97 μm	186.95 μm
45.00 MeV	6.684E−03	2.490E−06	2.32 mm	131.42 μm	226.21 μm
50.00 MeV	6.181E−03	2.268E−06	2.78 mm	156.98 μm	268.53 μm
55.00 MeV	5.760E−03	2.084E−06	3.27 mm	182.97 μm	313.81 μm
60.00 MeV	5.403E−03	1.929E−06	3.80 mm	209.52 μm	361.93 μm
65.00 MeV	5.095E−03	1.797E−06	4.37 mm	236.71 μm	412.81 μm
70.00 MeV	4.827E−03	1.682E−06	4.96 mm	264.59 μm	466.34 μm
80.00 MeV	4.384E−03	1.494E−06	6.25 mm	346.11 μm	581.01 μm
90.00 MeV	4.032E−03	1.345E−06	7.66 mm	425.93 μm	705.28 μm
100.00 MeV	3.745E−03	1.224E−06	9.18 mm	505.86 μm	838.50 μm
110.00 MeV	3.507E−03	1.124E−06	10.82 mm	586.68 μm	980.13 μm
120.00 MeV	3.306E−03	1.040E−06	12.56 mm	668.72 μm	1.13 mm
130.00 MeV	3.134E−03	9.676E−07	14.40 mm	752.14 μm	1.29 mm
140.00 MeV	2.985E−03	9.053E−07	16.34 mm	836.98 μm	1.45 mm
150.00 MeV	2.855E−03	8.509E−07	18.37 mm	923.23 μm	1.62 mm
160.00 MeV	2.740E−03	8.030E−07	20.49 mm	1.01 mm	1.80 mm
170.00 MeV	2.638E−03	7.604E−07	22.69 mm	1.10 mm	1.98 mm
180.00 MeV	2.547E−03	7.222E−07	24.98 mm	1.19 mm	2.17 mm
200.00 MeV	2.390E−03	6.568E−07	29.79 mm	1.45 mm	2.56 mm
220.00 MeV	2.261E−03	6.027E−07	34.90 mm	1.71 mm	2.97 mm
240.00 MeV	2.153E−03	5.572E−07	40.29 mm	1.96 mm	3.40 mm
260.00 MeV	2.060E−03	5.183E−07	45.93 mm	2.20 mm	3.84 mm
280.00 MeV	1.981E−03	4.847E−07	51.81 mm	2.45 mm	4.29 mm
300.00 MeV	1.912E−03	4.554E−07	57.92 mm	2.69 mm	4.76 mm
330.00 MeV	1.823E−03	4.177E−07	67.48 mm	3.16 mm	5.48 mm
360.00 MeV	1.749E−03	3.860E−07	77.47 mm	3.60 mm	6.22 mm
400.00 MeV	1.668E−03	3.508E−07	91.41 mm	4.29 mm	7.24 mm
450.00 MeV	1.587E−03	3.152E−07	109.70 mm	5.23 mm	8.54 mm
500.00 MeV	1.522E−03	2.864E−07	128.87 mm	6.12 mm	9.87 mm
550.00 MeV	1.470E−03	2.626E−07	148.79 mm	6.96 mm	11.22 mm
600.00 MeV	1.426E−03	2.426E−07	169.37 mm	7.78 mm	12.59 mm
650.00 MeV	1.390E−03	2.255E−07	190.55 mm	8.57 mm	13.96 mm
700.00 MeV	1.360E−03	2.108E−07	212.24 mm	9.34 mm	15.34 mm
800.00 MeV	1.312E−03	1.866E−07	256.93 mm	11.68 mm	18.11 mm
900.00 MeV	1.276E−03	1.676E−07	303.09 mm	13.78 mm	20.86 mm
1.00 GeV	1.249E−03	1.522E−07	350.41 mm	15.71 mm	23.58 mm

Appendix H1

Particle = Phosphorus. Target = Silicon. E = energy. $(dE/dx)e$ = electron stopping power (MeV-cm^2/mg). $(dE/dx)n$ = nuclear stopping power (MeV-cm^2/mg). R = Range. $L1$ = Longitudinal Straggling. $L2$ = Lateral Straggling.

E	$(dE/dx)e$	$(dE/dx)n$	R	$L1$	$L2$
20.00 keV	6.622E−01	1.899E+00	289 Å	128 Å	92 Å
22.00 keV	6.863E−01	1.891E+00	314 Å	137 Å	99 Å
24.00 keV	7.090E−01	1.882E+00	339 Å	147 Å	105 Å
26.00 keV	7.306E−01	1.871E+00	364 Å	156 Å	112 Å
28.00 keV	7.512E−01	1.859E+00	389 Å	165 Å	118 Å
30.00 keV	7.709E−01	1.846E+00	414 Å	175 Å	125 Å
33.00 keV	7.990E−01	1.825E+00	452 Å	188 Å	134 Å
36.00 keV	8.255E−01	1.804E+00	490 Å	202 Å	144 Å
40.00 keV	8.587E−01	1.775E+00	541 Å	220 Å	156 Å
45.00 keV	8.975E−01	1.738E+00	606 Å	242 Å	172 Å
50.00 keV	9.337E−01	1.702E+00	670 Å	263 Å	187 Å
55.00 keV	9.677E−01	1.667E+00	735 Å	284 Å	202 Å
60.00 keV	9.998E−01	1.633E+00	801 Å	305 Å	218 Å
65.00 keV	1.030E+00	1.600E+00	866 Å	326 Å	233 Å
70.00 keV	1.059E+00	1.569E+00	933 Å	347 Å	248 Å
80.00 keV	1.114E+00	1.509E+00	1067 Å	387 Å	278 Å
90.00 keV	1.164E+00	1.454E+00	1202 Å	426 Å	308 Å
100.00 keV	1.211E+00	1.403E+00	1338 Å	464 Å	338 Å
110.00 keV	1.255E+00	1.356E+00	1475 Å	502 Å	367 Å
120.00 keV	1.297E+00	1.313E+00	1614 Å	538 Å	397 Å
130.00 keV	1.336E+00	1.272E+00	1753 Å	574 Å	426 Å
140.00 keV	1.374E+00	1.235E+00	1892 Å	609 Å	455 Å
150.00 keV	1.410E+00	1.199E+00	2033 Å	644 Å	485 Å
160.00 keV	1.444E+00	1.167E+00	2174 Å	677 Å	513 Å
170.00 keV	1.477E+00	1.136E+00	2315 Å	710 Å	542 Å
180.00 keV	1.509E+00	1.107E+00	2457 Å	742 Å	571 Å
200.00 keV	1.570E+00	1.054E+00	2741 Å	805 Å	628 Å
220.00 keV	1.627E+00	1.007E+00	3026 Å	866 Å	684 Å
240.00 keV	1.681E+00	9.644E−01	3311 Å	924 Å	739 Å
260.00 keV	1.733E+00	9.259E−01	3597 Å	980 Å	793 Å
280.00 keV	1.781E+00	8.908E−01	3882 Å	1034 Å	847 Å
300.00 keV	1.828E+00	8.588E−01	4167 Å	1087 Å	899 Å
330.00 keV	1.895E+00	8.155E−01	4594 Å	1162 Å	977 Å
360.00 keV	1.957E+00	7.770E−01	5019 Å	1234 Å	1052 Å
400.00 keV	2.036E+00	7.318E−01	5583 Å	1326 Å	1150 Å
450.00 keV	2.172E+00	6.832E−01	6278 Å	1432 Å	1268 Å
500.00 keV	2.376E+00	6.415E−01	6946 Å	1526 Å	1379 Å
550.00 keV	2.579E+00	6.053E−01	7583 Å	1609 Å	1483 Å
600.00 keV	2.783E+00	5.736E−01	8189 Å	1682 Å	1579 Å
650.00 keV	2.986E+00	5.454E−01	8767 Å	1746 Å	1668 Å

Particle = Phosphorus. Target = Silicon. (Continued)

E	$(dE/dx)e$	$(dE/dx)n$	R	$L1$	$L2$
700.00 keV	3.188E+00	5.203E−01	9320 Å	1804 Å	1751 Å
800.00 keV	3.590E+00	4.772E−01	1.04 μm	1905 Å	1898 Å
900.00 keV	3.986E+00	4.415E−01	1.13 μm	1987 Å	2024 Å
1.00 MeV	4.374E+00	4.113E−01	1.22 μm	2056 Å	2135 Å
1.10 MeV	4.761E+00	3.855E−01	1.30 μm	2113 Å	2231 Å
1.20 MeV	5.162E+00	3.632E−01	1.38 μm	2162 Å	2316 Å
1.30 MeV	5.551E+00	3.435E−01	1.45 μm	2204 Å	2391 Å
1.40 MeV	5.927E+00	3.261E−01	1.52 μm	2240 Å	2458 Å
1.50 MeV	6.289E+00	3.106E−01	1.58 μm	2271 Å	2518 Å
1.60 MeV	6.635E+00	2.967E−01	1.65 μm	2299 Å	2573 Å
1.70 MeV	6.965E+00	2.841E−01	1.70 μm	2324 Å	2623 Å
1.80 MeV	7.280E+00	2.726E−01	1.76 μm	2347 Å	2668 Å
2.00 MeV	7.861E+00	2.526E−01	1.87 μm	2388 Å	2748 Å

APPENDIX H2

Particle = Boron. Target = Silicon. E = energy. $(dE/dx)e$ = electron stopping power (MeV-cm^2/mg). $(dE/dx)n$ = nuclear stopping power (MeV-cm^2/mg). R = Range. $L1$ = Longitudinal Straggling. $L2$ = Lateral Straggling

E	$(dE/dx)e$	$(dE/dx)n$	R	$L1$	$L2$
20.00 keV	6.047E−01	2.809E−01	728 Å	339 Å	264 Å
22.00 keV	6.267E−01	2.714E−01	799 Å	363 Å	284 Å
24.00 keV	6.475E−01	2.626E−01	869 Å	386 Å	305 Å
26.00 keV	6.672E−01	2.544E−01	940 Å	408 Å	324 Å
28.00 keV	6.860E−01	2.468E−01	1010 Å	430 Å	344 Å
30.00 keV	7.040E−01	2.397E−01	1081 Å	450 Å	363 Å
33.00 keV	7.296E−01	2.300E−01	1186 Å	480 Å	391 Å
36.00 keV	7.538E−01	2.211E−01	1290 Å	509 Å	418 Å
40.00 keV	7.842E−01	2.105E−01	1429 Å	545 Å	453 Å
45.00 keV	8.196E−01	1.988E−01	1601 Å	587 Å	495 Å
50.00 keV	8.527E−01	1.885E−01	1771 Å	627 Å	536 Å
55.00 keV	8.837E−01	1.794E−01	1939 Å	664 Å	575 Å
60.00 keV	9.130E−01	1.712E−01	2106 Å	700 Å	612 Å
65.00 keV	9.408E−01	1.639E−01	2271 Å	733 Å	649 Å
70.00 keV	9.673E−01	1.573E−01	2435 Å	765 Å	683 Å
80.00 keV	1.017E+00	1.458E−01	2757 Å	823 Å	750 Å
90.00 keV	1.063E+00	1.361E−01	3073 Å	877 Å	812 Å
100.00 keV	1.106E+00	1.277E−01	3384 Å	926 Å	871 Å
110.00 keV	1.146E+00	1.205E−01	3688 Å	971 Å	926 Å
120.00 keV	1.184E+00	1.142E−01	3987 Å	1013 Å	979 Å
130.00 keV	1.220E+00	1.085E−01	4282 Å	1052 Å	1029 Å
140.00 keV	1.254E+00	1.035E−01	4571 Å	1089 Å	1076 Å
150.00 keV	1.287E+00	9.899E−02	4856 Å	1123 Å	1121 Å
160.00 keV	1.342E+00	9.491E−02	5134 Å	1154 Å	1165 Å
170.00 keV	1.405E+00	9.119E−02	5403 Å	1183 Å	1205 Å
180.00 keV	1.468E+00	8.779E−02	5663 Å	1210 Å	1244 Å
200.00 keV	1.593E+00	8.178E−02	6156 Å	1258 Å	1314 Å
220.00 keV	1.715E+00	7.664E−02	6619 Å	1299 Å	1377 Å
240.00 keV	1.834E+00	7.218E−02	7055 Å	1334 Å	1433 Å
260.00 keV	1.951E+00	6.827E−02	7468 Å	1365 Å	1484 Å
280.00 keV	2.065E+00	6.480E−02	7861 Å	1392 Å	1529 Å
300.00 keV	2.177E+00	6.172E−02	8235 Å	1416 Å	1571 Å
330.00 keV	2.339E+00	5.766E−02	8766 Å	1448 Å	1627 Å
360.00 keV	2.494E+00	5.415E−02	9266 Å	1475 Å	1676 Å
400.00 keV	2.673E+00	5.016E−02	9895 Å	1507 Å	1734 Å
450.00 keV	2.877E+00	4.600E−02	1.06 μm	1540 Å	1796 Å
500.00 keV	3.060E+00	4.254E−02	1.13 μm	1569 Å	1850 Å
550.00 keV	3.225E+00	3.962E−02	1.20 μm	1593 Å	1897 Å
600.00 keV	3.371E+00	3.710E−02	1.26 μm	1614 Å	1939 Å
650.00 keV	3.499E+00	3.492E−02	1.32 μm	1633 Å	1977 Å

Particle = Boron. Target = Silicon. (Continued)

E	$(dE/dx)e$	$(dE/dx)n$	R	$L1$	$L2$
700.00 keV	3.612E+00	3.300E−02	1.38 μm	1650 Å	2011 Å
800.00 keV	3.795E+00	2.979E−02	1.49 μm	1683 Å	2072 Å
900.00 keV	3.929E+00	2.719E−02	1.60 μm	1711 Å	2126 Å
1.00 MeV	4.026E+00	2.505E−02	1.71 μm	1737 Å	2173 Å
1.10 MeV	4.094E+00	2.324E−02	1.81 μm	1760 Å	2217 Å
1.20 MeV	4.141E+00	2.170E−02	1.92 μm	1782 Å	2257 Å
1.30 MeV	4.171E+00	2.037E−02	2.02 μm	1802 Å	2294 Å
1.40 MeV	4.199E+00	1.920E−02	2.12 μm	1821 Å	2329 Å
1.50 MeV	4.221E+00	1.818E−02	2.22 μm	1839 Å	2363 Å
1.60 MeV	4.236E+00	1.726E−02	2.32 μm	1857 Å	2395 Å
1.70 MeV	4.244E+00	1.644E−02	2.42 μm	1873 Å	2426 Å
1.80 MeV	4.248E+00	1.570E−02	2.52 μm	1890 Å	2455 Å
2.00 MeV	4.246E+00	1.442E−02	2.72 μm	1932 Å	2512 Å

Appendix H3

Particle = Arsenic. Target = Silicon. E = energy. $(dE/dx)e$ = electron stopping power (MeV-cm^2/mg). $(dE/dx)n$ = nuclear stopping power (MeV-cm^2/mg). R = Range. $L1$ = Longitudinal Straggling. $L2$ = Lateral Straggling

E	$(dE/dx)e$	$(dE/dx)n$	R	$L1$	$L2$
20.00 keV	3.426E−01	2.754E−03	2946 Å	943 Å	946 Å
22.00 keV	3.632E−01	2.587E−03	3158 Å	968 Å	982 Å
24.00 keV	3.827E−01	2.441E−03	3361 Å	989 Å	1014 Å
26.00 keV	4.010E−01	2.312E−03	3556 Å	1009 Å	1044 Å
20.00 keV	5.515E−01	4.710E+00	192 Å	67 Å	51 Å
22.00 keV	5.768E−01	4.787E+00	205 Å	70 Å	54 Å
24.00 keV	6.009E−01	4.853E+00	218 Å	74 Å	57 Å
26.00 keV	6.239E−01	4.910E+00	231 Å	78 Å	60 Å
28.00 keV	6.460E−01	4.961E+00	244 Å	82 Å	63 Å
30.00 keV	6.673E−01	5.005E+00	256 Å	85 Å	66 Å
33.00 keV	6.979E−01	5.062E+00	275 Å	91 Å	70 Å
36.00 keV	7.270E−01	5.108E+00	293 Å	96 Å	74 Å
40.00 keV	7.639E−01	5.158E+00	318 Å	103 Å	80 Å
45.00 keV	8.074E−01	5.205E+00	348 Å	111 Å	86 Å
50.00 keV	8.484E−01	5.238E+00	378 Å	119 Å	93 Å
55.00 keV	8.873E−01	5.260E+00	408 Å	127 Å	99 Å
60.00 keV	9.243E−01	5.274E+00	437 Å	135 Å	105 Å
65.00 keV	9.597E−01	5.281E+00	466 Å	143 Å	111 Å
70.00 keV	9.938E−01	5.283E+00	496 Å	151 Å	117 Å
80.00 keV	1.058E+00	5.274E+00	554 Å	166 Å	129 Å
90.00 keV	1.118E+00	5.253E+00	612 Å	181 Å	140 Å
100.00 keV	1.175E+00	5.223E+00	670 Å	196 Å	152 Å
110.00 keV	1.229E+00	5.187E+00	728 Å	210 Å	163 Å
120.00 keV	1.280E+00	5.147E+00	786 Å	225 Å	174 Å
130.00 keV	1.329E+00	5.103E+00	844 Å	239 Å	185 Å
140.00 keV	1.376E+00	5.058E+00	903 Å	253 Å	195 Å
150.00 keV	1.422E+00	5.012E+00	961 Å	267 Å	206 Å
160.00 keV	1.466E+00	4.964E+00	1020 Å	281 Å	217 Å
170.00 keV	1.508E+00	4.916E+00	1079 Å	295 Å	227 Å
180.00 keV	1.549E+00	4.868E+00	1138 Å	308 Å	238 Å
200.00 keV	1.628E+00	4.773E+00	1256 Å	335 Å	259 Å
220.00 keV	1.702E+00	4.680E+00	1375 Å	362 Å	280 Å
240.00 keV	1.773E+00	4.588E+00	1495 Å	389 Å	301 Å
260.00 keV	1.841E+00	4.500E+00	1616 Å	415 Å	321 Å
280.00 keV	1.907E+00	4.414E+00	1738 Å	440 Å	342 Å
300.00 keV	1.969E+00	4.332E+00	1860 Å	466 Å	362 Å
330.00 keV	2.060E+00	4.214E+00	2044 Å	504 Å	393 Å
360.00 keV	2.146E+00	4.102E+00	2230 Å	541 Å	423 Å
400.00 keV	2.255E+00	3.963E+00	2480 Å	590 Å	464 Å
450.00 keV	2.383E+00	3.804E+00	2795 Å	650 Å	514 Å

Particle = Arsenic. Target = Silicon. (Continued)

E	$(dE/dx)e$	$(dE/dx)n$	R	$L1$	$L2$
500.00 keV	2.504E+00	3.658E+00	3113 Å	708 Å	564 Å
550.00 keV	2.619E+00	3.525E+00	3432 Å	765 Å	614 Å
600.00 keV	2.728E+00	3.402E+00	3754 Å	820 Å	664 Å
650.00 keV	2.832E+00	3.290E+00	4077 Å	874 Å	713 Å
700.00 keV	2.933E+00	3.185E+00	4401 Å	926 Å	762 Å
800.00 keV	3.123E+00	2.998E+00	5052 Å	1029 Å	859 Å
900.00 keV	3.301E+00	2.835E+00	5704 Å	1126 Å	954 Å
1.00 MeV	3.468E+00	2.692E+00	6356 Å	1219 Å	1047 Å
1.10 MeV	3.627E+00	2.565E+00	7008 Å	1308 Å	1139 Å
1.20 MeV	3.778E+00	2.451E+00	7658 Å	1392 Å	1228 Å
1.30 MeV	3.923E+00	2.348E+00	8305 Å	1473 Å	1316 Å
1.40 MeV	4.062E+00	2.255E+00	8949 Å	1550 Å	1402 Å
1.50 MeV	4.196E+00	2.170E+00	9590 Å	1625 Å	1485 Å
1.60 MeV	4.325E+00	2.093E+00	1.02 μm	1696 Å	1567 Å
1.70 MeV	4.450E+00	2.021E+00	1.09 μm	1764 Å	1648 Å
1.80 MeV	4.572E+00	1.956E+00	1.15 μm	1830 Å	1726 Å
2.00 MeV	4.804E+00	1.838E+00	1.27 μm	1957 Å	1878 Å

Glossary of CCD Terms

A-center
Also called an *oxygen vacancy* or *O-V center*. A trap created when a silicon vacancy interacts with an oxygen atom to form a stable trap defect.

active load
A MOSFET that serves as the load for the first-stage output amplifier.

acceptor material
A p-type semiconductor that contains impurities (boron or gallium) to increase hole concentration.

absorption length
The distance into the CCD where 63% $(1 - e^{-1})$ of the photons are absorbed at a specified wavelength.

adsorption
The condensation of a gas on the surface of a solid.

aliasing
A sampled-image artifact in which under-sampled frequencies in the scene give rise to low-frequency modulation in the image.

alpha particle
One of the particles that can be emitted by a radioactive nucleus. Equivalent with the nucleus of a helium atom consisting of two protons and two neutrons.

amplifier sensitivity
Sensitivity of the output amplifier from electrons to volts (V/e^-).

analog-to-digital converter (ADC)
An electronic device used to convert each pixel voltage to a digital number (DN).

antiblooming (AB) structure
A structure built into the pixel to prevent signal charge above full-well from blooming into adjacent pixels.

antireflection (AR) coatings
A coating employed on backside-illuminated CCDs to reduce reflection loss.

array transfer gate
The last vertical phase which is independently clocked to dump charge into the horizontal register.

Auger recombination
The process by which an electron recombines with a hole in a band-to-band transition and gives off the resulting energy to another electron or hole.

backside illumination
A CCD technology where incident photons enter the back of a sensor to achieve the highest QE possible.

backside well
A small potential well that develops at the surface of a backside-illuminated CCD after thinning.

backside charging
A QE accumulation technique used to induce negative charge on the surface of a backside-illuminated CCD.

backside dark current
Dark current generated on the rear surface of a backside-illuminated CCD.

backside accumulation
A surface passivation technique required by backside-illuminated CCDs to achieve high and stable QE sensitivity.

band gap
The minimum energy that a valence electron must acquire to jump into the conduction band.

barrier phase
A phase or region that exhibits the lowest potential to confine charge to a collecting phase or region.

baseline stabilization
A circuit that samples extended pixels and generates an offset level that is automatically subtracted at the input to the ADC.

beating
A sampled-image artifact in which interference between scene frequencies and the pixel Nyquist frequency results in low-frequency modulation in the image. Unlike aliasing, beating occurs for scene frequencies above and below Nyquist.

beta particle
A particle that can be emitted by a radioactive nucleus. Identical to an electron.

biased flash gate (BFG)
A QE accumulation technique that externally biases the surface of a backside-illuminated CCD into the QE pinned state.

bird's beak
The interface between the gate oxide over the signal channel and the thicker field oxide over the channel stop.

blem spillover
The process of charge spilling from a blemished column into the horizontal register.

blocked channel
A column that contains a defect pixel that allows a finite amount of charge to transfer through it, creating a dark column in a flat-field exposure.

bloomed full well (BFW)
A bias condition applied to the vertical registers to cause blooming when full well is exceeded.

blooming
A full-well condition where charge escapes a collecting region through a barrier region.

Bragg peak
The peak in the energy exchange versus depth for an ion traveling through the CCD.

breakdown
See **silicon breakdown**.

bulk damage
The damage associated with the displacement of silicon atoms from the lattice structure.

bulk traps
Defects or impurities in the silicon wafer which trap signal electrons.

bulk dark current
Dark current that is thermally generated in the bulk silicon.

buried-channel CCD
A CCD technology where signal carriers collect and transfer below the Si–SiO$_2$ interface.

bussing
Metal or poly silicon lines used to connect horizontal or vertical phases that are clocked together.

CCD gates
Conductive electrodes that define pixel boundaries and are clocked to collect and transfer signal charge.

Cm^{244}
An isotope of curium that generates 5.9-MeV alpha particles.

Co^{60}
An isotope of cobalt; standard gamma-ray source used for total dose (ionization) damage testing of CCDs.

calibrated photodiode
A National Bureau of Standards calibrated diode used in absolute QE measurements.

camera gain constant
A conversion constant that converts relative DN units to electrons (expressed in e^-/DN).

capture time constant
A statistical measure of the time required for an empty trap to capture an electron; after this amount of time, the trap has a 63% probability of being filled.

channel
See **buried-channel CCD**.

channel drive
The processing temperature and time used to diffuse the buried-channel implant into silicon; determines channel depth.

channel stop
Implanted regions of zero potential that confine charge laterally to the signal channel.

channel stop bounce
A resistive voltage bounce developed across the channel stops as holes leave the device when gates are clocked.

charge capacity
The amount of charge that can be held by a pixel before blooming (BFW) or surface interaction (SFW) occurs.

charge collection efficiency (CCE)
The collection efficiency of maintaining signal charge in the target pixel after charge is generated.

charge generation efficiency (CGE)
The efficiency resulting from when the CCD intercepts incoming photons and generates electron-hole pairs; this efficiency is quantified by the quantum efficiency (QE).

charge injection
Injection of charge from the output diode (sense node) into the horizontal register controlled by the output transfer gate and output summing well; also called *punch-through*.

charge transfer efficiency (CTE)
The fraction of charge successfully transferred per pixel transfer.

chemisorption charging
A backside accumulation technique in which backside charging occurs through the disassociation of oxygen and other ambient gases on a thin metal film.

chemical vapor deposition (CVD) accumulation
A backside accumulation technique in which doped oxide is grown on the surface of the CCD using a boron-based gas, followed by a high-temperature drive.

clocked antiblooming
A clocking technique used to eliminate blooming in multiphase CCDs that do not have built-in antiblooming structures.

clock feed-through
Electrical transients that accompany the CCD output signal induced by feed-through pulses by the vertical, horizontal, and reset clocks.

clock jitter noise
Phase variations in the camera master clock that translate into random CDS processing times and noise in the video signal.

clock overlap
The minimum time period given to transfer charge from phase to phase.

clock propagation
On-chip clock wave-shaping caused by poly gate resistance and pixel capacitance affecting full well in the center of the chip.

coincident-edge clocking
A clocking technique where the rising and falling clock edges are simultaneous to each other, thereby reducing the substrate bounce problem.

collecting phase
A high-potential phase that collects signal electrons.

column trace
An analog trace of pixels taken in the vertical direction.

column blemish
An isolated cosmetic problem that generates unwanted charge from a defect pixel.

Compton effect
The process of photon scattering off loosely bound electrons in which the photon imparts a portion of its energy to the electron; the magnitude of the energy transfer depends on the scattering angle.

Compton electron
The secondary electron created in the Compton scattering process, e.g., during the interaction of gamma rays with silicon atoms.

conduction band electron
An electron that is free to diffuse within the silicon lattice.

contact noise
Fluctuating conductivity by an improper bond wire connection between the output MOSFET amplifier and the CCD package.

contrast transfer function (CTF)
A CCE measurement technique used to characterize spatial resolution performance using a square-wave target.

contrast
Scene modulation superimposed on an average signal level.

correlated double sampler (CDS)
A popular CCD signal processing sampling technique used to achieve optimum S/N performance.

correlated dual slope (CDS)
A popular CCD signal processing integrating technique used to achieve optimum S/N performance.

cosmetic yield
The fraction of devices manufactured that satisfy both functional and cosmetic requirements.

cosmic rays
High-energy ions originating from interstellar space that transit through the CCD and generate ionization trials and interference events.

Coulomb scattering
Elastic scattering of an ion from the coulomb potential presented by the target ion.

critical absorption distance
For a given electric field, photons must be absorbed at a depth greater than this from the backside surface to prevent photoelectrons from recombining.

critical electric field
The field strength required to cancel the diffusion velocity and stop a photoelectron from reaching the backside surface.

cross section
A measure of the probability of a given particle interaction, expressed in units of area.

Curie
The quantity of any radioactive nuclide which undergoes 3.7×10^{10} disintegrations/sec.

dangling bond
An unsatisfied bonding site at the Si–SiO_2 interface which creates traps.

dark current
Carriers that are thermally generated under totally dark conditions.

dark current figure of merit
A dark current specification given by the manufacturer in units of nA/cm^2 at 300 K.

dark current nonuniformity
The differences in dark current from pixel to pixel, specified as a percentage of the average signal.

dark spikes
Isolated pixels that thermally generate dark current at a greater rate than the average dark current floor.

dark spike activation energy
The temperature required to anneal a dark spike.

dark spike efficiency factor
The number of dark spikes generated per incident particle (e.g., proton).

Debye length
The theoretical length that describes the maximum separation at which an electron will be influenced by an electric field generated by the presence of ions.

deep traps
Trap types whose emission time constant is longer than the shortest clock overlap time period.

deep-depletion CCD
A custom CCD fabricated on high-resistivity bulk silicon to extend the spectral range in the near-IR and hard x-ray regimes.

defect cluster
A region where several radiation damage defects are closely grouped.

defect-impurity complex
A stable defect in the silicon lattice formed when a silicon vacancy combines with an impurity atom.

deferred charge
Charge that lags behind the target pixel during charge transfer.

delta doping
An accumulation technique that grows a very thin highly doped epitaxial layer on the back surface of a CCD to obtain the QE-pinned condition.

depletion dark current
Dark current generated thermally in the depletion region.

depletion region
The region in the CCD where dopant atoms are ionized by applied gate and channel voltages that generate a potential well.

depth of field
The variation in depth of the focal plane where the image stays in focus.

design traps
CTE traps that result when the CCD is designed.

despiking
The process of removing dark spikes, cosmic-ray events and other transients from a CCD image.

dielectric
A material having a relatively low electrical conductivity: an insulator; a substance that contains few or no free electrons (e.g., silicon dioxide and silicon nitride).

diffusion
See **thermal diffusion**.

diffusion dark current
Dark current carriers that diffuse from regions outside the depletion region.

diffusion length
The length of travel through a material after which $(1 - e^{-1})$ carriers have recombined.

diffusion luminescence
In the drain region, when excessively high bias (V_{DD}) promotes avalanche and light emission.

diffusion MTF
MTF associated with the diffusion of signal charge from the target pixel into neighboring pixels caused by field-free material.

digital number (DN)
A digital number generated by an analog-to-digital converter for each pixel.

DN offset
A fixed DN number that accompanies each pixel value that is subtracted to obtain the true signal level.

displacement energy
The energy required to displace a silicon atom.

displacement damage
See **bulk damage**.

divacancy
A trap defect caused by two adjacent vacancies in the silicon lattice.

dominant time constant
The longest RC time constant that defines the bandwidth in a CCD camera system.

donor material
N-type semiconductor that contain impurities (phosphorus or arsenic) to increase electron concentration.

dose
See **radiation absorbed dose**.

dose enhancement
The process by which charge generated in the silicon by ionizing radiation finds its way into the oxide and enhances the dose.

dosimeter
An instrument used to measure total radiation absorbed dose.

drain pinch-off
A MOSFET bias condition that fully depletes the drain region into the saturation region.

drive capacitance
The pixel capacitance which includes oxide, depletion, overlap, field, and sidewall capacitances; given in F/pixel/phase.

dual gate insulator
A two-layer gate dielectric, usually silicon nitride deposited on top of silicon dioxide.

dual-stage amplifier
A high-speed output amplifier configuration that allows pixels rates greater than 22 Mpixels/sec.

dynamic range
The ratio of full well to read noise.

E-center
See **P-V center**.

effective threshold
The channel potential (V_{max}) under a gate when the gate is held at ground potential.

elastic interaction
An elastic scattering interaction of a high-energy ion and the coulomb potential presented by the target ion that can result in displacement damage.

electron
The signal carrier generated, collected, transferred, and measured in n-channel CCDs.

electron confinement range
The physical width in depth where a charge packet is confined in the signal channel.

electron-hole pair
A carrier pair produced when a photon or particle photoelectrically interacts with a silicon atom.

electrostatic discharge (ESD)
The rapid discharge of static electricity from one material to another.

emission time constant
The time period for a signal electron to escape from a trap thermally.

energy resolution
The minimum photon or particle energy difference that can be discriminated by the CCD.

epitaxial layer
A high-quality layer of silicon grown on a substrate where are all CCD functions take place.

epitaxial silicon
Silicon wafer type used in making high-performance CCDs.

escape peak
Charge generated by a x-ray event when a silicon x ray escapes a pixel.

excess charge
Charge leakage through the gate dielectric caused by a high electric field generated beyond inversion.

exit energy
The energy of an ion after passing through a shield.

extended pixels
Empty pixels used for offset purposes that are generated before and after the video pixels are read out.

extended pixel edge response (EPER)
A CTE measurement technique based on the deferred charge measured in the extended pixel region.

extreme ultraviolet (EUV)
The spectral range that covers 100–1000 Å.

Fano-factor
An empirical constant used to determine the variation in charge generated when an x-ray photon or particles interact.

Fano-noise
The noise associated with the uncertainty in signal generated by photons and particles (*see* **Fano-factor**).

Fano-noise-limited
An S/N condition where the output amplifier noise is lower than the Fano-noise.

fat-zero
A small signal injected into the CCD before an image is taken to fill unoccupied traps for CTE improvement.

Fe^{55}
A standard x-ray source used to measure absolute CTE performance.

field region
Within the depletion region where no charge diffusion takes place.

field-free region
The region outside the depletion regions where charge diffusion can take place.

field-assisted emission
A process by which electrons are emitted from a trap and accelerated by an electric field.

fill factor
The fractional area of a pixel that is light sensitive.

first pixel edge response
An absolute CTE measurement technique in which a flat-field image is clocked through an empty storage region.

fixed pattern noise
A image noise that results from sensitivity differences between pixels (also called *pixel nonuniformity*).

flash gate
An accumulation technique that uses a high-work-function metal to induce negative charge on the surface of a backside-illuminated CCD.

flash oxide
A technique for quickly growing a thin low-temperature oxide on the thinned back surface of a CCD in which water is the reactive species.

flat band shift
A shift in the clock potential caused by ionization damage.

flat bottom well effect
A fixed CTE loss effect where the number of traps seen by a charge packet is independent of signal level.

flat field
A light stimulus where each pixel is exposed to the same amount of light.

flicker noise
Low-frequency surface-state noise generated by the output MOSFET amplifier (also called $1/f$ noise).

floating diffusion
See **sense node**.

floating gate amplifier
An output amplifier whose gate is directly over the signal channel and used for nondestructive pixel readout.

fluence
The total concentration of particles that impinge on a CCD; given in particles/cm^2.

fluorescence
The property of emitting radiation as the result of absorption of radiation from some other source. The fluorescence radiation has a longer wavelength than that of the absorbed radiation.

forming gas
A gas mixture of hydrogen and nitrogen introduced during CCD processing to tie up dangling bonds at the Si–SiO$_2$ interface.

four-phase CCD
A popular CCD technology in which phases are defined with two levels of poly and clocked with four clocks.

fractional yield
The fraction of holes that remain in the gate dielectric after initial recombination and electron migration after being generated by ionizing radiation.

frame-transfer
A popular electronic shuttering CCD architecture that includes two sets of vertical registers that are clocked independently.

Frenkel pair
The atom-vacancy pair formed when an atom is dislocated from its position in a lattice by high-energy particles.

freeze out
A condition in which a CCD is cooled to the point where the dopant atoms do not ionize and form potential wells; the CCD ceases to work.

fringe
The locus of maximum constructive interference or destructive interference internal to the CCD where two or more coherent waves intersect.

fringing field
The electric field that is generated at the edge of a phase by a neighboring phase.

frontside illumination
A CCD technology where incident photons enter on the gate side of the sensor.

functional diagram
A diagram showing the order of the phases through which charge is clocked for a given CCD design.

full well
The maximum charge level that a pixel can hold and transfer (also referred to as well capacity).

full well transfer
A transfer curve showing full-well signal as a function of the positive vertical clock voltage.

full inversion clocking (FI)
A clocking mode where both the barrier and collecting phases are clocked into inversion during integration and readout.

gamma rays
Photons emitted from radioactive substances (e.g., C^{60}) that are higher in energy than x rays and able to traverse several centimeters of lead.

gate pinch-off
A MOSFET bias condition in which the gate region is fully depleted independent of source and drain voltage.

gate QE loss
Sensitivity loss that results when photons are absorbed in the gates of the CCD.

Gray total dose
A unit of dose equal to an absorbed energy of 1 Joule per kilogram of silicon. 100 rad(Si) = 1 Gy.

guard columns
Columns located on each side of an array that sink dark charge from the neutral bulk material outside the array (also called *dummy columns*).

half life
The average time required for half of the atoms in a radioactive sample to decay.

hard CCD
A CCD in which custom fabrication process steps are used to reduce the effects of radiation damage.

hermetic
A totally sealed packaging technique to avoid the leakage of gas in or out of a CCD package.

high-speed erasure
Clocking scheme in which unwanted lines are clocked from the array continuously without pausing for horizontal register readout.

hole
The signal carrier generated, collected, transferred and measured in p-channel CCDs.

hopping conduction
A dark current generation process where valence electrons transit through interface and bulk states into the conduction band.

horizontal x ray transfer
A x-ray stacking plot that measures absolute horizontal CTE performance characteristics.

horizontal shift register
The register responsible in shifting signal charge horizontally to the output amplifier (also referred to as the serial register).

horizontal stacking response
Several analog line trace responses superimposed upon one another.

hydrogen passivation
The process by which hydrogen atoms are introduced to a CCD during processing and tie up dangling bonds at the $Si–SiO_2$ interface. *See also* **forming gas**.

impact ionization
A process in which a high-energy electron interacts with the silicon lattice, breaking Si–Si covalent bonds and generating electron-hole pairs.

inelastic interaction
An interaction typically involving a high-energy particle and a silicon nucleus that results in displacement damage.

integration MTF
The MTF associated with the pixel pitch and fill factor.

interface state
Midband energy states found at the $Si–SiO_2$ interface that are responsible for dark current generation and charge trapping (also referred to as *interface traps*).

interlevel short
A short that occurs between two phases that are different poly layers.

interacting quantum efficiency
The ratio of the number of interacting photons to the number of incident photons.

interference filter
A narrow-band filter used to measure absolute QE performance.

interference pattern
The pattern of light and dark fringes produced when two or more coherent waves interfere or intersect.

interlacing
Image display scheme in which the image is updated in two sequential steps: first the odd- numbered lines, then the even-numbered lines.

interline transfer CCD
A CCD architecture that allows electronic exposure in which each column of pixels has its own dedicated transport storage register.

interstitial
The atom in the atom-vacancy pair produced when an atom is dislocated from its position in a lattice; *see also* **Frenkel pair**.

intralevel shorts
A short between two phases that are the same poly gate layer.

inversion
A clocked bias state where the surface potential at the Si–SiO$_2$ interface is equal to substrate potential.

ion
An energetic atom that carries a positive electric charge.

ion implantation
A process used to introduce dopants into silicon.

ionization
The process by which neutral atoms become electrically charged, either positively or negatively, by the loss or gain of electrons.

ionization damage
Damage caused by the generation of electron-hole pairs within the gate dielectric.

ionizing energy loss (IEL)
The energy lost by a particle traveling through matter as it creates electron-hole pairs.

ionizing radiation
Radiation sources that produce ionizing particles that damage the gate dielectric of the CCD.

isotropic source
A light source that generates photons uniformly in all directions.

Johnson noise
The noise associated with the erratic movement of carriers in a material induced by electron collisions, phonon collisions, and lattice vibrations. Also called *white noise* or *thermal noise*.

knife edge response
CCD response to an optical stimulus that exhibits an abrupt light-to-dark transition.

lateral antiblooming
Antiblooming structure in which charge levels over full well leave the pixel laterally to a n$^+$ drain region.

lightly doped drain (LDD)
An output amplifier design in which the source and drain regions are isolated from the gate and field regions.

light shield
A metal layer deposited over regions of the CCD to shield them from incoming light.

line drop out
Localized CTE problem caused by a single intralevel short.

line transfer time
The minimum time necessary to transfer a line vertically at full well.

line trace
An analog trace of pixels taken in the horizontal direction.

linearity residuals
Percentage deviation from ideal linearity over dynamic range.

linearity transfer
A transfer curve that plots signal as a function of exposure time.

load resistor
Off-chip resistor used to load the source of the output amplifier.

LOCOS process
LOCal Oxidation of Silicon, the semiconductor processing technology on which many CCDs are based.

luminescence
A light emission process seen in the array and output MOSFET amplifier.

lumogen
A phosphor coating applied to the CCD to extend the QE response into the UV and EUV spectral regimes.

mean free path
A statistical measure of the average distance a particle travels in a material before interacting.

mobility
A measure of the movement of a carrier in a given material (expressed in $cm^2/V\text{-sec}$).

modulation transfer function (MTF)
A transfer curve used to measure the spatial frequency response when stimulated with a sinusoidal input light source.

moiré patterns
A pattern resulting from interference beams between two sets of periodic strictures in an image (*see also* **beating**).

monochromatic source
A source of radiation that produces photons with exactly the same wavelength; *quasi-chromatic* is used to mean nearly the same wavelength.

mosaic
More than one CCD tiled together to create a larger detector area.

multi-pinned phase (MPP)
A multiphase CCD technology that suppresses surface dark current generation.

n-channel CCD
A buried-channel CCD in which signal electrons are transferred.

neutron
A neutral elementary particle with mass slightly heavier than a proton that is unstable with respect to beta-decay and with a half-life of about 12 minutes.

n-MOS process
A process technology used to fabricate n-channel CCD devices.

near infrared (NIR)
The spectral range that covers 7,000–10,000 Å.

noninversion clocking (NI)
A clocking mode where phases are not clocked into inversion.

nonionizing energy loss (NIEL)
The energy lost to interactions other than the creation of electron-hole pairs as an ion travels through a material.

notch channel
A small channel of higher potential within the main signal channel that isolates small charge packets so that they encounter less traps during charge transfer.

Nyquist frequency
The highest spatial frequency that can be properly sampled by the CCD.

O-V center
See **A-center**.

offset
A non-zero fixed electronic level that accompanies the video signal which must be removed for a true signal value.

Output amplifier
A MOSFET amplifier that provides an output voltage for each pixel.

open-pinned-phase (OPP)
A high-QE frontside-illuminated CCD technology that has a portion of the pixel open to incident photons.

operating window
A bias or clock voltage operating range where the CCD stays within performance specification.

output summing well
The last clocked gate of the horizontal register used to perform pixel binning (summation) and charge injection.

output transfer gate (OTG)
The last gate of the horizontal register used for charge injection and clock isolation between the sense node and output summing well.

output transfer gate transfer
The transfer curve used to measure the effective threshold and inversion breakpoint for a CCD.

overflow drain
A drain on the opposite edge of the device that keeps thermally generated charge from entering the device; used for CCDs with only one horizontal register.

overly thinned
Backside-illumination thinning that results where all substrate material is removed.

p+ donuts
A local thinning nonuniformity seen on self-accumulated CCDs where a ring-shaped region produces maximum QE response to a flat-field exposure.

p-channel CCD
A CCD in which holes are transferred in the signal channel.

P-V center
A phosphorus-vacancy center that traps a single electron induced when the CCD is exposed to high-energy particles.

pair production
A process in which a high-energy photon collides with a target atom and creates an electron-positron pair.

parallel readout
A readout scheme in which several output amplifiers are read out simultaneously, each reading a different portion of the image (also called *multiamplifier readout*).

partial events
X-ray events where some signal charge is lost to recombination.

partial inversion clocking (PI)
A clocking mode where the barrier phase is clocked into inversion and the collecting phase into noninversion.

phosphorescence
Emission of light that continues after the exciting mechanism has ceased.

photon
A elementary particle, or quantum, of radiant energy that stimulates the CCD and generates electron-hole pairs.

photodiode mode
A QE measuring technique where the CCD is configured into a photodiode without applied clocks or output amplifier bias.

photoelectric effect
A process in which valence band electrons are injected into the conduction band by photon interaction.

photoelectron
An electron that has been ejected from its parent atom by interaction between that atom and a photon.

photoemission
The process by which interacting photons stimulate the emission of electrons from a material into another material.

photolithography
The process of defining the poly silicon and gate dielectric layers and implants for the CCD.

photon transfer
An important transfer curve used to specify many CCD and camera performance parameters, including e^-/DN, read noise, fixed-pattern noise, dynamic range and full well.

pinch-off luminescence
The drain region of the output MOSFET amplifier that can luminesce.

pinning
A bias condition that occurs when the signal channel is driven into inversion and pins the Si–SiO_2 surface potential to substrate potential.

pinning implant
A highly concentrated, shallow boron p-doped layer that pins the surface potential to substrate potential.

pixel transfer time
The minimum time necessary to transfer a vertical full-well signal in the horizontal register one pixel.

pixel binning
A clocking process that sums lines and columns together on-chip (also referred to as *pixel summation*).

pixel nonuniformity
Variations in pixel sensitivity to incident photons.

pixel pitch
The center-to-center distance between pixels.

pixel
A picture element.

plasma etch
A process used in which a plasma supplies reactive ions to etch a surface.

pocket pumping
A clocking technique used to measure and characterize individual traps.

point spread function
A CCE response that measures the charge contained in the target pixel compared to its neighbors when stimulated with a point source of light or x rays.

poly silicon
Highly doped, semitransparent, semiconductive, noncrystalline silicon used to form the gates of a CCD.

poly stringers
Unintended unetched strands of poly connecting the same poly-layer together, creating a short which can affect CTE performance.

Poisson distribution
A statistical frequency distribution of n events that occur in a time interval x, provided that these events are individually independent and that the number in a subinterval does not influence the number occurring in any other nonoverlapping subinterval.

popcorn noise
Noise generated in the diffusion regions of a MOSFET, usually associated with premature breakdown caused by aluminum metal diffusing into the n-p junction.

potato chip factor
The peak-to-peak variation in flatness for a backside-illuminated CCD that is unsupported in the thinning process (i.e., window-thinned).

potential well
The potential distribution in the signal channel responsible for charge collection and transfer.

potential maximum
The highest potential within the potential well ($V_{\max}$).

potential pockets and bumps
Potential barriers that induce large traps through improper design or processing.

process traps
CTE traps that result as the CCD is processed.

progressive scan
A readout scheme in which each image line is read out progressively, i.e., not interlaced.

proton
A positively charged subatomic particle slightly less than the mass of a neutron but about 1,836 times heavier than an electron.

proton transfer
A CCD radiation transfer curve that plots displacement damage as a function of proton energy.

punch-through effect
A MOSFET effect where the drain and source depletion regions merge under the channel, causing premature breakdown; also called the *short channel effect*.

QE transfer
A transfer curve that characterizes interacting QE as a function of wavelength.

QE-pinned
An important backside-illuminated condition that achieves 100% internal QE.

quantizing noise
The pixel encoding noise related to the analog-to-digital converter.

quantum yield
The number of electrons generated per interacting photon.

quantum efficiency (QE)
The number of electrons generated per incident photon.

quantum efficiency hysteresis (QEH)
Instability related to QE performance.

RTS dark spike
A dark spike that generates charge erratically, generating a random telegraph signal.

radiation absorbed dose (RAD)
The standard unit of radiation dose, equivalent to the deposition of 100 ergs of energy per gram of material in the form of electron-hole pairs.

radiation events
Charge generated by energetic ions interacting with the CCD (also referred to as *interference events*).

radiation shield
A shield, usually made of aluminum or tantalum, placed near a CCD to reduce the dose received by the CCD.

radiation trap inefficiency factor (RTI)
The measured proportionality constant between the damage measured in terms of CTE and the theoretical CTE based on the number of vacancies predicted by NIEL and SRIM.

radiation traps
CTE traps induced when the CCD is exposed to high-energy particles.

radiation transfer
A transfer method used to transfer radiation damage data from one particle type to another.

range
As in the range of an ion in matter; the average distance traveled before absorption.

read noise
The random noise measured at the output of a camera system under dark conditions.

reflection QE loss
Sensitivity loss that results when photons are reflected away from the CCD surface.

reset MOSFET
Transistor used to reset the sense node before signal charge is transferred.

reset feed through
A large pulse that accompanies video at the time the sense node is reset.

reset noise
The uncertainty generated on the sense node during the reset operation (also called *kTC noise*).

reset remnant noise
Noise related to the finite time required to discharge the post-amp filter capacitor after the reset pulse.

residual bulk image (RBI)
A residual image caused by trapped charge escaping the epitaxial/substrate interface.

residual surface image (SRI)
A residual image caused by trapped charge escaping the $Si-SiO_2$ interface.

responsivity
Absolute QE sensitivity given in units of amps/watt.

reverse annealing
The process by which an irradiated CCD's dark current and flat-band shift continue to increase long after the initial exposure.

SRIM
Stopping and Range of Ions in Matter; software by IBM Research used to calculate stopping powers and ranges of ions in matter.

sample time
The time given to process a pixel after charge is transferred to the sense node.

sandbox
A process strategy for minimizing the cost of custom CCD development in which several customers share the space on the wafer and share the cost of the lot run.

sample-to-sample time
The time period between the first sample and second sample for a CDS processor.

self accumulation
Backside technique in which the accumulation is accomplished by leaving a thin layer of p^+ substrate material on the back surface.

self-induced drift
A self-repulsion field between electrons that is responsible in transferring large charge packets.

self-induced emission
The emission of electrons from the potential well caused by their mutual self-repulsion.

sense node
The region where signal charge is dumped from the horizontal register, allowing the measurement of the charge packet size as a voltage; also called the floating diffusion or output diode.

sense node sensitivity
Sensitivity of the sense node given in units of V/e^-.

serial shift register
See **horizontal shift register**.

shallow traps
Traps type whose emission time constant is shorter than the shortest clock overlap time period.

short channel effect
See **punch-through effect**.

shorts yield
The number of sensors that do not exhibit substrate, interlevel or intralevel shorts.

shot noise limited
The optimum S/N that can be achieved by a CCD camera system.

shot noise
The variance in signal generated by the random arrival of photons to individual pixels.

shutterless photon transfer
A photon transfer technique in which the CCD is illuminated nonuniformly and signal and noise statistics are taken from each column of a single image frame.

signal-to-noise ratio
A standard measure of image quality; the ratio of the signal level to the rms noise.

silicon breakdown
A high-field condition in which thermally generated electrons produce additional electrons through impact ionization.

single pixel events
X-ray events that occupy a single pixel.

seam
The interface in an image between two regions read out with different output amplifiers and gain and offset.

skipper CCD
A CCD employing a skipper amplifier, in which read noise is reduced through multiple sampling by shifting the charge packet back and forth between two electrodes, one of which is a floating gate connected to the output amplifier.

soft CCD
A CCD in which no special effort has been made in the design and fabrication to lessen the effects of radiation damage on the performance.

soft x-ray range
The spectral range that covers 1–100 Å.

South Atlantic Anomaly
A region of high proton fluence created where the Earth's magnetic field lines dip into the atmosphere because of the misalignment of the Earth's magnetic and rotational axes.

spatial frequency
The frequency of a pattern in an image, given in cycles per millimeter.

split frame transfer
A CCD frame transfer architecture that utilizes upper and lower storage regions.

split events
X-ray events that occupy more than one pixel.

spurious charge
An unwanted source of charge that is generated by impact ionization when a CCD is clocked into inversion.

square rooter
A signal processing technique used to optimally encode shot noise limited sensors.

step-and-repeat noise
Noise associated with the pixel nonuniformity caused by a step-and-repeat reticle fabrication process problem.

stiching
A step and repeat lithography process used to make large CCD arrays by projection printing.

stopping power
For an ion traveling through matter, the energy loss per unit of distance.

straggling
The variance of the lateral range on ions in matter.

strapping
Metal lines that run across the array to provide a low-impedance path for clock signals or ground returns.

substrate bounce
A change in substrate potential caused by holes entering and leaving the CCD as phases are clocked in and out of inversion.

substrate short
An electrical short between the substrate and any CCD gate.

supernotch
A processing technique that leaves low-doped p material between n-signal channel and p^+ field regions.

surface full well (SFW)
A bias condition that promotes signal interaction with the $Si-SiO_2$ interface when full well is exceeded.

surface-channel CCD
An early CCD technology where charge is collected and transferred at the $Si-SiO_2$ interface.

surface dark current
Thermally generated dark charge produced at the $Si-SiO_2$ interface.

target pixel
The pixel that should collect all signal charge for photons that interact above it.

thermal anneal
A process by which radiation traps are thermally annealed.

thermal diffusion
A process where signal charge randomly moves in field-free regions of the CCD.

thermionic emission
Emission of electrons from a potential well caused by the thermal excitation of the electrons.

Time Delay Integration (TDI)
An integration scheme in which rows of pixels are synchronized and transferred at the same motion and rate as the scene.

thin gate
A frontside-illuminated CCD technology that uses an ultrathin ($< 400\,\text{Å}$) gate as one of the phases for high QE response.

three-phase CCD
A popular CCD technology whose phases are defined with three levels of poly silicon.

total dose
The ionizing radiation dose received by a device. *See also* **radiation absorbed dose**.

transfer gate tunneling
An effect where charge escapes into the vertical registers as the horizontal register is read out.

transient event
A radiation-induced event that deposits electron-hole pairs in the device by ionization.

transmission QE loss
Sensitivity loss that results when photons are transmitted through a CCD's epitaxial layer.

transparent gate
High frontside QE technology where one of the polygate electrodes is replaced by an optically transparent conducting gate material.

trap
Undesired region in the signal channel where electrons can become deferred.

trap map
A CCD image that shows the locations of individual trapping sites.

trilevel clocking
A clocking technique to eliminate spurious charge generation.

tunnel anneal
A process that anneals traps in which electrons from the silicon tunnel into the oxide and recombine with the hole trap.

two-phase CCD
A popular CCD technology whose phases are defined with two levels of poly silicon and a barrier implant.

UV flooding
A backside charging technique used to accumulate and charge the surface of a backside-illuminated CCD negatively.

ultraviolet (UV)
The spectral range that covers $1000\text{--}4000\,\text{Å}$.

vacancy
What is produced when a silicon atom is displaced from its position in the lattice.

vacancy recombination
The recombination of a silicon atom vacancy with an interstitial before becoming a stable defect.

vertical antiblooming
Antiblooming structure in which charge greater than full well leaves the pixel vertically through the substrate.

vertical stacking response
Several column trace responses superimposed upon one another.

vertical shift register
The register responsible in shifting signal charge vertically to the horizontal shift register (also referred to as *parallel register*).

vertical x-ray transfer
An x-ray stacking plot that measures absolute vertical CTE characteristics.

virtual phase
A popular single-phase frontside-illumination technology that exhibits high UV sensitivity.

visible range
The spectral range that covers 4000–7000 Å.

white noise
Thermal or Johnson noise generated by the output MOSFET amplifier and off-chip electronics.

window clocking
A clocking scheme in which unwanted lines are clocked out quickly to reach a region of interest which is then clocked out normally.

window thinning
A backside-illuminated thinning method where only the imaging pixels are thinned, leaving a thick border of silicon around the membrane.

x ray
A type of radiation that comes from radioactive sources whose frequency is greater than UV photons but lower than gamma rays.

x-ray histogram
A plot of x-ray signal event occurrences as a function of signal.

x-ray point spread
The average point spread response generated by stacking many x-ray events.

yield
The percentage of good devices fabricated at a CCD manufacturer that pass shorts and cosmetic tests.

zebra effect
An interference fringing pattern in a CCD image produced by light reflecting between the back CCD surface and the glass support to which it is epoxied.

Index

A-center, 820
Absorption
  length, 729
  silicon, 170
Accumulation, 62
  active, 200, 256
  backside, 199
  passive, 200, 227
ADC
  bit weighting, 691
  transfer curve, 686
Aliasing, 373
Alpha-particle, 136, 284, 474, 782
Amplifier doping, 51
Annealing, 800
  thermal anneal, 756
  tunnel anneal, 756
Antiblooming, 51, 293, 300
Antireflection coating, 28, 169, 233, 246, 260, 266
Array transfer gate, 42
Auger
  lifetime, 617
  process, 133
  recombination, 617
Avalanche breakdown, 275

Back lapping, 195
Backside
  accumulation theory, 214
  dark current, 214, 232, 623
  electric field, 208, 219, 225
  illumination, 8, 28, 33, 168, 195, 408
  well, 197, 211, 258
Backside charging, 227
  chlorine, 232
  corona, 232
  NO, 231
  UV, 227
Bandgap, 26
  temperature dependence, 622
Bandwidth
  3-dB, 504, 536
  equivalent noise bandwidth, 536
Barrier phase, 24
Baseline stabilization, 561
Beating, 373
Biased flash gate, 249

Bird's beak, 761
Blem spillover, 679
Blocked channel, 57, 679
Bloomed full well, 275
Blooming, 102, 275, 397
Bond pad noise, 330
Bragg peak, 730, 740, 745
Bucket brigade device, 5
Bulk
  state limited, 453
  states, 553, 610
  traps, 35
Bulk damage, 723
  Co^{60}, 797
  equivalence, 812
  experiments, 782
Bulk damage remedy
  annealing, 836
  charge packet density, 824
  clocking, 824
  design and processing, 828
  fat-zero, 836
  operating temperature, 816
  shielding, 829
Buried channel, 8, 62
  doping, 274
  drive, 278
  implant, 278
  model, 84
Buried-channel
  box distribution, 85
  Gaussian distribution, 85
Burn-in, 668, 840

Camera gain
  constant, 98, 105, 132, 561, 690
  histogram, 108
  uncertainty, 110
Capacitance
  accumulation, 62
  depletion, 64
  drive, 483
  oxide, 62
Capacitors, decoupling, 701
Capture time constant, 825
Catastrophic blem, 275
Catastrophic pixels, 668

CCD
  architecture, 37
  Camera Calibration, 97
  Camera Performance, 96
  clock and bias optimization, 156
  design, 42
  fabrication, 51
  history, 3
  mosaic, 32
  operation, 22
  process, 52
  spectral range, 27
  theory, 61
  transfer curves, 95
  yield, 30, 38, 54
CCD Camera
  4-shooter, 15
  Cassini, 19
  Chandra, 22
  Deep Space, 22
  Galileo, 16
  Huygens Probe, 19
  MISR, 19
  MOC, 19
  Pathfinder, 22
  SXT, 16
  Traveling, 11
  WF/PC, 15
CCD performance, 25
  specifications, 36
  standards and units, 95
CCE, frontside illumination, 370
Cd^{109}, 151
Channel
  doping, 71
  drive, 71
  potential plots, 278
  resistance, 713
  stop, 24, 42
  stop bounce, 413, 417
Charge
  diffusion, 254, 273, 332, 738, 745
  generation, 25
  generation efficiency, 25
  injection, 79, 433, 655
  measurement, 35, 489
Charge capacity, 69
  optimum clock voltage, 277
Charge collection, 28
  efficiency, 200, 273, 369
Charge transfer, 34
  efficiency, 34, 387
  efficiency, diffusion drift, 393
  efficiency, fringing field drift, 396
  efficiency hysteresis, 449

  efficiency self-induced drift, 395
  inefficiency, 387
  mechanisms, 390
Chemical vapor deposition accumulation, 266
Chemisorption charging, 246
Cherenov events, 736
Cinema CCD, 60
Clock
  driver, 392
  propagation, 405
  reversal, 421
  wave-shaping, 392, 411
Clocked antiblooming, 293
Clocking modes, 280
Cm^{244}, 135, 782
Co^{60}, 732
Collecting phase, 23
Column blemish, 57, 678
Compton
  effect, 734, 749
  electrons, 673
Constant current diode, 503
Contacts, 46
Contrast transfer function, 338, 354
Cooling, 7, 35, 630
Coronene, 178
Correlated double sampler
  $1/f$ noise limited, 571
  baseline stabilization, 559
  camera gain constant, 561
  circuit, 559
  circuit elements, 557
  preamplifier circuit, 558
  timing diagram, 557
  transfer function, 563
  white noise limited, 565
Correlated double sampling, 540, 556
Cosmetic defects, 678
Cosmetic yield, 57
Cosmetics, 38, 54
Cosmic rays, 35, 670, 724, 736
Coulomb scattering, 736
Coulombic interaction, 773
Critical absorption distance, 223
Critical electric field, 222
Cross talk, 706, 711
Cross-section, 728, 777
CTE measurement techniques, 418
Curie, 135

Damage
  electrical, 837
  ESD, 839
  thermal, 838
Dangling bonds, 756, 773

INDEX

Dark current, 7, 35, 105, 312, 605, 623
    backside dark current, 622
    depletion dark current, 610, 616, 620
    diffusion dark current, 613
    equation, 622
    figure of merit, 622
    multi-pinned phase, 632
    nonuniformity, 629
    pixel nonuniformity, 105
    shot noise, 626
    spikes, 635
    substrate dark current, 617
    surface dark current, 618
Dark spike, 426, 789, 800
    activation energy, 800
    despiking, 646
    efficiency factor, 792
    virtual-phase CCD, 644
Debye length, 218
Decibel, 504
Decoupling, 699
Deep depletion, 251
    CCDs, 252
Defect clusters, 774
Defect-impurity complexes, 774
Deferred charge, 34, 132, 387, 418, 582
Delta doping, 258
Depletion, 63, 73
    depth, 67, 74, 87
Diffusion
    constant, 614
    length, 613, 615, 617
Digital number, 98
Displacement damage, *see* bulk damage, 723
Displacements, 777
Divacancy, 774, 820
Dominant time constant, 563
Dose enhancement, 736
Dosimeter, 727
Dual dielectric, 666
Dual slope processor, 578, 585
    transfer function, 579
Dual-gate insulator, 62
Dynamic range, 7, 33, 113, 690

E-center, 774
Edge response, 356, 817
Effective quantum yield, 369
Effective storage capacitance, 90
Effective threshold, 73, 78, 80, 278
    amplifier, 157
    array, 157
Elastic interactions, 773
Electromagnetic interference
    conductor shielding, 702
    structure shielding, 705

Electron
    concentration, 606
    confinement range, 827
    cross-section, 453
    detection, 265
    lifetime, 26, 612, 615
    mobility, 478
    thermal velocity, 453
Electron-hole
    capture and emission, 609
    pair, 26
Emission time constant, 544
Epitaxial silicon, 51, 605
Escape peak, 134, 143
ESD
    damage, 668
    event, 839
    protection diodes, 670
Etch stop, 196
Excess charge, 674, 771
Exit energy, 744
Extended pixel, 42, 104, 559, 817
    edge response, 423, 582

Fano-factor, 147, 599
Fano-noise, 599
Fano-noise limited performance, 149
Fat-zero, 469, 654
Fe^{55}, 132, 418, 732
Ferrite beads, 702
Field-assisted emission, 639
Field free
    material, 332
    region, 33, 132, 141, 201, 256
Fill factor, 24
First light, 49
First pixel edge response, 429
Fixed CTE loss, 466
Fixed pattern noise, 102, 318
Fixed pattern noise sources, 325
Flash gate, 242
Flash oxide, 234, 237
Flat bottom well effect, 466
Flatband, 86
    shift, 722
    voltage, 755
Flat-fielding, 321
Float gate amplifier, 585
Floating gate, 193
    amplifier, 58
Fluence, 730
Forming gas, 756, 771
Four-phase CCD, 56, 449
Fractional yield, 753, 760
Frame transfer, 37, 51
Freeze-out, 73, 624, 820
Frenkel pair, 773

Fringing, 204
  fields, 278, 335, 358
Frontside illumination, 167, 178
Full factor, 339
Full frame, 8
  array, 37
  CCD, 340
Full well, 102, 273, 760
  transfer, 280, 400
  optimum, 157
Full width half maximum, 147

Galileo, 11
Gamma, 117
Generation/recombination, 609
Germanium CCD, 27
Ghost image, 711
Global CTE, 439
Gray total dose, 726
Grounding, 706, 837
  daisy chain, 706
  diagram, 710
  multipoint, 707
  single point, 706
Guard columns, 424

Hard CCDs, 723
Hermeticity, 736
High-resistivity silicon, 252
High-speed channel stop (HSCS), 51
High-speed erasure, 305
High-speed operation, 392
High-speed straps, 51
Hole concentration, 606
Hole transit time, 255
Hopping conduction, 609, 618
Horizontal register, 24, 36, 38
Hot electrons, 666
Hubble
  lightpipe, 231
  Space Telescope, 6, 15
Hydrogen passivation, 756

I_D transfer, 507
Imaging registers, 37
Impact ionization, 275, 524, 649
Implant activation, 257
Impurities, 52
Index of refraction, silicon characteristics, 205
Inductive reactance, 707
Inelastic interactions, 773
Interface states, 196, 235, 244, 544, 756
Interference fringing, 204, 238, 329
Interlacing, 38
Interlevel short, 55
Interline CCD, 340
Interline transfer, 8

Interstitial, 773
Intralevel short, 55, 449
Intrinsic carrier concentration, 606
Intrinsic Fermi level, 610
Inversion, 84, 157, 278, 415
Inverted clocking, 280
Ion implantation, 256
Ion range, 257
Ionization damage, 722, 750
  equivalence, 750
  e-h generation, 750
  hole transport, 754
  hole trapping, 755
  interface state creation, 756
  process-induced, 771
  recombination, 753
  thick/thin oxides, 761
  UV, 750
Ionizing damage, minimum damage, 759
Ionizing energy, 73
Ionizing energy loss (IEL), 722
ITO gate, 194, 253

Junction diode, 73

Knife edge, 339
  response, 356

Lateral antiblooming
  gate, 300
  implant, 300
Light flood, 664
Light leak, 684
Light shield, 46
Lightly doped drain (LDD), 50, 520
Line drop out, 449
Linear array, 37
Linearity, 7, 117
  residuals, 117
  transfer, 117
Load resistor, 533
  active, 529
  channel resistor, 530
  passive, 530
  poly resistor, 530
Local CTE, 441
LOCOS process, 52
Lumigen, 178, 249
Luminescence, 665
  clocking luminescence, 666
  diffusion, 523
  diode luminescence, 669
  pinch-off, 523
  pixel luminescence, 668
Lyman-α, 183

Magnetosphere, 724
Masks, 42
MBE, 258

INDEX
903

Mean free path, 729, 777–778
Midband traps, 610, 636
Mobility, 75, 479
   electrons, 753
   holes, 753
Modulation transfer function, 33, 273, 338
Moiré effect, 377
MOS capacitor, 5, 22, 61
MOSFET
   bias, 507
   buried channel, 547
   depletion mode, 495
   effective threshold, 494
   enhancement mode, 495
   linear region, 490
   load line, 500
   mobility effect, 498
   p-channel, 495
   saturation region, 493
   subthreshold current, 494
   surface-channel, 495, 547
   threshold voltage, 494
   transconductance, 496
MPP
   CCD, 50, 232, 310, 828
   clocking, 280
MTF
   backside illumination, 345, 353
   CTE, 339, 351
   diffusion, 339, 344
   frontside illumination, 344
   integration, 339
Multi-pinned phase, 612, 620, 632
   CCD, 404, 764

n-channel CCD, 4
Native oxide, 234, 237
Neutron
   damage, 774, 802, 829
   half-life, 802
Noise, 36
   charge injection, 655
   clock-jitter, 697
   contact noise, 555
   correlation, 541
   coupling, capacitive and inductive, 702
   dark current, 626, 629
   electromagnetic interference, 699
   fat-zero, 654
   flicker noise, 543
   Johnson noise, 533
   kTC, 538
   off-chip, 684
   on-chip, 605
   partition noise, 541
   peak-to-peak, 535
   popcorn noise, 555
   preamplifier, 684
   quantizing, 686
   reset, 537
   root-mean-square, 534
   RTS, 551
   shot noise, 553
   spectral, 505, 527
   spurious charge, 650
   subelectron, 588
   transfer, 656
   white noise, 533, 541
Noise-reduction techniques, 714
Noninverted clocking, 280
Nonionizing energy loss (NIEL), 722, 804, 809
Nonlinear encoding, 691
Nonlinearity, 117
   floating diffusion, 517
   MOSFET, 518
Notch, 828
   channel, 50, 83, 472
Nyquist frequency, 339

O-V
   center, 820
   trap, 774
Offset, 104
Open pinned phase, 188
Operating window, 157, 761
Orange peel, 196
Output amplifier, 24, 35, 37–38, 42
   bias, 531
   breakdown characteristics, 521
   frequency response, 504
   linearity, 516
   loading, 499
   luminescence, 523
   multistage, 527
   noise, 532
   noise equation, 555
   operation, 490
   output impedance, 503
   power consumption, 531
   schematic, 489
   SEM image, 489
   sensitivity, 157, 509
   temperature characteristics, 519
   video time constant, 504, 528
   voltage gain, 498
Output diode, 35
Output summing well, 42
Output transfer gate, 42
   transfer, 80

p^+ donuts, 211
P_b center, 756

p-channel CCD, 3, 829
p-i-n diode, 252
P-V trap, 774
   emission time constant, 816
Pads, 47
Pair production, 749
Parallel readout, 32, 60
Parallel register, 23
Partial events, 141, 362
Partially inverted clocking, 280
Particle transients, 737
Phosphor coatings, 15, 27, 178
Photodiode standard, 151
Photoelectric effect, 26, 74, 167, 749
Photoemission, 227, 750
Photographic film, 5
Photolithography, 42
Photometric accuracy, 34
Photon transfer, 97, 339, 371, 693
   curve, 101
   derivation, 98
   generation, high-speed, 125
   simulation, 130
Photon transients, 749
Pinch-off, 159, 492
   drain, 495
   gate, 494
Pinning, 84
   implant, 185, 188, 767
Pixel
   binning, 43, 125
   cross talk, 33
   nonuniformity, 102, 273, 318, 321
   number, 28
   summation, 43
Plasma etching, 771
Pocket pumping, 430, 461, 783, 786, 802, 821
Point spread, 339
   response, 356, 361
Poisson's equation, 65, 85
Poly gaps, 593
Poly hole gate, 194
Poly overlap, 56
Poly silicon gate, 11
Poly stringers, 56, 449
Polysilicon, 56
Potential bump, 405
Potential well, 274
   buried-channel, 70
   surface-channel, 66
Power dissipation, 476
   charge motion, 478
   epitaxial, 480
   inversion, 481
   potential, 482
   reactive, 482

Power, frame transfer, 485
Primary knock-on atom, 773, 779
Progressive scan, 38
Proportional CTE loss, 466
Proton
   cross-section, 780
   damage, 784
   detection, 265
   displacement energy, 778
   displacement threshold energy, 778
   energy resolution, 747
   IEL, 777
   NIEL, 777
   QE, 745
   range, 738
   stopping power, 777
   straggling, 738
   transfer, 804
   vacancies, 777
Punch-through effect, 496

QE
   hysteresis, 198, 230, 261
   photodiode mode, 154
   transfer, 151, 169
   UV response, 182, 261
QE loss
   absorption, 167
   channel stop recombination, 175
   gate absorption, 27
   poly overlap, 174
   reflection, 28, 167
   substrate, 174
   temperature, 175
   transmission, 167
QE model
   backside illumination, 170
   frontside illumination, 173
   Monte Carlo, 177
QE-pinning, 200, 236
Quantum efficiency, 151, 167
   hysteresis, 663
Quantum yield, 27, 99, 106, 108, 167, 369
Quasi-Fermi level, 78
Quiet times, 714

Ra^{225}, 736
Radiation
   environment, 723
   interference, 673
   shields, 673 732
   transfer, 814
   trap inefficiency factor, 805
   units, 726
Radiation absorbed dose (rad), 726
Radiation damage, 6, 35, 136, 426, 632, 636
   x-ray, 263
Range, 730

INDEX

Read noise, 101
Reciprocity failure, 7
Reflection, silicon, 170, 261
Remnant
    noise, 583
    signal, 582
Reset clock feed-through, 82
Reset MOSFET, dual gate, 519
Reset switch, 42
Residual image, 284, 310
    residual bulk image, 659
    residual surface image, 657
Resistivity
    channel stop, 413
    epitaxial, 413
    metal, 408
Responsivity, 153
Reticles, 42
Reverse annealing, 757, 760
Rms definition, 687
RTI factor, 809
RTS dark spikes, 793
Rutherford scattering, 736

Sample-to-sample time, 563
Sandbox, 58
Seam factor, 682
Seam noise, 680
Self-accumulation, 201, 225
Self-induced emission, 289
Sense node, 35, 42
    floating diffusion capacitance, 510
    MOSFET capacitance, 514
Serial register, 24
Shield transfer, 833
Shielding, 829
Shockley Equation, 610
Short channel effect, 496
Shorts, 10, 33, 54
    latent, 668
    yield, 57
Shot noise, 101
Shutterless photon transfer, 127
Sidewall capacitance, 512
Signal contrast, 123
Signal-to-noise
    contrast, 121, 319
    flat-field, 120
Silicon
    breakdown, 275, 524
    resistivity, 75
    wafers, 51, 453, 460
Single pixel events, 141, 362
Skin depth, 705
Skipper, 193
    CCD, 58, 586
    dynamic range, 596

    signal-to-noise (extended images), 595
    signal-to-noise (point images), 599
Slant bar target, 354
Smear, 406
Soft CCDs, 723
Solar flares, 723
South Atlantic Anomaly, 673, 724
Spatial resolution, 28
Spiking, 555
Split events, 141, 362
Split frame transfer, 38, 51, 60
Spreading resistance, 71, 201
Spurious charge, 187, 649
    shot noise, 650
Spurious potential pockets, 439
Square rooter, 691
SRIM, 738, 777
Step-and-repeat noise, 329
Steam oxide, 240
Stitching, 42
Stopping power, 730
Storage registers, 37
Straggling, 257
Strapping, 408, 417
Substrate, 51
    bounce, 408, 670, 707
    contact, 50, 409
    short, 55
Super notch, 51, 192, 511, 765
    channel, 472
Super pixels, 33
Surface-channel, 8, 62
    model, 65
Surface
    full well, 277
    potential, 87
    recombination velocity, 619, 621
    states, 619, 657

Target pixel, 28, 132
Thermionic emission, 83, 289
Thermoelectric coolers, 630
Thin gate, 190
Thinning, 195
    spot, 238
Three-phase CCD, 23
    invention, 3
Three-phase functional diagram, 41
Three-phase timing diagram, 24, 41
Total dose, 726, 730
    sources, 732
Transfer curves, 37
    set point, 160
Transfer gate tunneling, 669
Transfer noise, 656
Transient event, 736

Transparent gate, 194
Trap map, 449, 464
Traps, 34, 187, 433
   backwards, 431
   bulk, 453
   deep, 454
   design, 434
   emission time constant, 453, 657
   forward, 431
   local, 441
   process, 439
   radiation-induced, 466
   shallow, 454
   statistical, 430, 462
Triboelectric charging, 839
Trilevel clocking, 651
Two-phase CCD, 5, 56, 183, 405, 417

UV flood, 721, 767

V_{DD} transfer, 521
Vacancy, 773
   recombination, 774
Van Allen belts, 724

Vertical
   antiblooming, 304, 415
   register, 23, 36, 38
Virtual-phase CCD, 183, 405, 417, 722, 767

Wave-shaping, 708
Well capacity, 28, 33, 68, 274
Window clocking, 129, 310
Window thinning, 201

X-ray
   absorption depth, 134
   characteristics, 131
   events, 362
   event cloud, 335
   event cloud diameter, 361
   fluorescence table, 135
   histogram, 143
   image, 139
   point spread, 361
   QE, 187, 192
   transfer, 131, 141, 418

Zebra effect, 329

**James Janesick** is currently a member of the technical staff at Conexant Systems, Personal Imaging Systems, developing high-performance CMOS imaging arrays. Previously he was Vice President and Chief Scientific Director for Advanced Systems Division of Pixel Vision, Inc., developing high-speed backside-illuminated CCDs for scientific and cinematographic applications. He was with the Jet Propulsion Laboratory for 22 years, where as group leader of the Advanced CCD Sensors Development Group he designed scientific CCDs and support electronics utilized in many NASA spaceborne and ground-based imaging systems. Flight CCDs and electronics included the Hubble Space Telescope, Galileo at Jupiter, Solar X-ray Telescope, Mars Pathfinder and Orbital Cameras, Multi-imaging Spectral Radiometer (MISR), Deep Space I and Cassini, a spacecraft en route to Saturn. Janesick has authored 75 publications on CCDs and has contributed to many NASA Technical Briefs and patents for various CCD innovations. He received NASA medals for Exceptional Engineering Achievement in 1982 and 1992. For his pioneering work on the CCD, Asteroid 4558 Janesick was named after him. He also has given several courses on CCD/CMOS technology and camera electronics at UCLA and through SPIE.